Elementary Linear Algebra
A Matrix Approach

Elementary Linear Algebra
A Matrix Approach

Lawrence E. Spence

Arnold J. Insel

Stephen H. Friedberg
Illinois State University

PRENTICE HALL
Upper Saddle River, New Jersey 07458

Library of Congress Cataloging-in-Publication Data

Spence, Lawrence E.
    Elementary linear algebra : a matrix approach / Lawrence E.
Spence, Arnold J. Insel, Stephen H. Friedberg.
      p.   cm.
    Includes bibliographical references and index.
    ISBN 0-13-716722-9
    1. Algebra, Linear.   I. Insel, Arnold J.   II. Friedberg, Stephen
H. III. Title.
QA184.S68   2000
$512^0.5$—DC21   99-23843
                                                         CIP

Acquisitions Editor: George Lobell
Editor-in-Chief: Jerome Grant
Production Editor: Brittney Corrigan-McElroy
Senior Managing Editor: Linda Mihatov Behrens
Executive Managing Editor: Kathleen Schiaparelli
Assistant Vice President of Production and Manufacturing: David W. Riccardi
Director of Marketing: Melody Marcus
Marketing Assistant: Vince Jansen
Manufacturing Buyer: Alan Fischer
Manufacturing Manager: Trudy Pisciotti
Supplements Editor/Editorial Assistant: Gale Epps
Art Director: Maureen Eide
Associate Creative Director: Amy Rosen
Director of Creative Services: Paul Belfonti
Assistant to Art Director: John Christiana
Art Manager: Gus Vibal
Art Editor: Grace Hazeldine
Interior/Cover Designer: John Christiana
Cover Photo: Joyce Lopez, "A Pre/Cautionary Tale." Owned by University of Illinois,
Champaign/Urbana. Photographer Mark Belter.

ⓒ 2000 by Prentice-Hall, Inc.
Upper Saddle River, New Jersey 07458

Printed in the United States of America

10 9 8 7 6 5 4 3 2 1

0-13-716722-9

Prentice-Hall International (UK) Limited, London
Prentice-Hall of Australia Pty. Limited, Sydney
Prentice-Hall Canada Inc., Toronto
Prentice-Hall Hispanoamericana, S.A., Mexico
Prentice-Hall of India Private Limited, New Delhi
Prentice-Hall of Japan, Inc., Tokyo
Prentice-Hall (Singapore) Pte. Ltd.
Editora Prentice-Hall do Brasil, Ltda., Rio de Janeiro

To our families:
Linda, Stephen, and Alison
Barbara, Thomas, Sara, and Max
Ruth Ann, Rachel, Jessica, and Jeremy

Contents

Preface   ix

To the Student   xiii

1   Matrices, Vectors, and Systems of Linear Equations   1

1.1   Matrices and Vectors   1
1.2   Linear Combinations, Matrix–Vector Products,
        and Special Matrices   11
1.3   Systems of Linear Equations   23
1.4   Gaussian Elimination   35
1.5* Applications of Systems of Linear Equations   50
1.6   The Span of a Set of Vectors   59
1.7   Linear Dependence and Linear Independence   68
        Chapter 1 Review Exercises   79

2   Matrices and Linear Transformations   83

2.1   Matrix Multiplication   83
2.2* Applications of Matrix Multiplication   98
2.3   Invertibility and Elementary Matrices   113
2.4   The Inverse of a Matrix   126
2.5* The LU Decomposition of a Matrix   136
2.6   Linear Transformations and Matrices   150
2.7   Composition and Invertibility of Linear Transformations   160
        Chapter 2 Review Exercises   172

3   Determinants   175

3.1   Cofactor Expansion   175
3.2   Properties of Determinants   185
        Chapter 3 Review Exercises   196

*Sections marked with an asterisk can be omitted wothout loss of continuity.

4 Subspaces and Their Properties   199

4.1 Subspaces 199
4.2 Basis and Dimension 211
4.3 The Dimensions of Subspaces Associated with a Matrix 221
4.4 Coordinate Systems 228
4.5 Matrix Representations of Linear Operators 239
     Chapter 4 Review Exercises 248

5 Eigenvalues, Eigenvectors, and Diagonalization   251

5.1 Eigenvalues and Eigenvectors 251
5.2 The Characteristic Polynomial 259
5.3 Diagonalization of Matrices 270
5.4 Diagonalization of Linear Operators 284
5.5* Applications of Eigenvalues 291
     Chapter 5 Review Exercises 308

6 Orthogonality   311

6.1 The Geometry of Vectors 311
6.2 Orthogonal Vectors 321
6.3 Least-Squares Approximation
     and Orthogonal Projection Matrices 335
6.4 Orthogonal Matrices and Operators 345
6.5 Symmetric Matrices 355
6.6 Singular Value Decomposition 364
6.7 Rotations of $\mathcal{R}^3$ and Computer Graphics 378
     Chapter 6 Review Exercises 394

7 Vector Spaces   397

7.1 Vector Spaces and Their Subspaces 397
7.2 Dimension and Isomorphism 406
7.3 Linear Transformations and Matrix Representations 418
7.4 Inner Product Spaces 430
     Chapter 7 Review Exercises 444

Appendix: Complex Numbers   447

References   451

Answers to Odd-Numbered Exercises   453

Index   475

Preface

That linear algebra plays a central role in mathematics is no longer debatable. In addition, it is increasingly used to create quantitative models in the physical and social sciences. The advent of computers (indeed, of supercalculators) has made matrices an indispensable tool in virtually all the sciences that require numerical computations. For these reasons, we have written a text that is matrix-oriented, and so will serve the needs of a wide variety of students. The presence of so diverse an audience, as well as the importance to mathematics majors of applications, has influenced our decision to include a wide variety of significant applications. Rather than postpone these applications to the end of the text, we have made an effort to introduce them as the necessary background is developed.

Prerequisites and Length

Although there is no use of calculus until the introduction of vector spaces in Chapter 7, the material is aimed at students who have had the maturity of one year of calculus. The core topics can be comfortably covered within one semester (however, there is adequate material for a two-quarter course).

Approach

We have adopted the recommendations of the Linear Algebra Curriculum Study Group.[1] Matrix operations are defined through column operations. For example, the matrix–vector product $A\mathbf{x}$ is defined as a linear combination of the columns of A with the components of $\mathbf{x}$ as the coefficients. This approach provides a gentle introduction to such difficult topics as linear combinations and linear independence, all of which are encountered in the setting of Euclidean spaces. As advocated by the Study Group, we build on the student's experience with the Euclidean plane and 3-space by presenting many topics (especially those involving linear transformations) with a geometric emphasis. Moreover, we attempt to keep the conceptual level of the text uniform throughout the book. To do so, we introduce some important conceptual ideas (such as linear combinations and spanning sets) early in the book in the context of matrix operations and systems of linear equations. Such an approach is very natural because, for example, the consistency of the system of linear equations $A\mathbf{x} = \mathbf{b}$ depends on whether $\mathbf{b}$ is a linear combination of the columns of A, which in turn depends on whether $\mathbf{b}$ belongs to the span of the columns of A. By introducing key ideas early and revisiting them in different contexts, the text enables students to interpret these ideas in many ways throughout the entire

---

[1] David Carlson, Charles R. Johnson, David C. Lay, and A. Duane Porter. The Linear Algebra Curriculum Study Group Recommendations for the First Course in Linear Algebra. *The College Mathematics Journal,* 24 (1993): 41–46.

book, rather than being forced to contend with a large number of difficult concepts in a short period of time.

Linear transformations between Euclidean spaces are introduced in Chapter 2 and appear repeatedly thereafter. Unlike abstract vector spaces, functions have been the mainstay of previous calculus courses and so are much more familiar to students. They provide a natural means of discussing the geometric transformations, and go hand in hand with matrices. For example, students will learn that the system $A\mathbf{x} = \mathbf{0}$ has a unique solution if and only if the associated linear transformation is one-to-one, which in turn is equivalent to the columns of A being linearly independent. Finally, the problem of finding simple matrix representations of a linear transformation leads directly to the topic of eigenvalues and eigenvectors. This is a more natural approach than motivating eigenvalues and eigenvectors as the solutions to the equation $\det(A - \lambda I) = 0$ and the system $A\mathbf{x} = \lambda\mathbf{x}$, respectively.

Core Topics

In Chapters 1 and 2, students are introduced to the basic properties of matrices, vectors, and linear transformations. By the time students have finished Chapter 1, they have been working with linear combinations, spans, systems of linear equations, and linear independence. In Chapter 2, the remaining properties of matrix multiplication are given, along with matrix inverses and linear transformations. Because we use determinants mainly in the context of eigenvalues, we provide a short but complete treatment in Chapter 3. The important topics of subspaces, bases and dimension, coordinate systems, and matrix representations of linear transformations are covered in Chapter 4. Although Chapters 3 and 4 can be interchanged, we prefer to cover determinants before subspaces so that there is no delay between the discussions of change of coordinates and diagonalization in Chapter 5. Chapter 5 contains the important results about eigenvalues, eigenvectors, and diagonalization of matrices and linear operators.

With the introduction of orthogonality, Chapter 6 provides a geometrical context for the interplay between vectors, matrices, and linear transformations. The important application of least-squares lines illustrates the power of this interplay. We continue with orthogonal matrices and operators and a study of symmetric matrices through the spectral theorem. The chapter concludes with the singular value decomposition and computer graphics.

Chapter 7 introduces abstract vector spaces. Because of the careful foundation that has been developed in Euclidean spaces, most of the concepts in earlier chapters—such as span, linear independence, and subspace—are easily generalized. We focus mainly on function and matrix spaces. For example, a nice application in this context arises when Lagrange interpolating polynomials are used to find a basis for a polynomial space. Differential operators and integrals are now explored as special cases of linear transformations. Finally, inner product spaces are introduced. The Gram–Schmidt process is applied to produce the Legendre polynomials. Least-squares theory and trigonometric polynomials are used to explore periodic motion in the setting of musical notes.

Other Topics

Users will likely have their own favorite topic that extends the core content. We have made an effort to include a number of optional topics, for example, the

LU decomposition, the singular value decomposition, the spectral decomposition for symmetric matrices, Lagrange interpolating polynomials, complex numbers, block multiplication, the Moore–Penrose generalized inverse, and quadratic forms.

Exercises and Examples

Exercises are the means by which students discover to what extent they understand the material and its implications. Examples not only illustrate but motivate theoretical results and definitions. For this reason we have included roughly 2200 exercises and more than 200 examples. The exercises range from simple numerical computations to problems of a theoretical nature. Every section has true/false exercises and practice problems with their solutions, and most exercise sets contain several problems that require technology. Finally, each chapter contains review problems.

Applications

As mentioned, we include a wide variety of applications, and they are given when the necessary prerequisites have been introduced. There are applications to economics (the Leontief input–output model), electrical networks (current flow through an electrical network), population change (the Leslie matrix), Markov chains, traffic flow, scheduling (adjacency matrices), anthropology, counting problems (difference equations), predator–prey models and harmonic motion (systems of differential equations), least-squares approximation, computer graphics, and music (using trigonometric polynomials). Although it is not assumed that any particular applications will be covered, it is our opinion that the core material is greatly enhanced when it is applied to many diverse areas.

Geometry

Because of its visual and historical importance, we view geometry as a significant component of the study of linear algebra. Although we have inserted geometric ideas throughout the text (for example, projections, reflections, and rotations), there are a number of places where we have given special emphasis to geometry. Specifically, we have devoted attention to orthogonal projections, area and volume, conic sections, rigid motion in the plane, distance in the plane and 3-space, and rotation and computer graphics in 3-space.

Technology

With the Linear Algebra Curriculum Study Group, we believe that students in this course should use technology, whether it be a supercalculator or computer software such as MATLAB. Throughout the text, we make appropriate comments about the use of technology. Yet, in order to be as flexible as possible, the text does not presume that any specific software or calculators will be used.

Acknowledgments

This text has been greatly improved by the comments of students at Illinois State University who have studied from the manuscript over a period of three semesters. We also appreciate the insightful suggestions of our colleagues,

Thomas Shilgalis and Charles Vanden Eynden, and of Beverly Hartter (Eureka College), who have used the text in their classes. In addition, we would like to express our thanks for the comments of Bohumil Cenkl (Northeastern University), Jane M. Day (San Jose State University), Caren L. Diefenderfer (Hollins University), John Gimbel (University of Alaska at Fairbanks), John Douglas Moore (University of California at Santa Barbara), Kathryn Turner (Utah State University), Bryan Smith (University of Puget Sound), and Cathleen Zucco-Teveloff (State University of New York at New Paltz), who carefully reviewed the manuscript.

We appreciate as well the helpful suggestions and encouragement of George Lobell, the Mathematics Acquisition Editor at Prentice Hall, and Brittney Corrigan-McElroy, project manager at Interactive Composition Corporation.

To find the latest information about this book, consult our home page on the World Wide Web. In addition, Prentice Hall is offering, free to all adopters of the text, a student website that has a number of activites to enhance learning. A link to this website can be found on our home page.

We encourage comments, which can be sent to us by e-mail or ordinary post. Our home page and e-mail addresses are listed below.

home page: **http://www.cas.ilstu.edu/math/matrix**
e-mail: **matrix@math.ilstu.edu**

Lawrence E. Spence
Arnold J. Insel
Stephen H. Friedberg

To the Student

Linear algebra is concerned with *vectors* and *matrices,* and with special functions called *linear transformations* that are defined on vectors. Since vectors arise in a wide variety of settings, linear algebra can be applied to many disciplines. In fact, it is one of the most important tools of applied mathematics. In this book, we present applications of linear algebra to economics, physics, biology, statistics, computer graphics, and other fields.

Like most areas of mathematics, linear algebra has its own terminology and notation. To be successful in your study of linear algebra, you must be able to solve problems and communicate ideas using its language and symbolism. Developing these abilities requires more than just attending lectures or casually reading this book—it requires active involvement with the subject matter. One of the best ways to be involved with mathematics is by working exercises. The exercise sets in this book begin with true/false questions that test your understanding of important ideas in the section. Problems that practice basic computations follow. Then come exercises that ask for conjectures, explanations, or justifications. Finally, most exercise sets end with questions that use technology. These different types of exercises help you learn different aspects of linear algebra.

Not only do the exercises help you to check your understanding of important concepts, but they also provide an opportunity to practice the vocabulary and symbolism that you are learning. For this reason, regular work on exercises is essential for success.

In mathematics courses prior to linear algebra, the emphasis is often on performing calculations. In linear algebra, however, you will be expected to understand the concepts and facts that are the basis for computations. In order to learn these concepts, you must be able to use the terminology and notation of linear algebra; so begin by learning the definitions of important terms and illustrate them with examples. The key results in the text are usually found in theorems or statements enclosed in a box. Pay particular attention to these, being certain that you understand thoroughly what is being said. Then try to express the result in your own words. If you can't communicate an idea in writing, then you probably don't understand it well.

In addition, we offer four specific suggestions that will enable you to get the most from your study of linear algebra.

- **Carefully read each section *before* the classroom discussion occurs**
  Some students use a textbook only as a source of examples when working exercises. This approach does not allow them to get the full benefit from either the textbook or the classroom discussions. By reading the text before a discussion occurs, you get an overview of the material to be discussed and know what you understand and what you don't. This enables you to be a more active participant in the classroom discussion.

Moreover, you do not learn much by imitating an example from the text. In fact, this practice usually deceives you into believing that you understand the problem better than you really do. Reading the text before the classroom discussion helps you to learn the material more quickly and prepares you to do the exercises.

- **Prepare regularly for each class**

You cannot expect to learn to play the piano merely by attending lessons once a week—long and careful practice between lessons is necessary. So it is with linear algebra. At the least, you should study the material presented in class. Often, however, the current lesson builds on previous material that must be reviewed. Usually there are exercises to work that deal with the material presented in class. In addition, you should prepare for the next class by reading ahead.

Each new lesson usually introduces several important concepts or definitions that must be learned in order for subsequent sections to be understood. As a result, falling behind in your study by even a single day prevents you from understanding the new material that follows. Many of the concepts of linear algebra are deep; to understand these ideas fully requires time. It is simply not possible to absorb an entire chapter's worth of material in a single day. To be successful, you must learn the new material as it arises and not wait to study until you are less busy or until an exam is imminent.

- **Ask questions of yourself and others**

Mathematics is not a spectator sport. It can only be learned by the interaction of study and questioning. Certain natural questions arise when a new topic is introduced: What purpose does it serve? How is the new topic related to previous ones? What examples illustrate the topic? For a new theorem, one might also ask: Why is it true? How does it relate to previous theorems? Why is it useful? Be sure you don't accept something as true unless you believe it. If you are not convinced that a statement is correct, you should ask for further details.

- **Review often**

As you attempt to understand new material, you may become aware that you have forgotten some previous material or that you haven't understood it well enough. By relearning such material, you not only gain a deeper understanding of previous topics, but you also enable yourself to learn new ideas more quickly and more deeply. When a new concept is introduced, search for related concepts or results and write them on paper. This enables you to see more easily the connections between new ideas and previous ones. Moreover, expressing ideas in writing helps you to learn because you must think carefully about the material as you write. A good test of your understanding of a section of the textbook is to ask yourself if, with the book closed, you can explain in writing what the section is about. If not, you will benefit from reading the section again.

We hope that your study of linear algebra is successful and that you take from the subject concepts and techniques that are useful in future courses and in your career.

CHAPTER 1

Matrices, Vectors, and Systems of Linear Equations

The most common use of linear algebra is to solve systems of linear equations, which arise in applications to such diverse disciplines as physics, biology, economics, engineering, and sociology. In this chapter, we describe the most efficient algorithm for solving systems of linear equations, *Gaussian elimination.* This procedure, or some variation of it, is used by most commercial software (such as MATLAB).

Systems of linear equations can be compactly written using arrays called *matrices* and *vectors.* More importantly, the arithmetic properties of these arrays allow us to compute solutions to such systems or to determine if no solutions exist. For this reason, the study of matrices and vectors is the principal emphasis in this text. This chapter begins by developing their basic properties. In Sections 1.6 and 1.7, we begin the study of two other important concepts, spanning sets and linear independence, which provide information about the existence and uniqueness of solutions to systems of linear equations.

1.1 Matrices and Vectors

Many situations occur where data are stored in two-dimensional arrays of numerical information. Examples arise in areas as diverse as sociology, business, communication theory, physics, and statistics; the reader may have seen such arrays in the use of spreadsheets.

Suppose that a company owns two bookstores, each of which sells newspapers, magazines, and (paperback) books. Assume that the sales (in hundreds of dollars) of the two bookstores for the months of July and August can be represented by the following arrays.

July				August		
Store	1	2		Store	1	2
Newspapers	6	8		Newspapers	7	9
Magazines	15	20	and	Magazines	18	31
Books	45	64		Books	52	68

For example, store 1 sold $1500 worth of magazines and $4500 worth of books during July. It is more convenient notationally to represent the information on July sales as

$$\begin{bmatrix} 6 & 8 \\ 15 & 20 \\ 45 & 64 \end{bmatrix}.$$

Such a rectangular array of real numbers is called a *matrix*.[1] It is customary to refer to real numbers as **scalars** (originally from the word *scale*) when working with a matrix. We denote the set of real numbers by $\mathcal{R}$.

Definitions.   A **matrix** (plural, matrices) is a rectangular array of scalars. If a matrix has m rows and n columns, we say that the **size** of the matrix is $\boldsymbol{m}$ **by** $\boldsymbol{n}$, written $m \times n$. The matrix is **square** if $m = n$. The scalar in the ith row and jth column is called the $(\boldsymbol{i}, \boldsymbol{j})$-**entry** of the matrix.

We use a capital letter to represent a matrix, and the corresponding lower case letter with subscript ij to denote the (i, j)-entry of the matrix. So if A is the name of a matrix, we denote its (i, j)-entry by a_{ij}. We say that two matrices A and B are **equal** if they have the same size and have equal corresponding entries, that is, $a_{ij} = b_{ij}$ for all i and j. Symbolically, we write $A = B$.

In our bookstore example, we may let

$$B = \begin{bmatrix} 6 & 8 \\ 15 & 20 \\ 45 & 64 \end{bmatrix} \quad \text{and} \quad C = \begin{bmatrix} 7 & 9 \\ 18 & 31 \\ 52 & 68 \end{bmatrix}.$$

Note that $b_{12} = 8$ and $c_{12} = 9$, so $B \neq C$. Both B and C are 3×2 matrices. Because of the context in which these matrices arise, they are called *inventory matrices*.

Other examples of matrices are

$$\begin{bmatrix} \frac{2}{3} & -4 & 0 \\ \pi & 1 & 6 \end{bmatrix}, \qquad \begin{bmatrix} 3 \\ 8 \\ 4 \end{bmatrix}, \qquad \text{and} \qquad \begin{bmatrix} -2 & 0 & 1 & 1 \end{bmatrix}.$$

The first matrix has size 2×3, the second has size 3×1, and the third has size 1×4.

Sometimes we are only interested in a subset of the information contained in a matrix. For example, suppose that we are only interested in magazine and book sales in July. Then the relevant information is contained in the last two rows of B; that is, in the matrix E defined by

$$E = \begin{bmatrix} 15 & 20 \\ 45 & 64 \end{bmatrix}.$$

In general, a **submatrix** of a matrix M is obtained by deleting from M entire rows, entire columns, or both. It is permissible, when forming a submatrix of M, to delete none of the rows or none of the columns of M. So E is a submatrix of B. As another example, if we delete the first row and the

---

[1] James Joseph Sylvester (1814–1897) coined the term *matrix* in the 1850s.

second column of B, we obtain the submatrix

$$\begin{bmatrix} 15 \\ 45 \end{bmatrix}.$$

Notice that we could have recorded the information about July sales in the form

Store	Newspapers	Magazines	Books
1	6	15	45
2	8	20	64

This representation produces the matrix

$$\begin{bmatrix} 6 & 15 & 45 \\ 8 & 20 & 64 \end{bmatrix}.$$

Compare this to

$$B = \begin{bmatrix} 6 & 8 \\ 15 & 20 \\ 45 & 64 \end{bmatrix}.$$

The rows of the first matrix are the columns of B, and the columns of the first matrix are the rows of B. This new matrix is called the *transpose* of B.

Definition. The **transpose** of an $m \times n$ matrix A is the $n \times m$ matrix A^T whose (i, j)-entry is the (j, i)-entry of A.

The matrix C in our bookstore example and its transpose are

$$C = \begin{bmatrix} 7 & 9 \\ 18 & 31 \\ 52 & 68 \end{bmatrix} \quad \text{and} \quad C^T = \begin{bmatrix} 7 & 18 & 52 \\ 9 & 31 & 68 \end{bmatrix}.$$

Matrices are more than convenient devices for storing information. Their usefulness lies in their *arithmetic*.

Returning to the bookstore example, suppose that we want to know the total numbers of newspapers, magazines, and books sold by both stores over the two-month period. It is natural to form one matrix whose entries are the sum of the corresponding entries of the matrices B and C, namely,

Store	1	2
Newspapers	13	17
Magazines	33	51
Books	97	132

This computation motivates the operation of *addition* for matrices.

Definition. Let A and B be $m \times n$ matrices. We define the **sum** of A and B, denoted $A + B$, to be the $m \times n$ matrix obtained by adding the corresponding entries of A and B; that is, the $m \times n$ matrix whose (i, j)-entry is $a_{ij} + b_{ij}$.

Notice that the matrices A and B must have the same size for their sum to be defined.

Suppose that in our bookstore example we focus on the July sales. If we are interested in the situation where the sales double in all categories, then the matrix of interest is

$$\begin{bmatrix} 12 & 16 \\ 30 & 40 \\ 90 & 128 \end{bmatrix}.$$

A reasonable notation for this matrix is $2B$.

Definition. Let A be an $m \times n$ matrix and c be a scalar. The **scalar multiple** cA of the matrix A is defined to be the $m \times n$ matrix whose (i, j)-entry is ca_{ij}. We define the matrix $-A$ to be $(-1)A$.

Example 1   Compute the matrices $A + B$, $3A$, $-A$, and $3A + 4B$, where

$$A = \begin{bmatrix} 3 & 4 & 2 \\ 2 & -3 & 0 \end{bmatrix} \quad \text{and} \quad B = \begin{bmatrix} -4 & 1 & 0 \\ 5 & -6 & 1 \end{bmatrix}.$$

Solution   We have

$$A + B = \begin{bmatrix} -1 & 5 & 2 \\ 7 & -9 & 1 \end{bmatrix}, \quad 3A = \begin{bmatrix} 9 & 12 & 6 \\ 6 & -9 & 0 \end{bmatrix}, \quad -A = \begin{bmatrix} -3 & -4 & -2 \\ -2 & 3 & 0 \end{bmatrix},$$

and

$$3A + 4B = \begin{bmatrix} 9 & 12 & 6 \\ 6 & -9 & 0 \end{bmatrix} + \begin{bmatrix} -16 & 4 & 0 \\ 20 & -24 & 4 \end{bmatrix} = \begin{bmatrix} -7 & 16 & 6 \\ 26 & -33 & 4 \end{bmatrix}.$$

Just as we have defined addition of matrices, we can also define **subtraction**. For any matrices A and B of the same size, we define $A - B$ to be the matrix $A + (-B)$. Thus the entries of $A - B$ are obtained by subtracting each entry of B from the corresponding entry of A. The $m \times n$ **zero matrix** is the $m \times n$ matrix all of whose entries are zero. We denote the $m \times n$ zero matrix by O. Notice that $A - A = O$ for all matrices A.

If, as in Example 1, we have

$$A = \begin{bmatrix} 3 & 4 & 2 \\ 2 & -3 & 0 \end{bmatrix}, \quad B = \begin{bmatrix} -4 & 1 & 0 \\ 5 & -6 & 1 \end{bmatrix}, \quad \text{and} \quad O = \begin{bmatrix} 0 & 0 & 0 \\ 0 & 0 & 0 \end{bmatrix},$$

then

$$-B = \begin{bmatrix} 4 & -1 & 0 \\ -5 & 6 & -1 \end{bmatrix}, \quad A - B = \begin{bmatrix} 7 & 3 & 2 \\ -3 & 3 & -1 \end{bmatrix},$$

and

$$A - O = \begin{bmatrix} 3 & 4 & 2 \\ 2 & -3 & 0 \end{bmatrix}.$$

We have now defined three operations with matrices: transposition (forming the transpose), addition, and scalar multiplication. The power of linear algebra lies in the natural relations between these operations, which are described in our first theorem.

Theorem 1.1 (Properties of matrix addition and scalar multiplication)

Let A, B, and C be $m \times n$ matrices and let s and t be any scalars. Then

(a) $A + B = B + A$ (commutative law of matrix addition)

(b) $(A + B) + C = A + (B + C)$ (associative law of matrix addition)

(c) $A + O = A$

(d) $A + (-A) = O$

(e) $1A = A$

(f) $0A = O$

(g) $(st)A = s(tA)$

(h) $s(A + B) = sA + sB$

(i) $(s + t)A = sA + tA$

(j) $(A + B)^T = A^T + B^T$

(k) $(sA)^T = sA^T$

(l) $(A^T)^T = A$

Proof We prove parts (b), (h), and (j). The rest are left as exercises.

(b) Clearly the matrices on each side of the equation are $m \times n$ matrices. We must show that each entry of $(A + B) + C$ is the same as the corresponding entry of $A + (B + C)$. Consider the (i, j)-entries. Because of the definition of matrix addition, the (i, j)-entry of $(A + B) + C$ is the sum of the (i, j)-entry of $A + B$, which is $a_{ij} + b_{ij}$, and the (i, j)-entry of C, which is c_{ij}. But this sum equals $(a_{ij} + b_{ij}) + c_{ij}$. Because the associative law holds for scalar addition, $(a_{ij} + b_{ij}) + c_{ij} = a_{ij} + (b_{ij} + c_{ij})$, which is the (i, j)-entry of $A + (B + C)$.

(h) Clearly the matrices on each side of the equation are $m \times n$ matrices. As in the proof of (b), we consider the (i, j)-entries of each side. The (i, j)-entry of $s(A + B)$ is defined to be the product of s and the (i, j)-entry of $A + B$, which is $a_{ij} + b_{ij}$. This product equals $s(a_{ij} + b_{ij})$. The (i, j)-entry of $sA + sB$ is the sum of the (i, j)-entry of sA, which is sa_{ij}, and the (i, j)-entry of sB, which is sb_{ij}. This sum is $sa_{ij} + sb_{ij}$. So (h) is proved.

(j) Clearly the matrices on each side of the equation are $n \times m$ matrices. So we show that the (i, j)-entry of $(A + B)^T$ equals the (i, j)-entry of $A^T + B^T$. By the definition of transpose, the (i, j)-entry of $(A + B)^T$ equals the (j, i)-entry of $A + B$, which is $a_{ji} + b_{ji}$. On the other hand, the (i, j)-entry of $A^T + B^T$ equals the sum of the (i, j)-entry of A^T and the (i, j)-entry of B^T, which is $a_{ji} + b_{ji}$.   ❑

Because of the associative law of matrix addition, sums of three or more matrices can be written unambiguously without parentheses. For example, we may write $A + B + C$ instead of either $(A + B) + C$ or $A + (B + C)$.

Vectors

In many situations, we are concerned with matrices that have exactly one column. Such matrices are called *vectors*.

Definitions. A **vector** is a matrix having a single column or a single row, that is, either an $n \times 1$ matrix or a $1 \times n$ matrix. The entries of the vector are called **components**. The set of all vectors with n components is denoted by $\mathcal{R}^n$.

Example 2

Consider the experiment of tossing a fair coin twice and recording the results: no heads, one head, or two heads. It can be shown that the probabilities of these events occurring are: $\frac{1}{4}$, $\frac{1}{2}$, and $\frac{1}{4}$, respectively. These probabilities can be recorded in a matrix $\begin{bmatrix} \frac{1}{4} \\ \frac{1}{2} \\ \frac{1}{4} \end{bmatrix}$, called a *probability vector* because its components are nonnegative and have a sum of 1. If we toss the coin n times, we arrive at the probability vector $\begin{bmatrix} p_0 \\ p_1 \\ \vdots \\ p_n \end{bmatrix}$, where p_i denotes the probability of obtaining exactly i heads in n tosses.

In this book, vectors are generally denoted by boldface lower case letters such as $\mathbf{u}$ and $\mathbf{v}$, and are usually written as $n \times 1$ matrices. The ith component of the vector $\mathbf{u}$ is denoted by u_i. So $\mathbf{u} = \begin{bmatrix} u_1 \\ u_2 \\ \vdots \\ u_n \end{bmatrix}$.

Because vectors are special cases of matrices, we may define the same three operations for vectors. In this case, we call the two arithmetic operations on vectors **vector addition** and **scalar multiplication**. The three vector operations satisfy the properties listed in Theorem 1.1. In particular, the vector in $\mathcal{R}^n$ with all zero components is denoted by $\mathbf{0}$ and is called the **zero vector**, and hence $\mathbf{u} + \mathbf{0} = \mathbf{u}$ and $0\mathbf{u} = \mathbf{0}$ for every $\mathbf{u}$ in $\mathcal{R}^n$.

On occasion, we identify a vector $\mathbf{u}$ in $\mathcal{R}^n$ with an ordered **n-tuple**, written $(u_1, u_2, \ldots, u_n)$. This is especially the case for $n = 2$ and $n = 3$, when we identify vectors with points in the xy-plane or in xyz-space, respectively.

In the special case that $n = 1$, we identify $\mathcal{R}^1$ with the set $\mathcal{R}$ of all scalars.

Example 3

Let $\mathbf{u} = \begin{bmatrix} 2 \\ -4 \\ 7 \end{bmatrix}$ and $\mathbf{v} = \begin{bmatrix} 5 \\ 3 \\ 0 \end{bmatrix}$. Then $u_2 = -4$ and $v_1 = 5$. Also

$$\mathbf{u} + \mathbf{v} = \begin{bmatrix} 7 \\ -1 \\ 7 \end{bmatrix}, \qquad \mathbf{u} - \mathbf{v} = \begin{bmatrix} -3 \\ -7 \\ 7 \end{bmatrix}, \qquad \text{and} \qquad 5\mathbf{v} = \begin{bmatrix} 25 \\ 15 \\ 0 \end{bmatrix}.$$

For a given matrix, it is often advantageous to consider its rows and columns as vectors. For example, if $A = \begin{bmatrix} 2 & 4 & 3 \\ 0 & 1 & -2 \end{bmatrix}$, then its **rows** are $\begin{bmatrix} 2 & 4 & 3 \end{bmatrix}$ and $\begin{bmatrix} 0 & 1 & -2 \end{bmatrix}$. Its **columns** are $\begin{bmatrix} 2 \\ 0 \end{bmatrix}$, $\begin{bmatrix} 4 \\ 1 \end{bmatrix}$, and $\begin{bmatrix} 3 \\ -2 \end{bmatrix}$.

Because the columns of a matrix play a more important role than the rows, we introduce a special notation. When a capital letter denotes a matrix, we use the corresponding lower case letter in boldface with a subscript j to represent the jth column of that matrix. So if C is an $m \times n$ matrix, its jth column is

$$\mathbf{c}_j = \begin{bmatrix} c_{1j} \\ c_{2j} \\ \vdots \\ c_{mj} \end{bmatrix}.$$

The importance of vectors in physics was recognized late in the nineteenth century.[2] In what follows, we concern ourselves briefly with vector geometry in the plane and in three-dimensional space. We return to additional geometric concepts later.

Vectors can be used to study geometry in $\mathcal{R}^2$. In this context, it is common to identify an ordered pair (a, b) in the xy-plane with the vector $\begin{bmatrix} a \\ b \end{bmatrix}$ in $\mathcal{R}^2$.

The geometry of vectors is realized by depicting a vector $\mathbf{v} = \begin{bmatrix} a \\ b \end{bmatrix}$ as an arrow emanating from the origin with the point (a, b) as its endpoint (see Figure 1.1). The power of the arithmetic follows, in part, from the geometric interpretation of the operations of addition and multiplication by scalars.

The addition of two vectors can be interpreted geometrically as a result about parallelograms called the *parallelogram law*. For nonzero vectors $\mathbf{u}$ and $\mathbf{v}$, form a parallelogram with adjacent sides $\mathbf{u}$ and $\mathbf{v}$. The sum $\mathbf{u} + \mathbf{v}$ is the arrow represented by the diagonal as directed in Figure 1.2. A justification of the parallelogram law by Heron of Alexandria (first century CE) appears in his *Mechanics*. Another justification is based on congruent triangles, as indicated in Figure 1.2.

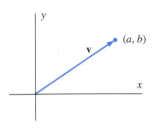

FIGURE 1.1

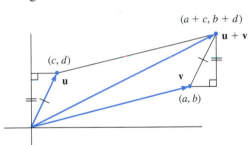

FIGURE 1.2

The velocity of an object is determined by its speed and direction. We may represent the velocity of a moving object by an arrow that points in the direction of the object's motion and whose length represents the speed of the object. Thus we can represent velocities as vectors.

Forces or velocities can be combined by adding vectors that represent them, according to the rules previously discussed. For example, consider a boat that cruises at 20 mph in still water. Now suppose that the boat is cruising on a river so that its bow (front) is pointing to the northeast, while the river is flowing toward the east at 7 mph. Then the actual velocity of the boat with respect to the shore results from adding the velocity of the river, which is carrying the boat eastward, to the velocity of the boat relative to the river. We can represent the velocity of the boat (with respect to the river) and the velocity of the river as vectors $\mathbf{u}$ and $\mathbf{v}$, respectively, in the xy-plane, where the positive direction

[2]The algebra of vectors, developed by Oliver Heaviside (1850–1925) and Josiah Willard Gibbs (1839–1903), won out over the algebra of quaternions to become the language of physicists.

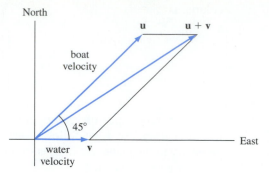

FIGURE 1.3

of the x-axis points east and the positive direction of the y-axis points north (see Figure 1.3).

The vector **u** can be represented as an arrow of length 20 making an angle of 45° with the positive direction of the x-axis. Therefore, if $\mathbf{u} = \begin{bmatrix} u_1 \\ u_2 \end{bmatrix}$, then

$$u_1 = 20\cos 45° = 10\sqrt{2} \quad \text{and} \quad u_2 = 20\sin 45° = 10\sqrt{2}.$$

Thus $\mathbf{u} = \begin{bmatrix} 10\sqrt{2} \\ 10\sqrt{2} \end{bmatrix}$. Similarly, it is easy to verify that $\mathbf{v} = \begin{bmatrix} 7 \\ 0 \end{bmatrix}$. It follows that the boat's velocity with respect to the shore, as shown in Figure 1.3, is given by the vector

$$\mathbf{u} + \mathbf{v} = \begin{bmatrix} 10\sqrt{2} + 7 \\ 10\sqrt{2} \end{bmatrix}.$$

To find the speed of the boat, we appeal to the Pythagorean theorem, which allows us to conclude that the length of a vector with endpoint (p, q) is $\sqrt{p^2 + q^2}$. Using the fact that the components of $\mathbf{u} + \mathbf{v}$ are $p = 10\sqrt{2} + 7$ and $q = 10\sqrt{2}$, respectively, it follows that the speed of the sailboat is

$$\sqrt{p^2 + q^2} \approx 25.44 \text{ mph.}$$

The motivation for defining $c \begin{bmatrix} a \\ b \end{bmatrix} = \begin{bmatrix} ca \\ cb \end{bmatrix}$ can be seen using similar triangles. See Figure 1.4(a) for the case that the scalar is a positive number. Figure 1.4(b) depicts the relative positions of a vector **v** and a negative multiple of **v**, in this case $-2\mathbf{v}$. From Figure 1.4 it makes sense to define two vectors as **parallel** if one of them is a scalar multiple of the other. It is easy to see that the vector $c\mathbf{v}$ has length equal to $|c|$ times the length of **v**.

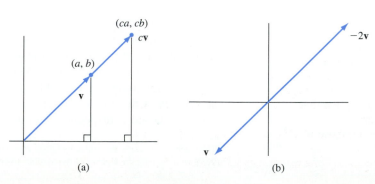

FIGURE 1.4
(a)   (b)

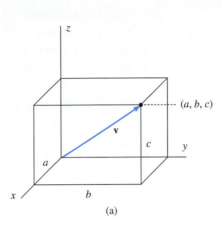

FIGURE 1.5

If we identify $\mathcal{R}^3$ as the set of all ordered triples, then the same geometric ideas that hold in $\mathcal{R}^2$ are also true in $\mathcal{R}^3$. We may depict a vector $\mathbf{v} = \begin{bmatrix} a \\ b \\ c \end{bmatrix}$ in $\mathcal{R}^3$ as an arrow emanating from the origin of the xyz-coordinate system with the point (a, b, c) as its endpoint (see Figure 1.5(a)). As is the case in $\mathcal{R}^2$, two nonzero vectors in $\mathcal{R}^3$ can be viewed as adjacent sides of a parallelogram, and their addition can be interpreted using the parallelogram law (see Figure 1.5(b)). In real life, motion takes place in 3-dimensional space, and quantities such as velocities and forces can be depicted as vectors in $\mathcal{R}^3$.

Practice Problems

1. Let $A = \begin{bmatrix} 4 & 2 \\ 1 & 3 \end{bmatrix}$.

(a) What is the $(1, 2)$-entry of A?
(b) What is the $(2, 2)$-entry of A?

2. Let $A = \begin{bmatrix} 4 & 2 \\ 1 & 3 \end{bmatrix}$ and $B = \begin{bmatrix} 1 & -1 \\ 2 & 3 \end{bmatrix}$. Compute the following matrices.

(a) $A + B$

(b) $2A$
(c) $A + 3B$

3. Let $A = \begin{bmatrix} 2 & -1 & 1 \\ 3 & 0 & -2 \end{bmatrix}$ and $B = \begin{bmatrix} 1 & 3 & 0 \\ 2 & -1 & 4 \end{bmatrix}$.
Compute the following matrices.

(a) A^T
(b) $(3B)^T$
(c) $(A + B)^T$

Exercises

1. Determine if the following statements are true or false.

(a) Matrices must be of the same size for their sum to be defined.
(b) The transpose of a sum of two matrices is the sum of the transposed matrices.
(c) Every vector is a matrix.
(d) A scalar multiple of the zero matrix is the zero scalar.
(e) The transpose of a matrix is a matrix of the same size.
(f) A submatrix of a matrix may be a vector.

(g) If B is a 3×4 matrix, then its rows are 4×1 vectors.

In Exercises 2–8, compute the indicated matrices, where

$$A = \begin{bmatrix} 2 & -1 & 5 \\ 3 & 4 & 1 \end{bmatrix} \quad \text{and} \quad B = \begin{bmatrix} 1 & 0 & -2 \\ 2 & 3 & 4 \end{bmatrix}.$$

2. $-A$ **3.** $4A$ **4.** $3A + 2B$
5. $4A - 2B$ **6.** $A^T + 2B^T$ **7.** $(2B)^T$
8. $(A + 2B)^T$

In Exercises 9–12, assume that $A = \begin{bmatrix} 3 & -2 \\ 0 & 1.6 \\ 2\pi & 5 \end{bmatrix}$.

9. Determine a_{12}.

10. Determine a_{21}.

11. Determine $\mathbf{a}_1$.

12. Determine $\mathbf{a}_2$.

In Exercises 13–16, assume $C = \begin{bmatrix} 2 & -3 & 0.4 \\ 2e & 12 & 0 \end{bmatrix}$.

13. Determine $\mathbf{c}_1$.

14. Determine $\mathbf{c}_3$.

15. Determine the first row of C.

16. Determine the second row of C.

17. An airplane is flying with a ground speed of 300 mph at an angle of 30° east of due north (see Figure 1.6). In addition, the airplane is climbing at a rate of 10 mph. Determine the vector in $\mathcal{R}^3$ that represents the velocity of the airplane.

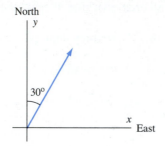

FIGURE 1.6

18. A swimmer is swimming northeast at 2 mph in still water.

(a) Give the velocity of the swimmer. Include a sketch.

(b) A current in a northerly direction at 1 mph affects the velocity of the swimmer. Give the new velocity and speed of the swimmer. Include a sketch.

19. A pilot keeps her airplane pointed in a northeasterly direction while maintaining an airspeed (speed relative to the surrounding air) of 300 mph. A wind from the west blows eastward at 50 mph.

(a) Find the velocity vector of the airplane relative to the ground.

(b) What is the speed of the airplane relative to the ground?

20. Suppose that in a medical study of 20 people, for each i, $1 \le i \le 20$, the 3×1 vector $\mathbf{u}_i$ is defined so that its components respectively represent the blood pressure, pulse rate, and cholesterol reading of the ith person. Provide an interpretation of the vector $\frac{1}{20}(\mathbf{u}_1 + \mathbf{u}_2 + \cdots + \mathbf{u}_{20})$.

21. Let A and B be matrices of the same size.

(a) Prove that the jth column of $A + B$ is $\mathbf{a}_j + \mathbf{b}_j$.

(b) Prove that for any scalar c, the jth column of cA is $c\mathbf{a}_j$.

22. Prove (a) of Theorem 1.1.

23. Prove (c) of Theorem 1.1.

24. Prove (d) of Theorem 1.1.

25. Prove (e) of Theorem 1.1.

26. Prove (f) of Theorem 1.1.

27. Prove (g) of Theorem 1.1.

28. Prove (i) of Theorem 1.1.

29. Prove (k) of Theorem 1.1.

30. Prove (l) of Theorem 1.1.

A square matrix A is called a **diagonal matrix** if $a_{ij} = 0$ whenever $i \ne j$. Exercises 31–34 are concerned with diagonal matrices.

31. Prove that a square zero matrix is a diagonal matrix.

32. Prove that if B is a diagonal matrix, then cB is a diagonal matrix for any scalar c.

33. Prove that if B is a diagonal matrix, then B^T is a diagonal matrix.

34. Prove that if B and C are diagonal matrices of the same size, then $B + C$ is a diagonal matrix.

A (square) matrix A is said to be **symmetric** if $A = A^T$. Exercises 35–42 are concerned with symmetric matrices.

35. Give examples of 2×2 and 3×3 symmetric matrices.

36. Prove that the (i, j)-entry of a symmetric matrix equals the (j, i)-entry.

37. Prove that a square zero matrix is symmetric.

38. Prove that if B is a symmetric matrix, then so is cB for any scalar c.

39. Prove that if B is a square matrix, then $B + B^T$ is symmetric.

40. Prove that if B and C are $n \times n$ symmetric matrices, then so is $B + C$.

41. Is a square submatrix of a symmetric matrix necessarily a symmetric matrix? Justify your answer.

42. Prove that a diagonal matrix is symmetric.

A (square) matrix A is called **skew-symmetric** if $A^T = -A$. Exercises 43–45 deal with skew-symmetric matrices.

43. What must be true about the (i, i)-entries of a skew-symmetric matrix? Justify your answer.

44. Give an example of a nonzero 2×2 skew-symmetric matrix B. Now show that every 2×2 skew-symmetric matrix is a scalar multiple of B.

45. Show that every 3×3 matrix can be written as the sum of a symmetric matrix and a skew-symmetric matrix.

46.[3] The **trace** of an $n \times n$ matrix A, written trace A, is defined as the sum

$$\text{trace } A = a_{11} + a_{22} + \cdots + a_{nn}.$$

---

[3]The result of this exercise are used in Sections 2.2, 7.1, and 7.4 (on pages 106, 403, and 431, respectively).

Prove that for any $n \times n$ matrices A and B and scalar c, the following are true.

(a) trace $(A + B) =$ trace $A +$ trace B
(b) trace $cA = c \cdot$ trace A
(c) trace $A^T =$ trace A

47. Suppose that $\mathbf{p}$ and $\mathbf{q}$ are probability vectors (see Example 2); that is, the components of each of the vectors are nonnegative and have a sum of 1. Assume that $\mathbf{p}$ and $\mathbf{q}$ have the same size. Show that $a\mathbf{p} + b\mathbf{q}$ is also a probability vector if a and b are nonnegative scalars with $a + b = 1$.

In the following exercise, use a calculator with matrix capabilities or computer software such as MATLAB *to solve the problem.*

48. Consider the matrices

$$A = \begin{bmatrix} 1.3 & 2.1 & -3.3 & 6.0 \\ 5.2 & 2.3 & -1.1 & 3.4 \\ 3.2 & -2.6 & 1.1 & -4.0 \\ 0.8 & -1.3 & -12.1 & 5.7 \\ -1.4 & 3.2 & 0.7 & 4.4 \end{bmatrix} \quad \text{and}$$

$$B = \begin{bmatrix} 2.6 & -1.3 & 0.7 & -4.4 \\ 2.2 & -2.6 & 1.3 & -3.2 \\ 7.1 & 1.5 & -8.3 & 4.6 \\ -0.9 & -1.2 & 2.4 & 5.9 \\ 3.3 & -0.9 & 1.4 & 6.2 \end{bmatrix}.$$

(a) Compute $A + 2B$.
(b) Compute $A - B$.
(c) Compute $A^T + B^T$.

Solutions to the Practice Problems

1. (a) The $(1, 2)$-entry of A is 2.
   (b) The $(2, 2)$-entry of A is 3.

2. (a) $A + B = \begin{bmatrix} 4 & 2 \\ 1 & 3 \end{bmatrix} + \begin{bmatrix} 1 & -1 \\ 2 & 3 \end{bmatrix} = \begin{bmatrix} 5 & 1 \\ 3 & 6 \end{bmatrix}$

   (b) $2\begin{bmatrix} 4 & 2 \\ 1 & 3 \end{bmatrix} = \begin{bmatrix} 8 & 4 \\ 2 & 6 \end{bmatrix}$

   (c) $A + 3B = \begin{bmatrix} 4 & 2 \\ 1 & 3 \end{bmatrix} + 3\begin{bmatrix} 1 & -1 \\ 2 & 3 \end{bmatrix}$

   $= \begin{bmatrix} 4 & 2 \\ 1 & 3 \end{bmatrix} + \begin{bmatrix} 3 & -3 \\ 6 & 9 \end{bmatrix} = \begin{bmatrix} 7 & -1 \\ 7 & 12 \end{bmatrix}$

3. (a) $A^T = \begin{bmatrix} 2 & 3 \\ -1 & 0 \\ 1 & -2 \end{bmatrix}$

   (b) $(3B)^T = \begin{bmatrix} 3 & 9 & 0 \\ 6 & -3 & 12 \end{bmatrix}^T = \begin{bmatrix} 3 & 6 \\ 9 & -3 \\ 0 & 12 \end{bmatrix}$

   (c) $(A + B)^T = \begin{bmatrix} 3 & 2 & 1 \\ 5 & -1 & 2 \end{bmatrix}^T = \begin{bmatrix} 3 & 5 \\ 2 & -1 \\ 1 & 2 \end{bmatrix}$

1.2 Linear Combinations, Matrix–Vector Products, and Special Matrices

In this section, we explore some applications involving matrix operations, and introduce the product of a matrix and a vector.

Suppose, during a linear algebra course with 20 students, two tests, a quiz, and a final exam are given. Let $\mathbf{u} = \begin{bmatrix} u_1 \\ u_2 \\ \vdots \\ u_{20} \end{bmatrix}$, where u_i denotes the score of the ith student on the first test. Likewise, let $\mathbf{v}$, $\mathbf{w}$, and $\mathbf{z}$ be similarly defined for the second test, quiz, and final exam, respectively. Assume that the instructor has decided that a student's course average is computed by counting each test score twice that of a quiz score, and that the final exam score counts three times as much as a test score. Thus the *weights* for the tests, quiz, and final exam score are, respectively, 2/11, 2/11, 1/11, 6/11 (the weights must sum to one). Now consider the vector

$$\mathbf{y} = \frac{2}{11}\mathbf{u} + \frac{2}{11}\mathbf{v} + \frac{1}{11}\mathbf{w} + \frac{6}{11}\mathbf{z}.$$

The first component y_1 represents the first student's course average, the second component y_2 represents the second student's course average, and so on. Notice that $\mathbf{y}$ is a sum of scalar multiples of $\mathbf{u}$, $\mathbf{v}$, $\mathbf{w}$, and $\mathbf{z}$. This form of vector sum is so important that it merits its own definition.

Definitions.   A **linear combination** of vectors is a sum of scalar multiples of the vectors; that is, a linear combination of vectors $\mathbf{u}_1, \mathbf{u}_2, \ldots, \mathbf{u}_k$ is a vector of the form

$$c_1\mathbf{u}_1 + c_2\mathbf{u}_2 + \cdots + c_k\mathbf{u}_k,$$

where $c_1, c_2, \ldots, c_k$ are scalars. These scalars are called the **coefficients** of the linear combination.

Note that a linear combination of one vector is simply a scalar multiple of that vector.

In the previous example, the vector $\mathbf{y}$ of the students' course averages is a linear combination of the vectors $\mathbf{u}$, $\mathbf{v}$, $\mathbf{w}$, and $\mathbf{z}$. The coefficients are the weights. Indeed, any weighted average produces a linear combination of the scores.

Notice that

$$\begin{bmatrix} 2 \\ 8 \end{bmatrix} = -3\begin{bmatrix} 1 \\ 1 \end{bmatrix} + 4\begin{bmatrix} 1 \\ 3 \end{bmatrix} + 1\begin{bmatrix} 1 \\ -1 \end{bmatrix}.$$

Thus $\begin{bmatrix} 2 \\ 8 \end{bmatrix}$ is a linear combination of $\begin{bmatrix} 1 \\ 1 \end{bmatrix}$, $\begin{bmatrix} 1 \\ 3 \end{bmatrix}$, and $\begin{bmatrix} 1 \\ -1 \end{bmatrix}$, with coefficients -3, 4, and 1. We can also write

$$\begin{bmatrix} 2 \\ 8 \end{bmatrix} = \begin{bmatrix} 1 \\ 1 \end{bmatrix} + 2\begin{bmatrix} 1 \\ 3 \end{bmatrix} - 1\begin{bmatrix} 1 \\ -1 \end{bmatrix}.$$

This equation also expresses $\begin{bmatrix} 2 \\ 8 \end{bmatrix}$ as a linear combination of $\begin{bmatrix} 1 \\ 1 \end{bmatrix}$, $\begin{bmatrix} 1 \\ 3 \end{bmatrix}$, and $\begin{bmatrix} 1 \\ -1 \end{bmatrix}$, but now the coefficients are 1, 2, and -1. So the set of coefficients that express one vector as a linear combination of the others need not be unique.

Example 1

(a) Determine if $\begin{bmatrix} 4 \\ -1 \end{bmatrix}$ is a linear combination of $\begin{bmatrix} 2 \\ 3 \end{bmatrix}$ and $\begin{bmatrix} 3 \\ 1 \end{bmatrix}$.

(b) Determine if $\begin{bmatrix} -4 \\ -2 \end{bmatrix}$ is a linear combination of $\begin{bmatrix} 6 \\ 3 \end{bmatrix}$ and $\begin{bmatrix} 2 \\ 1 \end{bmatrix}$.

(c) Determine if $\begin{bmatrix} 3 \\ 4 \end{bmatrix}$ is a linear combination of $\begin{bmatrix} 3 \\ 2 \end{bmatrix}$ and $\begin{bmatrix} 6 \\ 4 \end{bmatrix}$.

Solution   (a) We seek scalars x_1 and x_2 such that

$$\begin{bmatrix} 4 \\ -1 \end{bmatrix} = x_1\begin{bmatrix} 2 \\ 3 \end{bmatrix} + x_2\begin{bmatrix} 3 \\ 1 \end{bmatrix} = \begin{bmatrix} 2x_1 \\ 3x_1 \end{bmatrix} + \begin{bmatrix} 3x_2 \\ 1x_2 \end{bmatrix} = \begin{bmatrix} 2x_1 + 3x_2 \\ 3x_1 + x_2 \end{bmatrix}.$$

That is, we seek a solution to the following system of equations:

$$\begin{aligned} 2x_1 + 3x_2 &= 4 \\ 3x_1 + x_2 &= -1. \end{aligned}$$

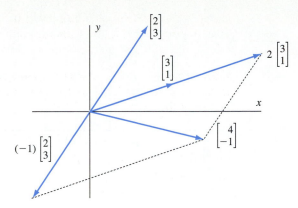

FIGURE 1.7

Because these equations represent nonparallel lines in the plane, there is exactly one solution, namely, $x_1 = -1$ and $x_2 = 2$. Therefore, $\begin{bmatrix} 4 \\ -1 \end{bmatrix}$ is a (unique) linear combination of the vectors $\begin{bmatrix} 2 \\ 3 \end{bmatrix}$ and $\begin{bmatrix} 3 \\ 1 \end{bmatrix}$, namely,

$$\begin{bmatrix} 4 \\ -1 \end{bmatrix} = -1 \begin{bmatrix} 2 \\ 3 \end{bmatrix} + 2 \begin{bmatrix} 3 \\ 1 \end{bmatrix}$$

(see Figure 1.7).

(b) To determine if $\begin{bmatrix} -4 \\ -2 \end{bmatrix}$ is a linear combination of $\begin{bmatrix} 6 \\ 3 \end{bmatrix}$ and $\begin{bmatrix} 2 \\ 1 \end{bmatrix}$, we perform a similar computation and produce the set of equations

$$6x_1 + 2x_2 = -4$$
$$3x_1 + \ x_2 = -2.$$

Since the first equation is twice the second, we need only solve $3x_1 + x_2 = -2$. This equation represents a line in the plane, and the coordinates of any point on the line give a solution. For example, we can let $x_1 = -2$ and $x_2 = 4$. In this case we have

$$\begin{bmatrix} -4 \\ -2 \end{bmatrix} = (-2) \begin{bmatrix} 6 \\ 3 \end{bmatrix} + 4 \begin{bmatrix} 2 \\ 1 \end{bmatrix}.$$

There are infinitely many solutions (see Figure 1.8).

(c) To determine if $\begin{bmatrix} 3 \\ 4 \end{bmatrix}$ is a linear combination of $\begin{bmatrix} 3 \\ 2 \end{bmatrix}$ and $\begin{bmatrix} 6 \\ 4 \end{bmatrix}$, we must solve the equations

$$3x_1 + 6x_2 = 3$$
$$2x_1 + 4x_2 = 4.$$

If we add $-\frac{2}{3}$ times the first equation to the second, we obtain $0 = 2$, an equation with no solutions. Indeed the two original equations represent parallel lines in the plane. So the original system has no solutions. We conclude that $\begin{bmatrix} 3 \\ 4 \end{bmatrix}$ is not a linear combination of $\begin{bmatrix} 3 \\ 2 \end{bmatrix}$ and $\begin{bmatrix} 6 \\ 4 \end{bmatrix}$ (see Figure 1.9).

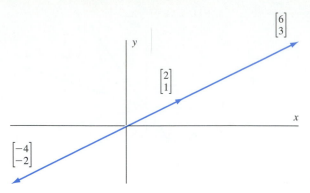

FIGURE 1.8

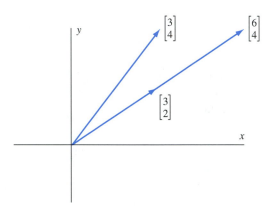

FIGURE 1.9

Example 2   Given vectors $\mathbf{u}_1$, $\mathbf{u}_2$, and $\mathbf{u}_3$, prove that the sum of any two linear combinations of these vectors is also a linear combination of these vectors.

Solution   Suppose that $\mathbf{w}$ and $\mathbf{z}$ are linear combinations of $\mathbf{u}_1$, $\mathbf{u}_2$, and $\mathbf{u}_3$. Then we may write

$$\mathbf{w} = a\mathbf{u}_1 + b\mathbf{u}_2 + c\mathbf{u}_3 \qquad \text{and} \qquad \mathbf{z} = a'\mathbf{u}_1 + b'\mathbf{u}_2 + c'\mathbf{u}_3,$$

where a, b, c, a', b', c' are scalars. Then

$$\mathbf{w} + \mathbf{z} = (a + a')\mathbf{u}_1 + (b + b')\mathbf{u}_2 + (c + c')\mathbf{u}_3,$$

which is also a linear combination of $\mathbf{u}_1$, $\mathbf{u}_2$, and $\mathbf{u}_3$.   ○

For any vector $\begin{bmatrix} a \\ b \end{bmatrix}$ in $\mathcal{R}^2$, we have that $\begin{bmatrix} a \\ b \end{bmatrix} = a \begin{bmatrix} 1 \\ 0 \end{bmatrix} + b \begin{bmatrix} 0 \\ 1 \end{bmatrix}$. That is, every vector in $\mathcal{R}^2$ is a linear combination of $\begin{bmatrix} 1 \\ 0 \end{bmatrix}$ and $\begin{bmatrix} 0 \\ 1 \end{bmatrix}$. Likewise, in $\mathcal{R}^3$ we have

$$\begin{bmatrix} a \\ b \\ c \end{bmatrix} = a \begin{bmatrix} 1 \\ 0 \\ 0 \end{bmatrix} + b \begin{bmatrix} 0 \\ 1 \\ 0 \end{bmatrix} + c \begin{bmatrix} 0 \\ 0 \\ 1 \end{bmatrix}.$$

These vectors with a single component equal to 1 and other components equal to zero are of such importance that they merit their own notation. We define the **standard vectors** of $\mathcal{R}^n$ by

$$\mathbf{e}_1 = \begin{bmatrix} 1 \\ 0 \\ \vdots \\ 0 \end{bmatrix}, \mathbf{e}_2 = \begin{bmatrix} 0 \\ 1 \\ \vdots \\ 0 \end{bmatrix}, \ldots, \mathbf{e}_n = \begin{bmatrix} 0 \\ 0 \\ \vdots \\ 1 \end{bmatrix}.$$

For example, the standard vectors in $\mathcal{R}^3$ are $\mathbf{e}_1 = \begin{bmatrix} 1 \\ 0 \\ 0 \end{bmatrix}$, $\mathbf{e}_2 = \begin{bmatrix} 0 \\ 1 \\ 0 \end{bmatrix}$, and

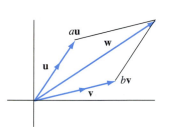

$$\mathbf{e}_3 = \begin{bmatrix} 0 \\ 0 \\ 1 \end{bmatrix}.$$

From the equations above, it is easy to see that every vector in $\mathcal{R}^n$ is a linear combination of the standard vectors of $\mathcal{R}^n$.

Now let $\mathbf{u}$ and $\mathbf{v}$ be nonparallel vectors, and let $\mathbf{w}$ be any vector in $\mathcal{R}^2$ (see Figure 1.10). Begin with the endpoint of $\mathbf{w}$ and create a parallelogram with sides $a\mathbf{u}$ and $b\mathbf{v}$, so that $\mathbf{w}$ is its diagonal. It follows that $\mathbf{w} = a\mathbf{u} + b\mathbf{v}$, that is, $\mathbf{w}$ is a linear combination of the vectors $\mathbf{u}$ and $\mathbf{v}$. More generally, the following statement is true.

FIGURE 1.10

> If $\mathbf{u}$ and $\mathbf{v}$ are any nonparallel vectors in $\mathcal{R}^2$, then every vector in $\mathcal{R}^2$ is a linear combination of $\mathbf{u}$ and $\mathbf{v}$.

Consider the following situation: A garden supply store sells three mixtures of grass seed. The deluxe mixture is 80% bluegrass and 20% rye, the standard mixture is 60% bluegrass and 40% rye, and the economy mixture is 40% bluegrass and 60% rye. One way to record this information is as the 2×3 matrix

$$B = \begin{array}{c} \\ \end{array} \begin{matrix} \text{deluxe} & \text{standard} & \text{economy} \\ \begin{bmatrix} .80 & .60 & .40 \\ .20 & .40 & .60 \end{bmatrix} & \begin{matrix} \text{bluegrass} \\ \text{rye} \end{matrix} \end{matrix}.$$

A customer wants to purchase a blend of grass seed containing 5 lb of blue grass and 3 lb of rye. There are two natural questions that arise. Is it possible to do this without any surplus for the customer; and, how much of each mixture should be blended? To find the amounts of each of the mixtures to be blended, let x_1, x_2, and x_3 denote the number of pounds of deluxe, standard, and economy mixtures, respectively. Then we have

$$.80x_1 + .60x_2 + .40x_3 = 5$$
$$.20x_1 + .40x_2 + .60x_3 = 3.$$

This is a *system of two linear equations in three unknowns.* Finding a solution to this system is equivalent to answering our second question. The technique for solving general systems of linear equations is explored in great detail in Sections 1.3 and 1.4.

Using matrix notation, we may rewrite these equations in the form

$$\begin{bmatrix} .80x_1 + .60x_2 + .40x_3 \\ .20x_1 + .40x_2 + .60x_3 \end{bmatrix} = \begin{bmatrix} 5 \\ 3 \end{bmatrix}.$$

Now we use matrix operations to rewrite this matrix equation using the columns of B as

$$x_1 \begin{bmatrix} .80 \\ .20 \end{bmatrix} + x_2 \begin{bmatrix} .60 \\ .40 \end{bmatrix} + x_3 \begin{bmatrix} .40 \\ .60 \end{bmatrix} = \begin{bmatrix} 5 \\ 3 \end{bmatrix}.$$

Thus, we can rephrase the first question as: Is $\begin{bmatrix} 5 \\ 3 \end{bmatrix}$ a linear combination of $\begin{bmatrix} .80 \\ .20 \end{bmatrix}$, $\begin{bmatrix} .60 \\ .40 \end{bmatrix}$, and $\begin{bmatrix} .40 \\ .60 \end{bmatrix}$, the columns of B? Our preceding boxed result provides an affirmative answer. Because no two of the three vectors are parallel, $\begin{bmatrix} 5 \\ 3 \end{bmatrix}$ is a linear combination of any pair of these vectors.

Now suppose we introduce the vector $\mathbf{x} = \begin{bmatrix} x_1 \\ x_2 \\ x_3 \end{bmatrix}$. We define the *matrix–vector product* $B\mathbf{x}$ to be the linear combination above; that is,

$$B\mathbf{x} = \begin{bmatrix} .80 & .60 & .40 \\ .20 & .40 & .60 \end{bmatrix} \begin{bmatrix} x_1 \\ x_2 \\ x_3 \end{bmatrix} = x_1 \begin{bmatrix} .80 \\ .20 \end{bmatrix} + x_2 \begin{bmatrix} .60 \\ .40 \end{bmatrix} + x_3 \begin{bmatrix} .40 \\ .60 \end{bmatrix}.$$

Consequently, one of our questions can be rephrased as: Does the vector $\begin{bmatrix} 5 \\ 3 \end{bmatrix}$ equal $B\mathbf{x}$ for some vector $\mathbf{x}$? Notice that for the matrix–vector product to make sense, the number of columns of B must equal the number of components in $\mathbf{x}$. Our definition extends to the more general case that follows.

Definition.   Let A be an $m \times n$ matrix and $\mathbf{v}$ be an $n \times 1$ vector. We define the **matrix–vector product** of A and $\mathbf{v}$, denoted $A\mathbf{v}$, to be the linear combination of the columns of A whose coefficients are the corresponding components of $\mathbf{v}$. That is,

$$A\mathbf{v} = v_1 \mathbf{a}_1 + v_2 \mathbf{a}_2 + \cdots + v_n \mathbf{a}_n.$$

As we noted above, the number of columns of A must equal the number of components of $\mathbf{v}$. For example, suppose that

$$A = \begin{bmatrix} 1 & 2 \\ 3 & 4 \\ 5 & 6 \end{bmatrix} \quad \text{and} \quad \mathbf{v} = \begin{bmatrix} 7 \\ 8 \end{bmatrix}.$$

Notice that A has two columns and $\mathbf{v}$ has two components. Then

$$A\mathbf{v} = \begin{bmatrix} 1 & 2 \\ 3 & 4 \\ 5 & 6 \end{bmatrix} \begin{bmatrix} 7 \\ 8 \end{bmatrix} = 7 \begin{bmatrix} 1 \\ 3 \\ 5 \end{bmatrix} + 8 \begin{bmatrix} 2 \\ 4 \\ 6 \end{bmatrix} = \begin{bmatrix} 7 \\ 21 \\ 35 \end{bmatrix} + \begin{bmatrix} 16 \\ 32 \\ 48 \end{bmatrix} = \begin{bmatrix} 23 \\ 53 \\ 83 \end{bmatrix}.$$

It is easy to show that for any $m \times n$ matrix A, $A\mathbf{0} = \mathbf{0}'$, where $\mathbf{0}$ is the $n \times 1$ zero vector and $\mathbf{0}'$ is the $m \times 1$ zero vector. Furthermore, for the $m \times n$ zero matrix O, $O\mathbf{v} = \mathbf{0}'$ for any $n \times 1$ vector $\mathbf{v}$ (see (f) and (g) of Theorem 1.2).

Returning to our garden supply store example, suppose that the store has 140 lb of seed in stock: 60 lb of the deluxe mixture, 50 lb of the standard mixture, and 30 lb of the economy mixture. We let $\mathbf{w} = \begin{bmatrix} 60 \\ 50 \\ 30 \end{bmatrix}$ represent this information. Now the matrix–vector product

$$B\mathbf{w} = \begin{bmatrix} .80 & .60 & .40 \\ .20 & .40 & .60 \end{bmatrix} \begin{bmatrix} 60 \\ 50 \\ 30 \end{bmatrix}$$

$$= 60 \begin{bmatrix} .80 \\ .20 \end{bmatrix} + 50 \begin{bmatrix} .60 \\ .40 \end{bmatrix} + 30 \begin{bmatrix} .40 \\ .60 \end{bmatrix}$$

$$\begin{array}{c} \text{seed (lb)} \\ = \begin{bmatrix} 90 \\ 50 \end{bmatrix} \quad \begin{array}{l} \text{bluegrass} \\ \text{rye} \end{array} \end{array}$$

gives the number of pounds of each type of seed contained in the 140 lb that the garden supply store has in stock. For example, there are 90 lb of bluegrass, because $90 = .80(60) + .60(50) + .40(30)$.

Suppose we let $I_2 = \begin{bmatrix} 1 & 0 \\ 0 & 1 \end{bmatrix}$ and $\mathbf{v}$ be any vector in $\mathcal{R}^2$. Then

$$I_2\mathbf{v} = \begin{bmatrix} 1 & 0 \\ 0 & 1 \end{bmatrix} \begin{bmatrix} v_1 \\ v_2 \end{bmatrix} = v_1 \begin{bmatrix} 1 \\ 0 \end{bmatrix} + v_2 \begin{bmatrix} 0 \\ 1 \end{bmatrix} = \begin{bmatrix} v_1 \\ 0 \end{bmatrix} + \begin{bmatrix} 0 \\ v_2 \end{bmatrix} = \begin{bmatrix} v_1 \\ v_2 \end{bmatrix} = \mathbf{v}.$$

So multiplication by I_2 leaves every vector $\mathbf{v}$ in $\mathcal{R}^2$ unchanged. The same property holds in a more general context.

Definition.   For each positive integer n, the $n \times n$ **identity matrix** I_n is the $n \times n$ matrix whose respective columns are the standard vectors $\mathbf{e}_1, \mathbf{e}_2, \ldots, \mathbf{e}_n$ in $\mathcal{R}^n$.

For example,

$$I_2 = \begin{bmatrix} 1 & 0 \\ 0 & 1 \end{bmatrix} \quad \text{and} \quad I_3 = \begin{bmatrix} 1 & 0 & 0 \\ 0 & 1 & 0 \\ 0 & 0 & 1 \end{bmatrix}.$$

Because the columns of I_n are the standard vectors of $\mathcal{R}^n$, it follows easily that $I_n\mathbf{v} = \mathbf{v}$ for any $\mathbf{v}$ in $\mathcal{R}^n$.

There is another approach to computing the matrix–vector product that relies more on the entries of A than its columns. Consider the following example.

$$A\mathbf{v} = \begin{bmatrix} a_{11} & a_{12} & a_{13} \\ a_{21} & a_{22} & a_{23} \end{bmatrix} \begin{bmatrix} v_1 \\ v_2 \\ v_3 \end{bmatrix}$$

$$= v_1 \begin{bmatrix} a_{11} \\ a_{21} \end{bmatrix} + v_2 \begin{bmatrix} a_{12} \\ a_{22} \end{bmatrix} + v_3 \begin{bmatrix} a_{13} \\ a_{23} \end{bmatrix}$$

$$= \begin{bmatrix} a_{11}v_1 + a_{12}v_2 + a_{13}v_3 \\ a_{21}v_1 + a_{22}v_2 + a_{23}v_3 \end{bmatrix}.$$

Notice that the first component of the vector $A\mathbf{v}$ is the sum of products of the corresponding entries of the first row of A and the components of $\mathbf{v}$. Likewise, the second component of $A\mathbf{v}$ is the sum of products of the corresponding entries of the second row of A and the components of $\mathbf{v}$. With this approach to computing a matrix–vector product, we can omit the intermediate step in the preceding illustration. For example, suppose

$$A = \begin{bmatrix} 2 & 3 & 1 \\ 1 & -2 & 3 \end{bmatrix} \quad \text{and} \quad \mathbf{v} = \begin{bmatrix} -1 \\ 1 \\ 3 \end{bmatrix};$$

then

$$A\mathbf{v} = \begin{bmatrix} 2 & 3 & 1 \\ 1 & -2 & 3 \end{bmatrix} \begin{bmatrix} -1 \\ 1 \\ 3 \end{bmatrix} = \begin{bmatrix} (2)(-1) + (3)(1) + (1)(3) \\ (1)(-1) + (-2)(1) + (3)(3) \end{bmatrix} = \begin{bmatrix} 4 \\ 6 \end{bmatrix}.$$

In general, this technique can be applied to computing $A\mathbf{v}$, where A is an $m \times n$ matrix and $\mathbf{v}$ is a vector in $\mathcal{R}^n$. In this case, the ith component of $A\mathbf{v}$ can be computed from the matrix–vector product

$$[a_{i1} \ a_{i2} \ \cdots \ a_{in}] \begin{bmatrix} v_1 \\ v_2 \\ \vdots \\ v_n \end{bmatrix};$$

that is, the ith component of $A\mathbf{v}$ is

$$a_{i1}v_1 + a_{i2}v_2 + \cdots + a_{in}v_n.$$

Henceforth, we will identify a 1×1 matrix with its single entry. So the ith component of $A\mathbf{v}$ is the matrix–vector product of the ith row of A and $\mathbf{v}$. Thus

$$A\mathbf{v} = \begin{bmatrix} a_{11} & a_{12} & \cdots & a_{1n} \\ a_{21} & a_{22} & \cdots & a_{2n} \\ \vdots & \vdots & & \vdots \\ a_{m1} & a_{m2} & \cdots & a_{mn} \end{bmatrix} \begin{bmatrix} v_1 \\ v_2 \\ \vdots \\ v_n \end{bmatrix} = \begin{bmatrix} a_{11}v_1 + a_{12}v_2 + \cdots + a_{1n}v_n \\ a_{21}v_1 + a_{22}v_2 + \cdots + a_{2n}v_n \\ \vdots \\ a_{m1}v_1 + a_{m2}v_2 + \cdots + a_{mn}v_n \end{bmatrix}.$$

Example 3   A sociologist is interested in studying the population changes within a metropolitan area as people move between the city and the suburbs. From empirical evidence, she has discovered that in any given year, 15% of those living in the city will move to the suburbs and 3% of those living in the suburbs will move to the city. For simplicity, we assume that the metropolitan population remains stable. This information may be represented by the following matrix.

$$
\text{To} \quad
\begin{matrix}
 & \text{From} & \\
 & \text{City} \quad \text{Suburbs} & \\
\text{City} & \begin{bmatrix} .85 & .03 \\ .15 & .97 \end{bmatrix} & = A \\
\text{Suburbs} & &
\end{matrix}
$$

Notice that the entries of A are nonnegative and that the entries of each column sum to 1. Such a matrix is called a **stochastic matrix**. Suppose that

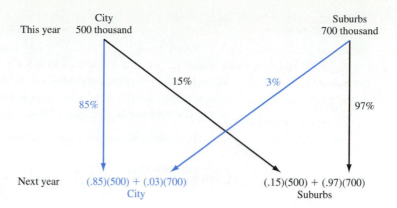

FIGURE 1.11

there are now 500 thousand people living in the city and 700 thousand people living in the suburbs. The sociologist would like to know how many people will be living in each of the two areas next year. Figure 1.11 describes the changes of population from one year to the next. It follows that the number of people (in thousands) who will be living in the city next year is $(.85)(500) + (.03)(700) = 446$ thousand, and the number of people living in the suburbs will be $(.15)(500) + (.97)(700) = 754$ thousand.

If we let **p** represent the vector of current populations of the city and suburbs, we have

$$\mathbf{p} = \begin{bmatrix} 500 \\ 700 \end{bmatrix}.$$

So the populations in the next year can be found by computing the matrix–vector product

$$A\mathbf{p} = \begin{bmatrix} .85 & .03 \\ .15 & .97 \end{bmatrix} \begin{bmatrix} 500 \\ 700 \end{bmatrix} = \begin{bmatrix} (.85)(500) + (.03)(700) \\ (.15)(500) + (.97)(700) \end{bmatrix} = \begin{bmatrix} 446 \\ 754 \end{bmatrix}.$$

If we want to determine the populations in the following year, we need only compute $A(A\mathbf{p})$.   ○

Rotation Matrices

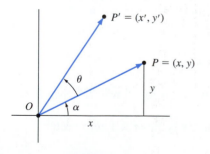

FIGURE 1.12

Consider a point $P = (x, y)$ in $\mathcal{R}^2$ with polar coordinates (r, α), where $r \geq 0$ and α is the angle between the segment $\overline{OP}$ and the positive x-axis (see Figure 1.12). Then $x = r \cos\alpha$ and $y = r \sin\alpha$. Suppose that $\overline{OP}$ is rotated by an angle θ to the segment $\overline{OP'}$, where $P' = (x', y')$. Then $(r, \alpha + \theta)$ represents the polar coordinates for P', and hence

$$x' = r\cos(\alpha + \theta)$$
$$= r(\cos\alpha\cos\theta - \sin\alpha\sin\theta)$$
$$= (r\cos\alpha)\cos\theta - (r\sin\alpha)\sin\theta$$
$$= x\cos\theta - y\sin\theta.$$

Similarly, $y' = x\sin\theta + y\cos\theta$. These equations can be expressed as a matrix equation using a matrix–vector product. If we define A_θ by

$$A_\theta = \begin{bmatrix} \cos\theta & -\sin\theta \\ \sin\theta & \cos\theta \end{bmatrix},$$

then

$$A_\theta \begin{bmatrix} x \\ y \end{bmatrix} = \begin{bmatrix} \cos\theta & -\sin\theta \\ \sin\theta & \cos\theta \end{bmatrix} \begin{bmatrix} x \\ y \end{bmatrix} = \begin{bmatrix} x\cos\theta - y\sin\theta \\ x\sin\theta + y\cos\theta \end{bmatrix} = \begin{bmatrix} x' \\ y' \end{bmatrix}.$$

We call A_θ the θ-**rotation matrix** or, more simply, a **rotation matrix**. For any vector $\mathbf{u}$, the vector $A_\theta\mathbf{u}$ is the vector obtained by rotating $\mathbf{u}$ by an angle θ, where the rotation is counterclockwise if $\theta > 0$ and clockwise if $\theta < 0$.

For example, suppose that we want to rotate the vector $\begin{bmatrix} 3 \\ 4 \end{bmatrix}$ by $30°$. We must compute $A_{30°}\begin{bmatrix} 3 \\ 4 \end{bmatrix}$; that is,

$$\begin{bmatrix} \cos 30° & -\sin 30° \\ \sin 30° & \cos 30° \end{bmatrix}\begin{bmatrix} 3 \\ 4 \end{bmatrix} = \begin{bmatrix} \dfrac{\sqrt{3}}{2} & -\dfrac{1}{2} \\ \dfrac{1}{2} & \dfrac{\sqrt{3}}{2} \end{bmatrix}\begin{bmatrix} 3 \\ 4 \end{bmatrix} = \begin{bmatrix} \dfrac{3\sqrt{3}}{2} - \dfrac{4}{2} \\ \dfrac{3}{2} + \dfrac{4\sqrt{3}}{2} \end{bmatrix}$$

$$= \frac{1}{2}\begin{bmatrix} 3\sqrt{3} - 4 \\ 3 + 4\sqrt{3} \end{bmatrix}.$$

It is interesting to observe that the zero-degree rotation matrix $A_{0°}$, which leaves a vector unchanged, is given by $A_{0°} = I_2$. This is quite reasonable because multiplication by I_2 also leaves vectors unchanged.

Besides rotations, other geometric transformations (such as reflections and projections) can be described as matrix–vector products. Examples are found in the exercises.

It is useful to note that the columns of a matrix can be represented as matrix–vector products of the matrix with the standard vectors. Suppose that $A = \begin{bmatrix} 2 & 4 \\ 3 & 6 \end{bmatrix}$. Then

$$A\mathbf{e}_1 = \begin{bmatrix} 2 & 4 \\ 3 & 6 \end{bmatrix}\begin{bmatrix} 1 \\ 0 \end{bmatrix} = \begin{bmatrix} 2 \\ 3 \end{bmatrix} \quad \text{and} \quad A\mathbf{e}_2 = \begin{bmatrix} 2 & 4 \\ 3 & 6 \end{bmatrix}\begin{bmatrix} 0 \\ 1 \end{bmatrix} = \begin{bmatrix} 4 \\ 6 \end{bmatrix}.$$

The general result is stated as (a) of the following theorem.

Theorem 1.2   (Properties of matrix–vector products)
Let A and B be $m \times n$ matrices, and let $\mathbf{u}$ and $\mathbf{v}$ be vectors in $\mathcal{R}^n$. Then

(a) $A\mathbf{e}_j = \mathbf{a}_j$ for $j = 1, 2, \ldots, n$, where $\mathbf{e}_j$ is the jth standard vector in $\mathcal{R}^n$.
(b) $A(\mathbf{u} + \mathbf{v}) = A\mathbf{u} + A\mathbf{v}$.
(c) $A(c\mathbf{u}) = c(A\mathbf{u}) = (cA)\mathbf{u}$ for every scalar c.
(d) $(A + B)\mathbf{u} = A\mathbf{u} + B\mathbf{u}$.
(e) If B is an $m \times n$ matrix such that $B\mathbf{w} = A\mathbf{w}$ for all $\mathbf{w}$ in $\mathcal{R}^n$, then $B = A$.
(f) $A\mathbf{0}$ is the $m \times 1$ zero vector.
(g) If O is the $m \times n$ zero matrix, then $O\mathbf{v}$ is the $m \times 1$ zero vector.
(h) $I_n\mathbf{v} = \mathbf{v}$.

Proof We prove (b) and leave the rest for the exercises.
(b) Because the ith component of $\mathbf{u} + \mathbf{v}$ is $u_i + v_i$, we have

$$A(\mathbf{u} + \mathbf{v}) = (u_1 + v_1)\mathbf{a}_1 + (u_2 + v_2)\mathbf{a}_2 + \cdots + (u_n + v_n)\mathbf{a}_n$$

$$= (u_1\mathbf{a}_1 + u_2\mathbf{a}_2 + \cdots + u_n\mathbf{a}_n) + (v_1\mathbf{a}_1 + v_2\mathbf{a}_2 + \cdots + v_n\mathbf{a}_n)$$

$$= A\mathbf{u} + A\mathbf{v}. \qquad \square$$

It follows by repeated applications of (b) and (c) of Theorem 1.2 that

$$A(c_1\mathbf{u}_1 + c_2\mathbf{u}_2 + \cdots + c_k\mathbf{u}_k) = c_1A\mathbf{u}_1 + c_2A\mathbf{u}_2 + \cdots + c_kA\mathbf{u}_k$$

for any scalars $c_1, c_2, \ldots, c_k$ and any vectors $\mathbf{u}_1, \mathbf{u}_2, \ldots, \mathbf{u}_k$ in $\mathcal{R}^n$.

Practice Problems

1. Let $A = \begin{bmatrix} 2 & -1 & 1 \\ 3 & 0 & -2 \end{bmatrix}$ and $\mathbf{v} = \begin{bmatrix} 3 \\ 1 \\ -1 \end{bmatrix}$. Compute the following matrices.

   (a) $A\mathbf{v}$
   (b) $(A\mathbf{v})^T$

2. Let $\mathbf{w} = \begin{bmatrix} -1 \\ 10 \end{bmatrix}$ and $\mathcal{S} = \left\{ \begin{bmatrix} 2 \\ 1 \end{bmatrix}, \begin{bmatrix} 3 \\ -2 \end{bmatrix} \right\}$.

   (a) Without doing any calculations, explain why $\mathbf{w}$ can be written as a linear combination of the vectors in $\mathcal{S}$.
   (b) Express $\mathbf{w}$ as a linear combination of the vectors in $\mathcal{S}$.

Exercises

1. Determine if the following statements are true or false.

   (a) A linear combination of vectors is a sum of scalar multiples of the vectors.
   (b) The coefficients in a linear combination can always be chosen to be positive scalars.
   (c) Every vector in $\mathcal{R}^2$ can be written as a linear combination of the standard vectors of $\mathcal{R}^2$.
   (d) Every vector in $\mathcal{R}^2$ is a linear combination of any two nonparallel vectors.
   (e) The zero vector is a linear combination of any nonempty set of vectors.
   (f) The matrix–vector product of a 2×3 matrix and a 3×1 vector is a 3×1 vector.
   (g) The matrix–vector product of a 2×3 matrix and a 3×1 vector equals a linear combination of the rows of the matrix.

In Exercises 2–13, compute the following matrix–vector products.

2. $\begin{bmatrix} 1 & -3 \\ 0 & 2 \\ -1 & 4 \end{bmatrix} \begin{bmatrix} 1 \\ 2 \end{bmatrix}$

3. $\begin{bmatrix} 3 & -2 & 1 \\ 4 & 0 & 2 \end{bmatrix} \begin{bmatrix} 1 \\ -2 \\ 5 \end{bmatrix}$

4. $\begin{bmatrix} 4 & 2 \\ 7 & -3 \end{bmatrix} \begin{bmatrix} 5 \\ 1 \end{bmatrix}$

5. $\begin{bmatrix} 2 & -1 & 3 \\ 1 & 0 & -1 \\ 0 & 2 & 4 \end{bmatrix} \begin{bmatrix} 2 \\ 1 \\ 2 \end{bmatrix}$

6. $\begin{bmatrix} 2 & 1 & 3 \end{bmatrix} \begin{bmatrix} -2 \\ 4 \\ 6 \end{bmatrix}$

7. $\begin{bmatrix} 1 & 0 \\ 0 & 1 \end{bmatrix} \begin{bmatrix} a \\ b \end{bmatrix}$

8. $\begin{bmatrix} 1 & 0 & 0 \\ 0 & 1 & 0 \\ 0 & 0 & 1 \end{bmatrix} \begin{bmatrix} a \\ b \\ c \end{bmatrix}$

9. $\begin{bmatrix} 3 & 0 \\ 2 & 1 \end{bmatrix}^T \begin{bmatrix} 4 \\ 5 \end{bmatrix}$

10. $\begin{bmatrix} 4 \\ 2 \\ -3 \end{bmatrix}^T \begin{bmatrix} 2 \\ -1 \\ 0 \end{bmatrix}$

11. $\begin{bmatrix} s & 0 & 0 \\ 0 & t & 0 \\ 0 & 0 & u \end{bmatrix} \begin{bmatrix} a \\ b \\ c \end{bmatrix}$

12. $\left(\begin{bmatrix} 3 & 0 \\ -2 & 4 \end{bmatrix} + \begin{bmatrix} 1 & 2 \\ 3 & -3 \end{bmatrix} \right) \begin{bmatrix} 4 \\ 5 \end{bmatrix}$

13. $\left(\begin{bmatrix} 3 & 0 \\ -2 & 4 \end{bmatrix}^T + \begin{bmatrix} 1 & 2 \\ 3 & -3 \end{bmatrix}^T \right) \begin{bmatrix} 4 \\ 5 \end{bmatrix}$

In Exercises 14–19, an angle θ and a vector $\mathbf{u}$ are given. Write the corresponding rotation matrix, and compute the vector found by rotating $\mathbf{u}$ by the angle θ. Draw a sketch and simplify your answers.

14. $\theta = 0°$, $\mathbf{u} = \mathbf{e}_1$

15. $\theta = 45°$, $\mathbf{u} = \mathbf{e}_2$

16. $\theta = 30°$, $\mathbf{u} = \begin{bmatrix} 1 \\ 2 \end{bmatrix}$

17. $\theta = 60°$, $\mathbf{u} = \begin{bmatrix} 3 \\ 1 \end{bmatrix}$

18. $\theta = 135°$, $\mathbf{u} = \begin{bmatrix} 2 \\ -1 \end{bmatrix}$

19. $\theta = 210°$, $\mathbf{u} = \begin{bmatrix} -1 \\ -3 \end{bmatrix}$

20. Use a matrix–vector product to show that if $\theta = 0°$, then $A_\theta\mathbf{v} = \mathbf{v}$ for all $\mathbf{v}$ in $\mathcal{R}^2$.

21. Use matrix–vector products to show that, for any angles θ and β and any vector $\mathbf{v}$ in $\mathcal{R}^2$, $A_\theta(A_\beta\mathbf{v}) = A_{\theta+\beta}\mathbf{v}$.

22. Compute $A_\theta^T(A_\theta\mathbf{u})$ and $A_\theta(A_\theta^T\mathbf{u})$ for any vector $\mathbf{u}$ in $\mathcal{R}^2$ and any angle θ.

In Exercises 23–32, a vector $\mathbf{u}$ and a set $\mathcal{S}$ are given. If possible, write $\mathbf{u}$ as a linear combination of the vectors in $\mathcal{S}$.

23. $\mathbf{u} = \begin{bmatrix} 1 \\ 1 \end{bmatrix}$, $\mathcal{S} = \left\{ \begin{bmatrix} 1 \\ 0 \end{bmatrix}, \begin{bmatrix} 0 \\ 1 \end{bmatrix} \right\}$

24. $\mathbf{u} = \begin{bmatrix} 1 \\ -1 \end{bmatrix}$, $\mathcal{S} = \left\{ \begin{bmatrix} 4 \\ -4 \end{bmatrix} \right\}$

25. $\mathbf{u} = \begin{bmatrix} 1 \\ -1 \end{bmatrix}$, $\mathcal{S} = \left\{ \begin{bmatrix} 4 \\ 4 \end{bmatrix} \right\}$

26. $\mathbf{u} = \begin{bmatrix} 1 \\ 1 \end{bmatrix}$, $\mathcal{S} = \left\{ \begin{bmatrix} 1 \\ 0 \end{bmatrix}, \begin{bmatrix} 0 \\ 1 \end{bmatrix} \right\}$

27. $\mathbf{u} = \begin{bmatrix} 1 \\ 1 \\ 2 \end{bmatrix}$, $S = \left\{ \begin{bmatrix} 1 \\ 0 \\ 1 \end{bmatrix}, \begin{bmatrix} 1 \\ 0 \\ -1 \end{bmatrix} \right\}$

28. $\mathbf{u} = \begin{bmatrix} 1 \\ 1 \end{bmatrix}$, $S = \left\{ \begin{bmatrix} 1 \\ 0 \end{bmatrix}, \begin{bmatrix} 0 \\ -1 \end{bmatrix}, \begin{bmatrix} 0 \\ 0 \end{bmatrix} \right\}$

29. $\mathbf{u} = \begin{bmatrix} -1 \\ 11 \end{bmatrix}$, $S = \left\{ \begin{bmatrix} 1 \\ 3 \end{bmatrix}, \begin{bmatrix} 2 \\ -1 \end{bmatrix} \right\}$

30. $\mathbf{u} = \begin{bmatrix} 1 \\ 1 \end{bmatrix}$, $S = \left\{ \begin{bmatrix} 1 \\ 0 \end{bmatrix}, \begin{bmatrix} 0 \\ -1 \end{bmatrix}, \begin{bmatrix} 1 \\ 1 \end{bmatrix} \right\}$

31. $\mathbf{u} = \begin{bmatrix} 3 \\ 8 \end{bmatrix}$, $S = \left\{ \begin{bmatrix} 1 \\ 2 \end{bmatrix}, \begin{bmatrix} 2 \\ 3 \end{bmatrix}, \begin{bmatrix} -2 \\ -5 \end{bmatrix} \right\}$

32. $\mathbf{u} = \begin{bmatrix} a \\ b \end{bmatrix}$, $S = \left\{ \begin{bmatrix} 1 \\ 1 \end{bmatrix}, \begin{bmatrix} 2 \\ -1 \end{bmatrix} \right\}$

33. Suppose that in a metropolitan area there are 400 thousand people living in the city and 300 thousand people living in the suburbs. Use the stochastic matrix in Example 3 to determine:

    (a) the number of people living in the city and suburbs after one year;
    (b) the number of people living in the city and suburbs after two years.

34. Let $A = \begin{bmatrix} 1 & 2 & 3 \\ 4 & 5 & 6 \\ 7 & 8 & 9 \end{bmatrix}$ and $\mathbf{u} = \begin{bmatrix} a \\ b \\ c \end{bmatrix}$. Represent $A\mathbf{u}$ as a linear combination of the columns of A.

In Exercises 35–38, let $A = \begin{bmatrix} -1 & 0 \\ 0 & 1 \end{bmatrix}$ and $\mathbf{u} = \begin{bmatrix} a \\ b \end{bmatrix}$.

35. Show that $A\mathbf{u}$ is the reflection of $\mathbf{u}$ about the y-axis.

36. Prove that $A(A\mathbf{u}) = \mathbf{u}$.

37. Modify the matrix A to obtain a matrix B, so that $B\mathbf{u}$ is the reflection of $\mathbf{u}$ about the x-axis.

38. Let C denote the rotation matrix that corresponds to $\theta = 180°$.

    (a) Find C.
    (b) Use the matrix B in Exercise 37 to show that

    $$A(C\mathbf{u}) = C(A\mathbf{u}) = B\mathbf{u}$$

    and

    $$B(C\mathbf{u}) = C(B\mathbf{u}) = A\mathbf{u}.$$

    (c) Interpret these equations in terms of reflections and rotations.

In Exercises 39–43, let $A = \begin{bmatrix} 1 & 0 \\ 0 & 0 \end{bmatrix}$ and $\mathbf{u} = \begin{bmatrix} a \\ b \end{bmatrix}$.

39. Show that $A\mathbf{u}$ is the projection of $\mathbf{u}$ onto the x-axis.

40. Prove that $A(A\mathbf{u}) = A\mathbf{u}$.

41. Show that if $\mathbf{v}$ is any vector whose endpoint lies on the x-axis, then $A\mathbf{v} = \mathbf{v}$.

42. Modify the matrix A to obtain a matrix B, so that $B\mathbf{u}$ is the projection of $\mathbf{u}$ onto the y-axis.

43. Let C denote the rotation matrix that corresponds to $\theta = 180°$ (see Exercise 38(a)).

    (a) Prove that $A(C\mathbf{u}) = C(A\mathbf{u})$.
    (b) Interpret the result in (a) geometrically.

44. Let $\mathbf{u}_1$ and $\mathbf{u}_2$ be vectors in $\mathcal{R}^n$. Prove that the sum of two linear combinations of these vectors is also a linear combination of these vectors.

45. Let $\mathbf{u}_1$ and $\mathbf{u}_2$ be vectors in $\mathcal{R}^n$. Let $\mathbf{v}$ and $\mathbf{w}$ be linear combinations of $\mathbf{u}_1$ and $\mathbf{u}_2$. Prove that any linear combination of $\mathbf{v}$ and $\mathbf{w}$ is also a linear combination of $\mathbf{u}_1$ and $\mathbf{u}_2$.

46. Let $\mathbf{u}_1$ and $\mathbf{u}_2$ be vectors in $\mathcal{R}^n$. Prove that a scalar multiple of a linear combination of these vectors is also a linear combination of these vectors.

47. Prove (a) of Theorem 1.2.

48. Prove (c) of Theorem 1.2.

49. Prove (d) of Theorem 1.2.

50. Prove (e) of Theorem 1.2.

51. Prove (f) of Theorem 1.2.

52. Prove (g) of Theorem 1.2.

53. Prove (h) of Theorem 1.2.

In Exercises 54 and 55, use a calculator with matrix capabilities or computer software such as MATLAB to solve the problem.

54. In reference to Exercise 33, determine the number of people living in the city and suburbs after ten years.

55. Consider the matrices

$$A = \begin{bmatrix} 2.1 & 1.3 & -0.1 & 6.0 \\ 1.3 & -9.9 & 4.5 & 6.2 \\ 4.4 & -2.2 & 5.7 & 2.0 \\ 0.2 & 9.8 & 1.1 & -8.5 \end{bmatrix}$$

and

$$B = \begin{bmatrix} 4.4 & 1.1 & 3.0 & 9.9 \\ -1.2 & 4.8 & 2.4 & 6.0 \\ 1.3 & 2.4 & -5.8 & 2.8 \\ 6.0 & -2.1 & -5.3 & 8.2 \end{bmatrix}$$

and the vectors

$$\mathbf{u} = \begin{bmatrix} 1 \\ -1 \\ 2 \\ 4 \end{bmatrix} \quad \text{and} \quad \mathbf{v} = \begin{bmatrix} 7 \\ -1 \\ 2 \\ 5 \end{bmatrix}.$$

    (a) Compute $A\mathbf{u}$.
    (b) Compute $B(\mathbf{u} + \mathbf{v})$.
    (c) Compute $(A + B)\mathbf{v}$.
    (d) Compute $A(B\mathbf{v})$.

Solutions to the Practice Problems

1. (a) $A\mathbf{v} = \begin{bmatrix} 2 & -1 & 1 \\ 3 & 0 & -2 \end{bmatrix} \begin{bmatrix} 3 \\ 1 \\ -1 \end{bmatrix} = \begin{bmatrix} 4 \\ 11 \end{bmatrix}$

(b) $(A\mathbf{v})^T = \begin{bmatrix} 4 \\ 11 \end{bmatrix}^T = \begin{bmatrix} 4 & 11 \end{bmatrix}$

2. (a) The vectors in S are nonparallel vectors in $\mathcal{R}^2$.

(b) To express $\mathbf{w}$ as a linear combination of the vectors in S, we must find scalars x_1 and x_2 such that

$$\begin{bmatrix} -1 \\ 10 \end{bmatrix} = x_1 \begin{bmatrix} 2 \\ 1 \end{bmatrix} + x_2 \begin{bmatrix} 3 \\ -2 \end{bmatrix} = \begin{bmatrix} 2x_1 + 3x_2 \\ x_1 - 2x_2 \end{bmatrix}.$$

That is, we must solve the following system:

$$2x_1 + 3x_2 = -1$$
$$x_1 - 2x_2 = 10.$$

Using elementary algebra, we see that $x_1 = 4$ and $x_2 = -3$. So

$$\begin{bmatrix} -1 \\ 10 \end{bmatrix} = 4 \begin{bmatrix} 2 \\ 1 \end{bmatrix} - 3 \begin{bmatrix} 3 \\ -2 \end{bmatrix}.$$

1.3   Systems of Linear Equations

By a **linear equation** in the variables (unknowns) $x_1, x_2, \ldots, x_n$, we mean an equation that can be written in the form

$$a_1x_1 + a_2x_2 + \cdots + a_nx_n = b,$$

where $a_1, a_2, \ldots, a_n$ and b are real numbers. The numbers $a_1, a_2, \ldots, a_n$ are called the **coefficients**, and **b** is called the **constant term** of the equation. For example, $3x_1 - 7x_2 + x_3 = 19$ is a linear equation in the variables x_1, x_2, and x_3, with coefficients 3, -7, and 1, and with constant term 19. The equation $8x_2 - 12x_5 = 4x_1 - 9x_3 + 6$ is also a linear equation because it can be written as

$$-4x_1 + 8x_2 + 9x_3 + 0x_4 - 12x_5 = 6.$$

On the other hand, the equations

$$x_1 + 5x_2x_3 = 7, \qquad 2x_1 - 7x_2 + x_3^2 = -3, \qquad \text{and} \qquad 4\sqrt{x_1} - 3x_2 = 15$$

are *not* linear equations because they contain terms involving a product, a square, or a square root of a variable.

By a **system of linear equations**, we mean a set of m linear equations in the same n variables, where m and n are positive integers. Such a system has the form

$$a_{11}x_1 + a_{12}x_2 + \cdots + a_{1n}x_n = b_1$$
$$a_{21}x_1 + a_{22}x_2 + \cdots + a_{2n}x_n = b_2$$
$$\vdots$$
$$a_{m1}x_1 + a_{m2}x_2 + \cdots + a_{mn}x_n = b_m.$$

For example, the problem involving the blending of grass seed that is described in Section 1.2 led to the following system of two linear equations in the variables x_1, x_2, and x_3:

$$.80x_1 + .60x_2 + .40x_3 = 5$$
$$.20x_1 + .40x_2 + .60x_3 = 3. \tag{1}$$

A **solution** to a system of linear equations in the variables $x_1, x_2, \ldots, x_n$ is an element $\begin{bmatrix} s_1 \\ s_2 \\ \vdots \\ s_n \end{bmatrix}$ of $\mathcal{R}^n$ such that every equation in the system is satisfied when each x_i is replaced by s_i. For example, $\begin{bmatrix} 2 \\ 5 \\ 1 \end{bmatrix}$ is a solution to (1) because

$$.80(2) + .60(5) + .40(1) = 5 \qquad \text{and} \qquad .20(2) + .40(5) + .60(1) = 3.$$

The set of all solutions to a system of linear equations is called the **solution set** of that system.

Systems of Two Linear Equations in Two Variables

A linear equation in the variables x and y has the form $ax + by = c$. When at least one of a and b is nonzero, this is the general form of an equation of a line in the xy-plane. Thus, except for degenerate cases where both coefficients are zero, a system of two linear equations in the variables x and y consists of a pair of equations, each of which describes some line in this plane. Geometrically, a solution to such a system corresponds to a point lying on each of these lines $\mathcal{L}_1$ and $\mathcal{L}_2$. There are three different situations that can arise, as follows.

If the lines are different and parallel, then they have no point in common. Hence, in this case, the system of equations has no solution (see Figure 1.13).

If the lines are different and nonparallel, then they have a unique point of intersection. In this event, the system of equations has exactly one solution (see Figure 1.14).

Finally, if the two lines coincide, then every point on $\mathcal{L}_1$ and $\mathcal{L}_2$ satisfies both of the equations in the system, and so every point on $\mathcal{L}_1$ and $\mathcal{L}_2$ is a solution to the system. Thus, there are infinitely many solutions in this case (see Figure 1.15).

As we soon see, it turns out that no matter how many equations and variables a system has, there are exactly three possibilities for its solution set.

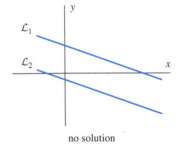

no solution

FIGURE 1.13

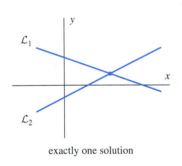

exactly one solution

FIGURE 1.14

> Every system of linear equations has no solution, exactly one solution, or infinitely many solutions.

A system of linear equations that has one or more solutions is called **consistent**; otherwise the system is called **inconsistent**.

Elementary Row Operations

To find the solution set of a system of linear equations, we replace it by a system with the same solutions that is easily solved. Two systems of linear equations that have exactly the same solutions are called **equivalent**. Thus we desire to replace the original system of linear equations by an easily solved equivalent system.

A basic technique for solving a system of linear equations is taught in high school algebra classes. We illustrate this procedure by solving the following

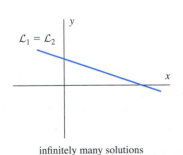

infinitely many solutions

FIGURE 1.15

system of three linear equations in the variables x_1, x_2, and x_3.

$$\begin{aligned} x_1 - 2x_2 - x_3 &= 3 \\ 3x_1 - 6x_2 - 5x_3 &= 3 \\ 2x_1 - x_2 + x_3 &= 0 \end{aligned} \qquad (2)$$

We begin the simplification by eliminating x_1 from every equation but the first. To do so, we add appropriate multiples of the first equation to the second and third equations so that the coefficient of x_1 becomes zero in these equations. Adding -3 times the first equation to the second makes the coefficient of x_1 zero in the result. The effect of these calculations is as follows.

$$\begin{aligned} -3x_1 + 6x_2 + 3x_3 &= -9 \quad \text{(-3 times first equation)} \\ 3x_1 - 6x_2 - 5x_3 &= 3 \quad \text{(second equation)} \\ \hline -2x_3 &= -6 \end{aligned}$$

Likewise, adding -2 times the first equation to the third makes the coefficient of x_1 zero in the new third equation.

$$\begin{aligned} -2x_1 + 4x_2 + 2x_3 &= -6 \quad \text{(-2 times first equation)} \\ 2x_1 - x_2 + x_3 &= 0 \quad \text{(third equation)} \\ \hline 3x_2 + 3x_3 &= -6 \end{aligned}$$

These changes transform (2) into the following system:

$$\begin{aligned} x_1 - 2x_2 - x_3 &= 3 \\ -2x_3 &= -6 \\ 3x_2 + 3x_3 &= -6. \end{aligned}$$

In this case, it happens that the calculation that made the coefficient of x_1 zero in the new second equation also made the coefficient of x_2 zero. (This does not always happen, as we saw when making the coefficient of x_1 zero in the new third equation.) If we now interchange the second and third equations in this system, we obtain the following system:

$$\begin{aligned} x_1 - 2x_2 - x_3 &= 3 \\ 3x_2 + 3x_3 &= -6 \\ -2x_3 &= -6. \end{aligned} \qquad (3)$$

The third equation can now be solved for x_3 by multiplying both sides by $-\frac{1}{2}$ (or equivalently, dividing both sides by -2). This produces

$$\begin{aligned} x_1 - 2x_2 - x_3 &= 3 \\ 3x_2 + 3x_3 &= -6 \\ x_3 &= 3. \end{aligned}$$

By adding appropriate multiples of the third equation to the first and second, we can eliminate x_3 from every equation but the third. Thus we add the third equation to the first and add -3 times the third equation to the second, to obtain

$$\begin{aligned} x_1 - 2x_2 \phantom{{}- x_3} &= 6 \\ 3x_2 \phantom{{}+ 3x_3} &= -15 \\ x_3 &= 3. \end{aligned}$$

Now we solve for x_2 by multiplying the second equation by $\frac{1}{3}$. The result is

$$
\begin{aligned}
x_1 - 2x_2 &= 6 \\
x_2 &= -5 \\
x_3 &= 3.
\end{aligned}
$$

Finally, adding 2 times the second equation to the first produces the solution

$$
\begin{aligned}
x_1 &= -4 \\
x_2 &= -5 \\
x_3 &= 3.
\end{aligned}
$$

The reader should check that replacing x_1 by -4, x_2 by -5, and x_3 by 3 makes each equation in (2) true; hence $\begin{bmatrix} -4 \\ -5 \\ 3 \end{bmatrix}$ is a solution to (2).

In the process of reducing the original system to a simpler form, the names of the variables played no essential role. Thus the manipulations required to solve the original system can be performed on matrices. In fact, the original system

$$
\begin{aligned}
x_1 - 2x_2 - x_3 &= 3 \\
3x_1 - 6x_2 - 5x_3 &= 3 \\
2x_1 - x_2 + x_3 &= 0
\end{aligned}
\tag{2}
$$

can be expressed as the matrix equation $A\mathbf{x} = \mathbf{b}$, where

$$
A = \begin{bmatrix} 1 & -2 & -1 \\ 3 & -6 & -5 \\ 2 & -1 & 1 \end{bmatrix}, \qquad \mathbf{x} = \begin{bmatrix} x_1 \\ x_2 \\ x_3 \end{bmatrix}, \qquad \text{and} \qquad \mathbf{b} = \begin{bmatrix} 3 \\ 3 \\ 0 \end{bmatrix}.
$$

Note that the columns of A contain the coefficients of x_1, x_2, and x_3 from (2). For this reason, A is called the **coefficient matrix** (or the **matrix of coefficients**) of (2). All the information about this system that is needed to find its solution set is contained in the matrix

$$
\begin{bmatrix} 1 & -2 & -1 & 3 \\ 3 & -6 & -5 & 3 \\ 2 & -1 & 1 & 0 \end{bmatrix},
$$

which is called the **augmented matrix** of the system. This matrix is formed by augmenting the coefficient matrix A to include the vector $\mathbf{b}$, and it is often convenient to denote this matrix by writing $[A \ \ \mathbf{b}]$.

If A is an $m \times n$ matrix, then a vector $\mathbf{u}$ in $\mathcal{R}^n$ is a solution to $A\mathbf{x} = \mathbf{b}$ if and only if $A\mathbf{u} = \mathbf{b}$. Thus $\begin{bmatrix} -4 \\ -5 \\ 3 \end{bmatrix}$ is a solution to (2) because

$$
A\mathbf{u} = \begin{bmatrix} 1 & -2 & -1 \\ 3 & -6 & -5 \\ 2 & -1 & 1 \end{bmatrix} \begin{bmatrix} -4 \\ -5 \\ 3 \end{bmatrix} = \begin{bmatrix} 3 \\ 3 \\ 0 \end{bmatrix} = \mathbf{b}.
$$

Example 1   For the system of linear equations

$$\begin{aligned} x_1 +   5x_3 - x_4 &= 7 \\ 2x_1 - x_2 + 6x_3 &= -8, \end{aligned}$$

the coefficient matrix and the augmented matrix are

$$\begin{bmatrix} 1 & 0 & 5 & -1 \\ 2 & -1 & 6 & 0 \end{bmatrix} \quad \text{and} \quad \begin{bmatrix} 1 & 0 & 5 & -1 & 7 \\ 2 & -1 & 6 & 0 & 8 \end{bmatrix},$$

respectively. Note that the variable x_2 is missing from the first equation and x_4 is missing from the second equation in the system (that is, the coefficients of x_2 in the first equation and x_4 in the second equation are zero). As a result, the $(1, 2)$- and $(2, 4)$-entries of the coefficient and augmented matrices of the system are zero. ○

The calculations that were performed to obtain the solution to (2) involved three types of operations: interchanging the position of two equations in a system, multiplying an equation in the system by a nonzero scalar, and adding a multiple of one equation in the system to another. The analogous operations that can be performed on the augmented matrix of a system are given in the following definition.

Definition.   Any one of the following three operations performed on a matrix is called an **elementary row operation**.

1. Interchange any two rows of the matrix (interchange operation).
2. Multiply every entry of some row of the matrix by the same nonzero constant (scaling operation).
3. Add a multiple of one row of the matrix to another row (row addition operation).

Every elementary row operation can be reversed. That is, if an elementary row operation is performed on a matrix A to produce a new matrix B, then there is an elementary row operation that can be performed on B to obtain A. If, for example, B is obtained by interchanging two rows of A, then interchanging the same rows of B yields A. Also, if B is obtained by multiplying some row of A by the nonzero constant c, then multiplying the same row of B by $\frac{1}{c}$ yields A. Finally, if B is obtained by adding c times row i of A to row j, then adding $-c$ times row i of B to row j results in A.

Suppose that an elementary row operation is performed on an augmented matrix $[A \ \mathbf{b}]$ to obtain a new matrix $[A' \ \mathbf{b}']$. The reversibility of the elementary row operations assures us that the solutions of $A\mathbf{x} = \mathbf{b}$ are the same as those of $A'\mathbf{x} = \mathbf{b}'$. *Thus performing an elementary row operation on the augmented matrix of a system of linear equations produces the augmented matrix of an equivalent system of linear equations.* (A formal proof of this important fact is given in Section 2.3.)

Reduced Row Echelon Form

Elementary row operations can be used to solve any system of linear equations. What must be done is to represent the system by its augmented matrix, and then use elementary row operations to transform the augmented matrix into a matrix having a special form. The system of linear equations whose augmented matrix has this special form is equivalent to the original system and is easily solved.

We now define this special form of a matrix. In the following discussion, we call a row of a matrix a *zero row* if all its entries are zero and a *nonzero row* otherwise. The leftmost nonzero entry of a nonzero row is called its **leading entry**.

Definitions.   A matrix is said to be in **row echelon form** if it satisfies the following three conditions.

1. Each nonzero row lies above every zero row.
2. The leading entry of a nonzero row lies in a column to the right of the column containing the leading entry of any preceding row.
3. If a column contains the leading entry of some row, then all entries of that column below the leading entry are zero.[4]

If a matrix also satisfies the following two additional conditions, then it is said to be in **reduced row echelon form**.[5]

4. If a column contains the leading entry of some row, then all the other entries of that column are zero.
5. The leading entry of each nonzero row is 1.

A matrix having either of the following forms is in reduced row echelon form. In these diagrams a $*$ denotes an arbitrary entry (that may or may not be zero).

$$\begin{bmatrix} 1 & * & 0 & 0 & * \\ 0 & 0 & 1 & 0 & * \\ 0 & 0 & 0 & 1 & * \\ 0 & 0 & 0 & 0 & 0 \end{bmatrix} \qquad \begin{bmatrix} 1 & 0 & * & * & 0 & 0 & * \\ 0 & 1 & * & * & 0 & 0 & * \\ 0 & 0 & 0 & 0 & 1 & 0 & * \\ 0 & 0 & 0 & 0 & 0 & 1 & * \\ 0 & 0 & 0 & 0 & 0 & 0 & 0 \end{bmatrix}$$

Notice that the leading entries (which must be 1s by condition 5) are located in a pattern suggestive of a flight of stairs. Moreover, these leading entries of 1 are the only nonzero entries in their columns. Also, each nonzero row lies above all of the zero rows.

---

[4]Condition 3 is a direct consequence of condition 2. We include it in this definition for emphasis, as is the common practice when defining the row echelon form.

[5]Inexpensive calculators are available that can compute the reduced row echelon form of a matrix. On such a calculator, or in computer software, the reduced row echelon form is usually obtained by using the command *rref*.

Example 2 The following matrices are *not* in reduced row echelon form:

$$A = \begin{bmatrix} 1 & 0 & 0 & 6 & 3 & 0 \\ 0 & 0 & 1 & 5 & 7 & 0 \\ 0 & 1 & 0 & 2 & 4 & 0 \\ 0 & 0 & 0 & 0 & 0 & 1 \end{bmatrix} \qquad B = \begin{bmatrix} 1 & 7 & 2 & -3 & 9 & 4 \\ 0 & 0 & 1 & 4 & 6 & 8 \\ 0 & 0 & 0 & 2 & 3 & 5 \\ 0 & 0 & 0 & 0 & 0 & 0 \\ 0 & 0 & 0 & 0 & 0 & 0 \end{bmatrix}.$$

Matrix A fails to be in reduced row echelon form because the leading entry of the third row does not lie to the right of the leading entry of the second row. Notice, however, that the matrix obtained by interchanging the second and third rows of A is in reduced row echelon form.

Matrix B is not in reduced row echelon form for two reasons. The first nonzero entry of the third row is not 1, and the leading entries in the second and third rows are not the only nonzero entries in their columns. That is, the third column of B contains the first nonzero entry in row 2, but the $(2, 3)$-entry of B is not the only nonzero entry in column 3. Notice, however, that although B *is* not in *reduced* row echelon form, it *is* in row echelon form.

A system of linear equations can be solved easily if its augmented matrix is in reduced row echelon form. For example, the system

$$\begin{aligned} x_1 & & & = -4 \\ & x_2 & & = -5 \\ & & x_3 & = 3, \end{aligned}$$

which was obtained while solving (2), has a solution that is immediately evident.

If a system of equations has infinitely many solutions, then obtaining the solution is somewhat more complicated. Consider, for example, the system of linear equations

$$\begin{aligned} x_1 - 3x_2 & + 2x_4 & = 7 \\ x_3 + 6x_4 & = 9 \\ x_5 & = 2 \\ 0 & = 0. \end{aligned} \tag{4}$$

The augmented matrix of this system is

$$\begin{bmatrix} 1 & -3 & 0 & 2 & 0 & 7 \\ 0 & 0 & 1 & 6 & 0 & 9 \\ 0 & 0 & 0 & 0 & 1 & 2 \\ 0 & 0 & 0 & 0 & 0 & 0 \end{bmatrix},$$

which is in reduced row echelon form.

Since the equation $0 = 0$ in (4) provides no useful information, we can disregard it. In (4), it is not possible to find a unique value for each variable because the system has infinitely many solutions. Instead, we can solve for some of the variables in terms of the others. Because the augmented matrix of this system is in reduced row echelon form, it is easy to solve for the **basic variables**, those corresponding to the columns of the reduced row echelon form that contain the leading entries of the nonzero rows. In this case, the leading entry of the first row is in the first column, the leading entry of the second row is in the third column, and the leading entry of the third row is in the fifth column. So the basic variables are x_1, x_3, and x_5, and they can be written easily in terms of the other variables (x_2 and x_4), which are called the **free variables**. In fact, doing so requires only that the terms involving the free variables be moved from the left side of the equations in (4) to the right.

The resulting equations,

$$
\begin{aligned}
x_1 &= 7 + 3x_2 - 2x_4 \\
x_2 &\quad \text{free} \\
x_3 &= 9 - 6x_4 \\
x_4 &\quad \text{free} \\
x_5 &= 2,
\end{aligned}
$$

provide a **general solution** to (4). This means that for every choice of values of the free variables, these equations give the corresponding values of x_1, x_3, and x_5 in a solution to the system and, furthermore, every solution of the system has this form for some values of the free variables. For example, choosing $x_2 = 0$ and $x_4 = 0$ gives the solution $\begin{bmatrix} 7 \\ 0 \\ 9 \\ 0 \\ 2 \end{bmatrix}$, and taking $x_2 = -2$ and $x_4 = 1$

produces the solution $\begin{bmatrix} -1 \\ -2 \\ 3 \\ 1 \\ 2 \end{bmatrix}$.

The general solution can also be written in vector form as

$$
\begin{bmatrix} x_1 \\ x_2 \\ x_3 \\ x_4 \\ x_5 \\ x_6 \end{bmatrix} = \begin{bmatrix} 7 + 3x_2 - 2x_4 \\ x_2 \\ 9 - x_4 \\ x_4 \\ 2 \end{bmatrix} = \begin{bmatrix} 7 \\ 0 \\ 9 \\ 0 \\ 2 \end{bmatrix} + x_2 \begin{bmatrix} 3 \\ 1 \\ 0 \\ 0 \\ 0 \end{bmatrix} + x_4 \begin{bmatrix} -2 \\ 0 \\ -1 \\ 1 \\ 0 \end{bmatrix}.
$$

In vector form, it is apparent that every solution to the system is the sum of $\begin{bmatrix} 7 \\ 0 \\ 9 \\ 0 \\ 2 \end{bmatrix}$ and an arbitrary linear combination of the vectors $\begin{bmatrix} 3 \\ 1 \\ 0 \\ 0 \\ 0 \end{bmatrix}$ and $\begin{bmatrix} -2 \\ 0 \\ -1 \\ 1 \\ 0 \end{bmatrix}$, with the coefficients being the free variables x_2 and x_4, respectively.

Example 3

Find a general solution to the system of linear equations

$$
\begin{aligned}
x_1 \qquad\quad + 2x_4 &= 7 \\
x_2 \qquad - 3x_4 &= 8 \\
x_3 + 6x_4 &= 9.
\end{aligned}
$$

Solution   Since the augmented matrix of this system is in reduced row echelon form, the solution can be obtained by solving for the basic variables in terms of the other variables. In this case, the basic variables are x_1, x_2, and x_3, and so we solve for x_1, x_2, and x_3 in terms of x_4. The resulting general solution is

$$x_1 = 7 - 2x_4$$
$$x_2 = 8 + 3x_4$$
$$x_3 = 9 - 6x_4$$
$$x_4 \quad \text{free.}$$

The general solution can be written in vector form as

$$\begin{bmatrix} x_1 \\ x_2 \\ x_3 \\ x_4 \end{bmatrix} = \begin{bmatrix} 7 \\ 8 \\ 9 \\ 0 \end{bmatrix} + x_4 \begin{bmatrix} -2 \\ 3 \\ -6 \\ 1 \end{bmatrix}.$$

There is one other case to consider. Suppose that the augmented matrix of a system of linear equations contains a row in which the only nonzero entry is in the last column, for example,

$$\begin{bmatrix} 1 & 0 & -3 & 5 \\ 0 & 1 & 2 & 4 \\ 0 & 0 & 0 & 1 \\ 0 & 0 & 0 & 0 \end{bmatrix}.$$

The system of linear equations corresponding to this matrix is

$$x_1 \qquad\quad - 3x_3 = 5$$
$$x_2 + 2x_3 = 4$$
$$0x_1 + 0x_2 + 0x_3 = 1$$
$$0x_1 + 0x_2 + 0x_3 = 0.$$

Clearly, there are no values of the variables that satisfy the third equation. Because a solution to the system must satisfy every equation in the system, it follows that this system of equations is inconsistent. More generally, the following statement is true.

> Whenever an augmented matrix contains a row in which the only nonzero entry lies in the last column, the corresponding system of linear equations has no solution.

Solving Systems of Linear Equations

So far, we have learned the following facts.

1. A system of linear equations can be represented by its augmented matrix.
2. A system of linear equations whose augmented matrix is in reduced row echelon form is easily solved.

The existence of a reduced row echelon form is established in Section 1.4, and its uniqueness is proved in Section 2.3. For the present, we state these results in the following theorem.

Theorem 1.3

Every matrix can be transformed into one and only one matrix in reduced row echelon form by means of a sequence of elementary row operations.

In fact, Section 1.4 describes an explicit procedure for performing this transformation. If there is a sequence of elementary row operations that transforms a matrix A into a matrix R in reduced row echelon form, then we call R the **reduced row echelon form of** A. Using the reduced row echelon form of the augmented matrix of a system of linear equations $A\mathbf{x} = \mathbf{b}$, we can solve the system as follows.

1. Write the augmented matrix $[A \ \ \mathbf{b}]$ of the system.
2. Find the reduced row echelon form $[R \ \ \mathbf{c}]$ of $[A \ \ \mathbf{b}]$.
3. If $[R \ \ \mathbf{c}]$ contains a row in which the only nonzero entry lies in the last column, then $A\mathbf{x} = \mathbf{b}$ has no solution. Otherwise, the system has at least one solution. Write a system of linear equations corresponding to the matrix $[R \ \ \mathbf{c}]$, and solve this system for the basic variables in terms of the free variables to obtain a general solution of $A\mathbf{x} = \mathbf{b}$.

Example 4   Solve the following system of linear equations:

$$
\begin{aligned}
x_1 + 2x_2 - x_3 + 2x_4 + x_5 &= 2 \\
-x_1 - 2x_2 + x_3 + 2x_4 + 3x_5 &= 6 \\
2x_1 + 4x_2 - 3x_3 + 2x_4 &= 3 \\
-3x_1 - 6x_2 + 2x_3 + 3x_5 &= 9.
\end{aligned}
$$

Solution   The augmented matrix of this system is

$$
\begin{bmatrix}
1 & 2 & -1 & 2 & 1 & 2 \\
-1 & -2 & 1 & 2 & 3 & 6 \\
2 & 4 & -3 & 2 & 0 & 3 \\
-3 & -6 & 2 & 0 & 3 & 9
\end{bmatrix}.
$$

In Section 1.4 we show that the reduced row echelon form of this matrix is

$$
\begin{bmatrix}
1 & 2 & 0 & 0 & -1 & -5 \\
0 & 0 & 1 & 0 & 0 & -3 \\
0 & 0 & 0 & 1 & 1 & 2 \\
0 & 0 & 0 & 0 & 0 & 0
\end{bmatrix}.
$$

Because there is no row in this matrix in which the only nonzero entry lies in the last column, the original system is consistent. This matrix corresponds to the system of linear equations

$$
\begin{aligned}
x_1 + 2x_2 \quad - x_5 &= -5 \\
x_3 \quad &= -3 \\
x_4 + x_5 &= 2.
\end{aligned}
$$

In this system, the basic variables are x_1, x_3, and x_4, and the free variables are x_2 and x_5. When we solve for the basic variables in terms of the free variables, we obtain the following general solution:

$$x_1 = -5 - 2x_2 + x_5$$
$$x_2 \quad \text{free}$$
$$x_3 = -3$$
$$x_4 = \quad 2 \qquad - x_5$$
$$x_5 \quad \text{free.}$$

This is a general solution to the original system of linear equations. ○

Practice Problems

1. Determine if (a) $\mathbf{u} = \begin{bmatrix} -2 \\ 3 \\ 2 \\ 1 \end{bmatrix}$ and (b) $\mathbf{v} = \begin{bmatrix} 5 \\ 8 \\ 1 \\ 3 \end{bmatrix}$ are solutions to the system of linear equations

$$x_1 + \quad 5x_3 - x_4 = 7$$
$$2x_1 - x_2 + 6x_3 \quad = 8.$$

2. The augmented matrix of a system of linear equations has

$$\begin{bmatrix} 0 & 1 & -4 & 0 & 3 & 5 \\ 0 & 0 & 0 & 1 & -2 & 6 \\ 0 & 0 & 0 & 0 & 0 & 1 \end{bmatrix}$$

as its reduced row echelon form. Determine if this system of linear equations is consistent and, if so, find its general solution.

3. The augmented matrix of a system of linear equations has

$$\begin{bmatrix} 0 & 1 & -3 & 0 & 2 & 4 \\ 0 & 0 & 0 & 1 & -1 & 5 \\ 0 & 0 & 0 & 0 & 0 & 0 \end{bmatrix}$$

as its reduced row echelon form. Write the corresponding system of linear equations, determine if it is consistent and, if so, find its general solution.

Exercises

1. Determine if the following statements are true or false.

(a) Every system of linear equations has at least one solution.

(b) Some systems of linear equations have exactly two solutions.

(c) If a matrix A can be transformed into a matrix B by performing an elementary row operation, then B can be transformed into A by performing an elementary row operation.

(d) If a matrix is in row echelon form, then the leading entry of each nonzero row must be 1.

(e) If a matrix is in reduced row echelon form, then the leading entry of each nonzero row is 1.

(f) Every matrix can be transformed into one in reduced row echelon form by a sequence of elementary row operations.

(g) Every matrix can be transformed into a unique matrix in row echelon form by a sequence of elementary row operations.

(h) Every matrix can be transformed into a unique matrix in reduced row echelon form by a sequence of elementary row operations.

(i) Performing an elementary row operation on the augmented matrix of a system of linear equations produces the augmented matrix of an equivalent system of linear equations.

(j) If the reduced row echelon form of the augmented matrix of a system of linear equations contains a zero row, then the system is consistent.

(k) If the only nonzero entry in some row of an augmented matrix of a system of linear equations lies in the last column, then the system is inconsistent.

In Exercises 2–6, write (a) the coefficient matrix, and (b) the augmented matrix of the given system.

2. $2x_1 - x_2 + 3x_3 = 4$

3. $\quad -x_2 + 2x_3 = \quad 0$
$\quad x_1 + 3x_2 \quad = -1$

4. $x_1 \quad + 2x_3 - x_4 = 3$
$2x_1 - x_2 + \quad x_4 = 0$

5. $\quad x_1 + 2x_2 = 3$
$-x_1 + 3x_2 = 2$
$-3x_1 + 4x_2 = 1$

6. $\quad 2x_2 - 3x_3 = \quad 4$
$-x_1 + x_2 + 2x_3 = -6$
$2x_1 + \quad x_3 = \quad 0$

In Exercises 7–12, perform the indicated elementary row operation on

$$\begin{bmatrix} 1 & -1 & 0 & 2 & -3 \\ -2 & 6 & 3 & -1 & 1 \\ 0 & 2 & -4 & 4 & 2 \end{bmatrix}.$$

7. Interchange rows 1 and 3.   8. Multiply row 1 by -3.

9. Add 2 times row 1 to row 2.

10. Interchange rows 1 and 2.

11. Multiply row 3 by $\frac{1}{2}$.

12. Add -3 times row 3 to row 2.

In Exercises 13–18, perform the indicated elementary row operation on

$$\begin{bmatrix} 1 & -2 & 0 \\ -1 & 1 & -1 \\ 2 & -4 & 6 \\ -3 & 2 & 1 \end{bmatrix}.$$

13. Multiply row 1 by -2.   14. Multiply row 2 by $\frac{1}{2}$.

15. Add -2 times row 1 to row 3.

16. Add 3 times row 1 to row 4.

17. Interchange rows 2 and 3.

18. Interchange rows 2 and 4.

In Exercises 19–22, determine if the given vector is a solution to the system

$$\begin{aligned} x_1 - 4x_2 \quad + 3x_4 &= 6 \\ x_3 - 2x_4 &= -3. \end{aligned}$$

19. $\begin{bmatrix} 1 \\ -2 \\ -5 \\ -1 \end{bmatrix}$ 20. $\begin{bmatrix} 2 \\ 0 \\ -1 \\ 1 \end{bmatrix}$ 21. $\begin{bmatrix} 3 \\ 0 \\ 2 \\ 1 \end{bmatrix}$ 22. $\begin{bmatrix} 4 \\ 1 \\ 1 \\ 2 \end{bmatrix}$

In Exercises 23–38, the reduced row echelon form of the augmented matrix of a system of linear equations is given. Determine if this system of linear equations is consistent and, if so, find its general solution. In Exercises 33–38, write the solution in vector form.

23. $\begin{bmatrix} 1 & -1 & 2 \end{bmatrix}$ 24. $\begin{bmatrix} 1 & 0 & -4 \\ 0 & 1 & 5 \end{bmatrix}$

25. $\begin{bmatrix} 1 & -2 & 6 \\ 0 & 0 & 0 \end{bmatrix}$ 26. $\begin{bmatrix} 1 & -4 & 5 \\ 0 & 0 & 0 \\ 0 & 0 & 0 \end{bmatrix}$

27. $\begin{bmatrix} 1 & -3 & 0 \\ 0 & 0 & 1 \\ 0 & 0 & 0 \end{bmatrix}$ 28. $\begin{bmatrix} 1 & 0 & -6 \\ 0 & 1 & 3 \\ 0 & 0 & 0 \end{bmatrix}$

29. $\begin{bmatrix} 1 & -2 & 0 & 4 \\ 0 & 0 & 1 & 3 \\ 0 & 0 & 0 & 0 \end{bmatrix}$ 30. $\begin{bmatrix} 1 & -2 & 0 & 0 \\ 0 & 0 & 1 & 0 \\ 0 & 0 & 0 & 1 \end{bmatrix}$

31. $\begin{bmatrix} 1 & 0 & 0 & -3 & 0 \\ 0 & 1 & 0 & -4 & 0 \\ 0 & 0 & 1 & 5 & 0 \end{bmatrix}$ 32. $\begin{bmatrix} 1 & 0 & -1 & 3 & 9 \\ 0 & 1 & 2 & -5 & 8 \\ 0 & 0 & 0 & 0 & 0 \end{bmatrix}$

33. $\begin{bmatrix} 0 & 1 & 0 & 0 & -3 \\ 0 & 0 & 1 & 0 & -4 \\ 0 & 0 & 0 & 1 & 5 \end{bmatrix}$ 34. $\begin{bmatrix} 1 & -2 & 0 & 0 & -3 \\ 0 & 0 & 1 & 0 & -4 \\ 0 & 0 & 0 & 1 & 5 \end{bmatrix}$

35. $\begin{bmatrix} 1 & 3 & 0 & -2 & 6 \\ 0 & 0 & 1 & 4 & 7 \\ 0 & 0 & 0 & 0 & 0 \end{bmatrix}$ 36. $\begin{bmatrix} 0 & 1 & 0 & 3 & -4 \\ 0 & 0 & 1 & 2 & 9 \\ 0 & 0 & 0 & 0 & 0 \end{bmatrix}$

37. $\begin{bmatrix} 1 & -3 & 2 & 0 & 4 & 0 \\ 0 & 0 & 0 & 0 & 0 & 1 \\ 0 & 0 & 0 & 0 & 0 & 0 \end{bmatrix}$

38. $\begin{bmatrix} 0 & 0 & 1 & -3 & 0 & 2 & 0 \\ 0 & 0 & 0 & 0 & 1 & -1 & 0 \\ 0 & 0 & 0 & 0 & 0 & 0 & 0 \end{bmatrix}$

39. Suppose that the general solution to a system of m linear equations in n variables contains k free variables. How many basic variables does it have? Explain your answer.

40. Suppose that R is a matrix in reduced row echelon form. If row 4 of R is nonzero and has its leading entry in column 5, describe column 5.

41.[6] Let $[A \ \mathbf{b}]$ be the augmented matrix of a system of linear equations. Prove that if its reduced row echelon form is $[R \ \mathbf{c}]$, then R is the reduced row echelon form of A.

42. Prove that if R is the reduced row echelon form of a matrix A, then $[R \ \mathbf{0}]$ is the reduced row echelon form of $[A \ \mathbf{0}]$.

43. Prove that for any $m \times n$ matrix A, the equation $A\mathbf{x} = \mathbf{0}$ is consistent, where $\mathbf{0}$ is the zero vector in $\mathcal{R}^m$.

44. Let A be an $m \times n$ matrix whose reduced row echelon form contains no zero rows. Prove that $A\mathbf{x} = \mathbf{b}$ is consistent for every $\mathbf{b}$ in $\mathcal{R}^m$.

45. In a matrix in reduced row echelon form, there are three types of entries: the leading entries of nonzero rows are required to be 1s, certain other entries are required to be 0s, and the remaining entries are arbitrary. Suppose that these arbitrary entries are denoted by asterisks. For example,

$$\begin{bmatrix} 0 & 1 & * & 0 & * & 0 & * \\ 0 & 0 & 0 & 1 & * & 0 & * \\ 0 & 0 & 0 & 0 & 0 & 1 & * \end{bmatrix}$$

is a possible reduced row echelon form for a 3×7 matrix. How many different such forms for a reduced row echelon matrix are possible if the matrix is 2×3?

46. Repeat Exercise 45 for a 2×4 matrix.

47. Suppose that B is obtained by performing one elementary row operation on matrix A. Prove that the same type of elementary operation (namely, an interchange, scaling, or row addition operation) that transforms A into B also transforms B into A.

48. Show that if an equation in a system of linear equations is multiplied by zero, the resulting system need not be equivalent to the original one.

---

[6] The result of this exercise is used in Section 1.6 (page 63).

49. Let S denote the following system of linear equations:

$$a_{11}x_1 + a_{12}x_2 + a_{13}x_3 = b_1$$
$$a_{21}x_1 + a_{22}x_2 + a_{23}x_3 = b_2$$
$$a_{31}x_1 + a_{32}x_2 + a_{33}x_3 = b_3.$$

Show that if the second equation of S is multiplied by a nonzero scalar c, then the resulting system is equivalent to S.

50. Let S be the system of linear equations in Exercise 49. Show that if k times the first equation of S is added to the third equation, then the resulting system is equivalent to S.

Solutions to the Practice Problems

1. (a) Since $2(-2) - 3 + 6(2) = 5$, $\mathbf{u}$ is not a solution to the second equation in the given system of equations. Therefore $\mathbf{u}$ is not a solution to the system. Another method for solving this problem is to represent the given system as a matrix equation $A\mathbf{x} = \mathbf{b}$, where

$$A = \begin{bmatrix} 1 & 0 & 5 & -1 \\ 2 & -1 & 6 & 0 \end{bmatrix} \quad \text{and} \quad \mathbf{b} = \begin{bmatrix} 7 \\ 8 \end{bmatrix}.$$

Because

$$A\mathbf{u} = \begin{bmatrix} 1 & 0 & 5 & -1 \\ 2 & -1 & 6 & 0 \end{bmatrix} \begin{bmatrix} -2 \\ 3 \\ 2 \\ 1 \end{bmatrix} = \begin{bmatrix} 7 \\ 5 \end{bmatrix} \neq \mathbf{b},$$

$\mathbf{u}$ is not a solution to the given system.

(b) Since $5 + 5(1) - 3 = 7$ and $2(5) - 8 + 6(1) = 8$, $\mathbf{v}$ satisfies both of the equations in the given system. Hence $\mathbf{v}$ is a solution to the system. Alternatively, using the matrix equation $A\mathbf{x} = \mathbf{b}$, we see that $\mathbf{v}$ is a solution because

$$A\mathbf{v} = \begin{bmatrix} 1 & 0 & 5 & -1 \\ 2 & -1 & 6 & 0 \end{bmatrix} \begin{bmatrix} 5 \\ 8 \\ 1 \\ 3 \end{bmatrix} = \begin{bmatrix} 7 \\ 8 \end{bmatrix} = \mathbf{b}.$$

2. In the given matrix, the only nonzero entry in the third row lies in the last column. Hence the system of linear equations corresponding to this matrix is not consistent.

3. The corresponding system of linear equations is

$$x_2 - 3x_3 \qquad + 2x_5 = 4$$
$$x_4 - \quad x_5 = 5.$$

Since the given matrix contains no row whose only nonzero entry lies in the last column, this system is consistent. The general solution to this system is

$$\begin{aligned} x_1 \quad &\text{free} \\ x_2 = \quad &4 \ + 3x_3 - 2x_5 \\ x_3 \quad &\text{free} \\ x_4 = \quad &5 \ + \ x_5 \\ x_5 \quad &\text{free.} \end{aligned}$$

Note that x_1, which is not a basic variable, is therefore a free variable.

1.4   Gaussian Elimination

In Section 1.3, we learned how to solve a system of linear equations for which the augmented matrix is in reduced row echelon form. Our primary concern in this section is to learn a procedure that can be used to transform any matrix into this form. In Section 2.3, we show that a given matrix A can be transformed into only one matrix in reduced row echelon form.

Suppose that R is the reduced row echelon form of a matrix A. Recall that the first nonzero entry in a nonzero row of R is called the *leading entry* of that row. Since R is uniquely determined by A, the positions that contain the leading entries of the nonzero rows of R are also uniquely determined by A. These positions are called the **pivot positions** of A, and a column of A that contains some pivot position of A is called a **pivot column** of A. For example, later in this section the reduced row echelon form of

$$A = \begin{bmatrix} 1 & 2 & -1 & 2 & 1 & 2 \\ -1 & -2 & 1 & 2 & 3 & 6 \\ 2 & 4 & -3 & 2 & 0 & 3 \\ -3 & -6 & 2 & 0 & 3 & 9 \end{bmatrix}$$

is shown to be

$$R = \begin{bmatrix} 1 & 2 & 0 & 0 & -1 & 5 \\ 0 & 0 & 1 & 0 & 0 & -3 \\ 0 & 0 & 0 & 1 & 1 & 2 \\ 0 & 0 & 0 & 0 & 0 & 0 \end{bmatrix}.$$

Here, the first three rows of R are its nonzero rows, and so A has three pivot positions. The first pivot position is row 1, column 1, because the leading entry in the first row of R lies in column 1. The second pivot position is row 2, column 3, because the leading entry in the second row of R lies in column 3. Finally, the third pivot position is row 3, column 4, because the leading entry in the third row of R lies in column 4. Hence the pivot columns of A are columns 1, 3, and 4 (see Figure 1.16).

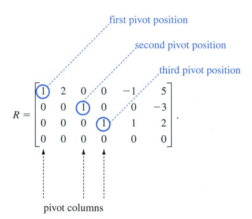

FIGURE 1.16

The pivot positions and pivot columns are easily determined from the reduced row echelon form of a matrix. Our problem, however, is to determine the pivot positions *as we are computing the reduced row echelon form*. The algorithm that we use to obtain the reduced row echelon form of a matrix is usually called **Gaussian elimination**.[7] This algorithm basically consists of locating the pivot positions, and then making certain entries of the matrix zero by means of elementary row operations. We will assume that the matrix is nonzero, because the reduced row echelon form of a zero matrix is the same zero matrix. Our procedure can be used to find the reduced row echelon form of any nonzero matrix. To illustrate the algorithm, we find the reduced row echelon form of the matrix

$$A = \begin{bmatrix} 0 & 0 & 2 & -4 & -5 & 2 & 5 \\ 0 & 1 & -1 & 1 & 3 & 1 & -1 \\ 0 & 6 & 0 & -6 & 5 & 16 & 7 \end{bmatrix}.$$

---

[7] This method is named after Carl Friedrich Gauss (1777–1855), whom many consider to be the greatest mathematician of all time. Gauss described this procedure in a paper that presented his calculations to determine the orbit of the asteroid Pallas. However, a similar method for solving systems of linear equations was known to the Chinese around 250 BC.

During steps 1–4 of the algorithm, certain rows of the matrix may be temporarily ignored. We depict such rows by shading them. At the beginning of the algorithm, no rows are ignored.

> **STEP 1** Determine the leftmost nonzero column. This is a pivot column, and the topmost position in this column is a pivot position.

Initially, no rows of A are ignored, and so we begin with the entire matrix. Since the second column of A is the leftmost nonzero column, it is the first pivot column. The topmost position in this column lies in row 1, and so the first pivot position is the row 1, column 2 position:

pivot position

$$A = \begin{bmatrix} 0 & 0 & 2 & -4 & -5 & 2 & 5 \\ 0 & 1 & -1 & 1 & 3 & 1 & -1 \\ 0 & 6 & 0 & -6 & 5 & 16 & 7 \end{bmatrix}.$$

pivot column

> **STEP 2** In the pivot column, choose any nonzero[8] entry in a row that is not ignored, and perform the appropriate row interchange to bring this entry into the pivot position.

Because the entry in the pivot position is zero, a row interchange is required. We must select a nonzero entry in the pivot column; suppose that we select the entry 1. By interchanging rows 1 and 2, we bring this entry into the pivot position. Thus our matrix now has the form

$$\begin{bmatrix} 0 & 1 & -1 & 1 & 3 & 1 & -1 \\ 0 & 0 & 2 & -4 & -5 & 2 & 5 \\ 0 & 6 & 0 & -6 & 5 & 16 & 7 \end{bmatrix}.$$

> **STEP 3** Add an appropriate multiple of the row containing the pivot position to each lower row in order to change each entry below the pivot position into zero.

In step 3, we must add multiples of row 1 of A to rows 2 and 3 so that the pivot column entries in rows 2 and 3 are changed into zero. In this case, the entry in row 2 is already zero, and so we need only change the row 3 entry. Thus we add -6 times row 1 to row 3. This calculation is usually done mentally,

[8] When performing calculations by hand, it may be advantageous to choose an entry of the pivot column that is ± 1, if possible, in order to simplify subsequent calculations.

but we show it below for the sake of completeness.

$$
\begin{array}{rrrrrrr}
0 & -6 & 6 & -6 & -18 & -6 & 6 \\
0 & 6 & 0 & -6 & 5 & 16 & 7 \\
\hline
0 & 0 & 6 & -12 & -13 & 10 & 13
\end{array}
\quad
\begin{array}{l}
(-6 \text{ times row } 1) \\
(\text{row } 3)
\end{array}
$$

The effect of this row operation is to transform A into the matrix

$$
\begin{bmatrix}
0 & 1 & -1 & 1 & 3 & 1 & -1 \\
0 & 0 & 2 & -4 & -5 & 2 & 5 \\
0 & 0 & 6 & -12 & -13 & 10 & 13
\end{bmatrix}.
$$

_____ **STEP 4** Ignore all the rows that contain previous pivot positions. If every row of the matrix has been ignored, or if the rows that are not ignored contain only zero entries, begin step 5 using the last nonzero row of the matrix. Otherwise, repeat steps 1–4 on the submatrix consisting of the rows that are not ignored.

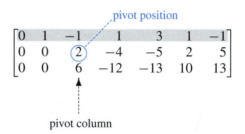

At this stage, row 1 contains the only pivot position, and therefore row 1 is ignored. Because the remaining submatrix contains nonzero entries, we return to step 1. The leftmost nonzero column of this submatrix is column 3, and so column 3 becomes the second pivot column. Since the topmost position in the second column of the submatrix lies in row 2 of the entire matrix, the second pivot position is the row 2, column 3 position.

Because the entry in the current pivot position is nonzero, no row interchange is required in step 2. Thus we continue to step 3, where we must add an appropriate multiple of row 2 to row 3 in order to create a zero in row 3, column 3. The addition of -3 times row 2 to row 3 is shown below.

$$
\begin{array}{rrrrrrr}
0 & 0 & -6 & 12 & 15 & -6 & -15 \\
0 & 0 & 6 & -12 & -13 & 10 & 13 \\
\hline
0 & 0 & 0 & 0 & 2 & 4 & -2
\end{array}
\quad
\begin{array}{l}
(-3 \text{ times row } 2) \\
(\text{row } 3)
\end{array}
$$

Thus the new matrix is

$$
\begin{bmatrix}
0 & 1 & -1 & 1 & 3 & 1 & -1 \\
0 & 0 & 2 & -4 & -5 & 2 & 5 \\
0 & 0 & 0 & 0 & 2 & 4 & -2
\end{bmatrix}.
$$

Since rows 1 and 2 both contain previous pivot positions, in step 4 we ignore rows 1 and 2. Because the remaining submatrix

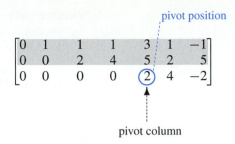

pivot column

contains nonzero entries, we repeat steps 1–4. At this stage, column 5 is the leftmost nonzero column of the submatrix. Hence column 5 is the third pivot column, and the row 3, column 5 position is the next pivot position. Because the entry in the pivot position is nonzero, no row interchange is needed in step 2. Moreover, there are no rows below the row containing the present pivot position, and so no operations are required in step 3. When step 4 is performed this time, rows 1, 2, and 3 are ignored, and so every row of the matrix has been ignored. Thus we proceed to step 5, which we perform on the last nonzero row of the matrix.

Notice that the preceding matrix is in row echelon form. Steps 5 and 6 will transform it into reduced row echelon form.

_____ STEP 5 If the leading entry of the row is not 1, perform the appropriate scaling operation to make it 1. Then add an appropriate multiple of this row to every preceding row to change each entry above the pivot position into zero.

Looking at the last nonzero row of the entire matrix, we see that the leading entry (which is the $(3, 5)$-entry) is not 1. To make it 1, we multiply the third row by $\frac{1}{2}$. This produces the following matrix:

$$\begin{bmatrix} 0 & 1 & -1 & 1 & 3 & 1 & -1 \\ 0 & 0 & 2 & -4 & -5 & 2 & 5 \\ 0 & 0 & 0 & 0 & 1 & 2 & -1 \end{bmatrix}.$$

Now we add appropriate multiples of the third row to every preceding row to change each entry above the leading entry into zero. Thus we add -3 times row 3 to row 1, and 5 times row 3 to row 2. The resulting matrix is

$$\begin{bmatrix} 0 & 1 & -1 & 1 & 0 & -5 & 2 \\ 0 & 0 & 2 & -4 & 0 & 12 & 0 \\ 0 & 0 & 0 & 0 & 1 & 2 & -1 \end{bmatrix}.$$

_____ STEP 6 If step 5 was performed using the first row, stop. Otherwise, repeat step 5 on the preceding row.

Since we just performed step 5 using the third row, we now repeat step 5 using the second row. To make the leading entry in this row a 1, we must multiply row 2 by $\frac{1}{2}$. This yields

$$\begin{bmatrix} 0 & 1 & -1 & 1 & 0 & -5 & 2 \\ 0 & 0 & 1 & -2 & 0 & 6 & 0 \\ 0 & 0 & 0 & 0 & 1 & 2 & -1 \end{bmatrix}.$$

Now the entry above the leading entry in row 2 must be changed to zero. We do this by adding row 2 to row 1. The resulting matrix is

$$\begin{bmatrix} 0 & 1 & 0 & -1 & 0 & 1 & 2 \\ 0 & 0 & 1 & -2 & 0 & 6 & 0 \\ 0 & 0 & 0 & 0 & 1 & 2 & -1 \end{bmatrix}.$$

Step 5 was just performed using the second row, and so we repeat it again using the first row. This time the leading entry in the row is 1. Consequently no scaling operation is needed in step 5. Moreover, because there are no rows above this row, no other operations are needed. Since we just worked with the first row, step 6 directs us to stop. We see that the preceding matrix is in reduced row echelon form.

Steps 1–4 of the algorithm above are called the *forward pass*. When the forward pass is complete, the original matrix has been transformed into a matrix in row echelon form. This matrix is further transformed into a matrix in reduced row echelon form by steps 5 and 6 of the algorithm, which are called the *backward pass*.

Example 1   Solve the following system of linear equations:

$$\begin{aligned} x_1 + 2x_2 - x_3 + 2x_4 + x_5 &= 2 \\ -x_1 - 2x_2 + x_3 + 2x_4 + 3x_5 &= 6 \\ 2x_1 + 4x_2 - 3x_3 + 2x_4 \phantom{{}+3x_5} &= 3 \\ -3x_1 - 6x_2 + 2x_3 + \phantom{2x_4 {}+{}} 3x_5 &= 9. \end{aligned}$$

Solution   The augmented matrix of this system is

$$\begin{bmatrix} 1 & 2 & -1 & 2 & 1 & 2 \\ -1 & -2 & 1 & 2 & 3 & 6 \\ 2 & 4 & -3 & 2 & 0 & 3 \\ -3 & -6 & 2 & 0 & 3 & 9 \end{bmatrix}.$$

We apply the Gaussian elimination algorithm to transform this matrix into one in reduced row echelon form. The first pivot position is row 1, column 1. When we add appropriate multiples of row 1 to change the entries below the pivot position to zero, the resulting matrix is

$$\begin{bmatrix} 1 & 2 & -1 & 2 & 1 & 2 \\ 0 & 0 & 0 & 4 & 4 & 8 \\ 0 & 0 & -1 & -2 & -2 & -1 \\ 0 & 0 & -1 & 6 & 6 & 15 \end{bmatrix}.$$

The second pivot position is row 2, column 3. Since the entry in this position is presently 0, we interchange rows 2 and 3 to obtain

$$\begin{bmatrix} 1 & 2 & -1 & 2 & 1 & 2 \\ 0 & 0 & -1 & -2 & -2 & -1 \\ 0 & 0 & 0 & 4 & 4 & 8 \\ 0 & 0 & -1 & 6 & 6 & 15 \end{bmatrix}.$$

Next we add -1 times row 2 to row 4. Our matrix now has the form

$$\begin{bmatrix} 1 & 2 & -1 & 2 & 1 & 2 \\ 0 & 0 & -1 & -2 & -2 & -1 \\ 0 & 0 & 0 & 4 & 4 & 8 \\ 0 & 0 & 0 & 8 & 8 & 16 \end{bmatrix}.$$

The third pivot position is row 3, column 4. Adding -2 times row 3 to row 4 produces

$$\begin{bmatrix} 1 & 2 & -1 & 2 & 1 & 2 \\ 0 & 0 & -1 & -2 & -2 & -1 \\ 0 & 0 & 0 & 4 & 4 & 8 \\ 0 & 0 & 0 & 0 & 0 & 0 \end{bmatrix}.$$

At this stage, step 4 is complete, and so we continue by performing step 5 on the third row. First, we multiply row 3 by $\frac{1}{4}$ to obtain

$$\begin{bmatrix} 1 & 2 & -1 & 2 & 1 & 2 \\ 0 & 0 & -1 & -2 & -2 & -1 \\ 0 & 0 & 0 & 1 & 1 & 2 \\ 0 & 0 & 0 & 0 & 0 & 0 \end{bmatrix}.$$

Then we add -2 times row 3 to row 1 and 2 times row 3 to row 2. The resulting matrix is

$$\begin{bmatrix} 1 & 2 & -1 & 0 & -1 & -2 \\ 0 & 0 & -1 & 0 & 0 & 3 \\ 0 & 0 & 0 & 1 & 1 & 2 \\ 0 & 0 & 0 & 0 & 0 & 0 \end{bmatrix}.$$

Now we must perform step 5 using row 2. This requires that we multiply row 2 by -1, which yields

$$\begin{bmatrix} 1 & 2 & -1 & 0 & -1 & -2 \\ 0 & 0 & 1 & 0 & 0 & -3 \\ 0 & 0 & 0 & 1 & 1 & 2 \\ 0 & 0 & 0 & 0 & 0 & 0 \end{bmatrix}.$$

Then we add row 2 to row 1 to obtain

$$\begin{bmatrix} 1 & 2 & 0 & 0 & -1 & -5 \\ 0 & 0 & 1 & 0 & 0 & -3 \\ 0 & 0 & 0 & 1 & 1 & 2 \\ 0 & 0 & 0 & 0 & 0 & 0 \end{bmatrix}.$$

Performing step 5 with row 1 produces no changes, and so the matrix above is the reduced row echelon form of the augmented matrix of the given system.

This matrix corresponds to the system of linear equations

$$
\begin{aligned}
x_1 + 2x_2 \quad\quad - x_5 &= -5 \\
x_3 \quad\quad\quad &= -3 \\
x_4 + x_5 &= \;\;\; 2,
\end{aligned}
$$

which (as we saw in Example 4 of Section 1.3) yields the following general solution to the original system of linear equations:

$$
\begin{aligned}
x_1 &= -5 - 2x_2 + x_5 \\
x_2 &\quad \text{free} \\
x_3 &= -3 \\
x_4 &= \;\;\; 2 \quad\quad - x_5 \\
x_5 &\quad \text{free.}
\end{aligned}
$$

The Rank and Nullity of a Matrix

It is clear that the Gaussian elimination algorithm transforms any matrix into one in reduced row echelon form. This establishes the existence of a reduced row echelon form for any matrix. As previously noted, the uniqueness of this reduced row echelon form is shown in Section 2.3. For the present, we assume the uniqueness of the reduced row echelon form of a matrix in order to discuss two significant numbers that are associated with a matrix.

Definitions.   The **rank** of an $m \times n$ matrix A, denoted rank A, is defined as the number of nonzero rows in the reduced row echelon form of A. The **nullity** of A, denoted nullity A, is defined as $n -$ rank A.

Example 2   For the matrix

$$
\begin{bmatrix}
1 & 2 & -1 & 2 & 1 & 2 \\
-1 & -2 & 1 & 2 & 3 & 6 \\
2 & 4 & -3 & 2 & 0 & 3 \\
-3 & -6 & 2 & 0 & 3 & 9
\end{bmatrix}
$$

in Example 1, the reduced row echelon form is

$$
\begin{bmatrix}
1 & 2 & 0 & 0 & -1 & -5 \\
0 & 0 & 1 & 0 & 0 & -3 \\
0 & 0 & 0 & 1 & 1 & 2 \\
0 & 0 & 0 & 0 & 0 & 0
\end{bmatrix}.
$$

Since the reduced row echelon form has three nonzero rows, the rank of the matrix is 3. The nullity of the matrix is found by subtracting its rank from the number of columns, and so the nullity of the matrix is $6 - 3 = 3$.

Example 3   The reduced row echelon form of the matrix

$$B = \begin{bmatrix} 2 & 3 & 1 & 5 & 2 \\ 0 & 1 & 1 & 3 & 2 \\ 4 & 5 & 1 & 7 & 2 \\ 2 & 1 & -1 & -1 & -2 \end{bmatrix}$$

is

$$\begin{bmatrix} 1 & 0 & -1 & -2 & -2 \\ 0 & 1 & 1 & 3 & 2 \\ 0 & 0 & 0 & 0 & 0 \\ 0 & 0 & 0 & 0 & 0 \end{bmatrix}.$$

Since the latter matrix has two nonzero rows, the rank of B is 2. The nullity of B is $5 - 2 = 3$.

Consider the system of linear equations in Example 1 as a matrix equation $A\mathbf{x} = \mathbf{b}$. Our method for solving the system is to find the reduced row echelon form $[R \ \mathbf{c}]$ of the augmented matrix $[A \ \mathbf{b}]$. The system $R\mathbf{x} = \mathbf{c}$ is then equivalent to $A\mathbf{x} = \mathbf{b}$, and $R\mathbf{x} = \mathbf{c}$ is easily solved, because $[R \ \mathbf{c}]$ is in reduced row echelon form. Note that each nontrivial equation in the system of linear equations $R\mathbf{x} = \mathbf{c}$ arises from a nonzero row of $[R \ \mathbf{c}]$. Moreover, each nontrivial equation in this system contains exactly one basic variable, and different equations contain different basic variables. It follows that the number of basic variables in a general solution to $R\mathbf{x} = \mathbf{c}$ equals the number of nonzero rows in $[R \ \mathbf{c}]$. Thus the number of basic variables in a general solution to $A\mathbf{x} = \mathbf{b}$ equals the rank of A. In addition, every variable that is not a basic variable must be a free variable, and so the number of free variables in the solution to $A\mathbf{x} = \mathbf{b}$ is the nullity of A.

These comments about the system of linear equations in Example 1 are true as well for any consistent system.

If $A\mathbf{x} = \mathbf{b}$ is the matrix form of a consistent system of linear equations, then:

(a) the number of basic variables in a general solution to the system equals the rank of A; and

(b) the number of free variables in a general solution to the system equals the nullity of A.

Note that the system of linear equations in Example 1 is a system of four equations in five variables. However, it is equivalent to a system of three equations in five variables: namely, the system corresponding to the reduced row echelon form of the augmented matrix. Thus the rank of the augmented matrix $[A \ \mathbf{b}]$ tells us the number of *nonredundant* equations in the original system. In Example 1, for instance, there are three nonredundant equations because the fourth equation in the original system is a combination of the first three equations. Specifically, it is the sum of -3 times the first equation, 2 times the second equation, and the third equation.

Suppose that Gaussian elimination is used to obtain the reduced row echelon form of a matrix. Each nonzero row of the reduced row echelon form of the matrix contains exactly one of the pivot positions that occur when using Gaussian elimination. Hence *the rank of A equals the number of pivot columns of A.* Actually, the preceding comments apply as well to any row echelon form of A. That is, each nonzero row of any row echelon form of A contains exactly one pivot position; so *the rank of A equals the number of nonzero rows in any row echelon form of A.*

The following theorem provides several conditions that are equivalent to the existence of solutions for a system of linear equations.

Theorem 1.4   (Test for consistency)
The following conditions are equivalent.
  (a) The matrix equation $A\mathbf{x} = \mathbf{b}$ is consistent.
  (b) Vector $\mathbf{b}$ is a linear combination of the columns of A.
  (c) The reduced row echelon form of $[A \ \ \mathbf{b}]$ has no row in which the only nonzero entry lies in the last column.[9]
  (d) The rank of $[A \ \ \mathbf{b}]$ equals the rank of A.

Proof Let A be an $m \times n$ matrix, and let $\mathbf{b} \in \mathcal{R}^n$. By the definition of a matrix–vector product, there exists a vector

$$\mathbf{v} = \begin{bmatrix} v_1 \\ v_2 \\ \vdots \\ v_n \end{bmatrix} \in \mathcal{R}^n$$

such that $A\mathbf{v} = \mathbf{b}$ if and only if

$$v_1\mathbf{a}_1 + v_2\mathbf{a}_2 + \cdots + v_n\mathbf{a}_n = \mathbf{b}.$$

Thus $A\mathbf{x} = \mathbf{b}$ is consistent if and only if $\mathbf{b}$ is a linear combination of the columns of A. Thus (a) is equivalent to (b).

We leave the proof that (a) is equivalent to (c) to the reader.

Finally, we show that (d) is equivalent to (c). Let $[R \ \ \mathbf{c}]$ be the reduced row echelon form of $[A \ \ \mathbf{b}]$. Then R is the reduced row echelon form of A. If the rank of $[A \ \ \mathbf{b}]$ equals the rank of A, then every nonzero entry in $\mathbf{c}$ occurs in the same row as some nonzero entry in R. Thus $[R \ \ \mathbf{c}]$, the reduced row echelon form of $[A \ \ \mathbf{b}]$, has no row in which the only nonzero number lies in the last column. Conversely, if the rank of $[R \ \ \mathbf{c}]$ does not equal the rank of A, there must be a nonzero row in $[R \ \ \mathbf{c}]$ for which the corresponding row of R consists entirely of zeros. Therefore (d) is equivalent to (c). ❑

Technological Considerations*

Gaussian elimination is the most efficient procedure for reducing a matrix to its reduced row echelon form. Nevertheless, it requires many tedious computations. It can be shown, in fact, that the number of arithmetic operations to obtain the reduced row echelon form of an $n \times (n+1)$ matrix is $\frac{2}{3}n^3 + \frac{1}{2}n^2 - \frac{7}{6}n$. Yet the algorithm is easily programmed on a computer or programmable calcu-

---

[9]Theorem 1.4 remains true if (c) is changed to: Every row echelon form of $[A \ \ \mathbf{b}]$ has no row in which the only nonzero entry lies in the last column.

*The remainder of this section may be omitted without loss of continuity.

lator, and so technology can be used to obtain the reduced row echelon form of a matrix. The use of technology, however, introduces other concerns. Although our statements pertain to the TI-85 calculator, it is important to understand that the types of problems we are going to describe can arise whenever technology is used. That is, every calculator or computer software package that uses floating-point arithmetic is subject to these errors.

For the matrix

$$\begin{bmatrix} 1 & -1 & 2 & 3 & 1 & -1 \\ 3 & -1 & 2 & 4 & 1 & 2 \\ 7 & -2 & 4 & 8 & 1 & 6 \end{bmatrix},$$

the TI-85 calculator gives the reduced row echelon form as

$$\begin{bmatrix} 1 & 0 & -1\text{E}-14 & 0 & -.999999999999 & 2 \\ 0 & 1 & -2 & 0 & 4 & 1.3\text{E}-12 \\ 0 & 0 & 0 & 1 & 2 & -1 \end{bmatrix}.$$

(The notation $a\text{E}b$ represents $a \times 10^b$.) It can be checked, however, that the third, fifth, and sixth columns should be

$$\begin{bmatrix} 0 \\ -2 \\ 0 \end{bmatrix}, \quad \begin{bmatrix} -1 \\ 4 \\ 2 \end{bmatrix}, \quad \text{and} \quad \begin{bmatrix} 2 \\ 0 \\ -1 \end{bmatrix},$$

respectively. On the TI-85 calculator, numbers are stored using 14 digits, and so a number containing more than 14 significant digits is stored inexactly in the calculator. Subsequent calculations with that number can accumulate errors to such a degree that the final result of the calculation is highly inaccurate. These types of errors are usually called *roundoff errors*. In our calculation of the reduced row echelon form of the matrix A, roundoff errors have affected the $(1, 3)$-entry, the $(1, 5)$-entry, and the $(2, 6)$-entry. In this instance, none of the affected entries is greatly changed, and it is reasonable to expect that the true entries in these positions should be 0, -1, and 0, respectively. But can we be absolutely sure? (We will learn a way of checking these entries in Section 2.3.)

It is not always so obvious that roundoff errors have occurred. Consider the system of linear equations

$$kx_1 + (k-1)x_2 = 1$$
$$(k+1)x_1 + kx_2 = 2.$$

By subtracting the first equation from the second, this system is easily solved, and the solution is found to be $x_1 = 2 - k$ and $x_2 = k - 1$. But for sufficiently large values of k, roundoff errors can cause problems. Taking $k = 4{,}935{,}937$, for example, the TI-85 calculator gives the reduced row echelon form of the augmented matrix

$$\begin{bmatrix} 4935937 & 4935936 & 1 \\ 4935938 & 4935937 & 2 \end{bmatrix}$$

as

$$\begin{bmatrix} 1 & .999999797404 & 0 \\ 0 & 0 & 1 \end{bmatrix}.$$

Since the last row of the reduced row echelon form has its only nonzero entry in the last column, we would incorrectly deduce from this that the original system is inconsistent!

The analysis of roundoff errors and related matters is a serious mathematical subject that is inappropriate for this book. (It is studied in the branch of mathematics called *numerical analysis*.) We encourage the use of technology whenever possible to perform the tedious calculations associated with matrices (such as those required to obtain the reduced row echelon form of a matrix). Nevertheless, a certain amount of skepticism is healthy when technology is used. Just because the calculations are performed using a calculator or computer is no guarantee that the result is correct. In this book, however, the examples and exercises usually involve simple numbers (often one- or two-digit integers), and so the likelihood of serious errors resulting from the use of technology is small.

Practice Problems

1. Consider the following system of linear equations:

$$x_1 + x_2 + x_3 = 1$$
$$x_1 + 3x_3 = -2 + s$$
$$x_1 - x_2 + rx_3 = 3.$$

(a) For what values of r and s is this system of linear equations inconsistent?

(b) For what values of r and s does this system of linear equations have infinitely many solutions?

(c) For what values of r and s does this system of linear equations have a unique solution?

2. Find the general solution to

$$x_1 - x_2 - 3x_3 + x_4 - x_5 = -2$$
$$-2x_1 + 2x_2 + 6x_3 - 6x_5 = -6$$
$$3x_1 - 2x_2 - 8x_3 + 3x_4 - 5x_5 = -7.$$

Exercises

1. Determine if the following statements are true or false.

(a) A column of a matrix A is a pivot column if the corresponding column in the reduced row echelon form of A contains the leading entry of some nonzero row.

(b) There is a unique sequence of elementary row operations that transforms a matrix into its reduced row echelon form.

(c) When the forward pass of Gaussian elimination is complete, the original matrix has been transformed into one in row echelon form.

(d) No scaling operations are required in the forward pass of Gaussian elimination.

(e) The rank of a matrix equals the number of pivot columns in the matrix.

(f) If $A\mathbf{x} = \mathbf{b}$ is consistent, then the nullity of A equals the number of free variables in the general solution to $A\mathbf{x} = \mathbf{b}$.

(g) There exists a 5×8 matrix with rank 3 and nullity 2.

(h) If a system of m linear equations in n variables is equivalent to a system of p linear equations in q variables, then $m = p$.

(i) If a system of m linear equations in n variables is equivalent to a system of p linear equations in q variables, then $n = q$.

(j) The equation $A\mathbf{x} = \mathbf{b}$ is consistent if and only if $\mathbf{b}$ is a linear combination of the columns of A.

(k) The equation $A\mathbf{x} = \mathbf{b}$ is consistent if and only if the rank of $[A \ \ \mathbf{b}]$ equals the rank of A.

(l) If the reduced row echelon form of $[A \ \ \mathbf{b}]$ contains a zero row, then $A\mathbf{x} = \mathbf{b}$ must have infinitely many solutions.

(m) If the reduced row echelon form of $[A \ \ \mathbf{b}]$ contains a zero row, then $A\mathbf{x} = \mathbf{b}$ must be consistent.

2. What is the solution set of $A\mathbf{x} = \mathbf{0}$ if A is the $m \times n$ zero matrix?

In Exercises 3–18, determine if the given system is consistent and, if so, find its general solution.

3. $2x_1 + 6x_2 = -4$

4. $x_1 - x_2 = 3$
$ -2x_1 + 2x_2 = -6$

5. $x_1 - 2x_2 = -6$
$ -2x_1 + 3x_2 = 7$

6. $x_1 - x_2 - 3x_3 = 3$
$2x_1 + x_2 - 3x_3 = 0$

7. $2x_1 - 2x_2 + 4x_3 = 1$
$ -4x_1 + 4x_2 - 8x_3 = -3$

8.
$$x_1 - 2x_2 - x_3 = 3$$
$$-2x_1 + 4x_2 + 2x_3 = -6$$
$$3x_1 - 6x_2 - 3x_3 = 9$$

9.
$$x_1 - 2x_2 - x_3 = -3$$
$$2x_1 - 4x_2 + 2x_3 = 2$$

10.
$$x_1 + x_2 - x_3 - x_4 = -2$$
$$2x_2 - 3x_3 - 12x_4 = -3$$
$$x_1 + x_3 + 6x_4 = 0$$

11.
$$x_1 - x_2 - 3x_3 + x_4 = 0$$
$$-2x_1 + x_2 + 5x_3 = -4$$
$$4x_1 - 2x_2 - 10x_3 + x_4 = 5$$

12.
$$x_1 - 3x_2 + x_3 + x_4 = 0$$
$$-3x_1 + 9x_2 - 2x_3 - 5x_4 = 1$$
$$2x_1 - 6x_2 - x_3 + 8x_4 = -2$$

13.
$$x_1 + 3x_2 + x_3 + x_4 = -1$$
$$-2x_1 - 6x_2 - x_3 = 5$$
$$x_1 + 3x_2 + 2x_3 + 3x_4 = 2$$

14.
$$x_1 + x_2 + x_3 = -1$$
$$2x_1 + x_2 - x_3 = 2$$
$$x_1 - 2x_3 = 3$$
$$-3x_1 - 2x_2 = -1$$

15.
$$x_1 + 2x_2 + x_3 = 1$$
$$-2x_1 - 4x_2 - x_3 = 0$$
$$5x_1 + 10x_2 + 3x_3 = 2$$
$$3x_1 + 6x_2 + 3x_3 = 4$$

16.
$$x_1 - x_2 + x_3 = 7$$
$$x_1 - 2x_2 - x_3 = 8$$
$$2x_1 - x_3 = 10$$
$$-x_1 - 4x_2 - x_3 = 2$$

17.
$$x_1 - x_3 - 2x_4 - 8x_5 = -3$$
$$-2x_1 + x_3 + 2x_4 + 9x_5 = 5$$
$$3x_1 - 2x_3 - 3x_4 - 15x_5 = -9$$

18.
$$x_1 - x_2 + x_4 = -4$$
$$x_1 - x_2 + 2x_4 + 2x_5 = -5$$
$$3x_1 - 3x_2 + 2x_4 - 2x_5 = -11$$

In Exercises 19–24, determine the values of r, if any, for which the given system of linear equations is inconsistent.

19.
$$-x_1 + 4x_2 = 3$$
$$3x_1 + rx_2 = 2$$

20.
$$3x_1 + rx_2 = -2$$
$$-x_1 + 4x_2 = 6$$

21.
$$x_1 - 2x_2 = 0$$
$$4x_1 - 8x_2 = r$$

22.
$$x_1 + rx_2 = -3$$
$$2x_1 = -6$$

23.
$$x_1 - 3x_2 = -2$$
$$2x_1 + rx_2 = -4$$

24.
$$-2x_1 + x_2 = 5$$
$$rx_1 + 4x_2 = 3$$

In Exercises 25–26, determine the values of r and s for which the given system of linear equations has (a) no solutions, (b) exactly one solution, and (c) infinitely many solutions.

25.
$$x_1 + rx_2 = 5$$
$$3x_1 + 6x_2 = s$$

26.
$$-x_1 + 4x_2 = s$$
$$2x_1 + rx_2 = 6$$

In Exercises 27–28, determine the values of r for which the given system of linear equations has (a) no solutions, (b) exactly one solution, and (c) infinitely many solutions.

27.
$$-x_1 + rx_2 = 2$$
$$rx_1 - 9x_2 = 6$$

28.
$$x_1 + rx_2 = 2$$
$$rx_1 + 16x_2 = 8$$

In Exercises 29–32, find the rank and nullity of the given matrix.

29.
$$\begin{bmatrix} 1 & -1 & -1 & 0 \\ 2 & -1 & -2 & 1 \\ 1 & -2 & -2 & 2 \\ -4 & 2 & 3 & 1 \\ 1 & -1 & -2 & 3 \end{bmatrix}$$

30.
$$\begin{bmatrix} 1 & -3 & -1 & 2 \\ -2 & 6 & 2 & -4 \\ 3 & -9 & 2 & 1 \\ 1 & -3 & 4 & -3 \\ -1 & 3 & -9 & 8 \end{bmatrix}$$

31.
$$\begin{bmatrix} -2 & 2 & 1 & 1 & -2 \\ 1 & -1 & -1 & -3 & 3 \\ -1 & 1 & -1 & -7 & 5 \end{bmatrix}$$

32.
$$\begin{bmatrix} 1 & 0 & -2 & -1 & 0 & -1 \\ 2 & -1 & -6 & -2 & 0 & -4 \\ 0 & 1 & 2 & 1 & 1 & 1 \\ -1 & 2 & 6 & 3 & 1 & 2 \end{bmatrix}$$

33. A mining company operates three mines that each produce three grades of ore. The daily yield of each mine is shown in the following table.

	Daily Yield		
	Mine 1	Mine 2	Mine 3
High-grade ore	1 ton	1 ton	2 tons
Medium-grade ore	1 ton	2 tons	2 tons
Low-grade ore	2 tons	1 ton	0 tons

(a) Can the company supply exactly 80 tons of high-grade, 100 tons of medium-grade, and 40 tons of low-grade ore? If so, how many days should each mine operate to fill this order?

(b) Can the company supply exactly 40 tons of high-grade, 100 tons of medium-grade, and 80 tons of low-grade ore? If so, how many days should each mine operate to fill this order?

34. A company makes three types of fertilizer. The first type contains 10% nitrogen and 3% phosphates by weight, the second contains 8% nitrogen and 6% phosphates, and the third contains 6% nitrogen and 1% phosphates.

    (a) Can the company mix these three types of fertilizers to supply exactly 600 lb of fertilizer containing 7.5% nitrogen and 5% phosphates? If so, how?
    (b) Can the company mix these three types of fertilizers to supply exactly 600 lb of fertilizer containing 9% nitrogen and 3.5% phosphates? If so, how?

35. A patient needs to consume exactly 660 mg of magnesium, 820 IU of vitamin D, and 750 mcg of folate per day. Three food supplements can be mixed to provide these nutrients. The amounts of the three nutrients provided by each of the supplements is given in the following table.

	Food Supplement		
	1	2	3
Magnesium (mg)	10	15	36
Vitamin D (IU)	10	20	44
Folate (mcg)	15	15	42

    (a) What is the maximum amount of supplement 3 that can be used to provide exactly the required amounts of the three nutrients?
    (b) Can the three supplements be mixed to provide exactly 720 mg of magnesium, 800 IU of vitamin D, and 750 mcg of folate? If so, how?

36. Three grades of crude oil are to be blended to obtain 100 barrels of oil costing $35 per barrel and containing 50 gm of sulfur per barrel. The cost and sulfur content of the three grades of oil are given in the following table.

	Grade		
	A	B	C
Cost per barrel	$40	$32	$24
Sulfur per barrel	30 gm	62 gm	94 gm

    (a) Find the amounts of each grade to be blended that use the least oil of grade C.
    (b) Find the amounts of each grade to be blended that use the most oil of grade C.

37. Find a polynomial function $f(x) = ax^2 + bx + c$ whose graph passes through the points $(-1, 14)$, $(1, 4)$, and $(3, 10)$.

38. Find a polynomial function $f(x) = ax^2 + bx + c$ whose graph passes through the points $(-2, -33)$, $(2, -1)$, and $(3, -8)$.

39. Find a polynomial function $f(x) = ax^3 + bx^2 + cx + d$ whose graph passes through the points $(-2, 32)$, $(-1, 13)$, $(2, 4)$, and $(3, 17)$.

40. Find a polynomial function $f(x) = ax^3 + bx^2 + cx + d$ whose graph passes through the points $(-2, 12)$, $(-1, -9)$, $(1, -3)$, and $(3, 27)$.

41. If the third pivot position of a matrix A is in column j, what can be said about column j of the reduced row echelon form of A? Explain your answer.

42. Suppose that the fourth pivot position of a matrix is in row i and column j. Say as much as possible about i and j. Explain your answer.

43. Describe an $m \times n$ matrix with rank 0.

44. What is the smallest possible rank of a 4×7 matrix? Explain your answer.

45. What is the largest possible rank of a 4×7 matrix? Explain your answer.

46. What is the largest possible rank of a 7×4 matrix? Explain your answer.

47. What is the smallest possible nullity of a 4×7 matrix? Explain your answer.

48. What is the smallest possible nullity of a 7×4 matrix? Explain your answer.

49. What is the largest possible rank of an $m \times n$ matrix? Explain your answer.

50. What is the smallest possible nullity of an $m \times n$ matrix? Explain your answer.

51. Let A be a 4×3 matrix. Is it possible that $Ax = b$ is consistent for every b in $\mathcal{R}^4$? Explain your answer.

52. Let A be an $m \times n$ matrix and b be a vector in $\mathcal{R}^m$. What must be true about the rank of A if $Ax = b$ has a unique solution? Justify your answer.

53. A system of linear equations is called *underdetermined* if it has fewer equations than variables. What can be said about the number of solutions to an underdetermined system?

54. A system of linear equations is called *overdetermined* if it has more equations than variables. Give examples of overdetermined systems that have

    (a) no solutions,
    (b) exactly one solution, and
    (c) infinitely many solutions.

55. Let A be an $m \times n$ matrix and b be a vector in $\mathcal{R}^m$ such that $Ax = b$ is consistent. Prove that $Ax = cb$ is consistent for every scalar c.

56. Let A be an $m \times n$ matrix and b_1 and b_2 be vectors in $\mathcal{R}^m$, such that both $Ax = b_1$ and $Ax = b_2$ are consistent. Prove that $Ax = b_1 + b_2$ is consistent.

57. Prove that if A is an $m \times n$ matrix with rank m, then $Ax = b$ is consistent for every b in $\mathcal{R}^m$.

58. Let u be a solution to $Ax = 0$, where A is an $m \times n$ matrix. Must cu be a solution to $Ax = 0$ for every scalar c? Justify your answer.

59. Let $\mathbf{u}$ and $\mathbf{v}$ be solutions to $A\mathbf{x} = \mathbf{b}$, where A is an $m \times n$ matrix and $\mathbf{b}$ is a vector in $\mathcal{R}^m$. Must $\mathbf{u} + \mathbf{v}$ be a solution to $A\mathbf{x} = \mathbf{b}$? Justify your answer.

60. Let $\mathbf{u}$ and $\mathbf{v}$ be solutions to $A\mathbf{x} = \mathbf{0}$, where A is an $m \times n$ matrix. Must $\mathbf{u} + \mathbf{v}$ be a solution to $A\mathbf{x} = \mathbf{0}$? Justify your answer.

61. Let $\mathbf{u}$ be a solution to $A\mathbf{x} = \mathbf{b}$ and $\mathbf{v}$ be a solution to $A\mathbf{x} = \mathbf{0}$, where A is an $m \times n$ matrix and $\mathbf{b}$ is a vector in $\mathcal{R}^m$. Prove that $\mathbf{u} + \mathbf{v}$ is a solution to $A\mathbf{x} = \mathbf{b}$.

62. Let $\mathbf{u}$ and $\mathbf{v}$ be solutions to $A\mathbf{x} = \mathbf{b}$, where A is an $m \times n$ matrix and $\mathbf{b}$ is a vector in $\mathcal{R}^m$. Prove that $\mathbf{u} - \mathbf{v}$ is a solution to $A\mathbf{x} = \mathbf{0}$.

63. Calculate the reduced row echelon form of $\begin{bmatrix} O & I_m \\ I_n & O' \end{bmatrix}$ if O and O' are zero matrices.

In Exercises 64–69, use a calculator with matrix capabilities or computer software such as MATLAB *to solve the problem.*

In Exercises 64–66, use Gaussian elimination on the augmented matrix of the system of linear equations to test for consistency, and to find the general solution.

64.
$$1.3x_1 + 0.5x_2 - 1.1x_3 + 2.7x_4 - 2.1x_5 = 12.9$$
$$2.2x_1 - 4.5x_2 + 3.1x_3 - 5.1x_4 + 3.2x_5 = -29.2$$
$$1.4x_1 - 2.1x_2 + 1.5x_3 - 3.1x_4 - 2.5x_5 = -11.9$$

65.
$$x_1 - x_2 + 3x_3 - x_4 + 2x_5 = 5$$
$$2x_1 + x_2 + 4x_3 + x_4 - x_5 = 7$$
$$3x_1 - x_2 + 2x_3 - 2x_4 + 2x_5 = 3$$
$$2x_1 - 4x_2 - x_3 - 4x_4 + 5x_5 = 6$$

66.
$$4x_1 - x_2 + 5x_3 - 2x_4 + x_5 = 0$$
$$7x_1 - 6x_2 + 3x_4 + 8x_5 = 15$$
$$9x_1 - 5x_2 + 4x_3 - 7x_4 + x_5 = 6$$
$$6x_1 + 9x_3 - 12x_4 - 6x_5 = 11$$

In Exercises 67–69, find the rank and the nullity of the matrix.

67.
$$\begin{bmatrix} 1.2 & 2.3 & -1.1 & 1.0 & 2.1 \\ 3.1 & 1.2 & -2.1 & 1.4 & 2.4 \\ -2.1 & 4.1 & 2.3 & -1.2 & 0.5 \\ 3.4 & 9.9 & -2.0 & 2.2 & 7.1 \end{bmatrix}$$

68.
$$\begin{bmatrix} 2.7 & 1.3 & 1.6 & 1.5 & -1.0 \\ 1.7 & 2.3 & -1.2 & 2.1 & 2.2 \\ 3.1 & -1.8 & 4.2 & 3.1 & 1.4 \\ 4.1 & -1.1 & 2.1 & 1.2 & 0.0 \\ 6.2 & -1.7 & 3.4 & 1.5 & 2.0 \end{bmatrix}$$

69.
$$\begin{bmatrix} 3 & -11 & 2 & 4 & -8 \\ 5 & 1 & 0 & 8 & 5 \\ 11 & 2 & -9 & 3 & -4 \\ 3 & 14 & -11 & 7 & 9 \\ 0 & 2 & 0 & 16 & 10 \end{bmatrix}$$

Solutions to the Practice Problems

1. Apply the Gaussian elimination algorithm to the augmented matrix of the given system to transform the matrix into one in row echelon form:

$$\begin{bmatrix} 1 & 1 & 1 & 1 \\ 1 & 0 & 3 & -2+s \\ 1 & -1 & r & 3 \end{bmatrix} \longrightarrow \begin{bmatrix} 1 & 1 & 1 & 1 \\ 0 & -1 & 2 & -3+s \\ 0 & -2 & r-1 & 2 \end{bmatrix}$$

$$\longrightarrow \begin{bmatrix} 1 & 1 & 1 & 1 \\ 0 & -1 & 2 & -3+s \\ 0 & 0 & r-5 & 8-2s \end{bmatrix}.$$

(a) The original system is inconsistent whenever there is a row whose only nonzero entry lies in the last column. Thus the original system is inconsistent whenever $r - 5 = 0$ and $8 - 2s \neq 0$, that is, when $r = 5$ and $s \neq 4$.

(b) The original system has infinitely many solutions whenever the system is consistent and there is a free variable in the general solution. In order to have a free variable, we must have $r - 5 = 0$, and in order for the system also to be consistent, we must have $8 - 2s = 0$. Thus the original system has infinitely many solutions if $r = 5$ and $s = 4$.

(c) The original system has a unique solution when the conditions in (a) and (b) do not hold; that is, when $r \neq 5$.

2. The augmented matrix of the given system is

$$\begin{bmatrix} 1 & -1 & -3 & 1 & -1 & -2 \\ -2 & 2 & 6 & 0 & -6 & -6 \\ 3 & -2 & -8 & 3 & -5 & -7 \end{bmatrix}.$$

Apply the Gaussian elimination algorithm to the augmented matrix of the given system to transform the matrix into one in row echelon form:

$$\begin{bmatrix} 1 & -1 & -3 & 1 & -1 & -2 \\ -2 & 2 & 6 & 0 & -6 & -6 \\ 3 & -2 & -8 & 3 & -5 & -7 \end{bmatrix}$$

$$\longrightarrow \begin{bmatrix} 1 & -1 & -3 & 1 & -1 & -2 \\ 0 & 0 & 0 & 2 & -8 & -10 \\ 0 & 1 & 1 & 0 & -2 & -1 \end{bmatrix}$$

$$\longrightarrow \begin{bmatrix} 1 & -1 & -3 & 1 & -1 & -2 \\ 0 & 1 & 1 & 0 & -2 & -1 \\ 0 & 0 & 0 & 2 & -8 & -10 \end{bmatrix}$$

$$\longrightarrow \begin{bmatrix} 1 & -1 & -3 & 1 & -1 & -2 \\ 0 & 1 & 1 & 0 & -2 & -1 \\ 0 & 0 & 0 & 1 & -4 & -5 \end{bmatrix}$$

$$\longrightarrow \begin{bmatrix} 1 & -1 & -3 & 0 & 3 & 3 \\ 0 & 1 & 1 & 0 & -2 & -1 \\ 0 & 0 & 0 & 1 & -4 & -5 \end{bmatrix}$$

$$\longrightarrow \begin{bmatrix} 1 & 0 & -2 & 0 & 1 & 2 \\ 0 & 1 & 1 & 0 & -2 & -1 \\ 0 & 0 & 0 & 1 & -4 & -5 \end{bmatrix}.$$

The final matrix corresponds to the following system of linear equations:

$$
\begin{aligned}
x_1 &\quad - 2x_3 \quad + x_5 = 2 \\
&x_2 + x_3 \quad - 2x_5 = -1 \\
&\qquad\qquad x_4 - 4x_5 = -5.
\end{aligned}
$$

The general solution to this system is

$$
\begin{aligned}
x_1 &= 2 + 2x_3 - x_5 \\
x_2 &= -1 - x_3 + 2x_5 \\
x_3 &\quad \text{free} \\
x_4 &= -5 + 4x_5 \\
x_5 &\quad \text{free.}
\end{aligned}
$$

1.5*   Applications of Systems of Linear Equations

Systems of linear equations arise in many applications of mathematics. In this section we present two such applications. They are independent of one another, and so they may be studied in either order.

The Leontief Input–Output Model

In a modern industrialized country, there are hundreds of different industries that supply goods and services needed for production. These industries are often mutually dependent. The agricultural industry, for instance, requires farm machinery to plant and harvest crops, whereas the makers of farm machinery need food produced by the agricultural industry. This interaction of industries is vital to a healthy economy, and yet it significantly complicates economic planning. A strike in one industry, for example, may have a devastating effect on another.

While a student in Berlin in the 1920s, the Russian-born economist Wassily Leontief developed a mathematical model, called the *input–output model*, for analyzing an economy. After arriving in the United States in 1931 to be a professor of economics at Harvard University, Leontief began to collect the data that would enable him to implement his ideas. Finally, after the end of World War II, he succeeded in extracting from government statistics the data necessary to create a model of the U.S. economy. This model proved to be highly accurate in predicting the behavior of the postwar U.S. economy, and earned for Leontief the 1973 Nobel Prize for Economics.

The input–output model divides an economy into sectors that provide products and services. In his 1947 model of the U.S. economy, Leontief grouped approximately 500 industries into 42 sectors, such as *electrical machinery*. Suppose that an economy is divided into n sectors, and that sector i produces some commodity or service S_i ($i = 1, 2, \ldots, n$). To facilitate comparisons among the sectors, we measure amounts of commodities and services in monetary units and hold their prices constant. For example, the amount of output of the steel industry could be measured in millions of dollars of steel produced rather than in thousands of tons of steel produced.

For each i and j, let c_{ij} denote the amount of S_i needed to produce one unit of S_j. Then the $n \times n$ matrix C whose (i, j)-entry is c_{ij} is called the **input–output matrix.** (or the **consumption matrix**) for the economy. To avoid complicated matrix calculations, we primarily consider 2×2 and 3×3 input–output matrices in our examples and exercises. Of course, no real economy could be meaningfully studied using such small matrices.

---

*This section can be omitted without loss of continuity.

To illustrate these ideas, consider an economy that is divided into three sectors: agriculture, manufacturing, and services. Suppose that each dollar's worth of agricultural output requires inputs of $0.10 from the agricultural sector, $0.20 from the manufacturing sector, and $0.30 from the services sector; each dollar's worth of manufacturing output requires inputs of $0.20 from the agricultural sector, $0.40 from the manufacturing sector, and $0.10 from the services sector; and each dollar's worth of services output requires inputs of $0.10 from the agricultural sector, $0.20 from the manufacturing sector, and $0.10 from the services sector.

From this information, we can form the following input–output matrix.

$$
\begin{array}{ccc}
\text{Agr.} & \text{Man.} & \text{Ser.}
\end{array}
$$

$$
C = \begin{bmatrix} .1 & .2 & .1 \\ .2 & .4 & .2 \\ .3 & .1 & .1 \end{bmatrix} \begin{array}{l} \text{Agriculture} \\ \text{Manufacturing} \\ \text{Services} \end{array}
$$

Note that the (i, j)-entry of the matrix represents the amount of input from sector i needed to produce a dollars' worth of output from sector j. Now let x_1, x_2, and x_3 denote the total output of the agriculture, manufacturing, and services sectors, respectively. Since x_1 dollars' worth of agricultural products are being produced, the first column of the input–output matrix shows that an input of $.1x_1$ is required from the agriculture sector, an input of $.2x_1$ is required from the manufacturing sector, and an input of $.3x_1$ is required from the services sector. Similar statements can be made for the manufacturing and services sectors. Thus Figure 1.17 shows the total amount of money flowing among the three sectors.

Note that in Figure 1.17 the three arcs leaving the agriculture sector give the total amount of agricultural output that is used as input during the production process. That is, $.1x_1 + .2x_2 + .1x_3$ represents the amount of agricultural output that is consumed during the production process. Similar statements

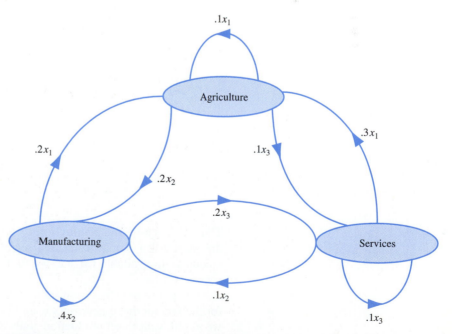

FIGURE 1.17

apply to the other two sectors. Hence

$$\begin{bmatrix} .1x_1 + .2x_2 + .1x_3 \\ .2x_1 + .4x_2 + .2x_3 \\ .3x_1 + .1x_2 + .1x_3 \end{bmatrix}$$

gives the amount of the total output of the economy that is consumed during the production process. This vector is just the matrix–vector product $C\mathbf{x}$, where $\mathbf{x}$ is the **gross production vector**

$$\mathbf{x} = \begin{bmatrix} x_1 \\ x_2 \\ x_3 \end{bmatrix}.$$

> For an economy with input–output matrix C and gross production vector $\mathbf{x}$, the total output of the economy that is consumed during the production process is $C\mathbf{x}$.

Example 1   Suppose that, in the economy previously described, the total outputs of the agriculture, manufacturing, and services sectors are \$100 million, \$150 million, and \$80 million, respectively. Then

$$C\mathbf{x} = \begin{bmatrix} .1 & .2 & .1 \\ .2 & .4 & .2 \\ .3 & .1 & .1 \end{bmatrix} \begin{bmatrix} 100 \\ 150 \\ 80 \end{bmatrix} = \begin{bmatrix} 48 \\ 96 \\ 53 \end{bmatrix},$$

and so the portion of the gross production that is consumed during the production process is \$48 million of agriculture, \$96 million of manufacturing, and \$53 million of services.   ○

Since in Example 1 the amount of the gross production consumed during the production process is

$$C\mathbf{x} = \begin{bmatrix} 48 \\ 96 \\ 53 \end{bmatrix},$$

the amount of the gross production that is not consumed during the production process is

$$\mathbf{x} - C\mathbf{x} = \begin{bmatrix} 100 \\ 150 \\ 80 \end{bmatrix} - \begin{bmatrix} 48 \\ 96 \\ 53 \end{bmatrix} = \begin{bmatrix} 52 \\ 54 \\ 27 \end{bmatrix}.$$

Thus $\mathbf{x} - C\mathbf{x}$ is the **net production** (or **surplus**) vector; its components indicate the amounts of output from each sector that remain after production. These amounts are available for sale within the economy or for export outside the economy.

Suppose now that we want to determine the amount of gross production for each sector that is necessary to yield a specific net production. For example, we might want to set production goals for the various sectors so that we have specific quantities available for export. Let $\mathbf{d}$ denote the **demand vector** whose components are the quantities required from each sector. In order to have exactly these amounts available after the production process is completed, the demand vector must equal the net production vector, that is, $\mathbf{d} = \mathbf{x} - C\mathbf{x}$. Using the algebra of matrices and vectors, we can rewrite this equation as follows.

$$\mathbf{x} - C\mathbf{x} = \mathbf{d}$$

$$I_3\mathbf{x} - C\mathbf{x} = \mathbf{d}$$

$$(I_3 - C)\mathbf{x} = \mathbf{d}$$

Thus the required gross production is a solution to the equation $(I_3 - C)\mathbf{x} = \mathbf{d}$.

> For an economy with $n \times n$ input–output matrix C, the gross production necessary to satisfy exactly a demand $\mathbf{d}$ is a solution to $(I_n - C)\mathbf{x} = \mathbf{d}$.

Example 2   For the economy described in Example 1, determine the gross production level needed to meet a consumer demand for \$90 million of agriculture, \$80 million of manufacturing, and \$60 million of services.

Solution   We must solve the matrix equation $(I_3 - C)\mathbf{x} = \mathbf{d}$, where C is the input output matrix and

$$\mathbf{d} = \begin{bmatrix} 90 \\ 80 \\ 60 \end{bmatrix}$$

is the demand vector. Since

$$I_3 - C = \begin{bmatrix} 1 & 0 & 0 \\ 0 & 1 & 0 \\ 0 & 0 & 1 \end{bmatrix} - \begin{bmatrix} .1 & .2 & .1 \\ .2 & .4 & .2 \\ .3 & .1 & .1 \end{bmatrix} = \begin{bmatrix} .9 & -.2 & -.1 \\ -.2 & .6 & -.2 \\ -.3 & -.1 & .9 \end{bmatrix},$$

the augmented matrix of the system to be solved is

$$\begin{bmatrix} .9 & -.2 & -.1 & 90 \\ -.2 & .6 & -.2 & 80 \\ -.3 & -.1 & .9 & 60 \end{bmatrix}.$$

Thus the solution to $(I_3 - C)\mathbf{x} = \mathbf{d}$ is

$$\begin{bmatrix} 170 \\ 240 \\ 150 \end{bmatrix},$$

and so the gross production needed to meet the demand is \$170 million of agriculture, \$240 million of manufacturing, and \$150 million of services.

Current Flow in Electrical Circuits

When a battery is connected in an electrical circuit, a current flows through the circuit. If the current passes through a resistor (a device that uses electricity), a drop in voltage occurs. These voltage drops obey *Ohm's law*, which states that

$$V = RI,$$

where V is the voltage drop across the resistor (measured in volts), R is the resistance (measured in ohms), and I is the current (measured in amperes).[10]

Figure 1.18 shows a simple electrical network consisting of a 20-volt battery (indicated by $|\!|$), and two resistors (indicated by ‑ww‑) with resistances of 3 ohms and 2 ohms. In Figure 1.18, the current flow is taken to be clockwise, but this is arbitrary. If the value of I is positive, then the flow is from the positive terminal of the battery (indicated by the longer side of the battery) to the negative terminal (indicated by the shorter side). In order to determine the value of I, we must utilize *Kirchoff's voltage law*.[11]

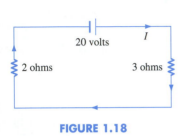

FIGURE 1.18

Kirchoff's Voltage Law

In a closed path within an electrical circuit, the algebraic sum of the voltage drops in any one direction equals the algebraic sum of the voltage sources in the same direction.

In the circuit shown in Figure 1.18, there are two voltage drops, one of $3I$ and the other of $2I$. Their sum equals 20, the voltage supplied by the single battery. Hence

$$3I + 2I = 20$$

$$5I = 20$$

$$I = 4.$$

Thus the current flow through the network is 4 amperes.

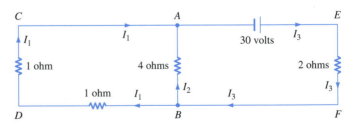

FIGURE 1.19

A more complicated circuit is shown in Figure 1.19. Here the junctions at A and B (indicated by the dots) create three branches in the circuit, each with its own current flow. Starting at B and applying Kirchoff's voltage law to

---

[10]Georg Simon Ohm (1787–1854) was a German physicist whose pamphlet *Die galvanische Kette mathematisch bearbeitet* greatly influenced the development of the theory of electricity.

[11]Gustav Robert Kirchoff (1824–1887) was a German physicist who made significant contributions to the fields of electricity and electromagnetic radiation.

the closed path $BDCAB$, we obtain $1I_1 + 1I_1 - 4I_2 = 0$; that is,

$$2I_1 - 4I_2 = 0. \tag{5}$$

Note that since we are proceeding around the closed path in a clockwise direction, the flow from A to B is opposite to the direction indicated for I_2. Thus the voltage drop at the 4-ohm resistor is $4(-I_2)$. Moreover, because there is no voltage source in this closed path, the algebraic sum of the three voltage drops is 0.

Similarly, from the closed path $BAEFB$, we obtain the equation

$$4I_2 + 2I_3 = 30, \tag{6}$$

and from the closed path $BDCAEFB$, we obtain the equation

$$2I_1 + 2I_3 = 30. \tag{7}$$

Note that, in this case, (7) is the sum of (5) and (6). Hence (7) provides no information not already given by (5) and (6), and so we can discard it. A similar situation occurs in all of the networks that we consider, and so we may ignore any closed paths that contain only currents that are accounted for in other equations obtained from Kirchoff's voltage law.

At this point we have two equations, (5) and (6), in three variables, and so another equation is required if we are to obtain a unique solution for I_1, I_2, and I_3. It is provided by another of Kirchoff's laws.

Kirchoff's Current Law

The current flow into any junction equals the current flow out of the junction.

In the context of Figure 1.19, Kirchoff's current law states that the flow into junction A, which is $I_1 + I_2$, equals the current flow out of A, which is I_3. Hence we obtain the equation $I_1 + I_2 = I_3$, or

$$I_1 + I_2 - I_3 = 0. \tag{8}$$

Notice that the current law also applies at junction B, where it yields the equation $I_3 = I_1 + I_2$. This equation is equivalent to (8), however, and so we ignore it. In general, if the current law is applied at each junction, then any one of the resulting equations is redundant and can be ignored.

Thus a system of equations that determines the current flows in the circuit in in Figure 1.19 is

$$2I_1 - 4I_2 \qquad = 0 \tag{5}$$

$$4I_2 + 2I_3 = 30 \tag{6}$$

$$I_1 + I_2 - I_3 = 0. \tag{8}$$

Solving this system by Gaussian elimination, we see that $I_1 = 6$, $I_2 = 3$, and $I_3 = 9$, and so the branch currents are 6 amperes, 3 amperes, and 9 amperes, respectively.

Practice Problems

1. An island's economy is divided into three sectors—tourism, transportation, and services. Suppose that each dollar's worth of tourism output requires inputs of $0.30 from the tourism sector, $0.10 from the transportation sector, and $0.30 from the services sector; each dollar's worth of transportation output requires inputs of $0.20 from the tourism sector, $0.40 from the transportation sector, and $0.20 from the services sector; and each dollar's worth of services output requires inputs of $0.05 from the tourism sector, $0.05 from the transportation sector, and $0.15 from the services sector.
   (a) Write the input–output matrix for this economy.
   (b) If the gross production for this economy is $10 million of tourism, $15 million of transportation, and $20 million of services, how much input from the tourism sector is required by the services sector?
   (c) If the gross production for this economy is $10 million of tourism, $15 million of transportation, and $20 million of services, what is the total value of the inputs consumed by each sector during the production process?
   (d) If the total outputs of the tourism, transportation, and services sectors are $70 million, $50 million, and $60 million, respectively, what is the net production of each sector?

   (e) What gross production is required to satisfy exactly a demand for $30 million of tourism, $50 million of transportation, and $40 million of services?
2. Determine the current through the following electrical circuit.

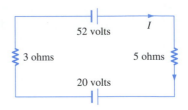

3. Determine the currents in each branch of the following electrical circuit.

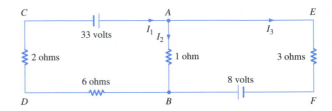

Exercises

1. Determine if the following statements are true or false.
   (a) If C is the input–output matrix for an economy with gross production vector $\mathbf{x}$, then $C\mathbf{x}$ is the net production vector.
   (b) For an economy with $n \times n$ input–output matrix C, the gross production necessary to satisfy exactly a demand $\mathbf{d}$ is a solution to $(I_n - C)\mathbf{x} = \mathbf{d}$.
   (c) The voltage drop at each resistor in an electrical network equals the product of the resistance and the amount of current through the resistor.
   (d) In any closed path within an electrical circuit, the algebraic sum of all the voltage drops in the same direction equals 0.
   (e) At every junction in an electrical circuit, the current flow into the junction equals the current flow out of the junction.

In Exercises 2–11, suppose that an economy is divided into four sectors (agriculture, manufacturing, services, and entertainment) with the following input–output matrix.

$$C = \begin{bmatrix} .12 & .11 & .15 & .18 \\ .20 & .08 & .24 & .07 \\ .18 & .16 & .06 & .22 \\ .09 & .07 & .12 & .05 \end{bmatrix} \begin{matrix} \text{Agriculture} \\ \text{Manufacturing} \\ \text{Services} \\ \text{Entertainment} \end{matrix}$$

with columns labeled Agr., Man., Ser., Ent.

2. What amount of input from the manufacturing sector is needed for a gross production of $100 million by the agriculture sector?
3. What amount of input from the services sector is needed for a gross production of $50 million by the entertainment sector?
4. Which sector is most dependent on services?
5. Which sector is least dependent on services?
6. On which sector is agriculture most dependent?
7. On which sector is agriculture least dependent?
8. If the gross production for this economy is $20 million of agriculture, $30 million of manufacturing, $20 million of services, and $10 million of entertainment, what is the total value of the inputs from each sector consumed during the production process?
9. If the gross production for this economy is $30 million of agriculture, $40 million of manufacturing, $30 million of services, and $20 million of entertainment, what is the total value of the inputs from each sector consumed during the production process?
10. If the gross production for this economy is $20 million of agriculture, $30 million of manufacturing, $20 million of services, and $10 million of entertainment, what is the net production of each sector?

11. If the gross production for this economy is $30 million of agriculture, $40 million of manufacturing, $30 million of services, and $20 million of entertainment, what is the net production of each sector?

12. The input–output matrix for an economy with sectors of metals, nonmetals, and services follows.

$$
\begin{array}{ccc}
\text{Met.} & \text{Non.} & \text{Ser.}
\end{array}
$$
$$
\begin{bmatrix}
.2 & .2 & .1 \\
.4 & .4 & .2 \\
.2 & .2 & .1
\end{bmatrix}
\begin{array}{l}
\text{Metals} \\
\text{Nonmetals} \\
\text{Services}
\end{array}
$$

(a) What is the net production corresponding to a gross production of $50 million of metals, $60 million of nonmetals, and $40 million of services?

(b) What gross production is required to satisfy exactly a demand for $120 million of metals, $180 million of nonmetals, and $150 million of services?

13. The input–output matrix for an economy producing transportation, food, and oil follows.

$$
\begin{array}{ccc}
\text{Tran.} & \text{Food} & \text{Oil}
\end{array}
$$
$$
\begin{bmatrix}
.2 & .20 & .3 \\
.4 & .30 & .1 \\
.2 & .25 & .3
\end{bmatrix}
\begin{array}{l}
\text{Transportation} \\
\text{Food} \\
\text{Oil}
\end{array}
$$

(a) What is the net production corresponding to a gross production of $40 million of transportation, $30 million of food, and $35 million of oil?

(b) What gross production is required to satisfy exactly a demand for $32 million of transportation, $48 million of food, and $24 million of oil?

14. Suppose that an economy is divided into two sectors, nongovernment and government. Each dollar's worth of nongovernment output requires $0.10 in nongovernment input and $0.10 in government input, and each dollar's worth of government output requires $0.20 in nongovernment input and $0.70 in government input.

(a) Write the input–output matrix for this economy.

(b) What is the net production corresponding to a gross production of $20 million in nongovernment and $30 million in government?

(c) What gross production is needed to satisfy exactly a demand for $45 million in nongovernment and $50 million in government?

15. Suppose that a nation's energy production is divided into two sectors, electricity and oil. Each dollar's worth of electricity output requires $0.10 of electricity input and $0.30 of oil input, and each dollar's worth of oil output requires $0.40 of electricity input and $0.20 of oil input.

(a) Write the input–output matrix for this economy.

(b) What is the net production corresponding to a gross production of $60 million of electricity and $50 million of oil?

(c) What gross production is needed to satisfy exactly a demand for $60 million of electricity and $72 million of oil?

16. Consider an economy that is divided into three sectors: agriculture, manufacturing, and services. Suppose that each dollar's worth of agricultural output requires inputs of $0.10 from the agricultural sector, $0.15 from the manufacturing sector, and $0.30 from the services sector; each dollar's worth of manufacturing output requires inputs of $0.20 from the agricultural sector, $0.25 from the manufacturing sector, and $0.10 from the services sector; and each dollar's worth of services output requires inputs of $0.20 from the agricultural sector, $0.35 from the manufacturing sector, and $0.10 from the services sector.

(a) What is the net production corresponding to a gross production of $40 million of agriculture, $50 million of manufacturing, and $30 million of services?

(b) What gross production is needed to satisfy exactly a demand for $90 million of agriculture, $72 million of manufacturing, and $96 million of services?

17. Consider an economy that is divided into three sectors: finance, goods, and services. Suppose that each dollar's worth of financial output requires inputs of $0.10 from the finance sector, $0.20 from the goods sector, and $0.20 from the services sector; each dollar's worth of goods requires inputs of $0.10 from the finance sector, $0.40 from the goods sector, and $0.20 from the services sector; and each dollar's worth of services requires inputs of $0.15 from the finance sector, $0.10 from the goods sector, and $0.30 from the services sector.

(a) What is the net production corresponding to a gross production of $70 million of finance, $50 million of goods, and $60 million of services?

(b) What is the gross production corresponding to a net production of $40 million of finance, $50 million of goods, and $30 million of services?

(c) What gross production is needed to satisfy exactly a demand for $40 million of finance, $36 million of goods, and $44 million of services?

18. Suppose that the columns of the input–output matrix

$$
\begin{array}{ccc}
\text{Agr.} & \text{Man.} & \text{Ser.}
\end{array}
$$
$$
C = \begin{bmatrix}
.1 & .2 & .1 \\
.2 & .4 & .2 \\
.3 & .1 & .1
\end{bmatrix}
\begin{array}{l}
\text{Agriculture} \\
\text{Manufacturing} \\
\text{Services}
\end{array}
$$

measure the amount (in tons) of each input needed to produce one ton of output from each sector. Let p_1, p_2, and p_3 denote the prices per ton of agricultural products, minerals, and textiles, respectively.

(a) Interpret the vector $C^T \mathbf{p}$, where $\mathbf{p} = \begin{bmatrix} p_1 \\ p_2 \\ p_3 \end{bmatrix}$.

(b) Interpret $\mathbf{p} - C^T \mathbf{p}$.

19. Let C be the input–output matrix for an economy, $\mathbf{x}$ be the gross production vector, $\mathbf{d}$ be the demand vector, and $\mathbf{p}$ be the vector whose components are the unit prices of the products or services produced by each sector. Economists call the vector $\mathbf{v} = \mathbf{p} - C^T\mathbf{p}$ the *value-added vector*. Show that $\mathbf{p}^T\mathbf{d} = \mathbf{v}^T\mathbf{x}$. (The single entry in the 1×1 matrix $\mathbf{p}^T\mathbf{d}$ represents the gross domestic product of the economy.) *Hint:* Compute $\mathbf{p}^T\mathbf{x}$ in two different ways. First, replace $\mathbf{p}$ by $\mathbf{v} + C^T\mathbf{p}$, and then replace $\mathbf{x}$ by $C\mathbf{x} + \mathbf{d}$.

In Exercises 20–24, determine the currents in each branch of the given circuit.

20.

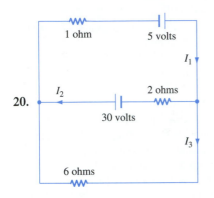

21.

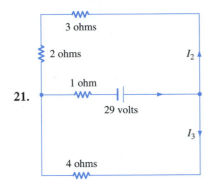

22.

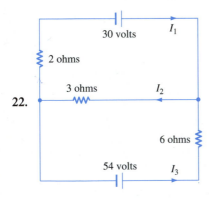

23.

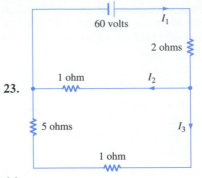

24.

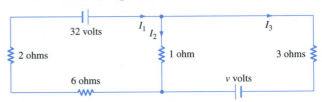

25. In the following electrical network, determine the value of v that makes $I_2 = 0$.

In the following exercise, use a calculator with matrix capabilities or computer software such as MATLAB *to solve the problem.*

26. Let

$$C = \begin{bmatrix} .12 & .03 & .20 & .10 & .05 & .09 \\ .21 & .11 & .06 & .11 & .07 & .07 \\ .05 & .21 & .11 & .15 & .11 & .06 \\ .11 & .18 & .13 & .22 & .03 & .18 \\ .16 & .15 & .07 & .12 & .19 & .14 \\ .07 & .23 & .06 & .05 & .15 & .19 \end{bmatrix} \quad \text{and}$$

$$\mathbf{d} = \begin{bmatrix} 100 \\ 150 \\ 200 \\ 125 \\ 300 \\ 180 \end{bmatrix},$$

where C is the input–output matrix for an economy that has been divided into six sectors, and $\mathbf{d}$ is the net production for this economy (where units are in millions of dollars). Find the gross production vector required to produce $\mathbf{d}$.

Solutions to the Practice Problems

1. (a) The input–output matrix is as follows.

$$C = \begin{bmatrix} .3 & .2 & .05 \\ .1 & .4 & .05 \\ .3 & .2 & .15 \end{bmatrix} \begin{matrix} \text{Tourism} \\ \text{Transportation} \\ \text{Services} \end{matrix}$$

with column headings Tour. Tran. Serv.

(b) Each dollar's worth of output from the services sector requires $.05 in input from the tourism sector. Hence a gross output of $20 million from the services sector requires $20(\$.05) = \$1$ million in input from the tourism sector.

(c) The total value of the inputs consumed by each sector during the production process is given by

$$C \begin{bmatrix} 10 \\ 15 \\ 20 \end{bmatrix} = \begin{bmatrix} .3 & .2 & .05 \\ .1 & .4 & .05 \\ .3 & .2 & .15 \end{bmatrix} \begin{bmatrix} 10 \\ 15 \\ 20 \end{bmatrix} = \begin{bmatrix} 7 \\ 8 \\ 9 \end{bmatrix}.$$

Hence during the production process, $7 million in inputs is consumed by the tourism sector, $8 million by the transportation sector, and $9 million by the services sector.

(d) The gross production vector is

$$\mathbf{x} = \begin{bmatrix} 70 \\ 50 \\ 60 \end{bmatrix},$$

and so the net production vector is

$$\mathbf{x} - C\mathbf{x} = \begin{bmatrix} 70 \\ 50 \\ 60 \end{bmatrix} - \begin{bmatrix} .3 & .2 & .05 \\ .1 & .4 & .05 \\ .3 & .2 & .15 \end{bmatrix} \begin{bmatrix} 70 \\ 50 \\ 60 \end{bmatrix}$$

$$= \begin{bmatrix} 70 \\ 50 \\ 60 \end{bmatrix} - \begin{bmatrix} 34 \\ 30 \\ 40 \end{bmatrix} = \begin{bmatrix} 36 \\ 20 \\ 20 \end{bmatrix}.$$

Hence the net productions of the tourism, transportation, and services sectors are $36 million, $20 million, and $20 million, respectively.

(e) To meet a demand

$$\mathbf{d} = \begin{bmatrix} 30 \\ 50 \\ 40 \end{bmatrix},$$

the gross production vector must be a solution to the equation $(I_3 - C)\mathbf{x} = \mathbf{d}$. The augmented matrix of the system is

$$\begin{bmatrix} .7 & -.2 & -.05 & 30 \\ -.1 & .6 & -.05 & 50 \\ -.3 & -.2 & .85 & 40 \end{bmatrix},$$

and so the gross production vector is

$$\begin{bmatrix} 80 \\ 105 \\ 100 \end{bmatrix}.$$

Thus the gross productions of the tourism, transportation, and services sectors are $80 million, $105 million, and $100 million, respectively.

2. The algebraic sum of the voltage drops around the circuit is $5I + 3I = 8I$. Since the current flow from the 20-volt battery is in the direction opposite to I, the algebraic sum of the voltage sources around the circuit is $52 + (-20) = 32$. Hence we have the equation $8I = 32$, so that $I = 4$ amperes.

3. There are two junctions in the given circuit, namely A and B in the following figure.

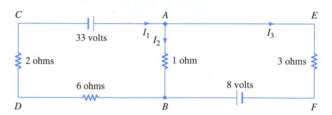

Applying Kirchoff's current law to junction A or junction B gives $I_1 = I_2 + I_3$. Applying Kirchoff's voltage law to the closed path $ABDCA$ yields

$$1I_2 + 6I_1 + 2I_1 = 33.$$

Similarly, from the closed path $AEFBA$, we obtain

$$3I_3 + 1(-I_2) = 8.$$

Hence the system of equations describing the current flows is

$$\begin{aligned} I_1 - I_2 - I_3 &= 0 \\ 8I_1 + I_2 &= 33 \\ -I_2 + 3I_3 &= 8. \end{aligned}$$

Solving this system gives $I_1 = 4$, $I_2 = 1$, and $I_3 = 3$. Hence the branch currents are 4 amperes, 1 ampere, and 3 amperes, respectively.

1.6 The Span of a Set of Vectors

In Section 1.2, we defined a linear combination of vectors $\mathbf{u}_1, \mathbf{u}_2, \ldots, \mathbf{u}_k$ in $\mathcal{R}^n$ to be a vector of the form $c_1\mathbf{u}_1 + c_2\mathbf{u}_2 + \ldots + c_k\mathbf{u}_k$, where $c_1, c_2, \ldots, c_k$ are scalars. For a given subset $\mathcal{S} = \{\mathbf{u}_1, \mathbf{u}_2, \ldots, \mathbf{u}_k\}$ of $\mathcal{R}^n$, the set of all the linear combinations of $\mathbf{u}_1, \mathbf{u}_2, \ldots, \mathbf{u}_k$ is often of interest. For example, if A is an $n \times p$ matrix, then the set of vectors $\mathbf{v} \in \mathcal{R}^n$ such that $A\mathbf{x} = \mathbf{v}$ is consistent is precisely the set of all the linear combinations of the columns of A. We now name such a set of linear combinations.

Definition.   For a nonempty set $\mathcal{S} = \{\mathbf{u}_1, \mathbf{u}_2, \ldots, \mathbf{u}_k\}$ of vectors in $\mathcal{R}^n$, we define the **span of** $\mathcal{S}$ to be the set of all linear combinations of $\mathbf{u}_1, \mathbf{u}_2, \ldots, \mathbf{u}_k$ in $\mathcal{R}^n$. This set is denoted Span $\mathcal{S}$ or Span $\{\mathbf{u}_1, \mathbf{u}_2, \ldots, \mathbf{u}_k\}$.

Since a linear combination of a single vector is just a multiple of that vector, if $\mathbf{u}$ is in $\mathcal{S}$, then every multiple of $\mathbf{u}$ is in Span $\mathcal{S}$. Thus the span of $\{\mathbf{u}\}$ consists of all multiples of $\mathbf{u}$. In particular, the span of $\{\mathbf{0}\}$ is $\{\mathbf{0}\}$. Note that, if $\mathcal{S}$ contains a nonzero vector, then Span $\mathcal{S}$ contains infinitely many vectors. Other examples of the span of a set follow.

Example 1   Describe the spans of the following subsets of $\mathcal{R}^2$:

$$\mathcal{S}_1 = \left\{ \begin{bmatrix} 1 \\ -1 \end{bmatrix} \right\}, \quad \mathcal{S}_2 = \left\{ \begin{bmatrix} 1 \\ -1 \end{bmatrix}, \begin{bmatrix} -2 \\ 2 \end{bmatrix} \right\}, \quad \mathcal{S}_3 = \left\{ \begin{bmatrix} 1 \\ -1 \end{bmatrix}, \begin{bmatrix} -2 \\ 2 \end{bmatrix}, \begin{bmatrix} 2 \\ 1 \end{bmatrix} \right\},$$

and

$$\mathcal{S}_4 = \left\{ \begin{bmatrix} 1 \\ -1 \end{bmatrix}, \begin{bmatrix} -2 \\ 2 \end{bmatrix}, \begin{bmatrix} 2 \\ 1 \end{bmatrix}, \begin{bmatrix} -1 \\ 3 \end{bmatrix} \right\}.$$

Solution   The span of $\mathcal{S}_1$ consists of all linear combinations of the vectors in $\mathcal{S}_1$. Since a linear combination of a single vector is just a multiple of that vector, the span of $\mathcal{S}_1$ consists of all multiples of $\begin{bmatrix} 1 \\ -1 \end{bmatrix}$; that is, all vectors of the form $\begin{bmatrix} c \\ -c \end{bmatrix}$ for some scalar c. These vectors all lie along the line with equation $y = -x$, as pictured in Figure 1.20.

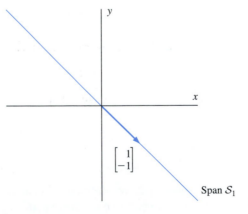

Span $\mathcal{S}_1$

FIGURE 1.20

The span of $\mathcal{S}_2$ consists of all linear combinations of the vectors $\begin{bmatrix} 1 \\ -1 \end{bmatrix}$ and $\begin{bmatrix} -2 \\ 2 \end{bmatrix}$. Such vectors have the form

$$a \begin{bmatrix} 1 \\ -1 \end{bmatrix} + b \begin{bmatrix} -2 \\ 2 \end{bmatrix} = a \begin{bmatrix} 1 \\ -1 \end{bmatrix} - 2b \begin{bmatrix} 1 \\ -1 \end{bmatrix} = (a - 2b) \begin{bmatrix} 1 \\ -1 \end{bmatrix},$$

where a and b are arbitrary scalars. Taking $c = a - 2b$, we see that these are the same vectors as those in the span of $\mathcal{S}_1$. Hence Span $\mathcal{S}_2$ = Span $\mathcal{S}_1$ (see Figure 1.21).

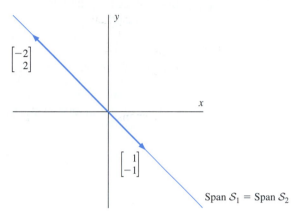

Span $\mathcal{S}_1$ = Span $\mathcal{S}_2$

FIGURE 1.21

The span of $\mathcal{S}_3$ consists of all linear combinations of the vectors $\begin{bmatrix} 1 \\ -1 \end{bmatrix}$, $\begin{bmatrix} -2 \\ 2 \end{bmatrix}$, and $\begin{bmatrix} 2 \\ 1 \end{bmatrix}$. Note that the vectors $\begin{bmatrix} 1 \\ -1 \end{bmatrix}$ and $\begin{bmatrix} 2 \\ 1 \end{bmatrix}$ are not parallel. Hence an arbitrary vector $\mathbf{v}$ in $\mathcal{R}^2$ is a linear combination of these two vectors, as we learned in Section 1.2. Suppose that $\mathbf{v} = a \begin{bmatrix} 1 \\ -1 \end{bmatrix} + b \begin{bmatrix} 2 \\ 1 \end{bmatrix}$ for some scalars a and b. Then

$$\mathbf{v} = a \begin{bmatrix} 1 \\ -1 \end{bmatrix} + 0 \begin{bmatrix} -2 \\ 2 \end{bmatrix} + b \begin{bmatrix} 2 \\ 1 \end{bmatrix},$$

and so every vector in $\mathcal{R}^2$ is a linear combination of the vectors in $\mathcal{S}_3$. It follows that the span of $\mathcal{S}_3$ is $\mathcal{R}^2$.

Finally, since every vector in $\mathcal{R}^2$ is a linear combination of the nonparallel vectors $\begin{bmatrix} 1 \\ -1 \end{bmatrix}$ and $\begin{bmatrix} 2 \\ 1 \end{bmatrix}$, every vector in $\mathcal{R}^2$ is also a linear combination of the vectors in $\mathcal{S}_4$. Therefore the span of $\mathcal{S}_4$ is again $\mathcal{R}^2$. ○

Example 2   For the standard vectors

$$\mathbf{e}_1 = \begin{bmatrix} 1 \\ 0 \\ 0 \end{bmatrix}, \qquad \mathbf{e}_2 = \begin{bmatrix} 0 \\ 1 \\ 0 \end{bmatrix}, \qquad \text{and} \qquad \mathbf{e}_3 = \begin{bmatrix} 0 \\ 0 \\ 1 \end{bmatrix}$$

in $\mathcal{R}^3$, we see that the span of $\{\mathbf{e}_1, \mathbf{e}_2\}$ is the set of vectors of the form

$$a\mathbf{e}_1 + b\mathbf{e}_2 = a \begin{bmatrix} 1 \\ 0 \\ 0 \end{bmatrix} + b \begin{bmatrix} 0 \\ 1 \\ 0 \end{bmatrix} = \begin{bmatrix} a \\ b \\ 0 \end{bmatrix}.$$

Thus Span $\{\mathbf{e}_1, \mathbf{e}_2\}$ can be visualized as the set of vectors in the xy-plane of $\mathcal{R}^3$. Similarly, Span $\{\mathbf{e}_3\}$ can be visualized as the set of vectors that lie along the z-axis in $\mathcal{R}^3$. ○

From the preceding examples, we see that saying "$\mathbf{v}$ belongs to the span of $S = \{\mathbf{u}_1, \mathbf{u}_2, \dots, \mathbf{u}_k\}$" means exactly the same as saying "$\mathbf{v}$ is a linear combination of the vectors $\mathbf{u}_1, \mathbf{u}_2, \dots, \mathbf{u}_k$." So our comment at the beginning of the section can be rephrased as follows.

> Let $S = \{\mathbf{u}_1, \mathbf{u}_2, \dots, \mathbf{u}_k\}$ be a set of vectors from $\mathcal{R}^n$, and let $A = \begin{bmatrix} \mathbf{u}_1 & \mathbf{u}_2 & \cdots & \mathbf{u}_k \end{bmatrix}$. Then a vector $\mathbf{v}$ in $\mathcal{R}^n$ belongs to the span of S (that is, $\mathbf{v}$ is a linear combination of $\mathbf{u}_1, \mathbf{u}_2, \dots, \mathbf{u}_k$) if and only if the equation $A\mathbf{x} = \mathbf{v}$ is consistent.

Example 3   Determine whether

$$\mathbf{v} = \begin{bmatrix} 3 \\ 0 \\ 5 \\ -1 \end{bmatrix} \quad \text{or} \quad \mathbf{w} = \begin{bmatrix} 2 \\ 1 \\ 3 \\ -1 \end{bmatrix}$$

is an element of the span of

$$S = \left\{ \begin{bmatrix} 1 \\ 2 \\ 1 \\ 1 \end{bmatrix}, \begin{bmatrix} -1 \\ 1 \\ -2 \\ 1 \end{bmatrix}, \begin{bmatrix} 1 \\ 8 \\ -1 \\ 5 \end{bmatrix} \right\}.$$

If so, express it as a linear combination of the vectors in S.

Solution   Let A be the matrix whose columns are the vectors in S. The vector $\mathbf{v}$ belongs to the span of S if and only if $A\mathbf{x} = \mathbf{v}$ is consistent. Since the reduced row echelon form of $[A \ \mathbf{v}]$ is

$$\begin{bmatrix} 1 & 0 & 3 & 1 \\ 0 & 1 & 2 & -2 \\ 0 & 0 & 0 & 0 \\ 0 & 0 & 0 & 0 \end{bmatrix},$$

$A\mathbf{x} = \mathbf{v}$ is consistent, by Theorem 1.4(c). Hence $\mathbf{v}$ belongs to the span of S.

We have argued that $\mathbf{v}$ belongs to the span of S by showing that $A\mathbf{x} = \mathbf{v}$ is consistent. Since we also need to express $\mathbf{v}$ as a linear combination of the vectors in S, it is necessary to find the actual solution to $A\mathbf{x} = \mathbf{v}$. Using the reduced row echelon form of $[A \ \mathbf{v}]$, we see that the general solution to this equation is

$$x_1 = 1 - 3x_3$$
$$x_2 = -2 - 2x_3$$
$$x_3 \quad \text{free.}$$

For example, by taking $x_3 = 0$, we find that

$$1\begin{bmatrix}1\\2\\1\\1\end{bmatrix} - 2\begin{bmatrix}-1\\1\\-2\\1\end{bmatrix} + 0\begin{bmatrix}1\\8\\-1\\5\end{bmatrix} = \begin{bmatrix}3\\0\\5\\-1\end{bmatrix} = \mathbf{v}.$$

In the same manner, $\mathbf{w}$ belongs to the span of S if and only if $A\mathbf{x} = \mathbf{w}$ is consistent. Because the reduced row echelon form of $[A \ \mathbf{w}]$ is

$$\begin{bmatrix}1&0&3&0\\0&1&2&0\\0&0&0&1\\0&0&0&0\end{bmatrix},$$

Theorem 1.4(c) shows that $A\mathbf{x} = \mathbf{w}$ is not consistent. Thus $\mathbf{w}$ does not belong to the span of S. ○

In our examples so far, we have started with a subset S of $\mathcal{R}^n$ and tried to describe the span of S. There are other important situations where we start with the span of S instead. If V is a subset of $\mathcal{R}^n$ and Span $S = V$, then we say that S is a **spanning set** for V or that S **spans** V. Note that because S is contained in its span (as shown in Theorem 1.6), a spanning set for V is necessarily a subset of V.

Example 4 Show that the set

$$S = \left\{\begin{bmatrix}1\\0\\0\end{bmatrix}, \begin{bmatrix}1\\1\\0\end{bmatrix}, \begin{bmatrix}1\\1\\1\end{bmatrix}, \begin{bmatrix}1\\-2\\-1\end{bmatrix}\right\}$$

is a spanning set for $\mathcal{R}^3$.

Solution Because S is contained in $\mathcal{R}^3$, it follows that Span S is contained in $\mathcal{R}^3$. Thus, in order to show that Span $S = \mathcal{R}^3$, we need only show that an arbitrary vector $\mathbf{v}$ in $\mathcal{R}^3$ belongs to Span S. Thus we must show that $A\mathbf{x} = \mathbf{v}$ is consistent, where

$$A = \begin{bmatrix}1&1&1&1\\0&1&1&-2\\0&0&1&-1\end{bmatrix}.$$

Let $[R \ \mathbf{c}]$ be the reduced row echelon form of $[A \ \mathbf{v}]$. Then R is the reduced row echelon form of A by Exercise 41 of Section 1.3. Since

$$R = \begin{bmatrix}1&0&0&3\\0&1&0&-1\\0&0&1&-1\end{bmatrix}$$

has no zero row, there can be no row in $[R \ \mathbf{c}]$ in which the only nonzero entry lies in the last column. Thus $A\mathbf{x} = \mathbf{v}$ is consistent by Theorem 1.4, and so $\mathbf{v}$ belongs to Span $\mathcal{S}$. Since $\mathbf{v}$ is an arbitrary vector in $\mathcal{R}^3$, it follows that Span $\mathcal{S} = \mathcal{R}^3$.   ○

The following theorem guarantees that the technique used in Example 4 can be applied to test whether any subset of $\mathcal{R}^m$ is a spanning set for $\mathcal{R}^m$.

Theorem 1.5

The following statements about an $m \times n$ matrix A are equivalent.

  (a) The span of the columns of A is $\mathcal{R}^m$.

  (b) The equation $A\mathbf{x} = \mathbf{b}$ has at least one solution (that is, $A\mathbf{x} = \mathbf{b}$ is consistent) for each $\mathbf{b}$ in $\mathcal{R}^m$.

  (c) The rank of A is m.

  (d) The reduced row echelon form of A has no zero rows.

Proof Since, by Theorem 1.4, the equation $A\mathbf{x} = \mathbf{b}$ is consistent precisely when $\mathbf{b}$ is a linear combination of the columns of A, statements (a) and (b) are equivalent. Also, because A is an $m \times n$ matrix, statements (c) and (d) are equivalent. The details of these arguments are left to the reader.

We now prove that statements (b) and (c) are equivalent. Let R denote the reduced row echelon form of A. There is a sequence of elementary row operations that transforms A into R. Since each of these elementary row operations is reversible, there is also a sequence of elementary row operations that transforms R into A. Apply the latter sequence of operations to $[R \ \mathbf{e}_m]$ to obtain a matrix $[A \ \mathbf{d}]$ for some $\mathbf{d}$ in $\mathcal{R}^m$. Then the system $A\mathbf{x} = \mathbf{d}$ is equivalent to the system $R\mathbf{x} = \mathbf{e}_m$.

If (b) is true, then $A\mathbf{x} = \mathbf{d}$, and hence $R\mathbf{x} = \mathbf{e}_m$, must be consistent. But then Theorem 1.4 implies that the last row of R cannot be a zero row, for otherwise $[R \ \mathbf{e}_m]$ would have a row in which the only nonzero entry lies in the last column. Because R is in reduced row echelon form, R must have no nonzero rows. It follows that the rank of A is m, establishing (c).

Conversely, assume that (c) is true. Then there is a $\mathbf{c}$ in $\mathcal{R}^m$ such that the reduced row echelon form of the $m \times (n+1)$ matrix $[A \ \mathbf{b}]$ is $[R \ \mathbf{c}]$. Since A has rank m, both R and $[R \ \mathbf{c}]$ have m nonzero rows. Therefore the ranks of both A and $[A \ \mathbf{b}]$ are m, and so (b) follows from Theorem 1.4.   ❑

Properties of the Span

In subsequent chapters, the concept of the span of a set becomes very important. We present here some of the elementary properties of this concept.

Let $\mathcal{S} = \{\mathbf{u}_1, \mathbf{u}_2\}$ be a set of vectors in $\mathcal{R}^n$. Clearly $\mathbf{u}_1 = 1\mathbf{u}_1 + 0\mathbf{u}_1$ and $\mathbf{u}_2 = 0\mathbf{u}_1 + 1\mathbf{u}_2$. Thus both $\mathbf{u}_1$ and $\mathbf{u}_2$ belong to the span of $\mathcal{S}$, and so $\mathcal{S}$ is a subset of its span.

Consider the vectors $\mathbf{v}_1 = 3\mathbf{u}_1 - 2\mathbf{u}_2$ and $\mathbf{v}_2 = -\mathbf{u}_1 + 4\mathbf{u}_2$. A vector $\mathbf{w}$ in the span of $\mathcal{S}' = \{\mathbf{v}_1, \mathbf{v}_2\}$ has the form $\mathbf{w} = a\mathbf{v}_1 + b\mathbf{v}_2$ for some scalars a and b. Since

$$\mathbf{w} = a\mathbf{v}_1 + b\mathbf{v}_2 = a(3\mathbf{u}_1 - 2\mathbf{u}_2) + b(-\mathbf{u}_1 + 4\mathbf{u}_2) = (3a - b)\mathbf{u}_1 + (-2a + 4b)\mathbf{u}_2,$$

we see that $\mathbf{w}$ belongs to the span of $\mathcal{S} = \{\mathbf{u}_1, \mathbf{u}_2\}$. Hence the span of $\mathcal{S}'$ is contained in the span of $\mathcal{S}$.

The following theorem generalizes the two properties discussed above and establishes a useful property that enables us to reduce the size of a spanning set in certain cases.

Theorem 1.6

For any finite subset S of $\mathcal{R}^n$, the following statements are true.

  (a) S is contained in the span of S.
  (b) If a finite set S' is contained in the span of S, then the span of S' is also contained in the span of S.
  (c) For any vector $\mathbf{z}$ in $\mathcal{R}^n$, the spans of S and $S \cup \{\mathbf{z}\}$ are equal if and only if $\mathbf{z}$ belongs to the span of S.

Proof We leave the proof of (a) to the reader.

(b) Suppose that $S' = \{\mathbf{v}_1, \mathbf{v}_2, \ldots, \mathbf{v}_m\}$ is a subset of Span S, where $S = \{\mathbf{u}_1, \mathbf{u}_2, \ldots, \mathbf{u}_k\}$, and that $\mathbf{w}$ belongs to the span of S'. Then $\mathbf{w}$ is a linear combination of $\mathbf{v}_1, \mathbf{v}_2, \ldots, \mathbf{v}_m$. Since each $\mathbf{v}_i$ $(i = 1, 2, \ldots, m)$ is a linear combination of $\mathbf{u}_1, \mathbf{u}_2, \ldots, \mathbf{u}_k$, we can therefore express $\mathbf{w}$ as a linear combination of $\mathbf{u}_1, \mathbf{u}_2, \ldots, \mathbf{u}_k$. Hence $\mathbf{w}$ belongs to the span of S, and therefore the span of S' is contained in the span of S.

(c) Let $\mathbf{z}$ be a vector in $\mathcal{R}^n$. Then $S \subseteq S \cup \{\mathbf{z}\} \subseteq \text{Span } (S \cup \{\mathbf{z}\})$ by (a). Hence the span of S is contained in the span of $S \cup \{\mathbf{z}\}$ by (b). Now suppose that $\mathbf{z}$ is in the span of S. Taking $S' = S \cup \{\mathbf{z}\}$ in (b), we see that the span of $S \cup \{\mathbf{z}\}$ is contained in the span of S. Since each of Span $(S \cup \{\mathbf{z}\})$ and Span S is contained in the other, the two spans are equal. Conversely, suppose that $\mathbf{z}$ does not belong to the span of S. Because $\mathbf{z}$ belongs to the span of $S \cup \{\mathbf{z}\}$ but not to the span of S, we have Span $S \neq$ Span $(S \cup \{\mathbf{z}\})$. ◻

If $S = \{\mathbf{u}_1, \mathbf{u}_2, \ldots, \mathbf{u}_k\}$ is a set of vectors in $\mathcal{R}^m$, then determining if $\mathbf{w}$ belongs to the span of S requires the solution to a system of m linear equations in k variables. (Example 3 illustrates this for the case that $k = 3$ and $m = 4$.) Thus the labor involved in answering this question would be reduced if the set S contained fewer vectors (so that the system of linear equations would have fewer variables). This is one of the reasons that we desire spanning sets to contain as few vectors as possible.

Theorem 1.6(c) provides a method for reducing the size of a spanning set. If one of the vectors in S is a linear combination of the others, it can be removed from S to obtain a smaller set having the same span as S. For example, in the set S of Example 3, we have

$$\begin{bmatrix} 1 \\ 8 \\ -1 \\ 5 \end{bmatrix} = 3 \begin{bmatrix} 1 \\ 2 \\ 1 \\ 1 \end{bmatrix} + 2 \begin{bmatrix} -1 \\ 1 \\ -2 \\ 1 \end{bmatrix}.$$

Hence the span of S is the same as the span of the smaller set

$$\left\{ \begin{bmatrix} 1 \\ 2 \\ 1 \\ 1 \end{bmatrix}, \begin{bmatrix} -1 \\ 1 \\ -2 \\ 1 \end{bmatrix} \right\}.$$

Practice Problems

1. Are $\mathbf{u} = \begin{bmatrix} -1 \\ 3 \\ 1 \end{bmatrix}$ and $\mathbf{v} = \begin{bmatrix} 1 \\ 1 \\ 2 \end{bmatrix}$ in the span of $S = \left\{ \begin{bmatrix} 2 \\ -1 \\ 1 \end{bmatrix}, \begin{bmatrix} -1 \\ 1 \\ 0 \end{bmatrix} \right\}$?

2. Is $S = \left\{ \begin{bmatrix} 1 \\ 0 \\ 1 \end{bmatrix}, \begin{bmatrix} -1 \\ 1 \\ 2 \end{bmatrix}, \begin{bmatrix} 1 \\ 3 \\ 9 \end{bmatrix}, \begin{bmatrix} 2 \\ -1 \\ 1 \end{bmatrix} \right\}$ a spanning set for $\mathcal{R}^3$?

3. Find the smallest subset of the set S in problem 2 that spans $\mathcal{R}^3$.

Exercises

1. Determine if the following statements are true or false. Justify each answer.

   (a) Let $S = \{\mathbf{u}_1, \mathbf{u}_2, \ldots, \mathbf{u}_k\}$ be a nonempty set of vectors in $\mathcal{R}^n$. A vector $\mathbf{v}$ belongs to the span of S if and only if $\mathbf{v} = c_1\mathbf{u}_1 + c_2\mathbf{u}_2 + \cdots + c_k\mathbf{u}_k$ for some scalars $c_1, c_2, \ldots, c_k$.

   (b) The span of $\{\mathbf{0}\}$ is $\{\mathbf{0}\}$.

   (c) If $A = [\mathbf{u}_1 \ \mathbf{u}_2 \ \ldots \ \mathbf{u}_k]$ and the matrix equation $A\mathbf{x} = \mathbf{v}$ is inconsistent, then $\mathbf{v}$ does not belong to the span of $\{\mathbf{u}_1, \mathbf{u}_2, \ldots, \mathbf{u}_k\}$.

   (d) If A is an $m \times n$ matrix, then $A\mathbf{x} = \mathbf{b}$ is consistent for every $\mathbf{b}$ in $\mathcal{R}^m$ if and only if the rank of A is n.

   (e) Let $S = \{\mathbf{u}_1, \mathbf{u}_2, \ldots, \mathbf{u}_k\}$ be a subset of $\mathcal{R}^n$. Then the span of S is $\mathcal{R}^n$ if and only if the rank of $[\mathbf{u}_1 \ \mathbf{u}_2 \ \ldots \ \mathbf{u}_k]$ is n.

   (f) Every finite subset of $\mathcal{R}^n$ is contained in its span.

   (g) If S_1 and S_2 are finite subsets of $\mathcal{R}^n$ such that S_1 is contained in Span S_2, then Span S_1 is contained in Span S_2.

   (h) If S_1 and S_2 are finite subsets of $\mathcal{R}^n$ having equal spans, then $S_1 = S_2$.

   (i) If S_1 and S_2 are finite subsets of $\mathcal{R}^n$ having equal spans, then S_1 and S_2 contain the same number of elements.

   (j) Let S be a nonempty set of vectors in $\mathcal{R}^n$, and let $\mathbf{v}$ be in $\mathcal{R}^n$. The spans of S and $S \cup \{\mathbf{v}\}$ are equal if and only if $\mathbf{v}$ is in S.

In Exercises 2–8, determine if the given vector is in Span $\left\{ \begin{bmatrix} 1 \\ 0 \\ 1 \end{bmatrix}, \begin{bmatrix} -1 \\ 1 \\ 1 \end{bmatrix}, \begin{bmatrix} 1 \\ 1 \\ 3 \end{bmatrix} \right\}$.

2. $\begin{bmatrix} 0 \\ 0 \\ 1 \end{bmatrix}$ 3. $\begin{bmatrix} -1 \\ 4 \\ 7 \end{bmatrix}$ 4. $\begin{bmatrix} 2 \\ -1 \\ 3 \end{bmatrix}$ 5. $\begin{bmatrix} 0 \\ 5 \\ 2 \end{bmatrix}$

6. $\begin{bmatrix} -3 \\ 2 \\ 1 \end{bmatrix}$ 7. $\begin{bmatrix} -1 \\ 1 \\ 1 \end{bmatrix}$ 8. $\begin{bmatrix} 1 \\ 1 \\ -1 \end{bmatrix}$

In Exercises 9–12, determine the values of r for which $\mathbf{v}$ *is in the span of S.*

9. $S = \left\{ \begin{bmatrix} 1 \\ 0 \\ -1 \end{bmatrix}, \begin{bmatrix} -1 \\ 3 \\ 2 \end{bmatrix} \right\}, \mathbf{v} = \begin{bmatrix} 2 \\ r \\ -1 \end{bmatrix}$

10. $S = \left\{ \begin{bmatrix} 1 \\ 2 \\ -1 \end{bmatrix}, \begin{bmatrix} -1 \\ -2 \\ 2 \end{bmatrix} \right\}, \mathbf{v} = \begin{bmatrix} 1 \\ r \\ 2 \end{bmatrix}$

11. $S = \left\{ \begin{bmatrix} -1 \\ 2 \\ 2 \end{bmatrix}, \begin{bmatrix} 1 \\ -1 \\ 0 \end{bmatrix} \right\}, \mathbf{v} = \begin{bmatrix} 2 \\ r \\ -8 \end{bmatrix}$

12. $S = \left\{ \begin{bmatrix} -1 \\ 1 \\ 1 \end{bmatrix}, \begin{bmatrix} 2 \\ -3 \\ 1 \end{bmatrix} \right\}, \mathbf{v} = \begin{bmatrix} r \\ 4 \\ 0 \end{bmatrix}$

In Exercises 13–20, a set of vectors in $\mathcal{R}^n$ is given. Determine if this set spans $\mathcal{R}^n$.

13. $\left\{ \begin{bmatrix} 1 \\ -1 \end{bmatrix}, \begin{bmatrix} -2 \\ 2 \end{bmatrix} \right\}$ 14. $\left\{ \begin{bmatrix} 1 \\ -2 \end{bmatrix}, \begin{bmatrix} -2 \\ 1 \end{bmatrix} \right\}$

15. $\left\{ \begin{bmatrix} 1 \\ -4 \end{bmatrix}, \begin{bmatrix} 3 \\ 2 \end{bmatrix}, \begin{bmatrix} -2 \\ 8 \end{bmatrix} \right\}$ 16. $\left\{ \begin{bmatrix} -2 \\ 4 \end{bmatrix}, \begin{bmatrix} 1 \\ -2 \end{bmatrix}, \begin{bmatrix} -3 \\ 6 \end{bmatrix} \right\}$

17. $\left\{ \begin{bmatrix} 1 \\ 0 \\ -2 \end{bmatrix}, \begin{bmatrix} -1 \\ 1 \\ 4 \end{bmatrix}, \begin{bmatrix} 1 \\ 2 \\ -2 \end{bmatrix} \right\}$

18. $\left\{ \begin{bmatrix} -1 \\ 2 \\ 1 \end{bmatrix}, \begin{bmatrix} -1 \\ 1 \\ 3 \end{bmatrix}, \begin{bmatrix} 1 \\ -3 \\ 1 \end{bmatrix} \right\}$

19. $\left\{ \begin{bmatrix} -1 \\ 1 \\ 2 \end{bmatrix}, \begin{bmatrix} 0 \\ -1 \\ 2 \end{bmatrix}, \begin{bmatrix} 3 \\ -7 \\ 2 \end{bmatrix}, \begin{bmatrix} -5 \\ 7 \\ 6 \end{bmatrix} \right\}$

20. $\left\{ \begin{bmatrix} -1 \\ 3 \\ 0 \end{bmatrix}, \begin{bmatrix} 0 \\ 1 \\ 1 \end{bmatrix}, \begin{bmatrix} 2 \\ -1 \\ 5 \end{bmatrix}, \begin{bmatrix} 2 \\ -1 \\ 1 \end{bmatrix} \right\}$

In Exercises 21–28, an $m \times n$ matrix A is given. Determine if the equation $A\mathbf{x} = \mathbf{b}$ is consistent for every $\mathbf{b}$ in $\mathcal{R}^m$.

21. $\begin{bmatrix} 1 & 0 \\ -2 & 1 \end{bmatrix}$ 22. $\begin{bmatrix} 1 & -2 \\ 2 & -4 \end{bmatrix}$

23. $\begin{bmatrix} 1 & 0 & -3 \\ -1 & 0 & 3 \end{bmatrix}$ 24. $\begin{bmatrix} 1 & 1 & 2 \\ -1 & -3 & 4 \end{bmatrix}$

25. $\begin{bmatrix} 1 & -1 \\ 0 & 1 \\ -2 & 2 \end{bmatrix}$ 26. $\begin{bmatrix} 1 & 0 & -1 \\ 2 & -1 & 1 \\ 0 & 3 & -2 \\ 1 & 1 & -3 \end{bmatrix}$

27. $\begin{bmatrix} 1 & 2 & 3 \\ 2 & 3 & 4 \\ 3 & 4 & 6 \end{bmatrix}$ 28. $\begin{bmatrix} 1 & 0 & 2 & 1 \\ 2 & 1 & 3 & 2 \\ 3 & 4 & 4 & 5 \end{bmatrix}$

In Exercises 29–36, a set S of vectors in $\mathcal{R}^n$ is given. Find a subset of S with the same span as S that is as small as possible.

29. $\left\{ \begin{bmatrix} 1 \\ 3 \end{bmatrix}, \begin{bmatrix} 0 \\ 1 \end{bmatrix} \right\}$ 30. $\left\{ \begin{bmatrix} -1 \\ 1 \end{bmatrix}, \begin{bmatrix} 2 \\ -2 \end{bmatrix}, \begin{bmatrix} 1 \\ 0 \end{bmatrix} \right\}$

31. $\left\{ \begin{bmatrix} 1 \\ 0 \\ -1 \end{bmatrix}, \begin{bmatrix} -2 \\ 0 \\ 2 \end{bmatrix}, \begin{bmatrix} 0 \\ 1 \\ 0 \end{bmatrix} \right\}$ 32. $\left\{ \begin{bmatrix} 1 \\ -1 \\ 2 \end{bmatrix}, \begin{bmatrix} 2 \\ -3 \\ 0 \end{bmatrix}, \begin{bmatrix} 0 \\ 0 \\ 0 \end{bmatrix} \right\}$

33. $\left\{ \begin{bmatrix} 1 \\ -2 \\ 1 \end{bmatrix}, \begin{bmatrix} -2 \\ 4 \\ -2 \end{bmatrix}, \begin{bmatrix} 0 \\ 0 \\ 0 \end{bmatrix} \right\}$ 34. $\left\{ \begin{bmatrix} 1 \\ 0 \\ 1 \end{bmatrix}, \begin{bmatrix} 1 \\ 1 \\ 0 \end{bmatrix}, \begin{bmatrix} 0 \\ 1 \\ 1 \end{bmatrix} \right\}$

35. $\left\{ \begin{bmatrix} -1 \\ 0 \\ 1 \end{bmatrix}, \begin{bmatrix} 0 \\ 1 \\ 2 \end{bmatrix}, \begin{bmatrix} 1 \\ 2 \\ 3 \end{bmatrix} \right\}$ 36. $\left\{ \begin{bmatrix} 1 \\ 0 \\ 0 \end{bmatrix}, \begin{bmatrix} 1 \\ 1 \\ 0 \end{bmatrix}, \begin{bmatrix} 1 \\ 1 \\ 1 \end{bmatrix}, \begin{bmatrix} 0 \\ 0 \\ 1 \end{bmatrix} \right\}$

37. Let $\mathbf{u}_1 = \begin{bmatrix} -1 \\ 3 \end{bmatrix}$ and $\mathbf{u}_2 = \begin{bmatrix} 1 \\ -2 \end{bmatrix}$.

    (a) How many vectors are in $\{\mathbf{u}_1, \mathbf{u}_2\}$?
    (b) How many vectors are in the span of $\{\mathbf{u}_1, \mathbf{u}_2\}$?

38. Give three different spanning sets for the set of vectors that lie in the xy-plane of $\mathcal{R}^3$.

39. Let A be an $m \times n$ matrix with $m > n$. Explain why $A\mathbf{x} = \mathbf{b}$ is inconsistent for some $\mathbf{b}$ in $\mathcal{R}^m$.

40. What can be said about the number of vectors in a spanning set for $\mathcal{R}^m$? Explain your answer.

41. Let $\mathcal{S}_1$ and $\mathcal{S}_2$ be finite subsets of $\mathcal{R}^n$, such that $\mathcal{S}_1$ is contained in $\mathcal{S}_2$. Prove that if $\mathcal{S}_1$ spans $\mathcal{R}^n$, then so does $\mathcal{S}_2$.

42. Let $\mathbf{u}$ and $\mathbf{v}$ be any vectors in $\mathcal{R}^n$. Prove that the spans of $\{\mathbf{u}, \mathbf{v}\}$ and $\{\mathbf{u} + \mathbf{v}, \mathbf{u} - \mathbf{v}\}$ are equal.

43. Let $\mathbf{u}_1, \mathbf{u}_2, \ldots, \mathbf{u}_k$ be vectors in $\mathcal{R}^n$ and $c_1, c_2, \ldots, c_k$

be nonzero scalars. Prove that Span $\{\mathbf{u}_1, \mathbf{u}_2, \ldots, \mathbf{u}_k\} =$ Span $\{c_1\mathbf{u}_1, c_2\mathbf{u}_2, \ldots, c_k\mathbf{u}_k\}$.

44. Let $\mathbf{u}_1, \mathbf{u}_2, \ldots, \mathbf{u}_k$ be vectors in $\mathcal{R}^n$ and c be a scalar. Prove that the span of $\{\mathbf{u}_1, \mathbf{u}_2, \ldots, \mathbf{u}_k\}$ is equal to the span of $\{\mathbf{u}_1 + c\mathbf{u}_2, \mathbf{u}_2, \ldots, \mathbf{u}_k\}$.

45. Let R be the reduced row echelon form of an $m \times n$ matrix A. Is the span of the columns of R equal to the span of the columns of A? Justify your answer.

46. Let $\mathcal{S}_1$ and $\mathcal{S}_2$ be finite subsets of $\mathcal{R}^n$ such that $\mathcal{S}_1$ is contained in $\mathcal{S}_2$. Use only Theorem 1.6(a) and (b) to prove that Span $\mathcal{S}_1$ is contained in Span $\mathcal{S}_2$.

47. Let $\mathcal{S}$ be a finite subset of $\mathcal{R}^n$. Prove that if $\mathbf{u}$ and $\mathbf{v}$ are in the span of $\mathcal{S}$, then so is $\mathbf{u} + c\mathbf{v}$ for any scalar c.

48. Let V be the span of a finite subset of $\mathcal{R}^n$. Show that either $V = \{\mathbf{0}\}$ or V contains infinitely many vectors.

49. Let B be a matrix obtained from A by performing a single elementary row operation on A. Prove that the span of the rows of B equals the span of the rows of A. *Hint:* Use Exercises 43 and 44.

50. Prove that every linear combination of the rows of A can be written as a linear combination of the rows of the reduced row echelon form of A. *Hint:* Use Exercise 49.

In Exercises 51–54, use a calculator with matrix capabilities or computer software such as MATLAB to determine if the given vector is in the span of

$$\left\{ \begin{bmatrix} 1.2 \\ -0.1 \\ 2.3 \\ 3.1 \\ -1.1 \\ -1.9 \end{bmatrix}, \begin{bmatrix} 3.4 \\ -1.7 \\ 0.0 \\ 2.4 \\ 1.7 \\ 2.6 \end{bmatrix}, \begin{bmatrix} -3.1 \\ 0.0 \\ 2.5 \\ 1.6 \\ -3.2 \\ 1.7 \end{bmatrix}, \begin{bmatrix} 7.7 \\ -1.8 \\ -0.2 \\ 3.9 \\ 3.8 \\ -1.0 \end{bmatrix} \right\}.$$

51. $\begin{bmatrix} 1.0 \\ -1.5 \\ -4.6 \\ -3.8 \\ 3.9 \\ 6.4 \end{bmatrix}$ 52. $\begin{bmatrix} -2.6 \\ -1.8 \\ 7.3 \\ 8.7 \\ -5.8 \\ 4.1 \end{bmatrix}$ 53. $\begin{bmatrix} 1.5 \\ -1.6 \\ 2.4 \\ 4.0 \\ -1.5 \\ 4.3 \end{bmatrix}$ 54. $\begin{bmatrix} -4.1 \\ 1.5 \\ 7.1 \\ 5.4 \\ -7.1 \\ -4.7 \end{bmatrix}$

Solutions to the Practice Problems

1. Let A be the matrix whose columns are the vectors of $\mathcal{S}$. Then $\mathbf{u}$ is in the span of $\mathcal{S}$ if and only if $A\mathbf{x} = \mathbf{u}$ is consistent. Because the reduced row echelon form of $[A \ \ \mathbf{u}]$ is I_3, this system is inconsistent. Hence $\mathbf{u}$ is not in the span of $\mathcal{S}$. On the other hand, the reduced row echelon form of $[A \ \ \mathbf{v}]$ is

   $$\begin{bmatrix} 1 & 0 & 2 \\ 0 & 1 & 3 \\ 0 & 0 & 0 \end{bmatrix}.$$

   Thus $A\mathbf{x} = \mathbf{v}$ is consistent, and so $\mathbf{v}$ is in the span of $\mathcal{S}$. In

fact, the reduced row echelon form of $[A \ \ \mathbf{v}]$ shows that

$$2\begin{bmatrix} 2 \\ -1 \\ 1 \end{bmatrix} + 3\begin{bmatrix} -1 \\ 1 \\ 0 \end{bmatrix} = \begin{bmatrix} 1 \\ 1 \\ 2 \end{bmatrix}.$$

2. Let A be the matrix whose columns are the vectors in $\mathcal{S}$. The reduced row echelon form of A is

$$\begin{bmatrix} 1 & 0 & 0 & 9 \\ 0 & 1 & 0 & 5 \\ 0 & 0 & 1 & -2 \end{bmatrix}.$$

Thus the rank of A is 3, and so S is a spanning set for $\mathcal{R}^3$, by Theorem 1.5.

3. From the reduced row echelon form of A in problem 2, we see that the last column of A is a linear combination of the first three columns. Thus the vector $\begin{bmatrix} 2 \\ -1 \\ 1 \end{bmatrix}$ can be removed from S without changing its span. So

$$S' = \left\{ \begin{bmatrix} 1 \\ 0 \\ 1 \end{bmatrix}, \begin{bmatrix} -1 \\ 1 \\ 2 \end{bmatrix}, \begin{bmatrix} 1 \\ 3 \\ 9 \end{bmatrix} \right\}$$

is a subset of S that is a spanning set for $\mathcal{R}^3$. Moreover, this set is the smallest spanning set possible because removing any vector from S' leaves a set of only two vectors. Since the matrix whose columns are the vectors in S' is a 3×2 matrix, it cannot have rank 3 and so cannot span $\mathcal{R}^3$, by Theorem 1.5.

1.7 Linear Dependence and Linear Independence

In Section 1.6, we saw that it is possible to reduce the size of a spanning set if some vector in the spanning set is a linear combination of the others. In fact, by Theorem 1.6(c), this vector can be removed without affecting the span. In this section, we consider the problem of recognizing when a spanning set cannot be made smaller. Consider, for example, the set $S = \{\mathbf{u}_1, \mathbf{u}_2, \mathbf{u}_3, \mathbf{u}_4\}$, where

$$\mathbf{u}_1 = \begin{bmatrix} 1 \\ -1 \\ 2 \\ 1 \end{bmatrix}, \quad \mathbf{u}_2 = \begin{bmatrix} 2 \\ 1 \\ -1 \\ -1 \end{bmatrix}, \quad \mathbf{u}_3 = \begin{bmatrix} -1 \\ -8 \\ 13 \\ 8 \end{bmatrix}, \quad \text{and} \quad \mathbf{u}_4 = \begin{bmatrix} 0 \\ 1 \\ -2 \\ 1 \end{bmatrix}.$$

In this case, the reader should check that $\mathbf{u}_4$ is not a linear combination of vectors $\mathbf{u}_1$, $\mathbf{u}_2$, and $\mathbf{u}_3$. However, this does *not* mean that we cannot find a smaller set having the same span as S, because it is possible that one of $\mathbf{u}_1$, $\mathbf{u}_2$, and $\mathbf{u}_3$ might be a linear combination of the other vectors in S. In fact, this is precisely the situation, because

$$\mathbf{u}_3 = 5\mathbf{u}_1 - 3\mathbf{u}_2 + 0\mathbf{u}_4.$$

Thus checking if one of the vectors in a spanning set is a linear combination of the others could require us to solve many systems of linear equations. Fortunately, a better method is available.

In the example above, in order not to guess which of $\mathbf{u}_1$, $\mathbf{u}_2$, $\mathbf{u}_3$, and $\mathbf{u}_4$ can be expressed as a linear combination of the others, let us formulate the problem differently. Note that because $\mathbf{u}_3 = 5\mathbf{u}_1 - 3\mathbf{u}_2 + 0\mathbf{u}_4$, we must have

$$-5\mathbf{u}_1 + 3\mathbf{u}_2 + \mathbf{u}_3 - 0\mathbf{u}_4 = \mathbf{0}.$$

Thus instead of trying to write some $\mathbf{u}_i$ as a linear combination of the others, we can try to write $\mathbf{0}$ as a linear combination of $\mathbf{u}_1$, $\mathbf{u}_2$, $\mathbf{u}_3$, and $\mathbf{u}_4$. Of course, this is always possible by taking each coefficient in the linear combination to be 0. But *if there is a linear combination of $\mathbf{u}_1$, $\mathbf{u}_2$, $\mathbf{u}_3$, and $\mathbf{u}_4$ that equals $\mathbf{0}$ in which not all of the coefficients are 0, then we can express one of the $\mathbf{u}_i$ as a linear combination of the others.* For example, the preceding equation $-5\mathbf{u}_1 + 3\mathbf{u}_2 + \mathbf{u}_3 - 0\mathbf{u}_4 = \mathbf{0}$ enables us to express any one of $\mathbf{u}_1$, $\mathbf{u}_2$, and $\mathbf{u}_3$ (but *not* $\mathbf{u}_4$) as a linear combination of the others. Solving for $\mathbf{u}_1$ in the equation $-5\mathbf{u}_1 + 3\mathbf{u}_2 + \mathbf{u}_3 - 0\mathbf{u}_4 = \mathbf{0}$, we have

$$-5\mathbf{u}_1 = -3\mathbf{u}_2 - \mathbf{u}_3 + 0\mathbf{u}_4$$

$$\mathbf{u}_1 = \frac{3}{5}\mathbf{u}_2 + \frac{1}{5}\mathbf{u}_3 + 0\mathbf{u}_4.$$

We see that at least one of the vectors *depends* on (is a linear combination of) the others. This idea motivates the following definitions.

Definitions.   A set of vectors $\{\mathbf{u}_1, \mathbf{u}_2, \ldots, \mathbf{u}_k\}$ in $\mathcal{R}^n$ is called **linearly dependent** if there exist scalars $c_1, c_2, \ldots, c_k$, not all zero, such that

$$c_1 \mathbf{u}_1 + c_2 \mathbf{u}_2 + \cdots + c_k \mathbf{u}_k = \mathbf{0}.$$

In this case we also say that **the vectors $\mathbf{u}_1, \mathbf{u}_1, \ldots, \mathbf{u}_k$ are linearly dependent**.

A set of vectors $\{\mathbf{u}_1, \mathbf{u}_2, \ldots, \mathbf{u}_k\}$ is called **linearly independent** if the only scalars $c_1, c_2, \ldots, c_k$, such that

$$c_1 \mathbf{u}_1 + c_2 \mathbf{u}_2 + \cdots + c_k \mathbf{u}_k = \mathbf{0}$$

are $c_1 = c_2 = \cdots = c_k = 0$. Again in this case, we also say that **the vectors $\mathbf{u}_1, \mathbf{u}_2, \ldots, \mathbf{u}_k$ are linearly independent**.

A set is linearly independent if and only if it is not linearly dependent. Note that *any finite subset $\mathcal{S} = \{\mathbf{0}, \mathbf{u}_1, \mathbf{u}_2, \ldots, \mathbf{u}_k\}$ of $\mathcal{R}^n$ that contains the zero vector is linearly dependent*, because

$$1 \cdot \mathbf{0} + 0\mathbf{u}_1 + 0\mathbf{u}_2 + \cdots + 0\mathbf{u}_k = \mathbf{0}$$

is a linear combination of the vectors in $\mathcal{S}$ in which at least one coefficient is nonzero. However, the zero vector is a linear combination of the vectors in any nonempty, finite subset $\{\mathbf{u}_1, \mathbf{u}_2, \ldots, \mathbf{u}_k\}$ of $\mathcal{R}^n$, namely,

$$0\mathbf{u}_1 + 0\mathbf{u}_2 + \cdots + 0\mathbf{u}_k = \mathbf{0}.$$

Thus this equation tells us *nothing* about the linear dependence or linear independence of $\{\mathbf{u}_1, \mathbf{u}_2, \ldots, \mathbf{u}_k\}$.

For example, suppose that we wish to determine whether or not

$$\mathcal{S} = \left\{ \begin{bmatrix} 2 \\ 3 \end{bmatrix}, \begin{bmatrix} 5 \\ 8 \end{bmatrix}, \begin{bmatrix} 1 \\ 2 \end{bmatrix} \right\}$$

is linearly dependent. As noted above, the equation

$$c_1 \begin{bmatrix} 2 \\ 3 \end{bmatrix} + c_2 \begin{bmatrix} 5 \\ 8 \end{bmatrix} + c_3 \begin{bmatrix} 1 \\ 2 \end{bmatrix} = \begin{bmatrix} 0 \\ 0 \end{bmatrix}$$

is true for $c_1 = c_2 = c_3 = 0$. Since a similar statement is true for *any* set of vectors, we cannot conclude anything about the linear independence or linear dependence of $\mathcal{S}$ from this equation. In this case, $\mathcal{S}$ is linearly dependent because there exists *another choice of the coefficients c_1, c_2, and c_3* that expresses $\mathbf{0}$ as a linear combination of the vectors in $\mathcal{S}$. For example, taking $c_1 = 2$, $c_2 = -1$, and $c_3 = 1$, we see that

$$2 \begin{bmatrix} 2 \\ 3 \end{bmatrix} + (-1) \begin{bmatrix} 5 \\ 8 \end{bmatrix} + 1 \begin{bmatrix} 1 \\ 2 \end{bmatrix} = \begin{bmatrix} 0 \\ 0 \end{bmatrix}.$$

Since the equation $c_1\mathbf{u}_1 + c_2\mathbf{u}_2 + \cdots + c_k\mathbf{u}_k = \mathbf{0}$ can be written as a matrix–vector product

$$[\mathbf{u}_1 \ \ \mathbf{u}_2 \ \ \cdots \ \ \mathbf{u}_k]\begin{bmatrix} c_1 \\ c_2 \\ \vdots \\ c_k \end{bmatrix} = \mathbf{0},$$

we have the following useful observation.

> The set $\{\mathbf{u}_1, \mathbf{u}_2, \ldots, \mathbf{u}_k\}$ is linearly dependent if and only if $A\mathbf{x} = \mathbf{0}$ has a nonzero solution, where $A = [\mathbf{u}_1 \ \ \mathbf{u}_2 \ \ \ldots \ \ \mathbf{u}_k]$.

Example 1   Determine whether the set

$$S = \left\{ \begin{bmatrix} 1 \\ 2 \\ 1 \end{bmatrix}, \begin{bmatrix} 1 \\ 0 \\ 1 \end{bmatrix}, \begin{bmatrix} 1 \\ 4 \\ 1 \end{bmatrix}, \begin{bmatrix} 1 \\ 2 \\ 3 \end{bmatrix} \right\}$$

is linearly dependent or linearly independent.

Solution   We must determine if $A\mathbf{x} = \mathbf{0}$ has a nonzero solution, where

$$A = \begin{bmatrix} 1 & 1 & 1 & 1 \\ 2 & 0 & 4 & 2 \\ 1 & 1 & 1 & 3 \end{bmatrix}$$

is the matrix whose columns are the vectors in S. The augmented matrix of $A\mathbf{x} = \mathbf{0}$ is

$$\begin{bmatrix} 1 & 1 & 1 & 1 & 0 \\ 2 & 0 & 4 & 2 & 0 \\ 1 & 1 & 1 & 3 & 0 \end{bmatrix},$$

and its reduced row echelon form is

$$\begin{bmatrix} 1 & 0 & 2 & 0 & 0 \\ 0 & 1 & -1 & 0 & 0 \\ 0 & 0 & 0 & 1 & 0 \end{bmatrix}.$$

Hence the general solution to this system is

$$\begin{aligned} x_1 &= -2x_3 \\ x_2 &= x_3 \\ x_3 &\quad \text{free} \\ x_4 &= 0. \end{aligned}$$

Because the solution to $A\mathbf{x} = \mathbf{0}$ contains a free variable, this system of linear equations has infinitely many solutions, and we can obtain a nonzero solution by choosing any nonzero value of the free variable. Taking $\mathbf{x}_3 = 1$, for instance, we see that

$$\begin{bmatrix} x_1 \\ x_2 \\ x_3 \\ x_4 \end{bmatrix} = \begin{bmatrix} -2 \\ 1 \\ 1 \\ 0 \end{bmatrix}$$

is a nonzero solution to $A\mathbf{x} = \mathbf{0}$. Thus $\mathcal{S}$ is a linearly dependent subset of $\mathcal{R}^3$. Observe that

$$-2 \begin{bmatrix} 1 \\ 2 \\ 1 \end{bmatrix} + 1 \begin{bmatrix} 1 \\ 0 \\ 1 \end{bmatrix} + 1 \begin{bmatrix} 1 \\ 4 \\ 1 \end{bmatrix} + 0 \begin{bmatrix} 1 \\ 2 \\ 3 \end{bmatrix} = \begin{bmatrix} 0 \\ 0 \\ 0 \end{bmatrix}$$

is a representation of $\mathbf{0}$ as a linear combination of the vectors in $\mathcal{S}$.

Example 2   Determine whether the set

$$\mathcal{S} = \left\{ \begin{bmatrix} 1 \\ 2 \\ 1 \end{bmatrix}, \begin{bmatrix} 2 \\ 2 \\ 3 \end{bmatrix}, \begin{bmatrix} 1 \\ 0 \\ 1 \end{bmatrix} \right\}$$

is linearly dependent or linearly independent.

Solution   As in Example 1, we must check if $A\mathbf{x} = \mathbf{0}$ has nonzero solutions, where

$$A = \begin{bmatrix} 1 & 2 & 1 \\ 2 & 2 & 0 \\ 1 & 3 & 1 \end{bmatrix}.$$

There is a way to do this without actually solving $A\mathbf{x} = \mathbf{0}$ (as we did in Example 1). Note that the system $A\mathbf{x} = \mathbf{0}$ has nonzero solutions if and only if its general solution contains a free variable. Since the reduced row echelon form of A is

$$\begin{bmatrix} 1 & 0 & 0 \\ 0 & 1 & 0 \\ 0 & 0 & 1 \end{bmatrix},$$

the rank of A is 3, and the nullity of A is $3 - 3 = 0$. Thus the general solution of $A\mathbf{x} = \mathbf{0}$ has no free variables. Hence $A\mathbf{x} = \mathbf{0}$ has no nonzero solutions, and $\mathcal{S}$ is linearly independent.

In Example 2, we showed that a particular set $\mathcal{S}$ is linearly independent without actually solving a system of linear equations. Our next theorem shows that a similar technique can be used for any set whatsoever. Note the relationship between this theorem and Theorem 1.5.

Theorem 1.7
The following statements about an $m \times n$ matrix A are equivalent.
 (a) The columns of A are linearly independent.
 (b) The equation $A\mathbf{x} = \mathbf{b}$ has at most one solution for each $\mathbf{b}$ in $\mathcal{R}^m$.
 (c) The nullity of A is zero.
 (d) The rank of A is n.

(e) The columns of the reduced row echelon form of A are distinct standard vectors in $\mathcal{R}^m$.

(f) The only solution to $A\mathbf{x} = \mathbf{0}$ is $\mathbf{0}$.

Proof We have already noted that (a) and (f) are equivalent. To complete the proof, we show that (b) implies (c), (c) implies (d), (d) implies (e), (e) implies (f), and (f) implies (b).

(b) *implies* (c) Since $\mathbf{0}$ is a solution to $A\mathbf{x} = \mathbf{0}$, (b) implies that $A\mathbf{x} = \mathbf{0}$ has no nonzero solutions. Thus the general solution to $A\mathbf{x} = \mathbf{0}$ has no free variables. Since the number of free variables is the nullity of A, we see that the nullity of A is zero.

(c) *implies* (d) Because rank $A + $ nullity $A = n$, (d) follows immediately from (c).

(d) *implies* (e) If the rank of A is n, then every column of A is a pivot column, and therefore the reduced row echelon form of A consists entirely of standard vectors. These are necessarily distinct because each column contains the first nonzero entry in some row.

(e) *implies* (f) Let R be the reduced row echelon form of A. If the columns of R are distinct standard vectors in $\mathcal{R}^m$, then $R = [\mathbf{e}_1 \ \mathbf{e}_2 \ \ldots \ \mathbf{e}_n]$. Clearly the only solution to $R\mathbf{x} = \mathbf{0}$ is $\mathbf{0}$, and since $A\mathbf{x} = \mathbf{0}$ is equivalent to $R\mathbf{x} = \mathbf{0}$, it follows that the only solution to $A\mathbf{x} = \mathbf{0}$ is $\mathbf{0}$.

(f) *implies* (b) Let $\mathbf{b}$ be any vector in $\mathcal{R}^m$. To show that $A\mathbf{x} = \mathbf{b}$ has at most one solution, we assume that $\mathbf{u}$ and $\mathbf{v}$ are both solutions to $A\mathbf{x} = \mathbf{b}$ and prove that $\mathbf{u} = \mathbf{v}$. Since $\mathbf{u}$ and $\mathbf{v}$ are solutions to $A\mathbf{x} = \mathbf{b}$, we have

$$A(\mathbf{u} - \mathbf{v}) = A\mathbf{u} - A\mathbf{v} = \mathbf{b} - \mathbf{b} = \mathbf{0}.$$

So $\mathbf{u} - \mathbf{v}$ is a solution to $A\mathbf{x} = \mathbf{0}$. Thus (f) implies that $\mathbf{u} - \mathbf{v} = \mathbf{0}$, that is, $\mathbf{u} = \mathbf{v}$. It follows that $A\mathbf{x} = \mathbf{b}$ has at most one solution. $\qquad \square$

The equation $A\mathbf{x} = \mathbf{b}$ is called **homogeneous** if $\mathbf{b} = \mathbf{0}$. As Examples 1 and 2 illustrate, in checking if a subset is linearly independent, we are led to a homogeneous equation. Note that, unlike an arbitrary equation, a homogeneous equation must be consistent because $\mathbf{0}$ is a solution to $A\mathbf{x} = \mathbf{0}$. As a result, the important question concerning a homogeneous equation is not *if* it has solutions, but rather *how many* solutions it has. For example, the general solution to a homogeneous system of linear equations with more variables than equations must have free variables. Hence *a homogeneous system of linear equations with more variables than equations has infinitely many solutions.* According to Theorem 1.7, the number of solutions to $A\mathbf{x} = \mathbf{0}$ determines the linear dependence or independence of the columns of A.

In order to investigate some other properties of the homogeneous equation $A\mathbf{x} = \mathbf{0}$, let us consider this equation for the matrix

$$A = \begin{bmatrix} 1 & -4 & 2 & -1 & 2 \\ 2 & -8 & 3 & 2 & -1 \end{bmatrix}.$$

Since the reduced row echelon form of $[A \ \ \mathbf{0}]$ is

$$\begin{bmatrix} 1 & -4 & 0 & 7 & -8 & 0 \\ 0 & 0 & 1 & -4 & 5 & 0 \end{bmatrix},$$

the general solution to $A\mathbf{x} = \mathbf{0}$ is

$$
\begin{aligned}
x_1 &= 4x_2 \quad - 7x_4 + 8x_5 \\
x_2 &\quad \text{free} \\
x_3 &= \qquad\qquad 4x_4 - 5x_5 \\
x_4 &\quad \text{free} \\
x_5 &\quad \text{free.}
\end{aligned}
$$

Expressing the solutions of $A\mathbf{x} = \mathbf{0}$ in vector form yields

$$
\begin{bmatrix} x_1 \\ x_2 \\ x_3 \\ x_4 \\ x_5 \end{bmatrix} = \begin{bmatrix} 4x_2 - 7x_4 + 8x_5 \\ x_2 \\ 4x_4 - 5x_5 \\ x_4 \\ x_5 \end{bmatrix} = x_2 \begin{bmatrix} 4 \\ 1 \\ 0 \\ 0 \\ 0 \end{bmatrix} + x_4 \begin{bmatrix} -7 \\ 0 \\ 4 \\ 1 \\ 0 \end{bmatrix} + x_5 \begin{bmatrix} 8 \\ 0 \\ -5 \\ 0 \\ 1 \end{bmatrix}. \quad (9)
$$

Thus the solution to $A\mathbf{x} = \mathbf{0}$ is the span of

$$
\mathcal{S} = \left\{ \begin{bmatrix} 4 \\ 1 \\ 0 \\ 0 \\ 0 \end{bmatrix}, \begin{bmatrix} -7 \\ 0 \\ 4 \\ 1 \\ 0 \end{bmatrix}, \begin{bmatrix} 8 \\ 0 \\ -5 \\ 0 \\ 1 \end{bmatrix} \right\}.
$$

In a similar manner, for a matrix A, any solution to $A\mathbf{x} = \mathbf{0}$ can be expressed as a linear combination of vectors in which the coefficients are the free variables in the general solution. We call such a representation a **parametric representation** of the general solution to $A\mathbf{x} = \mathbf{0}$. The solution set of this equation equals the span of the set of vectors that appear in a parametric representation of its general solution.

For the set $\mathcal{S}$ above, we see from (9) that the only linear combination of vectors in $\mathcal{S}$ equal to $\mathbf{0}$ is the one in which all of the coefficients are zero. Hence $\mathcal{S}$ is linearly independent. More generally, the following result is true.

> When a parametric representation of the general solution to $A\mathbf{x} = \mathbf{0}$ is obtained by the method described in Section 1.3, the vectors that appear in the parametric representation are linearly independent.

Linearly Dependent and Linearly Independent Sets

The following result provides a useful characterization of linearly dependent sets.

Theorem 1.8
Vectors $\mathbf{u}_1, \mathbf{u}_2, \dots, \mathbf{u}_k$ in $\mathcal{R}^n$ are linearly dependent if and only if $\mathbf{u}_1 = \mathbf{0}$ or some $\mathbf{u}_i$ ($i = 2, 3, \dots, k$) is linear combination of the preceding vectors.

Proof Suppose first that the vectors $\mathbf{u}_1, \mathbf{u}_2, \dots, \mathbf{u}_k$ in $\mathcal{R}^n$ are linearly dependent. If $\mathbf{u}_1 = \mathbf{0}$, then we are finished; so suppose $\mathbf{u}_1 \neq \mathbf{0}$. There exist scalars

$c_1, c_2, \ldots, c_k$, not all zero, such that

$$c_1\mathbf{u}_1 + c_2\mathbf{u}_2 + \cdots + c_k\mathbf{u}_k = \mathbf{0}.$$

Let i denote the largest index such that $c_i \neq 0$. Note that $i \geq 2$, for otherwise the preceding equation would reduce to $c_1\mathbf{u}_1 = \mathbf{0}$—which is false because $c_1 \neq 0$ and $\mathbf{u}_1 \neq \mathbf{0}$. Hence the preceding equation becomes

$$c_1\mathbf{u}_1 + c_2\mathbf{u}_2 + \cdots + c_i\mathbf{u}_i = \mathbf{0},$$

where $c_i \neq 0$. Solving this equation for $\mathbf{u}_i$, we obtain

$$c_i\mathbf{u}_i = -c_1\mathbf{u}_1 - c_2\mathbf{u}_2 - \cdots - c_{i-1}\mathbf{u}_{i-1}$$

$$\mathbf{u}_i = \frac{-c_1}{c_i}\mathbf{u}_1 - \frac{c_2}{c_i}\mathbf{u}_2 - \cdots - \frac{c_{i-1}}{c_i}\mathbf{u}_{i-1}.$$

Thus $\mathbf{u}_i$ is a linear combination of $\mathbf{u}_1, \mathbf{u}_2, \ldots, \mathbf{u}_{i-1}$.

We leave the proof of the converse as an exercise.   ❑

The following properties follow easily from Theorem 1.8.

Properties of Linearly Dependent and Independent Sets

1. A set consisting of a single nonzero vector is linearly independent, but $\{\mathbf{0}\}$ is linearly dependent.

2. A set of two vectors $\{\mathbf{u}_1, \mathbf{u}_2\}$ is linearly dependent if and only if $\mathbf{u}_1 = \mathbf{0}$ or $\mathbf{u}_2$ is in the span of $\{\mathbf{u}_1\}$; that is, if and only if $\mathbf{u}_1 = \mathbf{0}$ or $\mathbf{u}_2$ is a multiple of $\mathbf{u}_1$. Hence *a set of two vectors is linearly dependent if and only if one of the vectors is a multiple of the other.*

3. Let $\mathcal{S}$ be a linearly independent subset of $\mathcal{R}^n$, and let $\mathbf{v}$ be in $\mathcal{R}^n$. If $\mathbf{v}$ does not belong to the span of $\mathcal{S}$, then $\mathcal{S} \cup \{\mathbf{v}\}$ is linearly independent. For if $\mathcal{S} = \{\mathbf{u}_1, \mathbf{u}_2, \ldots, \mathbf{u}_k\}$ is linearly independent, then, by Theorem 1.8, $\mathbf{u}_1 \neq \mathbf{0}$ and, for $i = 2, 3, \ldots, k$, no $\mathbf{u}_i$ is in the span of $\{\mathbf{u}_1, \mathbf{u}_2, \ldots, \mathbf{u}_{i-1}\}$. Since $\mathbf{v}$ does not belong to the span of $\mathcal{S}$, the vectors $\mathbf{u}_1, \mathbf{u}_2, \ldots, \mathbf{u}_k, \mathbf{v}$ are also linearly independent, by Theorem 1.8.

Example 3   Determine by inspection if the following sets are linearly dependent or linearly independent.

$$\mathcal{S}_1 = \left\{ \begin{bmatrix} 3 \\ -1 \\ 7 \end{bmatrix}, \begin{bmatrix} 0 \\ 0 \\ 0 \end{bmatrix}, \begin{bmatrix} -2 \\ 5 \\ 1 \end{bmatrix} \right\}, \qquad \mathcal{S}_2 = \left\{ \begin{bmatrix} -4 \\ 12 \\ 6 \end{bmatrix}, \begin{bmatrix} -10 \\ 30 \\ 15 \end{bmatrix} \right\},$$

and

$$\mathcal{S}_3 = \left\{ \begin{bmatrix} -3 \\ 7 \\ 0 \end{bmatrix}, \begin{bmatrix} 2 \\ 9 \\ 0 \end{bmatrix}, \begin{bmatrix} -1 \\ 0 \\ 2 \end{bmatrix} \right\}.$$

Solution   Since $\mathcal{S}_1$ contains the zero vector, it is linearly dependent. To determine if $\mathcal{S}_2$, a set of two vectors, is linearly dependent or linearly independent, we need only check if either of the vectors in $\mathcal{S}_2$ is a multiple of the other.

Because

$$\frac{5}{2}\begin{bmatrix} -4 \\ 12 \\ 6 \end{bmatrix} = \begin{bmatrix} -10 \\ 30 \\ 15 \end{bmatrix},$$

we see that S_2 is linearly dependent. To see if S_3 is linearly independent, consider the subset

$$S = \left\{ \begin{bmatrix} -3 \\ 7 \\ 0 \end{bmatrix}, \begin{bmatrix} 2 \\ 9 \\ 0 \end{bmatrix} \right\}.$$

Because S is a set of two vectors, neither of which is a multiple of the other, S is linearly independent. Vectors in the span of S are linear combinations of the vectors in S and, therefore, must have 0 as their third component. Since

$$\mathbf{v} = \begin{bmatrix} -1 \\ 0 \\ 2 \end{bmatrix}$$

has a nonzero third component, it does not belong to the span of S. Hence by property 3 in the preceding list, $S_3 = S \cup \{\mathbf{v}\}$ is linearly independent.

Consider a subset $\{\mathbf{u}_1, \mathbf{u}_2, \ldots, \mathbf{u}_k\}$ of $\mathcal{R}^n$ that contains k vectors, where $k > n$. The $n \times k$ matrix $[\mathbf{u}_1 \ \mathbf{u}_2 \ \cdots \ \mathbf{u}_k]$ cannot have rank k, because it has only n rows. Thus the set $\{\mathbf{u}_1, \mathbf{u}_2, \ldots, \mathbf{u}_k\}$ is linearly dependent, by Theorem 1.7. In other words, *every subset of $\mathcal{R}^n$ containing more than n vectors must be linearly dependent.* However, Example 3 illustrates that subsets of $\mathcal{R}^n$ containing no more than n vectors may be either linearly dependent or linearly independent.

Although the concepts of linear independence and spanning sets appear unrelated, there is a connection suggested by the preceding paragraph. Note that $\mathcal{R}^n = \text{Span}\{\mathbf{e}_1, \mathbf{e}_2, \ldots, \mathbf{e}_n\}$, and any finite subset of $\mathcal{R}^n$ that contains more than n vectors must be linearly dependent. Our next theorem generalizes this fact; it plays a crucial role in Chapter 4.

Theorem 1.9
Let $V = \text{Span}\{\mathbf{u}_1, \mathbf{u}_2, \ldots, \mathbf{u}_k\}$, where $\mathbf{u}_1, \mathbf{u}_2, \ldots, \mathbf{u}_k$ are vectors in $\mathcal{R}^n$. Every subset of V containing more than k vectors is linearly dependent.

Proof Let $V = \text{Span}\{\mathbf{u}_1, \mathbf{u}_2, \ldots, \mathbf{u}_k\}$, and let $\{\mathbf{v}_1, \mathbf{v}_2, \ldots, \mathbf{v}_p\}$ be a subset of V containing p vectors, where $p > k$. Define $B = [\mathbf{u}_1 \ \mathbf{u}_2 \ \cdots \ \mathbf{u}_k]$. Because $\{\mathbf{u}_1, \mathbf{u}_2, \ldots, \mathbf{u}_k\}$ is a spanning set for V, there are vectors $\mathbf{a}_i \in \mathcal{R}^k$ for $i = 1, 2, \ldots, p$ such that $B\mathbf{a}_i = \mathbf{v}_i$. Now $\{\mathbf{a}_1, \mathbf{a}_2, \ldots, \mathbf{a}_p\}$ is a subset of $\mathcal{R}^k$ containing more than k vectors, and so it is linearly dependent. Hence there exist scalars $c_1, c_2, \ldots, c_p$, not all zero, such that $c_1\mathbf{a}_1 + c_2\mathbf{a}_2 + \cdots + c_p\mathbf{a}_p = \mathbf{0}$. Then

$$c_1\mathbf{v}_1 + c_2\mathbf{v}_2 + \cdots + c_p\mathbf{v}_p = c_1(B\mathbf{a}_1) + c_2(B\mathbf{a}_2) + \cdots + c_p(B\mathbf{a}_p)$$
$$= B(c_1\mathbf{a}_1) + B(c_2\mathbf{a}_2) + \cdots + B(c_p\mathbf{a}_p)$$
$$= B(c_1\mathbf{a}_1 + c_2\mathbf{a}_2 + \cdots + c_p\mathbf{a}_p)$$
$$= B\mathbf{0}$$
$$= \mathbf{0}.$$

Since not all of $c_1, c_2, \ldots, c_p$ are zero, $\{\mathbf{v}_1, \mathbf{v}_2, \ldots, \mathbf{v}_k\}$ is linearly dependent. $\qquad\blacksquare$

In this chapter, we introduced matrices and vectors and learned some of their fundamental properties. Since a system of linear equations can be written as an equation involving a matrix and vectors, we can use these arrays to solve any system of linear equations. It is surprising that the number of solutions to the equation $A\mathbf{x} = \mathbf{b}$ is related to both the simple concept of the rank of a matrix and also to the complex concepts of spanning sets and linearly independent sets. Yet this is exactly the case, as Theorems 1.5 and 1.7 show. To conclude this chapter, we present the following table, which summarizes the relationships among these ideas that were established in Sections 1.6 and 1.7. We assume that A is an $m \times n$ matrix with reduced row echelon form R. *Properties listed in the same row of the table are equivalent.*

Rank of A	Number of solutions to $A\mathbf{x} = \mathbf{b}$	Columns of A	Reduced row echelon form of A
rank $A = m$	$A\mathbf{x} = \mathbf{b}$ has at least one solution for every $\mathbf{b}$ in $\mathcal{R}^m$	The columns of A are a spanning set for $\mathcal{R}^m$	Every row of R contains a pivot position
rank $A = n$	$A\mathbf{x} = \mathbf{b}$ has at most one solution for every $\mathbf{b}$ in $\mathcal{R}^m$	The columns of A are linearly independent	Every column of R contains a pivot position

Practice Problems

1. Is some vector in the set

$$S = \left\{ \begin{bmatrix} -1 \\ 0 \\ 2 \\ 1 \end{bmatrix}, \begin{bmatrix} 1 \\ 1 \\ -1 \\ -1 \end{bmatrix}, \begin{bmatrix} 0 \\ 2 \\ -1 \\ 1 \end{bmatrix}, \begin{bmatrix} -1 \\ 3 \\ 1 \\ 2 \end{bmatrix} \right\}$$

a linear combination of the others?

2. Determine a parametric representation for the general solution to

$$x_1 - 3x_2 - x_3 + x_4 - x_5 = 0$$
$$2x_1 - 6x_2 + x_3 - 3x_4 - 9x_5 = 0$$
$$-2x_1 + 6x_2 + 3x_3 + 2x_4 + 11x_5 = 0.$$

Exercises

1. Determine if the following statements are true or false.

(a) If $\mathcal{S}$ is linearly independent, then no vector in $\mathcal{S}$ is a linear combination of the others.

(b) If the only solution to $A\mathbf{x} = \mathbf{0}$ is $\mathbf{0}$, then the rows of A are linearly independent.

(c) If the nullity of A is 0, then the columns of A are linearly dependent.

(d) If the columns of the reduced row echelon form of A are distinct standard vectors, then the only solution to $A\mathbf{x} = \mathbf{0}$ is $\mathbf{0}$.

(e) If A is an $m \times n$ matrix with rank n, then the columns of A are linearly independent.

(f) A homogeneous equation is always consistent.

(g) A homogeneous equation always has infinitely many solutions.

(h) If a parametric representation of the general solution to $A\mathbf{x} = \mathbf{0}$ is obtained by the method described in Section 1.3, then the vectors that appear in the parametric representation are linearly independent.

(i) For any vector $\mathbf{v}$, $\{\mathbf{v}\}$ is linearly dependent.

(j) A set of vectors in $\mathcal{R}^n$ is linearly dependent if and only if one of the vectors is a multiple of one of the others.

(k) If a subset of $\mathcal{R}^n$ is linearly dependent, then it must contain at least n vectors.

(l) If the columns of a 3×4 matrix are distinct, then they are linearly dependent.

(m) No subset of Span $\{\mathbf{u}_1, \mathbf{u}_2, \ldots, \mathbf{u}_k\}$ that contains more than k vectors is linearly independent.

In Exercises 2–6, determine by inspection if the given sets are linearly dependent.

2. $\left\{ \begin{bmatrix} 2 \\ -1 \end{bmatrix}, \begin{bmatrix} -1 \\ 2 \end{bmatrix} \right\}$

3. $\left\{ \begin{bmatrix} 1 \\ -3 \\ 0 \end{bmatrix}, \begin{bmatrix} -2 \\ 6 \\ 0 \end{bmatrix} \right\}$

4. $\left\{ \begin{bmatrix} 3 \\ -1 \\ 2 \end{bmatrix}, \begin{bmatrix} 0 \\ 0 \\ 0 \end{bmatrix}, \begin{bmatrix} -2 \\ 5 \\ 1 \end{bmatrix} \right\}$

5. $\left\{ \begin{bmatrix} 0 \\ 0 \\ -1 \end{bmatrix}, \begin{bmatrix} 0 \\ 2 \\ 1 \end{bmatrix}, \begin{bmatrix} -3 \\ 7 \\ 2 \end{bmatrix} \right\}$

6. $\left\{ \begin{bmatrix} 1 \\ -4 \end{bmatrix}, \begin{bmatrix} 2 \\ 3 \end{bmatrix}, \begin{bmatrix} -5 \\ 6 \end{bmatrix} \right\}$

In Exercises 7–10, a set S is given. Determine by inspection a smallest subset of S having the same span as S.

7. $\left\{ \begin{bmatrix} 1 \\ -2 \\ 3 \end{bmatrix}, \begin{bmatrix} -2 \\ 4 \\ -6 \end{bmatrix} \right\}$

8. $\left\{ \begin{bmatrix} 1 \\ 0 \\ 2 \end{bmatrix}, \begin{bmatrix} 3 \\ -1 \\ 1 \end{bmatrix} \right\}$

9. $\left\{ \begin{bmatrix} -3 \\ 2 \\ 0 \end{bmatrix}, \begin{bmatrix} 1 \\ 6 \\ 0 \end{bmatrix}, \begin{bmatrix} 0 \\ 0 \\ 0 \end{bmatrix} \right\}$

10. $\left\{ \begin{bmatrix} 0 \\ 0 \\ 1 \end{bmatrix}, \begin{bmatrix} 0 \\ 1 \\ 2 \end{bmatrix}, \begin{bmatrix} 1 \\ 2 \\ 3 \end{bmatrix}, \begin{bmatrix} 2 \\ 3 \\ 4 \end{bmatrix} \right\}$

In Exercises 11–18, determine if the given set is linearly independent.

11. $\left\{ \begin{bmatrix} 1 \\ -1 \\ -2 \end{bmatrix}, \begin{bmatrix} -1 \\ 0 \\ 1 \end{bmatrix}, \begin{bmatrix} 1 \\ 2 \\ 1 \end{bmatrix} \right\}$

12. $\left\{ \begin{bmatrix} 1 \\ -1 \\ 1 \end{bmatrix}, \begin{bmatrix} -1 \\ 0 \\ 2 \end{bmatrix}, \begin{bmatrix} 2 \\ 1 \\ 1 \end{bmatrix} \right\}$

13. $\left\{ \begin{bmatrix} 1 \\ 2 \\ 0 \\ -1 \end{bmatrix}, \begin{bmatrix} 1 \\ -3 \\ 1 \\ -2 \end{bmatrix}, \begin{bmatrix} 1 \\ 2 \\ -2 \\ 3 \end{bmatrix} \right\}$

14. $\left\{ \begin{bmatrix} -1 \\ 0 \\ 1 \\ 2 \end{bmatrix}, \begin{bmatrix} -2 \\ 1 \\ 1 \\ -3 \end{bmatrix}, \begin{bmatrix} -4 \\ 1 \\ 3 \\ 1 \end{bmatrix} \right\}$

15. $\left\{ \begin{bmatrix} 1 \\ 0 \\ 0 \\ -2 \end{bmatrix}, \begin{bmatrix} 0 \\ 1 \\ -1 \\ 0 \end{bmatrix}, \begin{bmatrix} 1 \\ 0 \\ 1 \\ 1 \end{bmatrix}, \begin{bmatrix} 0 \\ 1 \\ 0 \\ 1 \end{bmatrix} \right\}$

16. $\left\{ \begin{bmatrix} 1 \\ 0 \\ 1 \\ 0 \end{bmatrix}, \begin{bmatrix} -1 \\ 1 \\ 0 \\ 1 \end{bmatrix}, \begin{bmatrix} 1 \\ -1 \\ 1 \\ 0 \end{bmatrix}, \begin{bmatrix} 3 \\ -1 \\ 0 \\ -3 \end{bmatrix} \right\}$

17. $\left\{ \begin{bmatrix} 1 \\ -1 \\ -1 \\ 2 \end{bmatrix}, \begin{bmatrix} -1 \\ 0 \\ 1 \\ -1 \end{bmatrix}, \begin{bmatrix} -1 \\ -4 \\ 1 \\ 3 \end{bmatrix}, \begin{bmatrix} 0 \\ 1 \\ -2 \\ 1 \end{bmatrix} \right\}$

18. $\left\{ \begin{bmatrix} -1 \\ 0 \\ 1 \\ -1 \end{bmatrix}, \begin{bmatrix} 1 \\ 0 \\ -2 \\ 0 \end{bmatrix}, \begin{bmatrix} 0 \\ -2 \\ 1 \\ 2 \end{bmatrix}, \begin{bmatrix} 1 \\ -1 \\ -1 \\ 2 \end{bmatrix} \right\}$

In Exercises 19–22, a linearly dependent set S is given. Write some vector in S as a linear combination of the others.

19. $\left\{ \begin{bmatrix} -1 \\ 1 \\ 2 \end{bmatrix}, \begin{bmatrix} 3 \\ -3 \\ -6 \end{bmatrix}, \begin{bmatrix} 0 \\ 1 \\ 2 \end{bmatrix} \right\}$

20. $\left\{ \begin{bmatrix} 0 \\ 0 \\ 0 \end{bmatrix}, \begin{bmatrix} -2 \\ 3 \\ -4 \end{bmatrix}, \begin{bmatrix} 4 \\ -3 \\ 2 \end{bmatrix} \right\}$

21. $\left\{ \begin{bmatrix} 0 \\ 1 \\ 1 \end{bmatrix}, \begin{bmatrix} 1 \\ 0 \\ -1 \end{bmatrix}, \begin{bmatrix} 4 \\ 5 \\ 1 \end{bmatrix} \right\}$

22. $\left\{ \begin{bmatrix} 1 \\ 2 \\ -1 \end{bmatrix}, \begin{bmatrix} -1 \\ -3 \\ 2 \end{bmatrix}, \begin{bmatrix} 4 \\ 6 \\ -2 \end{bmatrix} \right\}$

In Exercises 23–26, determine, if possible, a value of r for which the given set is linearly dependent.

23. $\left\{ \begin{bmatrix} 1 \\ -1 \end{bmatrix}, \begin{bmatrix} -3 \\ 3 \end{bmatrix}, \begin{bmatrix} 4 \\ r \end{bmatrix} \right\}$

24. $\left\{ \begin{bmatrix} -2 \\ 0 \\ 1 \end{bmatrix}, \begin{bmatrix} 1 \\ 0 \\ -3 \end{bmatrix}, \begin{bmatrix} -1 \\ 1 \\ r \end{bmatrix} \right\}$

25. $\left\{ \begin{bmatrix} -2 \\ 0 \\ 1 \end{bmatrix}, \begin{bmatrix} 1 \\ 1 \\ -3 \end{bmatrix}, \begin{bmatrix} -1 \\ 1 \\ r \end{bmatrix} \right\}$

26. $\left\{ \begin{bmatrix} 1 \\ 0 \\ -1 \\ 1 \end{bmatrix}, \begin{bmatrix} 0 \\ -1 \\ 2 \\ 1 \end{bmatrix}, \begin{bmatrix} -1 \\ 1 \\ 1 \\ 0 \end{bmatrix}, \begin{bmatrix} -1 \\ 9 \\ r \\ -2 \end{bmatrix} \right\}$

In Exercises 27–34, write the parametric representation of the general solution to the given system of linear equations.

27. $x_1 - 4x_2 + 2x_3 = 0$

28. $x_1 \qquad + 5x_3 = 0$
$\qquad x_2 - 3x_3 = 0$

29. $x_1 + 3x_2 \qquad + 2x_4 = 0$
$\qquad\qquad x_3 - 6x_4 = 0$

30. $x_1 \qquad + 4x_4 = 0$
$\qquad x_2 - 2x_4 = 0$

31. $x_1 \qquad + 4x_3 - 2x_4 = 0$
$-x_1 + x_2 - 7x_3 + 7x_4 = 0$
$2x_1 + 3x_2 - x_3 + 11x_4 = 0$

32. $x_1 - 2x_2 - x_3 - 4x_4 = 0$
$2x_1 - 4x_2 + 3x_3 + 7x_4 = 0$
$-2x_1 + 4x_2 + x_3 + 5x_4 = 0$

33. $-x_1 + 2x_3 - 5x_4 + x_5 - x_6 = 0$
$x_1 - x_3 + 3x_4 - x_5 + 2x_6 = 0$
$x_1 + x_3 - x_4 + x_5 + 4x_6 = 0$

34. $-x_1 \qquad - 2x_3 - x_4 - 5x_5 = 0$
$\qquad - x_2 + 3x_3 + 2x_4 \qquad = 0$
$-2x_1 + x_2 + x_3 - x_4 + 8x_5 = 0$
$3x_1 - x_2 - 3x_3 - x_4 - 15x_5 = 0$

35. Find a 2×2 matrix A such that $\mathbf{0}$ is the only solution to $A\mathbf{x} = \mathbf{0}$.

36. Find a 2×2 matrix A such that $A\mathbf{x} = \mathbf{0}$ has infinitely many solutions.

37. State and prove the converse of property 3 of linearly dependent and independent sets.

38. Prove that if $S = \{\mathbf{u}_1, \mathbf{u}_2, \ldots, \mathbf{u}_k\}$ is linearly independent, then no spanning set for Span S can contain fewer than k vectors.

39. Let $\mathbf{u}$ and $\mathbf{v}$ be distinct vectors in $\mathcal{R}^n$. Prove that $\{\mathbf{u}, \mathbf{v}\}$ is linearly independent if and only if $\{\mathbf{u} + \mathbf{v}, \mathbf{u} - \mathbf{v}\}$ is linearly independent.

40. Let $\mathbf{u}$, $\mathbf{v}$, and $\mathbf{w}$ be distinct vectors in $\mathcal{R}^n$. Prove that $\{\mathbf{u}, \mathbf{v}, \mathbf{w}\}$ is linearly independent if and only if the set $\{\mathbf{u} + \mathbf{v}, \mathbf{u} + \mathbf{w}, \mathbf{v} + \mathbf{w}\}$ is linearly independent.

41. Prove that if $\{\mathbf{u}_1, \mathbf{u}_2, \ldots, \mathbf{u}_k\}$ is a linearly independent subset of $\mathcal{R}^n$ and $c_1, c_2, \ldots, c_k$ are nonzero scalars, then $\{c_1\mathbf{u}_1, c_2\mathbf{u}_2, \ldots, c_k\mathbf{u}_k\}$ is also linearly independent.

42. Complete the proof of Theorem 1.8 by showing that if $\mathbf{u}_1 = \mathbf{0}$ or $\mathbf{u}_i$ is in the span of $\{\mathbf{u}_1, \mathbf{u}_2, \ldots, \mathbf{u}_{i-1}\}$ for $i = 2, 3, \ldots, k$, then $\{\mathbf{u}_1, \mathbf{u}_2, \ldots, \mathbf{u}_k\}$ is linearly dependent. *Hint:* Separately consider the case in which $\mathbf{u}_1 = \mathbf{0}$ and the case in which vector $\mathbf{u}_i$ is in the span of $\{\mathbf{u}_1, \mathbf{u}_2, \ldots, \mathbf{u}_{i-1}\}$.

43.[12] Prove that any nonempty subset of a linearly independent subset of $\mathcal{R}^n$ is linearly independent.

44. Prove that if S_1 is a linearly dependent subset of $\mathcal{R}^n$ that is contained in a finite set S_2, then S_2 is linearly dependent.

45.[13] Let $S = \{\mathbf{u}_1, \mathbf{u}_2, \ldots, \mathbf{u}_k\}$ be a nonempty subset of $\mathcal{R}^n$. Prove that if S is linearly independent, then every vector in the span of S can be written in the form $c_1\mathbf{u}_1 + c_2\mathbf{u}_2 + \cdots + c_k\mathbf{u}_k$ for *unique* scalars $c_1, c_2, \ldots, c_k$.

46. State and prove the converse of Exercise 45.

47. Let $S = \{\mathbf{u}_1, \mathbf{u}_2, \ldots, \mathbf{u}_k\}$ be a nonempty subset of $\mathcal{R}^n$ and A be an $m \times n$ matrix. Prove that if S is linearly dependent, then so is $\{A\mathbf{u}_1, A\mathbf{u}_2, \ldots, A\mathbf{u}_k\}$.

48. Is the result of the preceding exercise true if *linearly dependent* is changed to *linearly independent*? Justify your answer.

49. Prove that if A is in row echelon form, then its nonzero rows are linearly independent.

50. Let $S = \{\mathbf{u}_1, \mathbf{u}_2, \ldots, \mathbf{u}_k\}$ be a nonempty subset of $\mathcal{R}^n$ and A be an $m \times n$ matrix with rank n. Prove that if S is a linearly independent set, then the set $\{A\mathbf{u}_1, A\mathbf{u}_2, \ldots, A\mathbf{u}_k\}$ is also linearly independent.

51. Let A and B be $m \times n$ matrices such that B can be obtained by performing a single elementary row operation on A. Prove that if the rows of A are linearly independent, then the rows of B are also linearly independent.

52. Prove that the rows of an $m \times n$ matrix A are linearly independent if and only if the rank of A is m.

In Exercises 53–56, use a calculator with matrix capabilities or computer software such as MATLAB *to determine if the given set is linearly dependent. In the case that the set is linearly dependent, write some vector in the set as a linear combination of the others.*

53. $\left\{ \begin{bmatrix} 1.1 \\ 2.3 \\ -1.4 \\ 2.7 \\ 3.6 \\ 0.0 \end{bmatrix}, \begin{bmatrix} -1.7 \\ 4.2 \\ 6.2 \\ 0.0 \\ 1.3 \\ -4.0 \end{bmatrix}, \begin{bmatrix} -5.7 \\ 8.1 \\ -4.3 \\ 7.2 \\ 10.5 \\ 2.9 \end{bmatrix}, \begin{bmatrix} -5.0 \\ 2.4 \\ 1.1 \\ 3.4 \\ 3.3 \\ 6.1 \end{bmatrix}, \begin{bmatrix} 2.9 \\ -1.1 \\ 2.6 \\ 1.6 \\ 0.0 \\ 3.2 \end{bmatrix} \right\}$

54. $\left\{ \begin{bmatrix} 1.2 \\ -5.4 \\ 3.7 \\ -2.6 \\ 0.3 \\ 1.4 \end{bmatrix}, \begin{bmatrix} -1.7 \\ 4.2 \\ 6.2 \\ 0.0 \\ 1.3 \\ -4.0 \end{bmatrix}, \begin{bmatrix} -5.0 \\ 2.4 \\ 1.1 \\ 3.4 \\ 3.3 \\ 6.1 \end{bmatrix}, \begin{bmatrix} -0.6 \\ 4.2 \\ 2.4 \\ -1.0 \\ 8.3 \\ -2.2 \end{bmatrix}, \begin{bmatrix} 2.4 \\ -1.4 \\ 0.0 \\ 5.6 \\ 2.3 \\ -1.0 \end{bmatrix} \right\}$

55. $\left\{ \begin{bmatrix} 21 \\ 25 \\ -15 \\ 42 \\ 17 \\ 10 \end{bmatrix}, \begin{bmatrix} 10 \\ -33 \\ 29 \\ 87 \\ -66 \\ 11 \end{bmatrix}, \begin{bmatrix} 32 \\ -21 \\ 15 \\ -11 \\ 25 \\ 16 \end{bmatrix}, \begin{bmatrix} 13 \\ 32 \\ -19 \\ 17 \\ -15 \\ 22 \end{bmatrix}, \begin{bmatrix} 26 \\ 18 \\ -37 \\ 0 \\ -7 \\ 22 \end{bmatrix}, \begin{bmatrix} 16 \\ 18 \\ 21 \\ 19 \\ -15 \\ 24 \end{bmatrix} \right\}$

56. $\left\{ \begin{bmatrix} 21 \\ 25 \\ -15 \\ 42 \\ 17 \\ 10 \end{bmatrix}, \begin{bmatrix} 10 \\ -33 \\ 29 \\ 87 \\ -66 \\ 11 \end{bmatrix}, \begin{bmatrix} -21 \\ 11 \\ 23 \\ -10 \\ 0 \\ 2 \end{bmatrix}, \begin{bmatrix} -14 \\ 3 \\ 15 \\ 0 \\ 45 \\ 15 \end{bmatrix}, \begin{bmatrix} 14 \\ 3 \\ -7 \\ 32 \\ -28 \\ -3 \end{bmatrix}, \begin{bmatrix} -8 \\ 21 \\ 30 \\ -17 \\ 34 \\ 7 \end{bmatrix} \right\}$

---

[12]The result of this exercise is used in Section 7.3 (page 412).
[13]The result of this exercise is used in Section 2.3 (page 122).

Solutions to the Practice Problems

1. Let A be the matrix whose columns are the vectors in S. Since the reduced row echelon form of A is I_4, the columns of A are linearly independent, by Theorem 1.7. Thus S is linearly independent, and so no vector in S is a linear combination of the others.

2. The augmented matrix of the given system is

$$\begin{bmatrix} 1 & -3 & -1 & 1 & -1 & 0 \\ 2 & -6 & 1 & -3 & -9 & 0 \\ -2 & 6 & 3 & 2 & 11 & 0 \end{bmatrix}.$$

Since the reduced row echelon form of this matrix is

$$\begin{bmatrix} 1 & -3 & 0 & 0 & -2 & 0 \\ 0 & 0 & 1 & 0 & 1 & 0 \\ 0 & 0 & 0 & 1 & 2 & 0 \end{bmatrix},$$

the general solution to the given system is

$$\begin{aligned} x_1 &= 3x_2 + 2x_5 \\ x_2 &\text{ free} \\ x_3 &= -x_5 \\ x_4 &= -2x_5 \\ x_5 &\text{ free.} \end{aligned}$$

To obtain its parametric representation, we express the general solution as a linear combination of vectors in which the coefficients are the free variables:

$$\begin{bmatrix} x_1 \\ x_2 \\ x_3 \\ x_4 \\ x_5 \end{bmatrix} = \begin{bmatrix} 3x_2 + 2x_5 \\ x_2 \\ -x_5 \\ -2x_5 \\ x_5 \end{bmatrix} = x_2 \begin{bmatrix} 3 \\ 1 \\ 0 \\ 0 \\ 0 \end{bmatrix} + x_5 \begin{bmatrix} 2 \\ 0 \\ -1 \\ -2 \\ 1 \end{bmatrix}.$$

Chapter 1 Review Exercises

1. Determine if the following statements are true or false.

   (a) If B is a 3×4 matrix, then its columns are 1×3 vectors.

   (b) Any scalar multiple of a vector $\mathbf{v}$ in $\mathcal{R}^n$ is a linear combination of $\mathbf{v}$.

   (c) If a vector $\mathbf{v}$ lies in the span of a finite subset S of $\mathcal{R}^n$, then $\mathbf{v}$ is a linear combination of the vectors in S.

   (d) The matrix–vector product of an $m \times n$ matrix A and a vector in $\mathcal{R}^n$ is a linear combination of the columns of A.

   (e) The rank of the coefficient matrix of a consistent system of linear equations is equal to the number of basic variables in the general solution to the system.

   (f) The nullity of the coefficient matrix of a consistent system of linear equations is equal to the number of free variables in the general solution to the system.

   (g) Every matrix can be transformed into one and only one matrix in reduced row echelon form by means of a sequence of elementary row operations.

   (h) If the last row of the reduced row echelon form of an augmented matrix of a system of linear equations has only one nonzero entry, then the system is inconsistent.

   (i) If the last row of the reduced row echelon form of an augmented matrix of a system of linear equations has only zero entries, then the system has infinitely many solutions.

   (j) The zero vector of $\mathcal{R}^n$ lies in the span of any finite subset of $\mathcal{R}^n$.

   (k) If the rank of an $m \times n$ matrix A is m, then the rows of A are linearly independent.

   (l) The columns of an $m \times n$ matrix A span $\mathcal{R}^m$ if and only if the rank of A is m.

   (m) If the columns of an $m \times n$ matrix are linearly dependent, then the rank of the matrix is less than m.

   (n) If S is a linearly independent subset of $\mathcal{R}^n$ and $\mathbf{v}$ is a vector in $\mathcal{R}^n$ such that $S \cup \{\mathbf{v}\}$ is linearly dependent, then $\mathbf{v}$ is in the span of S.

   (o) A subset of $\mathcal{R}^n$ containing more than n vectors must be linearly dependent.

   (p) A subset of $\mathcal{R}^n$ containing fewer than n vectors must be linearly independent.

   (q) A linearly dependent subset of $\mathcal{R}^n$ must contain more than n vectors.

2. Determine if each of the following phrases is a misuse of terminology. If so, explain what is wrong with each one.

   (a) an inconsistent matrix

   (b) the solution of a matrix

   (c) equivalent matrices

   (d) the nullity of a system of linear equations

   (e) the span of a matrix

   (f) a spanning set for a system of linear equations

   (g) a homogeneous matrix

   (h) a linearly independent matrix

3. (a) If A is an $m \times n$ matrix with rank n, what can be said about the number of solutions to $A\mathbf{x} = \mathbf{b}$ for every $\mathbf{b}$ in $\mathcal{R}^n$?

   (b) If A is an $m \times n$ matrix with rank m, what can be said about the number of solutions to $A\mathbf{x} = \mathbf{b}$ for every $\mathbf{b}$ in $\mathcal{R}^n$?

In Exercises 4–11, use the matrices below to compute the given expression, or give a reason why the expression is not defined.

$$A = \begin{bmatrix} 1 & 3 \\ -2 & 4 \\ 0 & 2 \end{bmatrix}, \qquad B = \begin{bmatrix} 2 & -1 \\ 0 & 3 \\ 4 & 1 \end{bmatrix},$$

$$C = \begin{bmatrix} 1 \\ 5 \end{bmatrix}, \qquad \text{and} \qquad D = [1 \ -1 \ 2].$$

4. $A + B^T$ 5. $A + B$ 6. BC 7. AD^T

8. $2A - 3B$ 9. $A^T D^T$ 10. $A^T - B$ 11. $C^T - 2D$

12. A boat is traveling on a river in a southwesterly direction at 10 mph while a passenger is walking from the port to the starboard (from the southeast side to the northwest side of the boat) at 2 mph. Find the velocity and the speed of the passenger with respect to the shore.

13. A supermarket chain has ten stores. For each i, $1 \le i \le$ 10, the 4×1 vector $\mathbf{v}_i$ is defined so that its respective components represent the total value of sales in produce, meats, dairy, and processed foods at store i during January of last year. Provide an interpretation of the vector $0.1(\mathbf{v}_1 + \mathbf{v}_2 + \cdots + \mathbf{v}_{10})$.

In Exercises 14–17, compute the following matrix–vector products.

14. $\begin{bmatrix} 3 & 1 \\ 0 & -1 \\ 1 & 2 \end{bmatrix} \begin{bmatrix} 4 \\ 1 \end{bmatrix}$

15. $\begin{bmatrix} 1 & 3 & 1 & 2 \\ 1 & -1 & 4 & 0 \end{bmatrix}^T \begin{bmatrix} -1 \\ 1 \end{bmatrix}$

16. $A_{45°} \begin{bmatrix} 2 \\ -1 \end{bmatrix}$

17. $A_{-30°} \begin{bmatrix} 2 \\ -1 \end{bmatrix}$

18. Suppose that

$$\mathbf{v}_1 = \begin{bmatrix} 2 \\ 1 \\ 3 \end{bmatrix} \quad \text{and} \quad \mathbf{v}_2 = \begin{bmatrix} -1 \\ 3 \\ 6 \end{bmatrix}.$$

Represent $3\mathbf{v}_1 - 4\mathbf{v}_2$ as the product of a 3×2 matrix and a vector in $\mathcal{R}^2$.

In Exercises 19–22, determine if the given vector $\mathbf{v}$ is in the span of

$$S = \left\{ \begin{bmatrix} -1 \\ 5 \\ 2 \end{bmatrix}, \begin{bmatrix} 1 \\ 3 \\ 4 \end{bmatrix}, \begin{bmatrix} 1 \\ -1 \\ 1 \end{bmatrix} \right\}.$$

If so, write $\mathbf{v}$ as a linear combination of the vectors in S.

19. $\mathbf{v} = \begin{bmatrix} 5 \\ 3 \\ 11 \end{bmatrix}$
20. $\mathbf{v} = \begin{bmatrix} 1 \\ 4 \\ 3 \end{bmatrix}$
21. $\mathbf{v} = \begin{bmatrix} 1 \\ 1 \\ 2 \end{bmatrix}$

22. $\mathbf{v} = \begin{bmatrix} 2 \\ 10 \\ 9 \end{bmatrix}$

In Exercises 23–28, determine if the given system is consistent; if so, find its general solution.

23. $x_1 + 2x_2 - x_3 = 1$

24. $\begin{aligned} x_1 + x_2 + x_3 &= 3 \\ -2x_1 + 4x_2 + 2x_3 &= 7 \\ 2x_1 - x_2 - 4x_3 &= 2 \end{aligned}$

25. $\begin{aligned} x_1 + 2x_2 + 3x_3 &= 1 \\ 2x_1 + x_2 + x_3 &= 2 \\ x_1 - 4x_2 - 7x_3 &= 4 \end{aligned}$

26. $\begin{aligned} x_1 + 3x_2 + 2x_3 + x_4 &= 2 \\ 2x_1 + x_2 + x_3 - x_4 &= 3 \\ x_1 - 2x_2 - x_3 - 2x_4 &= 4 \end{aligned}$

27. $\begin{aligned} x_1 + x_2 + 2x_3 + x_4 &= 2 \\ 2x_1 + 3x_2 + x_3 - x_4 &= -1 \end{aligned}$

28. $\begin{aligned} 2x_1 + 4x_2 - 2x_3 + 2x_4 &= 4 \\ 2x_1 + x_2 + 4x_3 + 2x_4 &= 1 \\ 4x_1 + 6x_2 + x_3 + 2x_4 &= 1 \end{aligned}$

In Exercises 29–32, find the rank and nullity of the given matrix.

29. $\begin{bmatrix} 1 & 2 & -3 & 0 & 1 \end{bmatrix}$

30. $\begin{bmatrix} 1 & 2 & -3 & 0 & 1 \\ 0 & 0 & 0 & 0 & 0 \end{bmatrix}$

31. $\begin{bmatrix} 1 & 2 & 1 & -1 & 2 \\ 2 & 1 & 0 & 1 & 3 \\ -1 & -3 & 1 & 2 & 4 \end{bmatrix}$
32. $\begin{bmatrix} 2 & 3 & 4 \\ 1 & 2 & 1 \\ -1 & 1 & 2 \\ 3 & 0 & 2 \end{bmatrix}$

33. A company that ships fruit has three kinds of fruit packs. The first pack consists of 10 oranges and 10 grapefruit, the second pack consists of 10 oranges, 15 grapefruit, and 10 apples, and the third pack consists of 5 oranges, 10 grapefruit and 5 apples. How many of each pack can be made from a stock of 500 oranges, 750 grapefruit, and 300 apples?

In Exercises 34–37, a set of vectors in $\mathcal{R}^n$ is given. Determine if the set spans $\mathcal{R}^n$.

34. $\left\{ \begin{bmatrix} 1 \\ 1 \\ -1 \\ 1 \end{bmatrix}, \begin{bmatrix} 1 \\ 0 \\ 0 \\ 2 \end{bmatrix}, \begin{bmatrix} 1 \\ 3 \\ -2 \\ 1 \end{bmatrix} \right\}$

35. $\left\{ \begin{bmatrix} -1 \\ 1 \\ 1 \end{bmatrix}, \begin{bmatrix} 1 \\ -1 \\ 1 \end{bmatrix}, \begin{bmatrix} 1 \\ 1 \\ -1 \end{bmatrix} \right\}$
36. $\left\{ \begin{bmatrix} 1 \\ 0 \\ 1 \end{bmatrix}, \begin{bmatrix} 1 \\ 1 \\ -1 \end{bmatrix}, \begin{bmatrix} 2 \\ 1 \\ 3 \end{bmatrix} \right\}$

37. $\left\{ \begin{bmatrix} 1 \\ 2 \\ 1 \end{bmatrix}, \begin{bmatrix} 1 \\ -1 \\ 1 \end{bmatrix}, \begin{bmatrix} 1 \\ 1 \\ 1 \end{bmatrix}, \begin{bmatrix} 0 \\ 1 \\ 0 \end{bmatrix} \right\}$

In Exercises 38–43, an $m \times n$ matrix A is given. Determine if the equation $A\mathbf{x} = \mathbf{b}$ is consistent for every $\mathbf{b}$ in $\mathcal{R}^n$.

38. $\begin{bmatrix} 1 & 2 \\ 3 & 6 \end{bmatrix}$
39. $\begin{bmatrix} 1 & 1 \\ 3 & 2 \end{bmatrix}$

40. $\begin{bmatrix} 1 & -1 & 1 \\ 2 & 0 & 1 \end{bmatrix}$
41. $\begin{bmatrix} -1 & 1 & 1 \\ 1 & -1 & 1 \\ 1 & 1 & -1 \end{bmatrix}$

42. $\begin{bmatrix} 1 & 2 & 1 \\ 3 & 0 & -3 \\ -1 & 1 & 2 \end{bmatrix}$
43. $\begin{bmatrix} 1 & 2 & 1 \\ 2 & -3 & 1 \\ -1 & 1 & 2 \\ 0 & 1 & 2 \end{bmatrix}$

In Exercises 44–47, determine if the given set is linearly dependent or linearly independent.

44. $\left\{ \begin{bmatrix} 1 \\ 3 \\ 2 \end{bmatrix}, \begin{bmatrix} 1 \\ -1 \\ 2 \end{bmatrix}, \begin{bmatrix} 3 \\ 1 \\ 6 \end{bmatrix} \right\}$
45. $\left\{ \begin{bmatrix} 1 \\ -1 \\ 2 \\ 0 \end{bmatrix}, \begin{bmatrix} 0 \\ 1 \\ 2 \\ 3 \end{bmatrix}, \begin{bmatrix} 1 \\ 0 \\ 1 \\ 1 \end{bmatrix} \right\}$

46. $\left\{ \begin{bmatrix} 2 \\ 3 \\ 5 \\ 7 \end{bmatrix}, \begin{bmatrix} 4 \\ 6 \\ 10 \\ 14 \end{bmatrix} \right\}$

47. $\left\{ \begin{bmatrix} 22.40 \\ 6.02 \\ 6.63 \end{bmatrix}, \begin{bmatrix} 9.11 \\ 1.76 \\ 9.27 \end{bmatrix}, \begin{bmatrix} 3.14 \\ 2.72 \\ 1.41 \end{bmatrix}, \begin{bmatrix} 31 \\ 37 \\ 41 \end{bmatrix} \right\}$

In Exercises 48–51, a linearly dependent set S is given. Write some vector in S as a linear combination of the others.

48. $\left\{ \begin{bmatrix} 1 \\ -1 \\ 3 \end{bmatrix}, \begin{bmatrix} 1 \\ 2 \\ 1 \end{bmatrix}, \begin{bmatrix} 2 \\ 4 \\ 2 \end{bmatrix} \right\}$

49. $\left\{ \begin{bmatrix} 1 \\ 2 \\ 3 \end{bmatrix}, \begin{bmatrix} 1 \\ -1 \\ 2 \end{bmatrix}, \begin{bmatrix} 3 \\ 3 \\ 8 \end{bmatrix} \right\}$

50. $\left\{ \begin{bmatrix} 3 \\ 1 \\ 4 \\ 1 \end{bmatrix}, \begin{bmatrix} 3 \\ 0 \\ 5 \\ 1 \end{bmatrix}, \begin{bmatrix} 3 \\ 3 \\ 2 \\ 1 \end{bmatrix} \right\}$

51. $\left\{ \begin{bmatrix} 1 \\ -1 \\ 1 \\ 2 \end{bmatrix}, \begin{bmatrix} 1 \\ 0 \\ 1 \\ 0 \end{bmatrix}, \begin{bmatrix} 1 \\ 1 \\ 1 \\ 1 \end{bmatrix}, \begin{bmatrix} 1 \\ -1 \\ 1 \\ -1 \end{bmatrix} \right\}$

In Exercises 52–55, write the parametric representation of the general solution to the given system of linear equations.

52. $x_1 + 2x_2 - x_3 + x_4 = 0$

53. $x_1 + 2x_2 - x_3 = 0$
    $x_1 + x_2 + x_3 = 0$
    $x_1 + 3x_2 - 3x_3 = 0$

54. $2x_1 + 5x_2 - x_3 + x_4 = 0$
    $x_1 + 3x_2 + 2x_3 - x_4 = 0$

55. $3x_1 + x_2 - x_3 + x_4 = 0$
    $2x_1 + 2x_2 + 4x_3 - 6x_4 = 0$
    $2x_1 + x_2 + 3x_3 - x_4 = 0$

56. Let A be an $m \times n$ matrix, let $\mathbf{b}$ be a vector in $\mathcal{R}^n$, and suppose that $\mathbf{v}$ is a solution to $A\mathbf{x} = \mathbf{b}$.

    (a) Prove that if $\mathbf{w}$ is a solution to $A\mathbf{x} = \mathbf{0}$, then $\mathbf{v} + \mathbf{w}$ is a solution to $A\mathbf{x} = \mathbf{b}$.

    (b) Prove that for any solution $\mathbf{u}$ to $A\mathbf{x} = \mathbf{b}$, there is a solution $\mathbf{w}$ to $A\mathbf{x} = \mathbf{0}$ such that $\mathbf{u} = \mathbf{v} + \mathbf{w}$.

57. Suppose that $\mathbf{w}_1$ and $\mathbf{w}_2$ are linear combinations of vectors $\mathbf{v}_1$ and $\mathbf{v}_2$ in $\mathcal{R}^n$ such that $\mathbf{w}_1$ and $\mathbf{w}_2$ are linearly independent. Prove that $\mathbf{v}_1$ and $\mathbf{v}_2$ are linearly independent.

CHAPTER 2

Matrices and Linear Transformations

In Section 1.2, we used matrix–vector products to describe processes in which vectors are transformed in various ways. For example, we saw how to use the matrix–vector product as a simple and elegant tool to describe such diverse processes as rotations in the plane, population shifts, and combinations of grass seed mixtures. These applications are possible because points in the plane, population distributions, and grass seed mixtures can be described by vectors, and the processes for transforming these vectors can be formulated as matrix–vector products.

In such situations, it is often useful to take the product of a matrix and a vector obtained from a previous matrix–vector product. This calculation leads naturally to an extension of the definition of multiplication to include products of matrices of various sizes.

In this chapter, we examine some of the elementary properties and uses of this extended definition of matrix multiplication (Sections 2.1–2.4). Later in the chapter, we study the matrix–vector product from the functional viewpoint of a rule of correspondence. This leads to the definition of *linear transformation* (Sections 2.6 and 2.7). There we see how the functional properties of linear transformations correspond to the properties of matrix multiplication studied earlier in this chapter.

2.1 Matrix Multiplication

There is a common situation in linear algebra in which a matrix–vector product $B\mathbf{v}$ is multiplied on the left by a matrix A to form a new matrix–vector product $A(B\mathbf{v})$. For example, at the end of Example 3 in Section 1.2, it is noted that the product $A(A\mathbf{p})$ can be used to predict the population distribution in two years. Two additional examples follow.

Example 1   In the seed example from Section 1.2, recall that

$$B = \begin{matrix} \text{deluxe} & \text{standard} & \text{economy} \\ \begin{bmatrix} .80 & .60 & .40 \\ .20 & .40 & .60 \end{bmatrix} & & \end{matrix} \begin{matrix} \text{bluegrass} \\ \text{rye} \end{matrix}$$

gives the proportions of bluegrass and rye in the deluxe, standard, and economy mixtures of grass seed, and that

$$\mathbf{w} = \begin{bmatrix} 60 \\ 50 \\ 30 \end{bmatrix}$$

gives the number of pounds of each mixture in stock. Suppose we have a seed manual with a table that gives us the germination rates of bluegrass and rye seeds under both wet and dry conditions. The table, given in the form of a matrix A, is

$$A = \begin{matrix} \text{bluegrass} & \text{rye} \\ \begin{bmatrix} .80 & .70 \\ .60 & .40 \end{bmatrix} & \begin{matrix} \text{wet} \\ \text{dry} \end{matrix} \end{matrix}$$

where, for example, the $(1, 1)$-entry, which is .80, signifies that 80% of the bluegrass seed germinates under wet conditions. Now suppose that we have a mixture of y_1 pounds of bluegrass seed and y_2 pounds of rye seed. Then $.80y_1 + .70y_2$ is the total weight (in pounds) of the seed that germinates under wet conditions and $.60y_1 + .40y_2$ is the total weight of the seed that germinates under dry conditions. Notice that these two expressions are the entries of the matrix–vector product $A\mathbf{v}$, where $\mathbf{v} = \begin{bmatrix} y_1 \\ y_2 \end{bmatrix}$. Let's combine this with the result in the example in Section 1.2. Using the notation of that example, $B\mathbf{w}$ is the vector whose components are the amounts of bluegrass and rye seed in a blend obtained by combining various amounts of seed mixtures. Then the components of the matrix–vector product

$$A(B\mathbf{w}) = \begin{bmatrix} .80 & .70 \\ .60 & .40 \end{bmatrix} \begin{bmatrix} 90 \\ 50 \end{bmatrix} = \begin{matrix} & \text{lb of seed} \\ & \text{germinated} \\ \begin{bmatrix} 107 \\ 74 \end{bmatrix} & \begin{matrix} \text{wet} \\ \text{dry} \end{matrix} \end{matrix}$$

are the amounts of seed that can be expected to germinate under each of the two types of weather conditions.   ○

Example 2   Recall the rotation matrix A_θ described in Section 1.2. Let $\mathbf{v}$ be a nonzero vector, and let α and β be angles. Then the expression $A_\beta(A_\alpha \mathbf{v})$ is the result of rotating $\mathbf{v}$ by α followed by β. From Figure 2.1, we see that this is the same as rotating $\mathbf{v}$ by β followed by α, which is also the same as rotating $\mathbf{v}$ by the angle $\alpha + \beta$. Thus

$$A_\beta(A_\alpha \mathbf{v}) = A_\alpha(A_\beta \mathbf{v}) = A_{\alpha+\beta}\mathbf{v}.$$ ○

In each of the examples above, a matrix–vector product is multiplied on the left by another matrix. An examination of this process leads to an extended definition of matrix multiplication.

Let A be an $m \times n$ matrix and B be an $n \times p$ matrix. Then for any $p \times 1$ vector $\mathbf{v}$, the product $B\mathbf{v}$ is an $n \times 1$ vector, and hence the new product $A(B\mathbf{v})$ is an $m \times 1$ vector. This raises the question: Is there is an $m \times p$ matrix C such that $A(B\mathbf{v}) = C\mathbf{v}$ for every $p \times 1$ vector $\mathbf{v}$? We have seen that this is the case for rotation matrices. Let's examine the general situation.

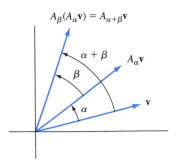

$A_\beta(A_\alpha\mathbf{v}) = A_{\alpha+\beta}\mathbf{v}$

FIGURE 2.1

By the definition of a matrix–vector product and Theorem 1.2, we have

$$A(B\mathbf{v}) = A(v_1\mathbf{b}_1 + v_2\mathbf{b}_2 + \cdots + v_p\mathbf{b}_p)$$
$$= A(v_1\mathbf{b}_1) + A(v_2\mathbf{b}_2) + \cdots + A(v_p\mathbf{b}_p)$$
$$= v_1 A\mathbf{b}_1 + v_2 A\mathbf{b}_2 + \cdots + v_p A\mathbf{b}_p$$
$$= [A\mathbf{b}_1 \quad A\mathbf{b}_2 \quad \cdots \quad A\mathbf{b}_p]\mathbf{v}.$$

Let C be the $m \times p$ matrix $[A\mathbf{b}_1 \quad A\mathbf{b}_2 \quad \cdots \quad A\mathbf{b}_p]$, that is, the matrix whose jth column is $\mathbf{c}_j = A\mathbf{b}_j$. Then $A(B\mathbf{v}) = C\mathbf{v}$.

Now that we have found a matrix that has the property $A(B\mathbf{v}) = C\mathbf{v}$, we observe that it is unique; that is, C is the only matrix that has this property. For let Q be any $m \times p$ matrix such that $Q\mathbf{v} = A(B\mathbf{v})$ for every $p \times 1$ vector $\mathbf{v}$. Then by Theorem 1.2, for any j we have

$$\mathbf{q}_j = Q\mathbf{e}_j = A(B\mathbf{e}_j) = A\mathbf{b}_j = \mathbf{c}_j.$$

Since the equation above tells us that each column of Q is equal to the corresponding column of C, we must have that $Q = C$. We summarize what we have shown in the following theorem.

Theorem 2.1

Let A be an $m \times n$ matrix and B be an $n \times p$ matrix. Then there is a unique $m \times p$ matrix C such that $C\mathbf{v} = A(B\mathbf{v})$ for every $p \times 1$ vector $\mathbf{v}$. Furthermore,

$$C = [A\mathbf{b}_1 \quad A\mathbf{b}_2 \quad \cdots \quad A\mathbf{b}_p].$$

This convenient method of combining the matrices A and B leads to the following definition.

Definition. Let A be an $m \times n$ matrix and B be an $n \times p$ matrix. We define the (**matrix**) **product** AB to be the unique $m \times p$ matrix whose jth column is $A\mathbf{b}_j$.

Notice that when the sizes of A and B are written side by side in the same order as the product, that is, $(m \times n)(n \times p)$, the inner dimensions must be equal, and the outer dimensions give the size of the product AB. Symbolically,

$$(m \times n)(n \times p) = (m \times p).$$

In particular, if A is an $m \times n$ matrix and B is an $n \times 1$ column vector, then the matrix product AB is defined and is the same as the matrix–vector product defined in Section 1.2.

Example 3 Let

$$A = \begin{bmatrix} 1 & 2 \\ 3 & 4 \\ 5 & 6 \end{bmatrix} \quad \text{and} \quad B = \begin{bmatrix} -1 & 1 \\ 3 & 2 \end{bmatrix}.$$

Notice that A has 2 columns and B has 2 rows. Then AB is the 3×2 matrix with first and second column

$$\begin{bmatrix} 1 & 2 \\ 3 & 4 \\ 5 & 6 \end{bmatrix} \begin{bmatrix} -1 \\ 3 \end{bmatrix} = \begin{bmatrix} 5 \\ 9 \\ 13 \end{bmatrix} \quad \text{and} \quad \begin{bmatrix} 1 & 2 \\ 3 & 4 \\ 5 & 6 \end{bmatrix} \begin{bmatrix} 1 \\ 2 \end{bmatrix} = \begin{bmatrix} 5 \\ 11 \\ 17 \end{bmatrix},$$

respectively. Thus

$$AB = \begin{bmatrix} 5 & 5 \\ 9 & 11 \\ 13 & 17 \end{bmatrix}.$$

In view of Theorem 2.1 and the definition of the matrix product, we have an *associative law* for the product of two matrices and a vector.

> For any $m \times n$ matrix A, any $n \times p$ matrix B, and any $p \times 1$ vector $\mathbf{v}$,
>
> $$(AB)\mathbf{v} = A(B\mathbf{v}).$$

Later in this section, we extend this associative law to the product of any three matrices of compatible sizes (see Theorem 2.2(b)).

Example 4   We return to Example 1 in this section and the matrix $A = \begin{bmatrix} .80 & .70 \\ .60 & .40 \end{bmatrix}$. Recall the matrix $B = \begin{bmatrix} .80 & .60 & .40 \\ .20 & .40 & .60 \end{bmatrix}$ and the vector $\mathbf{w} = \begin{bmatrix} 60 \\ 50 \\ 30 \end{bmatrix}$ from the related example in Section 1.2. Then

$$AB = \begin{bmatrix} .80 & .70 \\ .60 & .40 \end{bmatrix} \begin{bmatrix} .80 & .60 & .40 \\ .20 & .40 & .60 \end{bmatrix} = \begin{bmatrix} .78 & .76 & .74 \\ .56 & .52 & .48 \end{bmatrix},$$

and hence

$$(AB)\mathbf{w} = \begin{bmatrix} .78 & .76 & .74 \\ .56 & .52 & .48 \end{bmatrix} \begin{bmatrix} 60 \\ 50 \\ 30 \end{bmatrix} = \begin{bmatrix} 107 \\ 74 \end{bmatrix}.$$

This is the same result obtained in Example 1, where we computed $A(B\mathbf{w})$.

Example 5   As another example of this associative law, we can rewrite the equation in Example 2 as follows:

$$(A_\beta A_\alpha)\mathbf{v} = (A_\alpha A_\beta)\mathbf{v} = A_{\alpha+\beta}\mathbf{v}.$$

Since this equation is valid for every vector $\mathbf{v}$ in $\mathcal{R}^2$, we may apply Theorem 1.2(e) to obtain the refinement

$$A_\beta A_\alpha = A_\alpha A_\beta = A_{\alpha+\beta}.$$

Although $A_\beta A_\alpha = A_\alpha A_\beta$ in Example 5, in general the matrix products AB and BA are not equal. In fact, if A is an $m \times n$ matrix and B is a $n \times p$ matrix, then BA is undefined unless $p = m$. If $p = m$, then AB is $m \times m$ and BA is $n \times n$. Thus even if both products are defined, AB and BA need not be of the same size. But even if $m = n$, we are not guaranteed that the products are equal. For example, let

$$A = \begin{bmatrix} 1 & 2 \\ 1 & 1 \end{bmatrix} \quad \text{and} \quad B = \begin{bmatrix} 1 & 1 \\ 1 & -1 \end{bmatrix}.$$

Then

$$AB = \begin{bmatrix} 1 & 2 \\ 1 & 1 \end{bmatrix} \begin{bmatrix} 1 & 1 \\ 1 & -1 \end{bmatrix} = \begin{bmatrix} 3 & -1 \\ 2 & 0 \end{bmatrix}$$

and

$$BA = \begin{bmatrix} 1 & 1 \\ 1 & -1 \end{bmatrix} \begin{bmatrix} 1 & 2 \\ 1 & 1 \end{bmatrix} = \begin{bmatrix} 2 & 3 \\ 0 & 1 \end{bmatrix};$$

hence $AB \neq BA$.

Note that whenever the product AB is defined, it can be computed one column at a time as in Example 3. We can also compute AB one *entry* at a time. Observe that the (i, j)-entry of AB is the ith component of its jth column, $A\mathbf{b}_j$. But from our earlier observation, following the definition of a matrix–vector product, this entry equals

$$[a_{i1} \quad a_{i2} \quad \ldots \quad a_{in}] \begin{bmatrix} b_{1j} \\ b_{2j} \\ \vdots \\ b_{nj} \end{bmatrix} = a_{i1}b_{1j} + a_{i2}b_{2j} + \cdots + a_{in}b_{nj}.$$

This formula can be described by means of the following diagram.

(i, j)-entry of a Matrix Product

To compute the (i, j)-entry of the matrix product AB, locate the ith row of A and the jth column of B as in the preceding diagram. Moving across the ith row of A and down the jth column of B, multiply each entry of the row by the corresponding entry of the column. Then sum these products to obtain the (i, j)-entry of AB. In symbols, the (i, j)-entry of AB is

$$a_{i1}b_{1j} + a_{i2}b_{2j} + \cdots + a_{in}b_{nj}.$$

To illustrate this procedure, we compute the $(2, 1)$-entry of AB in Example 3. In this case we multiply each entry of the second row of A by the corresponding entry of the first column of B and sum the results:

$$[3 \quad 4] \begin{bmatrix} -1 \\ 3 \end{bmatrix} = (3)(-1) + (4)(3) = 9,$$

which is the $(2, 1)$-entry of AB.

Identity matrices, introduced in Section 1.2, leave matrices unchanged under matrix multiplication (see Theorem 2.2(e) and Exercise 46). For example, let $A = \begin{bmatrix} 1 & 2 & 3 \\ 4 & 5 & 6 \end{bmatrix}$. Then it is a simple matter to verify that $I_2A = AI_3 = A$.

Also, the product of a matrix and a zero matrix is a zero matrix, because each column of the product is a zero vector (see Theorem 1.2(f)).

The next theorem summarizes the various properties of matrix multiplication, and illustrates the interplay between matrix multiplication and the other matrix operations.

Theorem 2.2

Let A and B be $k \times m$ matrices, C be an $m \times n$ matrix, and P and Q be $n \times p$ matrices. Then the following statements are true.

(a) $s(AC) = (sA)C = A(sC)$ for any scalar s.
(b) $A(CP) = (AC)P$ (associative law of matrix multiplication).
(c) $(A + B)C = AC + BC$ (right distributive law).
(d) $C(P + Q) = CP + CQ$ (left distributive law).
(e) $I_kA = A = AI_m$.
(f) The product of any matrix and a zero matrix is a zero matrix.
(g) $(AC)^T = C^TA^T$.

Proof We prove (b), (c), and (g); the rest are left as exercises.

(b) First observe that both $A(CP)$ and $(AC)P$ are $k \times p$ matrices. Let $\mathbf{u}_j$ denote column j of CP. Since $\mathbf{u}_j = C\mathbf{p}_j$, column j of $A(CP)$ is $A\mathbf{u}_j = A(C\mathbf{p}_j)$. Furthermore, column j of $(AC)P$ is $(AC)\mathbf{p}_j = A(C\mathbf{p}_j)$. It follows that the corresponding columns of $A(CP)$ and $(AC)P$ are equal. Therefore $A(CP) = (AC)P$.

(c) Both $(A + B)C$ and $AC + BC$ are $k \times n$ matrices, and so we compare the corresponding columns of each matrix. For any j, $A\mathbf{c}_j$ and $B\mathbf{c}_j$ are the jth columns of AC and BC, respectively. But the jth column of $(A + B)C$ is

$$(A + B)\mathbf{c}_j = A\mathbf{c}_j + B\mathbf{c}_j,$$

by Theorem 1.2(d), and this is the jth column of $AC + BC$. This establishes (c).

(g) Both $(AC)^T$ and C^TA^T are $n \times k$ matrices, and so we compare the corresponding entries of each matrix. The (i, j)-entry of $(AC)^T$ is the (j, i)-entry of AC, which is

$$[a_{j1} \quad a_{j2} \quad \cdots \quad a_{jm}] \begin{bmatrix} c_{1i} \\ c_{2i} \\ \vdots \\ c_{mi} \end{bmatrix} = a_{j1}c_{1i} + a_{j2}c_{2i} + \cdots + a_{jm}c_{mi}.$$

Also, the (i, j)-entry of $C^T A^T$ is the product of row i of C^T and column j of A^T, which is

$$[c_{1i} \quad c_{2i} \quad \cdots \quad c_{mi}] \begin{bmatrix} a_{j1} \\ a_{j2} \\ \vdots \\ a_{jm} \end{bmatrix} = c_{1i}a_{j1} + c_{2i}a_{j2} + \cdots + c_{mi}a_{jm}.$$

Since the two displayed expressions are equal, the (i, j)-entry of $(AC)^T$ is equal to the (i, j)-entry of $C^T A^T$. This establishes (g). ❑

The associative law of matrix multiplication, Theorem 2.2(b), allows us to omit parentheses when writing products of matrices. For this reason, we usually write ABC for a product of the matrices A, B, and C.

If A is an $n \times n$ matrix, we can form products of A with itself any number of times. As with real numbers, we use the exponential notation A^k to denote the product of A with itself k times. By convention, $A^1 = A$ and $A^0 = I_n$.

Example 6   Recall that the stochastic matrix $A = \begin{bmatrix} .85 & .03 \\ .15 & .97 \end{bmatrix}$ in Example 3 of Section 1.2 can be used to study population shifts between the city and suburbs. In that example, if the components of $\mathbf{p}$ are the current populations of the city and suburbs, then the components of $A\mathbf{p}$ are the populations of the city and suburbs for the following year. Extending the argument of that example to subsequent years, we see that $A^2\mathbf{p} = A(A\mathbf{p})$ is the vector whose components are the population of the city and suburbs after two years, and in general, for any positive integer m, $A^m\mathbf{p}$ is the vector whose components are the population of the city and suburbs after m years. For example, if $\mathbf{p} = \begin{bmatrix} 500 \\ 700 \end{bmatrix}$ (as in Example 3 of Section 1.2), then the vector representing the city and suburban populations in ten years is given by

$$A^{10}\mathbf{p} \approx \begin{bmatrix} 241.2 \\ 958.8 \end{bmatrix}$$

(here, the entries have been rounded).

A sociologist may need to ascertain the long-range trend in population as a result of these annual shifts. In terms of matrix multiplication, this problem reduces to the study of the limit of $A^m\mathbf{p}$. A deeper understanding of matrices, which we acquire in Chapter 5, provides us with the tools to analyze this problem. ◯

Special Matrices

We briefly examine matrices with special properties that will be of interest later in this text. We begin with *diagonal matrices*, which are important because of their simplicity. The (i, j)-entry of a matrix A is called a **diagonal entry** if $i = j$. The set of diagonal entries is called the **diagonal** of A. A square matrix A is called a **diagonal matrix** if all its nondiagonal entries are zeros. For example, identity matrices and square zero matrices are diagonal matrices.

If A and B are $n \times n$ diagonal matrices, then AB is also an $n \times n$ diagonal matrix. Moreover, the diagonal entries of AB are the products of the corresponding diagonal entries of A and B (see Exercise 50). For example, suppose that

$$A = \begin{bmatrix} 1 & 0 & 0 \\ 0 & 2 & 0 \\ 0 & 0 & 3 \end{bmatrix} \quad \text{and} \quad B = \begin{bmatrix} 3 & 0 & 0 \\ 0 & -1 & 0 \\ 0 & 0 & 2 \end{bmatrix}.$$

Then

$$AB = \begin{bmatrix} 3 & 0 & 0 \\ 0 & -2 & 0 \\ 0 & 0 & 6 \end{bmatrix}.$$

The relationship between diagonal and nondiagonal square matrices is studied in depth in Chapter 5, where we see that in many circumstances, an ordinary square matrix can be replaced by a diagonal matrix, thus simplifying theoretical arguments as well as computations.

Symmetric matrices will be of use in Section 6.5. A (square) matrix A is called **symmetric** if $A^T = A$. For example, diagonal matrices are symmetric. Let $A = \begin{bmatrix} 1 & 2 & 4 \\ 2 & 3 & -1 \\ 4 & -1 & 5 \end{bmatrix}$ and $B = \begin{bmatrix} 1 & 2 \\ 3 & 4 \end{bmatrix}$. Then

$$A^T = \begin{bmatrix} 1 & 2 & 4 \\ 2 & 3 & -1 \\ 4 & -1 & 5 \end{bmatrix}^T = \begin{bmatrix} 1 & 2 & 4 \\ 2 & 3 & -1 \\ 4 & -1 & 5 \end{bmatrix} = A$$

and

$$B^T = \begin{bmatrix} 1 & 2 \\ 3 & 4 \end{bmatrix}^T = \begin{bmatrix} 1 & 3 \\ 2 & 4 \end{bmatrix} \neq B.$$

So A is symmetric but B is not. One way to visualize that A is symmetric is to observe that the entries of A below the diagonal are the "mirror images" of the entries above the diagonal.

For any matrix A, the matrix AA^T is square. Moreover, by Theorem 2.2(g),

$$(AA^T)^T = A^{TT}A^T = AA^T.$$

Hence AA^T is symmetric. Likewise $A^T A$ is symmetric.

Symmetric matrices arise in applications of linear algebra in which the notions of *distance* and *perpendicularity* are important. They were introduced in Exercises 35–42 of Section 1.1 and occur in several places in this book. They are studied in depth in Section 6.5.

*Partitioned Matrices and Block Multiplication**

Suppose that we wish to compute A^3, where

$$A = \begin{bmatrix} 1 & 0 & 0 & 0 \\ 0 & 1 & 0 & 0 \\ 6 & 8 & 5 & 0 \\ -7 & 9 & 0 & 5 \end{bmatrix}.$$

---

*The information in this and the following subsection is optional; the reader may skip to the numbered list at the end of the section (see p. 95).

Note that this calculation is tedious because A is a 4×4 matrix. There is another approach to matrix multiplication, however, that reduces the computations. Suppose that we write

$$A = \begin{bmatrix} I_2 & O \\ B & 5I_2 \end{bmatrix}, \qquad \text{where } B = \begin{bmatrix} 6 & 8 \\ -7 & 9 \end{bmatrix},$$

and then treat I_2, O, B, and $5I_2$ as if they were scalar entries of A. Then

$$A^2 = \begin{bmatrix} I_2 & O \\ B & 5I_2 \end{bmatrix} \begin{bmatrix} I_2 & O \\ B & 5I_2 \end{bmatrix} = \begin{bmatrix} I_2 & O \\ 6B & 5^2 I_2 \end{bmatrix}$$

and

$$A^3 = A^2 A = \begin{bmatrix} I_2 & O \\ 6B & 5^2 I_2 \end{bmatrix} \begin{bmatrix} I_2 & O \\ B & 5I_2 \end{bmatrix} = \begin{bmatrix} I_2 & O \\ 131B & 5^3 I_2 \end{bmatrix}.$$

This method for computing A^3 requires only that we multiply the 2×2 matrix B by a scalar.

This approach suggests that it may be useful to regard a matrix as an array of smaller matrices, which we call *blocks*. These blocks can be assembled from smaller matrices or they can be formed from an existing matrix by drawing horizontal and vertical lines. The resulting array is called a **partition** of the matrix, and the process of forming these blocks is called **partitioning**.

Naturally, there are many ways to partition a matrix, but a specific way is usually dictated by the application in which the matrix arises. For example, the matrix

$$A = \begin{bmatrix} 1 & 3 & 4 & 2 \\ 0 & 5 & -1 & 6 \\ 1 & 0 & 3 & -1 \end{bmatrix}$$

can be written as

$$A = \left[\begin{array}{cc|cc} 1 & 3 & 4 & 2 \\ 0 & 5 & -1 & 6 \\ \hline 1 & 0 & 3 & -1 \end{array} \right].$$

The horizontal and vertical lines partition A into an array of four blocks. The first row of the partition consists of the 2×2 matrices

$$\begin{bmatrix} 1 & 3 \\ 0 & 5 \end{bmatrix} \qquad \text{and} \qquad \begin{bmatrix} 4 & 2 \\ -1 & 6 \end{bmatrix},$$

and the second row consists of the 1×2 matrices $[1 \quad 0]$ and $[3 \quad -1]$. We can also partition A as

$$\left[\begin{array}{cc|cc} 1 & 3 & 4 & 2 \\ 0 & 5 & -1 & 6 \\ 1 & 0 & 3 & -1 \end{array} \right].$$

In this case, there is only one row and two columns. The blocks of this row are the 3×2 matrices

$$\begin{bmatrix} 1 & 3 \\ 0 & 5 \\ 1 & 0 \end{bmatrix} \qquad \text{and} \qquad \begin{bmatrix} 4 & 2 \\ -1 & 6 \\ 3 & -1 \end{bmatrix}.$$

As in the preceding example, matrices are partitioned into blocks to facilitate matrix multiplication. Two partitioned matrices can be multiplied by treating the blocks as if they were scalars, provided that the products of the individual blocks are defined.

Example 7   Let

$$A = \begin{bmatrix} 1 & 3 & 4 & 2 \\ 0 & 5 & -1 & 6 \\ 1 & 0 & 3 & -1 \end{bmatrix} \quad \text{and} \quad B = \begin{bmatrix} 1 & 0 & 3 \\ 1 & 2 & 0 \\ 2 & -1 & 2 \\ 0 & 3 & 1 \end{bmatrix}.$$

We can find the entries in the upper left block of AB using the given partition by computing

$$\begin{bmatrix} 1 & 3 \\ 0 & 5 \end{bmatrix} \begin{bmatrix} 1 & 0 \\ 1 & 2 \end{bmatrix} + \begin{bmatrix} 4 & 2 \\ -1 & 6 \end{bmatrix} \begin{bmatrix} 2 & -1 \\ 0 & 3 \end{bmatrix} = \begin{bmatrix} 4 & 6 \\ 5 & 10 \end{bmatrix} + \begin{bmatrix} 8 & 2 \\ -2 & 19 \end{bmatrix}$$

$$= \begin{bmatrix} 12 & 8 \\ 3 & 29 \end{bmatrix}.$$

Similarly, the upper right block of AB can be obtained by computing

$$\begin{bmatrix} 1 & 3 \\ 0 & 5 \end{bmatrix} \begin{bmatrix} 3 \\ 0 \end{bmatrix} + \begin{bmatrix} 4 & 2 \\ -1 & 6 \end{bmatrix} \begin{bmatrix} 2 \\ 1 \end{bmatrix} = \begin{bmatrix} 3 \\ 0 \end{bmatrix} + \begin{bmatrix} 10 \\ 4 \end{bmatrix} = \begin{bmatrix} 13 \\ 4 \end{bmatrix},$$

the lower left block of AB can be obtained by computing

$$\begin{bmatrix} 1 & 0 \end{bmatrix} \begin{bmatrix} 1 & 0 \\ 1 & 2 \end{bmatrix} + \begin{bmatrix} 3 & -1 \end{bmatrix} \begin{bmatrix} 2 & -1 \\ 0 & 3 \end{bmatrix} = \begin{bmatrix} 1 & 0 \end{bmatrix} + \begin{bmatrix} 6 & -6 \end{bmatrix} = \begin{bmatrix} 7 & -6 \end{bmatrix},$$

and the lower right block of AB can be obtained by computing

$$\begin{bmatrix} 1 & 0 \end{bmatrix} \begin{bmatrix} 3 \\ 0 \end{bmatrix} + \begin{bmatrix} 3 & -1 \end{bmatrix} \begin{bmatrix} 2 \\ 1 \end{bmatrix} = [3] + [5] = [8].$$

Putting these blocks together, we have

$$AB = \begin{bmatrix} 12 & 8 & 13 \\ 3 & 29 & 4 \\ 7 & -6 & 8 \end{bmatrix}.$$

As a general principle, we have the following rule.

> ### Block Multiplication
>
> If two matrices A and B are partitioned into blocks so that the number of blocks in each row of A is the same as the number of blocks in each column of B, then the matrices can be multiplied according to the usual rules for matrix multiplication by treating the blocks as if they were scalars, provided that the individual products are defined.

Four Interpretations of a Matrix Product

We are already familiar with the form of block multiplication that arises from the definition of a matrix product. Let A be an $m \times n$ matrix, and let B be an $n \times p$ matrix with columns $\mathbf{b}_1, \mathbf{b}_2, \ldots, \mathbf{b}_p$. Then we can regard A as a single block in a 1×1 array and B as a $1 \times p$ array of column vectors; that is,

$B = [\mathbf{b}_1 \quad \mathbf{b}_2 \quad \cdots \quad \mathbf{b}_p]$. It follows directly from the definition of the matrix product that

$$AB = A[\mathbf{b}_1 \quad \mathbf{b}_2 \quad \cdots \quad \mathbf{b}_p] = [A\mathbf{b}_1 \quad A\mathbf{b}_2 \quad \cdots \quad A\mathbf{b}_p],$$

which is how the rules of block multiplication apply in the context of these partitioned matrices.

We can extend this result to the situation

$$AB = A[B_1 \quad B_2 \quad \cdots \quad B_p] = [AB_1 \quad AB_2 \quad \cdots \quad AB_p],$$

where each B_j consists of a grouping of one or more adjacent columns of B. A special case of this, where $B = [C \quad D]$ and, hence, $AB = [AC \quad AD]$, will be useful later in this chapter. For example, suppose that

$$A = \begin{bmatrix} 1 & 2 \\ 3 & 4 \end{bmatrix} \quad \text{and} \quad B = \begin{bmatrix} 0 & 1 & 0 \\ 0 & 0 & 1 \end{bmatrix}.$$

Then we may partition $B = [C \quad D]$ with

$$C = \begin{bmatrix} 0 \\ 0 \end{bmatrix} = \mathbf{0} \quad \text{and} \quad D = \begin{bmatrix} 1 & 0 \\ 0 & 1 \end{bmatrix} = I_2.$$

Hence

$$AB = A[C \quad D] = [AC \quad AD] = [A\mathbf{0} \quad AI_2] = [\mathbf{0} \quad A] = \begin{bmatrix} 0 & 1 & 2 \\ 0 & 3 & 4 \end{bmatrix}.$$

Of course, we may eliminate the vertical line once we have computed AB.

A reverse approach, used less often, is to partition an $m \times n$ matrix A into an $m \times 1$ array of row vectors $\mathbf{a}_1', \mathbf{a}_2', \ldots, \mathbf{a}_m'$, and regard an $n \times p$ matrix B as a single block in a 1×1 array. In this case,

$$AB = \begin{bmatrix} \mathbf{a}_1' \\ \mathbf{a}_2' \\ \vdots \\ \mathbf{a}_m' \end{bmatrix} B = \begin{bmatrix} \mathbf{a}_1' B \\ \mathbf{a}_2' B \\ \vdots \\ \mathbf{a}_m' B \end{bmatrix}. \tag{1}$$

Thus the rows of AB are the products of the rows of A with B. More specifically, the ith row of AB is the matrix product of the ith row of A with B.

Yet another way to view matrix multiplication is in terms of matrix products involving a column of A and a row of B. Suppose that $\mathbf{a}_1, \mathbf{a}_2, \ldots, \mathbf{a}_n$ are the columns of an $m \times n$ matrix A, and $\mathbf{b}_1', \mathbf{b}_2', \ldots, \mathbf{b}_n'$ are the rows of an $n \times p$ matrix B. Then block multiplication gives

$$AB = [\mathbf{a}_1 \quad \mathbf{a}_2 \quad \cdots \quad \mathbf{a}_n] \begin{bmatrix} \mathbf{b}_1' \\ \mathbf{b}_2' \\ \vdots \\ \mathbf{b}_n' \end{bmatrix} = \mathbf{a}_1 \mathbf{b}_1' + \mathbf{a}_2 \mathbf{b}_2' + \cdots + \mathbf{a}_n \mathbf{b}_n'. \tag{2}$$

Thus AB is the sum of matrix products of each column of A with the corresponding row of B.

The terms $\mathbf{a}_i \mathbf{b}_i'$ in (2) are matrix products of two vectors, namely, column i of A and row i of B. Such products have an especially simple form. In order to present this result in a more standard notation, we consider the matrix product of $\mathbf{v}$ and $\mathbf{w}^T$, where

$$\mathbf{v} = \begin{bmatrix} v_1 \\ v_2 \\ \vdots \\ v_m \end{bmatrix} \quad \text{and} \quad \mathbf{w} = \begin{bmatrix} w_1 \\ w_2 \\ \vdots \\ w_n \end{bmatrix}.$$

It follows from (1) that

$$\mathbf{v}\mathbf{w}^T = \begin{bmatrix} v_1\mathbf{w}^T \\ v_2\mathbf{w}^T \\ \vdots \\ v_m\mathbf{w}^T \end{bmatrix}.$$

Thus the rows of the $m \times n$ matrix $\mathbf{v}\mathbf{w}^T$ are all multiples of $\mathbf{w}^T$. It follows (see Exercise 49) that the rank of the matrix $\mathbf{v}\mathbf{w}^T$ is 1.

Products of the form $\mathbf{v}\mathbf{w}^T$, where $\mathbf{v}$ is in $\mathcal{R}^m$ (and is regarded as an $m \times 1$ matrix) and $\mathbf{w}$ is in $\mathcal{R}^n$ (and is regarded as an $n \times 1$ matrix), are called **outer products**. In this terminology, (2) states that the product of an $m \times n$ matrix A and an $n \times p$ matrix B is the sum of n matrices of rank 1, namely the outer products of the columns of A with the corresponding rows of B.

One interesting example is the case that A is a $1 \times n$ matrix and, hence,

$$A = [a_1 \quad a_2 \quad \cdots \quad a_n]$$

is a row vector with the scalar entries a_i. In this case, the product (2) is the linear combination

$$AB = a_1\mathbf{b}_1' + a_2\mathbf{b}_2' + \cdots + a_n\mathbf{b}_n'$$

of the rows of B with the corresponding entries of A as the coefficients. For example,

$$[2 \quad 3]\begin{bmatrix} -1 & 4 \\ 5 & 0 \end{bmatrix} = 2[-1 \quad 4] + 3[5 \quad 0] = [13 \quad 8].$$

The following two examples illustrate our comments.

Example 8   Let

$$A = \begin{bmatrix} 1 & 2 & -1 \\ -1 & 1 & 3 \end{bmatrix} \quad \text{and} \quad B = \begin{bmatrix} -2 & 1 & 0 \\ 1 & -3 & 4 \\ 1 & -1 & -1 \end{bmatrix}.$$

Then

$$AB = \begin{bmatrix} -1 & -4 & 9 \\ 6 & -7 & 1 \end{bmatrix}.$$

Note that the rows of AB are

$$[1 \quad 2 \quad -1]\begin{bmatrix} -2 & 1 & 0 \\ 1 & -3 & 4 \\ 1 & -1 & -1 \end{bmatrix} = [-1 \quad -4 \quad 9]$$

and

$$[-1 \quad 1 \quad 3]\begin{bmatrix} -2 & 1 & 0 \\ 1 & -3 & 4 \\ 1 & -1 & -1 \end{bmatrix} = [6 \quad -7 \quad 1],$$

as guaranteed by (1).

Example 9   Use outer products to express the product AB in Example 8 as a sum of matrices of rank 1.

Solution   First form the outer products in (2) to obtain the following matrices of rank 1:

$$\begin{bmatrix} 1 \\ -1 \end{bmatrix} \begin{bmatrix} -2 & 1 & 0 \end{bmatrix} = \begin{bmatrix} -2 & 1 & 0 \\ 2 & -1 & 0 \end{bmatrix},$$

$$\begin{bmatrix} 2 \\ 1 \end{bmatrix} \begin{bmatrix} 1 & -3 & 4 \end{bmatrix} = \begin{bmatrix} 2 & -6 & 8 \\ 1 & -3 & 4 \end{bmatrix},$$

and

$$\begin{bmatrix} -1 \\ 3 \end{bmatrix} \begin{bmatrix} 1 & -1 & -1 \end{bmatrix} = \begin{bmatrix} -1 & 1 & 1 \\ 3 & -3 & -3 \end{bmatrix}.$$

Then

$$\begin{bmatrix} -2 & 1 & 0 \\ 2 & -1 & 0 \end{bmatrix} + \begin{bmatrix} 2 & -6 & 8 \\ 1 & -3 & 4 \end{bmatrix} + \begin{bmatrix} -1 & 1 & 1 \\ 3 & -3 & -3 \end{bmatrix} = \begin{bmatrix} -1 & -4 & 9 \\ 6 & -7 & 1 \end{bmatrix}$$

$$= AB,$$

as guaranteed by (2).   ○

In summary, there are four ways to think about a matrix product $C = AB$. We assume here that A is an $m \times n$ matrix with rows $\mathbf{a}'_1, \mathbf{a}'_2, \ldots, \mathbf{a}'_m$ and B is an $n \times p$ matrix with rows $\mathbf{b}'_1, \mathbf{b}'_2, \ldots, \mathbf{b}'_n$.

1. The jth column of AB is obtained by multiplying A by the jth column of B. Symbolically, $\mathbf{c}_j = A\mathbf{b}_j$ for all j.

2. The ith row of AB is obtained by multiplying the ith row of A by B; that is, $\mathbf{a}'_i B$.

3. The (i, j)-entry of AB is obtained by multiplying the ith row of A by the jth column of B; that is,

$$\mathbf{a}'_i \mathbf{b_j} = a_{i1}b_{1j} + a_{i2}b_{2j} + \cdots + a_{n1}b_{nj}.$$

4. The matrix AB is a sum of matrix products of each column of A with the corresponding row of B. Symbolically,

$$AB = \mathbf{a}_1 \mathbf{b}'_1 + \mathbf{a}_2 \mathbf{b}'_2 + \cdots + \mathbf{a}_n \mathbf{b}'_n.$$

Practice Problems

1. Suppose that A is a 2×4 matrix and B is a 2×3 matrix. Is the product BA^T defined? If so, what is its size?

2. For A and B as described in problem 1, is the product $A^T B$ defined? If so, what is its size?

3. For $A = \begin{bmatrix} 2 & -1 & 3 \\ 1 & 4 & -2 \end{bmatrix}$ and $B = \begin{bmatrix} -1 & 0 & 2 \\ 0 & -3 & 4 \\ 3 & 1 & -2 \end{bmatrix}$, compute AB.

Exercises

1. Determine if the following statements are true or false.

   (a) If A and B are $m \times n$ matrices, then AB is defined and is an $m \times n$ matrix.

   (b) If A is an $m \times n$ matrix, then A^2 is defined if and only if $m = n$.

   (c) If A and B are matrices, then both AB and BA are defined if and only if A and B are square matrices.

(d) If A is an $m \times n$ matrix and B is an $n \times p$ matrix, then $(AB)^T = A^T B^T$.

(e) If $\mathbf{v}$ and $\mathbf{w}$ are nonzero vectors in $\mathcal{R}^n$, then $\mathbf{vw}^T$ is an $n \times n$ matrix of rank 1.

(f) If A_α and A_β are both 2×2 rotation matrices, then $A_\alpha A_\beta$ is also a 2×2 rotation matrix.

(g) Let A and B be matrices such that AB is defined, and let A and B be partitioned into blocks so that the number of blocks in each row of A is the same as the number of blocks in each column of B. Then the matrices can be multiplied according to the usual rule for matrix multiplication by treating the blocks as if they were scalars.

In Exercises 2–4, decide if the matrix product AB is defined. If so, find its size.

2. A is a 2×4 matrix and B is a 4×6 matrix.

3. A is a 2×3 matrix and B^T is a 2×3 matrix.

4. A^T is a 3×3 matrix and B is a 2×3 matrix.

5. Give an example of matrices A and B such that BA is defined but AB is not.

In Exercises 6–21, use the following matrices to compute the given expression, or give a reason why the expression is not defined.

$$A = \begin{bmatrix} 1 & -2 \\ 3 & 4 \end{bmatrix} \quad B = \begin{bmatrix} 7 & 4 \\ 1 & 2 \end{bmatrix} \quad C = \begin{bmatrix} 3 & 8 & 1 \\ 2 & 0 & 4 \end{bmatrix}$$

$$\mathbf{x} = \begin{bmatrix} 2 \\ 3 \end{bmatrix} \quad \mathbf{y} = \begin{bmatrix} 1 \\ 3 \\ -5 \end{bmatrix} \quad \mathbf{z} = [7 \quad -1]$$

6. $B\mathbf{x}$
7. $C\mathbf{y}$
8. $B\mathbf{y}$
9. $\mathbf{xz}$

10. $A\mathbf{z}^T$
11. $AC\mathbf{x}$
12. AC
13. AB

14. BA
15. BC
16. CB
17. CB^T

18. A^2
19. A^3
20. B^2
21. C^2

In Exercises 22–24, use the matrices A, B, C, and $\mathbf{z}$ from Exercises 6–21.

22. Verify that $(AB)C = A(BC)$.

23. Verify that $(AB)^T = B^T A^T$.

24. Verify that $\mathbf{z}(AC) = (\mathbf{z}A)C$.

In Exercises 25–32, use the following matrices to compute the requested entry, row, or column of the matrix product.

$$A = \begin{bmatrix} 1 & 2 & 3 \\ 2 & -1 & 4 \\ -3 & -2 & 0 \end{bmatrix} \quad B = \begin{bmatrix} -1 & 0 \\ 4 & 1 \\ 3 & -2 \end{bmatrix}$$

$$C = \begin{bmatrix} 2 & 1 & -1 \\ 4 & 3 & -2 \end{bmatrix}$$

25. the (3, 2)-entry of AB
26. the (2, 1)-entry of BC

27. the (2, 3)-entry of CA
28. the (1, 1) entry of CB

29. column 2 of AB
30. column 3 of BC

31. row 1 of CA
32. row 2 of CB

In Exercises 33 and 34, use the matrices A, B, and C from Exercises 25–32.

33. Represent AB as the sum of three matrices of rank 1.

34. Represent BC as the sum of two matrices of rank 1.

In Exercises 35–39, use block multiplication to compute the matrix products.

35. $\begin{bmatrix} 1 & 1 & 2 & 1 \\ 1 & 0 & 0 & 0 \\ 0 & 1 & 0 & 0 \end{bmatrix} \begin{bmatrix} 1 & 0 & 1 & -1 \\ 0 & 1 & -1 & 1 \\ 0 & 0 & 1 & 0 \\ 0 & 0 & 0 & 1 \end{bmatrix}$

36. $\begin{bmatrix} 0 & 0 \\ 0 & 0 \\ 0 & 0 \\ 2 & 3 \end{bmatrix} \begin{bmatrix} 0 & 0 & 0 & 6 \\ 0 & 0 & 0 & -1 \end{bmatrix}$

37. $\begin{bmatrix} 3 & 0 \\ 0 & 3 \\ 2 & 0 \\ 0 & 2 \end{bmatrix} \begin{bmatrix} 1 & 2 \\ 3 & 4 \end{bmatrix}$

38. $\begin{bmatrix} 1 & 2 & 2 & -2 \\ 1 & 1 & -2 & 2 \end{bmatrix} \begin{bmatrix} 1 & 0 \\ 0 & 1 \\ 1 & 1 \\ 1 & 1 \end{bmatrix}$

39. $\begin{bmatrix} A_{20°} \\ A_{30°} \end{bmatrix} \begin{bmatrix} A_{40°} & | & A_{50°} \end{bmatrix}$

40. Suppose that A has the block form

$$\begin{bmatrix} I_m & O \\ O & B \end{bmatrix},$$

where B is an $m \times m$ matrix. Prove that for every positive integer k,

$$A^k = \begin{bmatrix} I_m & O \\ O & B^k \end{bmatrix}.$$

41. Let

$$\text{To} \quad \begin{matrix} & \text{From} \\ & \text{City} \quad \text{Suburbs} \\ \begin{matrix} \text{City} \\ \text{Suburbs} \end{matrix} & \begin{bmatrix} .85 & .03 \\ .15 & .97 \end{bmatrix} = A \end{matrix}$$

be the stochastic matrix used in Example 6 to predict population movement between the city and its suburbs. Suppose that 70% of city residents live in single unit houses (as opposed to multiple unit or apartment housing), and that 95% of suburban residents live in single unit houses.

(a) Find a 2×2 matrix B such that if v_1 people live in the city and v_2 people live in the suburbs, then

$B \begin{bmatrix} v_1 \\ v_2 \end{bmatrix} = \begin{bmatrix} u_1 \\ u_2 \end{bmatrix}$, where u_1 people live in single unit houses and u_2 people live in multiple unit houses.

(b) Explain the significance of $BA \begin{bmatrix} v_1 \\ v_2 \end{bmatrix}$.

42. Of those vehicle owners who live in the city, 60% drive cars, 30% drive vans, and 10% drive recreational vehicles. Of those vehicle owners who live in the suburbs, 30% drive cars, 50% drive vans, and 20% drive recreational vehicles. Of all of the vehicles (in the city and suburbs together), 60% of the cars, 40% of the vans, and 50% of the recreational vehicles are dark in color.

(a) Find a matrix B such that if v_1 vehicle owners live in the city and v_2 vehicle owners live in the suburbs, then $B \begin{bmatrix} v_1 \\ v_2 \end{bmatrix} = \begin{bmatrix} u_1 \\ u_2 \\ u_3 \end{bmatrix}$, where u_1 people drive cars, u_2 people drive vans, and u_3 people drive recreational vehicles.

(b) Find a matrix A such that $A \begin{bmatrix} u_1 \\ u_2 \\ u_3 \end{bmatrix} = \begin{bmatrix} w_1 \\ w_2 \end{bmatrix}$, where u_1, u_2, and u_3 are as in (a), w_1 is the number of people who drive dark vehicles, and w_2 is the number of people who drive light-colored vehicles.

(c) Find a matrix C such that $\begin{bmatrix} w_1 \\ w_2 \end{bmatrix} = C \begin{bmatrix} v_1 \\ v_2 \end{bmatrix}$, where v_1 and v_2 are as in (a) and w_1 and w_2 are as in (b).

43. In a certain elementary school, of those pupils who buy a hot lunch on a particular school day, 30% buy a hot lunch and 70% bring a bag lunch on the next school day. Furthermore, of those pupils who bring a bag lunch on a particular school day, 40% buy a hot lunch and 60% bring a bag lunch on the next school day.

(a) Find a matrix A such that if u_1 pupils buy a hot lunch and u_2 pupils bring a bag lunch on a particular day, then $A \begin{bmatrix} u_1 \\ u_2 \end{bmatrix} = \begin{bmatrix} v_1 \\ v_2 \end{bmatrix}$, where v_1 pupils buy a hot lunch and v_2 pupils bring a bag lunch on the next school day.

(b) Suppose that $u_1 = 100$ pupils buy a hot lunch and $u_2 = 200$ pupils bring a bag lunch on the first day of school. Compute $A \begin{bmatrix} u_1 \\ u_2 \end{bmatrix}$, $A^2 \begin{bmatrix} u_1 \\ u_2 \end{bmatrix}$, and $A^3 \begin{bmatrix} u_1 \\ u_2 \end{bmatrix}$. Explain the significance of each result.

(c) To do the following problem, you will need a calculator with matrix capabilities or access to computer software such as MATLAB. Using the notation of (b), compute $A^{100} \begin{bmatrix} u_1 \\ u_2 \end{bmatrix} = \begin{bmatrix} w_1 \\ w_2 \end{bmatrix}$. Explain the significance of this result. Now compute $A \begin{bmatrix} w_1 \\ w_2 \end{bmatrix}$, andcompare this result with $\begin{bmatrix} w_1 \\ w_2 \end{bmatrix}$. Explain.

44. Prove (a) of Theorem 2.2.

45. Prove (d) of Theorem 2.2.

46. Prove (e) of Theorem 2.2.

47. Prove (f) of Theorem 2.2.

48. Let $A = A_{180°}$, and let B be the matrix that reflects $\mathcal{R}^2$ about the x-axis; that is, $B = \begin{bmatrix} 1 & 0 \\ 0 & -1 \end{bmatrix}$. Compute BA, and describe how a vector $\mathbf{v}$ is affected by multiplication by BA.

49. Suppose $\mathbf{a}$ and $\mathbf{b}$ are nonzero vectors in $\mathcal{R}^m$ and $\mathcal{R}^n$, respectively. Prove that the outer product $\mathbf{ab}^T$ has rank 1.

50. Let A be an $n \times n$ matrix.

(a) Prove that A is a diagonal matrix if and only if its jth column equals $a_{jj}\mathbf{e}_j$.

(b) Use (a) to prove that if A and B are $n \times n$ diagonal matrices, then AB is a diagonal matrix whose jth column is $a_{jj}b_{jj}\mathbf{e}_j$.

51. A square matrix A is called **upper triangular** if the (i, j)-entry of A is zero whenever $i > j$. Prove that if A and B are both $n \times n$ upper triangular matrices, then AB is also an upper triangular matrix.

52. Let $A = \begin{bmatrix} 1 & -1 & 2 & -1 \\ -2 & 1 & -1 & 3 \\ -1 & -1 & 4 & 3 \\ -5 & 3 & -4 & 7 \end{bmatrix}$. Find a nonzero 4×2 matrix B with rank 2 such that $AB = O$.

53. Prove or disprove the following statement: If A and B are $n \times n$ matrices such that $AB = O$, then $BA = O$.

54. Let A and B be $n \times n$ matrices. Prove or disprove that the ranks of AB and BA are equal.

55. Recall the definition of the *trace* of a matrix given in Exercise 46 of Section 1.1. Prove that if A is an $m \times n$ matrix and B is an $n \times m$ matrix, then trace AB = trace BA.

56. Let $1 \le r, s \le n$ be integers, and let E be the $n \times n$ matrix with a 1 in the (r, s)-entry and 0s elsewhere. Let B be any $n \times n$ matrix. Describe EB in terms of the entries of B.

57. Prove that if A is a $k \times m$ matrix, B is an $m \times n$ matrix, and C is an $n \times p$ matrix, then $(ABC)^T = C^T B^T A^T$.

58. Let A and B be symmetric matrices of the same size. Prove that AB is symmetric if and only $AB = BA$.

In Exercises 59–63, use a calculator with matrix capabilities or computer software such as MATLAB *to solve the problem.*

59. Let A_θ denote a rotation matrix.

(a) For $\theta = \pi/2$, compute A_θ^2 by hand.

(b) For $\theta = \pi/3$, compute A_θ^3.

(c) For $\theta = \pi/8$, compute A_θ^8.

(d) Use the prevous results to make a conjecture about A_θ^k where $\theta = \pi/k$.

(e) Draw a sketch to support your conjecture in (d).

60. Let A, B, and C be randomly generated 4×4 matrices.

(a) Illustrate the distributive law $A(B+C) = AB+AC$.

(b) Check the validity of the equation $(A + B)^2 = A^2 + 2AB + B^2$.

(c) Make a conjecture about an expression that is equal to $(A + B)^2$ for arbitrary $n \times n$ matrices A and B. Justify your conjecture.

(d) Make a conjecture about the relationship that must hold between AB and BA for the equation in (b) to hold in general.

(e) Prove your conjecture in (d).

61. Suppose that A is a 4×4 matrix in the block form

$$A = \begin{bmatrix} B & C \\ O & D \end{bmatrix},$$

where the blocks are all 2×2 matrices.

(a) Use a randomly generated matrix for A to illustrate that $A^2 = \begin{bmatrix} B^2 & * \\ O & D^2 \end{bmatrix}$, where $*$ represents some 2×2 matrix.

(b) Use a randomly generated matrix for A to illustrate that $A^3 = \begin{bmatrix} B^3 & * \\ O & D^3 \end{bmatrix}$, where $*$ represents, some 2×2 matrix.

(c) Make a conjecture about the block form of A^k, where k is a positive integer.

(d) Prove your conjecture for $k = 3$.

62. Let A and B be randomly generated 4×4 matrices.

(a) Compute AB and its rank by finding the reduced row echelon form of AB.

(b) Compute BA and its rank by finding the reduced row echelon form of BA.

(c) Compare your answers with your solution to Exercise 54.

63. Let A be the stochastic matrix used in the population application in Example 6.

(a) Verify that A^{10} and $A^{10}\mathbf{p}$ are as given in the example.

(b) Determine the populations of the city and suburbs after 20 years.

(c) Determine the populations of the city and suburbs after 50 years.

(d) Make a conjecture about the eventual populations of the city and suburbs.

Solutions to the Practice Problems

1. Since B is a 2×3 matrix and A^T is a 4×2 matrix, the product BA^T is not defined.

2. Since A^T is a 4×2 matrix and B is a 2×3 matrix, the product $A^T B$ is defined. Its size is 4×3.

3. The matrix AB is a 2×3 matrix. Its first column is

$$\begin{bmatrix} 2 & -1 & 3 \\ 1 & 4 & -2 \end{bmatrix} \begin{bmatrix} -1 \\ 0 \\ 3 \end{bmatrix} = \begin{bmatrix} -2+0+9 \\ -1+0-6 \end{bmatrix} = \begin{bmatrix} 7 \\ -7 \end{bmatrix},$$

its second column is

$$\begin{bmatrix} 2 & -1 & 3 \\ 1 & 4 & -2 \end{bmatrix} \begin{bmatrix} 0 \\ -3 \\ 1 \end{bmatrix} = \begin{bmatrix} 0+ 3+3 \\ 0 - 12 - 2 \end{bmatrix} = \begin{bmatrix} 6 \\ -14 \end{bmatrix},$$

and its third column is

$$\begin{bmatrix} 2 & -1 & 3 \\ 1 & 4 & -2 \end{bmatrix} \begin{bmatrix} 2 \\ 4 \\ -2 \end{bmatrix} = \begin{bmatrix} 4 - 4 - 6 \\ 2 + 16 + 4 \end{bmatrix} = \begin{bmatrix} -6 \\ 22 \end{bmatrix}.$$

Hence $AB = \begin{bmatrix} 7 & 6 & -6 \\ -7 & -14 & 22 \end{bmatrix}$.

2.2*   Applications of Matrix Multiplication

In this section, we consider four applications of matrix multiplication. These are independent of one another, and so can be studied in any order.

The Leslie Matrix and Population Change

The population of an isolated colony of animals depends on the fertility and mortality rates for the various age groups of the colony. Fertility rates can be measured precisely if observations are limited to females (since they are the ones who give birth). Because a relationship between the number of males and the number of females can usually be established, this restriction is not serious.

The following hypothetical example illustrates these ideas. Suppose that the members of a certain colony of mammals have a life span of less than

---

*This section can be omitted without loss of continuity.

3 years. We divide the females into three age groups: those of age less than 1, those of age between 1 to 2, and those of age 2. Suppose that 40% of newborn females survive to age 1, and that 50% of these survive to age 2. In terms of fertility rates, suppose that the females under 1 year of age do not give birth, those of age between 1 and 2 have an average of two female offspring, and those of age 2 have an average of one female offspring. Let x_1, x_2, and x_3 be the number of females in the first, second, and third age groups, respectively, at the present time; and let y_1, y_2, and y_3 be the number of females in the corresponding groups for the next year. The changes from this year to next year are depicted by Figure 2.2.

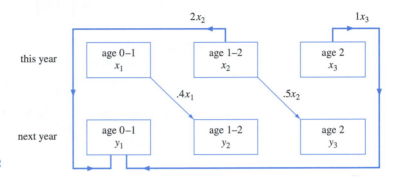

FIGURE 2.2

The vector $\mathbf{x} = \begin{bmatrix} x_1 \\ x_2 \\ x_3 \end{bmatrix}$ is called the **population distribution** for the female population of the colony. The preceding information can be used to predict the population distribution for the following year, $\mathbf{y} = \begin{bmatrix} y_1 \\ y_2 \\ y_3 \end{bmatrix}$. Referring to Figure 2.2, we see that y_1 is the total number of female offspring born during the year, and that these offspring come from females of the second and third age groups. Thus

$$y_1 = 2x_2 + x_3.$$

The number y_2 is the total number of females in the second age group for next year. Because these females are in the first age group this year, and because only 40% of them will survive to the next year, we have that $y_2 = 0.4x_1$. Similarly, $y_3 = 0.5x_2$. Collecting these three equations, we have

$$\begin{aligned} y_1 &= & 2.0x_2 &+ 1.0x_3 \\ y_2 &= 0.4x_1 & & \\ y_3 &= & 0.5x_2 & \end{aligned}$$

These three equations can be represented by the single matrix equation $\mathbf{y} = A\mathbf{x}$, where $\mathbf{x}$ and $\mathbf{y}$ are the population distributions as previously defined and A is the 3×3 matrix

$$A = \begin{bmatrix} 0.0 & 2.0 & 1.0 \\ 0.4 & 0.0 & 0.0 \\ 0.0 & 0.5 & 0.0 \end{bmatrix}.$$

For example, suppose that $\mathbf{x} = \begin{bmatrix} 1000 \\ 1000 \\ 1000 \end{bmatrix}$; that is, there are currently 1000 females in each age group. Then

$$\mathbf{y} = A\mathbf{x}$$

$$= \begin{bmatrix} 0.0 & 2.0 & 1.0 \\ 0.4 & 0.0 & 0.0 \\ 0.0 & 0.5 & 0.0 \end{bmatrix} \begin{bmatrix} 1000 \\ 1000 \\ 1000 \end{bmatrix}$$

$$= \begin{bmatrix} 3000 \\ 400 \\ 500 \end{bmatrix} .$$

So one year later there are 3000 females under 1 year of age, 400 females who are 1 year old, and 500 females who are 2 years old.

For each positive integer k, let $\mathbf{x}_k$ denote the population distribution after k years, and let $\mathbf{x}_0 = \mathbf{x} = \begin{bmatrix} x_1 \\ x_2 \\ x_3 \end{bmatrix}$ be the current population distribution. In the preceding example,

$$\mathbf{x}_0 = \begin{bmatrix} 1000 \\ 1000 \\ 1000 \end{bmatrix} \quad \text{and} \quad \mathbf{x}_1 = \mathbf{y} = \begin{bmatrix} 3000 \\ 400 \\ 500 \end{bmatrix} .$$

Then for any positive integer k, we have that $\mathbf{x}(k) = A\mathbf{x}(k-1)$. Thus

$$\mathbf{x}_k = A\mathbf{x}_{k-1} = A^2\mathbf{x}_{k-2} = \cdots = A^k\mathbf{x}_0.$$

In this way, we may predict population trends over the long term. For example, to predict the population distribution in ten years, we compute $\mathbf{x}(10) = A^{10}\mathbf{x}(0)$. Thus

$$\mathbf{x}_{10} = A^{10}\mathbf{x}_0$$

$$= \begin{bmatrix} 1987 \\ 851 \\ 387 \end{bmatrix} ,$$

where each entry is rounded off to the nearest whole number. If we continue this process in increments of ten years, we find that (rounding to whole numbers)

$$\mathbf{x}_{20} = \begin{bmatrix} 2043 \\ 819 \\ 408 \end{bmatrix} \quad \text{and} \quad \mathbf{x}_{30} = \mathbf{x}_{40} = \begin{bmatrix} 2045 \\ 818 \\ 409 \end{bmatrix} .$$

It appears that the population stabilizes after 30 years. In fact, for the vector

$$\mathbf{z} = \begin{bmatrix} 2045 \\ 818 \\ 409 \end{bmatrix} ,$$

we have that $A\mathbf{z} = \mathbf{z}$ precisely. Under this circumstance, the population distribution $\mathbf{z}$ is stable; that is, it does not change from year to year. Whether a population approaches a stable distribution for a colony depends on the nature of the survival and reproductive rates of the age groups (see, for example,

Exercises 4–7). Exercise 2 gives an example of a population model for which no nontrivial stable population distribution exists.

We can generalize this situation to an arbitrary colony of animals. Suppose that we divide the females of the colony into n age groups, where x_i is the number of members in the ith group. The duration of time in an individual age group need not be a year, but the various durations should be equal. Let

$$\mathbf{x} = \begin{bmatrix} x_1 \\ x_2 \\ \vdots \\ x_n \end{bmatrix}$$ be the population distribution of the females of the colony, p_i be the

portion of females in the ith group who survive to the $(i+1)$st group, and b_i be the average number of female offspring of a member of the ith age group.

If $\mathbf{y} = \begin{bmatrix} y_1 \\ y_2 \\ \vdots \\ y_n \end{bmatrix}$ is the population for the next time period, then

$$y_1 = b_1 x_1 + b_2 x_2 + \cdots + b_n x_n$$
$$y_2 = p_1 x_1$$
$$y_3 = \qquad p_2 x_2$$
$$\vdots$$
$$y_n = \qquad\qquad p_{n-1} x_{n-1}.$$

Therefore, for

$$A = \begin{bmatrix} b_1 & b_2 & \cdots & b_{n-1} & b_n \\ p_1 & 0 & \cdots & 0 & 0 \\ 0 & p_2 & \cdots & 0 & 0 \\ \vdots & \vdots & & \vdots & \vdots \\ 0 & 0 & \cdots & p_{n-1} & 0 \end{bmatrix},$$

we have that

$$\mathbf{y} = A\mathbf{x}.$$

The matrix A is called the **Leslie matrix** for the population. As in the earlier example, if $\mathbf{x}(k)$ is the population distribution after k time intervals, then

$$\mathbf{x}_k = A^k \mathbf{x}_0.$$

Analysis of Traffic Flow

Figure 2.3 represents the flow of traffic through a network of one-way streets. The arrows indicate the direction of traffic flow. The number on any street beyond an intersection is the portion of the traffic entering the street from that intersection. For example, 30% of the traffic leaving intersection P_1 goes to P_4, and the other 70% goes to P_2. Notice that all the traffic leaving P_5 must go to P_8.

Suppose that, on a particular day, x_1 cars enter the network from the left of P_1, and x_2 cars enter from the left of P_3. Let $w_1, w_2, w_3,$ and w_4 represent the number of cars leaving the network along the exits to the right (see Figure 2.3).

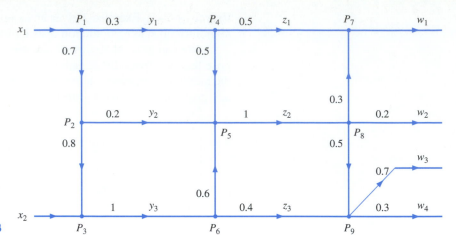

FIGURE 2.3

We wish to determine the values of the w_i. At first glance, this problem seems overwhelming since there are so many routes for the traffic. However, by decomposing the problem into several simpler ones, we can solve the simpler ones individually and combine their solutions to obtain the values of the w_i.

We begin with only the portion of the network involving intersections P_1, P_2, and P_3. Let y_1, y_2, and y_3 be the *expected* number of cars that exit along the three eastward routes. To find y_1, notice that 30% of all cars entering P_1 continue on to P_4. Therefore, $y_1 = 0.30x_1$. Also, $0.7x_1$ of the cars turn right at P_1, and of these, 20% enter the street between P_2 and P_5. Because these are the only cars to do so, it follows that $y_2 = (0.2)(0.7)x_1 = 0.14x_1$. Furthermore, since 80% of the cars entering P_2 continue on to P_3, the number of such cars is $(0.8)(0.7)x_1 = 0.56x_1$. Finally, all the cars entering P_3 from the left use the street between P_3 and P_6, so $y_3 = 0.56x_1 + x_2$. Summarizing, we have

$$y_1 = 0.30x_1$$
$$y_2 = 0.14x_1$$
$$y_3 = 0.56x_1 + x_2.$$

We can express this system of equations by the single matrix equation $\mathbf{y} = A\mathbf{x}$, where

$$\mathbf{y} = \begin{bmatrix} y_1 \\ y_2 \\ y_3 \end{bmatrix} \qquad A = \begin{bmatrix} 0.30 & 0 \\ 0.14 & 0 \\ 0.56 & 1 \end{bmatrix} \qquad \text{and} \qquad \mathbf{x} = \begin{bmatrix} x_1 \\ x_2 \end{bmatrix}.$$

Now consider the next set of intersections P_4, P_5, and P_6. If we let z_1, z_2, and z_3 represent the number of cars that exit from the right of P_4, P_5, and P_6, respectively, then by a similar analysis we have that

$$z_1 = 0.5y_1$$
$$z_2 = 0.5y_1 + y_2 + 0.6y_3$$
$$z_3 = \qquad\qquad 0.4y_3,$$

or $\mathbf{z} = B\mathbf{y}$, where

$$\mathbf{z} = \begin{bmatrix} z_1 \\ z_2 \\ z_3 \end{bmatrix} \qquad \text{and} \qquad B = \begin{bmatrix} 0.5 & 0 & 0 \\ 0.5 & 1 & 0.6 \\ 0 & 0 & 0.4 \end{bmatrix}.$$

Finally, if we set

$$
\mathbf{w} = \begin{bmatrix} w_1 \\ w_2 \\ w_3 \\ w_4 \end{bmatrix} \quad \text{and} \quad C = \begin{bmatrix} 1 & 0.30 & 0 \\ 0 & 0.20 & 0 \\ 0 & 0.35 & 0.7 \\ 0 & 0.15 & 0.3 \end{bmatrix},
$$

then by a similar argument, we have $\mathbf{w} = C\mathbf{z}$. It follows that

$$
\mathbf{w} = C\mathbf{z} = C(B\mathbf{y}) = (CB)A\mathbf{x} = (CBA)\mathbf{x}.
$$

Let $M = CBA$. Then

$$
M = \begin{bmatrix} 1 & 0.30 & 0 \\ 0 & 0.20 & 0 \\ 0 & 0.35 & 0.7 \\ 0 & 0.15 & 0.3 \end{bmatrix} \begin{bmatrix} 0.5 & 0 & 0 \\ 0.5 & 1 & 0.6 \\ 0 & 0 & 0.4 \end{bmatrix} \begin{bmatrix} 0.30 & 0 \\ 0.14 & 0 \\ 0.56 & 1 \end{bmatrix}
$$

$$
= \begin{bmatrix} 0.3378 & 0.18 \\ 0.1252 & 0.12 \\ 0.3759 & 0.49 \\ 0.1611 & 0.21 \end{bmatrix}.
$$

For example, if 1000 cars enter the traffic pattern at P_1 and 2000 enter at P_3, then for $\mathbf{x} = \begin{bmatrix} 1000 \\ 2000 \end{bmatrix}$, we have

$$
\mathbf{w} = M\mathbf{x}
$$

$$
= \begin{bmatrix} 0.3378 & 0.18 \\ 0.1252 & 0.12 \\ 0.3759 & 0.49 \\ 0.1611 & 0.21 \end{bmatrix} \begin{bmatrix} 1000 \\ 2000 \end{bmatrix}
$$

$$
= \begin{bmatrix} 697.8 \\ 365.2 \\ 1355.9 \\ 581.1 \end{bmatrix}.
$$

Naturally, the actual number of cars traveling on any path is a whole number, unlike the entries of $\mathbf{w}$. Since these calculations are based on probabilities, we cannot expect the answers to be exact. For example, approximately 698 cars exit the traffic pattern at P_7, and approximately 365 cars exit the pattern at P_8.

The same analysis can be applied if the quantities studied represent *rates* of traffic flow (for example, number of cars per hour), rather than the total number of cars.

Finally, this kind of analysis can be applied to other contexts, such as the flow of a fluid through a system of pipes or the movement of money in an economy. For other examples, see the exercises at the end of this section.

Adjacency Matrices

Matrices can be used to study certain relationships between objects. For example, suppose that there are five countries, and each of them maintains diplomatic relations with some of the others. To organize these relationships, we use a 5×5 matrix A defined as follows. For $1 \le i \le 5$ we have $a_{ii} = 0$, and for

$i \neq j$ we have

$$a_{ij} = \begin{cases} 1 & \text{if country } i \text{ maintains diplomatic relations with country } j \\ 0 & \text{otherwise.} \end{cases}$$

Note that all the entries of A are zeros and ones. Such matrices are called **adjacency matrices**, and they are worthy of study in their own right. For purposes of illustration, suppose that

$$A = \begin{bmatrix} 0 & 0 & 1 & 1 & 0 \\ 0 & 0 & 0 & 1 & 1 \\ 1 & 0 & 0 & 0 & 1 \\ 1 & 1 & 0 & 0 & 0 \\ 0 & 1 & 1 & 0 & 0 \end{bmatrix}.$$

Notice that $A = A^T$; that is, A is symmetric. The symmetry occurs here because the underlying relationship is symmetric (if country i maintains diplomatic relations with country j, then country j maintains diplomatic relations with country i). Such symmetry is true of many relationships of interest.

Let us consider the significance of the entries of the matrix $B = A^2$; for example,

$$b_{23} = a_{21}a_{13} + a_{22}a_{23} + a_{23}a_{33} + a_{24}a_{43} + a_{25}a_{53}.$$

A typical term on the right-hand side of the equation has the form $a_{2k}a_{k3}$. This term is 1 if and only if both factors are 1; that is, if and only if country 2 maintains diplomatic relations with another country that maintains diplomatic relations with country 3. Thus b_{23} gives the number of countries that *link* country 2 and country 3. To see all of these entries, we compute

$$B = A^2 = \begin{bmatrix} 2 & 1 & 0 & 0 & 1 \\ 1 & 2 & 1 & 0 & 0 \\ 0 & 1 & 2 & 1 & 0 \\ 0 & 0 & 1 & 2 & 1 \\ 1 & 0 & 0 & 1 & 2 \end{bmatrix}.$$

Since $b_{23} = 1$, there is exactly one country that links countries 2 and 3. A careful examination of the entries of A reveals that $a_{25} = a_{53} = 1$, and hence it is country 5 that serves as the link. Other deductions are left for the exercises.

By looking at other powers of A, additional information may be obtained. For example, it can be shown that if A is an $n \times n$ adjacency matrix and the (i, j)-entry of $A + A^2 + \cdots + A^{n-1}$ is nonzero, then there is a sequence of countries beginning with country i and ending with country j, such that every pair of consecutive countries in the sequence maintains diplomatic relations. Using such a sequence, it is possible for countries i and j to communicate by passing a message only from one country to another with which it has diplomatic relations. Conversely, if the (i, j)-entry of $A + A^2 + \cdots + A^{n-1}$ is zero, then such communication between countries i and j is impossible.

Example 1   Consider a set of three countries such that country 3 maintains diplomatic relations with both countries 1 and 2, and countries 1 and 2 do not maintain diplomatic relations with each other. These relationships can be described by the 3×3 adjacency matrix

$$A = \begin{bmatrix} 0 & 0 & 1 \\ 0 & 0 & 1 \\ 1 & 1 & 0 \end{bmatrix}.$$

In this case, we have

$$A + A^2 = \begin{bmatrix} 1 & 1 & 1 \\ 1 & 1 & 1 \\ 1 & 1 & 2 \end{bmatrix},$$

and so countries 1 and 2 can communicate even though they do not have diplomatic relations. Here the sequence linking them consists of countries 1, 3, and 2.   ○

Adjacency matrices are useful in resolving problems involving scheduling. Suppose, for example, that the administration of a small college with m students wants to plan the times for its n courses. The goal of such planning is to avoid scheduling popular courses at the same time. To minimize the number of time conflicts, the students are surveyed. Each student is asked which courses he or she would like to take during the following semester. The results of this survey may be put in matrix form. Define the $m \times n$ matrix A as follows:

$$a_{ij} = \begin{cases} 1 & \text{if student } i \text{ wants to take course } j \\ 0 & \text{otherwise.} \end{cases}$$

In this case, the matrix product $A^T A$ provides important information regarding the scheduling of course times. We begin with an interpretation of the elements of this matrix. Let $B = A^T$ and $C = A^T A = BA$. Then, for example,

$$c_{12} = b_{11}a_{12} + b_{12}a_{22} + \cdots + b_{1k}a_{k2} + \cdots + b_{1m}a_{m2}$$

$$= a_{11}a_{12} + a_{21}a_{22} + \cdots + a_{k1}a_{k2} + \cdots + a_{m1}a_{m2}.$$

A typical term on the right-hand side of the equation has the form $a_{k1}a_{k2}$. Now, $a_{k1}a_{k2} = 1$ if and only if $a_{k1} = 1$ and $a_{k2} = 1$; that is, if student k wants to take course 1 and course 2. So c_{12} represents the number of students who want to take both courses 1 and 2. In general, for $i \neq j$, c_{ij} is the number of students who want to take both course i and course j. In addition, c_{ii} represents the number of students who desire class i.

Example 2   Suppose that we have a group of ten students and five courses. The results of the survey concerning course preferences are as follows.

Student	Course Number 1	2	3	4	5
1	1	0	1	0	1
2	0	0	0	1	1
3	1	0	0	0	0
4	0	1	1	0	1
5	0	0	0	0	0
6	1	1	0	0	0
7	0	0	1	0	1
8	0	1	0	1	0
9	1	0	1	0	1
10	0	0	0	1	0

Let A be the 10×5 matrix with entries from the table above. Then

$$A^T A = \begin{bmatrix} 4 & 1 & 2 & 0 & 2 \\ 1 & 3 & 1 & 1 & 1 \\ 2 & 1 & 4 & 0 & 4 \\ 0 & 1 & 0 & 3 & 1 \\ 2 & 1 & 4 & 1 & 5 \end{bmatrix}.$$

From this matrix, we see that there are four students who want both course 3 and course 5. All other pairs of courses are wanted by at most two students. Furthermore, we see that four students prefer course 1, three students desire course 2, and so on. Thus, the trace (see Exercise 46 of Section 1.1) of $A^T A$ equals the total enrollment of all the courses (counting students as often as the number of courses they wish to take) if the courses are offered at different times. ○

As we saw in Section 2.1, the matrix $A^T A$ is symmetric. Hence we may save computational effort by computing only one of the (i, j)- and (j, i)-entries.

As a final comment, it should be pointed out that many of these facts about adjacency matrices can be adapted to apply to nonsymmetric relationships.

An Application to Anthropology

In this application we will see a fascinating use of matrix operations in the study of marriage laws of the Natchez Indians.[1]

Everyone in this tribe was a member of one of four classes: the Suns, the Nobles, the Honoreds, and the Stinkards. There were well-defined rules that determine class membership. The rules depended exclusively on the classes of the parents, and required that at least one of the parents be a Stinkard. Furthermore, the class of the child depended on the class of the other parent, as illustrated in Table 2.1.

TABLE 2.1:

Mother is a Stinkard		Father is a Stinkard	
Father	Child	Mother	Child
Sun	Noble	Sun	Sun
Noble	Honored	Noble	Noble
Honored	Stinkard	Honored	Honored
Stinkard	Stinkard	Stinkard	Stinkard

We are interested in determining the long-range distributions of these classes; that is, what the relative sizes of these classes will be in future generations. It is clear that there will be a problem of survival if the class of Stinkards becomes too small. To simplify matters, we make three assumptions.

1. In every generation, each class is divided equally between males and females.

2. Each adult marries exactly once.

3. Each pair of parents has exactly one son and one daughter.

---

[1] This example is taken from Samuel Goldberg, *Introduction to Difference Equations (with Illustrative Examples from Economics, Psychology, and Sociology)* (New York: Dover Publications, 1986), pp. 238–241.

Because of assumption 1, we need only keep track of the number of males in each class for every generation. To do this, we introduce the following notation:

$$s_k = \text{number of male Suns in the } k\text{th generation}$$

$$n_k = \text{number of male Nobles in the } k\text{th generation}$$

$$h_k = \text{number of male Honoreds in the } k\text{th generation}$$

$$st_k = \text{number of male Stinkards in the } k\text{th generation.}$$

Our immediate goal is to find a relationship between the numbers of members in each class of the kth and the $(k-1)$st generations. Since every Sun male must have a Sun mother (and vice versa), we obtain the equation

$$s_k = s_{k-1}. \tag{3}$$

The fact that every Noble male must have a Sun father or a Noble mother yields

$$n_k = s_{k-1} + n_{k-1}. \tag{4}$$

In addition, every Honored male must have a Noble father or an Honored mother. Thus,

$$h_k = n_{k-1} + h_{k-1}. \tag{5}$$

Finally, assumption 3 guarantees that the total number of males (and females) remains the same for each generation:

$$s_k + n_k + h_k + st_k = s_{k-1} + n_{k-1} + h_{k-1} + st_{k-1}. \tag{6}$$

Substituting the right sides of (3), (4), and (5) into (6), we obtain

$$s_{k-1} + (s_{k-1} + n_{k-1}) + (n_{k-1} + h_{k-1}) + st_k = s_{k-1} + n_{k-1} + h_{k-1} + st_{k-1},$$

which simplifies to

$$st_k = -s_{k-1} - n_{k-1} + st_{k-1}. \tag{7}$$

Equations (3), (4), (5), and (7) relate the number of males in a class in the kth generation to the number of males in the same class in the $(k-1)$st generation. If we let

$$\mathbf{x}_k = \begin{bmatrix} s_k \\ n_k \\ h_k \\ st_k \end{bmatrix} \quad \text{and} \quad A = \begin{bmatrix} 1 & 0 & 0 & 0 \\ 1 & 1 & 0 & 0 \\ 0 & 1 & 1 & 0 \\ -1 & -1 & 0 & 1 \end{bmatrix},$$

then we may represent all our relationships by the matrix equation

$$\mathbf{x}_k = A\mathbf{x}_{k-1}.$$

Because this equation must hold for all k, we have

$$\mathbf{x}_k = A\mathbf{x}_{k-1} = AA\mathbf{x}_{k-2} = \cdots = A^k \mathbf{x}_0.$$

To evaluate A^k, let $B = A - I_4$. We leave it to the exercises to show that for any positive integer k,

$$A^k = I + kB + \frac{k(k-1)}{2} B^2, \quad \text{for } k \geq 2$$

(see Exercise 14). Thus, carrying out the matrix multiplication, we obtain

$$\mathbf{x}_k = \begin{bmatrix} s_0 \\ n_0 + ks_0 \\ h_0 + kn_0 + \frac{k(k-1)}{2}s_0 \\ st_0 - kn_0 - \frac{k(k+1)}{2}s_0 \end{bmatrix}. \tag{8}$$

It is easy to see from (8) that if there are initially no Suns or Nobles (i.e., $n_0 = s_0 = 0$), the number of members in each class will remain the same from generation to generation. On the other hand, consider the last entry of $\mathbf{x}_k$. We can conclude that unless $n_0 = s_0 = 0$, the number of Stinkards will decrease to the point where there are insufficiently many of them to allow the other members to marry. At this point the social order ceases to exist.

Practice Problems

1. The life span of a certain species of mammal is at most 2 years, but only 25% of the females of this species survive to age 1. Suppose that, on average, the females under 1 year of age give birth to 0.5 females, and the females between 1 and 2 years of age give birth to 2 females.

   (a) Write the Leslie matrix for the population.
   (b) Suppose that this year there is a population of 200 females of age 1 and 200 females between 1 and 2 years of age. Find the population distribution for the next year and for 2 years from now.
   (c) Suppose this year's population distribution of females is given by the vector $\begin{bmatrix} 400 \\ 100 \end{bmatrix}$. What can you say about all future population distributions?
   (d) Suppose that the total population of females this year is 600. What should be the number of females in each age group so that the population distribution remains unchanged from year to year?

2. A supermarket chain in the Midwest imports soy sauce from Japan and South Korea. Of the soy sauce from Japan, 50% is shipped to Seattle and the rest is shipped to San Francisco. Of the soy sauce from South Korea, 60% is shipped to San Francisco and the rest is shipped to Los Angeles. All of the soy sauce shipped to Seattle is sent to Chicago, 30% of the soy sauce shipped to San Francisco is sent to Chicago and 70% to St. Louis, and all of the soy sauce shipped to Los Angeles is sent to St. Louis. Suppose that soy sauce is shipped from Japan and South Korea at the rates of 10,000 and 5,000 barrels a year, respectively. Find the number of barrels of soy sauce that are sent to Chicago and to St. Louis each year.

3. Suppose that we have four cities with airports. We define the 4×4 matrix A by

$$a_{ij} = \begin{cases} 1 & \text{if there is a nonstop commercial flight} \\ & \quad i \text{ from city to city } j \\ 0 & \text{if not.} \end{cases}$$

   (a) Prove that A is an adjacency matrix.
   (b) Interpret the $(2, 3)$-entry of A^2.
   (c) Suppose that

$$A = \begin{bmatrix} 0 & 1 & 1 & 0 \\ 1 & 0 & 0 & 1 \\ 1 & 0 & 0 & 0 \\ 0 & 1 & 0 & 0 \end{bmatrix}.$$

   Compute A^2.
   (d) How many flights are there with one layover (intermediate stop) from city 2 to city 3?
   (e) How many flights are there with two layovers from city 1 to city 2?
   (f) Is it possible to fly between each pair of cities?

Exercises

1. Determine if the following statements are true or false.

   (a) If A is a Leslie matrix and $\mathbf{v}$ is a population distribution, then each entry of $A\mathbf{v}$ must be greater than the corresponding entry of $\mathbf{v}$.
   (b) For any population distribution $\mathbf{v}$, if A is the Leslie matrix for the population, then as n grows, $A^n\mathbf{v}$ approaches a fixed population distribution.
   (c) In a Leslie matrix, the (i, j)-entry equals the average number of female offspring of a member of the ith age group.
   (d) In a Leslie matrix, the $(i + 1, i)$-entry equals the proportion of females in the ith age group who survive to the next age group.

(e) The application in this section on traffic flow relies on the associative law of matrix multiplication.

(f) If A and B are matrices and $\mathbf{x}$, $\mathbf{y}$, and $\mathbf{z}$ are vectors such that $\mathbf{y} = A\mathbf{x}$ and $\mathbf{z} = B\mathbf{y}$, then $\mathbf{z} = (AB)\mathbf{x}$.

(g) An adjacency matrix is a matrix with zeros and ones as its only entries.

(h) Every adjacency matrix is a symmetric matrix.

Exercises 2–8 deal with Leslie matrices.

2. By observing a certain colony of mice, it was found that all animals die within 3 years. Of those offspring that are females, 60% live for at least 1 year. Of these, 20% reach their second birthday. The females who are under 1 year of age have an average of three female offspring. Those females between 1 and 2 years of age have an average of two female offspring while they are in this age group. None of the females of age 2 give birth.

   (a) Construct the Leslie matrix that describes this situation.

   (b) Suppose that the current population distribution for females is given by the vector $\begin{bmatrix} 100 \\ 60 \\ 30 \end{bmatrix}$. Find the population distribution for the next year. Also, find the population distribution four years from now.

   (c) Show that there is no nonzero stable population distribution for the colony of mice. *Hint:* Let A be the Leslie matrix and suppose that $\mathbf{z}$ is a stable population distribution. Then $A\mathbf{z} = \mathbf{z}$. This is equivalent to $(A - I_3)\mathbf{z} = \mathbf{0}$. Solve this homogeneous system of linear equations.

3. Suppose that the females of a certain colony of animals are divided into two age groups, and suppose that the Leslie matrix for this population is
$$\begin{bmatrix} 0 & 1 \\ 1 & 0 \end{bmatrix}.$$

   (a) What proportion of the females of the first age group survive to the second age group?

   (b) How many female offspring do females of each age group average?

   (c) If $\mathbf{x} = \begin{bmatrix} a \\ b \end{bmatrix}$ is the current population distribution for the females of the colony, describe all future population distributions.

In Exercises 4–7, Use a calculator with matrix capabilities or computer software such as MATLAB.

4. A certain colony of lizards has a life span of less than 3 years. Suppose that the females are divided into three age groups: those of age less than 1, those of age 1, and those of age 2. Suppose further that the probability that a newborn female survives until age 1 is .5, and that the probability that a female of age 1 survives until age 2 is q. Assume also that females under age 1 do not give birth, those of age 1 have an average of 1.2 female offspring, and those of age 2 have an average of 1 female offspring.

Suppose there are initially 450 females of age less than 1,220 of age 1, and 70 of age 2.

   (a) Write a Leslie matrix A for this colony of lizards.

   (b) If $q = .3$, what will happen to the population in 50 years?

   (c) If $q = .9$, what will happen to the population in 50 years?

   (d) Find by trial and error a value of q for which the lizard population reaches a nonzero stable distribution. What is this stable distribution?

   (e) For the value of q found in (d), what happens to an initial population of 200 females of age less than 1, 360 of age 1, and 280 of age 2?

   (f) For what value of q does $(A - I_3)\mathbf{x} = \mathbf{0}$ have a nonzero solution?

   (g) For the value of q found in (f), find the general solution to $(A - I_3)\mathbf{x} = \mathbf{0}$. How does this relate to the stable distributions in (d) and (e)?

5. A certain colony of bats has a life span of less than 3 years. Suppose that the females are divided into three age groups: those of age less than 1, those of age 1, and those of age 2. Suppose further that the probability that a newborn female survives until age 1 is q and that the probability that a female of age 1 survives until age 2 is .5. Assume also that females under age 1 do not give birth, those of age 1 have an average of 2 female offspring, and those of age 2 have an average of 1 female offspring. Suppose there are initially 300 females of age less than 1,180 of age 1, and 130 of age 2.

   (a) Write a Leslie matrix A for this colony of bats.

   (b) If $q = .8$, what will happen to the population in 50 years?

   (c) If $q = .2$, what will happen to the population in 50 years?

   (d) Find by trial and error a value of q for which the bat population reaches a nonzero stable distribution. What is this stable distribution?

   (e) For the value of q found in (d), what happens to an initial population of 210 females of age less than 1,240 of age 1, and 180 of age 2?

   (f) For what value of q does $(A - I_3)\mathbf{x} = \mathbf{0}$ have a nonzero solution?

   (g) For the value of q found in (f), find the general solution to $(A - I_3)\mathbf{x} = \mathbf{0}$. How does this relate to the stable distributions in (d) and (e)?

6. A certain colony of voles has a life span of less than 3 years. Suppose that the females are divided into three age groups: those of age less than 1, those of age 1, and those of age 2. Suppose further that the probability that a newborn female survives until age 1 is .1 and that the probability that a female of age 1 survives until age 2 is .2. Assume also that females under age 1 do not give birth, those of age 1 have an average of b female offspring, and those of age 2 have an average of 10 female offspring. Suppose there are initially 150 females of age less than 1,300 of age 1, and 180 of age 2.

(a) Write a Leslie matrix A for this colony of voles.
(b) If $b = 10$, what will happen to the population in 50 years?
(c) If $b = 4$, what will happen to the population in 50 years?
(d) Find by trial and error a value of b for which the vole population reaches a nonzero stable distribution. What is this stable distribution?
(e) For the value of b found in (d), what happens to an initial population of 80 females of age less than 1,200 of age 1, and 397 of age 2?
(f) For what value of b does $(A - I_3)\mathbf{x} = \mathbf{0}$ have a nonzero solution?
(g) Let $\mathbf{p} = \begin{bmatrix} p_1 \\ p_2 \\ p_3 \end{bmatrix}$ be an arbitrary population vector and

   b have the value found in (f). Over time, $\mathbf{p}$ approaches a stable population vector $\mathbf{q}$. Express $\mathbf{q}$ in terms of p_1, p_2, and p_3.

7. A certain colony of squirrels has a life span of less than 3 years. Suppose that the females are divided into three age groups: those of age less than 1, those of age 1, and those of age 2. Suppose further that the probability that a newborn female survives until age 1 is .2 and that the probability that a female of age 1 survives until age 2 is .5. Assume also that females under age 1 do not give birth, those of age 1 have an average of 2 female offspring, and those of age 2 have an average of b female offspring. Suppose there are initially 240 females of age less than 1,400 of age 1, and 320 of age 2.

(a) Write a Leslie matrix A for this colony of squirrels.
(b) If $b = 3$, what will happen to the population in 50 years?
(c) If $b = 9$, what will happen to the population in 50 years?
(d) Find by trial and error a value of b for which the squirrel population reaches a nonzero stable distribution. What is this stable distribution?
(e) For the value of b found in (d), what happens to an initial population of 100 females of age less than 1, 280 of age 1, and 400 of age 2?
(f) For what value of b does $(A - I_3)\mathbf{x} = \mathbf{0}$ have a nonzero solution?
(g) Let $\mathbf{p} = \begin{bmatrix} p_1 \\ p_2 \\ p_3 \end{bmatrix}$ be an arbitrary population vector and

   b have the value found in (f). Over time, $\mathbf{p}$ approaches a stable population vector $\mathbf{q}$. Express $\mathbf{q}$ in terms of p_1, p_2, and p_3.

8. The maximum membership term for each member of the Service Club is 3 years. Each first-year and second-year member recruits one new person who begins the membership term in the following year. Of those in their first year of membership, 50% of the members resign, and of those in their second year of membership, 70% resign.

(a) Write a 3×3 matrix A so that if x_i is currently the number of Service Club members in their ith year of membership and y_i is the number in their ith year of membership a year from now, then $\mathbf{y} = A\mathbf{x}$.
(b) Suppose that there are 60 Service Club members in their first year, 20 members in their second year, and 40 members in their third year of membership. Find the distribution of members for next year and for two years from now.

Exercises 9 and 10 use the technique developed in the traffic flow application.

9. A certain medical foundation receives money from two sources: donations and interest earned on endowments. Of the donations received, 30% is used to defray the costs of raising funds; only 10% of the interest is used to defray the cost of managing the endowment funds. Of the rest of the money (the net income), 40% is used for research and 60% is used to maintain medical clinics. Of the three expenses (research, clinics, and fund raising), the portions going to materials and personnel are divided according to Table 2.2. Find a matrix M such that if p is the value of donations and q is the value of interest, then $M \begin{bmatrix} p \\ q \end{bmatrix} = \begin{bmatrix} m \\ f \end{bmatrix}$, where m and f are the material and personnel costs of the foundation, respectively.

Table 2.2

	Research	Clinics	Fund Raising
Material costs	80%	50%	70%
Personnel costs	20%	50%	30%

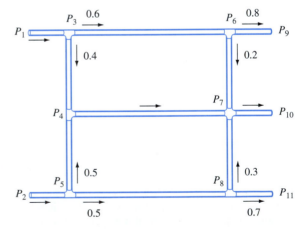

FIGURE 2.4

10. Water is pumped into a system of pipes at points P_1 and P_2, as shown in Figure 2.4. At each of the junctions P_3, P_4, P_5, P_6, P_7, and P_8, the pipes are split and water flows according to the proportions indicated in the diagram. Suppose that water flows into P_1 and P_2 at p and q

gallons per minute, respectively, and flows out of P_9, P_{10}, and P_{11} at the rates of a, b, and c gallons per minute, respectively. Find matrices A and B such that the vector of outputs is given by

$$\begin{bmatrix} a \\ b \\ c \end{bmatrix} = BA \begin{bmatrix} p \\ q \end{bmatrix}.$$

Exercises 11–15 deal with adjacency matrices.

11. Using the interpretation of an adjacency matrix in problem 3 of the Practice Problems, suppose that we have five cities with the associated matrix A given in the block form (see Section 2.1)

$$A = \begin{bmatrix} B & O_1 \\ O_2 & C \end{bmatrix},$$

where B is a 3×3 matrix, O_1 is the 3×2 zero matrix, O_2 is the 2×3 zero matrix, and C is a 2×2 matrix.

(a) What does the matrix A tell us about flight connections between the cities?
(b) Use block multiplication to obtain A^2, A^3, and A^k for any positive integer k.
(c) Interpret your result in (b) in terms of flights between the cities.

12. Recall the adjacency matrix

$$A = \begin{bmatrix} 0 & 0 & 1 & 1 & 0 \\ 0 & 0 & 0 & 1 & 1 \\ 1 & 0 & 0 & 0 & 1 \\ 1 & 1 & 0 & 0 & 0 \\ 0 & 1 & 1 & 0 & 0 \end{bmatrix}$$

in which the entries describe countries that maintain diplomatic relations with one another.

(a) Which pairs of countries maintain diplomatic relations?
(b) How many countries link country 1 with country 3?
(c) Give an interpretation of the $(1, 4)$-entry of A^3.

13. Suppose that there is a group of four people and an associated 4×4 adjacency matrix A, defined by

$$a_{ij} = \begin{cases} 1 & \text{if } i \neq j \text{ and person } i \text{ likes person } j \\ 0 & \text{otherwise.} \end{cases}$$

We say that persons i and j are *friends* if they like each other; that is, if $a_{ij} = a_{ji} = 1$. Suppose that A is given by

$$\begin{bmatrix} 0 & 1 & 0 & 1 \\ 1 & 0 & 1 & 0 \\ 0 & 1 & 0 & 1 \\ 1 & 1 & 1 & 0 \end{bmatrix}.$$

(a) List all pairs of friends.
(b) Give an interpretation of the entries of A^2.
(c) Let B be the 4×4 matrix defined by

$$b_{ij} = \begin{cases} 1 & \text{if persons } i \text{ and } j \text{ are friends} \\ 0 & \text{otherwise.} \end{cases}$$

Determine the matrix B. Is B a symmetric matrix?

(d) A *clique* is a set of three or more people, each of whom is friendly with all the other members of the set. Show that person i belongs to a clique if and only if the (i, i)-entry of B^3 is positive.
(e) Use computer software or a calculator that performs matrix arithmetic to count the cliques that exist among the four friends.

Exercises 14 and 15 refer to the scheduling example in this section.

14. Suppose that student preferences for a set of courses are given in the following table.

Student	Course Number 1	2	3	4	5
1	1	0	0	0	1
2	0	0	1	1	0
3	1	0	1	0	1
4	0	1	0	0	0
5	1	0	0	0	1
6	0	1	0	0	1
7	1	0	1	0	1
8	0	1	1	0	1
9	1	0	0	1	1
10	0	0	1	1	0

(a) Give the pair(s) of courses that are desired by the most students.
(b) Give the pair(s) of courses that are desired by the least students.
(c) Construct a matrix whose diagonal entries determine the number of students who prefer each course.

15. Let A be defined as in the scheduling example in the text.

(a) Justify the following interpretation: For $i \neq j$, the (i, j)-entry of AA^T is the number of classes that are desired by both students i and j.
(b) Show that the $(1, 2)$-entry of AA^T is 1 and the $(9, 1)$-entry of AA^T is 3.
(c) Interpret the answers to (b) in the context of (a) and the data in the scheduling example.
(d) Give an interpretation of the diagonal entries of AA^T.

Exercises 16 and 17 deal with the anthropology application in this section.

16. Recall the *binomial formula* for scalars a and b and any positive integer k:

$$(a+b)^k = a^k + ka^{k-1}b + \cdots + \frac{k!}{i!(k-i)!}a^{k-i}b^i + \cdots + b^k,$$

where $i!$ (i factorial) is given by $i! = 1 \cdot 2 \cdots (i-1) \cdot i$

(a) Suppose that A and B are $m \times m$ matrices that commute; that is, $AB = BA$. Prove that the binomial formula holds for A and B when $k = 2$ and $k = 3$. That is, prove that

$$(A + B)^2 = A^2 + 2AB + B^2$$

and
$$(A + B)^3 = A^3 + 3A^2B + 3AB^2 + B^3.$$

(b) Use mathematical induction to extend the results in part (a) to any positive integer k.

(c) For the matrices A and B at the end of this section, show that $B^3 = O$, and then use (b) to prove
$$A^k = I_4 + kB + \frac{k(k-1)}{2}B^2 \quad \text{for } k \geq 2.$$

In Exercises 17 and 18, use a calculator with matrix capabilities or computer software such as MATLAB to solve the problem.

17. In reference to the application in the text involving the Natchez Indians, suppose that initially there are 100 Sun males, 200 Noble males, 300 Honored males, and 8000 Stinkard males.

(a) How many males will there be in each class in $k = 1$, 2, and 3 generations?

(b) Use a computer to determine how many males there will be in each class in $k = 9$, 10, and 11 generations.

(c) What do your answers to (b) suggest for the future of the Natchez Indians if they hold to their current marriage laws?

(d) Produce an algebraic proof that for some k there will be insufficiently many Stinkards to allow the other members to marry.

18. Suppose that we have a group of six people, each of whom owns a communication device. We define a 6×6 matrix A as follows. For $1 \leq i \leq 6$ we have $a_{ii} = 0$, and for $i \neq j$ we have
$$a_{ij} = \begin{cases} 1 & \text{if person } i \text{ can send a message to person } j \\ 0 & \text{otherwise.} \end{cases}$$

(a) Show that A is an adjacency matrix.

(b) Give an interpretation of what it means for the term $a_{32}a_{21}$ to equal 1.

(c) Show that the (3, 1)-entry of A^2 represents the number of ways that person 3 can send a message to person 1 in two *stages*; that is, the number of people to whom person 3 can send a message and who in turn can send a message to person 1. *Hint:* Consider the number of terms in the expression $a_{31}a_{11} + a_{32}a_{21} + \cdots + a_{36}a_{61}$ that are not equal to zero.

(d) Generalize your result in (c) to the (i, j)-entry of A^2.

(e) Generalize your result in (d) to the (i, j)-entry of A^m.

Now suppose that

$$A = \begin{bmatrix} 0 & 0 & 0 & 1 & 0 & 1 \\ 1 & 0 & 1 & 1 & 0 & 0 \\ 0 & 1 & 0 & 1 & 0 & 0 \\ 1 & 0 & 1 & 0 & 0 & 0 \\ 1 & 1 & 1 & 0 & 0 & 1 \\ 0 & 0 & 1 & 1 & 0 & 0 \end{bmatrix}.$$

(f) Is there any person who cannot receive a message from anyone else in one stage? Justify your answer.

(g) How many ways can person 1 send a message to person 4 in 1, 2, 3, and 4 stages?

(h) The (i, j)-entry of $A + A^2 + \cdots + A^m$ can be shown to be the number of ways in which person i can send a message to person j in at most m stages. Use this result to determine the number of ways in which person 3 can send a message to person 4 in at most 4 stages.

Solutions to the Practice Problems

1. (a) $A = \begin{bmatrix} 0.50 & 2 \\ 0.25 & 0 \end{bmatrix}$.

(b) The population distribution of females for this year is given by the vector $\begin{bmatrix} 200 \\ 200 \end{bmatrix}$, and hence the population distribution of females for next year is
$$A \begin{bmatrix} 200 \\ 200 \end{bmatrix} = \begin{bmatrix} 0.50 & 2 \\ 0.25 & 0 \end{bmatrix} \begin{bmatrix} 200 \\ 200 \end{bmatrix} = \begin{bmatrix} 500 \\ 50 \end{bmatrix},$$
and the population distribution of females two years from now is
$$A \begin{bmatrix} 500 \\ 50 \end{bmatrix} = \begin{bmatrix} 0.50 & 2 \\ 0.25 & 0 \end{bmatrix} \begin{bmatrix} 500 \\ 50 \end{bmatrix} = \begin{bmatrix} 350 \\ 125 \end{bmatrix}.$$

(c) Since
$$A \begin{bmatrix} 400 \\ 100 \end{bmatrix} = \begin{bmatrix} 0.50 & 2 \\ 0.25 & 0 \end{bmatrix} \begin{bmatrix} 400 \\ 100 \end{bmatrix} = \begin{bmatrix} 400 \\ 100 \end{bmatrix},$$
the population distribution does not change from year to year.

(d) Let x_1 and x_2 be the number of females in the first and second age groups, respectively. Since next year's population distribution is the same as this year's, $Ax = x$, and hence
$$\begin{bmatrix} 0.50 & 2 \\ 0.25 & 0 \end{bmatrix} \begin{bmatrix} x_1 \\ x_2 \end{bmatrix} = \begin{bmatrix} 0.50x_1 + 2x_2 \\ 0.25x_1 \end{bmatrix} = \begin{bmatrix} x_1 \\ x_2 \end{bmatrix}.$$
Thus $x_1 = 4x_2$. But $x_1 + x_2 = 600$, and so $4x_2 + x_2 = 5x_2 = 600$, from which it follows that $x_2 = 120$. Finally, $x_1 = 4x_2 = 480$.

2. Let z_1 and z_2 be the rates at which soy sauce is shipped to Chicago and St. Louis, respectively. Then
$$\begin{bmatrix} z_1 \\ z_2 \end{bmatrix} = \begin{bmatrix} 1 & 0.3 & 0 \\ 0 & 0.7 & 1 \end{bmatrix} \begin{bmatrix} 0.5 & 0.0 \\ 0.5 & 0.6 \\ 0 & 0.4 \end{bmatrix} \begin{bmatrix} 10000 \\ 5000 \end{bmatrix}$$
$$= \begin{bmatrix} 7400 \\ 7600 \end{bmatrix}.$$

3. (a) Clearly every entry of A is either 0 or 1, so A is an adjacency matrix.
   (b) The (2, 3)-entry of A^2 is $a_{21}a_{13} + a_{22}a_{23} + a_{23}a_{33} + a_{24}a_{43}$. A typical term has the form $a_{2k}a_{k3}$, which equals 1 or 0. This term equals 1 if and only if $a_{2k} = 1$ and $a_{k3} = 1$. Consequently, this term equals 1 if and only if there is a nonstop flight between city 2 and city k as well as a nonstop flight between city k and city 3. That is, $a_{2k}a_{k3} = 1$ means that there is a flight with one layover (in city k) from city 2 to city 3. Therefore we may interpret the (2, 3)-entry of A^2 as the number of flights with one layover from city 2 to city 3.
   (c) $A^2 = \begin{bmatrix} 2 & 0 & 0 & 1 \\ 0 & 2 & 1 & 0 \\ 0 & 1 & 1 & 0 \\ 1 & 0 & 0 & 1 \end{bmatrix}.$
   (d) Because the (2, 3)-entry of A^2 is 1, there is one flight

with one layover from city 2 to city 3.
   (e) We compute A^3 to find the number of flights with two layovers from city 1 to city 2. We have
   $$A^3 = \begin{bmatrix} 0 & 3 & 2 & 0 \\ 3 & 0 & 0 & 2 \\ 2 & 0 & 0 & 1 \\ 0 & 2 & 1 & 0 \end{bmatrix}.$$
   Because the (1, 2)-entry of A^3 is 3, we see that there are three flights with two layovers from city 1 to city 2.
   (f) From the entries of A we see that there are nonstop flights between cities 1 and 2, cities 1 and 3, and cities 2 and 4. From A^2 we see that there are flights between cities 1 and 4 as well as cities 2 and 3. Finally, from A^3 we discover that there is a flight between cities 3 and 4. We conclude that there are flights between all pairs of cities.

2.3　Invertibility and Elementary Matrices

In this section, we introduce the concept of invertibility in the context of matrix multiplication, and examine special invertible matrices that are intimately associated with elementary row operations, the *elementary matrices*. We complete the section by applying what we have learned about invertible matrices and elementary matrices to prove that the reduced row echelon form of a matrix is unique.

For any real number $a \neq 0$ there is a unique real number b, called the (*multiplicative*) *inverse* of a, with the property that $ab = ba = 1$. For example, if $a = 2$, then $b = 0.5$. It is useful to consider the analogous situation in the context of matrices. Since the identity matrix I_n is analogous to the real number 1, we are interested in a matrix A for which there is a matrix B such that $AB = BA = I_n$. Notice that this last equation is possible only if both A and B are $n \times n$ matrices. This consideration motivates the following definition.

Definition. An $n \times n$ matrix A is called **invertible** if there is an $n \times n$ matrix B such that $AB = BA = I_n$. In this case B is called an **inverse** of A.

Because the roles of A and B are the same in this definition, it follows that if B is an inverse of A, then A is also an inverse of B.

Example 1　Let $A = \begin{bmatrix} 1 & 2 \\ 3 & 5 \end{bmatrix}$ and $B = \begin{bmatrix} -5 & 2 \\ 3 & -1 \end{bmatrix}$. Then

$$AB = \begin{bmatrix} 1 & 2 \\ 3 & 5 \end{bmatrix}\begin{bmatrix} -5 & 2 \\ 3 & -1 \end{bmatrix} = \begin{bmatrix} 1 & 0 \\ 0 & 1 \end{bmatrix} = I_2,$$

and

$$BA = \begin{bmatrix} -5 & 2 \\ 3 & -1 \end{bmatrix}\begin{bmatrix} 1 & 2 \\ 3 & 5 \end{bmatrix} = \begin{bmatrix} 1 & 0 \\ 0 & 1 \end{bmatrix} = I_2.$$

So A is invertible, and B is an inverse of A. Similarly, B is invertible, and A is an inverse of B.   ○

Just as the real number 0 has no multiplicative inverse, the $n \times n$ zero matrix O has no inverse because $OB = O \neq I_n$ for any $n \times n$ matrix B. But there are also other square matrices that are not invertible, for example, $A = \begin{bmatrix} 1 & 2 \\ 0 & 0 \end{bmatrix}$. For if $B = \begin{bmatrix} a & b \\ c & d \end{bmatrix}$ is any 2×2 matrix, then

$$AB = \begin{bmatrix} 1 & 2 \\ 0 & 0 \end{bmatrix} \begin{bmatrix} a & b \\ c & d \end{bmatrix} = \begin{bmatrix} a + 2c & b + 2d \\ 0 & 0 \end{bmatrix} \neq \begin{bmatrix} 1 & 0 \\ 0 & 1 \end{bmatrix}.$$

So A is not invertible.

Two questions naturally arise:

1. Which matrices are invertible?

2. If the matrix A is invertible, how can we find an inverse of A?

We postpone the answers to these questions until Section 2.4, where we delve more deeply into the subject of invertibility. Our task here is to establish some of the elementary properties of invertible matrices.

One elementary property of an invertible matrix is that it has a unique inverse. For suppose that both B and C are inverses of an $n \times n$ invertible matrix A. Then $AB = BA = I_n$ and $AC = CA = I_n$. So

$$B = BI_n = B(AC) = (BA)C = I_nC = C.$$

Hence we have the following result.

> If A is an invertible matrix, then A has exactly one inverse.

Because the inverse of an invertible matrix A is unique, we can denote it by the symbol A^{-1}. So $AA^{-1} = A^{-1}A = I_n$. Notice the similarity between this statement and the statement $2 \cdot 2^{-1} = 2^{-1} \cdot 2 = 1$, where 2^{-1} is the ordinary multiplicative inverse of the real number 2. Thus, in Example 1, we see that $A^{-1} = B$ and $B^{-1} = A$.

The inverse of a real number can be used to solve certain equations. For example, the equation $2x = 14$ can be solved by multiplying both sides of the equation by the inverse of 2:

$$2^{-1}(2x) = 2^{-1}(14)$$

$$(2^{-1}2)x = 7$$

$$1x = 7$$

$$x = 7.$$

In a similar manner, if A is an invertible $n \times n$ matrix, then A^{-1} can be used to solve matrix equations in which an unknown matrix is multiplied by A. For example, if A is invertible, then the matrix equation $A\mathbf{x} = \mathbf{b}$ can be solved as follows:[2]

$$A\mathbf{x} = \mathbf{b}$$
$$A^{-1}(A\mathbf{x}) = A^{-1}\mathbf{b}$$
$$(A^{-1}A)\mathbf{x} = A^{-1}\mathbf{b}$$
$$I_n\mathbf{x} = A^{-1}\mathbf{b}$$
$$\mathbf{x} = A^{-1}\mathbf{b}$$

Example 2   Use matrix inverses to solve the system of linear equations

$$x_1 + 2x_2 = 4$$
$$3x_1 + 5x_2 = 7.$$

Solution   The system above is the same as the matrix equation $A\mathbf{x} = \mathbf{b}$, where

$$A = \begin{bmatrix} 1 & 2 \\ 3 & 5 \end{bmatrix}, \qquad \mathbf{x} = \begin{bmatrix} x_1 \\ x_2 \end{bmatrix}, \qquad \text{and} \qquad \mathbf{b} = \begin{bmatrix} 4 \\ 7 \end{bmatrix}.$$

We saw in Example 1 that A is invertible. Hence we can solve this equation for $\mathbf{x}$ by multiplying both sides of the equation on the left by A^{-1}. Since

$$A^{-1} = \begin{bmatrix} -5 & 2 \\ 3 & -1 \end{bmatrix}$$

from Example 1, we have

$$\begin{bmatrix} x_1 \\ x_2 \end{bmatrix} = \mathbf{x} = A^{-1}\mathbf{b} = \begin{bmatrix} -5 & 2 \\ 3 & -1 \end{bmatrix}\begin{bmatrix} 4 \\ 7 \end{bmatrix} = \begin{bmatrix} -6 \\ 5 \end{bmatrix}.$$

Therefore $x_1 = -6$ and $x_2 = 5$ is the unique solution to the system.   ○

Example 3   Recall the rotation matrix

$$A_\theta = \begin{bmatrix} \cos\theta & -\sin\theta \\ \sin\theta & \cos\theta \end{bmatrix}$$

considered in Section 1.2, and Example 2 of Section 2.1. Notice that for $\theta = 0°$, $A_\theta = I_2$. Furthermore, for any angle α,

$$A_\alpha A_{-\alpha} = A_{\alpha+(-\alpha)} = A_{0°} = I_2.$$

Similarly, $A_{-\alpha} A_\alpha = I_2$. Hence A_α satisfies the definition of an invertible matrix, with inverse $A_{-\alpha}$. Therefore, $(A_\alpha)^{-1} = A_{-\alpha}$.   ○

The following theorem states some useful properties of matrix inverses.

Theorem 2.3
Let A and B be $n \times n$ matrices.
   (a) *If A is invertible, then A^{-1} is invertible and $(A^{-1})^{-1} = A$.*

---

[2]Although matrix inverses can be used to solve systems of linear equations whose coefficient matrices are invertible, the method of solving systems learned in Chapter 1 is far more efficient.

(b) If A and B are invertible, then AB is invertible and $(AB)^{-1} = B^{-1}A^{-1}$.

(c) If A is invertible,

Proof The proof of (a), which is omitted, is a simple consequence of the definition of a matrix inverse (see Exercise 20).

(b) Suppose that A and B are invertible. Then

$$(AB)(B^{-1}A^{-1}) = A(BB^{-1})A^{-1} = AI_nA^{-1} = AA^{-1} = I_n.$$

Similarly, $(B^{-1}A^{-1})(AB) = I_n$. Hence AB satisfies the definition of an invertible matrix with inverse $B^{-1}A^{-1}$; that is, $(AB)^{-1} = B^{-1}A^{-1}$.

(c) Suppose that A is invertible. Then $A^{-1}A = I_n$. Using Theorem 2.2(d), we obtain

$$A^T(A^{-1})^T = (A^{-1}A)^T = I_n^T = I_n.$$

Similarly, $(A^{-1})^T A^T = I_n$. Hence A^T satisfies the definition of an invertible matrix with the unique inverse $(A^{-1})^T$; that is, $(A^T)^{-1} = (A^{-1})^T$. $\square$

Part (b) of Theorem 2.3 can be easily extended to products of more than two matrices.

Let $A_1, A_2, \ldots, A_k$ be $n \times n$ invertible matrices. Then the product $A_1 A_2 \cdots A_k$ is invertible, and

$$(A_1 A_2 \cdots A_k)^{-1} = (A_k)^{-1}(A_{k-1})^{-1} \cdots (A_1)^{-1}.$$

Elementary Matrices

It is interesting that every elementary row operation can be performed by matrix multiplication. For example, multiplying row 2 of the matrix

$$A = \begin{bmatrix} a & b \\ c & d \end{bmatrix}$$

by the scalar k can be accomplished by computing the matrix product

$$\begin{bmatrix} 1 & 0 \\ 0 & k \end{bmatrix} \begin{bmatrix} a & b \\ c & d \end{bmatrix} = \begin{bmatrix} a & b \\ kc & kd \end{bmatrix}.$$

Also, rows 1 and 2 of A can be interchanged by the product

$$\begin{bmatrix} 0 & 1 \\ 1 & 0 \end{bmatrix} \begin{bmatrix} a & b \\ c & d \end{bmatrix} = \begin{bmatrix} c & d \\ a & b \end{bmatrix},$$

and adding k times row 1 of A to row 2 can be accomplished by computing

$$\begin{bmatrix} 1 & 0 \\ k & 1 \end{bmatrix} \begin{bmatrix} a & b \\ c & d \end{bmatrix} = \begin{bmatrix} a & b \\ ka+c & kb+d \end{bmatrix}.$$

The matrices

$$\begin{bmatrix} 1 & 0 \\ 0 & k \end{bmatrix}, \quad \begin{bmatrix} 0 & 1 \\ 1 & 0 \end{bmatrix}, \quad \text{and} \quad \begin{bmatrix} 1 & 0 \\ k & 1 \end{bmatrix}$$

are examples of *elementary matrices*. In general, an $n \times n$ matrix E is called an **elementary matrix** if E can be obtained from I_n by a single elementary row operation.

The matrix

$$E = \begin{bmatrix} 1 & 0 & 0 \\ 0 & 1 & 0 \\ 2 & 0 & 1 \end{bmatrix}$$

is an elementary matrix because E can be obtained from I_3 by the elementary row operation of adding two times the first row of I_3 to the third row of I_3. Note that if

$$A = \begin{bmatrix} 1 & 2 \\ 3 & 4 \\ 5 & 6 \end{bmatrix},$$

then

$$EA = \begin{bmatrix} 1 & 0 & 0 \\ 0 & 1 & 0 \\ 2 & 0 & 1 \end{bmatrix} \begin{bmatrix} 1 & 2 \\ 3 & 4 \\ 5 & 6 \end{bmatrix} = \begin{bmatrix} 1 & 2 \\ 3 & 4 \\ 7 & 10 \end{bmatrix}.$$

Hence EA can be obtained from A by adding two times the first row of A to the third row. This is the same elementary row operation that was applied to I_3 to produce E. A similar result holds for each of the three elementary row operations.

> Let A be an $m \times n$ matrix, and let E be an $m \times m$ elementary matrix obtained by performing an elementary row operation on I_m. Then the product EA can be obtained from A by performing the identical elementary row operation on A.

As noted in Section 1.3, any elementary row operation can be reversed. For example, a matrix obtained by adding 2 times the first row to the third row of A can be changed back into A by adding -2 times the first row to the third row of the new matrix. This concept of a *reverse* operation gives us a way of obtaining the inverse of an elementary matrix. To see how this is done, consider the matrix E above. Let

$$F = \begin{bmatrix} 1 & 0 & 0 \\ 0 & 1 & 0 \\ -2 & 0 & 1 \end{bmatrix}$$

be the elementary matrix obtained from I_3 by adding -2 times the first row of I_3 to the third row of I_3. If this elementary row operation is applied to E, then the result is I_3, and therefore $FE = I_3$. A similar argument can be used to show that $EF = I_3$. Hence E is invertible and $E^{-1} = F$. This method can be applied to any elementary matrix.

> Every elementary matrix is invertible. Furthermore, the inverse of an elementary matrix is also an elementary matrix.

Let A be an $m \times n$ matrix with reduced row echelon from R. Since A can be transformed into R by means of elementary row operations, this transformation can be executed by multiplying A on the left by a sequence of elementary matrices, one for each row operation. Thus there exist elementary matrices $E_1, E_2, \ldots, E_k$ such that

$$E_k E_{k-1} \cdots E_1 A = R.$$

Let $P = E_k E_{k-1} \cdots E_1$. Then P is a product of elementary matrices, and so P is an invertible matrix. Furthermore, $PA = R$. Thus we have established the following result which, although not used for practical computations, is important for its theoretical implications.

Theorem 2.4

Let A be an $m \times n$ matrix with reduced row echelon form R. Then there exists an invertible $m \times m$ matrix P such that $PA = R$.

We can immediately apply this result. In Section 1.4 we claimed that the general solution to the matrix equation $A\mathbf{x} = \mathbf{b}$ can be found by solving the equation $R\mathbf{x} = \mathbf{c}$, where $[R \quad \mathbf{c}]$ is the reduced row echelon form of the augmented matrix $[A \quad \mathbf{b}]$. We can now justify this claim by showing that the equation $A\mathbf{x} = \mathbf{b}$ has the same solutions as $R\mathbf{x} = \mathbf{c}$.

By Theorem 2.4, there is an invertible matrix P such that $P[A \quad \mathbf{b}] = [R \quad \mathbf{c}]$. Therefore

$$[PA \quad P\mathbf{b}] = P[A \quad \mathbf{b}] = [R \quad \mathbf{c}],$$

and hence $PA = R$ and $P\mathbf{b} = \mathbf{c}$. Because P is invertible, it follows that $A = P^{-1}R$ and $\mathbf{b} = P^{-1}\mathbf{c}$.

Suppose that $\mathbf{v}$ is a solution to $A\mathbf{x} = \mathbf{b}$. Then $A\mathbf{v} = \mathbf{b}$, and so

$$R\mathbf{v} = (PA)\mathbf{v} = P(A\mathbf{v}) = P\mathbf{b} = \mathbf{c},$$

and therefore $\mathbf{v}$ is a solution to $R\mathbf{x} = \mathbf{c}$. Conversely, suppose that $\mathbf{v}$ is a solution to $R\mathbf{x} = \mathbf{c}$. Then $R\mathbf{v} = \mathbf{c}$ and, hence,

$$A\mathbf{v} = (P^{-1}R)\mathbf{v} = P^{-1}(R\mathbf{v}) = P^{-1}\mathbf{c} = \mathbf{b}.$$

Therefore $\mathbf{v}$ is a solution to $A\mathbf{x} = \mathbf{b}$. Thus the equations $A\mathbf{x} = \mathbf{b}$ and $R\mathbf{x} = \mathbf{c}$ have the same solutions. As a special case of this, if $\mathbf{b} = \mathbf{0}$, then $\mathbf{c} = \mathbf{0}$; hence, $A\mathbf{x} = \mathbf{0}$ and $R\mathbf{x} = \mathbf{0}$ have the same solutions.

The Relation between a Matrix and Its Reduced Row Echelon Form

Suppose that we are given the set

$$S = \left\{ \begin{bmatrix} 1 \\ -1 \\ 2 \\ -3 \end{bmatrix}, \begin{bmatrix} 2 \\ -2 \\ 4 \\ -6 \end{bmatrix}, \begin{bmatrix} -1 \\ 1 \\ -3 \\ 2 \end{bmatrix}, \begin{bmatrix} 2 \\ 2 \\ 2 \\ 0 \end{bmatrix}, \begin{bmatrix} 1 \\ 3 \\ 0 \\ 3 \end{bmatrix}, \begin{bmatrix} 2 \\ 6 \\ 3 \\ 9 \end{bmatrix} \right\},$$

and asked to discover what relationships exist among the vectors of S. That is, we want to know which vectors are linear combinations of the others. We saw in Section 1.7 that at least one vector in a set is a linear combination of the other vectors precisely when $A\mathbf{x} = \mathbf{0}$ has nonzero solutions, where A is the matrix whose columns are the vectors in the set. In this case

$$A = \begin{bmatrix} 1 & 2 & -1 & 2 & 1 & 2 \\ -1 & -2 & 1 & 2 & 3 & 6 \\ 2 & 4 & -3 & 2 & 0 & 3 \\ -3 & -6 & 2 & 0 & 3 & 9 \end{bmatrix}.$$

But, as we have just proved, the solutions of $A\mathbf{x} = \mathbf{0}$ are the same as those of $R\mathbf{x} = \mathbf{0}$, where

$$R = \begin{bmatrix} 1 & 2 & 0 & 0 & -1 & -5 \\ 0 & 0 & 1 & 0 & 0 & -3 \\ 0 & 0 & 0 & 1 & 1 & 2 \\ 0 & 0 & 0 & 0 & 0 & 0 \end{bmatrix}$$

is the reduced row echelon form of A (obtained in Section 1.4). Since any solution to $A\mathbf{x} = \mathbf{0}$ gives the coefficients of a linear combination of columns of A that equals $\mathbf{0}$, it follows that the linear relationships among the vectors of S are the same as those among the vectors in the simpler set

$$S' = \left\{ \begin{bmatrix} 1 \\ 0 \\ 0 \\ 0 \end{bmatrix}, \begin{bmatrix} 2 \\ 0 \\ 0 \\ 0 \end{bmatrix}, \begin{bmatrix} 0 \\ 1 \\ 0 \\ 0 \end{bmatrix}, \begin{bmatrix} 0 \\ 0 \\ 1 \\ 0 \end{bmatrix}, \begin{bmatrix} -1 \\ 0 \\ 1 \\ 0 \end{bmatrix}, \begin{bmatrix} -5 \\ -3 \\ 2 \\ 0 \end{bmatrix} \right\},$$

whose vectors are the columns in R. For example, in R it is obvious that $\mathbf{r}_2 = 2\mathbf{r}_1$ and $\mathbf{r}_5 = -\mathbf{r}_1 + \mathbf{r}_4$. Although it is less obvious, similar relationships hold in A: $\mathbf{a}_2 = 2\mathbf{a}_1$ and $\mathbf{a}_5 = -\mathbf{a}_1 + \mathbf{a}_4$.

We show that the relationship between the matrices A and R above holds for *any* matrix and its reduced row echelon form. First, we make our terminology precise. For a fixed matrix A, an equation involving the columns of A in which each side of the equation is either $\mathbf{0}$ or a linear combination of one or more of the columns of A is called a *linear relationship* among the columns of A. For example, with regard to the preceding matrix A, it is easy to verify that

$$\mathbf{a}_5 + \mathbf{a}_6 = -6\mathbf{a}_1 - 3\mathbf{a}_2 + 3\mathbf{a}_3 \tag{9}$$

is a linear relationship among the columns of A. Any linear relationship can be rewritten in the form of a linear combination equal to $\mathbf{0}$. For example, the linear relationship (9) can be rewritten

$$6\mathbf{a}_1 + 3\mathbf{a}_2 - 3\mathbf{a}_3 + \mathbf{a}_5 + \mathbf{a}_6 = \mathbf{0}.$$

Notice that this last equation can be written as the matrix equation

$$A \begin{bmatrix} 6 \\ 3 \\ -3 \\ 0 \\ 1 \\ 1 \end{bmatrix} = \begin{bmatrix} 0 \\ 0 \\ 0 \\ 0 \\ 0 \\ 0 \end{bmatrix}. \tag{10}$$

In general, any linear relationship among the columns of an $m \times n$ matrix A can be written as a matrix equation of the form $A\mathbf{v} = \mathbf{0}$ for some vector $\mathbf{v}$ in $\mathcal{R}^n$.

Now let A be any $m \times n$ matrix and R be a reduced row echelon form of A. Then by the remarks following Theorem 2.4, the equations $A\mathbf{x} = \mathbf{0}$ and $R\mathbf{x} = \mathbf{0}$ have the same solutions. That is, for any vector $\mathbf{v}$ in $\mathcal{R}^n$, $A\mathbf{v} = \mathbf{0}$ if and only if $R\mathbf{v} = \mathbf{0}$. Thus a linear relationship among the columns of A, which can be written as a matrix equation of the form $A\mathbf{v} = \mathbf{0}$, is also equivalent to the matrix equation $R\mathbf{v} = \mathbf{0}$, which can be written as the corresponding linear relationship among the columns of R.

For example, (10) is equivalent to the matrix equation

$$R \begin{bmatrix} 6 \\ 3 \\ -3 \\ 0 \\ 1 \\ 1 \end{bmatrix} = \begin{bmatrix} 0 \\ 0 \\ 0 \\ 0 \\ 0 \\ 0 \end{bmatrix},$$

which can be written as the linear relationship

$$\mathbf{r}_5 + \mathbf{r}_6 = -6\mathbf{r}_1 - 3\mathbf{r}_2 + 3\mathbf{r}_3.$$

This last equation resembles (9), but with the columns of A replaced by the corresponding columns of R.

Thus we have a correspondence between linear relationships among the columns of a matrix and linear relationships among the columns of a reduced row echelon form of the matrix. Notice that this process is reversible. That is, a linear relationship among the columns of a reduced row echelon form of a matrix can be translated into a linear relationship among the columns of the matrix itself. We call this result the *linear correspondence property*.

Linear Correspondence Property

Any linear relationship among the columns of either a matrix or a reduced row echelon form of the matrix can be translated into a linear relationship among the corresponding columns of the other matrix, using the same coefficients.

In the following statements, let A be an $m \times n$ matrix with reduced row echelon form R. Then these statements are immediate consequences of the linear correspondence property.

Consequences of the Linear Correspondence Property

1. A column of A is $\mathbf{0}$ if and only if the corresponding column of R is $\mathbf{0}$. (For $\mathbf{a}_j = 0\mathbf{a}_j$ if and only if $\mathbf{r}_j = 0\mathbf{r}_j$. Therefore $\mathbf{a}_j = \mathbf{0}$ if and only if $\mathbf{r}_j = \mathbf{0}$.)

2. A set of columns of A is linearly independent if and only if the corresponding set of columns of R is linearly independent.

3. Column j of A is a linear combination of certain columns of A if and only if column j of R is a linear combination of the corresponding columns of R.

Our principal application of the linear correspondence property is to complete the proof of Theorem 1.3 (on page 32) by showing that the reduced row echelon form of a matrix is unique. This is important because such fundamental concepts as the *rank* and *nullity* of a matrix are defined in terms of the reduced row echelon form of the matrix.

Our strategy is to find a condition that characterizes the pivot columns of a matrix R in reduced row echelon form in terms of linear combinations of its columns, and then use the linear correspondence property to translate this condition into the corresponding condition about the columns of a matrix that has R as a reduced row echelon form.

For this purpose, consider the matrix

$$R = [\mathbf{r}_1 \ \mathbf{r}_2 \ \cdots \ \mathbf{r}_7] = \begin{bmatrix} 0 & 1 & 2 & 0 & 3 & 0 & 2 \\ 0 & 0 & 0 & 1 & 4 & 0 & 5 \\ 0 & 0 & 0 & 0 & 0 & 1 & 3 \\ 0 & 0 & 0 & 0 & 0 & 0 & 0 \end{bmatrix},$$

which is in reduced row echelon form. Notice that $\mathbf{r}_2$, $\mathbf{r}_4$, and $\mathbf{r}_6$ are the pivot columns of R. Notice, also, that these columns are the first three standard vectors of $\mathcal{R}^4$, and hence they are linearly independent. Also, the first pivot column, $\mathbf{r}_2$, is the first nonzero column of R, and no pivot column is a linear combination of the columns to its left. Except for the first column, which is the zero vector, any column of R that is not a pivot column is a linear combination of the preceding pivot columns. For example, the fifth column, which is not a pivot column, can be written $\mathbf{r}_5 = 3\mathbf{r}_2 + 4\mathbf{r}_4$. Finally, observe that every column of R is a linear combination of the pivot columns of R.

These properties are obviously true for any matrix in reduced row echelon form. We summarize them here.

Properties of a Matrix in Reduced Row Echelon Form

Let R be an $m \times n$ matrix in reduced row echelon form. Then the following statements are true.

(a) A column of R is a pivot column if and only if it is nonzero and not a linear combination of the preceding columns of R.

(b) The jth pivot column of R is $\mathbf{e}_j$, the jth standard vector of $\mathcal{R}^m$, and hence the pivot columns of R are linearly independent.

(c) Every column of R is a linear combination of the pivot columns of R.

We are now ready to complete the proof of Theorem 1.3.

Proof (Uniqueness of the reduced row echelon form of a matrix)   Let A be a matrix, and let R be a reduced row echelon form of A. By the preceding property (a), a column of R is a pivot column of R if and only if it is nonzero and it is not a linear combination of the preceding pivot columns of R. These two conditions can be combined with the first and third consequences of the linear correspondence property to produce a test for the pivot columns of A exclusively in terms of the columns of A: *A column of A is a pivot column if and only if it is nonzero and it is not a linear combination of the previous columns of A.* Thus the positions of the pivot columns of R are uniquely determined by the columns of A. Furthermore, since the jth pivot column of R is the jth standard vector of $\mathcal{R}^m$, the pivot columns of R are completely determined by the columns of A.

We show that the other columns of R are also determined by the columns of A. Suppose that $\mathbf{r}_j$ is not a pivot column of R. Then $\mathbf{r}_j$ is a linear combination of the pivot columns of R and hence $\mathbf{a}_j$ is a linear combination of the pivot columns of A by the consequences of the linear correspondence property. Furthermore, since pivot columns of R are linearly independent, the same can be said for the pivot columns of A. Suppose that $n_1, n_2, \ldots, n_k$ are the column numbers that correspond to the pivot columns of A, and hence of R. Then, by Exercise 45 of Section 1.7, there exist unique scalars $c_1, c_2, \ldots, c_k$ such that

$$\mathbf{a}_j = c_1 \mathbf{a}_{n_1} + c_2 \mathbf{a}_{n_2} + \cdots + c_k \mathbf{a}_{n_k}.$$

Hence, by the linear correspondence property,

$$\mathbf{r}_j = c_1 \mathbf{r}_{n_1} + c_2 \mathbf{r}_{n_2} + \cdots + c_k \mathbf{r}_{n_k} = c_1 \mathbf{e}_1 + c_2 \mathbf{e}_2 + \cdots + c_k \mathbf{e}_k.$$

So $\mathbf{r}_j$ is completely determined by A. We conclude that R is unique.   ∎

Two of the properties discussed in the proof above are important enough to list as a separate result.

Theorem 2.5
The following statements are true for any matrix A.

(a) The pivot columns of A are linearly independent.
(b) Every column of A is a linear combination of the pivot columns of A.

Example 4   The reduced row echelon form of a matrix A is

$$R = \begin{bmatrix} 1 & 2 & 0 & -1 \\ 0 & 0 & 1 & 1 \\ 0 & 0 & 0 & 0 \end{bmatrix}.$$

Determine A, given that the first and third columns of A are

$$\mathbf{a}_1 = \begin{bmatrix} 1 \\ 2 \\ 1 \end{bmatrix} \quad \text{and} \quad \mathbf{a}_3 = \begin{bmatrix} 2 \\ 2 \\ 3 \end{bmatrix}.$$

Solution   Since the first and third columns of R are the pivot columns, these are also the pivot columns of A. Now observe that the second column of R is $2\mathbf{r}_1$; hence, by the linear correspondence property, the second column of A is

$$\mathbf{a}_2 = 2\mathbf{a}_1 = 2 \begin{bmatrix} 1 \\ 2 \\ 1 \end{bmatrix} = \begin{bmatrix} 2 \\ 4 \\ 2 \end{bmatrix}.$$

Furthermore, the fourth column of R is $\mathbf{r}_4 = (-1)\mathbf{r}_1 + \mathbf{r}_3$, and so, again by the linear correspondence property,

$$\mathbf{a}_4 = (-1)\mathbf{a}_1 + \mathbf{a}_3 = (-1) \begin{bmatrix} 1 \\ 2 \\ 1 \end{bmatrix} + \begin{bmatrix} 2 \\ 2 \\ 3 \end{bmatrix} = \begin{bmatrix} 1 \\ 0 \\ 2 \end{bmatrix}.$$

Thus

$$A = \begin{bmatrix} 1 & 2 & 2 & 1 \\ 2 & 4 & 2 & 0 \\ 1 & 2 & 3 & 2 \end{bmatrix}.$$

Practice Problems

1. If $A = \begin{bmatrix} -1 & 0 & 1 \\ 1 & 2 & -2 \\ 2 & -1 & -1 \end{bmatrix}$ and $B = \begin{bmatrix} 4 & 1 & 2 \\ 3 & 1 & 1 \\ 5 & 1 & 2 \end{bmatrix}$, is $B = A^{-1}$?

2. Find an elementary matrix E such that $EA = B$ where

$$A = \begin{bmatrix} 3 & -4 & 1 \\ 2 & 5 & -1 \end{bmatrix} \quad \text{and} \quad B = \begin{bmatrix} 3 & -4 & 1 \\ -4 & 13 & -3 \end{bmatrix}.$$

3. Suppose that the reduced row echelon form of A is

$$R = \begin{bmatrix} 1 & -3 & 0 & 5 & 3 \\ 0 & 0 & 1 & 2 & -2 \\ 0 & 0 & 0 & 0 & 0 \end{bmatrix}.$$

Determine A, if the first and third columns of A are

$$\mathbf{a}_1 = \begin{bmatrix} 1 \\ -1 \\ 2 \end{bmatrix} \text{ and } \mathbf{a}_3 = \begin{bmatrix} 2 \\ 0 \\ -1 \end{bmatrix}, \text{ respectively.}$$

Exercises

1. Determine if the following statements are true or false.

   (a) Every square matrix is invertible.
   (b) Invertible matrices are square.
   (c) Elementary matrices are invertible.
   (d) If A and B are matrices such that $AB = I_n$ for some n, then both A and B are invertible.
   (e) If B and C are inverses of a matrix A, then $B = C$.
   (f) If A and B are invertible $n \times n$ matrices, then AB^T is invertible.

2.[3] Let A_α be the α-rotation matrix. Prove that $(A_\alpha)^T = (A_\alpha)^{-1}$.

For each of the matrices A and B in Exercises 3–6, determine if $B = A^{-1}$.

3. $A = \begin{bmatrix} 1 & 2 \\ 1 & -1 \end{bmatrix}$ and $B = \begin{bmatrix} 1 & 0.5 \\ 1 & -1 \end{bmatrix}$

4. $A = \begin{bmatrix} 1 & 2 \\ 3 & 5 \end{bmatrix}$ and $B = \begin{bmatrix} -5 & 2 \\ 3 & -1 \end{bmatrix}$

5. $A = \begin{bmatrix} 1 & 2 & 1 \\ 1 & 1 & 2 \\ 2 & 3 & 4 \end{bmatrix}$ and $B = \begin{bmatrix} 2 & 5 & -3 \\ 0 & -2 & 1 \\ -1 & -1 & 1 \end{bmatrix}$

6. $A = \begin{bmatrix} 1 & 1 & 2 \\ 0 & 1 & 1 \\ 0 & 0 & 1 \end{bmatrix}$ and $B = \begin{bmatrix} 1 & -1 & 1 \\ 1 & 2 & 1 \\ -1 & 0 & -1 \end{bmatrix}$

For Exercises 7–8, let A and B be the invertible 3×3 matrices such that

$$A^{-1} = \begin{bmatrix} 1 & 2 & 3 \\ 2 & 0 & 1 \\ 1 & 1 & -1 \end{bmatrix} \quad \text{and} \quad B^{-1} = \begin{bmatrix} 2 & -1 & 3 \\ 0 & 0 & 4 \\ 3 & -2 & 1 \end{bmatrix}.$$

7. Find $(AB^T)^{-1}$.

8. Find $((A^T)(B^T))^{-1}$.

9.[4] Let $A = \begin{bmatrix} a & b \\ c & d \end{bmatrix}$.

   (a) Prove that if $ad - bc \neq 0$, then A is invertible, and

   $$A^{-1} = \frac{1}{ad - bc} \begin{bmatrix} d & -b \\ -c & a \end{bmatrix}.$$

   (b) Prove the converse of (a): If A is invertible, then $ad - bc \neq 0$.

In Exercises 10–13, find the inverse of the given elementary matrices.

10. $\begin{bmatrix} 1 & 0 \\ 1 & 1 \end{bmatrix}$

11. $\begin{bmatrix} 1 & 0 & 0 \\ -2 & 1 & 0 \\ 0 & 0 & 1 \end{bmatrix}$

12. $\begin{bmatrix} 0 & 0 & 1 \\ 0 & 1 & 0 \\ 1 & 0 & 0 \end{bmatrix}$

13. $\begin{bmatrix} 1 & 0 & 0 & 0 \\ 0 & 4 & 0 & 0 \\ 0 & 0 & 1 & 0 \\ 0 & 0 & 0 & 1 \end{bmatrix}$

In Exercises 14–16, find an elementary matrix E such that $EA = B$.

14. $A = \begin{bmatrix} 1 & 2 \\ 3 & 4 \end{bmatrix}$ and $B = \begin{bmatrix} -1 & -2 \\ 3 & 4 \end{bmatrix}$

15. $A = \begin{bmatrix} 1 & 2 & 3 \\ 4 & 5 & 6 \\ 7 & 8 & 9 \end{bmatrix}$ and $B = \begin{bmatrix} 1 & 2 & 3 \\ 7 & 8 & 9 \\ 4 & 5 & 6 \end{bmatrix}$

---

[3]The result of this exercise is used in Section 4.4 (page 234).

[4]The result of this exercise is used in Section 3.1 (page 175).

16. $A = \begin{bmatrix} 1 & 2 & 3 & 4 \\ -1 & 1 & 3 & 2 \\ 2 & -1 & 0 & 4 \end{bmatrix}$ and

$B = \begin{bmatrix} 1 & 2 & 3 & 4 \\ -1 & 1 & 3 & 2 \\ 0 & -5 & -6 & -4 \end{bmatrix}$

17. Prove that the product of elementary matrices is invertible.

18. Let A be an invertible $n \times n$ matrix, and let $\mathbf{u}$ and $\mathbf{v}$ be vectors in $\mathcal{R}^n$ such that $\mathbf{u} \neq \mathbf{v}$. Prove that $A\mathbf{u} \neq A\mathbf{v}$.

19. Let Q be an invertible $n \times n$ matrix. Prove that the subset $\{\mathbf{u}_1, \mathbf{u}_2, \dots, \mathbf{u}_k\}$ of $\mathcal{R}^n$ is linearly independent if and only if $\{Q\mathbf{u}_1, Q\mathbf{u}_2, \dots, Q\mathbf{u}_k\}$ is linearly independent.

20. Prove Theorem 2.3(a).

21. Prove that if A, B, and C are invertible $n \times n$ matrices, then ABC is invertible and $(ABC)^{-1} = C^{-1}B^{-1}A^{-1}$.

22. Let A and B be $n \times n$ matrices such that both A and AB are invertible. Prove that B is invertible.

23. Let A and B be $n \times n$ matrices such that $AB = I_n$. Prove that the rank of A is n.

24. Prove that if A is an $m \times n$ matrix and B is an $n \times p$ matrix, then rank $AB \leq$ rank B. *Hint:* Prove that if column number k is not a pivot column of B, then column number k is not a pivot column of AB.

25. Prove that if B is an $n \times n$ matrix with rank n, then there exists an $n \times n$ matrix C such that $BC = I_n$.

26. Prove that if A and B are $n \times n$ matrices such that $AB = I_n$, then B is invertible and $A = B^{-1}$. *Hint:* Use Exercises 24 and 25.

27. Prove that if an $n \times n$ matrix has rank n, then it is invertible. *Hint:* Use Exercises 25 and 26.

28. Let $M = \begin{bmatrix} A & O_1 \\ O_2 & B \end{bmatrix}$, where A and B are square and O_1 and O_2 are zero matrices. Prove that M is invertible if and only if A and B are both invertible. If M is invertible, find M^{-1} in terms of A and B.

In Exercises 29–32, find the matrix A given the reduced row echelon form R of A and information about certain columns of A.

29. $R = \begin{bmatrix} 1 & 0 & 1 \\ 0 & 1 & 2 \end{bmatrix}$, $\mathbf{a}_1 = \begin{bmatrix} 3 \\ -1 \end{bmatrix}$,   and

$\mathbf{a}_2 = \begin{bmatrix} 2 \\ 5 \end{bmatrix}$

30. $R = \begin{bmatrix} 1 & 2 & 0 & -3 & 0 & 1 \\ 0 & 0 & 1 & 2 & 0 & 2 \\ 0 & 0 & 0 & 0 & 1 & 3 \\ 0 & 0 & 0 & 0 & 0 & 0 \end{bmatrix}$, $\mathbf{a}_1 = \begin{bmatrix} 2 \\ 0 \\ -1 \\ 1 \end{bmatrix}$,

$\mathbf{a}_3 = \begin{bmatrix} 1 \\ -1 \\ 2 \\ 0 \end{bmatrix}$,   and $\mathbf{a}_5 = \begin{bmatrix} 2 \\ 3 \\ 0 \\ 1 \end{bmatrix}$

31. $R = \begin{bmatrix} 1 & -1 & 0 & 0 & 1 \\ 0 & 0 & 1 & 0 & 2 \\ 0 & 0 & 0 & 1 & 3 \end{bmatrix}$, $\mathbf{a}_2 = \begin{bmatrix} 1 \\ -2 \\ 1 \end{bmatrix}$,

$\mathbf{a}_3 = \begin{bmatrix} 1 \\ -1 \\ 0 \end{bmatrix}$,   and $\mathbf{a}_4 = \begin{bmatrix} 4 \\ 1 \\ 3 \end{bmatrix}$.

32. $R = \begin{bmatrix} 1 & 2 & 0 & 1 & 0 & 1 \\ 0 & 0 & 1 & -1 & 0 & -1 \\ 0 & 0 & 0 & 0 & 1 & 1 \\ 0 & 0 & 0 & 0 & 0 & 0 \end{bmatrix}$, $\mathbf{a}_2 = \begin{bmatrix} 2 \\ 4 \\ 6 \\ -2 \end{bmatrix}$,

$\mathbf{a}_4 = \begin{bmatrix} 1 \\ 3 \\ -1 \\ 1 \end{bmatrix}$, $\mathbf{a}_6 = \begin{bmatrix} 2 \\ -1 \\ -1 \\ 2 \end{bmatrix}$

33. Suppose that $\mathbf{u}$ and $\mathbf{v}$ are linearly independent vectors in $\mathcal{R}^3$. Find the reduced row echelon form of $A = [\mathbf{a}_1 \ \mathbf{a}_2 \ \mathbf{a}_3 \ \mathbf{a}_4]$, given that

$$\mathbf{a}_1 = \mathbf{u}, \qquad \mathbf{a}_2 = 2\mathbf{u}, \qquad \mathbf{a}_3 = \mathbf{u} + \mathbf{v}, \qquad \text{and} \qquad \mathbf{a}_4 = \mathbf{v}.$$

34. Let A be an $n \times n$ invertible matrix, and let $\mathbf{e}_j$ be the jth standard vector of $\mathcal{R}^n$.

    (a) Prove that the jth column of A^{-1} is a solution to $A\mathbf{x} = \mathbf{e}_j$.
    (b) Why does (a) imply that A^{-1} is unique?
    (c) Why does (a) imply that rank $A = n$?

35. Prove items 2 and 3 in the list of consequences of the linear correspondence property (see page 120).

36. Let R be an $m \times n$ matrix in reduced row echelon form. Find a relationship between the columns of R^T and the columns of $R^T R$. Justify your answer.

37. Let R be an $m \times n$ matrix in reduced row echelon form. Describe the reduced row echelon form of R^T. Justify your answer.

38. Let A be an $m \times n$ matrix with reduced row echelon form R. Then there exists an invertible matrix P such that $PA = R$, and an invertible matrix Q such that QR^T is the reduced row echelon form of R^T. Describe the matrix PAQ^T in terms of A. Justify your answer.

39. Let A and B be $m \times n$ matrices. Prove that the following conditions are equivalent.

    (a) A and B have the same reduced row echelon form.
    (b) There is an invertible $m \times m$ matrix P such that $B = PA$.

40. Let A be an $n \times n$ matrix. Find a condition on A that is equivalent to the statement $AB = AC$ if and only if $B = C$. Justify your answer.

41. Prove the following assertion about the relation between elementary row operations and multiplication by elementary matrices: Let A be an $m \times n$ matrix, and let E be an elementary matrix obtained by performing an elementary row operation on I_m. Then the product EA can be obtained from A by performing the identical elementary

row operation on A. *Hint:* Prove this separately for each of the three kinds of elementary row operations.

42. Let A and B be $m \times n$ matrices. Prove that the following conditions are equivalent.

   (a) A and B have the same reduced row echelon form.
   (b) $A\mathbf{x} = \mathbf{0}$ if and only if $B\mathbf{x} = \mathbf{0}$.

We can define an elementary column operation in a manner analogous to the definition of an elementary row operation presented in Section 1.3. Each of the following operations on a matrix is called an **elementary column operation**: interchanging any two columns of the matrix, multiplying any column by a nonzero constant, and adding a multiple of one column of the matrix to another column of the matrix.

43. Let E be an $n \times n$ elementary matrix. Prove that E is an elementary matrix if and only if E can be obtained from I_n by a single elementary column operation.

44. Prove that if E is obtained from I_n by a single elementary column operation, then for any $m \times n$ matrix A, the product AE can be obtained from A by performing the identical elementary column operation on A.

In Exercises 45–50, use a calculator with matrix capabilities or computer software such as MATLAB *to solve the problem.*

45. Let

$$A = \begin{bmatrix} 1 & 1 & 0 & -1 \\ 0 & 1 & 1 & 2 \\ 2 & 1 & 0 & -3 \\ -1 & -1 & 1 & 1 \end{bmatrix}.$$

Let B be the matrix obtained by interchanging rows 1 and 3 of A, and let C be the matrix obtained by interchanging rows 2 and 4 of A.

   (a) Prove that A is invertible.
   (b) Prove that B and C are invertible.
   (c) Compare B^{-1} and C^{-1} to A^{-1}.
   (d) Now let A be any invertible $n \times n$ matrix, and let B be the matrix obtained by interchanging rows i and j of A, where $1 \le i < j \le n$. Make a conjecture about the relationship between B^{-1} and A^{-1}.
   (e) Prove your conjecture in (d).

46. Let

$$A = \begin{bmatrix} 15 & 30 & 17 & 31 \\ 30 & 66 & 36 & 61 \\ 17 & 36 & 20 & 35 \\ 31 & 61 & 35 & 65 \end{bmatrix}.$$

   (a) Show that A is symmetric and invertible.
   (b) Show that A^{-1} is symmetric and invertible.
   (c) Prove that the inverse of any symmetric and invertible matrix is also symmetric and invertible.

47. Let

$$A = \begin{bmatrix} 1 & 2 & 0 & 3 \\ 2 & 5 & -1 & 8 \\ 2 & 4 & 1 & 6 \\ 3 & 6 & 1 & 8 \end{bmatrix}.$$

   (a) Show that A and A^2 are invertible by using their reduced row echelon forms.
   (b) Compute the inverse of A^2, and show that it equals $(A^{-1})^2$.
   (c) State and verify a result similar to (b) for A^3.
   (d) Generalize and prove a result analogous to (c) for the nth power of any invertible matrix.

48. Consider the following system $A\mathbf{x} = \mathbf{b}$.

$$\begin{aligned} x_1 + 3x_2 + 2x_3 + x_4 &= 4 \\ x_1 + 2x_2 + 4x_3 &= -3 \\ 2x_1 + 6x_2 + 5x_3 + 2x_4 &= -1 \\ x_1 + 3x_2 + 2x_3 + 2x_4 &= 2 \end{aligned}$$

   (a) Compute the inverse of A and use it to solve the system.
   (b) Show that your solution is correct by showing that it satisfies the matrix equation $A\mathbf{x} = \mathbf{b}$.

49. It is generally easier to compute the inverse of several smaller matrices than one large matrix. Consider the matrix M partitioned in the following way.

$$M = \left[\begin{array}{cccc|cccc} 6 & -3 & 4 & 5 & 2 & 7 & -2 & 6 \\ -3 & 0 & 1 & -8 & 9 & 11 & 0 & -7 \\ 4 & 5 & -5 & 6 & 8 & 0 & -8 & 2 \\ 6 & 4 & 4 & -3 & 8 & 3 & 5 & 2 \\ \hline 0 & 0 & 0 & 0 & -6 & 9 & 8 & 8 \\ 0 & 0 & 0 & 0 & 2 & -1 & 1 & 5 \\ 0 & 0 & 0 & 0 & 6 & 5 & -4 & 4 \\ 0 & 0 & 0 & 0 & 9 & 5 & 1 & 9 \end{array}\right]$$

   (a) Compute the inverse of M.
   (b) Write M in the form $M = \begin{bmatrix} A & B \\ O & C \end{bmatrix}$, and compute the inverses of A, B, and C.
   (c) Use your answers to (b) to form the matrix

$$F = \begin{bmatrix} A^{-1} & -A^{-1}BC^{-1} \\ O & C^{-1} \end{bmatrix}.$$

   (d) Verify that the matrix F in (c) is the inverse of M by comparing your answers to (a) and (c).
   (e) Use your answers to the preceding parts to make and prove a conjecture about the inverse of a matrix in the form of the matrix M.

50. The purpose of this exercise is to illustrate the method of finding an inverse described in Exercise 34. In Exercise

47, you found the inverse of

$$A = \begin{bmatrix} 1 & 2 & 0 & 3 \\ 2 & 5 & -1 & 8 \\ 2 & 4 & 1 & 6 \\ 3 & 6 & 1 & 8 \end{bmatrix}.$$

(a) Solve the system $A\mathbf{x} = \mathbf{e}_1$, and compare your solution to the first column of A^{-1}.

(b) Repeat (a) for the vectors $\mathbf{e}_2$, $\mathbf{e}_3$, and $\mathbf{e}_4$.

Solutions to the Practice Problems

1. Since

$$AB = \begin{bmatrix} -1 & 0 & 1 \\ 1 & 2 & -2 \\ 2 & -1 & -1 \end{bmatrix} \begin{bmatrix} 4 & 1 & 2 \\ 3 & 1 & 1 \\ 5 & 1 & 2 \end{bmatrix} = \begin{bmatrix} 1 & 0 & 0 \\ 0 & 1 & 0 \\ 0 & 0 & 1 \end{bmatrix}$$

and

$$BA = \begin{bmatrix} 4 & 1 & 2 \\ 3 & 1 & 1 \\ 5 & 1 & 2 \end{bmatrix} \begin{bmatrix} -1 & 0 & 1 \\ 1 & 2 & -2 \\ 2 & -1 & -1 \end{bmatrix} = \begin{bmatrix} 1 & 0 & 0 \\ 0 & 1 & 0 \\ 0 & 0 & 1 \end{bmatrix},$$

$B = A^{-1}$.

2. The matrix B is obtained from A by adding -2 times row 1 of A to row 2. Adding -2 times row 1 of I_2 to row 2 produces the elementary matrix $E = \begin{bmatrix} 1 & 0 \\ -2 & 1 \end{bmatrix}$. This matrix has the property that $EA = B$.

3. Let $\mathbf{r}_1, \mathbf{r}_2, \mathbf{r}_3, \mathbf{r}_4,$ and $\mathbf{r}_5$ denote the columns of R. The pivot columns of R are columns 1 and 3, and so the pivot columns of A are columns 1 and 3. Every other column of R is a linear combination of its pivot columns. In fact,

$\mathbf{r}_2 = -3\mathbf{r}_1$, $\mathbf{r}_4 = 5\mathbf{r}_1 + 2\mathbf{r}_3$, and $\mathbf{r}_5 = 3\mathbf{r}_1 - 2\mathbf{r}_3$. Thus the linear correspondence property implies that column 2 of A is

$$\mathbf{a}_2 = -3\mathbf{a}_1 = -3 \begin{bmatrix} 1 \\ -1 \\ 2 \end{bmatrix} = \begin{bmatrix} -3 \\ 3 \\ -6 \end{bmatrix}.$$

Similarly, the fourth and fifth columns of A are

$$\mathbf{a}_4 = 5\mathbf{a}_1 + 2\mathbf{a}_3 = 5 \begin{bmatrix} 1 \\ -1 \\ 2 \end{bmatrix} + 2 \begin{bmatrix} 2 \\ 0 \\ -1 \end{bmatrix} = \begin{bmatrix} 9 \\ -5 \\ 8 \end{bmatrix},$$

and

$$\mathbf{a}_5 = 3\mathbf{a}_1 - 2\mathbf{a}_3 = 3 \begin{bmatrix} 1 \\ -1 \\ 2 \end{bmatrix} - 2 \begin{bmatrix} 2 \\ 0 \\ -1 \end{bmatrix} = \begin{bmatrix} -1 \\ -3 \\ 8 \end{bmatrix}.$$

Hence

$$A = \begin{bmatrix} 1 & -3 & 2 & 9 & -1 \\ -1 & 3 & 0 & -5 & -3 \\ 2 & -6 & -1 & 8 & 8 \end{bmatrix}.$$

2.4   The Inverse of a Matrix

In this section we apply what we have learned about invertible and elementary matrices to answer the two questions posed early in Section 2.3. Theorem 2.6 gives us a list of conditions equivalent to matrix invertibility. Some of these (for example, (j)), can be used as a test for matrix invertibility. We also devise a method for actually computing the inverse of an invertible matrix.

Theorem 2.6

Let A be an $n \times n$ matrix. The following conditions are equivalent.

(a) A is invertible.

(b) There exists an $n \times n$ matrix B such that $BA = I_n$.

(c) There exists an $n \times n$ matrix C such that $AC = I_n$.

(d) The columns of A span $\mathcal{R}^n$.

(e) For every $\mathbf{b} \in \mathcal{R}^n$, the system $A\mathbf{x} = \mathbf{b}$ is consistent.

(f) The rank of A is n.

(g) The columns of A are linearly independent.

(h) The only solution to $A\mathbf{x} = \mathbf{0}$ is the zero vector.

(i) The nullity of A is zero.

(j) The reduced row echelon form of A is I_n.

(k) A is a product of elementary matrices.

Proof By Theorem 1.4, (d), (e), and (f) are equivalent. By Theorem 1.7, (g), (h), and (i) are equivalent. Since nullity $A = n - \text{rank } A$, nullity $A = 0$ if and only if rank $A = n$, and hence (f) is equivalent to (i).

Next we prove the implications (i) $\Rightarrow$ (j) $\Rightarrow$ (k) $\Rightarrow$ (a) $\Rightarrow$ (b) $\Rightarrow$ (h), from which it follows that (j), (k), (a), and (b) are each equivalent to the items (d) through (i). Finally, we prove that (a) $\Rightarrow$ (c) $\Rightarrow$ (e), from which it follows that (c) is equivalent to each of the other items, thus establishing that all of the items in the list are equivalent.

(i) *implies* (j) Suppose that the nullity of A is zero. By Theorem 1.7, the columns of the reduced row echelon form of A are distinct standard vectors of $\mathcal{R}^n$. But this is the definition of I_n, and hence (j) follows.

(j) *implies* (k) Suppose that $R = I_n$ is the reduced row echelon form of A. By Theorem 2.4, there is an invertible matrix P such that $PA = R = I_n$. Notice that in the discussion preceding Theorem 2.4, P is a product of elementary matrices. We can multiply both sides of this equation by P^{-1} to obtain that $A = P^{-1}$, which is the product of the inverses of these elementary matrices (in the reverse order), each of which is also elementary. Therefore A is a product of elementary matrices.

(k) *implies* (a) Suppose that A is the product of elementary matrices. Since elementary matrices are invertible and the product of invertible matrices is invertible, A is invertible.

(a) *implies* (b) Suppose that A is invertible. Take $B = A^{-1}$.

(b) *implies* (h) Suppose that $BA = I_n$ for some $n \times n$ matrix B, and suppose that $A\mathbf{v} = \mathbf{0}$ for some vector $\mathbf{v}$. Then $\mathbf{v} = I_n\mathbf{v} = (BA)\mathbf{v} = B(A\mathbf{v}) = B\mathbf{0} = \mathbf{0}$.

(a) *implies* (c) Suppose that A is invertible. Take $C = A^{-1}$.

(c) *implies* (e) Suppose that $AC = I_n$ for some $n \times n$ matrix C. Let $\mathbf{b} \in \mathcal{R}^n$. Then $A(C\mathbf{b}) = (AC)\mathbf{b} = I_n\mathbf{b} = \mathbf{b}$. So $C\mathbf{b}$ is a solution to $A\mathbf{x} = \mathbf{b}$. $\quad\square$

Example 1 Use Theorem 2.6 to test the following matrices for invertibility:

$$A = \begin{bmatrix} 1 & 2 & 3 \\ 2 & 5 & 6 \\ 3 & 4 & 8 \end{bmatrix} \quad \text{and} \quad B = \begin{bmatrix} 1 & 1 & 2 \\ 2 & 1 & -1 \\ 1 & 0 & -1 \end{bmatrix}.$$

Solution The reader should check that the reduced row echelon form of A is I_3. Therefore part (j) of Theorem 2.6 is true, and hence A is invertible.

On the other hand, the reduced row echelon form of B is

$$\begin{bmatrix} 1 & 0 & -1 \\ 0 & 1 & 3 \\ 0 & 0 & 0 \end{bmatrix}.$$

Since part (j) of Theorem 2.6 is not satisfied, we conclude that B is not invertible. $\quad\circ$

Notice that conditions (b) and (c) of Theorem 2.6 are the two conditions given in the definition of invertibility. Thus to verify that a square matrix A is invertible, we need only verify one of these conditions. For example, suppose that for a given $n \times n$ matrix A, there is an $n \times n$ matrix B such that $AB = I_n$. Since A is invertible by Theorem 2.6, we may multiply both sides of this equation on the left by A^{-1} to obtain $A^{-1}(AB) = A^{-1}I_n$, which reduces to $B = A^{-1}$. Similarly, we can apply Theorem 2.6(b) to the case that $BA = I_n$ for some matrix B and obtain $B = A^{-1}$.

Theorem 2.6 stipulates that the matrices be square. Notice that there are nonsquare matrices A and B for which the product AB is an identity matrix (condition (c) of Theorem 2.6). For example, let

$$A = \begin{bmatrix} 1 & 1 & 0 \\ 1 & 2 & 1 \end{bmatrix} \quad \text{and} \quad B = \begin{bmatrix} 2 & 1 \\ -1 & -1 \\ 0 & 2 \end{bmatrix}.$$

Then

$$AB = \begin{bmatrix} 1 & 1 & 0 \\ 1 & 2 & 1 \end{bmatrix} \begin{bmatrix} 2 & 1 \\ -1 & -1 \\ 0 & 2 \end{bmatrix} = \begin{bmatrix} 1 & 0 \\ 0 & 1 \end{bmatrix} = I_2.$$

Theorem 2.6 not only provides a method for determining if a matrix is invertible, it also enables us to devise a method for actually calculating the inverse when it exists. We know that any $n \times n$ matrix A can be transformed into a matrix R in reduced row echelon form by means of elementary row operations. Applying the same operations to the $n \times 2n$ matrix $[A \ I_n]$ transforms this matrix into an $n \times 2n$ matrix $[R \ B]$ for some $n \times n$ matrix B. Hence there is an invertible matrix P such that $P[A \ I_n] = [R \ B]$. Thus

$$[R \ B] = P[A \ I_n] = [PA \ PI_n] = [PA \ P].$$

It follows that $PA = R$, and $P = B$. If $R \neq I_n$, then we know that A is not invertible, by Theorem 2.6. On the other hand, if $R = I_n$, then A is invertible. Furthermore, since $PA = I_n$ and $P = B$, it follows that $B = A^{-1}$. Thus we have the following algorithm for computing the inverse of a matrix.

Algorithm for Matrix Inversion

Let A be an $n \times n$ matrix, and suppose that the $n \times 2n$ partitioned matrix $[A \ I_n]$ is transformed by means of elementary row operations into the matrix $[R \ B]$, where R is a matrix in reduced row echelon form. Then either

(a) $R \neq I_n$, in which case A is not invertible, or

(b) $R = I_n$, in which case A is invertible and $B = A^{-1}$.

Example 2   We use the algorithm for matrix inversion to compute A^{-1} for the invertible matrix A of Example 1. This algorithm requires us to transform $[A \ I_3]$ into a matrix of the form $[I_3 \ B]$ by means of elementary row operations. For this purpose, we use the Gaussian elimination algorithm of Section 1.4 to transform A into its reduced row echelon form I_3, while applying each row operation to the entire row of the 3×6 matrix.

$$[A \ I_3] = \begin{bmatrix} 1 & 2 & 3 & | & 1 & 0 & 0 \\ 2 & 5 & 6 & | & 0 & 1 & 0 \\ 3 & 4 & 8 & | & 0 & 0 & 1 \end{bmatrix} \rightarrow \begin{bmatrix} 1 & 2 & 3 & | & 1 & 0 & 0 \\ 0 & 1 & 0 & | & -2 & 1 & 0 \\ 0 & -2 & -1 & | & -3 & 0 & 1 \end{bmatrix}$$

$$\rightarrow \begin{bmatrix} 1 & 2 & 3 & | & 1 & 0 & 0 \\ 0 & 1 & 0 & | & -2 & 1 & 0 \\ 0 & 0 & -1 & | & -7 & 2 & 1 \end{bmatrix} \rightarrow \begin{bmatrix} 1 & 2 & 3 & | & 1 & 0 & 0 \\ 0 & 1 & 0 & | & -2 & 1 & 0 \\ 0 & 0 & 1 & | & 7 & -2 & -1 \end{bmatrix}$$

$$\rightarrow \begin{bmatrix} 1 & 2 & 0 & | & -20 & 6 & 3 \\ 0 & 1 & 0 & | & -2 & 1 & 0 \\ 0 & 0 & 1 & | & 7 & -2 & -1 \end{bmatrix}$$

$$\rightarrow \begin{bmatrix} 1 & 0 & 0 & | & -16 & 4 & 3 \\ 0 & 1 & 0 & | & -2 & 1 & 0 \\ 0 & 0 & 1 & | & 7 & -2 & -1 \end{bmatrix}$$

$$= [I_3 \ B]$$

Thus

$$A^{-1} = B = \begin{bmatrix} -16 & 4 & 3 \\ -2 & 1 & 0 \\ 7 & -2 & -1 \end{bmatrix}.$$

In the context of the discussion above, if $[R \ B]$ is in reduced row echelon form, then so is R (see Exercise 34). This observation is useful if there is a calculator or computer available that can produce the reduced row echelon form of a matrix. In this case, we simply find the reduced row echelon form $[R \ C]$ of $[A \ I_n]$. Then, as before, either

(a) $R \neq I_n$, in which case A is not invertible, or
(b) $R = I_n$, in which case A is invertible and $C = A^{-1}$.

Example 3   To illustrate the previous paragraph, we test

$$A = \begin{bmatrix} 1 & 1 \\ 2 & 2 \end{bmatrix}$$

for invertibility. If we are performing calculations by hand, we transform $[A \ I_2]$ into a matrix $[R \ B]$ such that R is in reduced row echelon form:

$$[A \ I_2] = \begin{bmatrix} 1 & 1 & | & 1 & 0 \\ 2 & 2 & | & 0 & 1 \end{bmatrix} \rightarrow \begin{bmatrix} 1 & 1 & | & 1 & 0 \\ 0 & 0 & | & -2 & 1 \end{bmatrix} = [R \ B].$$

Since $R \neq I_2$, A is not invertible.

Note that, in this case, $[R \ B]$ is not in reduced row echelon form. If a computing device is available that can produce the reduced row echelon form of a matrix, we use it to find the reduced row echelon form of $[A \ I_2]$, which is

$$[R \ C] = \begin{bmatrix} 1 & 1 & | & 0 & 0.5 \\ 0 & 0 & | & 1 & -0.5 \end{bmatrix}.$$

Again we see that A is not invertible, because $R \neq I_2$.

When A is an invertible $n \times n$ matrix and B is any $n \times p$ matrix, we can extend the algorithm for matrix inversion to the computation of $A^{-1}B$. Consider the $n \times (n + p)$ matrix $[A \ B]$. Suppose this matrix is transformed by means of elementary row operations into the matrix $[I_n \ C]$. As in the discussion of the algorithm for matrix inversion, there is an $n \times n$ invertible matrix P such that

$$[I_n \ C] = P[A \ B] = [PA \ PB],$$

from which it follows that $PA = I_n$ and $C = PB$. Therefore, $P = A^{-1}$, and hence $C = A^{-1}B$. Thus we have the following algorithm.

Algorithm for Computing $A^{-1}B$

Let A be an invertible $n \times n$ matrix and B be an $n \times p$ matrix. Suppose that the $n \times (n + p)$ matrix $[A \ B]$ is transformed by means of elementary row operations into the matrix $[I_n \ C]$ in reduced row echelon form. Then $C = A^{-1}B$.

Example 4   Use the algorithm above to compute $A^{-1}B$ for

$$A = \begin{bmatrix} 1 & 2 & 1 \\ 2 & 5 & 1 \\ 2 & 4 & 1 \end{bmatrix} \quad \text{and} \quad B = \begin{bmatrix} 2 & -1 \\ 1 & 3 \\ 0 & 2 \end{bmatrix}.$$

Solution   We apply elementary row operations to transform $[A \ B]$ into its reduced row echelon form, $[I_3 \ A^{-1}B]$.

$$[A \ B] = \left[\begin{array}{ccc|cc} 1 & 2 & 1 & 2 & -1 \\ 2 & 5 & 1 & 1 & 3 \\ 2 & 4 & 1 & 0 & 2 \end{array} \right] \rightarrow \left[\begin{array}{ccc|cc} 1 & 2 & 1 & 2 & -1 \\ 0 & 1 & -1 & -3 & 5 \\ 0 & 0 & -1 & -4 & 4 \end{array} \right]$$

$$\rightarrow \left[\begin{array}{ccc|cc} 1 & 2 & 1 & 2 & -1 \\ 0 & 1 & -1 & -3 & 5 \\ 0 & 0 & 1 & 4 & -4 \end{array} \right] \rightarrow \left[\begin{array}{ccc|cc} 1 & 2 & 0 & -2 & 3 \\ 0 & 1 & 0 & 1 & 1 \\ 0 & 0 & 1 & 4 & -4 \end{array} \right]$$

$$\rightarrow \left[\begin{array}{ccc|cc} 1 & 0 & 0 & -4 & 1 \\ 0 & 1 & 0 & 1 & 1 \\ 0 & 0 & 1 & 4 & -4 \end{array} \right]$$

Therefore,

$$A^{-1}B = \begin{bmatrix} -4 & 1 \\ 1 & 1 \\ 4 & -4 \end{bmatrix}.$$

An Interpretation of the Inverse Matrix

Consider the system $A\mathbf{x} = \mathbf{b}$, where A is an invertible $n \times n$ matrix. It is often of interest to know how a solution to this system is affected by a change in $\mathbf{b}$. This situation may occur when $\mathbf{b}$ is altered by roundoff, or when, as we see below, $\mathbf{b}$ is replaced by a new demand vector in the Leontief input–output model. To examine what happens, let $P = A^{-1}$. Then

$$[\mathbf{e}_1 \ \mathbf{e}_2 \ \cdots \ \mathbf{e}_n] = I_n = AP = A[\mathbf{p}_1 \ \mathbf{p}_2 \ \cdots \ \mathbf{p}_n] = [A\mathbf{p}_1 \ A\mathbf{p}_2 \ \cdots \ A\mathbf{p}_n].$$

It follows that for each j, $A\mathbf{p}_j = \mathbf{e}_j$. Suppose that $\mathbf{u}$ is a solution to $A\mathbf{x} = \mathbf{b}$, and assume that we change the kth component of $\mathbf{b}$ from b_k to $b_k + d$, where d is some scalar. Thus we have replaced $\mathbf{b}$ by $\mathbf{b} + d\mathbf{e}_k$ to obtain the new system $A\mathbf{x} = \mathbf{b} + d\mathbf{e}_k$. Notice that

$$A(\mathbf{u} + d\mathbf{p}_k) = A\mathbf{u} + d\,A\mathbf{p}_k = \mathbf{b} + d\mathbf{e}_k;$$

so $\mathbf{u} + d\mathbf{p}_k$ is a solution to $A\mathbf{x} = \mathbf{b} + d\mathbf{e}_k$. This solution differs from the original solution by $d\mathbf{p}_k$; that is, $d\mathbf{p}_k$ measures the change in a solution to $A\mathbf{x} = \mathbf{b}$ when the kth component of $\mathbf{b}$ is increased by d.

One situation in which this type of information is useful occurs in the context of the Leontief input–output model (discussed in Section 1.5). When an economy is being planned, the gross production vectors corresponding to several different demand vectors may need to be calculated. For example, we may want to compare the effect of increasing the demand vector from

$$\mathbf{d}_1 = \begin{bmatrix} 90 \\ 80 \\ 60 \end{bmatrix} \quad \text{to} \quad \mathbf{d}_2 = \begin{bmatrix} 100 \\ 80 \\ 60 \end{bmatrix}.$$

If C is the input–output matrix for the economy and $I_3 - C$ is invertible, then such comparisons are easily made.[5] In this case, the solution to $(I_3 - C)\mathbf{x} = \mathbf{d}_1$ is $(I_3 - C)^{-1}\mathbf{d}_1$, as we saw in Section 2.3. Hence the gross production vector needed to meet the demand $\mathbf{d}_2$ is

$$(I_3 - C)^{-1}\mathbf{d}_1 + 10\mathbf{p}_1,$$

where $\mathbf{p}_1$ is the first column of $(I_3 - C)^{-1}$. For the economy in Example 2 of Section 1.5, we have

$$(I_3 - C)^{-1} = \begin{bmatrix} 1.3 & 0.475 & 0.25 \\ 0.6 & 1.950 & 0.50 \\ 0.5 & 0.375 & 1.25 \end{bmatrix} \quad \text{and} \quad (I_3 - C)^{-1}\mathbf{d}_1 = \begin{bmatrix} 170 \\ 240 \\ 150 \end{bmatrix}.$$

So the gross production vector needed to meet the demand $\mathbf{d}_2$ is

$$(I_3 - C)^{-1}\mathbf{d}_1 + 10\mathbf{p}_1 = \begin{bmatrix} 170 \\ 240 \\ 150 \end{bmatrix} + 10\begin{bmatrix} 1.3 \\ 0.6 \\ 0.5 \end{bmatrix} = \begin{bmatrix} 183 \\ 246 \\ 155 \end{bmatrix}.$$

Example 5 For the input–output matrix and demand in Example 2 of Section 1.5, determine the additional inputs needed to increase the demand for services from \$60 million to \$70 million.

Solution The additional inputs needed to increase demand for services by one unit are given by the third column of the preceding matrix $(I_3 - C)^{-1}$. Hence an increase

---

[5]In practice, the sums of the entries in each column of C are less than 1 because each dollar's worth of output normally requires inputs whose total value is less than one dollar. In this situation, it can be shown that $I_n - C$ is invertible and has nonnegative entries.

of $10 million in the demand for services requires additional inputs of

$$10 \begin{bmatrix} 0.25 \\ 0.50 \\ 1.25 \end{bmatrix} = \begin{bmatrix} 2.5 \\ 5.0 \\ 12.5 \end{bmatrix} ;$$

that is, additional inputs of $2.5 million of agriculture, $5 million of manufacturing, and $12.5 million of services.   ○

Invertible Matrices and Rank

We conclude this section with a result that tells us that multiplying a matrix by an invertible matrix preserves its rank.

Theorem 2.7
Let A be an $m \times n$ matrix. Then the following statements are true.
(a) *If P is an $m \times m$ invertible matrix, then rank $(PA) = $ rank A.*
(b) *If Q is an $n \times n$ invertible matrix, then rank $(AQ) = $ rank A.*

Proof (a) Suppose that P is invertible. Let $C = PA$, and let R be the reduced row echelon form of C. Then, as in Section 2.3, there is an invertible matrix B such that $R = BC = B(PA) = (BP)A$. Since B and P are invertible, BP is invertible. So by Theorem 2.6(k), BP is a product of elementary matrices, and hence A can be transformed into R by means of elementary row operations. It follows that R is the reduced row echelon form of A. Therefore C and A have the same rank, which is the number of nonzero rows of R.

(b) Let Q be an $n \times n$ matrix, A have rank r, and R be the reduced row echelon form of A. Then the last $m - r$ rows of R are zero, and hence the last $m - r$ rows of RQ are zero. It follows that the last $m - r$ rows of the reduced row echelon form of RQ are zero, and hence rank $RQ \leq r$. By Theorem 2.4 there is an invertible matrix P such that $PA = R$. Hence, by (a), we have

$$\text{rank } AQ = \text{rank } P(AQ) = \text{rank } (PA)Q = \text{rank } RQ \leq r = \text{rank } A.$$

Now suppose that Q is invertible. Then, applying the inequality above to $(AQ)Q^{-1}$, we have

$$\text{rank } A = \text{rank } A(QQ^{-1}) = \text{rank } (AQ)Q^{-1} \leq \text{rank } AQ.$$

Combining these two inequalities, we obtain rank $AQ = $ rank A.   ❑

Practice Problems

1. Determine if $A = \begin{bmatrix} 1 & -2 & 1 \\ 2 & -1 & -1 \\ -2 & -5 & 7 \end{bmatrix}$ is invertible. If so, find its inverse.

2. Determine if $B = \begin{bmatrix} 1 & 1 & 0 \\ 3 & 4 & 1 \\ -1 & 4 & 4 \end{bmatrix}$ is invertible. If so, find its inverse.

3. Consider the system of linear equations

$$\begin{aligned} x_1 - x_2 + 2x_3 &= 2 \\ x_1 + 2x_2 &= 3 \\ -x_2 + x_3 &= -1 . \end{aligned}$$

(a) Write the system in the form of a matrix equation $A\mathbf{x} = \mathbf{b}$.
(b) Show that A is invertible, and find A^{-1}.
(c) Solve the system using the answer to (b).

Exercises

1. Determine if the following statements are true or false.
   (a) For any two matrices A and B, if $AB = I_n$ for some positive integer n, then A is invertible.
   (b) For any two $n \times n$ matrices A and B, if $AB = I_n$, then $BA = I_n$.
   (c) For any two $n \times n$ matrices A and B, if $AB = I_n$, then A is invertible and $A^{-1} = B$.
   (d) If a square matrix has a column consisting of all zeros, then it is not invertible.
   (e) If a square matrix has a row consisting of all zeros, then it is not invertible.
   (f) If A and B are invertible $n \times n$ matrices, then $A + B$ is invertible.
   (g) If A is an $m \times n$ matrix, and B and C are invertible matrices such that BAC is defined, then A and BAC have the same rank.

In Exercises 2–10, determine if the given matrix is invertible. If so, find its inverse.

2. $\begin{bmatrix} 1 & 2 \\ 2 & 4 \end{bmatrix}$

3. $\begin{bmatrix} 1 & 3 \\ 1 & 2 \end{bmatrix}$

4. $\begin{bmatrix} 1 & 3 & 2 \\ 2 & 5 & 5 \\ 1 & 3 & 1 \end{bmatrix}$

5. $\begin{bmatrix} 1 & -2 & 1 \\ 1 & 0 & 1 \\ 1 & -1 & 1 \end{bmatrix}$

6. $\begin{bmatrix} 2 & -1 & 2 \\ 1 & 0 & 3 \\ 0 & 1 & 4 \end{bmatrix}$

7. $\begin{bmatrix} 1 & 1 & 2 \\ 2 & -1 & 1 \\ 2 & 3 & 4 \end{bmatrix}$

8. $\begin{bmatrix} 1 & 2 & 1 & -1 \\ 2 & 5 & 1 & -1 \\ 1 & 3 & 1 & 2 \\ 2 & 4 & 2 & -1 \end{bmatrix}$

9. $\begin{bmatrix} 1 & 0 & 0 & 1 \\ 0 & 1 & 1 & 0 \\ 1 & 0 & 1 & 0 \\ 0 & 1 & 0 & 1 \end{bmatrix}$

10. $\begin{bmatrix} 1 & 1 & 1 & 0 \\ 1 & 1 & 0 & 1 \\ 1 & 0 & 1 & 1 \\ 0 & 1 & 1 & 1 \end{bmatrix}$

In Exercises 11–16, use the algorithm for computing $A^{-1}B$.

11. $A = \begin{bmatrix} 1 & 2 \\ 2 & 3 \end{bmatrix}$ and $B = \begin{bmatrix} 1 & -1 & 2 \\ 1 & 0 & 1 \end{bmatrix}$

12. $A = \begin{bmatrix} -1 & 2 \\ 2 & -3 \end{bmatrix}$ and $B = \begin{bmatrix} 4 & -1 \\ 1 & 2 \end{bmatrix}$

13. $A = \begin{bmatrix} 2 & 2 \\ 2 & 1 \end{bmatrix}$ and $B = \begin{bmatrix} 2 & 4 & 2 & 6 \\ 0 & -2 & 8 & -4 \end{bmatrix}$

14. $A = \begin{bmatrix} 1 & -1 & 1 \\ 2 & -1 & 4 \\ 2 & -2 & 3 \end{bmatrix}$ and $B = \begin{bmatrix} 3 & -2 \\ 1 & -1 \\ 4 & 2 \end{bmatrix}$

15. $A = \begin{bmatrix} -2 & 3 & 7 \\ -1 & 1 & 2 \\ 1 & 1 & 2 \end{bmatrix}$
    and $B = \begin{bmatrix} 2 & 0 & 1 & -1 \\ 1 & 2 & -2 & 1 \\ 3 & 1 & 1 & 3 \end{bmatrix}$

16. $A = \begin{bmatrix} 1 & 0 & 1 & 1 \\ 0 & 1 & 1 & -1 \\ 0 & 0 & 1 & -1 \\ 0 & 0 & 0 & 1 \end{bmatrix}$ and
    $B = \begin{bmatrix} 2 & 1 & -1 \\ 0 & 1 & 1 \\ 1 & 0 & 1 \\ 3 & 1 & 2 \end{bmatrix}$

In Exercises 17–22, a matrix A is given. Determine (a) the reduced row echelon form R of A, and (b) an invertible matrix P such that $PA = R$.

17. $\begin{bmatrix} 1 & -1 & 2 \\ -2 & 1 & -1 \end{bmatrix}$

18. $\begin{bmatrix} 1 & 1 & -1 \\ 1 & -1 & 2 \\ 1 & 0 & 1 \end{bmatrix}$

19. $\begin{bmatrix} -1 & 0 & 2 & 1 \\ 0 & 1 & 1 & -1 \\ 2 & 3 & -1 & -5 \end{bmatrix}$

20. $\begin{bmatrix} 1 & -1 & 0 & -1 & 2 \\ -1 & 1 & 1 & -2 & 1 \\ 5 & -5 & -3 & 4 & 1 \end{bmatrix}$

21. $\begin{bmatrix} 2 & 1 & 0 & -2 \\ 0 & 1 & -1 & 0 \\ -1 & -2 & 2 & 1 \\ 1 & 3 & 1 & 0 \end{bmatrix}$

22. $\begin{bmatrix} 1 & 0 & 1 & 2 & 1 \\ 0 & 1 & -1 & -1 & 0 \\ 1 & 1 & -2 & 7 & 4 \\ 2 & 1 & 3 & -3 & -1 \end{bmatrix}$

23. Suppose that $A\mathbf{x} = \mathbf{b}$ is a system of linear equations such that A is an invertible $n \times n$ matrix.
    (a) Prove that a vector $\mathbf{v}$ is a solution to the system if and only if $\mathbf{v} = A^{-1}\mathbf{b}$.
    (b) Use the result of (a) to prove (e) and (h) of Theorem 2.6.

In Exercises 24–27, a system of linear equations is given. For each exercise:
    (a) Write the system as a matrix equation $A\mathbf{x} = \mathbf{b}$;
    (b) Show that A is invertible, and find A^{-1}; and
    (c) Use the method of Exercise 23 to solve the system.

24. $\begin{aligned} x_1 + 2x_2 &= 9 \\ 2x_1 + 3x_2 &= 3 \end{aligned}$

25. $\begin{aligned} -x_1 - 3x_2 &= -6 \\ 2x_1 + 5x_2 &= 4 \end{aligned}$

26.
$$x_1 + x_2 + x_3 = 4$$
$$2x_1 + x_2 + 4x_3 = 7$$
$$3x_1 + 2x_2 + 6x_3 = -1$$

27.
$$-x_1 + x_3 = -4$$
$$x_1 + 2x_2 - 2x_3 = 3$$
$$2x_1 - x_2 + x_3 = 1$$

28. Let A be an $n \times n$ matrix such that $A^2 = I_n$. Prove that A is invertible and $A^{-1} = A$.

29. Let A be an $n \times n$ matrix such that $A^k = I_n$ for some positive integer k.

(a) Prove that A is invertible.
(b) Express A^{-1} as a power of A.

30. Let $A = \begin{bmatrix} 1 & 1 \\ 1 & 2 \end{bmatrix}$.

(a) Verify that $A^2 - 3A + I_2 = O$.
(b) Use the result of (a) to prove that A is invertible and that $A^{-1} = 3I_2 - A$.

31. Let R be an $m \times n$ matrix in reduced row echelon form with rank $R = r$. Prove that the reduced row echelon form of R^T is the $n \times m$ matrix $[\mathbf{e}_1 \ \mathbf{e}_2 \ \cdots \ \mathbf{e}_r \ \mathbf{0} \ \cdots \ \mathbf{0}]$, where $\mathbf{e}_j$ is the jth standard vector of $\mathcal{R}^n$ for $1 \le j \le r$.

32. Prove that for any matrix A, rank A = rank A^T. *Hint:* Combine the results of Theorem 2.4, Exercise 31, and Theorem 2.7.

33. According to Theorem 2.4, if A is an $m \times n$ matrix with reduced row echelon form R, there exists an invertible $m \times m$ matrix P such that $PA = R$. Is P unique? Justify your answer.

34. Prove that for any $m \times n$ matrix A and any $n \times p$ matrix B, rank $AB \le$ rank A. *Hint:* Apply Theorem 2.4 to A, and then use Theorem 2.7.

35. Combine Exercises 32 and 34 to prove that for any $m \times n$ matrix A and any $n \times p$ matrix B, rank $AB \le$ rank B.

36. Use Theorem 2.6 to prove that for any subset $\mathcal{S}$ of n vectors in $\mathcal{R}^n$, the set $\mathcal{S}$ is linearly independent if and only if $\mathcal{S}$ spans $\mathcal{R}^n$.

37. Let R and S be matrices with the same number of rows, and suppose that the matrix $[R \ S]$ is in reduced row echelon form. Prove that R is in reduced echelon form.

38. Consider the system of linear equations $A\mathbf{x} = \mathbf{b}$, where
$$A = \begin{bmatrix} 1 & 2 & 3 \\ 2 & 3 & 4 \\ 3 & 4 & 5 \end{bmatrix} \quad \text{and} \quad \mathbf{b} = \begin{bmatrix} 20 \\ 30 \\ 40 \end{bmatrix}.$$

(a) Solve this matrix equation using Gaussian elimination.
(b) If this system is solved on a TI-85 calculator using the method described in Exercise 23, we obtain a solution to
$$A^{-1}\mathbf{b} = \begin{bmatrix} 8 \\ 10 \\ 4 \end{bmatrix}.$$

But this is *not* a solution to $A\mathbf{x} = \mathbf{b}$. Does this example contradict Exercise 23? Explain why or why not.

39. Repeat Exercise 38 with
$$A = \begin{bmatrix} 1 & 2 & 3 \\ 2 & 3 & 4 \\ 6 & 7 & 8 \end{bmatrix}, \quad \mathbf{b} = \begin{bmatrix} 5 \\ 6 \\ 10 \end{bmatrix}, \quad \text{and}$$
$$A^{-1}\mathbf{b} = \begin{bmatrix} 0 \\ -8 \\ 3 \end{bmatrix}.$$

40. Repeat Exercise 38 with
$$A = \begin{bmatrix} 1 & 2 & 3 \\ 4 & 5 & 6 \\ 7 & 8 & 9 \end{bmatrix}, \quad \mathbf{b} = \begin{bmatrix} 15 \\ 18 \\ 21 \end{bmatrix}, \quad \text{and} \quad A^{-1}\mathbf{b} = \begin{bmatrix} -9 \\ 8 \\ 4 \end{bmatrix}.$$

41. In Exercise 15(c) of Section 1.5, how much is required in additional input from each sector to increase the net production of oil by $3 million?

42. In Exercise 14(c) of Section 1.5, how much is required in additional input from each sector to increase net production in the nongovernment sector by $1 million?

43. In Exercise 17(b) of Section 1.5, how much is required in additional input from each sector to increase the net production of services by $40 million?

44. In Exercise 16(b) of Section 1.5, how much is required in additional input from each sector to increase the net production of manufacturing by $24 million?

45. Suppose that the input–output matrix C for an economy is such that $I_n - C$ is invertible and every entry of $(I_n - C)^{-1}$ is positive. If the net production of one particular sector of the economy must be increased, how does this affect the gross production of the economy?

46. Let A be an $m \times n$ matrix with reduced row echelon form R. There exists an invertible matrix P such that $PA = R$ and an invertible matrix Q such that QR^T is the reduced row echelon form of R^T (see Exercise 31). Describe the matrix PAQ^T in terms of A. Justify your answer.

47. Let A be an $m \times n$ matrix with reduced row echelon form R.

(a) Prove that if rank $A = m$, then there is a unique $m \times m$ matrix P such that $PA = R$. Furthermore, P is invertible. *Hint:* For each j, let $\mathbf{u}_j$ denote the jth pivot column of A. Prove that the $m \times m$ matrix $[\mathbf{u}_1 \ \mathbf{u}_2 \ \cdots \ \mathbf{u}_m]$ is invertible. Now let P be the inverse of this matrix.
(b) Prove that if rank $A < m$, then there is more than one invertible $m \times m$ matrix P such that $PA = R$. *Hint:* There is an elementary $m \times m$ matrix E distinct from I_m such that $ER = R$.

*Let A and B be $n \times n$ matrices. We say that A is **similar** to B if $B = P^{-1}AP$ for some invertible matrix P. Exercises 48–52 deal with the similarity relation.*

48. Let A, B, and C be $n \times n$ matrices. Prove the following.

   (a) A is similar to A.
   (b) If A is similar to B, then B is similar to A.
   (c) If A is similar to B and B is similar to C, then A is similar to C.

49. Let A be an $n \times n$ matrix.

   (a) Prove that if A is similar to I_n, then $A = I_n$.
   (b) Prove that if A is similar to O, the $n \times n$ zero matrix, then $A = O$.
   (c) Suppose that $B = cI_n$ for some scalar c. (The matrix B is called a *scalar matrix*.) What can you say about A if A is similar to B?

50. Suppose that A and B are $n \times n$ matrices such that A is similar to B. Prove that if A is invertible, then B is invertible, and A^{-1} is similar to B^{-1}.

51. Suppose that A and B are $n \times n$ matrices such that A is similar to B. Prove that A^T is similar to B^T.

52. Suppose that A and B are $n \times n$ matrices such that A is similar to B. Prove that rank A = rank B.

In Exercises 53–56, use a calculator with matrix capabilities or computer software such as MATLAB *to solve the problem.*

Exercises 53–55 refer to the matrix

$$A = \begin{bmatrix} 2 & 5 & 6 & 1 \\ 3 & 8 & 9 & 2 \\ 2 & 6 & 5 & 2 \\ 3 & 9 & 7 & 4 \end{bmatrix}.$$

53. Show that A is invertible by computing its reduced row echelon form and using Theorem 2.6.

54. Show that A is invertible by solving the system $A\mathbf{x} = \mathbf{0}$, and using Theorem 2.6.

55. Show that A is invertible by computing its rank, and using Theorem 2.6.

56. Show that the matrix

$$P = \begin{bmatrix} 1 & 2 & -1 & 3 \\ 2 & 3 & 2 & 8 \\ 2 & 4 & -1 & 4 \\ 3 & 6 & -2 & 8 \end{bmatrix}$$

is invertible. Illustrate Theorem 2.7 by creating several randomly generated 4×4 matrices A, and showing that rank PA = rank A.

Solutions to the Practice Problems

1. The reduced row echelon form of A is $\begin{bmatrix} 1 & 0 & -1 \\ 0 & 1 & -1 \\ 0 & 0 & 0 \end{bmatrix}$.

   Since this matrix is not I_3, Theorem 2.6(j) implies that A is not invertible.

2. The reduced row echelon form of B is I_3, and so Theorem 2.6(j) implies that B is invertible. To compute B^{-1}, we find the reduced row echelon form of $[B \ I_3]$:

$$\begin{bmatrix} 1 & 1 & 0 & 1 & 0 & 0 \\ 3 & 4 & 1 & 0 & 1 & 0 \\ -1 & 4 & 4 & 0 & 0 & 1 \end{bmatrix} \longrightarrow \begin{bmatrix} 1 & 1 & 0 & 1 & 0 & 0 \\ 0 & 1 & 1 & -3 & 1 & 0 \\ 0 & 5 & 4 & 1 & 0 & 1 \end{bmatrix}$$

$$\longrightarrow \begin{bmatrix} 1 & 1 & 0 & 1 & 0 & 0 \\ 0 & 1 & 1 & -3 & 1 & 0 \\ 0 & 0 & -1 & 16 & -5 & 1 \end{bmatrix}$$

$$\longrightarrow \begin{bmatrix} 1 & 1 & 0 & 1 & 0 & 0 \\ 0 & 1 & 1 & -3 & 1 & 0 \\ 0 & 0 & 1 & -16 & 5 & -1 \end{bmatrix}$$

$$\longrightarrow \begin{bmatrix} 1 & 1 & 0 & 1 & 0 & 0 \\ 0 & 1 & 0 & 13 & -4 & 1 \\ 0 & 0 & 1 & -16 & 5 & -1 \end{bmatrix}$$

$$\longrightarrow \begin{bmatrix} 1 & 0 & 0 & -12 & 4 & -1 \\ 0 & 1 & 0 & 13 & -4 & 1 \\ 0 & 0 & 1 & -16 & 5 & -1 \end{bmatrix}$$

   Thus $B^{-1} = \begin{bmatrix} -12 & 4 & -1 \\ 13 & -4 & 1 \\ -16 & 5 & -1 \end{bmatrix}$.

3. (a) The matrix form of the given system of linear equations is

$$\begin{bmatrix} 1 & -1 & 2 \\ 1 & 2 & 0 \\ 0 & -1 & 1 \end{bmatrix} \begin{bmatrix} x_1 \\ x_2 \\ x_3 \end{bmatrix} = \begin{bmatrix} 2 \\ 3 \\ -1 \end{bmatrix}.$$

   (b) Because the reduced row echelon of the 3×3 matrix in (a) is I_3, the matrix is invertible.

   (c) The solution to $A\mathbf{x} = \mathbf{b}$ is

$$\mathbf{x} = A^{-1}\mathbf{b} = \begin{bmatrix} 2 & -1 & -4 \\ -1 & 1 & 2 \\ -1 & 1 & 3 \end{bmatrix} \begin{bmatrix} 2 \\ 3 \\ -1 \end{bmatrix} = \begin{bmatrix} 5 \\ -1 \\ -2 \end{bmatrix}.$$

   Thus the unique solution to the given system of linear equations is $x_1 = 5$, $x_2 = -1$, and $x_3 = -2$.

2.5*   The LU Decomposition of a Matrix

Sections 1.3 and 1.4 describe how to use Gaussian elimination, an efficient method for solving a system of linear equations. In practice, it is often necessary to solve many systems of linear equations with the same coefficient matrix. In these situations, using Gaussian elimination on each system involves a great deal of duplication of effort since the augmented matrices for these systems are almost identical. In this section, we examine a method that avoids this duplication.

For now, suppose that an $m \times n$ matrix A can be transformed into a matrix U in row echelon form using steps 1, 3, and 4 *but not step 2* (interchanging rows) of the Gaussian elimination algorithm described in Section 1.4. Then U can be written as

$$U = E_k E_{k-1} \cdots E_1 A,$$

where $E_1, \ldots, E_{k-1}, E_k$ are the elementary matrices corresponding to the elementary row operations that transform A into U. Solving this equation for A, we obtain

$$A = (E_k E_{k-1} \cdots E_1)^{-1} U = E_1^{-1} E_2^{-1} \cdots E_k^{-1} U = LU,$$

where

$$L = E_1^{-1} E_2^{-1} \cdots E_k^{-1}.$$

Since each elementary row operation used in the process of transforming A into U is the result of adding a multiple of a row to a subsequent row of a matrix, the corresponding elementary matrix E_p and its inverse E_p^{-1} are of the forms

$$E_p = \begin{array}{c} \\ \\ \text{row } j \rightarrow \\ \\ \\ \text{row } i \rightarrow \\ \end{array} \begin{bmatrix} 1 & 0 & \cdots & & & & 0 \\ 0 & \ddots & & & & & \vdots \\ & & 1 & & & & \\ \vdots & & \vdots & \ddots & & & \\ & & c & & \ddots & & \\ 0 & & & & & & 1 \end{bmatrix}$$

$$\begin{array}{c} \uparrow \\ \text{column } j \end{array}$$

and

$$E_p^{-1} = \begin{array}{c} \\ \\ \text{row } j \rightarrow \\ \\ \\ \text{row } i \rightarrow \\ \end{array} \begin{bmatrix} 1 & 0 & \cdots & & & & 0 \\ 0 & \ddots & & & & & \vdots \\ & & 1 & & & & \\ \vdots & & \vdots & \ddots & & & \\ & & -c & & \ddots & & \\ 0 & & & & & & 1 \end{bmatrix}$$

$$\begin{array}{c} \uparrow \\ \text{column } j \end{array}$$

where c is the multiple, and $j < i$ are the rows. Notice that E_p and E_p^{-1} can be

---

*This section can be omitted without loss of continuity.

obtained from I_m by changing the (i, j)-entry from 0 to c and from 0 to $-c$, respectively.

We can make some observations about L and U. Since U is in row echelon form, the entries of U below and to the left of the diagonal entries are all zeros. Any matrix with this description is called an **upper triangular matrix**. The entries above and to the right of the diagonal entries of each E_p^{-1} are zeros. Any matrix with this description is called a **lower triangular matrix**. Furthermore, the diagonal entries of each E_p^{-1} are ones. A lower triangular matrix whose diagonal entries are all ones is called a **unit lower triangular matrix**. Since the product of unit lower triangular matrices is a unit lower triangular matrix (see Exercise 12), L is also unit lower triangular. Thus we can factor $A = LU$ into the product of a unit lower triangular matrix L and an upper triangular matrix U. Such a factorization is called an **LU decomposition of** A. It can be shown that if a matrix has an LU decomposition and is also invertible, then the LU decomposition is unique (see Exercise 14).

Example 1   Let

$$A = \begin{bmatrix} 1 & 0 & 0 \\ 0 & 2 & 0 \\ 3 & 4 & 3 \end{bmatrix}, \qquad B = \begin{bmatrix} 1 & 0 \\ 4 & 1 \end{bmatrix}, \qquad \text{and} \qquad C = \begin{bmatrix} 2 & 0 & 1 & -1 \\ 0 & 0 & 3 & 4 \\ 0 & 0 & 3 & 0 \end{bmatrix}.$$

Both A and B are lower triangular matrices, because the entries above and to the right of the diagonal entries are zeros. Both diagonal entries of B are ones; hence B is a unit lower triangular matrix, whereas A is not. The entries below and to the left of the diagonal entries of C are zeros; hence C is an upper triangular matrix.   ○

It should be noted that not every matrix can be transformed into a matrix in row echelon form without the use of row interchanges. The matrix $\begin{bmatrix} 0 & 1 \\ 1 & 0 \end{bmatrix}$ is a simple example. It can be shown that if a matrix cannot be transformed into a matrix in row echelon form without the use of row interchanges, then the matrix has no LU decomposition.

For the present, we consider an $m \times n$ matrix A that can be transformed into the $m \times n$ matrix U in row echelon form using elementary row operations without interchanging rows. We have two tasks before us. The first is to explain a practical way of finding an $m \times m$ unit lower triangular matrix L such that $A = LU$ in order to obtain an LU decomposition of A. The second is to show how to use the LU decomposition of A to solve a system of linear equations $A\mathbf{x} = \mathbf{b}$. Given the LU decomposition of A, the number of steps required to solve several systems of linear equations with coefficient matrix A is significantly less than the total number of steps required to use Gaussian elimination on each system separately.

Computing the LU Decomposition

Since $L = E_1^{-1} E_2^{-1} \cdots E_k^{-1} I_m$, we can compute L from I_m by applying the elementary row operations corresponding to the E_p^{-1}'s, starting with the last operation and working our way back to the first. If the elementary row operation corresponding to E_p involves adding c times row j to row i of a matrix, then the elementary matrix corresponding to E_p^{-1} involves adding $-c$ times row j to row i of that matrix. We illustrate this process in the following examples.

Example 2   Find the LU decomposition of the matrix

$$A = \begin{bmatrix} 1 & -1 & 2 \\ 3 & -1 & 7 \\ 2 & -4 & 5 \end{bmatrix}.$$

Solution   First, we use Gaussian elimination to transform A into an upper triangular matrix U in row echelon form without the use of row interchanges. This process consists of three elementary row operations performed in succession: adding -3 times row 1 to row 2 of A, adding -2 times row 1 to row 3 of the resulting matrix, and adding 1 times row 2 to row 3 of the previous matrix to obtain the final result, U. This is illustrated by the following diagram, where the label $(c)\mathbf{r}_j + \mathbf{r}_i$ above an arrow describes the elementary row operation in which c times row j is added to row i to transform the matrix before the arrow into the matrix after the arrow:

$$A = \begin{bmatrix} 1 & -1 & 2 \\ 3 & -1 & 7 \\ 2 & -4 & 5 \end{bmatrix} \xrightarrow{(-3)\mathbf{r}_1+\mathbf{r}_2} \begin{bmatrix} 1 & -1 & 2 \\ 0 & 2 & 1 \\ 2 & -4 & 5 \end{bmatrix} \xrightarrow{(-2)\mathbf{r}_1+\mathbf{r}_3} \begin{bmatrix} 1 & -1 & 2 \\ 0 & 2 & 1 \\ 0 & -2 & 1 \end{bmatrix}$$

$$\xrightarrow{(1)\mathbf{r}_2+\mathbf{r}_3} \begin{bmatrix} 1 & -1 & 2 \\ 0 & 2 & 1 \\ 0 & 0 & 2 \end{bmatrix} = U.$$

The reverse of the last operation, adding -1 times row 2 to row 3 of a matrix, is the first operation used in the transformation of I_3 into L. We continue to apply the reverse row operations in the opposite order to complete the transformation of I_3 into L:

$$I_3 = \begin{bmatrix} 1 & 0 & 0 \\ 0 & 1 & 0 \\ 0 & 0 & 1 \end{bmatrix} \xrightarrow{(-1)\mathbf{r}_2+\mathbf{r}_3} \begin{bmatrix} 1 & 0 & 0 \\ 0 & 1 & 0 \\ 0 & -1 & 1 \end{bmatrix} \xrightarrow{(2)\mathbf{r}_1+\mathbf{r}_3} \begin{bmatrix} 1 & 0 & 0 \\ 0 & 1 & 0 \\ 2 & -1 & 1 \end{bmatrix}$$

$$\xrightarrow{(3)\mathbf{r}_1+\mathbf{r}_2} \begin{bmatrix} 1 & 0 & 0 \\ 3 & 1 & 0 \\ 2 & -1 & 1 \end{bmatrix} = L.$$

In retrospect, we see that the entries of L below the diagonal can be obtained directly from the row operations used to transform A into U. In particular, the (i, j)-entry of L is $-c$, where c times row j is added to row i in one of the elementary row operations used to transform A into U. For example, in the first of the three elementary row operations used to transform A into U, -3 times row 1 is added to row 2. Thus the $(2, 1)$-entry of L is 3. Since -2 times row 1 is added to row 3 in the second operation, 2 is the $(3, 1)$-entry of L. Finally, 1 times row 2 is added to row 3 to complete the transformation of A to U, and hence -1 is the $(3, 2)$-entry of L. These entries below the diagonal of L are called **multipliers**. The following diagram shows how I_3 is transformed into L by changing the entries below the diagonal of I_3 into the multipliers obtained from the elementary row operations used to transform

A into *U*.

$$A = \begin{bmatrix} 1 & -1 & 2 \\ 3 & -1 & 7 \\ 2 & -4 & 5 \end{bmatrix} \qquad I_3 = \begin{bmatrix} 1 & 0 & 0 \\ 0 & 1 & 0 \\ 0 & 0 & 1 \end{bmatrix}$$

$$(-3)\mathbf{r}_1 + \mathbf{r}_2 \Bigg\downarrow \qquad\qquad\qquad\qquad \Bigg\downarrow$$

$$\begin{bmatrix} 1 & -1 & 2 \\ 0 & 2 & 1 \\ 2 & -4 & 5 \end{bmatrix} \qquad \begin{bmatrix} 1 & 0 & 0 \\ 3 & 1 & 0 \\ 0 & 0 & 1 \end{bmatrix}$$

$$(-2)\mathbf{r}_1 + \mathbf{r}_3 \Bigg\downarrow \qquad\qquad\qquad\qquad \Bigg\downarrow$$

$$\begin{bmatrix} 1 & -1 & 2 \\ 0 & 2 & 1 \\ 0 & -2 & 1 \end{bmatrix} \qquad \begin{bmatrix} 1 & 0 & 0 \\ 3 & 1 & 0 \\ 2 & 0 & 1 \end{bmatrix}$$

$$(1)\mathbf{r}_2 + \mathbf{r}_3 \Bigg\downarrow \qquad\qquad\qquad\qquad \Bigg\downarrow$$

$$U = \begin{bmatrix} 1 & -1 & 2 \\ 0 & 2 & 1 \\ 0 & 0 & 2 \end{bmatrix} \qquad \begin{bmatrix} 1 & 0 & 0 \\ 3 & 1 & 0 \\ 2 & -1 & 1 \end{bmatrix} = L$$

Example 3   Find an *LU* decomposition of the matrix

$$A = \begin{bmatrix} 2 & -2 & 2 & 4 \\ -2 & 4 & 2 & -1 \\ 6 & -2 & 4 & 14 \end{bmatrix}.$$

Solution   We continue using the notation in Example 2 to indicate the various row operations performed in transforming *A* into *U*. We have

$$A = \begin{bmatrix} 2 & -2 & 2 & 4 \\ -2 & 4 & 2 & -1 \\ 6 & -2 & 4 & 14 \end{bmatrix} \xrightarrow{(1)\mathbf{r}_1 + \mathbf{r}_2} \begin{bmatrix} 2 & -2 & 2 & 4 \\ 0 & 2 & 4 & 3 \\ 6 & -2 & 4 & 14 \end{bmatrix}$$

$$\xrightarrow{(-3)\mathbf{r}_1 + \mathbf{r}_3} \begin{bmatrix} 2 & -2 & 2 & 4 \\ 0 & 2 & 4 & 3 \\ 0 & 4 & -2 & 2 \end{bmatrix} \xrightarrow{(-2)\mathbf{r}_2 + \mathbf{r}_3} \begin{bmatrix} 2 & -2 & 2 & 4 \\ 0 & 2 & 4 & 3 \\ 0 & 0 & -10 & -4 \end{bmatrix} = U.$$

We are now prepared to obtain *L*. Of course, the diagonal entries of *L* are ones and the entries above the diagonal are zeros. The entries below the diagonal, the multipliers, can be obtained directly from the labels above the arrows in the transformation of *A* into *U*. A label of the form $(c)\mathbf{r}_j + \mathbf{r}_i$ indicates that the (i, j)-entry of *L* is $-c$. It follows that

$$L = \begin{bmatrix} 1 & 0 & 0 \\ -1 & 1 & 0 \\ 3 & 2 & 1 \end{bmatrix}.$$

We now summarize the process of obtaining the LU decomposition of a matrix.

The LU decomposition of an $m \times n$ matrix A

(a) Use steps 1, 3, and 4 of Gaussian elimination (as described in Section 1.4) to transform A into a matrix U in row echelon form by means of elementary row operations. If this is impossible, then A has no LU decomposition.

(b) While performing (a), create an $m \times m$ matrix L as follows.

   (i) Each diagonal entry of L is 1.

   (ii) If some elementary row operation in (a) adds c times row j of a matrix to row i, then $l_{ij} = -c$; otherwise, $l_{ij} = 0$.

Using an LU Decomposition to Solve a System of Linear Equations

Consider a system of m linear equations in n variables of the form $A\mathbf{x} = \mathbf{b}$. Suppose that the $m \times n$ coefficient matrix A has an LU decomposition $A = LU$. Then the system can be written as $LU\mathbf{x} = \mathbf{b}$. Letting $\mathbf{y} = U\mathbf{x}$, we can write the system as $L\mathbf{y} = \mathbf{b}$. Since L is a unit lower triangular square matrix, it is invertible. Hence the system $L\mathbf{y} = \mathbf{b}$ has a unique solution. Furthermore, the first equation of this system has only one variable, for which we can easily solve. The solution is substituted into the second equation, which reduces the number of variables in this equation to one. Solve for this variable. We continue this process, working our way downward from one equation to the next, substituting the values of the variables for which we have already solved, and then solving for the single remaining variable. In this manner, we can solve the system $L\mathbf{y} = \mathbf{b}$ for $\mathbf{y}$. This process of solving a system of linear equations with a lower triangular coefficient matrix is called **forward substitution**. Once $\mathbf{y}$ has been obtained, we use the system of equations $U\mathbf{x} = \mathbf{y}$ to solve for $\mathbf{x}$.

Now suppose that the coefficient matrix A is invertible. Then U is invertible, and hence the process of solving $U\mathbf{x} = \mathbf{y}$ for $\mathbf{x}$ is fairly simple. Since U is upper triangular and invertible, the last equation in the system has only one variable, for which we can easily solve. We then substitute this value in the equation above and solve it for the remaining variable. We continue this process, working our way upward just as we worked our way downward in solving $L\mathbf{y} = \mathbf{b}$ for $\mathbf{y}$. Since we work backwards, this process is called **back substitution**. We illustrate this procedure in the following example.

Example 4   Use LU decomposition to solve the system

$$\begin{aligned}
x_1 - x_2 + 2x_3 &= 2 \\
3x_1 - x_2 + 7x_3 &= 10 \\
2x_1 - 4x_2 + 5x_3 &= 4.
\end{aligned}$$

The coefficient matrix of this system is

$$A = \begin{bmatrix} 1 & -1 & 2 \\ 3 & -1 & 7 \\ 2 & -4 & 5 \end{bmatrix}.$$

The *LU* decomposition of A, which we obtained in Example 2, is given by

$$L = \begin{bmatrix} 1 & 0 & 0 \\ 3 & 1 & 0 \\ 2 & -1 & 1 \end{bmatrix} \quad \text{and} \quad U = \begin{bmatrix} 1 & -1 & 2 \\ 0 & 2 & 1 \\ 0 & 0 & 2 \end{bmatrix}.$$

We first solve the system $L\mathbf{y} = \mathbf{b}$, with

$$\mathbf{y} = \begin{bmatrix} y_1 \\ y_2 \\ y_3 \end{bmatrix} \quad \text{and} \quad \mathbf{b} = \begin{bmatrix} 2 \\ 10 \\ 4 \end{bmatrix},$$

which can be written

$$\begin{aligned} y_1 & & = 2 \\ 3y_1 + y_2 & & = 10 \\ 2y_1 - y_2 + y_3 & = & 4. \end{aligned}$$

Using forward substitution, we immediately obtain that $y_1 = 2$. Substituting this into the second equation, we solve for y_2 to obtain $y_2 = 10 - 3(2) = 4$. Substituting the values for y_1 and y_2 into the third equation, we obtain $y_3 = 4 - 2(2) + 4 = 4$. Thus

$$\mathbf{y} = \begin{bmatrix} 2 \\ 4 \\ 4 \end{bmatrix}.$$

We now use back substitution to solve the system $U\mathbf{x} = \mathbf{y}$, which can be written

$$\begin{aligned} x_1 - x_2 + 2x_3 &= 2 \\ 2x_2 + x_3 &= 4 \\ 2x_3 &= 4. \end{aligned}$$

Solving the third equation, we obtain $x_3 = 2$. Substituting this into the second equation and solving for x_2, we obtain $x_2 = (4-2)/2 = 1$. Finally, substituting the values for x_3 and x_2 into the first equation and solving for x_1 gives $x_1 = 2 + 1 - 2(2) = -1$. Thus

$$\mathbf{x} = \begin{bmatrix} x_1 \\ x_2 \\ x_3 \end{bmatrix} = \begin{bmatrix} -1 \\ 1 \\ 4 \end{bmatrix}.$$

If the coefficient matrix of a system of linear equations is not invertible—for example, if the matrix is not square—we can still solve the system using an *LU* decomposition. In this case, the process of back substitution is complicated by the presence of free variables. The following example illustrates this situation.

Example 5   Use *LU* decomposition to solve the system

$$\begin{aligned} 2x_1 - 2x_2 + 2x_3 + 4x_4 &= 6 \\ -2x_1 + 4x_2 + 2x_3 - x_4 &= 4 \\ 6x_1 - 2x_2 + 4x_3 + 14x_4 &= 20. \end{aligned}$$

The coefficient matrix of this system is

$$A = \begin{bmatrix} 2 & -2 & 2 & 4 \\ -2 & 4 & 2 & -1 \\ 6 & -2 & 4 & 14 \end{bmatrix}.$$

An LU decomposition of A was obtained in Example 3; it is given by

$$L = \begin{bmatrix} 1 & 0 & 0 \\ -1 & 1 & 0 \\ 3 & 2 & 1 \end{bmatrix} \quad \text{and} \quad U = \begin{bmatrix} 2 & -2 & 2 & 4 \\ 0 & 2 & 4 & 3 \\ 0 & 0 & -10 & -4 \end{bmatrix}.$$

We first use forward substitution to solve the system $L\mathbf{y} = \mathbf{b}$, where

$$\mathbf{y} = \begin{bmatrix} y_1 \\ y_2 \\ y_3 \end{bmatrix} \quad \text{and} \quad \mathbf{b} = \begin{bmatrix} 6 \\ 4 \\ 20 \end{bmatrix}.$$

As in Example 4, the unique solution to this system can be shown to be

$$\mathbf{y} = \begin{bmatrix} 6 \\ 10 \\ -18 \end{bmatrix}.$$

Next, we use back substitution to solve the system $U\mathbf{x} = \mathbf{y}$, which can be written

$$\begin{aligned} 2x_1 - 2x_2 + 2x_3 + 4x_4 &= 6 \\ 2x_2 + 4x_3 + 3x_4 &= 10 \\ -10x_3 - 4x_4 &= -18. \end{aligned}$$

We begin with the last equation. In this equation, we solve for the first variable x_3, treating x_4 as a free variable. This yields

$$x_3 = \frac{9}{5} - \frac{2}{5}x_4.$$

Working our way upwards, we substitute this solution into the second equation and solve for the first variable in this equation, x_2. Any other new variables that we encounter here are treated as free variables, but in this case there are none. Solving for x_2, we obtain

$$2x_2 = 10 - 4x_3 - 3x_4 = 10 - 4\left(\frac{9}{5} - \frac{2}{5}x_4\right) - 3x_4 = \frac{14}{5} - \frac{7}{5}x_4;$$

hence

$$x_2 = \frac{7}{5} - \frac{7}{10}x_4.$$

Finally, we solve for x_1 in the first equation, substituting the expressions we have already obtained in the previous equations. In this case, there are no new variables other than x_1, and hence no additional free variables. Thus we have

$$\begin{aligned} 2x_1 &= 6 + 2x_2 - 2x_3 - 4x_4 \\ &= 6 + 2\left(\frac{7}{5} - \frac{7}{10}x_4\right) - 2\left(\frac{9}{5} - \frac{2}{5}x_4\right) - 4x_4 \\ &= \frac{26}{5} - \frac{23}{5}x_4; \end{aligned}$$

hence

$$x_1 = \frac{13}{5} - \frac{23}{10}x_4.$$

○

Relative Efficiencies of Methods for Solving Systems of Linear Equations

Let $A\mathbf{x} = \mathbf{b}$ be the matrix form of a system of n linear equations in n variables. Suppose that A is invertible and has an LU decomposition. We have seen three different methods for solving this system, as follows.

1. Use Gaussian elimination to transform the augmented matrix $[A \quad \mathbf{b}]$ to reduced row echelon form.

2. Apply elementary row operations to the augmented matrix $[A \quad I_n]$ to compute A^{-1}, and then form $A^{-1}\mathbf{b}$.

3. Compute the LU decomposition of A, and then use the methods described in this section to solve the system $A\mathbf{x} = \mathbf{b}$.

The relative efficiencies of different methods for producing the same matrix computation can be compared by estimating the number of arithmetic operations (additions, subtractions, multiplications, and divisions) used for each method. In the context of an electronic computer, which is required for computations with matrices of substantial size, any arithmetic operation is called a **flop** (floating point operation). The total number of flops used to perform a matrix computation using a particular method is called the **flop count** for that method.

Typically, a flop count for a computation involving an $n \times n$ matrix is a polynomial in n. Since these counts are usually rough estimates and are significant only for large values of n, the terms in the polynomial of lower degree are usually ignored, and hence a flop count for a method is usually approximated as a multiple of a power of n.

The table below lists approximate flop counts for various matrix computations that can be used in solving a system of n equations in n unknowns.

Flop Counts for Various Procedures

In each case, the size of the matrix is $n \times n$ or the system consists of n equations and n variables.

Procedure	Approximate Flop Count
LU decomposition of an invertible matrix	$\dfrac{2n^3}{3}$
solution of a system using Gaussian elimination	$\dfrac{2n^3}{3}$
solution of a system given the LU decomposition of the coefficient matrix	$2n^2$
computation of the inverse of an invertible matrix	$2n^3$

We can use these approximations to compare the various methods of solving a system $A\mathbf{x} = \mathbf{b}$ of n linear equations in n variables, where A is invertible. Solving a system by computing the LU decomposition of the coefficient matrix requires approximately $\dfrac{2n^3}{3}$ flops for computing the LU decomposition of A and approximately $2n^2$ flops for solving the system using the LU decomposition of A. The sum, ignoring the term of lower degree, is approximately $\dfrac{2n^3}{3}$, which is the approximate number of flops used to solve the system by Gaussian elimination. Solving $A\mathbf{x} = \mathbf{b}$ using a matrix inverse requires approximately $2n^3$ flops to compute A^{-1}, followed by approximately $2n^2$ flops to compute the matrix product $\mathbf{x} = A^{-1}\mathbf{b}$ (see Exercise 23). Clearly, this compares unfavorably with either of the other two methods.

If several systems with the same coefficient matrix are to be solved, then the cost of using Gaussian elimination is approximately $\dfrac{2n^3}{3}$ flops to solve each system, whereas the cost of using LU decomposition involves an initial investment of approximately $\dfrac{2n^3}{3}$ flops for the LU decomposition followed by a much lower cost of approximately $2n^2$ flops to solve each system. For example, the cost of solving n different systems using Gaussian elimination is approximately

$$n\left(\frac{2n^3}{3}\right) = \frac{2n^4}{3}$$

flops, whereas the cost for solving the same n systems using an LU decomposition is approximately

$$\frac{2n^3}{3} + n \cdot 2n^2 = \frac{8n^3}{3}$$

flops.

What if a Matrix Has No LU Decomposition?

We have seen that not every matrix has an LU decomposition. Suppose that A is such a matrix. Then, by means of Gaussian elimination, A can be transformed into an upper triangular matrix U using elementary row operations that include one or more row interchanges. It can be shown (we omit the justification) that if these row interchanges are applied to A initially, then the resulting matrix, C, can be transformed into U by means of elementary row operations that do not include row interchanges. Consequently, there is a unit lower triangular matrix L such that $C = LU$. The matrix C has the same rows as A but in a different sequence. Thus there is a sequence of row interchanges that transforms A into C. Performing this same sequence of row interchanges on the appropriate identity matrix produces a matrix P such that $C = PA$. Any matrix P obtained by permuting the rows of an identity matrix is called a **permutation matrix**. So if A does not have an LU decomposition, there is a permutation matrix P such that PA has an LU decomposition.

In the next example, we illustrate how to find such a permutation matrix for a matrix having no LU decomposition.

Example 6   Let

$$A = \begin{bmatrix} 0 & 2 & 2 & 4 \\ 0 & 2 & 2 & 2 \\ 1 & 2 & 2 & 1 \\ 2 & 6 & 7 & 5 \end{bmatrix}.$$

Find a permutation matrix P such that $C = PA$ has an LU decomposition.

Solution   In this example, we denote the interchange of the pth and the qth rows of a matrix by $\mathbf{R}(p, q)$ and, as before, we use the notation $(c)\mathbf{r}_j + \mathbf{r}_i$ to denote the addition of c times row j to row i. We begin by transforming A into a matrix U in row echelon form, keeping track of the elementary row operations, as in Example 2:

$$A = \begin{bmatrix} 0 & 2 & 2 & 4 \\ 0 & 2 & 2 & 2 \\ 1 & 2 & 2 & 1 \\ 2 & 6 & 7 & 5 \end{bmatrix} \xrightarrow{\mathbf{R}(1,3)} \begin{bmatrix} 1 & 2 & 2 & 1 \\ 0 & 2 & 2 & 2 \\ 0 & 2 & 2 & 4 \\ 2 & 6 & 7 & 5 \end{bmatrix} \xrightarrow{(-2)\mathbf{r}_1+\mathbf{r}_4} \begin{bmatrix} 1 & 2 & 2 & 1 \\ 0 & 2 & 2 & 2 \\ 0 & 2 & 2 & 4 \\ 0 & 2 & 3 & 3 \end{bmatrix}$$

$$\xrightarrow{(-1)\mathbf{r}_2+\mathbf{r}_3} \begin{bmatrix} 1 & 2 & 2 & 1 \\ 0 & 2 & 2 & 2 \\ 0 & 0 & 0 & 2 \\ 0 & 2 & 3 & 3 \end{bmatrix} \xrightarrow{(-1)\mathbf{r}_2+\mathbf{r}_4} \begin{bmatrix} 1 & 2 & 2 & 1 \\ 0 & 2 & 2 & 2 \\ 0 & 0 & 0 & 2 \\ 0 & 0 & 1 & 1 \end{bmatrix}$$

$$\xrightarrow{\mathbf{R}(3,4)} \begin{bmatrix} 1 & 2 & 2 & 1 \\ 0 & 2 & 2 & 2 \\ 0 & 0 & 1 & 1 \\ 0 & 0 & 0 & 2 \end{bmatrix} = U.$$

In this computation, two row interchanges were performed. If we apply these directly to A, we obtain

$$A = \begin{bmatrix} 0 & 2 & 2 & 4 \\ 0 & 2 & 2 & 2 \\ 1 & 2 & 2 & 1 \\ 2 & 6 & 7 & 5 \end{bmatrix} \xrightarrow{\mathbf{R}(1,3)} \begin{bmatrix} 1 & 2 & 2 & 1 \\ 0 & 2 & 2 & 2 \\ 0 & 2 & 2 & 4 \\ 2 & 6 & 7 & 5 \end{bmatrix} \xrightarrow{\mathbf{R}(3,4)} \begin{bmatrix} 1 & 2 & 2 & 1 \\ 0 & 2 & 2 & 2 \\ 2 & 6 & 7 & 5 \\ 0 & 2 & 2 & 4 \end{bmatrix}$$

$$= C.$$

To find P, simply perform the preceding row interchanges on I_4 in the same order:

$$I_4 \xrightarrow{\mathbf{R}(1,3)} \begin{bmatrix} 0 & 0 & 1 & 0 \\ 0 & 1 & 0 & 0 \\ 1 & 0 & 0 & 0 \\ 0 & 0 & 0 & 1 \end{bmatrix} \xrightarrow{\mathbf{R}(3,4)} \begin{bmatrix} 0 & 0 & 1 & 0 \\ 0 & 1 & 0 & 0 \\ 0 & 0 & 0 & 1 \\ 1 & 0 & 0 & 0 \end{bmatrix} = P.$$

Then $C = PA$ has an LU decomposition.   ○

In the next example, we give a direct method for computing an LU decomposition of $C = PA$ as in Example 6.

Example 7   For the matrices A, P, and U in Example 6, find a unit lower triangular matrix L such that $LU = PA$.

Solution   Since we can transform PA into U using elementary row operations without interchanges, we can obtain L using the method described in Example 2. However, it is more efficient to obtain L from the elementary row operations used to transform A into U in Example 6. The method used here is similar to the method in Example 2, but with one complication. If, in the process of transforming A to U (in Example 6), row p is switched with row q as the result of an interchange, then the corresponding multipliers are switched in the same way. So, for example, in the process of transforming A into U in Example 6, -2 times row 1 is added to row 4. This results in the multiplier 2 in the $(4, 1)$-position. However, several steps later, row 4 is interchanged with row 3, and hence this multiplier moves to the $(3, 1)$-position. Since there are no more row interchanges, it follows that $l_{31} = 2$. A common way of keeping track of the entries of L is to place the multipliers in the appropriate positions of the intermediate matrices, replacing the actual entries, which are zeros. So as not to confuse these multipliers with the actual entries, we place each multiplier in parentheses:

$$A = \begin{bmatrix} 0 & 2 & 2 & 4 \\ 0 & 2 & 2 & 2 \\ 1 & 2 & 2 & 1 \\ 2 & 6 & 7 & 5 \end{bmatrix} \xrightarrow{\mathbf{R}(1,3)} \begin{bmatrix} 1 & 2 & 2 & 1 \\ 0 & 2 & 2 & 2 \\ 0 & 2 & 2 & 4 \\ 2 & 6 & 7 & 5 \end{bmatrix}$$

$$\xrightarrow{(-2)\mathbf{r}_1+\mathbf{r}_4} \begin{bmatrix} 1 & 2 & 2 & 1 \\ 0 & 2 & 2 & 2 \\ 0 & 2 & 2 & 4 \\ (2) & 2 & 3 & 3 \end{bmatrix}$$

$$\xrightarrow{(-1)\mathbf{r}_2+\mathbf{r}_3} \begin{bmatrix} 1 & 2 & 2 & 1 \\ 0 & 2 & 2 & 2 \\ 0 & (1) & 0 & 2 \\ (2) & 2 & 3 & 3 \end{bmatrix}$$

$$\xrightarrow{(-1)\mathbf{r}_2+\mathbf{r}_4} \begin{bmatrix} 1 & 2 & 2 & 1 \\ 0 & 2 & 2 & 2 \\ 0 & (1) & 0 & 2 \\ (2) & (1) & 1 & 1 \end{bmatrix}$$

$$\xrightarrow{\mathbf{R}(3,4)} \begin{bmatrix} 1 & 2 & 2 & 1 \\ 0 & 2 & 2 & 2 \\ (2) & (1) & 1 & 1 \\ 0 & (1) & 0 & 2 \end{bmatrix}.$$

Notice that if the entries in parentheses are replaced by zeros in the last matrix of the sequence above, we obtain U. Finally, we obtain L using the entries in parentheses in the last matrix of the sequence. The other nondiagonal entries of L are zeros. Thus

$$L = \begin{bmatrix} 1 & 0 & 0 & 0 \\ 0 & 1 & 0 & 0 \\ 2 & 1 & 1 & 0 \\ 0 & 1 & 0 & 1 \end{bmatrix}.$$

It is a simple matter to verify that $LU = PA$.

We now use these results to solve a system of linear equations.

Example 8   Use the results of Example 7 to solve the system of linear equations

$$
\begin{aligned}
2x_2 + 2x_3 + 4x_4 &= -6 \\
2x_2 + 2x_3 + 2x_4 &= -2 \\
x_1 + 2x_2 + 2x_3 + x_4 &= 3 \\
2x_1 + 6x_2 + 7x_3 + 5x_4 &= 2.
\end{aligned}
$$

Solution   This system can be written as the matrix equation $A\mathbf{x} = \mathbf{b}$, where

$$
A = \begin{bmatrix} 0 & 2 & 2 & 4 \\ 0 & 2 & 2 & 2 \\ 1 & 2 & 2 & 1 \\ 2 & 6 & 7 & 5 \end{bmatrix} \quad \text{and} \quad \mathbf{b} = \begin{bmatrix} -6 \\ -2 \\ 3 \\ 2 \end{bmatrix}.
$$

In Example 7, we found a permutation matrix P such that PA has an LU decomposition. We multiply both sides of the equation $A\mathbf{x} = \mathbf{b}$ from the left by P to obtain the equivalent equation $PA\mathbf{x} = P\mathbf{b}$. Then $PA = LU$ for the matrices L and U obtained in Example 7. Setting $\mathbf{b}' = P\mathbf{b}$, we have reduced the problem to solving the system $LU = \mathbf{b}'$, where

$$
\mathbf{b}' = P\mathbf{b} = \begin{bmatrix} 0 & 0 & 1 & 0 \\ 0 & 1 & 0 & 0 \\ 0 & 0 & 0 & 1 \\ 1 & 0 & 0 & 0 \end{bmatrix} \begin{bmatrix} -6 \\ -2 \\ 3 \\ 2 \end{bmatrix} = \begin{bmatrix} 3 \\ -2 \\ 2 \\ -6 \end{bmatrix}.
$$

As in Example 4, we set $\mathbf{y} = U\mathbf{x}$ and solve the system $L\mathbf{y} = \mathbf{b}'$, which has the form

$$
\begin{aligned}
y_1 &= 3 \\
y_2 &= -2 \\
2y_1 + y_2 + y_3 &= 2 \\
y_2 + y_4 &= -6.
\end{aligned}
$$

We solve this system using forward substitution to obtain

$$
\mathbf{y} = \begin{bmatrix} y_1 \\ y_2 \\ y_3 \\ y_4 \end{bmatrix} = \begin{bmatrix} 3 \\ -2 \\ -2 \\ -4 \end{bmatrix}.
$$

Finally, to obtain the solution to the original system, we solve $U\mathbf{x} = \mathbf{y}$, which has the form

$$
\begin{aligned}
x_1 + 2x_2 + 2x_3 + x_4 &= 3 \\
2x_2 + 2x_3 + 2x_4 &= -2 \\
x_3 + x_4 &= -2 \\
2x_4 &= -4.
\end{aligned}
$$

We solve this system using back substitution, to obtain the desired solution

$$
\mathbf{x} = \begin{bmatrix} x_1 \\ x_2 \\ x_3 \\ x_4 \end{bmatrix} = \begin{bmatrix} 3 \\ 1 \\ 0 \\ -2 \end{bmatrix}.
$$

Practice Problems

1. Find an LU decomposition of

$$A = \begin{bmatrix} 1 & -1 & -2 & -8 \\ -2 & 1 & 2 & 9 \\ 3 & 0 & 2 & 1 \end{bmatrix}.$$

2. Use your answer to problem 1 to solve $Ax = b$, where A is the matrix in problem 1 and

$$b = \begin{bmatrix} -3 \\ 5 \\ -8 \end{bmatrix}.$$

3. Find an LU decomposition of PA, where P is a permutation matrix and

$$A = \begin{bmatrix} 0 & 3 & -6 & 1 \\ -2 & -2 & 2 & 6 \\ 1 & 1 & -1 & -1 \\ 2 & -1 & 2 & -2 \end{bmatrix}.$$

4. Use your answer to problem 3 to solve $Ax = b$, where A is the matrix in problem 3 and

$$b = \begin{bmatrix} -13 \\ -6 \\ 1 \\ 8 \end{bmatrix}.$$

Exercises

1. Determine if the following statements are true or false.

(a) Every matrix has an LU decomposition.

(b) If a matrix A has an LU decomposition, then A can be transformed into a matrix in row echelon form without using any row interchanges.

In Exercises 2–5, find the LU decomposition of the given matrix.

2. $\begin{bmatrix} 2 & -1 & 1 \\ 4 & -1 & 4 \\ -2 & 1 & 2 \end{bmatrix}$

3. $\begin{bmatrix} 2 & 3 & 4 \\ 6 & 8 & 10 \\ -2 & -4 & -3 \end{bmatrix}$

4. $\begin{bmatrix} 3 & 1 & -1 & 1 \\ 6 & 4 & -1 & 4 \\ -3 & -1 & 2 & -1 \\ 3 & 5 & 0 & 3 \end{bmatrix}$

5. $\begin{bmatrix} -1 & 2 & 1 & -1 & 3 \\ 1 & -4 & 0 & 5 & -5 \\ -2 & 6 & -1 & -5 & 7 \\ -1 & -4 & 4 & 11 & -2 \end{bmatrix}$

In Exercises 6–9, use the results of Exercises 2–5 to solve the given system of linear equations.

6. $2x_1 - x_2 + x_3 = -1$
  $4x_1 - x_2 + 4x_3 = -2$
  $-2x_1 + x_2 + 2x_3 = -2$

7. $2x_1 + 3x_2 + 4x_3 = 1$
  $6x_1 + 8x_2 + 10x_3 = 4$
  $-2x_1 - 4x_2 - 3x_3 = 0$

8. $3x_1 + x_2 - x_3 + x_4 = 0$
  $6x_1 + 4x_2 - x_3 + 4x_4 = 15$
  $-3x_1 - x_2 + 2x_3 - x_4 = 1$
  $3x_1 + 5x_2 + 3x_4 = 21$

9. $-x_1 + 2x_2 + x_3 - x_4 + 3x_5 = 7$
  $x_1 - 4x_2 + 5x_4 - 5x_5 = -7$
  $-2x_1 + 6x_2 - x_3 - 5x_4 + 7x_5 = 6$
  $-x_1 - 4x_2 + 4x_3 + 11x_4 - 2x_5 = 11$

10. Prove that the product of two $n \times n$ upper triangular matrices is an upper triangular matrix.

11. Prove that if U is an invertible upper triangular matrix, then U^{-1} is also upper triangular.

12. Let A and B be $n \times n$ lower triangular matrices.

(a) Prove that AB is a lower triangular matrix.

(b) Prove that if both A and B are unit lower triangular matrices, then AB is a unit lower triangular matrix.

13. Prove that a square unit lower triangular matrix L is invertible, and that L^{-1} is also a unit lower triangular matrix.

14. Suppose that LU and $L'U'$ are two LU decompositions for an invertible matrix. Prove that $L = L'$ and $U = U'$. Thus the LU decomposition for an invertible matrix is unique. *Hint:* Use the results of Exercises 10–13.

In Exercises 15–18, for each matrix A find (a) a permutation matrix P such that PA has an LU decomposition, and (b) the LU decomposition of PA.

15. $\begin{bmatrix} 1 & -1 & 3 \\ 2 & -2 & 5 \\ -1 & 2 & -1 \end{bmatrix}$

16. $\begin{bmatrix} 0 & 2 & -1 \\ 2 & 6 & 0 \\ 1 & 3 & -1 \end{bmatrix}$

17. $\begin{bmatrix} 1 & 2 & 1 & -1 \\ 2 & 4 & 1 & 1 \\ 3 & 2 & -1 & -2 \\ 2 & 5 & 3 & 0 \end{bmatrix}$

18. $\begin{bmatrix} 2 & 4 & -6 & 0 \\ -2 & 1 & 3 & 2 \\ 2 & 9 & -9 & 1 \\ 4 & 3 & -3 & 0 \end{bmatrix}$

In Exercises 19–22, use the results of Exercises 15–18 to solve the given system of linear equations.

19. $\begin{aligned} x_1 - \ \ x_2 + 3x_3 &= 6 \\ 2x_1 - 2x_2 + 5x_3 &= 9 \\ -x_1 + 2x_2 - \ \ x_3 &= 1 \end{aligned}$

20. $\begin{aligned} 2x_2 - x_3 &= \ \ 2 \\ 2x_1 + 6x_2 \ \ \ \ \ \ \ &= -2 \\ x_1 + 3x_2 - x_3 &= -1 \end{aligned}$

21. $\begin{aligned} x_1 + 2x_2 + \ \ x_3 - \ \ x_4 &= \ \ 3 \\ 2x_1 + 4x_2 + \ \ x_3 + \ \ x_4 &= \ \ 2 \\ 3x_1 + 2x_2 - \ \ x_3 - 2x_4 &= -4 \\ 2x_1 + 5x_2 + 3x_3 \ \ \ \ \ \ \ &= \ \ 7 \end{aligned}$

22. $\begin{aligned} 2x_1 + 4x_2 - 6x_3 \ \ \ \ \ \ \ &= \ \ 2 \\ -2x_1 + \ \ x_2 + 3x_3 + 2x_4 &= \ \ 7 \\ 2x_1 + 9x_2 - 9x_3 + \ \ x_4 &= 11 \\ 4x_1 + 3x_2 - 3x_3 \ \ \ \ \ \ \ &= \ \ 7 \end{aligned}$

23. Let C be an $n \times n$ matrix and $\mathbf{b}$ be a vector in $\mathcal{R}^n$.

(a) Show that it requires n multiplications and $n - 1$ additions to compute each component of $C\mathbf{b}$.

(b) Show that the approximate flop count for computing $C\mathbf{b}$ is $2n^2$.

24. Suppose that we are given n systems of n linear equations in n variables, all of which have as their coefficient matrix the same invertible matrix A. Estimate the total flop count for solving all of these systems by first computing A^{-1}, and then computing the product of A^{-1} with the constant vector of each system.

25. Suppose that A is an $m \times n$ matrix and B is an $n \times p$ matrix. Find the exact flop count for computing the product AB.

26. Suppose that A is an $m \times n$ matrix, B is an $n \times p$ matrix, and C is a $p \times q$ matrix. The product ABC can be computed in two ways: (a) First compute AB and then multiply this (on the right) by C. (b) First compute BC

and then multiply this (on the left) by A. Use Exercise 25 to devise a strategy that compares the two ways, so that the more efficient one can be chosen.

In Exercises 27–30, use a calculator with matrix capabilities or computer software such as MATLAB *to solve the problem.*[6]

In Exercises 27 and 28, find the LU decomposition of the given matrix.

27. $\begin{bmatrix} 2 & -1 & 3 & 2 & 1 \\ -2 & 2 & -1 & 1 & 4 \\ 4 & 1 & 15 & 12 & 19 \\ 6 & -6 & 9 & -4 & 0 \\ 4 & -2 & 9 & 2 & 9 \end{bmatrix}$

28. $\begin{bmatrix} -3 & 1 & 0 & 2 & 1 \\ -6 & 0 & 1 & 3 & 5 \\ -15 & 7 & 4 & 1 & 12 \\ 0 & -4 & 2 & -6 & 8 \end{bmatrix}$

In Exercises 29 and 30, for each matrix A find (a) a permutation matrix P such that PA has an LU decomposition, and (b) the LU decomposition of PA.

29. $\begin{bmatrix} 0 & 1 & 2 & -1 & 1 \\ 2 & -2 & -1 & 3 & 4 \\ 1 & 1 & 2 & -1 & 2 \\ -1 & 0 & 3 & 0 & 1 \\ 3 & 4 & -1 & 2 & 4 \end{bmatrix}$

30. $\begin{bmatrix} 1 & 2 & -3 & 1 & 4 \\ 3 & 6 & -5 & 4 & 8 \\ 2 & 3 & -3 & 2 & 1 \\ -1 & 2 & 1 & 4 & 2 \\ 3 & 2 & 4 & -4 & 0 \end{bmatrix}$

---

[6]Caution! The MATLAB function `lu` does not compute the standard LU decomposition of a matrix.

Solutions to the Practice Problems

1. First we apply Gaussian elimination to transform A into an upper triangular matrix U in row echelon form without using row interchanges:

$$A = \begin{bmatrix} 1 & -1 & -2 & -8 \\ -2 & 1 & 2 & 9 \\ 3 & 0 & 2 & 1 \end{bmatrix}$$

$$\xrightarrow{(2)\mathbf{r}_1 + \mathbf{r}_2} \begin{bmatrix} 1 & -1 & -2 & -8 \\ 0 & -1 & -2 & -7 \\ 3 & 0 & 2 & 1 \end{bmatrix}$$

$$\xrightarrow{(-3)\mathbf{r}_1 + \mathbf{r}_3} \begin{bmatrix} 1 & -1 & -2 & -8 \\ 0 & -1 & -2 & -7 \\ 0 & 3 & 8 & 25 \end{bmatrix}$$

$$\xrightarrow{(3)\mathbf{r}_2 + \mathbf{r}_3} \begin{bmatrix} 1 & -1 & -2 & -8 \\ 0 & -1 & -2 & -7 \\ 0 & 0 & 2 & 4 \end{bmatrix} = U.$$

Then L is the unit lower triangular matrix whose (i, j)-entry, for $i > j$, is $-c$, where $(c)\mathbf{r}_j + \mathbf{r}_i$ is a label over an arrow in the preceding reduction process. Thus

$$L = \begin{bmatrix} 1 & 0 & 0 \\ -2 & 1 & 0 \\ 3 & -3 & 1 \end{bmatrix}.$$

2. Substituting the LU decomposition for A obtained in problem 1, we write the system as the matrix equation $LU\mathbf{x} = \mathbf{b}$. Setting $\mathbf{y} = U\mathbf{x}$, the system becomes $L\mathbf{y} = \mathbf{b}$, which can be written

$$\begin{aligned} y_1 \phantom{-2y_1+{}y_2} &= -3 \\ -2y_1 + y_2 \phantom{{}+y_3} &= 5 \\ 3y_1 - 3y_2 + y_3 &= -8. \end{aligned}$$

Using forward substitution, we solve this system to obtain

$$\mathbf{y} = \begin{bmatrix} y_1 \\ y_2 \\ y_3 \end{bmatrix} = \begin{bmatrix} -3 \\ -1 \\ -2 \end{bmatrix}.$$

Next we use back substitution to solve the system $U\mathbf{x} = \mathbf{y}$, which can be written

$$\begin{aligned} x_1 - x_2 - 2x_3 - 8x_4 &= -3 \\ -x_2 - 2x_3 - 7x_4 &= -1 \\ 2x_3 + 4x_4 &= -2. \end{aligned}$$

Treating x_4 as a free variable, we obtain the following general solution.

$$\begin{aligned} x_1 &= -2 + x_4 \\ x_2 &= 3 - 3x_4 \\ x_3 &= -1 - 2x_4 \\ x_4 & \quad \text{free.} \end{aligned}$$

3. Using the method of Example 7, we have

$$A = \begin{bmatrix} 0 & 3 & -6 & 1 \\ -2 & -2 & 2 & 6 \\ 1 & 1 & -1 & -1 \\ 2 & -1 & 2 & -2 \end{bmatrix}$$

$$\xrightarrow{\mathbf{R}(1,3)} \begin{bmatrix} 1 & 1 & -1 & -1 \\ -2 & -2 & 2 & 6 \\ 0 & 3 & -6 & 1 \\ 2 & -1 & 2 & -2 \end{bmatrix}$$

$$\xrightarrow{(2)\mathbf{r}_1+\mathbf{r}_2} \begin{bmatrix} 1 & 1 & -1 & -1 \\ (-2) & 0 & 0 & 4 \\ 0 & 3 & -6 & 1 \\ 2 & -1 & 2 & -2 \end{bmatrix}$$

$$\xrightarrow{(-2)\mathbf{r}_1+\mathbf{r}_4} \begin{bmatrix} 1 & 1 & -1 & -1 \\ (-2) & 0 & 0 & 4 \\ 0 & 3 & -6 & 1 \\ (2) & -3 & 4 & 0 \end{bmatrix}$$

$$\xrightarrow{\mathbf{R}(2,4)} \begin{bmatrix} 1 & 1 & -1 & -1 \\ (2) & -3 & 4 & 0 \\ 0 & 3 & -6 & 1 \\ (-2) & 0 & 0 & 4 \end{bmatrix}$$

$$\xrightarrow{(1)\mathbf{r}_2+\mathbf{r}_3} \begin{bmatrix} 1 & 1 & -1 & -1 \\ (2) & -3 & 4 & 0 \\ 0 & (-1) & -2 & 1 \\ (-2) & 0 & 0 & 4 \end{bmatrix}.$$

Thus

$$L = \begin{bmatrix} 1 & 0 & 0 & 0 \\ 2 & 1 & 0 & 0 \\ 0 & -1 & 1 & 0 \\ -2 & 0 & 0 & 1 \end{bmatrix}$$

and

$$U = \begin{bmatrix} 1 & 1 & -1 & -1 \\ 0 & -3 & 4 & 0 \\ 0 & 0 & -2 & 1 \\ 0 & 0 & 0 & 4 \end{bmatrix}.$$

Finally, we apply $\mathbf{R}(1, 3)$ and $\mathbf{R}(2, 4)$ to I_4 to obtain P:

$$I_4 \xrightarrow{\mathbf{R}(1,3)} \begin{bmatrix} 0 & 0 & 1 & 0 \\ 0 & 1 & 0 & 0 \\ 1 & 0 & 0 & 0 \\ 0 & 0 & 0 & 1 \end{bmatrix}$$

$$\xrightarrow{\mathbf{R}(2,4)} \begin{bmatrix} 0 & 0 & 1 & 0 \\ 0 & 0 & 0 & 1 \\ 1 & 0 & 0 & 0 \\ 0 & 1 & 0 & 0 \end{bmatrix} = P.$$

4. Using the permutation matrix P and the LU decomposition of PA obtained in problem 3, we transform the system of equations $A\mathbf{x} = \mathbf{b}$ into

$$LU\mathbf{x} = PA\mathbf{x} = P\mathbf{b} = \begin{bmatrix} 0 & 0 & 1 & 0 \\ 0 & 0 & 0 & 1 \\ 1 & 0 & 0 & 0 \\ 0 & 1 & 0 & 0 \end{bmatrix} \begin{bmatrix} -13 \\ -6 \\ 1 \\ 8 \end{bmatrix}$$

$$= \begin{bmatrix} 1 \\ 8 \\ -13 \\ -6 \end{bmatrix}.$$

This system can now be solved by the method used in problem 2 to obtain the unique solution

$$\mathbf{x} = \begin{bmatrix} x_1 \\ x_2 \\ x_3 \\ x_4 \end{bmatrix} = \begin{bmatrix} 1 \\ 2 \\ 3 \\ -1 \end{bmatrix}.$$

2.6   Linear Transformations and Matrices

In Section 1.2, we defined the matrix–vector product $A\mathbf{v}$, where A is an $m \times n$ matrix and $\mathbf{v} \in \mathcal{R}^n$. The correspondence that associates to each vector $\mathbf{v}$ in $\mathcal{R}^n$ the vector $A\mathbf{v}$ in $\mathcal{R}^m$ is an example of a function from $\mathcal{R}^n$ to $\mathcal{R}^m$. Most of the functions that we encounter in this book are defined from a subset of $\mathcal{R}^n$ to a subset of $\mathcal{R}^m$. We begin by recalling some definitions.

Definition.   Let $\mathcal{S}_1$ and $\mathcal{S}_2$ be subsets of $\mathcal{R}^n$ and $\mathcal{R}^m$, respectively. A **function** f from $\mathcal{S}_1$ to $\mathcal{S}_2$, written $f\colon \mathcal{S}_1 \to \mathcal{S}_2$, is a rule that assigns to each vector $\mathbf{x}$ in $\mathcal{S}_1$ a unique vector $f(\mathbf{x})$ in $\mathcal{S}_2$. The vector $f(\mathbf{x})$ is called the **image** of $\mathbf{x}$ (under f). The set $\mathcal{S}_1$ is called the **domain of a function** of f, and the set $\mathcal{S}_2$ is called the **codomain** of f. The **range** of f is defined to be the set of images $f(\mathbf{x})$ for all $\mathbf{x}$ in $\mathcal{S}_1$.

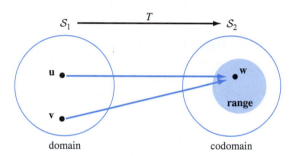

FIGURE 2.5

In Figure 2.5, we see that $\mathbf{u}$ and $\mathbf{v}$ both have $\mathbf{w}$ as their image. So $\mathbf{w} = f(\mathbf{u})$ and $\mathbf{w} = f(\mathbf{v})$.

Example 1   Define $f\colon \mathcal{R}^3 \to \mathcal{R}^2$ by the rule

$$f\left(\begin{bmatrix} x_1 \\ x_2 \\ x_3 \end{bmatrix}\right) = \begin{bmatrix} x_1 + x_2 + x_3 \\ x_1^2 \end{bmatrix}.$$

Then f is a function whose domain is $\mathcal{R}^3$ and codomain is $\mathcal{R}^2$. Notice that

$$f\left(\begin{bmatrix} 0 \\ 1 \\ 1 \end{bmatrix}\right) = \begin{bmatrix} 2 \\ 0 \end{bmatrix} \qquad \text{and} \qquad f\left(\begin{bmatrix} 0 \\ 3 \\ -1 \end{bmatrix}\right) = \begin{bmatrix} 2 \\ 0 \end{bmatrix}.$$

So $\begin{bmatrix} 2 \\ 0 \end{bmatrix}$ is the image of both $\begin{bmatrix} 0 \\ 1 \\ 1 \end{bmatrix}$ and $\begin{bmatrix} 0 \\ 3 \\ -1 \end{bmatrix}$. However, not every vector in $\mathcal{R}^2$ is an image of a vector in $\mathcal{R}^3$, because every image must have a nonnegative second component.

Example 2   Let A be the 3×2 matrix

$$A = \begin{bmatrix} 1 & 0 \\ 2 & 1 \\ 1 & -1 \end{bmatrix}.$$

Define the function $T_A\colon \mathcal{R}^2 \to \mathcal{R}^3$ by

$$T_A(\mathbf{x}) = A\mathbf{x}.$$

Notice that because A is a 3×2 matrix and $\mathbf{x}$ is a 2×1 vector, the vector $A\mathbf{x}$ has size 3×1. Also, observe that there is a reversal in the order of the size 3×2 of A and the "sizes" of the domain $\mathcal{R}^2$ and the codomain $\mathcal{R}^3$ of T_A.

We can easily obtain the same form for the rule of T_A as for f in Example 1, by computing

$$T_A \left(\begin{bmatrix} x_1 \\ x_2 \end{bmatrix} \right) = \begin{bmatrix} 1 & 0 \\ 2 & 1 \\ 1 & -1 \end{bmatrix} \begin{bmatrix} x_1 \\ x_2 \end{bmatrix} = \begin{bmatrix} x_1 \\ 2x_1 + x_2 \\ x_1 - x_2 \end{bmatrix}.$$

A definition similar to that in Example 2 can be given for any $m \times n$ matrix A, in which case we obtain a function T_A with domain $\mathcal{R}^n$ and codomain $\mathcal{R}^m$.

Definition.   Let A be an $m \times n$ matrix. The function $T_A : \mathcal{R}^n \to \mathcal{R}^m$ defined by $T_A(\mathbf{x}) = A\mathbf{x}$ for all $\mathbf{x}$ in $\mathcal{R}^n$ is called the **matrix transformation induced by** A.

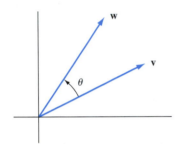

FIGURE 2.6

Later we show that the function f of Example 1 is not a matrix transformation. It is interesting to note that we have already seen an important example of a matrix transformation in Section 1.2, using the rotation matrix A_θ. Here $T_{A_\theta} : \mathcal{R}^2 \to \mathcal{R}^2$ represents the function that rotates a vector counterclockwise by θ. To show that the range of T_{A_θ} is all of $\mathcal{R}^2$, suppose that $\mathbf{w}$ is any vector in $\mathcal{R}^2$. If we let $\mathbf{v} = A_{-\theta}\mathbf{w}$ (see Figure 2.6), then, as in Example 2 of Section 2.1, we have

$$T_{A_\theta}(\mathbf{v}) = A_\theta(\mathbf{v}) = A_\theta A_{-\theta}\mathbf{w} = A_{\theta - \theta}\mathbf{w} = A_{0°}\mathbf{w} = \mathbf{w}.$$

So every vector $\mathbf{w}$ in $\mathcal{R}^2$ is in the range of T_{A_θ}.

Example 3   Let A be the matrix

$$\begin{bmatrix} 1 & 0 & 0 \\ 0 & 1 & 0 \\ 0 & 0 & 0 \end{bmatrix}.$$

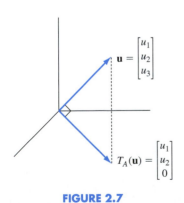

FIGURE 2.7

So $T_A : \mathcal{R}^3 \to \mathcal{R}^3$ is defined by

$$T_A \left(\begin{bmatrix} x_1 \\ x_2 \\ x_3 \end{bmatrix} \right) = \begin{bmatrix} 1 & 0 & 0 \\ 0 & 1 & 0 \\ 0 & 0 & 0 \end{bmatrix} \begin{bmatrix} x_1 \\ x_2 \\ x_3 \end{bmatrix} = \begin{bmatrix} x_1 \\ x_2 \\ 0 \end{bmatrix}.$$

We can see from Figure 2.7 that $T_A(\mathbf{u})$ is the *projection of* $\mathbf{u}$ *onto the xy-plane.* The range of T_A is the xy-plane in $\mathcal{R}^3$.

So far we have seen that rotations and projections are matrix transformations. In the exercises, we discover that other geometric transformations (namely, reflections, contractions, and dilations), are also matrix transformations. The next example introduces yet another geometric transformation.

Example 4   Let k be a scalar and $A = \begin{bmatrix} 1 & k \\ 0 & 1 \end{bmatrix}$. The function $T_A : \mathcal{R}^2 \to \mathcal{R}^2$ is defined by

$T_A \left(\begin{bmatrix} x_1 \\ x_2 \end{bmatrix} \right) = \begin{bmatrix} x_1 + kx_2 \\ x_2 \end{bmatrix}$, and is called a **shear transformation**. Notice the effect on the vector **u** in Figure 2.8(a). The head of the vector is moved to the right, but at the same height. In Figure 2.8(b), the letter "I" is centered on the y-axis. Notice the effect of the transformation T_A, where $k = 2$.   ○

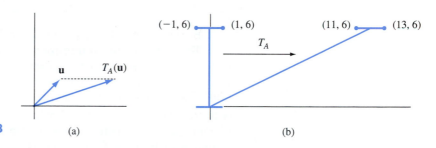

FIGURE 2.8   (a)   (b)

The next result follows immediately from Theorem 1.2. It is used throughout the remaining chapters.

Theorem 2.8

For any $m \times n$ matrix A and any vectors **u** *and* **v** *in* $\mathcal{R}^n$, *the following are true.*

(a) $T_A(\mathbf{u} + \mathbf{v}) = T_A(\mathbf{u}) + T_A(\mathbf{v})$.

(b) $T_A(c\mathbf{u}) = cT_A(\mathbf{u})$ for any scalar c.

We see from (a) and (b) that T_A preserves the two vector operations; that is, the image of a sum of two vectors is the sum of the images, and the image of a scalar multiple of a vector is the same scalar multiple of the image. We can easily see that the function f of Example 1 does not satisfy either of these properties. For example,

$$ f \left(\begin{bmatrix} 1 \\ 0 \\ 1 \end{bmatrix} + \begin{bmatrix} 2 \\ 0 \\ 0 \end{bmatrix} \right) = f \left(\begin{bmatrix} 3 \\ 0 \\ 1 \end{bmatrix} \right) = \begin{bmatrix} 4 \\ 9 \end{bmatrix} $$

but

$$ f \left(\begin{bmatrix} 1 \\ 0 \\ 1 \end{bmatrix} \right) + f \left(\begin{bmatrix} 2 \\ 0 \\ 0 \end{bmatrix} \right) = \begin{bmatrix} 2 \\ 1 \end{bmatrix} + \begin{bmatrix} 2 \\ 4 \end{bmatrix} = \begin{bmatrix} 4 \\ 5 \end{bmatrix} . $$

So (a) is not satisfied for all vectors. Also,

$$ f \left(2 \begin{bmatrix} 1 \\ 0 \\ 1 \end{bmatrix} \right) = f \left(\begin{bmatrix} 2 \\ 0 \\ 2 \end{bmatrix} \right) = \begin{bmatrix} 4 \\ 4 \end{bmatrix} \qquad \text{but} \qquad 2f \left(\begin{bmatrix} 1 \\ 0 \\ 1 \end{bmatrix} \right) = 2 \begin{bmatrix} 2 \\ 1 \end{bmatrix} = \begin{bmatrix} 4 \\ 2 \end{bmatrix} . $$

So (b) is not satisfied for all vectors and scalars.

Functions that do satisfy (a) and (b) of Theorem 2.8 merit their own definition.

Definition. A function T from $\mathcal{R}^n$ to $\mathcal{R}^m$, written $T: \mathcal{R}^n \to \mathcal{R}^m$, is called a **linear transformation** (or simply **linear**) if, for all vectors $\mathbf{u}$ and $\mathbf{v}$ in $\mathcal{R}^n$ and all scalars c, both of the following conditions hold.

(i) $T(\mathbf{u} + \mathbf{v}) = T(\mathbf{u}) + T(\mathbf{v})$. (In this case, we say that T **preserves vector addition**.)

(ii) $T(c\mathbf{u}) = cT(\mathbf{u})$. (In this case, we say that T **preserves scalar multiplication**.)

By Theorem 2.8, every matrix transformation is linear.

There are two linear transformations that deserve special attention. The first is the **identity transformation** $I: \mathcal{R}^n \to \mathcal{R}^n$, which is defined by $I(\mathbf{x}) = \mathbf{x}$ for all $\mathbf{x}$ in $\mathcal{R}^n$. By Exercise 35, $I = T_{I_n}$, and so I is linear. It is easy to show that the range of I is all of $\mathcal{R}^n$. The second transformation is the **zero transformation** $T_0: \mathcal{R}^n \to \mathcal{R}^m$, which is defined by $T_0(\mathbf{x}) = \mathbf{0}$ for all $\mathbf{x}$ in $\mathcal{R}^n$. Like the identity transformation, T_0 can be represented as a matrix transformation, namely, $T_0 = T_O$, where O is the $m \times n$ zero matrix (see Exercise 36). So T_0 is also linear. The range of T_0 consists precisely of the zero vector.

From properties (i) and (ii) there follow many significant results. In particular, we see that linear transformations preserve linear combinations.

Theorem 2.9

For any linear transformation $T: \mathcal{R}^n \to \mathcal{R}^m$, the following statements are true.

(a) $T(\mathbf{0}) = \mathbf{0}$.
(b) $T(-\mathbf{u}) = -T(\mathbf{u})$ for all vectors $\mathbf{u}$ in $\mathcal{R}^n$.
(c) $T(a\mathbf{u} + b\mathbf{v}) = aT(\mathbf{u}) + bT(\mathbf{v})$ for all vectors $\mathbf{u}$ and $\mathbf{v}$ in $\mathcal{R}^n$ and all scalars a and b.

Proof (a) By property (i) of the definition of a linear transformation, we have

$$T(\mathbf{0}) = T(\mathbf{0} + \mathbf{0}) = T(\mathbf{0}) + T(\mathbf{0}).$$

Subtracting $T(\mathbf{0})$ from both sides yields $\mathbf{0} = T(\mathbf{0})$.

(b) Let $\mathbf{u}$ be a vector in $\mathcal{R}^n$. Using property (ii), we have

$$T(-\mathbf{u}) = T((-1)\mathbf{u}) = (-1)T(\mathbf{u}) = -T(\mathbf{u}).$$

(c) Combining properties (i) and (ii) we have, for any vectors $\mathbf{u}$ and $\mathbf{v}$ in $\mathcal{R}^n$ and scalars a and b, that

$$T(a\mathbf{u} + b\mathbf{v}) = T(a\mathbf{u}) + T(b\mathbf{v}) = aT(\mathbf{u}) + bT(\mathbf{v}).$$

It follows from Theorem 2.9(c) that T preserves linear combinations.

Let $T: \mathcal{R}^n \to \mathcal{R}^m$ be a linear transformation. If $\mathbf{u}_1, \mathbf{u}_2, \ldots, \mathbf{u}_k$ are vectors in $\mathcal{R}^n$ and $a_1, a_2, \ldots, a_k$ are scalars, then

$$T(a_1\mathbf{u}_1 + a_2\mathbf{u}_2 + \cdots + a_k\mathbf{u}_k) = a_1T(\mathbf{u}_1) + a_2T(\mathbf{u}_2) + \cdots + a_kT(\mathbf{u}_k).$$

Theorem 2.9 can often be used to show that a function is not linear. For example, the function $T: \mathcal{R} \to \mathcal{R}$ defined by $T(x) = 2x + 3$ is not linear because $T(0) = 3 \neq 0$. On the other hand, the function f of Example 1 satisfies $f(0) = 0$, but we showed that f is not linear. The next example illustrates how the definition of a linear transformation is used to prove that a function is linear.

Example 5   Define $T: \mathcal{R}^2 \to \mathcal{R}^2$ by $T\left(\begin{bmatrix} x_1 \\ x_2 \end{bmatrix}\right) = \begin{bmatrix} 2x_1 - x_2 \\ x_1 \end{bmatrix}$. To prove that T is linear, let

$\mathbf{u}$ and $\mathbf{v}$ be vectors in $\mathcal{R}^2$. Then $\mathbf{u} = \begin{bmatrix} u_1 \\ u_2 \end{bmatrix}$, $\mathbf{v} = \begin{bmatrix} v_1 \\ v_2 \end{bmatrix}$, and $\mathbf{u} + \mathbf{v} = \begin{bmatrix} u_1 + v_1 \\ u_2 + v_2 \end{bmatrix}$.
So

$$T(\mathbf{u} + \mathbf{v}) = T\left(\begin{bmatrix} u_1 + v_1 \\ u_2 + v_2 \end{bmatrix}\right) = \begin{bmatrix} 2(u_1 + v_1) - (u_2 + v_2) \\ u_1 + v_1 \end{bmatrix}.$$

On the other hand,

$$T(\mathbf{u}) + T(\mathbf{v}) = T\left(\begin{bmatrix} u_1 \\ u_2 \end{bmatrix}\right) + T\left(\begin{bmatrix} v_1 \\ v_2 \end{bmatrix}\right) = \begin{bmatrix} 2u_1 - u_2 \\ u_1 \end{bmatrix} + \begin{bmatrix} 2v_1 - v_2 \\ v_1 \end{bmatrix}$$

$$= \begin{bmatrix} (2u_1 - u_2) + (2v_1 - v_2) \\ u_1 + v_1 \end{bmatrix} = \begin{bmatrix} 2(u_1 + v_1) - (u_2 + v_2) \\ u_1 + v_1 \end{bmatrix}.$$

So $T(\mathbf{u} + \mathbf{v}) = T(\mathbf{u}) + T(\mathbf{v})$.
Now suppose that c is any scalar. Then

$$T(c\mathbf{u}) = T\left(\begin{bmatrix} cu_1 \\ cu_2 \end{bmatrix}\right) = \begin{bmatrix} 2cu_1 - cu_2 \\ cu_1 \end{bmatrix}.$$

Also

$$cT(\mathbf{u}) = c\begin{bmatrix} 2u_1 - u_2 \\ u_1 \end{bmatrix} = \begin{bmatrix} 2cu_1 - cu_2 \\ cu_1 \end{bmatrix}.$$

Hence $T(c\mathbf{u}) = cT(\mathbf{u})$. Therefore T is linear.   ○

Another way to prove that the transformation T in Example 5 is linear is to find a matrix A such that $T = T_A$, and then appeal to Theorem 2.8. Suppose we let

$$A = \begin{bmatrix} 2 & -1 \\ 1 & 0 \end{bmatrix}.$$

Then

$$T_A\left(\begin{bmatrix} x_1 \\ x_2 \end{bmatrix}\right) = \begin{bmatrix} 2 & -1 \\ 1 & 0 \end{bmatrix}\begin{bmatrix} x_1 \\ x_2 \end{bmatrix} = \begin{bmatrix} 2x_1 - x_2 \\ x_1 \end{bmatrix} = T\left(\begin{bmatrix} x_1 \\ x_2 \end{bmatrix}\right).$$

So $T = T_A$.
Now we show that *every* linear transformation with domain $\mathcal{R}^n$ and codomain $\mathcal{R}^m$ is a matrix transformation. This means that if a transformation T is linear, we can produce a corresponding matrix A such that $T = T_A$. From this result we discover many connections between the linear transformation T and the matrix A. We begin with a preliminary result stating that a linear transformation is uniquely determined by its "action" on the standard vectors.

Theorem 2.10

Let $T, U: \mathcal{R}^n \to \mathcal{R}^m$ be linear transformations. Suppose that $T(\mathbf{e}_i) = U(\mathbf{e}_i)$ for every standard vector $\mathbf{e}_i$ in $\mathcal{R}^n$. Then $T = U$; that is, $T(\mathbf{v}) = U(\mathbf{v})$ for all $\mathbf{v}$ in $\mathcal{R}^n$.

Proof Let $\mathbf{v} \in \mathcal{R}^n$. We begin by writing $\mathbf{v} = v_1\mathbf{e}_1 + v_2\mathbf{e}_2 + \cdots + v_n\mathbf{e}_n$. Apply the preceding boxed result twice to obtain

$$
\begin{aligned}
T(\mathbf{v}) &= T(v_1\mathbf{e}_1 + v_2\mathbf{e}_2 + \cdots + v_n\mathbf{e}_n) \\
&= v_1 T(\mathbf{e}_1) + v_2 T(\mathbf{e}_2) + \cdots + v_n T(\mathbf{e}_n) \\
&= v_1 U(\mathbf{e}_1) + v_2 U(\mathbf{e}_2) + \cdots + v_n U(\mathbf{e}_n) \\
&= U(v_1\mathbf{e}_1 + v_2\mathbf{e}_2 + \cdots + v_n\mathbf{e}_n) \\
&= U(\mathbf{v}).
\end{aligned}
$$

With Theorem 2.10 in place, we now prove that every linear transformation with domain $\mathcal{R}^n$ and codomain $\mathcal{R}^m$ is a matrix transformation induced by a unique $m \times n$ matrix. (However, in Chapter 7 we extend the notion of linearity to a more general context, in which not all linear transformations are matrix transformations.) The uniqueness result (namely, $T_A = T_B$ if and only if $A = B$) allows us to link matrix attributes with those typically associated with functions, and vice versa.

Theorem 2.11

Let $T: \mathcal{R}^n \to \mathcal{R}^m$ be linear. Then there is a unique $m \times n$ matrix A such that $T(\mathbf{v}) = A\mathbf{v}$ for all $\mathbf{v}$ in $\mathcal{R}^n$. In fact, A is the $m \times n$ matrix whose columns are the images under T of the standard vectors for $\mathcal{R}^n$.

Proof Let A be the $m \times n$ matrix with columns $T(\mathbf{e}_1), T(\mathbf{e}_2), \ldots, T(\mathbf{e}_n)$. Recall from Theorem 1.2(a) that the jth column of A equals $A\mathbf{e}_j$. Hence, $T_A(\mathbf{e}_j) = A\mathbf{e}_j = T(\mathbf{e}_j)$ for every j. So T_A and T agree at each standard vector. By Theorem 2.10, it follows that $T_A = T$. Thus, for all $\mathbf{x}$ in $\mathcal{R}^n$, we have $T(\mathbf{v}) = T_A(\mathbf{v}) = A\mathbf{v}$.

It is left to show that A is unique in the sense that if, for some $m \times n$ matrix B, we also have $T(\mathbf{v}) = B\mathbf{v}$ for all $\mathbf{v}$ in $\mathcal{R}^n$, then $A = B$. Again using Theorem 1.2(a), we see that the jth columns of A and B are $A\mathbf{e}_j$ and $B\mathbf{e}_j$, respectively. Since

$$
A\mathbf{e}_j = T(\mathbf{e}_j) = B\mathbf{e}_j,
$$

the corresponding columns of A and B are equal, and hence $A = B$.

If $T: \mathcal{R}^n \to \mathcal{R}^m$ is a linear transformation, we call the unique $m \times n$ matrix A such that $T(\mathbf{v}) = A\mathbf{v}$ for all $\mathbf{v}$ in $\mathcal{R}^n$ the **standard matrix** of T. So Theorem 2.11 states that every linear transformation $T: \mathcal{R}^n \to \mathcal{R}^m$ has a unique standard matrix, which is $[T(\mathbf{e}_1) \quad T(\mathbf{e}_2) \quad \cdots \quad T(\mathbf{e}_n)]$. In other words, every linear transformation is a matrix transformation—namely, the matrix transformation induced by its standard matrix.

Example 6 Let $T: \mathcal{R}^3 \to \mathcal{R}^2$ be defined by $T\left(\begin{bmatrix} x_1 \\ x_2 \\ x_3 \end{bmatrix}\right) = \begin{bmatrix} 3x_1 - 4x_2 \\ 2x_1 + x_3 \end{bmatrix}$. It is easy to show that T is linear. To find the standard matrix of T, we compute its columns $T(\mathbf{e}_1)$,

$T(\mathbf{e}_2)$, and $T(\mathbf{e}_3)$. We have $T(\mathbf{e}_1) = \begin{bmatrix} 3 \\ 2 \end{bmatrix}$, $T(\mathbf{e}_2) = \begin{bmatrix} -4 \\ 0 \end{bmatrix}$, and $T(\mathbf{e}_3) = \begin{bmatrix} 0 \\ 1 \end{bmatrix}$. So the standard matrix of T is

$$\begin{bmatrix} 3 & -4 & 0 \\ 2 & 0 & 1 \end{bmatrix}.$$

Example 7 Let $T: \mathcal{R}^2 \rightarrow \mathcal{R}^2$ denote the *contraction* $T(\mathbf{x}) = \frac{1}{2}\mathbf{x}$. To find the standard matrix of T, we compute $T(\mathbf{e}_1) = \frac{1}{2}\mathbf{e}_1$ and $T(\mathbf{e}_2) = \frac{1}{2}\mathbf{e}_2$. So the standard matrix of T is

$$\begin{bmatrix} \frac{1}{2} & 0 \\ 0 & \frac{1}{2} \end{bmatrix}.$$

In the next section we illustrate the close relationship between a linear transformation and its standard matrix.

Practice Problems

1. Let $A = \begin{bmatrix} 1 & -2 \\ 3 & 1 \\ -1 & 4 \end{bmatrix}$.

   (a) What is the domain of T_A?
   (b) What is the codomain of T_A?
   (c) Compute $T_A\left(\begin{bmatrix} 4 \\ 3 \end{bmatrix} \right)$.

2. Determine the standard matrix of the linear transformation $T: \mathcal{R}^3 \rightarrow \mathcal{R}^2$ defined by

   $$T\left(\begin{bmatrix} x_1 \\ x_2 \\ x_3 \end{bmatrix} \right) = \begin{bmatrix} 2x_1 - 5x_3 \\ -3x_2 + 4x_3 \end{bmatrix}.$$

3. Suppose that $T: \mathcal{R}^2 \rightarrow \mathcal{R}^3$ is a linear transformation such that

   $$T\left(\begin{bmatrix} -1 \\ 0 \end{bmatrix} \right) = \begin{bmatrix} -2 \\ 1 \\ 3 \end{bmatrix} \quad \text{and} \quad T\left(\begin{bmatrix} 2 \\ 1 \end{bmatrix} \right) = \begin{bmatrix} 3 \\ -1 \\ -4 \end{bmatrix}.$$

   Determine $T\left(\begin{bmatrix} x_1 \\ x_2 \end{bmatrix} \right)$.

Exercises

1. Determine if the following statements are true or false.

   (a) Every function from $\mathcal{R}^n$ to $\mathcal{R}^m$ has a standard matrix.
   (b) Every matrix transformation is linear.
   (c) A function from $\mathcal{R}^n$ to $\mathcal{R}^m$ that preserves scalar multiplication is linear.
   (d) The image of the zero vector under any linear transformation is the zero vector.
   (e) If $T: \mathcal{R}^3 \rightarrow \mathcal{R}^2$ is linear, then its standard matrix has size 3×2.
   (f) The zero transformation is linear.
   (g) A function is uniquely determined by the images of the standard vectors in its domain.
   (h) The first column of the standard matrix of a linear transformation is the image of the first standard vector under the transformation.

Exercises 2–9 refer to the following matrices:

$$A = \begin{bmatrix} 2 & -3 & 1 \\ 4 & 0 & -2 \end{bmatrix}, \quad B = \begin{bmatrix} 1 & 5 & 0 \\ 2 & -1 & 3 \end{bmatrix}, \quad \text{and}$$

$$C = \begin{bmatrix} 1 & 2 \\ 0 & -2 \\ 4 & 1 \end{bmatrix}.$$

2. Give the domain and codomain of the matrix transformation T_A.

3. Give the domain and codomain of the matrix transformation T_B.

4. Give the domain and codomain of the matrix transformation T_C.

5. Compute $T_A\left(\begin{bmatrix} 3 \\ -1 \\ 2 \end{bmatrix} \right)$.

6. Compute $T_B \left(\begin{bmatrix} 1 \\ 0 \\ 1 \end{bmatrix} \right)$.

7. Compute $T_C \left(\begin{bmatrix} 2 \\ 3 \end{bmatrix} \right)$.

8. Compute $T_{(A+B)} \left(\begin{bmatrix} 2 \\ 1 \\ 1 \end{bmatrix} \right)$, and $T_A \left(\begin{bmatrix} 2 \\ 1 \\ 1 \end{bmatrix} \right) + T_B \left(\begin{bmatrix} 2 \\ 1 \\ 1 \end{bmatrix} \right)$.

9. Compute $T_A(\mathbf{e}_1)$ and $T_A(\mathbf{e}_3)$.

10. Suppose that $T:\mathcal{R}^n \to \mathcal{R}^m$ is defined by $T \left(\begin{bmatrix} x_1 \\ x_2 \\ x_3 \end{bmatrix} \right) = \begin{bmatrix} 2x_1 \\ x_1 - x_2 \end{bmatrix}$. What are the values of n and m?

11. Suppose that $T:\mathcal{R}^n \to \mathcal{R}^m$ is defined by $T \left(\begin{bmatrix} x_1 \\ x_2 \end{bmatrix} \right) = \begin{bmatrix} x_1 + x_2 \\ x_1 - x_2 \\ x_2 \end{bmatrix}$. What are the values of n and m?

12. If T is the identity transformation, what is true about the domain and the codomain of T?

13. Suppose that T is linear and $T \left(\begin{bmatrix} 8 \\ 2 \end{bmatrix} \right) = \begin{bmatrix} 2 \\ -4 \\ 6 \end{bmatrix}$.
Determine $T \left(\begin{bmatrix} 16 \\ 4 \end{bmatrix} \right)$ and $T \left(\begin{bmatrix} -4 \\ -1 \end{bmatrix} \right)$. Justify your answers.

14. Suppose that T is linear with $T \left(\begin{bmatrix} 2 \\ 3 \end{bmatrix} \right) = \begin{bmatrix} 1 \\ 2 \end{bmatrix}$ and $T \left(\begin{bmatrix} -4 \\ 0 \end{bmatrix} \right) = \begin{bmatrix} -5 \\ 1 \end{bmatrix}$. Determine $T \left(\begin{bmatrix} -2 \\ 3 \end{bmatrix} \right)$. Justify your answer.

15. Suppose that $T:\mathcal{R}^2 \to \mathcal{R}^2$ is a linear transformation such that
$$T \left(\begin{bmatrix} 1 \\ -2 \end{bmatrix} \right) = \begin{bmatrix} 2 \\ 1 \end{bmatrix} \quad \text{and} \quad T \left(\begin{bmatrix} -1 \\ 3 \end{bmatrix} \right) = \begin{bmatrix} 3 \\ 0 \end{bmatrix}.$$
Determine $T \left(\begin{bmatrix} x_1 \\ x_2 \end{bmatrix} \right)$.

16. Suppose that $T:\mathcal{R}^2 \to \mathcal{R}^3$ is a linear transformation such that
$$T \left(\begin{bmatrix} 3 \\ -5 \end{bmatrix} \right) = \begin{bmatrix} 1 \\ -1 \\ 2 \end{bmatrix} \quad \text{and} \quad T \left(\begin{bmatrix} -1 \\ 2 \end{bmatrix} \right) = \begin{bmatrix} 3 \\ 0 \\ -2 \end{bmatrix}.$$
Determine $T \left(\begin{bmatrix} x_1 \\ x_2 \end{bmatrix} \right)$.

In Exercises 17–23, a linear transformation is given. Compute its standard matrix.

17. $T:\mathcal{R}^2 \to \mathcal{R}^2$ defined by $T \left(\begin{bmatrix} x_1 \\ x_2 \end{bmatrix} \right) = \begin{bmatrix} x_2 \\ x_1 + x_2 \end{bmatrix}$

18. $T:\mathcal{R}^2 \to \mathcal{R}^2$ defined by $T \left(\begin{bmatrix} x_1 \\ x_2 \end{bmatrix} \right) = \begin{bmatrix} 2x_1 + 3x_2 \\ 4x_1 + 5x_2 \end{bmatrix}$

19. $T:\mathcal{R}^3 \to \mathcal{R}^2$ defined by $T \left(\begin{bmatrix} x_1 \\ x_2 \\ x_3 \end{bmatrix} \right) = \begin{bmatrix} x_1 + x_2 + x_3 \\ 2x_1 \end{bmatrix}$

20. $T:\mathcal{R}^2 \to \mathcal{R}^3$ defined by $T \left(\begin{bmatrix} x_1 \\ x_2 \end{bmatrix} \right) = \begin{bmatrix} 3x_2 \\ 2x_1 - x_2 \\ x_1 + x_2 \end{bmatrix}$

21. $T:\mathcal{R}^3 \to \mathcal{R}^3$ defined by $T(\mathbf{v}) = \mathbf{v}$ for all $\mathbf{v}$ in $\mathcal{R}^3$

22. $T:\mathcal{R}^3 \to \mathcal{R}^2$ defined by $T(\mathbf{v}) = \mathbf{0}$ for all $\mathbf{v}$ in $\mathcal{R}^3$

23. $T:\mathcal{R}^3 \to \mathcal{R}^3$ defined by $T(\mathbf{v}) = 4\mathbf{v}$ for all $\mathbf{v}$ in $\mathcal{R}^3$

24. A linear transformation $T:\mathcal{R}^2 \to \mathcal{R}^2$ has the property that $T(\mathbf{e}_1) = \begin{bmatrix} 2 \\ 3 \end{bmatrix}$ and $T(\mathbf{e}_2) = \begin{bmatrix} 4 \\ 1 \end{bmatrix}$. Determine $T \left(\begin{bmatrix} 5 \\ 6 \end{bmatrix} \right)$. Justify your answer.

25. A linear transformation $T:\mathcal{R}^2 \to \mathcal{R}^2$ has the property that $T(\mathbf{e}_1) = \begin{bmatrix} 2 \\ 3 \end{bmatrix}$ and $T(\mathbf{e}_2) = \begin{bmatrix} 4 \\ 1 \end{bmatrix}$. Determine $T \left(\begin{bmatrix} a \\ b \end{bmatrix} \right)$ for any $\begin{bmatrix} a \\ b \end{bmatrix}$ in $\mathcal{R}^2$. Justify your answer.

In Exercises 26–34, a function $T:\mathcal{R}^n \to \mathcal{R}^m$ is given. Either prove that T is linear, or explain why T is not linear.

26. $T:\mathcal{R}^2 \to \mathcal{R}^2$ defined by $T \left(\begin{bmatrix} x_1 \\ x_2 \end{bmatrix} \right) = \begin{bmatrix} 2x_1 \\ x_2^2 \end{bmatrix}$

27. $T:\mathcal{R}^2 \to \mathcal{R}^2$ defined by $T \left(\begin{bmatrix} x_1 \\ x_2 \end{bmatrix} \right) = \begin{bmatrix} 0 \\ 2x_1 \end{bmatrix}$

28. $T:\mathcal{R}^2 \to \mathcal{R}^2$ defined by $T \left(\begin{bmatrix} x_1 \\ x_2 \end{bmatrix} \right) = \begin{bmatrix} 1 \\ 2x_1 \end{bmatrix}$

29. $T:\mathcal{R}^3 \to \mathcal{R}$ defined by $T \left(\begin{bmatrix} x_1 \\ x_2 \\ x_3 \end{bmatrix} \right) = x_1 + x_2 + x_3 - 1$

30. $T:\mathcal{R}^3 \to \mathcal{R}$ defined by $T \left(\begin{bmatrix} x_1 \\ x_2 \\ x_3 \end{bmatrix} \right) = x_1 + x_2 + x_3$

31. $T:\mathcal{R}^2 \to \mathcal{R}^2$ defined by $T \left(\begin{bmatrix} x_1 \\ x_2 \end{bmatrix} \right) = \begin{bmatrix} x_1 + x_2 \\ 2x_1 - x_2 \end{bmatrix}$

32. $T:\mathcal{R}^2 \to \mathcal{R}^2$ defined by $T \left(\begin{bmatrix} x_1 \\ x_2 \end{bmatrix} \right) = \begin{bmatrix} x_2 \\ |x_1| \end{bmatrix}$

33. $T:\mathcal{R} \to \mathcal{R}^2$ defined by $T(x) = \begin{bmatrix} \sin x \\ x \end{bmatrix}$

34. $T:\mathcal{R}^2 \to \mathcal{R}^2$ defined by $T \left(\begin{bmatrix} x_1 \\ x_2 \end{bmatrix} \right) = \begin{bmatrix} ax_1 \\ bx_2 \end{bmatrix}$, where a and b are fixed scalars.

35. Prove that the identity transformation $I:\mathcal{R}^n \to \mathcal{R}^n$ equals T_{I_n} and, hence, is linear.

36. Prove that the zero transformation $T_0:\mathcal{R}^n \to \mathcal{R}^m$ equals T_O and, hence, is linear.

Let $T, U:\mathcal{R}^n \to \mathcal{R}^m$ be functions and c be a scalar. Define $(T + U):\mathcal{R}^n \to \mathcal{R}^m$ and $cT:\mathcal{R}^n \to \mathcal{R}^m$ by

$$(T + U)(\mathbf{x}) = T(\mathbf{x}) + U(\mathbf{x}) \quad \text{and} \quad (cT)(\mathbf{x}) = cT(\mathbf{x})$$

for all $\mathbf{x}$ in $\mathcal{R}^n$.

Exercises 37–40 assume the preceding definitions of $T + U$ and cT, where $T, U:\mathcal{R}^n \to \mathcal{R}^m$ are linear.

37. Prove that if T is linear and c is a scalar, then cT is linear.

38. Prove that if T and U are linear, then $T + U$ is linear.

39. Suppose c is a scalar. Use Exercise 37 to prove that if T is linear and has standard matrix A, then the standard matrix of cT is cA.

40. Use Exercise 38 to prove that if T and U are linear with standard matrices A and B, respectively, then the standard matrix of $T + U$ is $A + B$.

41. Let $T:\mathcal{R}^2 \to \mathcal{R}^2$ be a linear transformation. Prove that there exist unique scalars a, b, c, and d such that
$$T\left(\begin{bmatrix} x_1 \\ x_2 \end{bmatrix}\right) = \begin{bmatrix} ax_1 + bx_2 \\ cx_1 + dx_2 \end{bmatrix} \text{ for every vector } \begin{bmatrix} x_1 \\ x_2 \end{bmatrix} \text{ in } \mathcal{R}^2.$$
Hint: Use Theorem 2.11.

42. State and prove a generalization of Exercise 41.

43. Define $T:\mathcal{R}^2 \to \mathcal{R}^2$ by $T\left(\begin{bmatrix} x_1 \\ x_2 \end{bmatrix}\right) = \begin{bmatrix} x_1 \\ 0 \end{bmatrix}$. T represents the *orthogonal projection of $\mathcal{R}^2$ onto the x-axis.*

(a) Prove that T is linear.
(b) Find the standard matrix of T.
(c) Prove that $T(T(\mathbf{v})) = T(\mathbf{v})$ for every $\mathbf{v}$ in $\mathcal{R}^2$.

44. Define $T:\mathcal{R}^3 \to \mathcal{R}^3$ by $T\left(\begin{bmatrix} x_1 \\ x_2 \\ x_3 \end{bmatrix}\right) = \begin{bmatrix} 0 \\ x_2 \\ x_3 \end{bmatrix}$. T represents the *orthogonal projection of $\mathcal{R}^3$ onto the yz-plane.*

(a) Prove that T is linear.
(b) Find the standard matrix of T.
(c) Prove that $T(T(\mathbf{v})) = T(\mathbf{v})$ for every $\mathbf{v}$ in $\mathcal{R}^3$.

45. Define the linear transformation $T:\mathcal{R}^2 \to \mathcal{R}^2$ by $T\left(\begin{bmatrix} x_1 \\ x_2 \end{bmatrix}\right) = \begin{bmatrix} -x_1 \\ x_2 \end{bmatrix}$. T represents the *reflection of $\mathcal{R}^2$ about the y-axis.*

(a) Show that T is a matrix transformation.
(b) Determine the range of T.

46. Define the linear transformation $T:\mathcal{R}^3 \to \mathcal{R}^3$ by $T\left(\begin{bmatrix} x_1 \\ x_2 \\ x_3 \end{bmatrix}\right) = \begin{bmatrix} x_1 \\ x_2 \\ -x_3 \end{bmatrix}$. T represents the *reflection of $\mathcal{R}^3$ about the xy-plane.*

(a) Show that T is a matrix transformation.
(b) Determine the range of T.

47. A linear transformation $T:\mathcal{R}^n \to \mathcal{R}^n$ defined by $T(\mathbf{x}) = k\mathbf{x}$, where $0 < k < 1$, is called a *contraction.*

(a) Show that T is a matrix transformation.
(b) Determine the range of T.

48. A linear transformation $T:\mathcal{R}^n \to \mathcal{R}^n$ defined by $T(\mathbf{x}) = k\mathbf{x}$, where $k > 1$, is called a *dilation.*

(a) Show that T is a matrix transformation.
(b) Determine the range of T.

49. Let $T:\mathcal{R}^n \to \mathcal{R}^m$ be a linear transformation. Prove that $T(\mathbf{u}) = T(\mathbf{v})$ if and only if $T(\mathbf{u} - \mathbf{v}) = \mathbf{0}$.

50. Find functions $f:\mathcal{R}^2 \to \mathcal{R}^2$ and $g:\mathcal{R}^2 \to \mathcal{R}^2$ such that $f(\mathbf{e}_1) = g(\mathbf{e}_1)$ and $f(\mathbf{e}_2) = g(\mathbf{e}_2)$, but $f(\mathbf{v}) \neq g(\mathbf{v})$ for some $\mathbf{v}$ in $\mathcal{R}^2$.

51. Let A be an invertible $n \times n$ matrix. Determine $T_{A^{-1}}(T_A(\mathbf{v}))$ and $T_A(T_{A^{-1}}(\mathbf{v}))$ for all $\mathbf{v}$ in $\mathcal{R}^n$.

52. Let A be an $m \times n$ matrix and B an $n \times p$ matrix. Prove that $T_{AB}(\mathbf{v}) = T_A(T_B(\mathbf{v}))$ for all $\mathbf{v}$ in $\mathcal{R}^p$.

53. Let $T:\mathcal{R}^n \to \mathcal{R}^m$ be a linear transformation with standard matrix A. Prove that the range of T is spanned by the columns of A.

54. Prove that a linear transformation $T:\mathcal{R}^n \to \mathcal{R}^m$ has $\mathcal{R}^m$ as its range if and only if the rank of its standard matrix is m.

55. Let $T:\mathcal{R}^n \to \mathcal{R}^m$ be a linear transformation and $S = \{\mathbf{v}_1, \mathbf{v}_2, \ldots, \mathbf{v}_k\}$ be a subset of $\mathcal{R}^n$. Prove that if the set $\{T(\mathbf{v}_1), T(\mathbf{v}_2), \ldots, T(\mathbf{v}_k)\}$ is a linearly independent subset of $\mathcal{R}^m$, then S is a linearly independent subset of $\mathcal{R}^n$.

In Exercises 56 and 57, use a calculator with matrix capabilities or computer software such as MATLAB *to solve the problem.*

56. Suppose that $T:\mathcal{R}^4 \to \mathcal{R}^4$ is a linear transformation such that

$$T\left(\begin{bmatrix} 1 \\ 2 \\ 0 \\ -1 \end{bmatrix}\right) = \begin{bmatrix} 0 \\ 1 \\ 1 \\ 0 \end{bmatrix}, \quad T\left(\begin{bmatrix} 1 \\ 1 \\ 1 \\ -1 \end{bmatrix}\right) = \begin{bmatrix} -2 \\ 1 \\ 3 \\ 2 \end{bmatrix},$$

$$T\left(\begin{bmatrix} 0 \\ 1 \\ 0 \\ 1 \end{bmatrix}\right) = \begin{bmatrix} 4 \\ 6 \\ 0 \\ -3 \end{bmatrix},$$

and

$$T\left(\begin{bmatrix} -1 \\ 2 \\ -3 \\ 1 \end{bmatrix}\right) = \begin{bmatrix} 0 \\ 0 \\ 0 \\ 0 \end{bmatrix}.$$

(a) Find a rule for T.
(b) Is T uniquely determined by the four images above? Justify your answer.

57. Suppose that $T: \mathcal{R}^4 \to \mathcal{R}^4$ is the linear transformation defined by the rule

$$T\left(\begin{bmatrix} x_1 \\ x_2 \\ x_3 \\ x_4 \end{bmatrix}\right) = \begin{bmatrix} x_1 + x_2 + x_3 + 2x_4 \\ x_1 + 2x_2 - 3x_3 + 4x_4 \\ x_2 + 2x_4 \\ x_1 + 5x_2 - x_3 \end{bmatrix}.$$

Determine if the vector $\begin{bmatrix} 2 \\ -1 \\ 0 \\ 3 \end{bmatrix}$ is in the range of T.

Solutions to the Practice Problems

1. (a) Because A is a 3×2 matrix, the domain of T_A is $\mathcal{R}^2$.

(b) The codomain of T_A is $\mathcal{R}^3$.

(c) We have

$$T_A\left(\begin{bmatrix} 4 \\ 3 \end{bmatrix}\right) = A\begin{bmatrix} 4 \\ 3 \end{bmatrix} = \begin{bmatrix} 1 & -2 \\ 3 & 1 \\ -1 & 4 \end{bmatrix}\begin{bmatrix} 4 \\ 3 \end{bmatrix} = \begin{bmatrix} -2 \\ 15 \\ 8 \end{bmatrix}.$$

2. Since

$$T(\mathbf{e}_1) = \begin{bmatrix} 2 \\ 0 \end{bmatrix}, \qquad T(\mathbf{e}_2) = \begin{bmatrix} 0 \\ -3 \end{bmatrix}, \qquad \text{and}$$

$$T(\mathbf{e}_3) = \begin{bmatrix} -5 \\ 4 \end{bmatrix},$$

the standard matrix of T is

$$\begin{bmatrix} 2 & 0 & -5 \\ 0 & -3 & 4 \end{bmatrix}.$$

3. We begin by finding the standard matrix of T. This requires that we express each of the standard vectors for $\mathcal{R}^2$ as a linear combination of $\begin{bmatrix} -1 \\ 0 \end{bmatrix}$ and $\begin{bmatrix} 2 \\ 1 \end{bmatrix}$. Clearly

$$\mathbf{e}_1 = -\begin{bmatrix} -1 \\ 0 \end{bmatrix}, \text{ and also } \mathbf{e}_2 = \begin{bmatrix} 0 \\ 1 \end{bmatrix} = 2\begin{bmatrix} -1 \\ 0 \end{bmatrix} + \begin{bmatrix} 2 \\ 1 \end{bmatrix}.$$

Thus we have

$$T(\mathbf{e}_1) = T\left(-\begin{bmatrix} -1 \\ 0 \end{bmatrix}\right) = -T\left(\begin{bmatrix} -1 \\ 0 \end{bmatrix}\right)$$

$$= -\begin{bmatrix} -2 \\ 1 \\ 3 \end{bmatrix} = \begin{bmatrix} 2 \\ -1 \\ -3 \end{bmatrix},$$

and

$$T(\mathbf{e}_2) = T\left(2\begin{bmatrix} -1 \\ 0 \end{bmatrix} + \begin{bmatrix} 2 \\ 1 \end{bmatrix}\right) = 2T\left(\begin{bmatrix} -1 \\ 0 \end{bmatrix}\right) + T\left(\begin{bmatrix} 2 \\ 1 \end{bmatrix}\right)$$

$$= 2\begin{bmatrix} -2 \\ 1 \\ 3 \end{bmatrix} + \begin{bmatrix} 3 \\ -1 \\ -4 \end{bmatrix} = \begin{bmatrix} -1 \\ 1 \\ 2 \end{bmatrix}.$$

Hence the standard matrix of T is

$$A = \begin{bmatrix} 2 & -1 \\ -1 & 1 \\ -3 & 2 \end{bmatrix}.$$

Therefore

$$T\left(\begin{bmatrix} x_1 \\ x_2 \end{bmatrix}\right) = A\begin{bmatrix} x_1 \\ x_2 \end{bmatrix} = \begin{bmatrix} 2 & -1 \\ -1 & 1 \\ -3 & 2 \end{bmatrix}\begin{bmatrix} x_1 \\ x_2 \end{bmatrix}$$

$$= \begin{bmatrix} 2x_1 & - & x_2 \\ -x_1 & + & x_2 \\ -3x_1 & + & 2x_2 \end{bmatrix}.$$

2.7 Composition and Invertibility of Linear Transformations

In this section, we exploit the relationship between a linear transformation and its standard matrix. Properties that we normally associate with functions—for example, one-to-one and onto—have close relationships to properties of the columns of a matrix, which in turn are related to the consistency of particular systems of linear equations.

Once we have the standard matrix of a linear transformation T, it is easy to find a spanning set for the range of T. For example, suppose that $T: \mathcal{R}^3 \to \mathcal{R}^2$ is defined by $T\left(\begin{bmatrix} x_1 \\ x_2 \\ x_3 \end{bmatrix}\right) = \begin{bmatrix} 3x_1 - 4x_2 \\ 2x_1 + x_3 \end{bmatrix}$. In Example 6 of Section 2.6, we saw that the standard matrix of T is

$$A = [T(\mathbf{e}_1)\ \ T(\mathbf{e}_2)\ \ T(\mathbf{e}_3)] = \begin{bmatrix} 3 & -4 & 0 \\ 2 & 0 & 1 \end{bmatrix}.$$

Then $\mathbf{w}$ is in the range of T if and only if $\mathbf{w} = T(\mathbf{v})$ for some $\mathbf{v}$ in $\mathcal{R}^2$. Writing $\mathbf{v} = v_1\mathbf{e}_1 + v_2\mathbf{e}_2 + v_3\mathbf{e}_3$, we see that

$$\mathbf{w} = T(\mathbf{v}) = T(v_1\mathbf{e}_1 + v_2\mathbf{e}_2 + v_3\mathbf{e}_3) = v_1 T(\mathbf{e}_1) + v_2 T(\mathbf{e}_2) + v_3 T(\mathbf{e}_3),$$

which is a linear combination of the columns of A. Likewise, it is clear from the same computation that every linear combination of the columns of A is in the range of T. We conclude that the range of T is spanned by the vectors $\begin{bmatrix} 3 \\ 2 \end{bmatrix}$, $\begin{bmatrix} -4 \\ 0 \end{bmatrix}$, and $\begin{bmatrix} 0 \\ 1 \end{bmatrix}$. This argument can be generalized to prove the following result.

> The range of a linear transformation equals the span of the columns of its standard matrix.

In what follows, additional properties about a linear transformation are obtained from its standard matrix. First, however, we need to recall some definitions about functions.

Definition. A function $f: \mathcal{R}^n \to \mathcal{R}^m$ is said to be **onto** if its range is all of $\mathcal{R}^m$; that is, if every vector in $\mathcal{R}^m$ is an image.

So a function is onto if and only if its range equals its codomain. To show that a function $f: \mathcal{R}^n \to \mathcal{R}^m$ is onto, we need to show that for any vector $\mathbf{w}$ in $\mathcal{R}^m$, there exists a vector $\mathbf{u}$ in $\mathcal{R}^n$ such that $f(\mathbf{u}) = \mathbf{w}$; that is, that $\mathbf{w}$ is an image. For example, the contraction $T: \mathcal{R}^2 \to \mathcal{R}^2$ in Example 7 of Section 2.6 defined by $T(\mathbf{x}) = \frac{1}{2}\mathbf{x}$ is onto because every vector $\mathbf{w}$ in $\mathcal{R}^2$ can be written as $\mathbf{w} = \frac{1}{2}(2\mathbf{w}) = T(2\mathbf{w})$—that is, every vector $\mathbf{w}$ is an image.

From the preceding boxed statement, it follows that a linear transformation is onto if and only if the columns of its standard matrix span its codomain. In the case of the preceding contraction, to demonstrate that the columns $\frac{1}{2}\mathbf{e}_1, \frac{1}{2}\mathbf{e}_2$ of its standard matrix span $\mathcal{R}^2$, we observe that any vector $\mathbf{w}$ in $\mathcal{R}^2$ can be written as

$$\mathbf{w} = w_1\mathbf{e}_1 + w_2\mathbf{e}_2 = 2w_1\left(\frac{1}{2}\mathbf{e}_1\right) + 2w_2\left(\frac{1}{2}\mathbf{e}_2\right).$$

So $\mathbf{w}$ is a linear combination of the columns of A.

It is useful to note that we may use the rank of the standard matrix A to determine if T is onto. If A is an $m \times n$ matrix, then by Theorem 1.5 the columns of A span $\mathcal{R}^m$ if and only if rank $A = m$.

Example 1

Determine if the linear transformation $T: \mathcal{R}^3 \to \mathcal{R}^3$ defined by

$$T\left(\begin{bmatrix} x_1 \\ x_2 \\ x_3 \end{bmatrix}\right) = \begin{bmatrix} x_1 + 2x_2 + 4x_3 \\ x_1 + 3x_2 + 6x_3 \\ 2x_1 + 5x_2 + 10x_3 \end{bmatrix}$$

is onto.

Solution   First we compute the standard matrix A of T and its reduced row echelon form R:

$$A = \begin{bmatrix} 1 & 2 & 4 \\ 1 & 3 & 6 \\ 2 & 5 & 10 \end{bmatrix} \quad \text{and} \quad R = \begin{bmatrix} 1 & 0 & 0 \\ 0 & 1 & 2 \\ 0 & 0 & 0 \end{bmatrix}.$$

Clearly, rank $A = 2 \neq 3$. So T is not onto.   ○

The reader is invited to investigate if every vector $\mathbf{b}$ in $\mathcal{R}^3$ is an image under T, and compare this computation with the shorter one in Example 1.

We now state the first of many theorems that relate a linear transformation to its standard matrix. Its proof follows from our preceding observations and from Theorem 1.5.

Theorem 2.12

Let $T: \mathcal{R}^n \to \mathcal{R}^m$ be a linear transformation with standard matrix A. The following conditions are equivalent.

(a) *T is onto.*
(b) *The range of T is $\mathcal{R}^m$.*
(c) *The columns of A are a spanning set for $\mathcal{R}^m$.*
(d) *rank $A = m$.*

There is a close relationship between the range of a matrix transformation and the consistency of a system of linear equations. For example, consider the system

$$\begin{aligned} x_1 \phantom{{}+ x_2} &= 1 \\ 2x_1 + x_2 &= 3 \\ x_1 - x_2 &= 1. \end{aligned}$$

The system is equivalent to the matrix equation $A\mathbf{x} = \mathbf{b}$, where

$$A = \begin{bmatrix} 1 & 0 \\ 2 & 1 \\ 1 & -1 \end{bmatrix}, \quad \mathbf{x} = \begin{bmatrix} x_1 \\ x_2 \end{bmatrix}, \quad \text{and} \quad \mathbf{b} = \begin{bmatrix} 1 \\ 3 \\ 1 \end{bmatrix}.$$

We see that the matrix equation may be written as $T_A(\mathbf{x}) = \mathbf{b}$. We conclude that the system has a solution if and only if $\mathbf{b}$ is in the range of T_A.

Now suppose that we define $T: \mathcal{R}^3 \to \mathcal{R}^2$ by

$$T\left(\begin{bmatrix} x_1 \\ x_2 \\ x_3 \end{bmatrix} \right) = \begin{bmatrix} x_1 + x_2 + x_3 \\ x_1 + 3x_2 - x_3 \end{bmatrix}.$$

It is easy to see that $T = T_A$, where $A = \begin{bmatrix} 1 & 1 & 1 \\ 1 & 3 & -1 \end{bmatrix}$; so T is linear. Suppose that we want to determine if the vector $\mathbf{w} = \begin{bmatrix} 2 \\ 8 \end{bmatrix}$ is in the range of T. This question is equivalent to asking if there exists a vector $\mathbf{x}$ such that $T(\mathbf{x}) = \mathbf{w}$, or if the following system is consistent:

$$\begin{aligned} x_1 + x_2 + x_3 &= 2 \\ x_1 + 3x_2 - x_3 &= 8. \end{aligned}$$

Using Gaussian elimination, we obtain the general solution

$$x_1 = -2x_3 - 1$$
$$x_2 = x_3 + 3$$
$$x_3 \quad \text{free.}$$

We conclude that the system is consistent, and there are infinitely many vectors whose image is **w**. For example, for the cases $x_3 = 0$ and $x_3 = 1$, we obtain the vectors $\begin{bmatrix} -1 \\ 3 \\ 0 \end{bmatrix}$ and $\begin{bmatrix} -3 \\ 4 \\ 1 \end{bmatrix}$, respectively. Alternatively, we could have observed that A has rank 2 and then appealed to Theorem 2.12 to conclude that every vector in $\mathcal{R}^2$, including **w**, is in the range of T.

Another property of some functions is that of being *one-to-one*.

Definition. A function $f : \mathcal{R}^n \to \mathcal{R}^m$ is said to be **one-to-one** if every pair of distinct vectors in $\mathcal{R}^n$ have distinct images. That is, if **u** and **v** are distinct vectors in $\mathcal{R}^n$, then $f(\mathbf{u})$ and $f(\mathbf{v})$ are distinct.

In Figure 2.9(a), we see that distinct vectors **u** and **v** have distinct images, which is necessary for T to be one-to-one. In Figure 2.9(b), T is not one-to-one because there exist distinct vectors **u** and **v** that have the same image **w**.

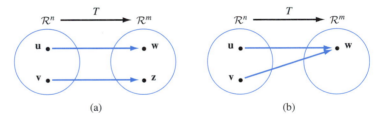

FIGURE 2.9   (a)   (b)

A common way of showing that f is one-to-one is to show that the relation $f(\mathbf{u}) = f(\mathbf{v})$ implies $\mathbf{u} = \mathbf{v}$. For example, the contraction $T : \mathcal{R}^2 \to \mathcal{R}^2$ defined by $T(\mathbf{x}) = \frac{1}{2}\mathbf{x}$ is one-to-one because if $T(\mathbf{u}) = T(\mathbf{v})$, then $\frac{1}{2}\mathbf{u} = \frac{1}{2}\mathbf{v}$; so $\mathbf{u} = \mathbf{v}$.

Now let $T : \mathcal{R}^n \to \mathcal{R}^m$ be a linear transformation. Suppose that T is one-to-one. If **w** is a vector in $\mathcal{R}^n$ such that $T(\mathbf{w}) = \mathbf{0}$, then $T(\mathbf{w}) = T(\mathbf{0}) = \mathbf{0}$ and, hence, $\mathbf{w} = \mathbf{0}$. So **0** is the only vector in $\mathcal{R}^n$ whose image under T is the zero vector of $\mathcal{R}^m$. It is interesting that this condition, which is a trivial consequence of T being one-to-one, also implies that T is one-to-one. To see this, assume that **0** is the only vector in $\mathcal{R}^n$ whose image under T is the zero vector of $\mathcal{R}^n$. Assume that **u** and **v** are vectors in $\mathcal{R}^n$ such that $T(\mathbf{u}) = T(\mathbf{v})$. Let $\mathbf{w} = \mathbf{u} - \mathbf{v}$. Then

$$T(\mathbf{w}) = T(\mathbf{u} - \mathbf{v}) = T(\mathbf{u}) - T(\mathbf{v}) = \mathbf{0},$$

hence $\mathbf{u} - \mathbf{v} = \mathbf{w} = \mathbf{0}$. Thus $\mathbf{u} = \mathbf{v}$, and we conclude that T is one-to-one. So the set of vectors whose image is the zero vector is of particular interest.

Definition. Let $T : \mathcal{R}^n \to \mathcal{R}^m$ be linear. The **null space** of T is the set of all **x** in $\mathcal{R}^n$ such that $T(\mathbf{x}) = \mathbf{0}$.

Note that if A is the standard matrix of a linear transformation T, then the null space of T is the set of solutions to $A\mathbf{x} = \mathbf{0}$.

Example 2   Suppose that $T: \mathcal{R}^3 \rightarrow \mathcal{R}^2$ is defined by

$$T\left(\begin{bmatrix} x_1 \\ x_2 \\ x_3 \end{bmatrix}\right) = \begin{bmatrix} x_1 - x_2 + 2x_3 \\ -x_1 + x_2 - 3x_3 \end{bmatrix}.$$

Find a spanning set for the null space of T.

Solution   If we let A be the standard matrix of T, then

$$A = \begin{bmatrix} 1 & -1 & 2 \\ -1 & 1 & -3 \end{bmatrix}.$$

Because $\mathbf{v}$ is in the null space of T if and only if $T(\mathbf{v}) = A\mathbf{v} = \mathbf{0}$, the null space of T is the set of solutions to $A\mathbf{x} = \mathbf{0}$. The reduced row echelon form of A is

$$\begin{bmatrix} 1 & -1 & 0 \\ 0 & 0 & 1 \end{bmatrix}.$$

This matrix corresponds to the system

$$\begin{aligned} x_1 - x_2 \quad &= 0 \\ x_3 &= 0. \end{aligned}$$

So every solution has the form

$$\begin{bmatrix} x_1 \\ x_2 \\ x_3 \end{bmatrix} = \begin{bmatrix} x_2 \\ x_2 \\ 0 \end{bmatrix} = x_2 \begin{bmatrix} 1 \\ 1 \\ 0 \end{bmatrix}.$$

Thus a spanning set for the null space of T is $\left\{ \begin{bmatrix} 1 \\ 1 \\ 0 \end{bmatrix} \right\}$.

A linear transformation $T: \mathcal{R}^n \rightarrow \mathcal{R}^n$ with standard matrix A is one-to-one if and only if the zero vector is the only solution to $A\mathbf{x} = \mathbf{0}$. Several other equivalent properties hold if T is one-to-one. First, the condition that the only solution to $A\mathbf{x} = \mathbf{0}$ is the zero solution is equivalent to the statement that the columns of A are linearly independent. It is also equivalent to the statement that the number of free variables in the general solution must be zero; that is, the nullity of A is 0. Since rank $A = n -$ nullity A, we have that T is one-to-one if and only if rank $A = n$. Using these observations and Theorem 1.7, we obtain the following theorem.

Theorem 2.13
Let $T: \mathcal{R}^n \rightarrow \mathcal{R}^m$ be a linear transformation with standard matrix A. Then the following statements are equivalent.

(a) *T is one-to-one.*
(b) *The null space of T consists of only the zero vector.*
(c) *The columns of A are linearly independent.*
(d) *rank $A = n$.*

Example 3   Determine if the linear transformation $T: \mathcal{R}^3 \rightarrow \mathcal{R}^3$ in Example 1 is one-to-one.

Solution   We saw in Example 1 that rank $A = 2$. So by (a) and (d) of Theorem 2.13, T is not one-to-one.   ○

Finally, we relate all three topics: linear transformations, matrices, and systems of linear equations. Suppose that we begin with the system $A\mathbf{x} = \mathbf{b}$, where A is an $m \times n$ matrix, $\mathbf{x}$ is a vector in $\mathcal{R}^n$, and $\mathbf{b}$ is a vector in $\mathcal{R}^m$. We may write the system in the equivalent form $T_A(\mathbf{x}) = \mathbf{b}$. For the system $A\mathbf{x} = \mathbf{b}$ to be consistent means that $A\mathbf{u} = \mathbf{b}$ for some vector $\mathbf{u}$. That is, $A\mathbf{x} = \mathbf{b}$ has a solution if and only if $\mathbf{b}$ is in the range of T_A, and so $A\mathbf{x} = \mathbf{b}$ has a solution for every $\mathbf{b}$ if and only if T_A is onto. Also, if T_A is one-to-one, then the solution $\mathbf{u}$ is unique. Conversely, if there is at most one solution to $A\mathbf{x} = \mathbf{b}$ for every $\mathbf{b}$, then T_A is one-to-one.

Example 4   Let

$$A = \begin{bmatrix} 0 & 0 & 1 & 3 & 3 \\ 2 & 3 & 1 & 5 & 2 \\ 4 & 6 & 1 & 6 & 2 \\ 4 & 6 & 1 & 7 & 1 \end{bmatrix}.$$

Is the system $A\mathbf{x} = \mathbf{b}$ consistent for every $\mathbf{b}$? If $A\mathbf{x} = \mathbf{b}$ is consistent for some $\mathbf{b}$, is the solution unique?

Solution   Let $T = T_A: \mathcal{R}^5 \rightarrow \mathcal{R}^4$. First, note that the reduced row echelon form of A is

$$R = \begin{bmatrix} 1 & 1.5 & 0 & 1 & 0 \\ 0 & 0 & 1 & 3 & 0 \\ 0 & 0 & 0 & 0 & 1 \\ 0 & 0 & 0 & 0 & 0 \end{bmatrix}.$$

Because rank $A = 3 \neq 4$, we see that T is not onto, by Theorem 2.12. So there exists a $\mathbf{b}$ in $\mathcal{R}^4$ that is not in the range of T. That is, $A\mathbf{x} = \mathbf{b}$ is not consistent. Also, Theorem 2.13 shows T is not one-to-one, so there exists a nonzero solution $\mathbf{u}$ to $A\mathbf{x} = \mathbf{0}$. Therefore, if for some $\mathbf{b}$ we have $A\mathbf{v} = \mathbf{b}$, then we also have

$$A(\mathbf{v} + \mathbf{u}) = A(\mathbf{v}) + A(\mathbf{u}) = \mathbf{b} + \mathbf{0} = \mathbf{b}.$$

So there is *never* a unique solution to $A\mathbf{x} = \mathbf{b}$.   ○

Composition of Linear Transformations

It is useful to observe that matrix multiplication has a parallel with function composition. Recall that if $f: \mathcal{S}_1 \rightarrow \mathcal{S}_2$ and $g: \mathcal{S}_2 \rightarrow \mathcal{S}_3$, then the *composite* $g \circ f: \mathcal{S}_1 \rightarrow \mathcal{S}_3$ is defined by $(g \circ f)(\mathbf{u}) = g(f(\mathbf{u}))$ for all $\mathbf{u}$ in $\mathcal{S}_1$ (see Figure 2.10).

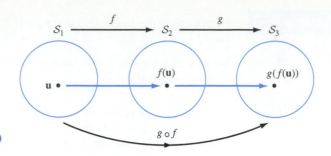

FIGURE 2.10

Example 5 Suppose $f: \mathcal{R}^2 \to \mathcal{R}^3$ and $g: \mathcal{R}^3 \to \mathcal{R}^2$ are the functions defined by

$$f\left(\begin{bmatrix} x_1 \\ x_2 \end{bmatrix}\right) = \begin{bmatrix} x_1^2 \\ x_1 x_2 \\ x_1 + x_2 \end{bmatrix} \qquad \text{and} \qquad g\left(\begin{bmatrix} x_1 \\ x_2 \\ x_3 \end{bmatrix}\right) = \begin{bmatrix} x_1 - x_3 \\ 3x_2 \end{bmatrix}.$$

Then $g \circ f: \mathcal{R}^2 \to \mathcal{R}^2$ is defined by

$$(g \circ f)\left(\begin{bmatrix} x_1 \\ x_2 \end{bmatrix}\right) = g\left(f\left(\begin{bmatrix} x_1 \\ x_2 \end{bmatrix}\right) \right) = g\left(\begin{bmatrix} x_1^2 \\ x_1 x_2 \\ x_1 + x_2 \end{bmatrix}\right)$$

$$= \begin{bmatrix} x_1^2 - (x_1 + x_2) \\ 3x_1 x_2 \end{bmatrix}.$$

In linear algebra, it is customary to drop the "circle" notation and write the composite of linear transformations $U: \mathcal{R}^m \to \mathcal{R}^p$ and $T: \mathcal{R}^n \to \mathcal{R}^m$ as UT rather than $U \circ T$. In this case, UT has domain $\mathcal{R}^n$ and codomain $\mathcal{R}^p$.

Now suppose that we have an $m \times n$ matrix A and an $n \times p$ matrix B, so that AB is an $m \times p$ matrix. The corresponding matrix transformations are $T_A: \mathcal{R}^n \to \mathcal{R}^m$, $T_B: \mathcal{R}^p \to \mathcal{R}^n$, and $T_{AB}: \mathcal{R}^p \to \mathcal{R}^m$. For any $\mathbf{v}$ in $\mathcal{R}^p$, we have

$$T_{AB}(\mathbf{v}) = (AB)\mathbf{v} = A(B\mathbf{v}) = A(T_B(\mathbf{v})) = T_A(T_B(\mathbf{v})).$$

It follows that the matrix transformation T_{AB} is the composite of T_A and T_B; that is,

$$T_A T_B = T_{AB}.$$

Example 6 In $\mathcal{R}^2$, show that a rotation by $180°$ followed by a reflection about the x-axis is equivalent to a reflection about the y-axis.

Solution Let U and T denote the given rotation and reflection, respectively. We want to show that TU is a reflection about the y-axis. Let B and A be the standard matrices of U and T, respectively. These matrices are given by

$$A = \begin{bmatrix} 1 & 0 \\ 0 & -1 \end{bmatrix} \qquad \text{and} \qquad B = \begin{bmatrix} -1 & 0 \\ 0 & -1 \end{bmatrix}.$$

Then $U = T_B$, $T = T_A$, and $AB = \begin{bmatrix} -1 & 0 \\ 0 & 1 \end{bmatrix}$. Thus, for any $\mathbf{u} = \begin{bmatrix} u_1 \\ u_2 \end{bmatrix}$, we have

$$(TU)\left(\begin{bmatrix} u_1 \\ u_2 \end{bmatrix}\right) = (T_A T_B)\left(\begin{bmatrix} u_1 \\ u_2 \end{bmatrix}\right) = T_{AB}\left(\begin{bmatrix} u_1 \\ u_2 \end{bmatrix}\right)$$

$$= (AB)\begin{bmatrix} u_1 \\ u_2 \end{bmatrix} = \begin{bmatrix} -1 & 0 \\ 0 & 1 \end{bmatrix}\begin{bmatrix} u_1 \\ u_2 \end{bmatrix} = \begin{bmatrix} -u_1 \\ u_2 \end{bmatrix},$$

which represents a reflection about the y-axis (see Figure 2.11).

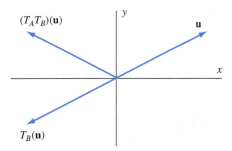

FIGURE 2.11

We can use the preceding result to obtain an interesting relationship between an invertible matrix and the corresponding matrix transformation. First, recall that a function $f: \mathcal{S}_1 \to \mathcal{S}_2$ is *invertible* if there exists a function $g: \mathcal{S}_2 \to \mathcal{S}_1$ such that $g(f(\mathbf{v})) = \mathbf{v}$ for all $\mathbf{v}$ in $\mathcal{S}_1$ and $f(g(\mathbf{v})) = \mathbf{v}$ for all $\mathbf{v}$ in $\mathcal{S}_2$. If f is invertible, then the function g is unique and is called the *inverse* of f; it is denoted by f^{-1}. It can be shown that a function is invertible if and only if it is one-to-one and onto.

Now suppose that A is an $n \times n$ invertible matrix. Then, for all $\mathbf{v}$ in $\mathcal{R}^n$, we have

$$T_A(T_{A^{-1}}(\mathbf{v})) = (T_A T_{A^{-1}})(\mathbf{v}) = T_{AA^{-1}}(\mathbf{v}) = T_{I_n}(\mathbf{v}) = I_n\mathbf{v} = \mathbf{v}.$$

Likewise, $T_{A^{-1}}(T_A(\mathbf{v})) = \mathbf{v}$. We conclude that T_A is invertible and

$$T_A^{-1} = T_{A^{-1}}.$$

It is easy to see, for example, that if T_A represents rotation in $\mathcal{R}^2$ by θ, then T_A^{-1} represents rotation by $-\theta$. Using the result above, this can also be shown by computing the rotation matrix $A_{-\theta}$ (see Example 3 of Section 2.3).

Example 7   Suppose that $A = \begin{bmatrix} 1 & 2 \\ 3 & 5 \end{bmatrix}$, so that $T_A\left(\begin{bmatrix} v_1 \\ v_2 \end{bmatrix}\right) = \begin{bmatrix} v_1 + 2v_2 \\ 3v_1 + 5v_2 \end{bmatrix}$. We saw in Example 1 of Section 2.3 that $A^{-1} = \begin{bmatrix} -5 & 2 \\ 3 & -1 \end{bmatrix}$. To compute the rule for T_A^{-1}, we have

$$T_A^{-1}\left(\begin{bmatrix} v_1 \\ v_2 \end{bmatrix}\right) = T_{A^{-1}}\left(\begin{bmatrix} v_1 \\ v_2 \end{bmatrix}\right) = \begin{bmatrix} -5 & 2 \\ 3 & -1 \end{bmatrix}\begin{bmatrix} v_1 \\ v_2 \end{bmatrix} = \begin{bmatrix} -5v_1 + 2v_2 \\ 3v_1 - v_2 \end{bmatrix}.$$

We can easily check that $T_A^{-1}(T_A(\mathbf{v})) = \mathbf{v}$ and $T_A(T_A^{-1}(\mathbf{v})) = \mathbf{v}$ for every $\mathbf{v}$ in $\mathcal{R}^2$. ○

If $T:\mathcal{R}^n \to \mathcal{R}^n$ is linear and invertible, then it is also one-to-one. By Theorem 2.13, its standard matrix A has rank n and, hence, is also invertible. Thus we have $T^{-1} = T_{A^{-1}}$. In particular, T^{-1} is a matrix transformation, and therefore linear.

The following theorem summarizes several of the preceding observations.

Theorem 2.14

Let $T:\mathcal{R}^n \to \mathcal{R}^m$ and $U:\mathcal{R}^m \to \mathcal{R}^p$ be linear transformations with standard matrices A and B, respectively.

(a) The composite function $UT:\mathcal{R}^n \to \mathcal{R}^p$ is also linear, and its standard matrix is BA.

(b) Suppose $m = n$. Then T is invertible if and only if A is invertible, in which case $T^{-1} = T_{A^{-1}}$. Thus T^{-1} is linear and its standard matrix is A^{-1}.

Proof By Theorem 2.11, we have that $T = T_A$ and $U = T_B$. By our previous observation, we have $UT = T_B T_A = T_{BA}$, which is a matrix transformation and hence is linear. So (a) is proven. Likewise, (b) follows from the preceding discussion. ❑

We list the highlights from this section in the following table. Assume that $T:\mathcal{R}^n \to \mathcal{R}^m$ is a linear transformation with standard matrix A, which has size $m \times n$. *Properties listed in the same row of the table are equivalent.*

Property of T	Number of solutions to $A\mathbf{x} = \mathbf{b}$	Property of the columns of A	Property of the rank of A
T is onto	$A\mathbf{x} = \mathbf{b}$ has at least one solution for every $\mathbf{b}$ in $\mathcal{R}^m$	The columns of A are a spanning set for $\mathcal{R}^m$	rank $A = m$
T is one-to-one	$A\mathbf{x} = \mathbf{b}$ has at most one solution for every $\mathbf{b}$ in $\mathcal{R}^m$	The columns of A are linearly independent	rank $A = n$
T is invertible	$A\mathbf{x} = \mathbf{b}$ has a unique solution for every $\mathbf{b}$ in $\mathcal{R}^m$	The columns of A are a linearly independent spanning set for $\mathcal{R}^m$	rank $A = m = n$

Practice Problems

1. Let $T:\mathcal{R}^2 \to \mathcal{R}^3$ be the linear transformation defined by

$$T\left(\begin{bmatrix} x_1 \\ x_2 \end{bmatrix}\right) = \begin{bmatrix} 3x_1 - x_2 \\ -x_1 + 2x_2 \\ 2x_1 \end{bmatrix}.$$

(a) Determine a spanning set for the range of T.
(b) Determine a spanning set for the null space of T.
(c) Is T onto?
(d) Is T one-to-one?

2. Let $U:\mathcal{R}^2 \to \mathcal{R}^2$ be the linear transformation defined by

$$U\left(\begin{bmatrix} x_1 \\ x_2 \\ x_3 \end{bmatrix}\right) = \begin{bmatrix} x_2 - 4x_3 \\ 2x_1 + 3x_3 \end{bmatrix}.$$

Determine $UT\left(\begin{bmatrix} x_1 \\ x_2 \end{bmatrix}\right)$, where T is as in problem 1.

3. Let $T:\mathcal{R}^2 \to \mathcal{R}^2$ be the linear transformation defined by

$$T\left(\begin{bmatrix} x_1 \\ x_2 \end{bmatrix}\right) = \begin{bmatrix} x_1 + 4x_2 \\ 2x_1 + 7x_2 \end{bmatrix}.$$

Show that T is invertible, and determine $T^{-1}\left(\begin{bmatrix} x_1 \\ x_2 \end{bmatrix}\right)$.

Exercises

1. Determine if the following statements are true or false.

   (a) A linear transformation with codomain $\mathcal{R}^m$ is onto if and only if the rank of its standard matrix equals m.

   (b) A linear transformation is onto if and only if the columns of its standard matrix span its range.

   (c) A linear transformation is onto if and only if the columns of its standard matrix are linearly independent.

   (d) A linear transformation is one-to-one if and only if every vector in its range is the image of a unique vector in its domain.

   (e) A linear transformation is one-to-one if and only if its null space consists of only the zero vector.

   (f) A linear transformation is invertible if and only if its standard matrix is invertible.

   (g) The system $A\mathbf{x} = \mathbf{b}$ is consistent for all $\mathbf{b}$ if and only if the transformation T_A is one-to-one.

   (h) Let A be an $m \times n$ matrix. The system $A\mathbf{x} = \mathbf{b}$ is consistent for all $\mathbf{b}$ in $\mathcal{R}^m$ if and only if the columns of A form a spanning set for $\mathcal{R}^m$.

In Exercises 2–12, find a spanning set for the range of the given linear transformation T.

2. $T : \mathcal{R}^2 \to \mathcal{R}^2$ defined by $T\left(\begin{bmatrix} x_1 \\ x_2 \end{bmatrix}\right) = \begin{bmatrix} x_2 \\ x_1 + x_2 \end{bmatrix}$

3. $T : \mathcal{R}^2 \to \mathcal{R}^2$ defined by $T\left(\begin{bmatrix} x_1 \\ x_2 \end{bmatrix}\right) = \begin{bmatrix} 2x_1 + 3x_2 \\ 4x_1 + 5x_2 \end{bmatrix}$

4. $T : \mathcal{R}^3 \to \mathcal{R}^2$ defined by
$$T\left(\begin{bmatrix} x_1 \\ x_2 \\ x_3 \end{bmatrix}\right) = \begin{bmatrix} x_1 + x_2 + x_3 \\ 2x_1 \end{bmatrix}$$

5. $T : \mathcal{R}^2 \to \mathcal{R}^3$ defined by $T\left(\begin{bmatrix} x_1 \\ x_2 \end{bmatrix}\right) = \begin{bmatrix} 3x_2 \\ 2x_1 - x_2 \\ x_1 + x_2 \end{bmatrix}$

6. $T : \mathcal{R}^3 \to \mathcal{R}^3$ defined by
$$T\left(\begin{bmatrix} x_1 \\ x_2 \\ x_3 \end{bmatrix}\right) = \begin{bmatrix} 5x_1 - 4x_2 + x_3 \\ x_1 - 2x_2 \\ x_1 + x_3 \end{bmatrix}$$

7. $T : \mathcal{R}^3 \to \mathcal{R}^3$ defined by
$$T\left(\begin{bmatrix} x_1 \\ x_2 \\ x_3 \end{bmatrix}\right) = \begin{bmatrix} 2x_1 + x_2 + x_3 \\ 2x_1 + 2x_2 + 3x_3 \\ 4x_1 + x_2 \end{bmatrix}$$

8. $T : \mathcal{R}^3 \to \mathcal{R}^3$ defined by $T(\mathbf{v}) = \mathbf{v}$

9. $T : \mathcal{R}^3 \to \mathcal{R}^2$ defined by $T(\mathbf{v}) = \mathbf{0}$

10. $T : \mathcal{R}^3 \to \mathcal{R}^3$ defined by $T(\mathbf{v}) = 4\mathbf{v}$

11. $T : \mathcal{R}^2 \to \mathcal{R}^2$ defined by $T\left(\begin{bmatrix} x_1 \\ x_2 \end{bmatrix}\right) = \begin{bmatrix} x_1 \\ 0 \end{bmatrix}$

12. $T : \mathcal{R}^3 \to \mathcal{R}^3$ defined by $T\left(\begin{bmatrix} x_1 \\ x_2 \\ x_3 \end{bmatrix}\right) = \begin{bmatrix} x_1 \\ x_2 \\ 0 \end{bmatrix}$

In Exercises 13–23, find a spanning set for the null space of the given linear transformation T, and use your answer to determine if T is one-to-one.

13. $T : \mathcal{R}^2 \to \mathcal{R}^2$ defined by $T\left(\begin{bmatrix} x_1 \\ x_2 \end{bmatrix}\right) = \begin{bmatrix} x_2 \\ x_1 + x_2 \end{bmatrix}$

14. $T : \mathcal{R}^2 \to \mathcal{R}^2$ defined by $T\left(\begin{bmatrix} x_1 \\ x_2 \end{bmatrix}\right) = \begin{bmatrix} 2x_1 + 3x_2 \\ 4x_1 + 5x_2 \end{bmatrix}$

15. $T : \mathcal{R}^3 \to \mathcal{R}^2$ defined by
$$T\left(\begin{bmatrix} x_1 \\ x_2 \\ x_3 \end{bmatrix}\right) = \begin{bmatrix} x_1 + x_2 + x_3 \\ 2x_1 \end{bmatrix}$$

16. $T : \mathcal{R}^2 \to \mathcal{R}^3$ defined by $T\left(\begin{bmatrix} x_1 \\ x_2 \end{bmatrix}\right) = \begin{bmatrix} 3x_2 \\ 2x_1 - x_2 \\ x_1 + x_2 \end{bmatrix}$

17. $T : \mathcal{R}^3 \to \mathcal{R}^3$ defined by
$$T\left(\begin{bmatrix} x_1 \\ x_2 \\ x_3 \end{bmatrix}\right) = \begin{bmatrix} x_1 + 2x_2 + x_3 \\ x_1 + 3x_2 + 2x_3 \\ 2x_1 + 5x_2 + 3x_3 \end{bmatrix}$$

18. $T : \mathcal{R}^3 \to \mathcal{R}^3$ defined by
$$T\left(\begin{bmatrix} x_1 \\ x_2 \\ x_3 \end{bmatrix}\right) = \begin{bmatrix} 2x_1 + 3x_2 \\ x_1 - x_3 \\ x_1 + x_2 + 4x_3 \end{bmatrix}$$

19. $T : \mathcal{R}^3 \to \mathcal{R}^3$ defined by $T(\mathbf{v}) = \mathbf{v}$

20. $T : \mathcal{R}^3 \to \mathcal{R}^2$ defined by $T(\mathbf{v}) = \mathbf{0}$

21. $T : \mathcal{R}^2 \to \mathcal{R}^2$ defined by $T\left(\begin{bmatrix} x_1 \\ x_2 \end{bmatrix}\right) = \begin{bmatrix} x_1 \\ 0 \end{bmatrix}$

22. $T : \mathcal{R}^3 \to \mathcal{R}^1$ defined by $T\left(\begin{bmatrix} x_1 \\ x_2 \\ x_3 \end{bmatrix}\right) = x_1 + 2x_3$

23. $T : \mathcal{R}^4 \to \mathcal{R}^3$ defined by
$$T\left(\begin{bmatrix} x_1 \\ x_2 \\ x_3 \\ x_4 \end{bmatrix}\right) = \begin{bmatrix} 2x_1 + x_2 + x_3 - x_4 \\ x_1 + x_2 + 2x_3 + 2x_4 \\ x_1 - x_3 - 3x_4 \end{bmatrix}$$

In Exercises 24–27, first find the standard matrix of the given linear transformation T. Then determine if the linear transformation is one-to-one, by checking if the columns of the standard matrix are linearly independent.

24. $T:\mathcal{R}^2 \to \mathcal{R}^2$ defined by $T\left(\begin{bmatrix} x_1 \\ x_2 \end{bmatrix}\right) = \begin{bmatrix} x_2 \\ x_1 + x_2 \end{bmatrix}$

25. $T:\mathcal{R}^2 \to \mathcal{R}^2$ defined by $T\left(\begin{bmatrix} x_1 \\ x_2 \end{bmatrix}\right) = \begin{bmatrix} 2x_1 + 3x_2 \\ 4x_1 + 5x_2 \end{bmatrix}$

26. $T:\mathcal{R}^3 \to \mathcal{R}^2$ defined by $T\left(\begin{bmatrix} x_1 \\ x_2 \\ x_3 \end{bmatrix}\right) = \begin{bmatrix} x_1 + x_2 + x_3 \\ 2x_1 \end{bmatrix}$

27. $T:\mathcal{R}^2 \to \mathcal{R}^3$ defined by $T\left(\begin{bmatrix} x_1 \\ x_2 \end{bmatrix}\right) = \begin{bmatrix} 3x_2 \\ 2x_1 - x_2 \\ x_1 + x_2 \end{bmatrix}$

In Exercises 28–32, find the standard matrix of the given linear transformation T. Then determine if the linear transformation is onto, by checking the rank of the standard matrix.

28. $T:\mathcal{R}^2 \to \mathcal{R}^2$ defined by $T\left(\begin{bmatrix} x_1 \\ x_2 \end{bmatrix}\right) = \begin{bmatrix} x_2 \\ x_1 + x_2 \end{bmatrix}$

29. $T:\mathcal{R}^2 \to \mathcal{R}^2$ defined by $T\left(\begin{bmatrix} x_1 \\ x_2 \end{bmatrix}\right) = \begin{bmatrix} 2x_1 + 3x_2 \\ 4x_1 + 5x_2 \end{bmatrix}$

30. $T:\mathcal{R}^3 \to \mathcal{R}^2$ defined by

$$T\left(\begin{bmatrix} x_1 \\ x_2 \\ x_3 \end{bmatrix}\right) = \begin{bmatrix} x_1 + x_2 + x_3 \\ 2x_1 \end{bmatrix}$$

31. $T:\mathcal{R}^2 \to \mathcal{R}^3$ defined by $T\left(\begin{bmatrix} x_1 \\ x_2 \end{bmatrix}\right) = \begin{bmatrix} 3x_2 \\ 2x_1 - x_2 \\ x_1 + x_2 \end{bmatrix}$

32. $T:\mathcal{R}^3 \to \mathcal{R}^3$ defined by

$$T\left(\begin{bmatrix} x_1 \\ x_2 \\ x_3 \end{bmatrix}\right) = \begin{bmatrix} x_2 - 2x_3 \\ x_1 - x_3 \\ -x_1 + 2x_2 - 3x_3 \end{bmatrix}$$

33. Suppose that $T:\mathcal{R}^2 \to \mathcal{R}^2$ is the linear transformation that rotates a vector by $90°$.

  (a) What is the null space of T?
  (b) Is T one-to-one?
  (c) What is the range of T?
  (d) Is T onto?

34. Suppose that $T:\mathcal{R}^2 \to \mathcal{R}^2$ is the reflection of $\mathcal{R}^2$ about the x-axis (see Exercise 37 of Section 1.2).

  (a) What is the null space of T?
  (b) Is T one-to-one?
  (c) What is the range of T?
  (d) Is T onto?

35. Define $T:\mathcal{R}^2 \to \mathcal{R}^2$ by $T\left(\begin{bmatrix} x_1 \\ x_2 \end{bmatrix}\right) = \begin{bmatrix} 0 \\ x_2 \end{bmatrix}$. (This vector is the projection of $\begin{bmatrix} x_1 \\ x_2 \end{bmatrix}$ onto the y-axis.)

  (a) What is the null space of T?
  (b) Is T one-to-one?
  (c) What is the range of T?
  (d) Is T onto?

36. Define $T:\mathcal{R}^3 \to \mathcal{R}^3$ by $T\left(\begin{bmatrix} x_1 \\ x_2 \\ x_3 \end{bmatrix}\right) = \begin{bmatrix} 0 \\ 0 \\ x_3 \end{bmatrix}$. (This vector is the projection of $\begin{bmatrix} x_1 \\ x_2 \\ x_3 \end{bmatrix}$ onto the z-axis.)

  (a) What is the null space of T?
  (b) Is T one-to-one?
  (c) What is the range of T?
  (d) Is T onto?

37. Define $T:\mathcal{R}^3 \to \mathcal{R}^3$ by $T\left(\begin{bmatrix} x_1 \\ x_2 \\ x_3 \end{bmatrix}\right) = \begin{bmatrix} x_1 \\ x_2 \\ 0 \end{bmatrix}$. (This vector is the projection of $\begin{bmatrix} x_1 \\ x_2 \\ x_3 \end{bmatrix}$ onto the xy-plane.)

  (a) What is the null space of T?
  (b) Is T one-to-one?
  (c) What is the range of T?
  (d) Is T onto?

38. Define $T:\mathcal{R}^3 \to \mathcal{R}^3$ by $T\left(\begin{bmatrix} x_1 \\ x_2 \\ x_3 \end{bmatrix}\right) = \begin{bmatrix} x_1 \\ x_2 \\ -x_3 \end{bmatrix}$. (See Exercise 46 in Section 2.6.)

  (a) What is the null space of T?
  (b) Is T one-to-one?
  (c) What is the range of T?
  (d) Is T onto?

39. Suppose that $T:\mathcal{R}^2 \to \mathcal{R}^2$ is linear and has the property that $T(\mathbf{e}_1) = \begin{bmatrix} 3 \\ 1 \end{bmatrix}$ and $T(\mathbf{e}_2) = \begin{bmatrix} 4 \\ 2 \end{bmatrix}$.

  (a) Determine if T is one-to-one.
  (b) Determine if T is onto.

40. Suppose that $T:\mathcal{R}^2 \to \mathcal{R}^2$ is linear and has the property that $T(\mathbf{e}_1) = \begin{bmatrix} 3 \\ 1 \end{bmatrix}$ and $T(\mathbf{e}_2) = \begin{bmatrix} 6 \\ 2 \end{bmatrix}$.

  (a) Determine if T is one-to-one.
  (b) Determine if T is onto.

Exercises 41–47 deal with the linear transformations $T:\mathcal{R}^2 \to \mathcal{R}^3$ and $U:\mathcal{R}^3 \to \mathcal{R}^2$ defined as

$$T\left(\begin{bmatrix} x_1 \\ x_2 \end{bmatrix}\right) = \begin{bmatrix} x_1 + x_2 \\ x_1 - 3x_2 \\ 4x_1 \end{bmatrix}$$

and

$$U\left(\begin{bmatrix} x_1 \\ x_2 \\ x_3 \end{bmatrix}\right) = \begin{bmatrix} x_1 - x_2 + 4x_3 \\ x_1 + 3x_2 \end{bmatrix}.$$

41. Determine the domain, the codomain, and the rule for UT.

42. Use the rule for UT obtained in Exercise 41 to find the standard matrix of UT.

43. Determine the standard matrices A and B of T and U, respectively.

44. Compute the product BA of the matrices found in Exercise 43, and illustrate Theorem 2.14 by comparing your answer to the result obtained in Exercise 42.

45. Determine the domain, the codomain, and the rule for TU.

46. Use the rule for TU obtained in Exercise 45 to find the standard matrix of TU.

47. Compute the product AB of the matrices found in Exercise 43, and illustrate Theorem 2.14 by comparing your answer to the result obtained in Exercise 46.

Exercises 48–54 deal with the linear transformations $T:\mathcal{R}^2 \to \mathcal{R}^2$ and $U:\mathcal{R}^2 \to \mathcal{R}^2$ defined as

$$T\left(\begin{bmatrix} x_1 \\ x_2 \end{bmatrix}\right) = \begin{bmatrix} x_1 + 2x_2 \\ 3x_1 - x_2 \end{bmatrix}$$

and

$$U\left(\begin{bmatrix} x_1 \\ x_2 \end{bmatrix}\right) = \begin{bmatrix} 2x_1 - x_2 \\ 5x_2 \end{bmatrix}.$$

48. Determine the domain, the codomain, and the rule for UT.

49. Use the rule for UT obtained in Exercise 48 to find the standard matrix of UT.

50. Determine the standard matrices A and B of T and U, respectively.

51. Compute the product BA of the matrices found in Exercise 50, and illustrate Theorem 2.14 by comparing your answer to the result obtained in Exercise 49.

52. Determine the domain, the codomain, and the rule for TU.

53. Use the rule for TU obtained in Exercise 52 to find the standard matrix of TU.

54. Compute the product AB of the matrices found in Exercise 50, and illustrate Theorem 2.14 by comparing your answer to the result obtained in Exercise 53.

In Exercises 55–58, an invertible linear transformation T is defined. Determine a similar definition for the inverse T^{-1} of the linear transformation.

55. $T:\mathcal{R}^2 \to \mathcal{R}^2$ defined by $T\left(\begin{bmatrix} x_1 \\ x_2 \end{bmatrix}\right) = \begin{bmatrix} 2x_1 - x_2 \\ x_1 + x_2 \end{bmatrix}$

56. $T:\mathcal{R}^2 \to \mathcal{R}^2$ defined by $T\left(\begin{bmatrix} x_1 \\ x_2 \end{bmatrix}\right) = \begin{bmatrix} x_1 + 3x_2 \\ 2x_1 + x_2 \end{bmatrix}$

57. $T:\mathcal{R}^3 \to \mathcal{R}^3$ defined by

$$T\left(\begin{bmatrix} x_1 \\ x_2 \\ x_3 \end{bmatrix}\right) = \begin{bmatrix} -x_1 + x_2 + 3x_3 \\ 2x_1 - x_3 \\ -x_1 + 2x_2 + 5x_3 \end{bmatrix}$$

58. $T:\mathcal{R}^3 \to \mathcal{R}^3$ defined by

$$T\left(\begin{bmatrix} x_1 \\ x_2 \\ x_3 \end{bmatrix}\right) = \begin{bmatrix} x_2 - 2x_3 \\ x_1 - x_3 \\ -x_1 + 2x_2 - 2x_3 \end{bmatrix}$$

59. Prove that the composition of two one-to-one linear transformations is one-to-one. Is the result true if the transformations are not linear? Justify your answer.

60. Prove that if two linear transformations are onto, then their composition is onto. Is the result true if the transformations are not linear? Justify your answer.

61. In $\mathcal{R}^2$, show that the composition of two reflections about the x-axis is the identity transformation.

62. In $\mathcal{R}^2$, show that a reflection about the y-axis followed by a rotation of $180°$ is equal to a reflection about the x-axis.

63. In $\mathcal{R}^2$, show that the composition of the projection on the x-axis followed by the reflection about the y-axis is equal to the composition of the reflection about the y-axis followed by the projection on the x-axis.

64. Prove that the composition of two shear transformations is a shear transformation (see Example 4 of Section 2.6).

65.[7] Suppose that $T:\mathcal{R}^n \to \mathcal{R}^m$ is linear and one-to-one. Let $\{\mathbf{v}_1, \mathbf{v}_2, \ldots, \mathbf{v}_k\}$ be a linearly independent subset of $\mathcal{R}^n$.
  (a) Prove that the set $\{T(\mathbf{v}_1), T(\mathbf{v}_2), \ldots, T(\mathbf{v}_k)\}$ is a linearly independent subset of $\mathcal{R}^m$.
  (b) Show by example that (a) is false if T is not one-to-one.

66. Use Theorem 2.14 to prove that matrix multiplication is associative.

In Exercises 67 and 68, use a calculator with matrix capabilities or computer software such as MATLAB to solve the problem.

67. The linear transformations $T, U:\mathcal{R}^4 \to \mathcal{R}^4$ are defined as follows:

$$T\left(\begin{bmatrix} x_1 \\ x_2 \\ x_3 \\ x_4 \end{bmatrix}\right) = \begin{bmatrix} x_1 + 3x_2 - 2x_3 + x_4 \\ 3x_1 + 4x_3 + x_4 \\ 2x_1 - x_2 + 2x_4 \\ x_3 + x_4 \end{bmatrix}$$

and

$$U\left(\begin{bmatrix} x_1 \\ x_2 \\ x_3 \\ x_4 \end{bmatrix}\right) = \begin{bmatrix} x_2 - 3x_4 \\ 2x_1 + x_3 - x_4 \\ x_1 - 2x_2 + 4x_4 \\ 5x_2 + x_3 \end{bmatrix}.$$

  (a) Compute the standard matrices A and B of T and U, respectively.
  (b) Compute the product AB.
  (c) Use your answer to (b) to write a rule for TU.

---

[7]The result of this exercise is used in Section 7.2 (page 411).

68. Define the linear transformation $T: \mathcal{R}^4 \to \mathcal{R}^4$ by the rule

$$T\left(\begin{bmatrix} x_1 \\ x_2 \\ x_3 \\ x_4 \end{bmatrix}\right) = \begin{bmatrix} 2x_1 + 4x_2 + x_3 + 6x_4 \\ 3x_1 + 7x_2 - x_3 + 11x_4 \\ x_1 + 2x_2 + 2x_4 \\ 2x_1 + 5x_2 - x_3 + 8x_4 \end{bmatrix}.$$

(a) Find the standard matrix A of T.
(b) Show that A is invertible and find its inverse.
(c) Use your answer to (b) to find the rule for T^{-1}.

Solutions to the Practice Problems

1. The standard matrix of T is $A = \begin{bmatrix} 3 & -1 \\ -1 & 2 \\ 2 & 0 \end{bmatrix}$.

(a) Since the columns of A span the range of T, the desired spanning set is

$$\left\{ \begin{bmatrix} 3 \\ -1 \\ 2 \end{bmatrix}, \begin{bmatrix} -1 \\ 2 \\ 0 \end{bmatrix} \right\}.$$

(b) The null space of A is the solution set of $A\mathbf{x} = \mathbf{0}$.
Since the reduced row echelon form of A is $\begin{bmatrix} 1 & 0 \\ 0 & 1 \\ 0 & 0 \end{bmatrix}$,
we see that the general solution of $A\mathbf{x} = \mathbf{0}$ is

$$x_1 = 0$$
$$x_2 = 0.$$

Hence the null space of T is $\{\mathbf{0}\}$, and so a spanning set for the null space of T is $\{\mathbf{0}\}$.

(c) From (b) we see that the rank of A is 2. Since Theorem 2.12 implies that T is onto if and only if the rank of A is 3, we see that T is not onto.

(d) From (b) we see that the null space of T is $\{\mathbf{0}\}$. Hence T is one-to-one, by Theorem 2.13.

2. The standard matrices of T and U are

$$A = \begin{bmatrix} 3 & -1 \\ -1 & 2 \\ 2 & 0 \end{bmatrix} \quad \text{and} \quad B = \begin{bmatrix} 0 & 1 & -4 \\ 2 & 0 & 3 \end{bmatrix},$$

respectively. Hence, by Theorem 2.14(a), the standard matrix of UT is

$$BA = \begin{bmatrix} -9 & 2 \\ 12 & -2 \end{bmatrix}.$$

Therefore

$$UT\left(\begin{bmatrix} x_1 \\ x_2 \end{bmatrix}\right) = \begin{bmatrix} -9 & 2 \\ 12 & -2 \end{bmatrix}\begin{bmatrix} x_1 \\ x_2 \end{bmatrix} = \begin{bmatrix} -9x_1 + 2x_2 \\ 12x_1 - 2x_2 \end{bmatrix}.$$

3. The standard matrix of T is $A = \begin{bmatrix} 1 & 4 \\ 2 & 7 \end{bmatrix}$. Because the rank of A is 2, A is invertible. In fact, $A^{-1} = \begin{bmatrix} -7 & 4 \\ 2 & -1 \end{bmatrix}$.
Hence, by Theorem 2.14, T is invertible, and

$$T^{-1}\left(\begin{bmatrix} x_1 \\ x_2 \end{bmatrix}\right) = \begin{bmatrix} -7 & 4 \\ 2 & -1 \end{bmatrix}\begin{bmatrix} x_1 \\ x_2 \end{bmatrix} = \begin{bmatrix} -7x_1 + 4x_2 \\ 2x_1 - x_2 \end{bmatrix}.$$

Chapter 2 Review Exercises

1. Determine if the following statements are true or false.

(a) A symmetric matrix equals its transpose.
(b) If a symmetric matrix is written in block form, then the blocks are also symmetric matrices.
(c) The product of square matrices is always defined.
(d) The transpose of an invertible matrix is invertible.
(e) It is possible for an invertible matrix to have two distinct inverses.
(f) The sum of an invertible matrix and its inverse is the zero matrix.
(g) The columns of an invertible matrix are linearly independent.
(h) If a matrix is invertible, then its rank equals the number of its rows.
(i) If A is an $n \times n$ matrix and the system $A\mathbf{x} = \mathbf{b}$ is consistent for some $\mathbf{b}$, then A is invertible.
(j) The range of a linear transformation is contained in the codomain of the linear transformation.
(k) The null space of a linear transformation is contained in the codomain of the linear transformation.
(l) Linear transformations preserve linear combinations.
(m) Linear transformations preserve linearly independent sets.
(n) Every linear transformation has a standard matrix.
(o) The zero transformation is the only linear transformation whose standard matrix is the zero matrix.
(p) If a linear transformation is one-to-one, then it is invertible.
(q) If a linear transformation is onto, then its range equals its codomain.
(r) If a linear transformation is one-to-one, then its range consists exactly of the zero vector.
(s) If a linear transformation is onto, then its rows span its codomain.
(t) If a linear transformation is one-to-one, then its columns form a linearly independent set.

2. Determine if each of the following phrases is a misuse of terminology. If so, explain what is wrong with each one.

(a) the range of a matrix
(b) the standard matrix of a function $f: \mathcal{R}^n \to \mathcal{R}^n$
(c) a spanning set for the range of a linear transformation
(d) the null space of a system of linear equations
(e) a one-to-one matrix

3. Let A be an $m \times n$ matrix and B be a $p \times q$ matrix.

(a) Under what conditions is the matrix product BA defined?
(b) If BA is defined, what size is it?

In Exercises 4–15, use the following matrices to compute the given expression, or give a reason why the expression is not defined.

$$A = \begin{bmatrix} 2 & 1 \\ 4 & -1 \end{bmatrix}, \qquad B = \begin{bmatrix} 2 & 3 \\ 4 & 6 \end{bmatrix}, \qquad C = \begin{bmatrix} 2 & -1 \\ 3 & 5 \\ 0 & 1 \end{bmatrix},$$

$$\mathbf{u} = \begin{bmatrix} 3 \\ 2 \\ -1 \end{bmatrix}, \qquad \mathbf{v} = \begin{bmatrix} 1 & -2 & 2 \end{bmatrix}, \quad \text{and} \quad \mathbf{w} = \begin{bmatrix} 3 \\ 4 \end{bmatrix}.$$

4. $A\mathbf{w}$ 5. ABA 6. $A\mathbf{u}$
7. $C\mathbf{w}$ 8. $\mathbf{v}C$ 9. $\mathbf{v}A$
10. $A^T B$ 11. $A^{-1}B^T$ 12. $B^{-1}\mathbf{w}$
13. $AC^T\mathbf{u}$ 14. B^3 15. $\mathbf{u}^2$

In Exercises 16 and 17, compute the product of the matrices in block form.

16. $\begin{bmatrix} 1 & 0 & 3 & 1 \\ 0 & 1 & 2 & 4 \\ 0 & 0 & 2 & 1 \\ 0 & 0 & -1 & 3 \end{bmatrix} \begin{bmatrix} 1 & -1 & 0 & 0 \\ 2 & 1 & 0 & 0 \\ 0 & 0 & 2 & 0 \\ 0 & 0 & 0 & 2 \end{bmatrix}$

17. $\begin{bmatrix} I_2 & -I_2 \end{bmatrix} \begin{bmatrix} 1 \\ 3 \\ -7 \\ -4 \end{bmatrix}$

In Exercises 18 and 19, determine if the given matrix is invertible. If so, find its inverse; if not, explain why.

18. $\begin{bmatrix} 1 & 0 & 2 \\ 2 & -1 & 3 \\ 4 & 1 & 8 \end{bmatrix}$ 19. $\begin{bmatrix} 2 & -1 & 3 \\ 1 & 2 & -4 \\ 4 & 3 & 5 \end{bmatrix}$

20. Let A and B be square matrices of the same size. Prove that if the first row of A is zero, then the first row of AB is zero.

21. Let A and B be square matrices of the same size. Prove that if the first column of B is zero, then the first column of AB is zero.

22. Give examples of 2×2 matrices A and B such that A and B are invertible, and $(A + B)^{-1} \neq A^{-1} + B^{-1}$.

In Exercises 23 and 24, systems of equations are given. First use the appropriate matrix inverse to solve the system, and

then use Gaussian elimination to check your answer.

23. $\begin{aligned} 2x_1 + x_2 &= 3 \\ x_1 + x_2 &= 5 \end{aligned}$ 24. $\begin{aligned} x_1 + x_2 + x_3 &= 3 \\ x_1 + 3x_2 + 4x_3 &= -1 \\ 2x_1 + 4x_2 + x_3 &= 2 \end{aligned}$

25. Suppose that the reduced row echelon form R and three columns of A are given by

$$R = \begin{bmatrix} 1 & 2 & 0 & 0 & -2 \\ 0 & 0 & 1 & 0 & 3 \\ 0 & 0 & 0 & 1 & 1 \end{bmatrix}, \quad \mathbf{a}_1 = \begin{bmatrix} 3 \\ 5 \\ 2 \end{bmatrix}, \quad \mathbf{a}_3 = \begin{bmatrix} 2 \\ 0 \\ -1 \end{bmatrix},$$

and

$$\mathbf{a}_4 = \begin{bmatrix} 2 \\ -1 \\ 3 \end{bmatrix}.$$

Determine the matrix A.

Exercises 26–29 refer to the following matrices:

$$A = \begin{bmatrix} 2 & -1 & 3 \\ 4 & 0 & -2 \end{bmatrix} \quad \text{and} \quad B = \begin{bmatrix} 4 & 2 \\ 1 & -3 \\ 0 & 1 \end{bmatrix}.$$

26. Find the range and codomain of the matrix transformation T_A.

27. Find the range and codomain of the matrix transformation T_B.

28. Compute $T_A \left(\begin{bmatrix} 2 \\ 0 \\ 3 \end{bmatrix} \right)$ 29. Compute $T_B \left(\begin{bmatrix} 4 \\ 2 \end{bmatrix} \right)$

In Exercises 30–33, a linear transformation is given. Compute its standard matrix.

30. $T: \mathcal{R}^2 \to \mathcal{R}^2$ defined by $T \left(\begin{bmatrix} x_1 \\ x_2 \end{bmatrix} \right) = \begin{bmatrix} 3x_1 - x_2 \\ 4x_1 \end{bmatrix}$

31. $T: \mathcal{R}^3 \to \mathcal{R}^2$ defined by $T \left(\begin{bmatrix} x_1 \\ x_2 \\ x_3 \end{bmatrix} \right) = \begin{bmatrix} 2x_1 - x_3 \\ 4x_1 \end{bmatrix}$

32. $T: \mathcal{R}^2 \to \mathcal{R}^2$ defined by $T(\mathbf{v}) = 6\mathbf{v}$

33. $T: \mathcal{R}^2 \to \mathcal{R}^2$ defined by $T(\mathbf{v}) = 2\mathbf{v} + U(\mathbf{v})$, where $U: \mathcal{R}^2 \to \mathcal{R}^2$ is the linear transformation defined by $U \left(\begin{bmatrix} x_1 \\ x_2 \end{bmatrix} \right) = \begin{bmatrix} 2x_1 + x_2 \\ 3x_1 \end{bmatrix}$

In Exercises 34–37, a function $T: \mathcal{R}^n \to \mathcal{R}^m$ is given. Either prove that T is linear or explain why T is not linear.

34. $T: \mathcal{R}^2 \to \mathcal{R}^2$ defined by $T \left(\begin{bmatrix} x_1 \\ x_2 \end{bmatrix} \right) = \begin{bmatrix} x_1 + 1 \\ x_2 \end{bmatrix}$

35. $T: \mathcal{R}^2 \to \mathcal{R}^2$ defined by $T \left(\begin{bmatrix} x_1 \\ x_2 \end{bmatrix} \right) = \begin{bmatrix} 2x_2 \\ x_1 \end{bmatrix}$

36. $T: \mathcal{R}^2 \to \mathcal{R}^2$ defined by $T \left(\begin{bmatrix} x_1 \\ x_2 \end{bmatrix} \right) = \begin{bmatrix} x_1 x_2 \\ x_1 \end{bmatrix}$

37. $T: \mathcal{R}^3 \to \mathcal{R}^2$ defined by $T\left(\begin{bmatrix} x_1 \\ x_2 \\ x_3 \end{bmatrix}\right) = \begin{bmatrix} x_1 + x_2 \\ x_3 \end{bmatrix}$

In Exercises 38 and 39, find a spanning set for the range of the given linear transformation T.

38. $T: \mathcal{R}^2 \to \mathcal{R}^3$ defined by $T\left(\begin{bmatrix} x_1 \\ x_2 \end{bmatrix}\right) = \begin{bmatrix} x_1 + x_2 \\ 0 \\ 2x_1 - x_2 \end{bmatrix}$

39. $T: \mathcal{R}^3 \to \mathcal{R}^2$ defined by $T\left(\begin{bmatrix} x_1 \\ x_2 \\ x_3 \end{bmatrix}\right) = \begin{bmatrix} x_1 + 2x_2 \\ x_2 - x_3 \end{bmatrix}$

In Exercises 40 and 41, find a spanning set for the null space of the given linear transformation T. Use your answer to determine if T is one-to-one.

40. $T: \mathcal{R}^2 \to \mathcal{R}^3$ defined by $T\left(\begin{bmatrix} x_1 \\ x_2 \end{bmatrix}\right) = \begin{bmatrix} x_1 + x_2 \\ 0 \\ 2x_1 - x_2 \end{bmatrix}$

41. $T: \mathcal{R}^3 \to \mathcal{R}^2$ defined by $T\left(\begin{bmatrix} x_1 \\ x_2 \\ x_3 \end{bmatrix}\right) = \begin{bmatrix} x_1 + 2x_2 \\ x_2 - x_3 \end{bmatrix}$

In Exercises 42 and 43, first find the standard matrix of the given linear transformation T. Then determine if the linear transformation is one-to-one, by checking if the columns of the standard matrix are linearly independent.

42. $T: \mathcal{R}^3 \to \mathcal{R}^2$ defined by $T\left(\begin{bmatrix} x_1 \\ x_2 \\ x_3 \end{bmatrix}\right) = \begin{bmatrix} x_1 + 2x_2 \\ x_2 - x_3 \end{bmatrix}$

43. $T: \mathcal{R}^2 \to \mathcal{R}^3$ defined by $T\left(\begin{bmatrix} x_1 \\ x_2 \end{bmatrix}\right) = \begin{bmatrix} x_1 + x_2 \\ 0 \\ 2x_1 - x_2 \end{bmatrix}$

In Exercises 44 and 45, first find the standard matrix of the given linear transformation T. Then determine if the linear transformation is onto, by checking the rank of its standard matrix.

44. $T: \mathcal{R}^3 \to \mathcal{R}^2$ defined by $T\left(\begin{bmatrix} x_1 \\ x_2 \\ x_3 \end{bmatrix}\right) = \begin{bmatrix} 2x_1 + x_3 \\ x_1 + x_2 - x_3 \end{bmatrix}$

45. $T: \mathcal{R}^2 \to \mathcal{R}^3$ defined by $T\left(\begin{bmatrix} x_1 \\ x_2 \end{bmatrix}\right) = \begin{bmatrix} 3x_1 - x_2 \\ x_2 \\ x_1 + x_2 \end{bmatrix}$

Exercises 46–52 deal with the linear transformations $T: \mathcal{R}^3 \to \mathcal{R}^2$ and $U: \mathcal{R}^2 \to \mathcal{R}^3$ defined as

$$T\left(\begin{bmatrix} x_1 \\ x_2 \\ x_3 \end{bmatrix}\right) = \begin{bmatrix} 2x_1 + x_3 \\ x_1 + x_2 - x_3 \end{bmatrix}$$

and

$$U\left(\begin{bmatrix} x_1 \\ x_2 \end{bmatrix}\right) = \begin{bmatrix} 3x_1 - x_2 \\ x_2 \\ x_1 + x_2 \end{bmatrix}.$$

46. Determine the domain, codomain, and the rule for UT.

47. Use the rule for UT obtained in Exercise 44 to find the standard matrix of UT.

48. Determine the standard matrices A and B of T and U, respectively.

49. Compute the product BA of the matrices found in Exercise 46, and illustrate Theorem 2.14 by comparing your answer to the result found in Exercise 45.

50. Determine the domain, codomain, and the rule for TU.

51. Use the rule for TU obtained in Exercise 48 to find the standard matrix of TU.

52. Compute the product AB of the matrices found in Exercise 46, and illustrate Theorem 2.14 by comparing your answer to the result found in Exercise 49.

In Exercises 53 and 54, an invertible linear transformation T is defined. Determine a similar definition for T^{-1}.

53. $T: \mathcal{R}^2 \to \mathcal{R}^2$ defined by $T\left(\begin{bmatrix} x_1 \\ x_2 \end{bmatrix}\right) = \begin{bmatrix} x_1 + 2x_2 \\ -x_1 + 3x_2 \end{bmatrix}$

54. $T: \mathcal{R}^3 \to \mathcal{R}^3$ defined by

$$T\left(\begin{bmatrix} x_1 \\ x_2 \\ x_3 \end{bmatrix}\right) = \begin{bmatrix} x_1 + x_2 + x_3 \\ x_1 + 3x_2 + 4x_3 \\ 2x_1 + 4x_2 + x_3 \end{bmatrix}$$

CHAPTER 3

Determinants

The *determinant*[1] of a square matrix is a scalar that provides information about the matrix, such as whether or not the matrix is invertible. Determinants were first considered in the late seventeenth century. For more than one hundred years thereafter, they were studied principally because of their connection to systems of linear equations. The best-known result involving determinants and systems of linear equations is *Cramer's rule*, presented in Section 3.2.

In recent years the use of determinants as a computational tool has greatly diminished. This is primarily because the size of the systems of linear equations that arise in applications has increased greatly, requiring the use of high-speed computers and efficient computational algorithms to obtain solutions. Since calculations with determinants are usually inefficient, they are normally avoided. Although determinants can be used to compute the areas and volumes of geometric objects, our principal use of determinants in this book is to determine the *eigenvalues* of a square matrix (discussed in Chapter 5).

3.1 Cofactor Expansion

In Exercise 9 of Section 2.3 we found that a 2×2 matrix

$$A = \begin{bmatrix} a & b \\ c & d \end{bmatrix}$$

is invertible if and only if $ad - bc \neq 0$. We call the scalar $ad - bc$ the **determinant** of A and denote it by det A or $|A|$.

Example 1 For

$$A = \begin{bmatrix} 1 & 2 \\ 3 & 4 \end{bmatrix} \qquad \text{and} \qquad B = \begin{bmatrix} 1 & 2 \\ 3 & 6 \end{bmatrix},$$

---

[1] Although work with determinants can be found in ancient China and in the writings of Gabriel Cramer in 1750, the study of determinants dates mainly from the early nineteenth century. In an 84-page paper presented to the French Institute in 1812, Augustin-Louis Cauchy (1789–1857) introduced the term *determinant* and proved many of the well-known results about determinants. He also used modern double-subscript notation for matrices, and showed how to evaluate a determinant by cofactor expansion.

the determinants are

$$\det A = 1 \cdot 4 - 2 \cdot 3 = -2 \qquad \text{and} \qquad \det B = 1 \cdot 6 - 2 \cdot 3 = 0.$$

Thus A is invertible, but B is not.   ○

Example 2   Determine the scalars c for which $A - cI_2$ is not invertible, where

$$A = \begin{bmatrix} 11 & 12 \\ -8 & -9 \end{bmatrix}.$$

Solution   The matrix $A - cI_2$ has the form

$$A - cI_2 = \begin{bmatrix} 11 & 12 \\ -8 & -9 \end{bmatrix} - c \begin{bmatrix} 1 & 0 \\ 0 & 1 \end{bmatrix} = \begin{bmatrix} 11 - c & 12 \\ -8 & -9 - c \end{bmatrix}.$$

Although we can use elementary row operations to determine the values of c for which this matrix is not invertible, the presence of the unknown scalar c makes the calculations difficult. By using the determinant instead, we obtain an easier computation. Since

$$\det (A - cI_2) = \det \begin{bmatrix} 11 - c & 12 \\ -8 & -9 - c \end{bmatrix}$$

$$= (11 - c)(-9 - c) - 12(-8)$$

$$= (c^2 - 2c - 99) + 96$$

$$= c^2 - 2c - 3$$

$$= (c + 1)(c - 3),$$

we see that $\det (A - cI_2) = 0$ if and only if $c = -1$ or $c = 3$. Thus $A - cI_2$ is not invertible when $c = -1$ or $c = 3$.   ○

Our principal use for determinants in this book is to calculate the scalars c for which a matrix $A - cI_n$ is not invertible, as in Example 2. In order to have such a test for $n \times n$ matrices, we must extend the definition of determinants to square matrices of any size so that a nonzero determinant is equivalent to invertibility. For 1×1 matrices, the appropriate definition is not hard to discover. Since the product of 1×1 matrices satisfies $[a][b] = [ab]$, we see that $[a]$ is invertible if and only if $a \neq 0$. Hence for a 1×1 matrix $[a]$, we define the **determinant** of $[a]$ by $\det [a] = a$.

Unfortunately, for $n \geq 3$, the determinant of an $n \times n$ matrix A is more complicated to define. To begin, we need some additional notation. We define the $(n - 1) \times (n - 1)$ matrix A_{ij} to be the matrix obtained from A by deleting row i and column j.

$$A_{ij} = \begin{bmatrix} a_{11} & \cdots & a_{1j} & \cdots & a_{1n} \\ \vdots & & \vdots & & \vdots \\ a_{i1} & \cdots & a_{ij} & \cdots & a_{in} \\ \vdots & & \vdots & & \vdots \\ a_{n1} & \cdots & a_{nj} & \cdots & a_{nn} \end{bmatrix} \longleftarrow \text{row } i$$

$$\uparrow$$
$$\text{column } j$$

Thus, for example, if

$$A = \begin{bmatrix} 1 & 2 & 3 \\ 4 & 5 & 6 \\ 7 & 9 & 8 \end{bmatrix},$$

then

$$A_{12} = \begin{bmatrix} 4 & 6 \\ 7 & 8 \end{bmatrix}, A_{21} = \begin{bmatrix} 2 & 3 \\ 9 & 8 \end{bmatrix}, A_{23} = \begin{bmatrix} 1 & 2 \\ 7 & 9 \end{bmatrix}, \text{ and } A_{33} = \begin{bmatrix} 1 & 2 \\ 4 & 5 \end{bmatrix}.$$

We can express the determinant of a 2×2 matrix using these matrices. For if

$$A = \begin{bmatrix} a & b \\ c & d \end{bmatrix},$$

then $A_{11} = [d] = d$ and $A_{12} = [c] = c$. Thus

$$\det A = ad - bc = a \cdot \det A_{11} - b \cdot \det A_{12}. \tag{1}$$

Notice that this representation expresses the determinant of the 2×2 matrix A in terms of the determinants of the 1×1 matrices A_{ij}.

Using 1 as a motivation, we define the **determinant** of an $n \times n$ matrix A for $n \geq 3$ by

$$\det A = a_{11} \cdot \det A_{11} - a_{12} \cdot \det A_{12} + \cdots + (-1)^{1+n} a_{1n} \cdot \det A_{1n}. \tag{2}$$

We denote the determinant of A by $\det A$ or $|A|$. Note that the expression on the right-hand side of (2) is an alternating sum of products of entries from the first row of A multiplied by the determinant of the corresponding matrix A_{1j}. If we let $c_{ij} = (-1)^{i+j} \cdot \det A_{ij}$, then our definition of the determinant of A can be written as

$$\det A = a_{11}c_{11} + a_{12}c_{12} + \cdots + a_{1n}c_{1n}. \tag{3}$$

The number c_{ij} is called the **(i, j)-cofactor** of A, and (3) is called the **cofactor expansion** of A along the first row.

Equations (2) and (3) define the determinant of an $n \times n$ matrix recursively. That is, the determinant of a matrix is defined in terms of determinants of smaller matrices. For example, if we want to compute the determinant of a 4×4 matrix A, (2) enables us to express the determinant of A in terms of determinants of 3×3 matrices. The determinants of these 3×3 matrices can then be expressed by (2) in terms of determinants of 2×2 matrices, and the determinants of these 2×2 matrices can finally be evaluated by (1).

Example 3

Evaluate the determinant of A using the cofactor expansion along the first row, if

$$A = \begin{bmatrix} 1 & 2 & 3 \\ 4 & 5 & 6 \\ 7 & 9 & 8 \end{bmatrix}.$$

Solution   The cofactor expansion of A along the first row yields

$$\det A = a_{11}c_{11} + a_{12}c_{12} + a_{13}c_{13}$$

$$= 1(-1)^{1+1} \det A_{11} + 2(-1)^{1+2} \det A_{12} + 3(-1)^{1+3} \det A_{13}$$

$$= 1(-1)^{1+1} \det \begin{bmatrix} 5 & 6 \\ 9 & 8 \end{bmatrix} + 2(-1)^{1+2} \det \begin{bmatrix} 4 & 6 \\ 7 & 8 \end{bmatrix} + 3(-1)^{1+3} \det \begin{bmatrix} 4 & 5 \\ 7 & 9 \end{bmatrix}$$

$$= 1(1)(5 \cdot 8 - 6 \cdot 9) + 2(-1)(4 \cdot 8 - 6 \cdot 7) + 3(1)(4 \cdot 9 - 5 \cdot 7)$$

$$= 1(1)(-14) + (2)(-1)(-10) + 3(1)(1)$$

$$= -14 + 20 + 3$$

$$= 9.$$

As we illustrate in Example 5, it is often more efficient to evaluate a determinant by cofactor expansion along a row other than the first row. The following important result enables us to do this. (For a proof of this theorem, see reference [1].)

Theorem 3.1

For any $i = 1, 2, \ldots, n$, *we have*

$$\det A = a_{i1}c_{i1} + a_{i2}c_{i2} + \cdots + a_{in}c_{in},$$

where c_{ij} *denotes the* (i, j)-*cofactor of* A.

The expression $a_{i1}c_{i1} + a_{i2}c_{i2} + \cdots + a_{in}c_{in}$ in Theorem 3.1 is called the **cofactor expansion** of A along row i. Thus the determinant of an $n \times n$ matrix can be evaluated using a cofactor expansion along *any* row.

Example 4   To illustrate Theorem 3.1, we compute the determinant of the matrix

$$A = \begin{bmatrix} 1 & 2 & 3 \\ 4 & 5 & 6 \\ 7 & 9 & 8 \end{bmatrix}$$

from Example 3 using the cofactor expansion along the second row.

Solution   Using the cofactor expansion of A along the second row, we have

$$\det A = a_{21}c_{21} + a_{22}c_{22} + a_{23}c_{23}$$

$$= 4(-1)^{2+1} \det A_{21} + 5(-1)^{2+2} \det A_{22} + 6(-1)^{2+3} \det A_{23}$$

$$= 4(-1)^{2+1} \det \begin{bmatrix} 2 & 3 \\ 9 & 8 \end{bmatrix} + 5(-1)^{2+2} \det \begin{bmatrix} 1 & 3 \\ 7 & 8 \end{bmatrix} + 6(-1)^{2+3} \det \begin{bmatrix} 1 & 2 \\ 7 & 9 \end{bmatrix}$$

$$= 4(-1)(2 \cdot 8 - 3 \cdot 9) + 5(1)(1 \cdot 8 - 3 \cdot 7) + 6(-1)(1 \cdot 9 - 2 \cdot 7)$$

$$= 4(-1)(-11) + 5(1)(-13) + (6)(-1)(-5)$$

$$= 44 - 65 + 30$$

$$= 9.$$

Note that we obtained the same value for $\det A$ as in Example 3.

Example 5   Let

$$M = \begin{bmatrix} 1 & 2 & 3 & 8 & 5 \\ 4 & 5 & 6 & 9 & 1 \\ 7 & 9 & 8 & 4 & 7 \\ 0 & 0 & 0 & 1 & 0 \\ 0 & 0 & 0 & 0 & 1 \end{bmatrix}.$$

Since the last row of M has only one nonzero entry, the cofactor expansion of M along the last row has only one nonzero term. Thus the cofactor expansion of M along the last row involves only one-fifth the work of the cofactor expansion along the first row. Using the last row, we obtain

$$\det M = 0 + 0 + 0 + 0 + 1(-1)^{5+5}\det M_{55} = \det M_{55} = \det \begin{bmatrix} 1 & 2 & 3 & 8 \\ 4 & 5 & 6 & 9 \\ 7 & 9 & 8 & 4 \\ 0 & 0 & 0 & 1 \end{bmatrix}.$$

Once again, we use the cofactor expansion along the last row to compute $\det M_{55}$.

$$\det M = \det \begin{bmatrix} 1 & 2 & 3 & 8 \\ 4 & 5 & 6 & 9 \\ 7 & 9 & 8 & 4 \\ 0 & 0 & 0 & 1 \end{bmatrix}$$

$$= 0 + 0 + 0 + 1(-1)^{4+4}\det \begin{bmatrix} 1 & 2 & 3 \\ 4 & 5 & 6 \\ 7 & 8 & 9 \end{bmatrix} = \det A,$$

where A is the matrix in Example 3. Thus $\det M = \det A = 9$.

Note that M has the form

$$M = \begin{bmatrix} A & B \\ O & I_2 \end{bmatrix},$$

where

$$A = \begin{bmatrix} 1 & 2 & 3 \\ 4 & 5 & 6 \\ 7 & 9 & 8 \end{bmatrix} \quad \text{and} \quad B = \begin{bmatrix} 8 & 5 \\ 9 & 1 \\ 4 & 7 \end{bmatrix}.$$

More generally, the approach used in the preceding paragraph can be used to show that for any $m \times m$ matrix A and $m \times n$ matrix B,

$$\det \begin{bmatrix} A & B \\ O & I_n \end{bmatrix} = \det A.$$

Evaluating the determinant of an arbitrary matrix by cofactor expansion is extremely inefficient. In fact, the cofactor expansion of an arbitrary $n \times n$ matrix requires approximately $e \cdot n!$ arithmetic operations, where e is the base of the natural logarithm. Suppose that we have a computer capable of performing 1 billion arithmetic operations per second and use it to evaluate the cofactor expansion of a 20 × 20 matrix (which, in applied problems, is a relatively small matrix). For such a computer, the amount of time required to perform this calculation is approximately

$$\frac{e \cdot 20!}{10^9} \text{ seconds} > 6.613 \cdot 10^9 \text{ seconds}$$

$$> 1.837 \cdot 10^6 \text{ hours}$$

$$> 76{,}542 \text{ days}$$

$$> 209 \text{ years.}$$

If determinants are to be of practical value, we must have an efficient method for evaluating them. The key to developing such a method is to observe that we can easily evaluate the determinant of a matrix such as

$$B = \begin{bmatrix} 3 & -4 & -7 & -5 \\ 0 & 8 & -2 & 6 \\ 0 & 0 & 9 & -1 \\ 0 & 0 & 0 & 4 \end{bmatrix}.$$

If we repeatedly evaluate the determinant by a cofactor expansion along the last row, we obtain

$$\det B = \det \begin{bmatrix} 3 & -4 & -7 & -5 \\ 0 & 8 & -2 & 6 \\ 0 & 0 & 9 & -1 \\ 0 & 0 & 0 & 4 \end{bmatrix} = 4(-1)^{4+4} \cdot \det \begin{bmatrix} 3 & -4 & -7 \\ 0 & 8 & -2 \\ 0 & 0 & 9 \end{bmatrix}$$

$$= 4 \cdot 9(-1)^{3+3} \cdot \det \begin{bmatrix} 3 & -4 \\ 0 & 8 \end{bmatrix}$$

$$= 4 \cdot 9 \cdot 8(-1)^{2+2} \cdot \det [3]$$

$$= 4 \cdot 9 \cdot 8 \cdot 3.$$

In Section 2.5, we defined a matrix to be **upper triangular** if all its entries to the left and below the diagonal entries are zero and to be **lower triangular** if all its entries to the right and above the diagonal entries are zero. The matrix B above is an upper triangular 4×4 matrix, and its determinant equals the product of its diagonal entries. The determinant of any such matrix can be computed in a similar fashion.

Theorem 3.2
The determinant of an upper triangular $n \times n$ matrix or a lower triangular $n \times n$ matrix equals the product of its diagonal entries.

An important consequence of Theorem 3.2 is that $\det I_n = 1$.

Example 6   Compute the determinants of

$$A = \begin{bmatrix} -2 & 0 & 0 & 0 \\ 8 & 7 & 0 & 0 \\ -6 & -1 & -3 & 0 \\ 4 & 3 & 9 & 5 \end{bmatrix} \quad \text{and} \quad B = \begin{bmatrix} 2 & 3 & 4 \\ 0 & 5 & 6 \\ 0 & 0 & 7 \end{bmatrix}.$$

Solution Since A is a lower triangular 4×4 matrix and B is an upper triangular 3×3 matrix, we have

$$\det A = (-2)(7)(-3)(5) = 210 \qquad \text{and} \qquad \det B = 2 \cdot 5 \cdot 7 = 70.$$

Geometric Applications of the Determinant*

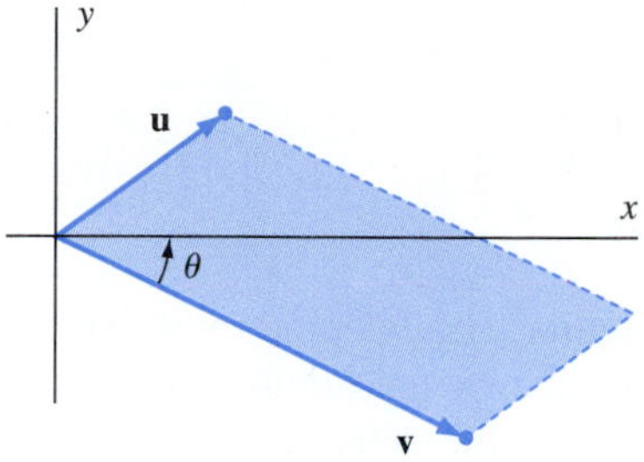

Two vectors $\mathbf{u}$ and $\mathbf{v}$ in $\mathcal{R}^2$ determine a parallelogram having $\mathbf{u}$ and $\mathbf{v}$ as adjacent sides (see Figure 3.1). We call this the **parallelogram determined by $\mathbf{u}$ and $\mathbf{v}$**. Note that if this parallelogram is rotated through the angle θ in Figure 3.1 we obtain the parallelogram determined by $\mathbf{u}'$ and $\mathbf{v}'$ in Figure 3.2.

Suppose that

$$\mathbf{u}' = \begin{bmatrix} u_1 \\ u_2 \end{bmatrix} \qquad \text{and} \qquad \mathbf{v}' = \begin{bmatrix} v_1 \\ 0 \end{bmatrix}.$$

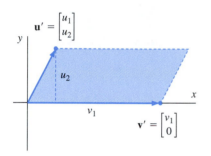

FIGURE 3.1

FIGURE 3.2

Then the parallelogram determined by $\mathbf{u}'$ and $\mathbf{v}'$ has base v_1 and height u_2, so that its area is

$$v_1 u_2 = \left| \det \begin{bmatrix} u_1 & v_1 \\ u_2 & 0 \end{bmatrix} \right| = |\det [\mathbf{u}' \ \mathbf{v}']|.$$

Because a rotation maps a parallelogram into a congruent parallelogram, the parallelogram determined by $\mathbf{u}$ and $\mathbf{v}$ has the same area as that determined by $\mathbf{u}'$ and $\mathbf{v}'$. Recall that multiplying by the rotation matrix A_θ rotates a vector by the angle θ. Using the facts that $\det AB = (\det A)(\det B)$ for any 2×2 matrices A and B (Exercise 39) and $\det A_\theta = 1$ (Exercise 33), we see that the area of the parallelogram determined by $\mathbf{u}$ and $\mathbf{v}$ is

$$|\det [\mathbf{u}' \ \mathbf{v}']| = |\det [A_\theta \mathbf{u} \ \ A_\theta \mathbf{v}]|$$
$$= |\det (A_\theta [\mathbf{u} \ \ \mathbf{v}])|$$
$$= |(\det A_\theta)(\det[\mathbf{u} \ \ \mathbf{v}])|$$
$$= |(1)(\det [\mathbf{u} \ \ \mathbf{v}])|$$
$$= |\det [\mathbf{u} \ \ \mathbf{v}]|.$$

The area of the parallelogram determined by $\mathbf{u}$ and $\mathbf{v}$ is $|\det [\mathbf{u} \ \ \mathbf{v}]|$.

Moreover, a corresponding result can be proved for $\mathcal{R}^n$ using the appropriate n-dimensional analog of area.

Example 7 The area of the parallelogram in $\mathcal{R}^2$ determined by the vectors

$$\begin{bmatrix} -2 \\ 3 \end{bmatrix} \qquad \text{and} \qquad \begin{bmatrix} 1 \\ 5 \end{bmatrix}$$

is

$$\left| \det \begin{bmatrix} -2 & 1 \\ 3 & 5 \end{bmatrix} \right| = |(-2)(5) - (1)(3)| = |-13| = 13.$$

---
*The remainder of this section may be omitted without loss of continuity.

Example 8   The volume of the parallelepiped in $\mathcal{R}^3$ determined by the vectors

$$\begin{bmatrix} 1 \\ 1 \\ 1 \end{bmatrix}, \qquad \begin{bmatrix} 1 \\ -2 \\ 1 \end{bmatrix}, \qquad \text{and} \qquad \begin{bmatrix} 1 \\ 0 \\ -1 \end{bmatrix}$$

is

$$\left| \det \begin{bmatrix} 1 & 1 & 1 \\ 1 & -2 & 0 \\ 1 & 1 & -1 \end{bmatrix} \right| = 6.$$

Note that the object in question is a rectangular parallelepiped with sides of length $\sqrt{3}$, $\sqrt{6}$, and $\sqrt{2}$ (see Figure 3.3). Hence by the familiar formula for volume, its volume should be $\sqrt{3} \cdot \sqrt{6} \cdot \sqrt{2}$, as the determinant calculation shows. ○

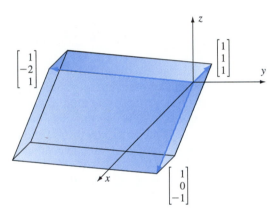

FIGURE 3.3

The points on or within the parallelogram in $\mathcal{R}^2$ determined by $\mathbf{u}$ and $\mathbf{v}$ can be written in the form $a\mathbf{u} + b\mathbf{v}$, where a and b are scalars such that $0 \le a \le 1$ and $0 \le b \le 1$ (see Figure 3.4). If T is a linear operator on $\mathcal{R}^2$, then

$$T(a\mathbf{u} + b\mathbf{v}) = aT(\mathbf{u}) + bT(\mathbf{v}).$$

Hence the image under T of the parallelogram determined by $\mathbf{u}$ and $\mathbf{v}$ is the parallelogram determined by $T(\mathbf{u})$ and $T(\mathbf{v})$. The area of this parallelogram is

$$|\det [T(\mathbf{u}) \ \ T(\mathbf{v})]| = |\det [A\mathbf{u} \ \ A\mathbf{v}]| |\det A[\mathbf{u} \ \ \mathbf{v}]| = |\det A| \cdot |\det [\mathbf{u} \ \ \mathbf{v}]|,$$

where A is the standard matrix of T. Thus the area of the image parallelogram is $|\det A|$ times larger than that of the parallelogram determined by $\mathbf{u}$ and $\mathbf{v}$. (If T is not invertible, then $\det A = 0$, and the parallelogram determined by $T(\mathbf{u})$ and $T(\mathbf{v})$ is degenerate.)

More generally, the area of any "sufficiently nice" region R in $\mathcal{R}^2$ can be approximated as the sum of the areas of rectangles. In fact, as the lengths of the sides of these rectangles approach zero, the sum of the areas of the rectangles approaches the area of R. Hence the area of the image of R under T equals $|\det A|$ times the area of R. A corresponding theorem can be proved about the volume of a sufficiently nice region in $\mathcal{R}^3$. In fact, using the appropriate n-dimensional analog of volume, the following result is true.

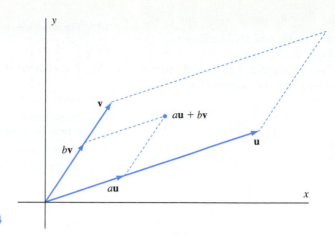

FIGURE 3.4

> If R is a "sufficiently nice" region in $\mathcal{R}^n$ and T is an invertible linear operator on $\mathcal{R}^n$ with standard matrix A, then the n-dimensional volume of the image of R under T equals $|\det A|$ times the n-dimensional volume of R.

This result plays a crucial role when making a change of variables in calculus.

Practice Problems

1. Evaluate the determinant of

$$\begin{bmatrix} 8 & 3 \\ -6 & 5 \end{bmatrix}.$$

2. Evaluate the determinant of

$$\begin{bmatrix} 1 & 3 & -3 \\ -3 & -5 & 2 \\ -4 & 4 & -6 \end{bmatrix}$$

using the cofactor expansion along the second row.

Exercises

1. Determine if the following statements are true or false.

   (a) $\det \begin{bmatrix} a & b \\ c & d \end{bmatrix} = ad + bc$.

   (b) For $n \geq 2$, the (i, j)-cofactor of an $n \times n$ matrix A is the determinant of the $(n - 1) \times (n - 1)$ matrix obtained by deleting row i and column j from A.

   (c) The determinant of an $n \times n$ matrix can be evaluated using a cofactor expansion along any row.

   (d) The determinant of an upper triangular $n \times n$ matrix or a lower triangular $n \times n$ matrix equals the sum of its diagonal entries.

   (e) The area of the parallelogram determined by $\mathbf{u}$ and $\mathbf{v}$ is $\det [\mathbf{u} \ \ \mathbf{v}]$.

In Exercises 2–6, compute the determinant of the given matrix.

2. $\begin{bmatrix} 4 & 5 \\ 3 & -7 \end{bmatrix}$ 3. $\begin{bmatrix} -2 & 9 \\ 1 & 8 \end{bmatrix}$ 4. $\begin{bmatrix} 9 & -2 \\ 3 & 4 \end{bmatrix}$

5. $\begin{bmatrix} -5 & -6 \\ 10 & 12 \end{bmatrix}$ 6. $\begin{bmatrix} -7 & 8 \\ 4 & -5 \end{bmatrix}$

In Exercises 7–10, compute the indicated cofactor of the matrix

$$A = \begin{bmatrix} 9 & -2 & 4 \\ -1 & 6 & 3 \\ 7 & 8 & -5 \end{bmatrix}.$$

7. the $(1, 2)$-cofactor   **8.** the $(2, 3)$-cofactor

9. the $(3, 1)$-cofactor   **10.** the $(3, 3)$-cofactor

In Exercises 11–18, compute the determinant of the matrix A by a cofactor expansion along the indicated row.

11. $\begin{bmatrix} 2 & -1 & 3 \\ 1 & 4 & -2 \\ -1 & 0 & 1 \end{bmatrix}$
along the first row

12. $\begin{bmatrix} 1 & -2 & 2 \\ 2 & -1 & 3 \\ 0 & 1 & -1 \end{bmatrix}$
along the second row

13. $\begin{bmatrix} 1 & -2 & 2 \\ 2 & -1 & 3 \\ 0 & 1 & -1 \end{bmatrix}$
along the third row

14. $\begin{bmatrix} 2 & -1 & 3 \\ 1 & 4 & -2 \\ -1 & 0 & 1 \end{bmatrix}$
along the third row

15. $\begin{bmatrix} 1 & 4 & -3 \\ 5 & 0 & 0 \\ 2 & 0 & -1 \end{bmatrix}$
along the second row

16. $\begin{bmatrix} 4 & 1 & 0 \\ 0 & 3 & -2 \\ 2 & 0 & 5 \end{bmatrix}$
along the first row

17. $\begin{bmatrix} 1 & 2 & 1 & -1 \\ 0 & -1 & 0 & 1 \\ 4 & -3 & 2 & -1 \\ 0 & 3 & 0 & -2 \end{bmatrix}$
along the second row

18. $\begin{bmatrix} 0 & -1 & 0 & 1 \\ -2 & 3 & 1 & 4 \\ 1 & -2 & 2 & 3 \\ 0 & 1 & 0 & -2 \end{bmatrix}$
along the fourth row

In Exercises 19–26, compute the determinant of A by any legitimate method.

19. $\begin{bmatrix} 4 & -1 & 2 \\ 0 & 3 & 7 \\ 0 & 0 & 5 \end{bmatrix}$

20. $\begin{bmatrix} 8 & 0 & 0 \\ -1 & -2 & 0 \\ 4 & 5 & 3 \end{bmatrix}$

21. $\begin{bmatrix} -6 & 0 & 0 \\ 7 & -3 & 2 \\ 2 & 9 & 4 \end{bmatrix}$

22. $\begin{bmatrix} 7 & 1 & 8 \\ 0 & -3 & 4 \\ 0 & 0 & -2 \end{bmatrix}$

23. $\begin{bmatrix} 2 & 3 & 4 \\ 5 & 6 & 1 \\ 7 & 0 & 0 \end{bmatrix}$

24. $\begin{bmatrix} 5 & 1 & 1 \\ 0 & 2 & 0 \\ 6 & -4 & 3 \end{bmatrix}$

25. $\begin{bmatrix} -2 & -1 & -5 & 1 \\ 0 & 0 & 0 & 4 \\ 0 & -2 & 0 & 5 \\ 3 & 1 & 6 & -2 \end{bmatrix}$

26. $\begin{bmatrix} 4 & 2 & 2 & -3 \\ 6 & -1 & 1 & 5 \\ 0 & -3 & 0 & 0 \\ 2 & -5 & 0 & 0 \end{bmatrix}$

*In Exercises 27–30, compute the area of the parallelogram determined by **u** and **v**.*

27. $\mathbf{u} = \begin{bmatrix} 3 \\ 5 \end{bmatrix}, \mathbf{v} = \begin{bmatrix} -2 \\ 7 \end{bmatrix}$

28. $\mathbf{u} = \begin{bmatrix} -3 \\ 6 \end{bmatrix}, \mathbf{v} = \begin{bmatrix} 8 \\ -5 \end{bmatrix}$

29. $\mathbf{u} = \begin{bmatrix} 6 \\ 4 \end{bmatrix}, \mathbf{v} = \begin{bmatrix} 3 \\ 2 \end{bmatrix}$

30. $\mathbf{u} = \begin{bmatrix} -1 \\ 2 \end{bmatrix}, \mathbf{v} = \begin{bmatrix} 4 \\ 5 \end{bmatrix}$

31. For what value of c is $\begin{bmatrix} 3 & 6 \\ c & 4 \end{bmatrix}$ not invertible?

32. For what value of c is $\begin{bmatrix} 9 & -18 \\ 4 & c \end{bmatrix}$ not invertible?

33. Show that the determinant of the rotation matrix A_θ is 1.

34. Show that if A is a 2×2 matrix in which every entry is 0 or 1, then the determinant of A equals 0, 1, or -1.

35. Show that the conclusion of Exercise 34 is false for 3×3 matrices by calculating
$$\det \begin{bmatrix} 1 & 0 & 1 \\ 1 & 1 & 0 \\ 0 & 1 & 1 \end{bmatrix}.$$

36. Prove that if a 2×2 matrix has identical rows, then its determinant is zero.

37. Prove that, for any 2×2 matrix A, $\det A^T = \det A$.

38. Let A be a 2×2 matrix and k be a scalar. How does $\det kA$ compare to $\det A$? Justify your answer.

39. Prove that, for any 2×2 matrices A and B, $\det AB = (\det A)(\det B)$.

40. What is the determinant of an $n \times n$ matrix with a zero row? Justify your answer.

For the elementary matrix E given in Exercises 41–44 and the matrix
$$A = \begin{bmatrix} a & b \\ c & d \end{bmatrix},$$
verify that $\det EA = (\det E)(\det A)$.

41. $\begin{bmatrix} 1 & 0 \\ 0 & k \end{bmatrix}$ **42.** $\begin{bmatrix} 0 & 1 \\ 1 & 0 \end{bmatrix}$ **43.** $\begin{bmatrix} 1 & 0 \\ k & 1 \end{bmatrix}$ **44.** $\begin{bmatrix} 1 & k \\ 0 & 1 \end{bmatrix}$

45. Prove that
$$\det \begin{bmatrix} a & b \\ c+kp & d+kq \end{bmatrix} = \det \begin{bmatrix} a & b \\ c & d \end{bmatrix} + k \cdot \det \begin{bmatrix} a & b \\ p & q \end{bmatrix}$$

46. The TI-85 calculator gives
$$\det \begin{bmatrix} 1 & 2 & 3 \\ 2 & 3 & 4 \\ 3 & 4 & 5 \end{bmatrix} = -4 \times 10^{-13}.$$

Why must this answer be wrong? *Hint:* State a general fact about the determinant of a square matrix in which all the entries are integers.

47. Use a determinant to express the area of the triangle in $\mathcal{R}^2$ having vertices $\mathbf{0}$, $\mathbf{u}$, and $\mathbf{v}$.

48. Calculate the determinant of $\begin{bmatrix} O & I_m \\ I_n & O' \end{bmatrix}$ if O and O' are zero matrices.

In Exercises 49–52, use a calculator with matrix capabilities or computer software such as MATLAB to solve the problem.

49. (a) Randomly generate 4×4 matrices A and B. Evaluate $\det A$, $\det B$, and $\det (A + B)$.
(b) Repeat (a) with randomly generated 5×5 matrices.

(c) Does it appear that $\det(A + B) = \det A + \det B$ for all $n \times n$ matrices A and B?

50. (a) Randomly generate 4×4 matrices A and B. Evaluate $\det A$, $\det B$, and $\det(AB)$.

(b) Repeat (a) with randomly generated 5×5 matrices.

(c) Does it appear to be true that $\det (AB) = (\det A)(\det B)$ for all $n \times n$ matrices A and B?

51. (a) Randomly generate a 4×4 matrix A. Evaluate $\det A$ and $\det A^T$.

(b) Repeat (a) with a randomly generated 5×5 matrix.

(c) Does it appear to be true that $\det A = \det A^T$ for all $n \times n$ matrices?

52. (a) Let

$$A = \begin{bmatrix} 0 & -1 & 2 & 2 \\ 1 & -1 & 0 & -2 \\ 2 & 1 & 0 & 1 \\ -1 & 1 & 2 & -1 \end{bmatrix}.$$

Show that A is invertible, and evaluate $\det A$ and $\det A^{-1}$.

(b) Repeat (a) with a randomly generated invertible 5×5 matrix.

(c) Make a conjecture about $\det A$ and $\det A^{-1}$ for any invertible matrix A.

Solutions to the Practice Problems

1. $\det \begin{bmatrix} 8 & 3 \\ -6 & 5 \end{bmatrix} = 8(5) - 3(-6) = 40 + 18 = 58.$

2. Let c_{ij} denote the (i, j)-cofactor of A. The cofactor expansion along the second row gives the following value for $\det A$:

$$\det A = (-3)c_{21} + (-5)c_{22} + 2c_{23}$$

$$= -3(-1)^{2+1} \cdot \det \begin{bmatrix} 3 & -3 \\ 4 & -6 \end{bmatrix}$$

$$+ (-5)(-1)^{2+2} \cdot \det \begin{bmatrix} 1 & -3 \\ -4 & -6 \end{bmatrix}$$

$$+ 2(-1)^{2+3} \cdot \det \begin{bmatrix} 1 & 3 \\ -4 & 4 \end{bmatrix}$$

$$= -3(-1)[3(-6) - (-3)(4)]$$
$$+ (-5)(1)[1(-6) - (-3)(-4)]$$
$$+ 2(-1)[1(4) - 3(-4)]$$
$$= 3(-6) + (-5)(-18) + (-2)(16)$$
$$= 40.$$

3.2   Properties of Determinants

We have seen that, for arbitrary matrices, the computation of determinants by cofactor expansion is quite inefficient. Theorem 3.2, however, provides a simple and efficient method for evaluating the determinant of an upper triangular matrix. Fortunately, the forward pass of the Gaussian elimination algorithm in Section 1.4 transforms any matrix into an upper triangular matrix using a sequence of elementary row operations. If we knew the effect of these elementary row operations on the determinant of a matrix, we could then evaluate the determinant efficiently by using Theorem 3.2.

For example, the following sequence of elementary row operations transforms

$$A = \begin{bmatrix} 1 & 2 & 3 \\ 4 & 5 & 6 \\ 7 & 9 & 8 \end{bmatrix}$$

into an upper triangular matrix U:

$$A = \begin{bmatrix} 1 & 2 & 3 \\ 4 & 5 & 6 \\ 7 & 9 & 8 \end{bmatrix} \longrightarrow \begin{bmatrix} 1 & 2 & 3 \\ 0 & -3 & -6 \\ 7 & 9 & 8 \end{bmatrix}$$

$$\longrightarrow \begin{bmatrix} 1 & 2 & 3 \\ 0 & -3 & -6 \\ 0 & -5 & -13 \end{bmatrix}$$

$$\longrightarrow \begin{bmatrix} 1 & 2 & 3 \\ 0 & -3 & -6 \\ 0 & 0 & -3 \end{bmatrix} = U.$$

The three elementary row operations used in this transformation are row addition operations (operations that add a multiple of some row to another). Theorem 3.3 shows that this type of elementary row operation leaves the determinant unchanged. Hence the determinant of each matrix in the above sequence is the same, and so

$$\det A = \det U = 1(-3)(-3) = 9.$$

This calculation is much more efficient than the one in Example 3 of Section 3.1.

The following theorem enables us to use elementary row operations to evaluate determinants.

Theorem 3.3

Let A be an n × n matrix.

(a) If B is a matrix obtained by interchanging two rows of A, then $\det B = -\det A$.

(b) If B is a matrix obtained by multiplying each entry of some row of A by a scalar k, then $\det B = k \cdot \det A$.

(c) If B is a matrix obtained by adding a multiple of some row of A to a different row, then $\det B = \det A$.

(d) For any $n \times n$ elementary matrix E, we have $\det EA = (\det E)(\det A)$.

Parts (a), (b), and (c) of Theorem 3.3 describe how the determinant of a matrix changes when an elementary row operation is performed on the matrix. (The proof of the theorem is found at the end of this section.) Note that if $A = I_n$ in Theorem 3.3, then (a), (b), and (c) give the value of the determinant of each type of elementary matrix. In particular, $\det E = 1$ if E performs a row addition operation and $\det E = -1$ if E performs a row interchange operation.

Suppose that an $n \times n$ matrix is transformed into an upper triangular matrix U by a sequence of elementary row operations other than scaling operations. (This can always be done by using steps 1–4 of the Gaussian elimination algorithm. The elementary row operations that occur are interchange operations in step 2 and row addition operations in step 3.) We saw in Section 2.3 that each of these elementary row operations can be implemented by multiplying by an elementary matrix. Thus there is a sequence of elementary matrices $E_1, E_2, \ldots, E_k$ such that

$$E_k \cdots E_2 E_1 A = U.$$

By Theorem 3.3(d) we have

$$(\det E_k) \cdots (\det E_2)(\det E_1)(\det A) = \det U.$$

Thus

$$(-1)^r (\det A) = \det U,$$

where r is the number of row interchange operations that occur in the transformation of A into U. Since U is an upper triangular matrix, its determinant is the product $u_{11} u_{22} \cdots u_{nn}$ of its diagonal entries, by Theorem 3.2. Hence we have the following important result, which provides an efficient method for evaluating a determinant.

> If an $n \times n$ matrix A is transformed into an upper triangular matrix U by using elementary row operations other than scaling operations, then
>
> $$\det A = (-1)^r u_{11} u_{22} \cdots u_{nn},$$
>
> where r is the number of row interchanges performed.

Example 1   Use elementary row operations to compute the determinant of

$$A = \begin{bmatrix} 0 & 1 & 3 & -3 \\ 0 & 0 & 4 & -2 \\ -2 & 0 & 4 & -7 \\ 4 & -4 & 4 & 15 \end{bmatrix}.$$

Solution   In order to transform A into an upper triangular matrix U, we must begin with a row interchange. Suppose that rows 1 and 3 are interchanged to produce the matrix

$$\begin{bmatrix} -2 & 0 & 4 & -7 \\ 0 & 0 & 4 & -2 \\ 0 & 1 & 3 & -3 \\ 4 & -4 & 4 & 15 \end{bmatrix}.$$

Adding 2 times row 1 to row 4 yields

$$\begin{bmatrix} -2 & 0 & 4 & -7 \\ 0 & 0 & 4 & -2 \\ 0 & 1 & 3 & -3 \\ 0 & -4 & 12 & 1 \end{bmatrix}.$$

At this point, another row interchange is required to make the $(2, 2)$-entry nonzero. Interchanging rows 2 and 3 produces

$$\begin{bmatrix} -2 & 0 & 4 & -7 \\ 0 & 1 & 3 & -3 \\ 0 & 0 & 4 & -2 \\ 0 & -4 & 12 & 1 \end{bmatrix}.$$

Now we add 4 times row 2 to row 4 to obtain

$$\begin{bmatrix} -2 & 0 & 4 & -7 \\ 0 & 1 & 3 & -3 \\ 0 & 0 & 4 & -2 \\ 0 & 0 & 24 & -11 \end{bmatrix}.$$

Adding -6 times row 3 to row 4 yields

$$\begin{bmatrix} -2 & 0 & 4 & -7 \\ 0 & 1 & 3 & -3 \\ 0 & 0 & 4 & -2 \\ 0 & 0 & 0 & 1 \end{bmatrix} = U.$$

Since U is an upper triangular matrix, we have $\det U = (-2) \cdot 1 \cdot 4 \cdot 1 = -8$. During the transformation of A into U, two row interchanges were performed. Thus

$$\det A = (-1)^2 \cdot \det U = -8.$$

In Section 3.1, we mentioned that the cofactor expansion of an arbitrary $n \times n$ matrix requires approximately $e \cdot n!$ arithmetic operations. By contrast, evaluating the determinant of an $n \times n$ matrix using elementary row operations requires only about $2n^3/3$ arithmetic operations. Thus a computer capable of performing 1 billion arithmetic operations per second could calculate the determinant of a 20×20 matrix in about 5 millionths of a second by using elementary row operations, as compared to the more than 209 years it needs to evaluate the determinant using a cofactor expansion.

Four Properties of Determinants

Several useful results about determinants can be proved from Theorem 3.3. Among these is the desired test for the invertibility of a matrix.

Theorem 3.4

Let A and B be square matrices of the same size. The following statements are true.

(a) A is invertible if and only if $\det A \neq 0$.

(b) $\det AB = (\det A)(\det B)$.

(c) $\det A^T = \det A$.

(d) If A is invertible, then $\det A^{-1} = \dfrac{1}{\det A}$.

Proof We first prove (a), (b), and (c) for an invertible $n \times n$ matrix A. If A is invertible, there are elementary matrices $E_1, E_2, \ldots, E_k$ such that $A = E_k \cdots E_2 E_1$ (Theorem 2.6). Hence by repeated applications of Theorem 3.3(d), we obtain

$$\det A = (\det E_k) \cdots (\det E_2)(\det E_1).$$

Since the determinant of an elementary matrix is nonzero, we have $\det A \neq 0$. This proves (a) for an invertible matrix. Moreover, for any $n \times n$ matrix B, repeated applications of Theorem 3.3(d) give

$$(\det A)(\det B) = (\det E_k) \cdots (\det E_2)(\det E_1)(\det B)$$
$$= (\det E_k) \cdots (\det E_2)(\det E_1 B)$$
$$\vdots$$
$$= \det(E_k \cdots E_2 E_1 B)$$
$$= \det AB.$$

This proves (b) when A is invertible. Furthermore, we also have

$$A^T = (E_k \cdots E_2 E_1)^T = E_1^T E_2^T \cdots E_k^T.$$

We leave as an exercise the proof that $\det E^T = \det E$ for every elementary matrix E. It follows that

$$\det A^T = \det \left(E_1^T E_2^T \cdots E_k^T \right)$$
$$= \left(\det E_1^T \right)\left(\det E_2^T \right) \cdots \left(\det E_k^T \right)$$
$$= (\det E_1)(\det E_2) \cdots (\det E_k)$$
$$= (\det E_k) \cdots (\det E_2)(\det E_1)$$
$$= \det(E_k \cdots E_2 E_1)$$
$$= \det A,$$

proving (c) for an invertible matrix.

Now we prove (a), (b), and (c) in the case that A is an $n \times n$ matrix that is not invertible. By Theorem 2.4, there exists an invertible matrix P such that $PA = R$, where R is the reduced row echelon form of A. Since the rank of A is not n (Theorem 2.6), the $n \times n$ matrix R must contain a row of zeros. Performing the cofactor expansion of R along this row yields $\det R = 0$. Because P^{-1} is invertible, (b) implies that

$$\det A = \det(P^{-1}R) = (\det P^{-1})(\det R) = (\det P^{-1}) \cdot 0 = 0,$$

completing the proof of (a). Moreover, since R contains a row of zeros, so does RB. Thus RB is not invertible, and so $\det RB = 0$ by (a). Applying (b) to the invertible matrix P, we obtain

$$(\det P)(\det AB) = \det P(AB) = \det(PA)B = \det RB = 0.$$

Since $\det P \neq 0$ by (a), it follows that

$$\det AB = 0 = 0 \cdot \det B = (\det A)(\det B),$$

completing the proof of (b). Finally, by Theorem 2.3, A^T is not invertible (for otherwise $(A^T)^T = A$ would be invertible). Hence $\det A^T = 0 = \det A$ by (a). This completes the proof of (c).

The proof of (d) follows from (b) and the fact that $\det I_n = 1$. We leave the details as an exercise. ❏

As we have said, our principal reason for studying determinants is that they provide a means for testing if a matrix is invertible, namely, the result of Theorem 3.4(a). This fact is essential to Chapter 5, where it is used to determine the eigenvalues of a matrix. The next example illustrates how this test can be used.

Example 2   For what scalar c is the matrix

$$A = \begin{bmatrix} 1 & -1 & 2 \\ -1 & 0 & c \\ 2 & 1 & 7 \end{bmatrix}$$

not invertible?

Solution   To answer this question, we compute the determinant of A. The following sequence of row addition operations transforms A into an upper triangular matrix:

$$\begin{bmatrix} 1 & -1 & 2 \\ -1 & 0 & c \\ 2 & 1 & 7 \end{bmatrix} \longrightarrow \begin{bmatrix} 1 & -1 & 2 \\ 0 & -1 & c+2 \\ 2 & 1 & 7 \end{bmatrix} \longrightarrow \begin{bmatrix} 1 & -1 & 2 \\ 0 & -1 & c+2 \\ 0 & 3 & 3 \end{bmatrix}$$

$$\longrightarrow \begin{bmatrix} 1 & -1 & 2 \\ 0 & -1 & c+2 \\ 0 & 0 & 3c+9 \end{bmatrix}.$$

Hence $\det A = 1(-1)(3c + 9) = -3c - 9$. Theorem 3.4(a) states that A is not invertible if and only if its determinant is 0. Thus A is not invertible if and only if $0 = -3c - 9$; that is, if and only if $c = -3$. ○

The following example illustrates how Theorem 3.4(b) can be used. The result of this example is needed in Section 5.2.

Example 3   Suppose that a matrix M can be partitioned as

$$\begin{bmatrix} A & B \\ O & C \end{bmatrix},$$

where A is a $m \times m$ matrix, C is an $n \times n$ matrix, and O is the $n \times m$ zero matrix. Show that $\det M = (\det A)(\det C)$.

Solution   Using block multiplication, we see that

$$\begin{bmatrix} I_m & O' \\ O & C \end{bmatrix} \begin{bmatrix} A & B \\ O & I_n \end{bmatrix} = \begin{bmatrix} A & B \\ O & C \end{bmatrix} = M,$$

where O' is the $m \times n$ zero matrix. Therefore, by Theorem 3.4,

$$\det \begin{bmatrix} I_m & O' \\ O & C \end{bmatrix} \cdot \det \begin{bmatrix} A & B \\ O & I_n \end{bmatrix} = \det M.$$

As in Example 5 of Section 3.1, it can be shown that

$$\det \begin{bmatrix} A & B \\ O & I_n \end{bmatrix} = \det A.$$

A similar argument (using cofactor expansions along the first row) yields

$$\det \begin{bmatrix} I_m & O' \\ O & C \end{bmatrix} = \det C.$$

Hence

$$\det M = \det \begin{bmatrix} I_m & O' \\ O & C \end{bmatrix} \cdot \det \begin{bmatrix} A & B \\ O & I_n \end{bmatrix} = (\det C)(\det A). \qquad \circ$$

Several important theoretical results follow from Theorem 3.4(c). For example, we can evaluate the determinant of a matrix A by computing the determinant of its transpose. Thus we can evaluate the determinant of A by a cofactor expansion along the rows of A^T. But the rows of A^T are the columns of A, and so the determinant of A can be evaluated by cofactor expansion along *any column* of A, as well as any row.

Let A and B be $n \times n$ matrices. Although $\det AB = (\det A)(\det B)$ by Theorem 3.4(b), the corresponding property for matrix addition is *not* true. Consider, for instance, the matrices

$$A = \begin{bmatrix} 1 & 0 \\ 0 & 0 \end{bmatrix} \quad \text{and} \quad B = \begin{bmatrix} 0 & 0 \\ 0 & 1 \end{bmatrix}.$$

Clearly $\det A = \det B = 0$, whereas $\det(A + B) = \det I_2 = 1$. Therefore

$$\det(A + B) \neq \det A + \det B,$$

and so the determinant of a sum of matrices need not be equal to the sum of their determinants.

Cramer's Rule*

One of the original motivations for studying determinants was that they provide a method for solving systems of linear equations having an invertible coefficient matrix. The following result was published in 1750 by the Swiss mathematician Gabriel Cramer (1704–1752).

Theorem 3.5 (Cramer's rule[3])

Let A be an invertible $n \times n$ matrix and $\mathbf{b} \in \mathcal{R}^n$. Let M_i be the matrix obtained from A by replacing column i of A by $\mathbf{b}$. Then $A\mathbf{x} = \mathbf{b}$ has a unique solution $\mathbf{u}$ in which the components are given by

$$u_i = \frac{\det M_i}{\det A} \text{ for } i = 1, 2, \ldots, n.$$

Proof Since A is invertible, $A\mathbf{x} = \mathbf{b}$ has the unique solution $\mathbf{u} = A^{-1}\mathbf{b}$, as we saw in Section 2.3. For each i, let U_i denote the matrix obtained by replacing column i of I_n by

$$\mathbf{u} = \begin{bmatrix} u_1 \\ u_2 \\ \vdots \\ u_n \end{bmatrix}.$$

Then

$$AU_i = A[\mathbf{e}_1 \ \cdots \ \mathbf{e}_{i-1} \ \mathbf{u} \ \mathbf{e}_{i+1} \ \cdots \ \mathbf{e}_n]$$
$$= [A\mathbf{e}_1 \ \cdots \ A\mathbf{e}_{i-1} \ A\mathbf{u} \ A\mathbf{e}_{i+1} \ \cdots \ A\mathbf{e}_n]$$
$$= [\mathbf{a}_1 \ \cdots \ \mathbf{a}_{i-1} \ \mathbf{b} \ \mathbf{a}_{i+1} \ \cdots \ \mathbf{a}_n]$$
$$= M_i.$$

Evaluating U_i by cofactor expansion along row i produces

$$\det U_i = u_i \cdot \det I_{n-1} = u_i.$$

Hence, by Theorem 3.4(b), we have

$$\det M_i = \det AU_i = (\det A) \cdot (\det U_i) = (\det A) \cdot u_i.$$

Because $\det A \neq 0$ by Theorem 3.4(a), it follows that

$$u_i = \frac{\det M_i}{\det A}. \qquad \square$$

Example 4 Use Cramer's rule to solve the system of equations

$$x_1 + 2x_2 + 3x_3 = 2$$
$$x_1 + \qquad\ x_3 = 3$$
$$x_1 + \ x_2 - \ x_3 = 1.$$

---

*The remainder of this section may be omitted without loss of continuity.

[3]This rule was first stated in its most general form in Cramer's 1750 paper, where it was used to find the equations of curves in the plane passing through given points.

Solution   The coefficient matrix of this system is

$$A = \begin{bmatrix} 1 & 2 & 3 \\ 1 & 0 & 1 \\ 1 & 1 & -1 \end{bmatrix}.$$

Since $\det A = 6$, A is invertible by Theorem 3.4(a), and hence Cramer's rule can be used. In the notation of Theorem 3.5, we have

$$M_1 = \begin{bmatrix} 2 & 2 & 3 \\ 3 & 0 & 1 \\ 1 & 1 & -1 \end{bmatrix}, \quad M_2 = \begin{bmatrix} 1 & 2 & 3 \\ 1 & 3 & 1 \\ 1 & 1 & -1 \end{bmatrix}, \quad \text{and} \quad M_3 = \begin{bmatrix} 1 & 2 & 2 \\ 1 & 0 & 3 \\ 1 & 1 & 1 \end{bmatrix}.$$

Therefore the unique solution to the given system is the vector with components

$$u_1 = \frac{\det M_1}{\det A} = \frac{15}{6} = \frac{5}{2}, u_2 = \frac{\det M_2}{\det A} = \frac{-6}{6} = -1, \text{ and}$$

$$u_3 = \frac{\det M_3}{\det A} = \frac{3}{6} = \frac{1}{2}.$$

It is readily checked that these values satisfy each of the equations in the given system. ○

We noted earlier that evaluating the determinant of an $n \times n$ matrix using elementary row operations requires about $2n^3/3$ arithmetic operations. On the other hand, we saw in Section 2.5 that an entire system of n linear equations in n unknowns can be solved by Gaussian elimination with about the same number of operations. Thus Cramer's rule is not an efficient method for solving systems of linear equations and, moreover, it can be used only in the special case where the coefficient matrix is invertible. Nevertheless, Cramer's rule is useful for certain theoretical purposes. It can be used, for example, to analyze how the solution to $A\mathbf{x} = \mathbf{b}$ is influenced by changes in $\mathbf{b}$.

We conclude this section with a proof of Theorem 3.3.

Proof (Theorem 3.3)   Let A be an $n \times n$ matrix with rows $\mathbf{a}_1', \mathbf{a}_2', \ldots, \mathbf{a}_n'$, respectively.

(a) Suppose that B is the matrix obtained from A by interchanging rows r and s, where $r < s$. We begin by establishing the result in the case that $s = r + 1$. In this case $a_{rj} = b_{sj}$ and $A_{rj} = B_{sj}$ for each j. Thus each cofactor in the cofactor expansion of B along row s is the negative of the corresponding cofactor in the cofactor expansion of A along row r. It follows that $\det B = -\det A$.

Now suppose that $s > r + 1$. Beginning with rows r and $r + 1$, we successively interchange $\mathbf{a}_r'$ with the following row until the rows of A are in the order

$$\mathbf{a}_1', \ldots, \mathbf{a}_{r-1}', \mathbf{a}_{r+1}', \ldots, \mathbf{a}_s', \mathbf{a}_r', \mathbf{a}_{s+1}', \ldots, \mathbf{a}_n'.$$

A total of $s - r$ interchanges are necessary to produce this ordering. Now we successively interchange $\mathbf{a}_s'$ with the preceding row until the rows are in the order

$$\mathbf{a}_1', \ldots, \mathbf{a}_{r-1}', \mathbf{a}_s', \mathbf{a}_{r+1}', \ldots, \mathbf{a}_{s-1}', \mathbf{a}_r', \mathbf{a}_{s+1}', \ldots, \mathbf{a}_n'.$$

This process requires another $s - r - 1$ interchanges of adjacent rows and produces the matrix B. Thus the preceding paragraph shows that

$$\det B = (-1)^{s-r}(-1)^{s-r-1} \cdot \det A = (-1)^{2(s-r)-1} \cdot \det A = -\det A.$$

(b) Suppose that B is a matrix obtained by multiplying each entry of row r of A by a scalar k. Comparing the cofactor expansion of B along row r to that of A, it is easy to see that $\det B = k \cdot \det A$. We leave the details to the reader.

(c) We first show that if C is an $n \times n$ matrix having two identical rows, then $\det C = 0$. Suppose that rows r and s of C are equal, and let M be obtained from C by interchanging rows r and s. Then $\det M = -\det C$ by (a). But since rows r and s of C are equal, we also have $C = M$. Thus $\det C = \det M$. Combining the two equations involving $\det M$, we obtain $\det C = -\det C$. Therefore $\det C = 0$.

Now suppose that B is obtained from A by adding k times row s of A to row r, where $r \neq s$. Let C be the $n \times n$ matrix obtained from A by replacing $\mathbf{a}'_r = (u_1, u_2, \ldots, u_n)$ by $\mathbf{a}'_s = (v_1, v_2, \ldots, v_n)$. Since A, B, and C differ only in row r, we have $A_{rj} = B_{rj} = C_{rj}$ for every j. Using the cofactor expansion of B along row r, we obtain

$$\det B = (u_1 + kv_1)(-1)^{r+1} \det B_{r1} + \cdots + (u_n + kv_n)(-1)^{r+n} \det B_{rn}$$

$$= \left(u_1(-1)^{r+1} \det B_{r1} + \cdots + u_n(-1)^{r+n} \det B_{rn} \right)$$

$$+ k \left(v_1(-1)^{r+1} \det B_{r1} + \cdots + v_n(-1)^{r+n} \det B_{rn} \right)$$

$$= \left[u_1(-1)^{r+1} \det A_{r1} + \cdots + u_n(-1)^{r+n} \det A_{rn} \right]$$

$$+ k \left[v_1(-1)^{r+1} \det C_{r1} + \cdots + v_n(-1)^{r+n} \det C_{rn} \right].$$

In this equation, the first expression in brackets is the cofactor expansion of A along row r, and the second is the cofactor expansion of C along row r. Thus we have

$$\det B = \det A + k \cdot \det C.$$

However, C is a matrix with two identical rows (namely, rows r and s, which are both equal to $\mathbf{a}'_s$). Since $\det C = 0$ by the preceding paragraph, it follows that $\det B = \det A$.

(d) Let E be an elementary matrix obtained by interchanging two rows of I_n. Then $\det EA = -\det A$ by (a). Since $\det E = -1$, we have $\det EA = (\det E)(\det A)$. Similar arguments establish (d) for the other two types of elementary matrices. ❏

Practice Problems

1. Use elementary row operations to evaluate the determinant of

$$A = \begin{bmatrix} 1 & 3 & -3 \\ -3 & -9 & 2 \\ -4 & 4 & -6 \end{bmatrix}.$$

2. For what value of c is

$$B = \begin{bmatrix} 1 & -1 & 2 \\ -1 & 0 & c \\ 2 & 1 & 4 \end{bmatrix}$$

not invertible?

Exercises

1. Determine if the following statements are true or false.

(a) The determinant of a square matrix equals the product of its diagonal entries.

(b) Performing a row addition operation on a square matrix does not change its determinant.

(c) Performing a scaling operation on a square matrix does not change its determinant.

(d) Performing an interchange operation on a square matrix changes its determinant by a factor of -1.

(e) For any $n \times n$ matrices A and B, $\det (A + B) = \det A + \det B$.

(f) For any $n \times n$ matrices A and B, $\det AB = (\det A)(\det B)$.

(g) If A is any invertible matrix, then $\det A = 0$.

(h) For any square matrix A, $\det A^T = -\det A$.

(i) The determinant of any square matrix can be evaluated using a cofactor expansion along any column.

(j) The determinant of any square matrix equals the product of the diagonal entries of its reduced row echelon form.

In Exercises 2–4, evaluate the determinant of the given matrix using a cofactor expansion along the indicated column.

2.
$\begin{bmatrix} 1 & -2 & 2 \\ 2 & -1 & 3 \\ 0 & 1 & -1 \end{bmatrix}$
first column

3.
$\begin{bmatrix} 2 & -1 & 3 \\ 1 & 4 & -2 \\ -1 & 0 & 1 \end{bmatrix}$
second column

4.
$\begin{bmatrix} -1 & 2 & 1 \\ 5 & -9 & -2 \\ 3 & -1 & 2 \end{bmatrix}$
third column

In Exercises 5–18, evaluate the determinant of the given matrix using elementary row operations.

5.
$\begin{bmatrix} 0 & 0 & 5 \\ 0 & 3 & 7 \\ 4 & -1 & -2 \end{bmatrix}$

6.
$\begin{bmatrix} -6 & 0 & 0 \\ 2 & 9 & 4 \\ 7 & -3 & 0 \end{bmatrix}$

7.
$\begin{bmatrix} 1 & -2 & 2 \\ 0 & 5 & -1 \\ 2 & -4 & 1 \end{bmatrix}$

8.
$\begin{bmatrix} -2 & 1 & -2 \\ 4 & -2 & -1 \\ 0 & 3 & 6 \end{bmatrix}$

9.
$\begin{bmatrix} 3 & -2 & 1 \\ 0 & 0 & 5 \\ -9 & 4 & 2 \end{bmatrix}$

10.
$\begin{bmatrix} -2 & 6 & 1 \\ 0 & 0 & 3 \\ 4 & -1 & 2 \end{bmatrix}$

11.
$\begin{bmatrix} 1 & 4 & 2 \\ 2 & -1 & 3 \\ -1 & 3 & 1 \end{bmatrix}$

12.
$\begin{bmatrix} -1 & 2 & 1 \\ 5 & -9 & -2 \\ 3 & -1 & 2 \end{bmatrix}$

13.
$\begin{bmatrix} 1 & 2 & 1 \\ 1 & 1 & 2 \\ 3 & 4 & 8 \end{bmatrix}$

14.
$\begin{bmatrix} 3 & 4 & 2 \\ 2 & -1 & 3 \\ -1 & 3 & 1 \end{bmatrix}$

15.
$\begin{bmatrix} 1 & -1 & 2 & 1 \\ 2 & -1 & -1 & 4 \\ -4 & 5 & -10 & -6 \\ 3 & -2 & 10 & -1 \end{bmatrix}$

16.
$\begin{bmatrix} 2 & 1 & 5 & 2 \\ 2 & 1 & 8 & 1 \\ 2 & -1 & 5 & 3 \\ 4 & -2 & 10 & 3 \end{bmatrix}$

17.
$\begin{bmatrix} 0 & 4 & -1 & 1 \\ -3 & 1 & 1 & 2 \\ 1 & 0 & -2 & 3 \\ 2 & 3 & 0 & 1 \end{bmatrix}$

18.
$\begin{bmatrix} 1 & -1 & 2 & -1 \\ 2 & -2 & -3 & 8 \\ -3 & 4 & 1 & -1 \\ -2 & 6 & -4 & 18 \end{bmatrix}$

For each of the matrices in Exercises 19–28, determine the value(s) of c for which the given matrix is not invertible.

19.
$\begin{bmatrix} 4 & c \\ 3 & -6 \end{bmatrix}$

20.
$\begin{bmatrix} 3 & 9 \\ 5 & c \end{bmatrix}$

21.
$\begin{bmatrix} c & 6 \\ 2 & c+4 \end{bmatrix}$

22.
$\begin{bmatrix} c & c-1 \\ -8 & c-6 \end{bmatrix}$

23.
$\begin{bmatrix} 1 & 2 & -1 \\ 0 & -1 & c \\ 3 & 4 & 7 \end{bmatrix}$

24.
$\begin{bmatrix} 1 & 2 & -6 \\ 2 & 4 & c \\ -3 & -5 & 7 \end{bmatrix}$

25.
$\begin{bmatrix} 1 & -1 & 2 \\ -1 & 0 & 4 \\ 2 & 1 & c \end{bmatrix}$

26.
$\begin{bmatrix} 1 & 2 & c \\ -2 & -2 & 4 \\ 1 & 6 & -12 \end{bmatrix}$

27.
$\begin{bmatrix} 1 & 2 & -1 \\ 2 & 3 & c \\ 0 & c & -15 \end{bmatrix}$

28.
$\begin{bmatrix} -1 & 1 & 1 \\ 3 & -2 & -c \\ 0 & c & -10 \end{bmatrix}$

In Exercises 29–36, solve the given system using Cramer's rule.

29. $\begin{aligned} x_1 + 2x_2 &= 6 \\ 3x_1 + 4x_2 &= -3 \end{aligned}$

30. $\begin{aligned} 2x_1 + 3x_2 &= 7 \\ 3x_1 + 4x_2 &= 6 \end{aligned}$

31. $\begin{aligned} 2x_1 + 4x_2 &= -2 \\ 7x_1 + 12x_2 &= 5 \end{aligned}$

32. $\begin{aligned} 3x_1 + 2x_2 &= -6 \\ 6x_1 + 5x_2 &= 9 \end{aligned}$

33. $\begin{aligned} x_1 \quad\quad - 2x_3 &= 6 \\ -x_1 + x_2 + 3x_3 &= -5 \\ 2x_2 + x_3 &= 4 \end{aligned}$

34. $\begin{aligned} -x_1 + 2x_2 + x_3 &= -3 \\ x_2 + 2x_3 &= -1 \\ x_1 - x_2 + 3x_3 &= 4 \end{aligned}$

35. $\begin{aligned} 2x_1 - x_2 + x_3 &= -5 \\ x_1 \quad\quad - x_3 &= 2 \\ -x_1 + 3x_2 + 2x_3 &= 1 \end{aligned}$

36. $\begin{aligned} -2x_1 + 3x_2 + x_3 &= -2 \\ 3x_1 + x_2 - x_3 &= 1 \\ -x_1 + 2x_2 + x_3 &= -1 \end{aligned}$

37. Give an example to show that $\det kA \neq k \cdot \det A$ for some matrix A and scalar k.

38. Evaluate $\det kA$ if A is an $n \times n$ matrix and k is a scalar. Justify your answer.

39. Prove that if A is an invertible matrix, then

$$\det A^{-1} = \frac{1}{\det A}.$$

40. Under what conditions is det $(-A) = -\det A$? Justify your answer.

41. Let A and B be $n \times n$ matrices such that B is invertible. Prove that $\det(B^{-1}AB) = \det A$.

42. An $n \times n$ matrix A is called *nilpotent* if, for some positive integer k, $A^k = O$, where O is the $n \times n$ zero matrix. Prove that if A is nilpotent, then $\det A = 0$.

43. An $n \times n$ matrix Q is called *orthogonal* if $QQ^T = I_n$. Prove that if Q is orthogonal, then $\det Q = \pm 1$.

44. A square matrix A is called *skew-symmetric* if $A^T = -A$. Prove that if A is a skew-symmetric $n \times n$ matrix and n is odd, then A is not invertible. What if n is even?

45. The matrix

$$A = \begin{bmatrix} 1 & a & a^2 \\ 1 & b & b^2 \\ 1 & c & c^2 \end{bmatrix}$$

is called a *Vandermonde matrix*. Show that

$$\det A = (b - a)(c - a)(c - b).$$

46. Use properties of determinants to show that the equation of the line in $\mathcal{R}^2$ passing through the points (x_1, y_1) and (x_2, y_2) can be written

$$\det \begin{bmatrix} 1 & x_1 & y_1 \\ 1 & x_2 & y_2 \\ 1 & x & y \end{bmatrix} = 0.$$

47. Let $\mathcal{B} = \{\mathbf{b}_1, \mathbf{b}_2, \cdots, \mathbf{b}_n\}$ be a subset of $\mathcal{R}^n$ containing n distinct vectors, and let $B = [\mathbf{b}_1 \ \mathbf{b}_2 \ \cdots \ \mathbf{b}_n]$. Prove that $\mathcal{B}$ is linearly independent if and only if $\det B \neq 0$.

48. Let A be an $n \times n$ matrix with rows $\mathbf{a}_1', \mathbf{a}_2', \ldots, \mathbf{a}_n'$ and B be the $n \times n$ matrix with rows $\mathbf{a}_n', \ldots, \mathbf{a}_2', \mathbf{a}_1'$. How are the determinants of A and B related? Justify your answer.

49. Complete the proof of Theorem 3.3(b).

50. Complete the proof of Theorem 3.3(d).

51. Prove that $\det E^T = \det E$ for every elementary matrix E.

52. Let A be an $n \times n$ matrix and c_{jk} denote the (k, j)-cofactor of A.

(a) Prove that if B is the matrix obtained from A by replacing column k by $\mathbf{e}_j$, then $\det B = c_{kj}$.

(b) Show that for each j we have

$$A \begin{bmatrix} c_{1j} \\ c_{2j} \\ \vdots \\ c_{nj} \end{bmatrix} = (\det A) \cdot \mathbf{e}_j.$$

Hint: Apply Cramer's rule to $A\mathbf{x} = \mathbf{e}_j$.

(c) Deduce that if C is the $n \times n$ matrix whose (i, j)-entry is c_{ij}, then $AC = (\det A)I_n$. This matrix C is called the *classical adjoint* of A.

(d) Show that if $\det A \neq 0$, then $A^{-1} = \dfrac{1}{\det A} C$.

In Exercises 53–55, use a calculator with matrix capabilities or computer software such as MATLAB *to solve the problem.*

53. (a) Use elementary row operations other than scaling operations to transform

$$A = \begin{bmatrix} 0.0 & -3.0 & -2 & -5 \\ 2.4 & 3.0 & -6 & 9 \\ -4.8 & 6.3 & 4 & -2 \\ 9.6 & 1.5 & 5 & 9 \end{bmatrix}$$

into an upper triangular matrix U.

(b) Use the boxed result on page 187 to compute $\det A$.

54. (a) Solve $A\mathbf{x} = \mathbf{b}$ using Cramer's rule, where

$$A = \begin{bmatrix} 0 & 1 & 2 & -1 \\ 1 & 2 & 1 & -2 \\ 2 & -1 & 0 & 3 \\ 3 & 0 & -3 & 1 \end{bmatrix} \quad \text{and} \quad \mathbf{b} = \begin{bmatrix} 24 \\ -16 \\ 8 \\ 10 \end{bmatrix}.$$

(b) How many determinants of 4×4 matrices are evaluated in (a)?

55. Compute the classical adjoint (as defined in Exercise 52) of the matrix A in Exercise 54.

Solutions to the Practice Problems

1. The following sequence of elementary row operations transforms A into an upper triangular matrix U.

$$\begin{bmatrix} 1 & 3 & -3 \\ -3 & -9 & 2 \\ -4 & 4 & -6 \end{bmatrix} \longrightarrow \begin{bmatrix} 1 & 3 & -3 \\ 0 & 0 & -7 \\ 0 & 16 & -18 \end{bmatrix}$$

$$\longrightarrow \begin{bmatrix} 1 & 3 & -3 \\ 0 & 16 & -18 \\ 0 & 0 & -7 \end{bmatrix} = U$$

Since one row interchange operation was performed, we have

$$\det A = (-1)^1 \cdot \det U = (-1)(1)(16)(-7) = 112.$$

2. The determinant of B is $-3(c + 2)$. Since a matrix is invertible if and only if its determinant is nonzero, the only value for which B is not invertible is $c = -2$.

Chapter 3 Review Exercises

1. Determine if the following statements are true or false.

   (a) $\det \begin{bmatrix} a & b \\ c & d \end{bmatrix} = bc - ad$.

   (b) For $n \geq 2$, the (i, j)-cofactor of an $n \times n$ matrix A is the determinant of the $(n-1) \times (n-1)$ matrix obtained by deleting row j and column i from A.

   (c) If A is an $n \times n$ matrix and c_{ij} denotes the (i, j)-cofactor of A, then $\det A = a_{i1}c_{i1} + a_{i2}c_{i2} + \cdots + a_{in}c_{in}$ for any $i = 1, 2, \ldots, n$.

   (d) For all $n \times n$ matrices A and B, $\det(A + B) = \det A + \det B$.

   (e) For all $n \times n$ matrices A and B, $\det AB = (\det A)(\det B)$.

   (f) If B is obtained by interchanging two rows of an $n \times n$ matrix A, then $\det B = \det A$.

   (g) An $n \times n$ matrix is invertible if and only if its determinant is 0.

   (h) For any square matrix A, $\det A^T = \det A$.

   (i) For any square matrix A, $\det A^{-1} = -\det A$.

   (j) For any square matrix A and scalar c, $\det cA = c \det A$.

In Exercises 2–5, compute the indicated cofactor of the matrix

$$\begin{bmatrix} 1 & -1 & 2 \\ -1 & 2 & -1 \\ 2 & 1 & 3 \end{bmatrix}.$$

2. the $(1, 2)$-cofactor   3. the $(2, 1)$-cofactor
4. the $(2, 3)$-cofactor   5. the $(3, 1)$-cofactor

In Exercises 6–9, compute the determinant of the matrix in Exercises 2–5 using a cofactor expansion along the indicated row or column.

6. row 1   7. row 3   8. column 2
9. column 1

In Exercises 10–13, evaluate the determinant of the given matrix by any legitimate method.

10. $\begin{bmatrix} 5 & 6 \\ 3 & 2 \end{bmatrix}$ 11. $\begin{bmatrix} -5.0 & 3.0 \\ 3.5 & -2.1 \end{bmatrix}$

12. $\begin{bmatrix} 1 & -1 & 2 \\ 2 & -1 & 3 \\ 3 & -1 & 4 \end{bmatrix}$ 13. $\begin{bmatrix} 1 & -3 & 1 \\ 4 & -2 & 1 \\ 2 & 5 & -1 \end{bmatrix}$

14. (a) Perform a sequence of elementary row operations on

$$A = \begin{bmatrix} 0 & 3 & -6 & 1 \\ -2 & -2 & 2 & 6 \\ 1 & 1 & -1 & -1 \\ 2 & -1 & 2 & -2 \end{bmatrix}$$

   to transform it into an upper triangular matrix.

   (b) Use your answer to (a) to compute $\det A$.

In Exercises 15–18, use a determinant to find all values of the scalar c for which the given matrix is not invertible.

15. $\begin{bmatrix} c - 17 & -13 \\ 20 & c + 16 \end{bmatrix}$ 16. $\begin{bmatrix} 1 & c + 1 \\ 2 & 3c + 4 \end{bmatrix}$

17. $\begin{bmatrix} c + 4 & -1 & c + 5 \\ -3 & 3 & -4 \\ c + 6 & -3 & c + 7 \end{bmatrix}$

18. $\begin{bmatrix} -1 & c - 1 & 1 - c \\ -c - 2 & 2c - 3 & 4 - c \\ -c - 2 & c - 1 & 2 \end{bmatrix}$

19. Compute the area of the parallelogram in $\mathcal{R}^2$ determined by the vectors $\begin{bmatrix} 3 \\ 7 \end{bmatrix}$ and $\begin{bmatrix} 4 \\ 1 \end{bmatrix}$.

20. Compute the volume of the parallelepiped in $\mathcal{R}^3$ determined by the vectors $\begin{bmatrix} 1 \\ 0 \\ 2 \end{bmatrix}$, $\begin{bmatrix} -1 \\ 2 \\ 1 \end{bmatrix}$, and $\begin{bmatrix} 3 \\ 1 \\ -1 \end{bmatrix}$.

In Exercises 21–22, solve the given system of linear equations using Cramer's rule.

21. $2x_1 + x_2 = 5$
    $-4x_1 + 3x_2 = -6$

22. $x_1 - x_2 + 2x_3 = 7$
    $-x_1 + 2x_2 - x_3 = -3$
    $2x_1 + x_2 + 2x_3 = 4$

Let A be a 3×3 matrix such that $\det A = 5$. Evaluate the determinant of the matrix given in Exercises 23–30.

23. A^T 24. A^{-1} 25. $2A$ 26. A^3

27. $\begin{bmatrix} a_{11} - 3a_{21} & a_{12} - 3a_{22} & a_{13} - 3a_{23} \\ a_{21} & a_{22} & a_{23} \\ a_{31} & a_{32} & a_{33} \end{bmatrix}$

28. $\begin{bmatrix} a_{11} & a_{12} & a_{13} \\ -2a_{21} & -2a_{22} & -2a_{23} \\ a_{31} & a_{32} & a_{33} \end{bmatrix}$

29. $\begin{bmatrix} a_{11} + 5a_{31} & a_{12} + 5a_{32} & a_{13} + 5a_{33} \\ 4a_{21} & 4a_{22} & 4a_{23} \\ a_{31} - 2a_{21} & a_{32} - 2a_{22} & a_{33} - 2a_{23} \end{bmatrix}$

30. $\begin{bmatrix} a_{31} & a_{32} & a_{33} \\ a_{21} & a_{22} & a_{23} \\ a_{11} & a_{12} & a_{13} \end{bmatrix}$

31. A square matrix B is called *idempotent* if $B^2 = B$. What can be said about the determinant of an idempotent matrix?

32. Suppose that an $n \times n$ matrix can be expressed in the form $A = PDP^{-1}$, where P is an invertible matrix and D is a diagonal matrix. Prove that the determinant of A equals the product of the diagonal entries of D.

33. Show that the equation

$$\det \begin{bmatrix} 1 & x & y \\ 1 & x_1 & y_1 \\ 0 & 1 & m \end{bmatrix} = 0$$

yields the equation of the line through the point (x_1, y_1) with slope m.

CHAPTER 4

Subspaces and Their Properties

Many of the sets that we have encountered so far possess two *closure* properties: The sum of any pair of vectors in the set lies in the set, and every scalar multiple of a vector in the set lies in the set. Among the sets with these properties are the solution set of a homogeneous system of linear equations, the span of a finite set of vectors, and both the null space and range of a linear transformation. Such sets are called *subspaces* and are studied in detail in this chapter.

The techniques learned in Chapter 1 enable us to find spanning sets for particular subspaces. A spanning set of the smallest size is especially useful in analyzing a subspace. Such smallest spanning sets are studied in Section 4.2. We see there that the size of a smallest spanning set is an intrinsic property of a subspace. This number is not only the size of a smallest spanning set but also the size of a largest linearly independent subset.

Moreover, a smallest spanning set for $\mathcal{R}^n$ provides a means for establishing a coordinate system on $\mathcal{R}^n$. In Sections 4.4 and 4.5, we investigate the process of establishing a coordinate system on $\mathcal{R}^n$ and describe several situations where it is advantageous to use a coordinate system other than the usual one.

4.1   Subspaces

When two vectors in $\mathcal{R}^n$ are added or a vector in $\mathcal{R}^n$ is multiplied by a scalar, the resulting vectors are also in $\mathcal{R}^n$. In other words, $\mathcal{R}^n$ is *closed* under the operations of vector addition and scalar multiplication. This fact underlies many of the calculations with vectors and matrices that we have done so far. For example, both the solution set of a homogeneous system of linear equations and the span of a finite, nonempty subset of $\mathcal{R}^n$ share these properties. In this section, we study the subsets of $\mathcal{R}^n$ that possess this type of closure.

Definition.   A set W of vectors in $\mathcal{R}^n$ is called a **subspace** of $\mathcal{R}^n$ if it has the following three properties.

1. The zero vector belongs to W.

2. Whenever $\mathbf{u}$ and $\mathbf{v}$ belong to W, then $\mathbf{u} + \mathbf{v}$ belongs to W. (In this case, we say that W is **closed under (vector) addition**.)

3. Whenever **u** belongs to W and c is a scalar, then $c\mathbf{u}$ belongs to W. (In this case, we say that W is **closed under scalar multiplication**.)

Example 1   Determine which properties of a subspace are possessed by the sets

$$S_1 = \left\{ \begin{bmatrix} w_1 \\ w_2 \end{bmatrix} \in \mathcal{R}^2 \colon w_1 \geq 0 \text{ and } w_2 \geq 0 \right\}$$

and

$$S_2 = \left\{ \begin{bmatrix} w_1 \\ w_2 \end{bmatrix} \in \mathcal{R}^2 \colon w_1^2 = w_2^2 \right\}.$$

Solution   The vectors in S_1 are those that lie in the first quadrant of $\mathcal{R}^2$ including the nonnegative parts of the x- and y-axes (see Figure 4.1(a)). Clearly $\mathbf{0} = \begin{bmatrix} 0 \\ 0 \end{bmatrix}$ is in S_1. Suppose that $\mathbf{u} = \begin{bmatrix} u_1 \\ u_2 \end{bmatrix}$ and $\mathbf{v} = \begin{bmatrix} v_1 \\ v_2 \end{bmatrix}$ are in S_1. Then $u_1 \geq 0$, $u_2 \geq 0$, $v_1 \geq 0$, and $v_2 \geq 0$. Hence $u_1 + v_1 \geq 0$ and $u_2 + v_2 \geq 0$, so that

$$\mathbf{u} + \mathbf{v} = \begin{bmatrix} u_1 + v_1 \\ u_2 + v_2 \end{bmatrix}$$

is in S_1. Thus S_1 is closed under vector addition. This conclusion can also be seen by using the parallelogram law. However, S_1 is *not* closed under scalar multiplication because $\mathbf{u} = \begin{bmatrix} 1 \\ 2 \end{bmatrix}$ belongs to S_1, but $(-1)\mathbf{u} = \begin{bmatrix} -1 \\ -2 \end{bmatrix}$ does not.

Consider a vector $\mathbf{u} = \begin{bmatrix} u_1 \\ u_2 \end{bmatrix}$ in S_2. Since $u_1^2 = u_2^2$, we have $u_2 = \pm u_1$. Hence the vector $\mathbf{u}$ lies along one of the lines $y = x$ or $y = -x$ (see Fig-

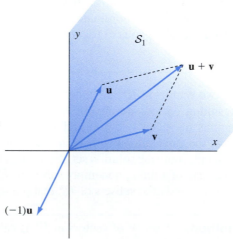

S_1 is closed under vector addition
but not under scalar multiplication

(a)

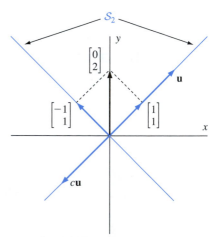

S_2 is closed under scalar multiplication but not under vector addition

(b)

FIGURE 4.1

ure 4.1(b)). Clearly $\mathbf{0} = \begin{bmatrix} 0 \\ 0 \end{bmatrix}$ belongs to $\mathcal{S}_2$. Moreover, if $\mathbf{u} = \begin{bmatrix} u_1 \\ u_2 \end{bmatrix}$ is in $\mathcal{S}_2$, then $u_1^2 = u_2^2$. So, for any scalar c, $c\mathbf{u} = \begin{bmatrix} cu_1 \\ cu_2 \end{bmatrix}$ is in $\mathcal{S}_2$ because $(cu_1)^2 = c^2 u_1^2 = c^2 u_2^2 = (cu_2)^2$. However, $\mathcal{S}_2$ is *not* closed under vector addition. For example, the vectors $\begin{bmatrix} 1 \\ 1 \end{bmatrix}$ and $\begin{bmatrix} -1 \\ 1 \end{bmatrix}$ belong to $\mathcal{S}_2$, but $\begin{bmatrix} 1 \\ 1 \end{bmatrix} + \begin{bmatrix} -1 \\ 1 \end{bmatrix} = \begin{bmatrix} 0 \\ 2 \end{bmatrix}$ does not.

The next two examples describe two special subspaces of $\mathcal{R}^n$.

Example 2 The set $\mathcal{R}^n$ is a subspace of itself because the zero vector belongs to $\mathcal{R}^n$, the sum of any two vectors in $\mathcal{R}^n$ is also in $\mathcal{R}^n$, and every scalar multiple of a vector in $\mathcal{R}^n$ belongs to $\mathcal{R}^n$.

Example 3 The set W consisting of only the zero vector in $\mathcal{R}^n$ is a subspace of $\mathcal{R}^n$ called the **zero subspace**. Clearly $\mathbf{0}$ is in W. Moreover, if $\mathbf{u}$ and $\mathbf{v}$ are vectors in W, then $\mathbf{u} = \mathbf{0}$ and $\mathbf{v} = \mathbf{0}$, so that $\mathbf{u} + \mathbf{v} = \mathbf{0} + \mathbf{0} = \mathbf{0}$. Hence $\mathbf{u} + \mathbf{v}$ is in W, so that W is closed under vector addition. Finally, if $\mathbf{u}$ is in W and c is a scalar, then $c\mathbf{u} = c\mathbf{0} = \mathbf{0}$, so that $c\mathbf{u}$ is in W. Hence W is also closed under scalar multiplication.

A subspace of $\mathcal{R}^n$ other than $\{\mathbf{0}\}$ is called a **nonzero subspace**. Examples 4 and 5 show how to verify that two nonzero subspaces of $\mathcal{R}^3$ satisfy the three properties in the definition of a subspace.

Example 4 The set $W = \left\{ \begin{bmatrix} w_1 \\ w_2 \\ w_3 \end{bmatrix} \in \mathcal{R}^3 \colon 6w_1 - 5w_2 + 4w_3 = 0 \right\}$ is a subspace of $\mathcal{R}^3$, as shown below.

1. Since $6(0) - 5(0) + 4(0) = 0$, the components of $\mathbf{0}$ satisfy the equation that defines W. Hence $\mathbf{0} = \begin{bmatrix} 0 \\ 0 \\ 0 \end{bmatrix}$ is in W.

2. Let $\mathbf{u} = \begin{bmatrix} u_1 \\ u_2 \\ u_3 \end{bmatrix}$ and $\mathbf{v} = \begin{bmatrix} v_1 \\ v_2 \\ v_3 \end{bmatrix}$ be in W. Then $6u_1 - 5u_2 + 4u_3 = 0$ and also $6v_1 - 5v_2 + 4v_3 = 0$. Now $\mathbf{u} + \mathbf{v} = \begin{bmatrix} u_1 + v_1 \\ u_2 + v_2 \\ u_3 + v_3 \end{bmatrix}$. Since

$$6(u_1 + v_1) - 5(u_2 + v_2) + 4(u_3 + v_3) = (6u_1 - 5u_2 + 4u_3)$$
$$+ (6v_1 - 5v_2 + 4v_3)$$
$$= 0 + 0$$
$$= 0,$$

we see that the components of $\mathbf{u} + \mathbf{v}$ satisfy the equation defining W. Therefore $\mathbf{u} + \mathbf{v}$ is in W, and so W is closed under vector addition.

3. Let $\mathbf{u} = \begin{bmatrix} u_1 \\ u_2 \\ u_3 \end{bmatrix}$ be in W. For any scalar c, we have $c\mathbf{u} = c \begin{bmatrix} u_1 \\ u_2 \\ u_3 \end{bmatrix}$

$= \begin{bmatrix} cu_1 \\ cu_2 \\ cu_3 \end{bmatrix}$. Because

$$6(cu_1) - 5(cu_2) + 4(cu_3) = c(6u_1 - 5u_2 + 4u_3) = c(0) = 0,$$

the components of $c\mathbf{u}$ satisfy the equation defining W. Therefore $c\mathbf{u}$ is in W, and so W is also closed under scalar multiplication.

Since W is a subset of $\mathcal{R}^3$ that contains the zero vector and is closed under both vector addition and scalar multiplication, W is a subspace of $\mathcal{R}^3$. Geometrically, W is a plane through the origin of $\mathcal{R}^3$.   ○

Example 5   Show that the set W of all multiples of a fixed nonzero vector $\mathbf{w}$ in $\mathcal{R}^3$ is a subspace of $\mathcal{R}^3$.

Solution   First, $\mathbf{0} = 0\mathbf{w}$ is in W. Next, let $\mathbf{u}$ and $\mathbf{v}$ be vectors in W. Then $\mathbf{u} = a\mathbf{w}$ and $\mathbf{v} = b\mathbf{w}$ for some scalars a and b. Since

$$\mathbf{u} + \mathbf{v} = a\mathbf{w} + b\mathbf{w} = (a + b)\mathbf{w},$$

we see that $\mathbf{u} + \mathbf{v}$ is a multiple of $\mathbf{w}$. Hence $\mathbf{u} + \mathbf{v}$ is in W, and so W is closed under vector addition. Finally, for any scalar c, $c\mathbf{u} = c(a\mathbf{w}) = (ca)\mathbf{w}$ is a multiple of $\mathbf{w}$. Thus $c\mathbf{u}$ is in W, so that W is also closed under scalar multiplication. Therefore W is a subspace of $\mathcal{R}^3$. Note that W can be depicted as a line in $\mathcal{R}^3$ through the origin (see Figure 4.2).   ○

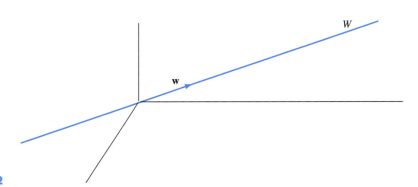

FIGURE 4.2

Example 5 shows that the set of vectors in $\mathcal{R}^3$ along a line through the origin is a subspace of $\mathcal{R}^3$. However, the set of vectors on a line in $\mathcal{R}^3$ that does not pass through the origin is not a subspace, for such a set does not contain the zero vector of $\mathcal{R}^3$.

Our first theorem generalizes Example 5.

Theorem 4.1

The span of a finite nonempty subset of $\mathcal{R}^n$ is a subspace of $\mathcal{R}^n$.

Proof Let $\mathcal{S} = \{\mathbf{w}_1, \mathbf{w}_2, \ldots, \mathbf{w}_k\}$. Since

$$0\mathbf{w}_1 + 0\mathbf{w}_2 + \cdots + 0\mathbf{w}_k = \mathbf{0},$$

we see that $\mathbf{0}$ belongs to the span of $\mathcal{S}$. Let $\mathbf{u}$ and $\mathbf{v}$ belong to the span of $\mathcal{S}$. Then

$$\mathbf{u} = a_1\mathbf{w}_1 + a_2\mathbf{w}_2 + \cdots + a_k\mathbf{w}_k \qquad \text{and} \qquad \mathbf{v} = b_1\mathbf{w}_1 + b_2\mathbf{w}_2 + \cdots + b_k\mathbf{w}_k$$

for some scalars $a_1, a_2, \ldots, a_k$ and $b_1, b_2, \ldots, b_k$. Since

$$\mathbf{u} + \mathbf{v} = (a_1\mathbf{w}_1 + a_2\mathbf{w}_2 + \cdots + a_k\mathbf{w}_k) + (b_1\mathbf{w}_1 + b_2\mathbf{w}_2 + \cdots + b_k\mathbf{w}_k)$$
$$= (a_1 + b_1)\mathbf{w}_1 + (a_2 + b_2)\mathbf{w}_2 + \cdots + (a_k + b_k)\mathbf{w}_k,$$

we see that $\mathbf{u} + \mathbf{v}$ belongs to the span of $\mathcal{S}$. Hence the span of $\mathcal{S}$ is closed under vector addition. Furthermore, for any scalar c,

$$c\mathbf{u} = c(a_1\mathbf{w}_1 + a_2\mathbf{w}_2 + \cdots + a_k\mathbf{w}_k)$$
$$= (c_1a_1)\mathbf{w}_1 + (c_2a_2)\mathbf{w}_2 + \cdots + (c_ka_k)\mathbf{w}_k,$$

and so $c\mathbf{u}$ belongs to the span of $\mathcal{S}$. Thus the span of $\mathcal{S}$ is also closed under scalar multiplication, and so the span of $\mathcal{S}$ is a subspace of $\mathcal{R}^n$. $\qquad \square$

Example 6

We can use Theorem 4.1 to show that the set of vectors of the form

$$W = \left\{ \begin{bmatrix} 2a - 3b \\ b \\ -a + 4b \end{bmatrix} \in \mathcal{R}^3 \colon a \in \mathcal{R} \text{ and } b \in \mathcal{R} \right\}$$

is a subspace of $\mathcal{R}^3$. Simply observe that

$$\begin{bmatrix} 2a - 3b \\ b \\ -a + 4b \end{bmatrix} = a \begin{bmatrix} 2 \\ 0 \\ -1 \end{bmatrix} + b \begin{bmatrix} -3 \\ 1 \\ 4 \end{bmatrix},$$

and so

$$W = \text{Span} \left\{ \begin{bmatrix} 2 \\ 0 \\ -1 \end{bmatrix}, \begin{bmatrix} -3 \\ 1 \\ 4 \end{bmatrix} \right\}.$$

Therefore W is a subspace of $\mathcal{R}^3$ by Theorem 4.1. $\qquad \circ$

Subspaces Associated with a Matrix

Several important subspaces are associated with a matrix. The first one that we consider is the *null space*.

Definition. The **null space** of a matrix A is the solution set of $A\mathbf{x} = \mathbf{0}$. It is denoted Null A.

For an $m \times n$ matrix A, the null space of A is the set

$$\text{Null } A = \{\mathbf{v} \in \mathcal{R}^n \colon A\mathbf{v} = \mathbf{0}\}.$$

For example, the null space of the matrix

$$\begin{bmatrix} 1 & -5 & 3 \\ 2 & -9 & -6 \end{bmatrix}$$

equals the solution set of the homogeneous system of linear equations

$$x_1 - 5x_2 + 3x_3 = 0$$
$$2x_1 - 9x_2 - 6x_3 = 0.$$

More generally, the solution set of any homogeneous system of linear equations equals the null space of the coefficient matrix of that system.

The set W in Example 4 is such a solution set. (In this case, it is the solution set of a single equation in 3 variables, $6x_1 - 5x_2 + 4x_3 = 0$.) We saw in Example 4 that W is a subspace of $\mathcal{R}^3$. Such sets are always subspaces, as the next theorem shows.

Theorem 4.2

If A is an $m \times n$ matrix, then Null A *is a subspace of* $\mathcal{R}^n$.

Proof Since A is an $m \times n$ matrix, the elements of Null A, which are the solutions to $A\mathbf{x} = \mathbf{0}$, belong to $\mathcal{R}^n$. Clearly $\mathbf{0}$ is in Null A because $A\mathbf{0} = \mathbf{0}$. Suppose that $\mathbf{u}$ and $\mathbf{v}$ belong to Null A. Then $A\mathbf{u} = \mathbf{0}$ and $A\mathbf{v} = \mathbf{0}$. Hence, by Theorem 1.2(b), we have

$$A(\mathbf{u} + \mathbf{v}) = A\mathbf{u} + A\mathbf{v} = \mathbf{0} + \mathbf{0} = \mathbf{0}.$$

Thus argument proves that $\mathbf{u} + \mathbf{v}$ is in Null A, and so Null A is closed under vector addition. Moreover, for any scalar c, we have by Theorem 1.2(c) that

$$A(c\mathbf{u}) = c(A\mathbf{u}) = c\mathbf{0} = \mathbf{0}.$$

Thus $c\mathbf{u}$ is in Null A, and so Null A is also closed under scalar multiplication. Therefore Null A is a subspace of $\mathcal{R}^n$.   ◻

Another important subspace associated with a matrix is its *column space*.

Definition.   The **column space** of a matrix A is the span of its columns. It is denoted Col A.

For example, if

$$A = \begin{bmatrix} 1 & -5 & 3 \\ 2 & -9 & -6 \end{bmatrix},$$

then

$$\text{Col } A = \text{Span} \left\{ \begin{bmatrix} 1 \\ 2 \end{bmatrix}, \begin{bmatrix} -5 \\ -9 \end{bmatrix}, \begin{bmatrix} 3 \\ -6 \end{bmatrix} \right\}.$$

Recall from Section 1.6 that a vector $\mathbf{b}$ is a linear combination of the columns of A if and only if the matrix equation $A\mathbf{x} = \mathbf{b}$ is consistent. Hence

$$\text{Col } A = \{A\mathbf{v} : \mathbf{v} \in \mathcal{R}^n\}.$$

It follows from Theorem 4.1 that the column space of an $m \times n$ matrix is a subspace of $\mathcal{R}^m$. Since the null space of A is a subspace of $\mathcal{R}^n$, the column space and null space of an $m \times n$ matrix are contained in different spaces if $m \neq n$. Even if $m = n$, however, these two subspaces are rarely equal.

Example 7 Find a spanning set for the column space of the matrix

$$A = \begin{bmatrix} 1 & 2 & 1 & -1 \\ 2 & 4 & 0 & -8 \\ 0 & 0 & 2 & 6 \end{bmatrix}.$$

Is $\mathbf{u} = \begin{bmatrix} 2 \\ 1 \\ 1 \end{bmatrix}$ in Col A? Is $\mathbf{v} = \begin{bmatrix} 2 \\ 1 \\ 3 \end{bmatrix}$ in Col A?

Solution The column space of A is the span of the columns of A. Hence one spanning set for Col A is

$$\left\{ \begin{bmatrix} 1 \\ 2 \\ 0 \end{bmatrix}, \begin{bmatrix} 2 \\ 4 \\ 0 \end{bmatrix}, \begin{bmatrix} 1 \\ 0 \\ 2 \end{bmatrix}, \begin{bmatrix} -1 \\ -8 \\ 6 \end{bmatrix} \right\}.$$

To see if the vector $\mathbf{u}$ lies in the column space of A, we must determine if $A\mathbf{x} = \mathbf{u}$ is consistent. Since the reduced row echelon form of $[A \quad \mathbf{u}]$ is

$$\begin{bmatrix} 1 & 2 & 0 & -4 & 0 \\ 0 & 0 & 1 & 3 & 0 \\ 0 & 0 & 0 & 0 & 1 \end{bmatrix},$$

we see that the system is inconsistent, and hence $\mathbf{u}$ is not in Col A. On the other hand, the reduced row echelon form of $[A \quad \mathbf{v}]$ is

$$\begin{bmatrix} 1 & 2 & 0 & -4 & 0.5 \\ 0 & 0 & 1 & 3 & 1.5 \\ 0 & 0 & 0 & 0 & 0 \end{bmatrix}.$$

Thus the system $A\mathbf{x} = \mathbf{v}$ is consistent, and so $\mathbf{v}$ is in Col A. ○

Example 8 Find a spanning set for the null space of the matrix A in Example 7. Is

$$\mathbf{u} = \begin{bmatrix} 2 \\ -3 \\ 3 \\ -1 \end{bmatrix} \text{ in Null } A? \text{ Is } \mathbf{v} = \begin{bmatrix} 5 \\ -3 \\ 2 \\ 1 \end{bmatrix} \text{ in Null } A?$$

Solution Unlike the calculation of a spanning set for the column space of A in Example 7, there is no easy way to obtain a spanning set for the null space of A directly from the entries of A. To describe these vectors, we must solve $A\mathbf{x} = \mathbf{0}$. Because the reduced row echelon form of A is

$$\begin{bmatrix} 1 & 2 & 0 & -4 \\ 0 & 0 & 1 & 3 \\ 0 & 0 & 0 & 0 \end{bmatrix},$$

the parametric form of the general solution to $A\mathbf{x} = \mathbf{0}$ is

$$\begin{bmatrix} x_1 \\ x_2 \\ x_3 \\ x_4 \end{bmatrix} = \begin{bmatrix} -2x_2 + 4x_4 \\ x_2 \\ -3x_4 \\ x_4 \end{bmatrix} = x_2 \begin{bmatrix} -2 \\ 1 \\ 0 \\ 0 \end{bmatrix} + x_4 \begin{bmatrix} 4 \\ 0 \\ -3 \\ 1 \end{bmatrix}.$$

It follows that

$$\text{Null } A = \text{Span} \left\{ \begin{bmatrix} -2 \\ 1 \\ 0 \\ 0 \end{bmatrix}, \begin{bmatrix} 4 \\ 0 \\ -3 \\ 1 \end{bmatrix} \right\}.$$

To see if the vector $\mathbf{u}$ lies in the null space of A, we must determine if $A\mathbf{u} = \mathbf{0}$. An easy calculation confirms this; so $\mathbf{u}$ belongs to Null A. On the other hand,

$$A\mathbf{v} = \begin{bmatrix} 1 & 2 & 1 & -1 \\ 2 & 4 & 0 & -8 \\ 0 & 0 & 2 & 6 \end{bmatrix} \begin{bmatrix} 5 \\ -3 \\ 2 \\ 1 \end{bmatrix} = \begin{bmatrix} 0 \\ -10 \\ 10 \end{bmatrix}.$$

Since $A\mathbf{v} \neq \mathbf{0}$, we see that $\mathbf{v}$ is not in Null A. ○

A subspace of $\mathcal{R}^n$ is usually defined as the span of a given set of vectors or as the solution set of a homogeneous system of linear equations. As Examples 7 and 8 illustrate, when a subspace is defined by giving a spanning set, there is no work involved in obtaining a spanning set, but a system of linear equations must be solved to check if a particular vector belongs to the subspace. On the other hand, if a subspace is defined as the solution set of a homogeneous system of linear equations, then a system of linear equations must be solved in order to find a spanning set, but a particular vector can be easily checked for membership in the subspace (by verifying that its components satisfy the linear system defining the subspace).

Like the column space of a matrix, we define the **row space** of a matrix to be the subspace spanned by its rows. The row space of a matrix A is denoted Row A. So for the matrix

$$\begin{bmatrix} 1 & 2 & 1 & -1 \\ 2 & 4 & 0 & -8 \\ 0 & 0 & 2 & 6 \end{bmatrix}$$

in Example 7, we have

$$\text{Row } A = \text{Span} \{[1 \quad 2 \quad 1 \quad -1], [2 \quad 4 \quad 0 \quad -8], [0 \quad 0 \quad 2 \quad 6]\}.$$

Recall that, for any matrix A, the rows of A are the columns of A^T. Hence Row $A = \text{Col } A^T$, and so the row space of an $m \times n$ matrix is a subspace of $\mathcal{R}^n$. In general, the three subspaces Null A, Col A, and Row A are all distinct.

Subspaces Associated with a Linear Transformation

In Section 2.7, we saw that the range of a linear transformation is the span of the columns of its standard matrix. We have just defined the span of the columns of a matrix to be its column space. Thus we can reformulate this statement from Section 2.7 as follows.

> The range of a linear transformation is the same as the column space of its standard matrix.

As a consequence of this result, we see that the range of a linear transformation $T: \mathcal{R}^n \to \mathcal{R}^m$ is a subspace of $\mathcal{R}^m$.

Example 9   Find a spanning set for the range of the linear transformation $T: \mathcal{R}^4 \to \mathcal{R}^3$ defined by

$$T\left(\begin{bmatrix} x_1 \\ x_2 \\ x_3 \\ x_4 \end{bmatrix}\right) = \begin{bmatrix} x_1 + 2x_2 + x_3 - x_4 \\ 2x_1 + 4x_2 - 8x_4 \\ 2x_3 + 6x_4 \end{bmatrix}.$$

Solution   The standard matrix of T is

$$A = \begin{bmatrix} 1 & 2 & 1 & -1 \\ 2 & 4 & 0 & -8 \\ 0 & 0 & 2 & 6 \end{bmatrix}.$$

Since the range of T is the same as the column space of A, a spanning set for the range of T is

$$\left\{ \begin{bmatrix} 1 \\ 2 \\ 0 \end{bmatrix}, \begin{bmatrix} 2 \\ 4 \\ 0 \end{bmatrix}, \begin{bmatrix} 1 \\ 0 \\ 2 \end{bmatrix}, \begin{bmatrix} -1 \\ -8 \\ -6 \end{bmatrix} \right\},$$

the set of columns of A.

We also learned in Section 2.7 that the null space of a linear transformation is the solution set of $A\mathbf{x} = \mathbf{0}$, where A is the standard matrix of T. Using the terminology of this section, we can reformulate this result as follows.

> The null space of a linear transformation is the same as the null space of its standard matrix.

Note that this result implies that the null space of a linear transformation $T: \mathcal{R}^n \to \mathcal{R}^m$ is a subspace of $\mathcal{R}^n$.

Example 10   Determine a spanning set for the null space of the linear transformation in Example 9.

Solution   The standard matrix of T is given in Example 9. Its reduced row echelon form is the matrix

$$\begin{bmatrix} 1 & 2 & 0 & -4 \\ 0 & 0 & 1 & 3 \\ 0 & 0 & 0 & 0 \end{bmatrix}$$

in Example 8. From the latter example, we see that

$$\text{Null } A = \text{Span} \left\{ \begin{bmatrix} -2 \\ 1 \\ 0 \\ 0 \end{bmatrix}, \begin{bmatrix} 4 \\ 0 \\ -3 \\ 1 \end{bmatrix} \right\}.$$

Practice Problems

1. Show that

$$V = \left\{ \begin{bmatrix} -s \\ 2t \\ 3s - t \end{bmatrix} \in \mathcal{R}^3 : s, t \in \mathcal{R} \right\}$$

is a subspace of $\mathcal{R}^3$ by finding a spanning set for V.

2. Find a spanning set for the column space and null space of

$$A = \begin{bmatrix} 1 & 2 & -1 \\ -1 & -3 & 4 \end{bmatrix}.$$

3. Find a spanning set for the null space and range of the linear transformation $T: \mathcal{R}^4 \to \mathcal{R}^3$ defined by

$$T\left(\begin{bmatrix} x_1 \\ x_2 \\ x_3 \\ x_4 \end{bmatrix} \right) = \begin{bmatrix} x_1 + x_3 + 2x_4 \\ -x_2 + x_3 + x_4 \\ 2x_1 + 3x_2 - x_3 + x_4 \end{bmatrix}.$$

Exercises

1. Determine if the following statements are true or false.

   (a) If V is a subspace of $\mathcal{R}^n$ and $\mathbf{v}$ is in V, then $c\mathbf{v}$ is in V for every scalar c.
   (b) The empty set is a subspace of $\mathcal{R}^n$.
   (c) The subspace $\{\mathbf{0}\}$ is called the null space.
   (d) The span of a finite nonempty subset of $\mathcal{R}^n$ is a subspace of $\mathcal{R}^n$.
   (e) The null space of an $m \times n$ matrix is contained in $\mathcal{R}^n$.
   (f) The column space of an $m \times n$ matrix is contained in $\mathcal{R}^n$.
   (g) The row space of an $m \times n$ matrix is contained in $\mathcal{R}^m$.
   (h) The row space of an $m \times n$ matrix is the set $\{A\mathbf{v} : \mathbf{v} \in R^n\}$.
   (i) For any matrix A, the row space of A^T equals the column space of A.
   (j) The null space of every linear transformation is a subspace.
   (k) The range of a linear transformation need not be a subspace.

In Exercises 2–10, find a spanning set for the given subspace.

2. $\left\{ \begin{bmatrix} 2s \\ -3s \end{bmatrix} \in \mathcal{R}^2 : s \in \mathcal{R} \right\}$ 3. $\left\{ \begin{bmatrix} 4s \\ -s \end{bmatrix} \in \mathcal{R}^2 : s \in \mathcal{R} \right\}$

4. $\left\{ \begin{bmatrix} 4t \\ s + t \\ -3s + t \end{bmatrix} \in \mathcal{R}^3 : s, t \in \mathcal{R} \right\}$

5. $\left\{ \begin{bmatrix} -s + t \\ 2s - t \\ s + 3t \end{bmatrix} \in \mathcal{R}^3 : s, t \in \mathcal{R} \right\}$

6. $\left\{ \begin{bmatrix} -r + 3s \\ 0 \\ s - t \\ r - 2t \end{bmatrix} \in \mathcal{R}^4 : r, s, t \in \mathcal{R} \right\}$

7. $\left\{ \begin{bmatrix} -r + s \\ 4s - 3t \\ 0 \\ 3r - t \end{bmatrix} \in \mathcal{R}^4 : r, s, t \in \mathcal{R} \right\}$

8. $\left\{ \begin{bmatrix} r - s + 3t \\ 2r - t \\ -r + 3s + 2t \\ -2r + s + t \end{bmatrix} \in \mathcal{R}^4 : r, s, t \in \mathcal{R} \right\}$

9. $\left\{ \begin{bmatrix} 2s - 5t \\ 3r + s - 2t \\ r - 4s + 3t \\ -r + 2s \end{bmatrix} \in \mathcal{R}^4 : r, s, t \in \mathcal{R} \right\}$

10. $\left\{ \begin{bmatrix} -r + 4t \\ r - s + 2t \\ 3t \\ r - t \end{bmatrix} \in \mathcal{R}^4 : r, s, t \in \mathcal{R} \right\}$

In Exercises 11–18, show that the given set is not a subspace of the appropriate $\mathcal{R}^n$.

11. $\left\{ \begin{bmatrix} u_1 \\ u_2 \end{bmatrix} \in \mathcal{R}^2 : u_1 u_2 = 0 \right\}$

12. $\left\{ \begin{bmatrix} u_1 \\ u_2 \end{bmatrix} \in \mathcal{R}^2 : 2u_1^2 + 3u_2^2 = 12 \right\}$

13. $\left\{ \begin{bmatrix} 3s - 2 \\ 2s + 4t \\ -t \end{bmatrix} \in \mathcal{R}^3 : s, t \in \mathcal{R} \right\}$

14. $\left\{ \begin{bmatrix} u_1 \\ u_2 \end{bmatrix} \in \mathcal{R}^2 : u_1^2 + u_2^2 \le 1 \right\}$

15. $\left\{ \begin{bmatrix} u_1 \\ u_2 \\ u_3 \end{bmatrix} \in \mathcal{R}^3 : u_1 > u_2 \text{ and } u_3 < 0 \right\}$

16. $\left\{ \begin{bmatrix} u_1 \\ u_2 \\ u_3 \end{bmatrix} \in \mathcal{R}^3 : u_1 \ge u_2 \ge u_3 \right\}$

17. $\left\{ \begin{bmatrix} u_1 \\ u_2 \\ u_3 \end{bmatrix} \in \mathcal{R}^3 \colon u_1 = u_2 u_3 \right\}$

18. $\left\{ \begin{bmatrix} u_1 \\ u_2 \\ u_3 \end{bmatrix} \in \mathcal{R}^3 \colon u_1 u_2 = u_3^2 \right\}$

In Exercises 19–22, determine if the given vector belongs to Null A, where

$$A = \begin{bmatrix} 1 & -2 & -1 & 0 \\ 0 & 1 & 3 & -2 \\ -2 & 3 & -1 & 2 \end{bmatrix}.$$

19. $\begin{bmatrix} 1 \\ 1 \\ -1 \\ -1 \end{bmatrix}$ 20. $\begin{bmatrix} 1 \\ 0 \\ 1 \\ 2 \end{bmatrix}$ 21. $\begin{bmatrix} -1 \\ 2 \\ -2 \\ -2 \end{bmatrix}$ 22. $\begin{bmatrix} 2 \\ 0 \\ 2 \\ 3 \end{bmatrix}$

In Exercises 23–26, determine if the given vector belongs to Col A, where A is the matrix used in Exercises 19–22.

23. $\begin{bmatrix} 2 \\ -1 \\ 3 \end{bmatrix}$ 24. $\begin{bmatrix} -1 \\ 3 \\ -1 \end{bmatrix}$ 25. $\begin{bmatrix} 1 \\ -4 \\ 2 \end{bmatrix}$ 26. $\begin{bmatrix} -1 \\ 2 \\ 1 \end{bmatrix}$

In Exercises 27–30, find a spanning set for the null space of the given matrix.

27. $\begin{bmatrix} -1 & 1 & 2 \\ 1 & -2 & 3 \end{bmatrix}$ 28. $\begin{bmatrix} 1 & 2 & 0 \\ 0 & -1 & 1 \\ 1 & 0 & 2 \end{bmatrix}$

29. $\begin{bmatrix} 1 & 1 & 0 & 2 & 1 \\ 3 & 2 & 1 & 6 & 3 \\ 0 & -1 & 1 & -1 & -1 \end{bmatrix}$

30. $\begin{bmatrix} 1 & -3 & 0 & 1 & -2 & -2 \\ 2 & -6 & -1 & 0 & 2 & 5 \\ -1 & 3 & 2 & 3 & -1 & 2 \end{bmatrix}$

31. Find a spanning set containing exactly two vectors for the column space of the matrix in Exercise 27.

32. Find a spanning set containing exactly two vectors for the column space of the matrix in Exercise 28.

33. Find a spanning set containing exactly four vectors for the column space of the matrix in Exercise 29.

34. Find a spanning set containing exactly four vectors for the column space of the matrix in Exercise 30.

35. Determine the null space, column space, and row space of the $m \times n$ zero matrix.

36. Let R be the reduced row echelon form of A. Is Null $A =$ Null R? Justify your answer.

37. Let R be the reduced row echelon form of A. Is Col $A =$ Col R? Justify your answer.

38. Let R be the reduced row echelon form of A. Prove that Row $A =$ Row R.

39. Give an example of a nonzero matrix for which the row space equals the column space.

40. Give an example of a matrix for which the null space equals the column space.

41. Prove that the intersection of two subspaces of $\mathcal{R}^n$ is a subspace of $\mathcal{R}^n$.

42. Let $V = \left\{ \begin{bmatrix} v_1 \\ v_2 \end{bmatrix} \in \mathcal{R}^2 \colon v_1 = 0 \right\}$ and

$W = \left\{ \begin{bmatrix} v_1 \\ v_2 \end{bmatrix} \in \mathcal{R}^2 \colon v_2 = 0 \right\}.$

(a) Prove that both V and W are subspaces of $\mathcal{R}^2$.

(b) Show that $V \cup W$ is *not* a subspace of $\mathcal{R}^2$.

43. Let S be a nonempty subset of $\mathcal{R}^n$. Prove that S is a subspace of $\mathcal{R}^n$ if and only if, for all vectors $\mathbf{u}$ and $\mathbf{v}$ in S and all scalars c, the vector $\mathbf{u} + c\mathbf{v}$ is in S.

44. Prove that if V is a subspace of $\mathcal{R}^n$ containing vectors $\mathbf{u}_1, \mathbf{u}_2, \dots, \mathbf{u}_k$, then V contains the span of $\{\mathbf{u}_1, \mathbf{u}_2, \dots, \mathbf{u}_k\}$. Hence the span of $\{\mathbf{u}_1, \mathbf{u}_2, \dots, \mathbf{u}_k\}$ is the smallest subspace of $\mathcal{R}^n$ containing the vectors $\mathbf{u}_1, \mathbf{u}_2, \dots, \mathbf{u}_k$.

In Exercises 45–50, find a spanning set for the range and null space of the given linear transformation.

45. $T\left(\begin{bmatrix} x_1 \\ x_2 \\ x_3 \end{bmatrix} \right) = [x_1 + 2x_2 - x_3]$

46. $T\left(\begin{bmatrix} x_1 \\ x_2 \end{bmatrix} \right) = \begin{bmatrix} x_1 + 2x_2 \\ 2x_1 + 4x_2 \end{bmatrix}$

47. $T\left(\begin{bmatrix} x_1 \\ x_2 \end{bmatrix} \right) = \begin{bmatrix} x_1 + x_2 \\ x_1 - x_2 \\ x_1 \\ x_2 \end{bmatrix}$

48. $T\left(\begin{bmatrix} x_1 \\ x_2 \\ x_3 \end{bmatrix} \right) = \begin{bmatrix} x_1 - 2x_2 + 3x_3 \\ -2x_1 + 4x_2 - 6x_3 \end{bmatrix}$

49. $T\left(\begin{bmatrix} x_1 \\ x_2 \\ x_3 \end{bmatrix} \right) = \begin{bmatrix} x_1 + x_2 - x_3 \\ 0 \\ 2x_1 - x_3 \end{bmatrix}$

50. $T\left(\begin{bmatrix} x_1 \\ x_2 \\ x_3 \end{bmatrix} \right) = \begin{bmatrix} x_1 + x_2 \\ x_2 + x_3 \\ x_1 - x_3 \\ x_1 + 2x_2 + x_3 \end{bmatrix}$

In Exercises 51–56, use the definition of a subspace as in Example 4 to prove that the given set is a subspace of the appropriate $\mathcal{R}^n$.

51. $\left\{ \begin{bmatrix} u_1 \\ u_2 \end{bmatrix} \in \mathcal{R}^2 \colon u_1 - 3u_2 = 0 \right\}$

52. $\left\{ \begin{bmatrix} u_1 \\ u_2 \end{bmatrix} \in \mathcal{R}^2 \colon 5u_1 + 4u_2 = 0 \right\}$

53. $\left\{ \begin{bmatrix} u_1 \\ u_2 \\ u_3 \end{bmatrix} \in \mathcal{R}^3 \colon 2u_1 + 5u_2 - 4u_3 = 0 \right\}$

54. $\left\{ \begin{bmatrix} u_1 \\ u_2 \\ u_3 \end{bmatrix} \in \mathcal{R}^3 \colon -u_1 + 7u_2 + 2u_3 = 0 \right\}$

55. $\left\{ \begin{bmatrix} u_1 \\ u_2 \\ u_3 \\ u_4 \end{bmatrix} \in \mathcal{R}^4 \colon 3u_1 - u_2 + 6u_4 = 0 \text{ and } u_3 = 0 \right\}$

56. $\left\{ \begin{bmatrix} u_1 \\ u_2 \\ u_3 \\ u_4 \end{bmatrix} \in \mathcal{R}^4 \colon u_1 + 5u_3 = 0 \text{ and } 4u_2 - 3u_4 = 0 \right\}$

57. Let $T \colon \mathcal{R}^n \to \mathcal{R}^m$ be a linear transformation. Use the definition of a subspace to prove that the null space of T is a subspace of $\mathcal{R}^n$.

58. Let $T \colon \mathcal{R}^n \to \mathcal{R}^m$ be a linear transformation. Use the definition of a subspace to prove that the range of T is a subspace of $\mathcal{R}^m$.

59. Let $T \colon \mathcal{R}^n \to \mathcal{R}^m$ be a linear transformation. Prove that if V is a subspace of $\mathcal{R}^n$, then $\{T(\mathbf{u}) \in \mathcal{R}^m \colon \mathbf{u} \in V\}$ is a subspace of $\mathcal{R}^m$.

60. Let $T \colon \mathcal{R}^n \to \mathcal{R}^m$ be a linear transformation. Prove that if W is a subspace of $\mathcal{R}^m$, then $\{\mathbf{u} \colon T(\mathbf{u}) \in W\}$ is a subspace of $\mathcal{R}^n$.

In Exercises 61–63, use a calculator with matrix capabilities or computer software such as MATLAB *to solve the problem.*

61. Let

$$A = \begin{bmatrix} -1 & 0 & 2 & 1 & 1 \\ 1 & 1 & 1 & 0 & 0 \\ 1 & -1 & -5 & 3 & -2 \\ 1 & 1 & 1 & -1 & 0 \\ 0 & 1 & 3 & -2 & 1 \end{bmatrix}, \quad \mathbf{u} = \begin{bmatrix} 3.0 \\ 1.8 \\ -10.3 \\ 2.3 \\ 6.3 \end{bmatrix},$$

and $\quad \mathbf{v} = \begin{bmatrix} -.6 \\ 1.4 \\ -1.6 \\ 1.2 \\ 1.8 \end{bmatrix}.$

(a) Is $\mathbf{u}$ in the column space of A?
(b) Is $\mathbf{v}$ in the column space of A?

62. Let A be the matrix in Exercise 61, and let

$$\mathbf{u} = \begin{bmatrix} 0.5 \\ -1.6 \\ -2.1 \\ 0.0 \\ 4.7 \end{bmatrix} \quad \text{and} \quad \mathbf{v} = \begin{bmatrix} 2.4 \\ -6.3 \\ 3.9 \\ 0.0 \\ -5.4 \end{bmatrix}.$$

(a) Is $\mathbf{u}$ in the null space of A?
(b) Is $\mathbf{v}$ in the null space of A?

63. Let A be the matrix in Exercise 61, and let

$$\mathbf{u} = \begin{bmatrix} -5.1 & -2.2 & 3.6 & 8.2 & 2.9 \end{bmatrix}$$

and $\quad \mathbf{v} = \begin{bmatrix} -5.6 & -1.4 & 3.5 & 2.9 & 4.2 \end{bmatrix}.$

(a) Is $\mathbf{u}$ in the row space of A?
(b) Is $\mathbf{v}$ in the row space of A?

Solutions to the Practice Problems

1. The vectors in V can be written in the form $s \begin{bmatrix} -1 \\ 0 \\ 3 \end{bmatrix} + t \begin{bmatrix} 0 \\ 2 \\ -1 \end{bmatrix}$. Hence

$$\left\{ \begin{bmatrix} -1 \\ 0 \\ 3 \end{bmatrix}, \begin{bmatrix} 0 \\ 2 \\ -1 \end{bmatrix} \right\}$$

is a spanning set for V.

2. The set

$$\left\{ \begin{bmatrix} 1 \\ -1 \end{bmatrix}, \begin{bmatrix} 2 \\ -3 \end{bmatrix}, \begin{bmatrix} -1 \\ 4 \end{bmatrix} \right\}$$

consisting of the columns of A is a spanning set for the column space of A. To find a spanning set for the null space of A, we must solve the equation $A\mathbf{x} = \mathbf{0}$. Since

the reduced row echelon form of A is

$$\begin{bmatrix} 1 & 0 & 5 \\ 0 & 1 & -3 \end{bmatrix},$$

the parametric form of the general solution is

$$\begin{bmatrix} x_1 \\ x_2 \\ x_3 \end{bmatrix} = \begin{bmatrix} -5x_3 \\ 3x_3 \\ x_3 \end{bmatrix} = x_3 \begin{bmatrix} -5 \\ 3 \\ 1 \end{bmatrix}.$$

Hence

$$\left\{ \begin{bmatrix} -5 \\ 3 \\ 1 \end{bmatrix} \right\}$$

is a spanning set for the null space of A.

3. The standard matrix of T is

$$A = \begin{bmatrix} 1 & 0 & 1 & 2 \\ 0 & -1 & 1 & 1 \\ 2 & 3 & -1 & 1 \end{bmatrix}.$$

The null space of T is the same as the null space of A, and so it is the solution set of $A\mathbf{x} = \mathbf{0}$. Since the reduced row echelon form of A is

$$\begin{bmatrix} 1 & 0 & 1 & 2 \\ 0 & 1 & -1 & -1 \\ 0 & 0 & 0 & 0 \end{bmatrix},$$

we see that the solutions of $A\mathbf{x} = \mathbf{0}$ can be written as

$$\begin{bmatrix} x_1 \\ x_2 \\ x_3 \\ x_4 \end{bmatrix} = \begin{bmatrix} -x_3 - 2x_4 \\ x_3 + x_4 \\ x_3 \\ x_4 \end{bmatrix} = x_3 \begin{bmatrix} -1 \\ 1 \\ 1 \\ 0 \end{bmatrix} + x_4 \begin{bmatrix} -2 \\ 1 \\ 0 \\ 1 \end{bmatrix}.$$

Thus

$$\left\{ \begin{bmatrix} -1 \\ 1 \\ 1 \\ 0 \end{bmatrix}, \begin{bmatrix} -2 \\ 1 \\ 0 \\ 1 \end{bmatrix} \right\}$$

is a spanning set for the null space of T.

The range of T is the same as the column space of A. Hence the set

$$\left\{ \begin{bmatrix} 1 \\ 0 \\ 2 \end{bmatrix}, \begin{bmatrix} 0 \\ -1 \\ 3 \end{bmatrix}, \begin{bmatrix} 1 \\ 1 \\ -1 \end{bmatrix}, \begin{bmatrix} 2 \\ 1 \\ 1 \end{bmatrix} \right\}$$

of columns of A is a spanning set for the range of T.

4.2   Basis and Dimension

We have seen that subspaces are often described by spanning sets. In such cases, it clearly desirable to use spanning sets that are as small as possible. Linearly independent sets are guaranteed to have no "redundancy," that is, no vector in the set can be written as a linear combination of the others. Thus it is advantageous to use spanning sets for a subspace that are also linearly independent. Sets with both these properties are so important that they deserve their own definition.

Definition. Let V be a subspace of $\mathcal{R}^n$. A **basis** for V is a linearly independent spanning set for V.

For example, the set of standard vectors $\{\mathbf{e}_1, \mathbf{e}_2, \ldots, \mathbf{e}_n\}$ in $\mathcal{R}^n$ is both a linearly independent set and a spanning set for $\mathcal{R}^n$. Hence $\{\mathbf{e}_1, \mathbf{e}_2, \ldots, \mathbf{e}_n\}$ is a basis for $\mathcal{R}^n$. We call this basis the **standard basis** for $\mathcal{R}^n$ and denote it by $\mathcal{E}$.

Some of our previous results about spanning sets and linearly independent sets can be reinterpreted using the concept of a basis. Recall, for instance, that Theorem 2.5 states that the pivot columns of a matrix form a spanning set for its column space and that the pivot columns of a matrix are also linearly independent. Thus we have the following useful result.

The pivot columns of a matrix form a basis for its column space.

The following example shows that we already know how to find a basis for certain subspaces.

Example 1   Find a basis for Col A if

$$A = \begin{bmatrix} 1 & 2 & -1 & 2 & 1 & 2 \\ -1 & -2 & 1 & 2 & 3 & 6 \\ 2 & 4 & -3 & 2 & 0 & 3 \\ -3 & -6 & 2 & 0 & 3 & 9 \end{bmatrix}.$$

Solution   In Example 1 of Section 1.4, we showed that the reduced row echelon form of A is

$$\begin{bmatrix} 1 & 2 & 0 & 0 & -1 & -5 \\ 0 & 0 & 1 & 0 & 0 & -3 \\ 0 & 0 & 0 & 1 & 1 & 2 \\ 0 & 0 & 0 & 0 & 0 & 0 \end{bmatrix}.$$

As noted above, the pivot columns of A are a basis for Col A. (Note that it is the pivot columns of A, and not those of the reduced row echelon form of A, that form a basis for Col A.) Thus the desired basis for Col A is

$$\left\{ \begin{bmatrix} 1 \\ -1 \\ 2 \\ -3 \end{bmatrix}, \begin{bmatrix} -1 \\ 1 \\ -3 \\ 2 \end{bmatrix}, \begin{bmatrix} 2 \\ 2 \\ 2 \\ 0 \end{bmatrix} \right\}.$$

Note that in Example 1 the column space of A is different from the column space of its reduced row echelon form. Moreover, the method for finding a basis in Example 1 is more general than it appears: it can be used to find a basis for any subspace if we know a finite spanning set for the subspace.

Theorem 4.3   (Reduction principle)
If S is a finite spanning set for a subspace V of $\mathcal{R}^n$, then there is a basis for V that is contained in S.

Proof Let V be a subspace of $\mathcal{R}^n$ and $S = \{\mathbf{u}_1, \mathbf{u}_2, \ldots, \mathbf{u}_k\}$ be a spanning set for V. If $A = [\mathbf{u}_1 \ \mathbf{u}_2 \ \cdots \ \mathbf{u}_k]$, then Col A = Span $\{\mathbf{u}_1, \mathbf{u}_2, \ldots, \mathbf{u}_k\} = V$. Since the pivot columns of A form a basis for Col A, the set consisting of the pivot columns of A is a basis for V. This basis is clearly contained in S. ◻

In Theorem 1.5, we showed that the columns of an $n \times k$ matrix are a spanning set for $\mathcal{R}^n$ if and only if the matrix has rank n; that is, if and only if the matrix has a pivot position in each row. Suppose that $\{\mathbf{u}_1, \mathbf{u}_2, \ldots, \mathbf{u}_k\}$ is a spanning set for $\mathcal{R}^n$. Then the matrix $[\mathbf{u}_1 \ \mathbf{u}_2 \ \cdots \ \mathbf{u}_k]$ must have a pivot position in each row. Since no two pivot positions can lie in the same column, it follows that this matrix must have at least n columns. Thus $k \geq n$, that is, *a spanning set for $\mathcal{R}^n$ must contain at least n vectors.*

Now suppose that $\{\mathbf{v}_1, \mathbf{v}_2, \ldots, \mathbf{v}_j\}$ is a linearly independent subset of $\mathcal{R}^n$. As noted in Section 1.7, every subset of $\mathcal{R}^n$ containing more than n vectors must be linearly dependent. Hence, in order that $\{\mathbf{v}_1, \mathbf{v}_2, \ldots, \mathbf{v}_j\}$ be linearly independent, we must have $j \leq n$. That is, *a linearly independent subset of $\mathcal{R}^n$ must contain no more than n vectors.*

Combining the observations in the two preceding paragraphs, we see that *every basis for $\mathcal{R}^n$ must contain exactly n vectors.* Our next theorem shows that every subspace of $\mathcal{R}^n$ has a basis and that the number of vectors in any basis is an intrinsic property of the subspace.

Theorem 4.4
Let V be a nonzero subspace of $\mathcal{R}^n$.

(a) **(Extension principle)** Every linearly independent subset of V is contained in a basis for V.
(b) V has a basis.
(c) Any two bases for V contain the same number of vectors.

Proof (a) Let $\mathcal{L}$ be a linearly independent subset of V. If the span of $\mathcal{L}$ is V, then $\mathcal{L}$ is a basis for V that contains $\mathcal{L}$, and we are done. Otherwise, there exists a vector $\mathbf{v}_1$ in V that is not in the span of $\mathcal{L}$. Property 3 on page 74 implies that $\mathcal{L} \cup \{\mathbf{v}_1\}$ is linearly independent. If the span of $\mathcal{L} \cup \{\mathbf{v}_1\}$ is V, then $\mathcal{L} \cup \{\mathbf{v}_1\}$ is a basis for V that contains $\mathcal{L}$, and again we are done. Otherwise, there exists a vector $\mathbf{v}_2$ in V that is not in the span of $\mathcal{L} \cup \{\mathbf{v}_1\}$. As before, $\mathcal{L} \cup \{\mathbf{v}_1, \mathbf{v}_2\}$ is linearly independent. We continue this process of selecting larger linearly independent subsets of V containing $\mathcal{L}$ until one of them is a spanning set for V (and hence a basis for V that contains $\mathcal{L}$). Note that this process must stop in at most n steps because every subset of $\mathcal{R}^n$ containing more than n vectors is linearly dependent.

(b) Let $\mathbf{u}$ be a nonzero vector in V. Applying the extension principle to $\mathcal{L} = \{\mathbf{u}\}$, which is a linearly independent subset of V (property 1 on page 74), we see that V has a basis.

(c) Let $\mathcal{A}$ be a basis for V containing exactly j vectors, and let $\mathcal{B}$ be a basis for V containing exactly k vectors. Applying Theorem 1.9 to the linearly independent set $\mathcal{A}$ and the spanning set $\mathcal{B}$ shows that $j \leq k$. Applying the same result to the linearly independent set $\mathcal{B}$ and the spanning set $\mathcal{A}$ also shows that $k \leq j$. Hence $j = k$; that is, $\mathcal{A}$ and $\mathcal{B}$ contain the same number of vectors. $\square$

The reduction and extension principles give two different characterizations of a basis. The first of these theorems implies that *a basis for a subspace is a smallest possible spanning set for the subspace,* whereas the second theorem implies that *a basis is a largest possible linearly independent subset of the subspace.*

According to Theorem 4.4(c), the size of every basis for a subspace is the same. This permits the following definition.

Definition.   The number of vectors in a basis for a subspace V of $\mathcal{R}^n$ is called the **dimension** of V and is denoted $\dim V$.

Since the standard basis for $\mathcal{R}^n$ contains n vectors, $\dim \mathcal{R}^n = n$. Section 4.3 is concerned with studying the dimension of several of the important subspaces that we have previously encountered.

Theorem 1.9 contains important information about the size of linearly independent subsets of a subspace. We restate it here using our new terminology.

Theorem 1.9′
Let V be a subspace of $\mathcal{R}^n$ with dimension k. Then any finite subset of V containing more than k vectors is linearly dependent or, equivalently, every linearly independent subset of V contains at most k vectors.

To find the dimension of a subspace, we must usually determine a basis for that subspace. In this book, a subspace is almost always defined in one of two ways:

(a)   as the span of a given set of vectors, or
(b)   as the solution set of a homogeneous system of linear equations.

Recall that a basis for a subspace defined by means of a finite spanning set can be found by the technique used in Example 1. When a subspace is defined as in (b), we can obtain a basis by solving a system of linear equations. The method is demonstrated in the following example.

Example 2   Find a basis for the subspace

$$V = \left\{ \begin{bmatrix} x_1 \\ x_2 \\ x_3 \\ x_4 \end{bmatrix} \in \mathcal{R}^4 : x_1 - 3x_2 + 5x_3 - 6x_4 = 0 \right\}$$

of $\mathcal{R}^4$, and determine the dimension of V. (Since V is defined as the set of solutions to a homogeneous system of linear equations, V is, in fact, a subspace.)

Solution   The vectors in V are solutions to $x_1 - 3x_2 + 5x_3 - 6x_4 = 0$, a system of 1 linear equation in 4 variables. To solve this system, we apply the technique described in Section 1.3. Since

$$x_1 = 3x_2 - 5x_3 + 6x_4,$$

the parametric representation of the general solution to this system is

$$\begin{bmatrix} x_1 \\ x_2 \\ x_3 \\ x_4 \end{bmatrix} = \begin{bmatrix} 3x_2 - 5x_3 + 6x_4 \\ x_2 \\ x_3 \\ x_4 \end{bmatrix} = x_2 \begin{bmatrix} 3 \\ 1 \\ 0 \\ 0 \end{bmatrix} + x_3 \begin{bmatrix} -5 \\ 0 \\ 1 \\ 0 \end{bmatrix} + x_4 \begin{bmatrix} 6 \\ 0 \\ 0 \\ 1 \end{bmatrix}.$$

As noted in Section 1.7, the set

$$\mathcal{S} = \left\{ \begin{bmatrix} 3 \\ 1 \\ 0 \\ 0 \end{bmatrix}, \begin{bmatrix} -5 \\ 0 \\ 1 \\ 0 \end{bmatrix}, \begin{bmatrix} 6 \\ 0 \\ 0 \\ 1 \end{bmatrix} \right\}$$

containing the vectors in the parametric representation is both a spanning set for V and a linearly independent set. Therefore $\mathcal{S}$ is a basis for V. Since $\mathcal{S}$ is a basis for V containing 3 vectors, $\dim V = 3$.   ○

Confirming that a Set Is a Basis for a Subspace

As we have just seen, our previous techniques enable us to find a basis for many common subspaces. Sometimes the basis obtained by these methods does not possess a property that we desire. It may be possible, however, to find another basis with the desired property. (Note that Exercise 35 shows that every nonzero subspace of $\mathcal{R}^n$ has infinitely many bases.)

Consider, for example, the subspace

$$V = \left\{ \begin{bmatrix} v_1 \\ v_2 \\ v_3 \end{bmatrix} \in \mathcal{R}^3 : v_1 - v_2 + 2v_3 = 0 \right\}$$

of $\mathcal{R}^3$. As in Example 2, a basis for V can be found by solving the equation $v_1 - v_2 + 2v_3 = 0$ that defines V. The resulting basis is

$$\mathcal{B} = \left\{ \begin{bmatrix} 1 \\ 1 \\ 0 \end{bmatrix}, \begin{bmatrix} -2 \\ 0 \\ 1 \end{bmatrix} \right\}.$$

There are circumstances where we want the vectors in a basis to be perpendicular to one another. This property is not possessed by $\mathcal{B}$. In Section 6.2, we will learn a technique for transforming $\mathcal{B}$ into a basis whose vectors are perpendicular. One such basis is

$$C = \left\{ \begin{bmatrix} 1 \\ 1 \\ 0 \end{bmatrix}, \begin{bmatrix} -1 \\ 1 \\ 1 \end{bmatrix} \right\}.$$

Our concern here is not with the method for obtaining $\mathcal{C}$, but rather with the question of how to show that $\mathcal{C}$ is actually a basis for V. By definition, $\mathcal{C}$ is a basis for V if $\mathcal{C}$ is both linearly independent and a spanning set for V. In this case, it is clear that $\mathcal{C}$ is linearly independent because $\mathcal{C}$ contains two vectors, neither of which is a multiple of the other. It remains to show only that $\mathcal{C}$ is a spanning set for V; that is, to show that $\mathcal{C}$ is a subset of V such that every vector in V is a linear combination of the vectors in $\mathcal{C}$. Checking that $\mathcal{C}$ is a subset of V is straightforward. Because $1 - 1 + 2(0) = 0$ and $-1 - 1 + 2(1) = 0$, the vectors in $\mathcal{C}$ satisfy the equation defining V. Hence we have

$$\begin{bmatrix} 1 \\ 1 \\ 0 \end{bmatrix} \in V \qquad \text{and} \qquad \begin{bmatrix} -1 \\ 1 \\ 1 \end{bmatrix} \in V.$$

Unfortunately, checking that every vector in V is a linear combination of the vectors in $\mathcal{C}$ is more tedious. The difficulty is that we must show that *every* vector in V is a linear combination of the vectors in $\mathcal{C}$. This requires us to show that, for every

$$\begin{bmatrix} v_1 \\ v_2 \\ v_3 \end{bmatrix} \in V,$$

there are scalars c_1 and c_2 such that

$$c_1 \begin{bmatrix} 1 \\ 1 \\ 0 \end{bmatrix} + c_2 \begin{bmatrix} -1 \\ 1 \\ 1 \end{bmatrix} = \begin{bmatrix} v_1 \\ v_2 \\ v_3 \end{bmatrix}.$$

The following result enables us to avoid this type of calculation.

Theorem 4.5

Let V be a subspace of $\mathcal{R}^n$ having dimension k. Then any two of the following conditions about a subset S of V imply that S is a basis for V.

(a) S is linearly independent.
(b) S is a spanning set for V.
(c) S contains exactly k vectors.

Proof Conditions (a) and (b) are the definition of a basis. Thus it is enough to show that (a) and (c) imply (b) and that (b) and (c) imply (a).

   We first show that (a) and (c) imply (b). Let S be a linearly independent subset of V that contains exactly k vectors. By the extension principle, there is a basis $\mathcal{B}$ for V that contains S, and $\mathcal{B}$ must contain exactly k vectors because V has dimension k. Therefore $\mathcal{B} = S$, and so S is a basis for V.

   We next show that (b) and (c) imply (a). Let S be a spanning set for V that contains exactly k vectors. By the reduction principle, there is a subset $\mathcal{B}$

of S that is a basis for V. As in the preceding paragraph, $\mathcal{B}$ contains exactly k vectors. Hence $\mathcal{B} = S$, and so S is a basis for V. ◻

This theorem gives us three straightforward steps to show that a given set $\mathcal{B}$ is a basis for a subspace V.

(i) Show that $\mathcal{B}$ is contained in V.
(ii) Show that $\mathcal{B}$ is linearly independent.
(iii) Compute the dimension of V, and confirm that the number of vectors in $\mathcal{B}$ equals dim V.

In the context of the preceding example, $\mathcal{C}$ was shown to be a linearly independent subset of V containing two vectors. Because dim $V = 2$, all three conditions are satisfied, and hence $\mathcal{C}$ is a basis for V. Therefore, we need not verify that $\mathcal{C}$ spans V.

Two more examples of this technique follow.

Example 3 Show that

$$\mathcal{B} = \left\{ \begin{bmatrix} 1 \\ -1 \\ 1 \\ 0 \end{bmatrix}, \begin{bmatrix} 1 \\ 0 \\ 1 \\ -1 \end{bmatrix}, \begin{bmatrix} 0 \\ 1 \\ 1 \\ -1 \end{bmatrix} \right\}$$

is a basis for

$$V = \left\{ \begin{bmatrix} v_1 \\ v_2 \\ v_3 \\ v_4 \end{bmatrix} \in \mathcal{R}^4 : v_1 + v_2 + v_4 = 0 \right\}.$$

Solution Clearly the components of the three vectors in $\mathcal{B}$ all satisfy the equation $v_1 + v_2 + v_4 = 0$. Thus $\mathcal{B}$ is a subset of V, and so (i) is satisfied.

Because the reduced row echelon form of

$$\begin{bmatrix} 1 & 1 & 0 \\ -1 & 0 & 1 \\ 1 & 1 & 1 \\ 0 & -1 & -1 \end{bmatrix} \quad \text{is} \quad \begin{bmatrix} 1 & 0 & 0 \\ 0 & 1 & 0 \\ 0 & 0 & 1 \\ 0 & 0 & 0 \end{bmatrix},$$

it follows that $\mathcal{B}$ is linearly independent, and so (ii) is satisfied.

As in Example 2, we find that

$$\left\{ \begin{bmatrix} -1 \\ 1 \\ 0 \\ 0 \end{bmatrix}, \begin{bmatrix} 0 \\ 0 \\ 1 \\ 0 \end{bmatrix}, \begin{bmatrix} -1 \\ 0 \\ 0 \\ 1 \end{bmatrix} \right\}$$

is a basis for V. Hence the dimension of V is 3. But $\mathcal{B}$ contains three vectors, so (iii) is satisfied, and hence $\mathcal{B}$ is a basis for V. ○

Example 4   Let W be the span of $\mathcal{S}$, where

$$\mathcal{S} = \left\{ \begin{bmatrix} 1 \\ 1 \\ 1 \\ 2 \end{bmatrix}, \begin{bmatrix} -1 \\ 3 \\ 1 \\ -1 \end{bmatrix}, \begin{bmatrix} 3 \\ 1 \\ -1 \\ 1 \end{bmatrix}, \begin{bmatrix} 1 \\ 1 \\ -1 \\ -1 \end{bmatrix} \right\}.$$

Show that a basis for W is

$$\mathcal{B} = \left\{ \begin{bmatrix} 1 \\ 2 \\ 0 \\ 0 \end{bmatrix}, \begin{bmatrix} 1 \\ 0 \\ 0 \\ 1 \end{bmatrix}, \begin{bmatrix} 0 \\ 1 \\ 1 \\ 1 \end{bmatrix} \right\}.$$

Solution   Let

$$B = \begin{bmatrix} 1 & 1 & 0 \\ 2 & 0 & 1 \\ 0 & 0 & 1 \\ 0 & 1 & 1 \end{bmatrix} \quad \text{and} \quad A = \begin{bmatrix} 1 & -1 & 3 & 1 \\ 1 & 3 & 1 & 1 \\ 1 & 1 & -1 & -1 \\ 2 & -1 & 1 & -1 \end{bmatrix}.$$

The reader should check that the equation $A\mathbf{x} = \mathbf{b}$ is consistent for each $\mathbf{b}$ in $\mathcal{B}$. Therefore $\mathcal{B}$ is a subset of W, and so (i) is satisfied.

We can easily verify that the reduced row echelon form of B is

$$\begin{bmatrix} 1 & 0 & 0 \\ 0 & 1 & 0 \\ 0 & 0 & 1 \\ 0 & 0 & 0 \end{bmatrix},$$

and so $\mathcal{B}$ is linearly independent. Thus (ii) is satisfied.

Since the reduced row echelon form of A is

$$\begin{bmatrix} 1 & 0 & 0 & -\frac{2}{3} \\ 0 & 1 & 0 & \frac{1}{3} \\ 0 & 0 & 1 & \frac{2}{3} \\ 0 & 0 & 0 & 0 \end{bmatrix},$$

we see that the first 3 columns of A are its pivot columns and hence are a basis for Col $A = W$. Thus dim $W = 3$, which equals the number of vectors contained in $\mathcal{B}$, and so (iii) is satisfied. Hence $\mathcal{B}$ is a basis for W.   ○

Practice Problems

1. Find a basis for the column space and null space of

$$\begin{bmatrix} -1 & 2 & 1 & -1 \\ 2 & -4 & -3 & 0 \\ 1 & -2 & 0 & 3 \end{bmatrix}.$$

Then find the dimension of each of these subspaces.

2. Show that

$$\left\{ \begin{bmatrix} -1 \\ 1 \\ -2 \\ 1 \end{bmatrix}, \begin{bmatrix} 0 \\ 3 \\ -4 \\ 2 \end{bmatrix} \right\}$$

is a basis for the null space of the matrix in problem 1.

3. Is $\left\{ \begin{bmatrix} 0 \\ -1 \\ 1 \end{bmatrix}, \begin{bmatrix} -1 \\ 1 \\ 2 \end{bmatrix} \right\}$ a basis for $\mathcal{R}^3$? Justify your answer.

Exercises

1. Determine if the following statements are true or false.

   (a) Every nonzero subspace of $\mathcal{R}^n$ has a unique basis.

   (b) Every nonzero subspace of $\mathcal{R}^n$ has a basis.

   (c) A basis for a subspace is a spanning set that is as large as possible.

   (d) If $\mathcal{S}$ is a linearly independent set such that every vector in a subspace V is a linear combination of the vectors in $\mathcal{S}$, then $\mathcal{S}$ is a basis for V.

   (e) Every finite spanning set for a subspace contains a basis for the subspace.

   (f) A basis for a subspace is a linearly independent subset of the subspace that is as large as possible.

   (g) Every basis for a particular subspace contains the same number of vectors.

   (h) The columns of any matrix form a basis for its column space.

   (i) The pivot columns of the reduced row echelon form of A form a basis for the column space of A.

   (j) The vectors in the parametric form of the general solution to $A\mathbf{x} = \mathbf{0}$ form a basis for the null space of A.

2. Explain why $\left\{ \begin{bmatrix} 1 \\ -1 \\ 2 \\ 1 \end{bmatrix}, \begin{bmatrix} 1 \\ 3 \\ -1 \\ 4 \end{bmatrix}, \begin{bmatrix} 2 \\ 1 \\ 5 \\ -3 \end{bmatrix} \right\}$ is not a spanning set for $\mathcal{R}^4$.

3. Explain why $\left\{ \begin{bmatrix} 1 \\ -3 \\ 4 \end{bmatrix}, \begin{bmatrix} -2 \\ 5 \\ 3 \end{bmatrix}, \begin{bmatrix} -1 \\ 6 \\ -4 \end{bmatrix}, \begin{bmatrix} 5 \\ 3 \\ -1 \end{bmatrix} \right\}$ is not linearly independent.

4. Explain why $\left\{ \begin{bmatrix} -4 \\ 6 \\ 2 \end{bmatrix}, \begin{bmatrix} 2 \\ -3 \\ 7 \end{bmatrix} \right\}$ is not a basis for $\mathcal{R}^3$.

In Exercises 5–14, find a basis for the given subspace.

5. $\left\{ \begin{bmatrix} s \\ -2s \end{bmatrix} \in \mathcal{R}^2 \colon s \in \mathcal{R} \right\}$

6. $\left\{ \begin{bmatrix} 2s \\ -s + 4t \\ s - 3t \end{bmatrix} \in \mathcal{R}^3 \colon s, t \in \mathcal{R} \right\}$

7. $\left\{ \begin{bmatrix} x_1 \\ x_2 \\ x_3 \end{bmatrix} \in \mathcal{R}^3 \colon x_1 - 3x_2 + 5x_3 = 0 \right\}$

8. $\left\{ \begin{bmatrix} x_1 \\ x_2 \\ x_3 \end{bmatrix} \in \mathcal{R}^3 \colon -x_1 + x_2 + 2x_3 = 0 \right.$

   $\left. \text{and} \quad 2x_1 - 3x_2 + 4x_3 = 0 \right\}$

9. Span $\left\{ \begin{bmatrix} 1 \\ 2 \\ 1 \end{bmatrix}, \begin{bmatrix} 2 \\ 1 \\ 3 \end{bmatrix}, \begin{bmatrix} 1 \\ -4 \\ 3 \end{bmatrix} \right\}$

10. Span $\left\{ \begin{bmatrix} 1 \\ 1 \\ -1 \end{bmatrix}, \begin{bmatrix} 2 \\ 2 \\ -2 \end{bmatrix}, \begin{bmatrix} 1 \\ 2 \\ 0 \end{bmatrix}, \begin{bmatrix} -1 \\ 1 \\ 3 \end{bmatrix} \right\}$

11. Span $\left\{ \begin{bmatrix} 1 \\ -1 \\ 3 \end{bmatrix}, \begin{bmatrix} 0 \\ -1 \\ 1 \end{bmatrix}, \begin{bmatrix} 2 \\ 3 \\ 1 \end{bmatrix}, \begin{bmatrix} 1 \\ -2 \\ 0 \end{bmatrix}, \begin{bmatrix} 4 \\ -7 \\ -9 \end{bmatrix} \right\}$

12. Span $\left\{ \begin{bmatrix} 2 \\ 3 \\ -5 \end{bmatrix}, \begin{bmatrix} 8 \\ -12 \\ 20 \end{bmatrix}, \begin{bmatrix} 1 \\ 0 \\ -2 \end{bmatrix}, \begin{bmatrix} 0 \\ 2 \\ -1 \end{bmatrix}, \begin{bmatrix} 7 \\ 2 \\ 0 \end{bmatrix} \right\}$

13. Span $\left\{ \begin{bmatrix} 1 \\ 0 \\ -1 \\ 2 \end{bmatrix}, \begin{bmatrix} 1 \\ 1 \\ -2 \\ 1 \end{bmatrix}, \begin{bmatrix} -2 \\ 3 \\ -1 \\ -7 \end{bmatrix}, \begin{bmatrix} 1 \\ -1 \\ 0 \\ 3 \end{bmatrix}, \begin{bmatrix} 0 \\ 1 \\ -1 \\ 2 \end{bmatrix} \right\}$

14. Span $\left\{ \begin{bmatrix} 0 \\ 2 \\ 3 \\ 1 \end{bmatrix}, \begin{bmatrix} 1 \\ 1 \\ 1 \\ 3 \end{bmatrix}, \begin{bmatrix} 3 \\ 1 \\ 0 \\ 8 \end{bmatrix}, \begin{bmatrix} 1 \\ 0 \\ 1 \\ -1 \end{bmatrix}, \begin{bmatrix} -6 \\ 2 \\ 3 \\ -7 \end{bmatrix} \right\}$

In Exercises 15–18, find a basis for (a) the column space and (b) the null space of the given matrix.

15. $\begin{bmatrix} 1 & -2 & 0 & 2 \\ -1 & 2 & 1 & -3 \\ 2 & -4 & 3 & 1 \end{bmatrix}$

16. $\begin{bmatrix} 1 & 1 & -1 & -2 \\ -1 & -2 & 1 & 3 \\ 2 & 3 & 1 & 4 \end{bmatrix}$

17. $\begin{bmatrix} -1 & 1 & 2 & 2 \\ 2 & 0 & -5 & 3 \\ 1 & -1 & -1 & -1 \\ 0 & 1 & -2 & 2 \end{bmatrix}$

18. $\begin{bmatrix} 1 & -1 & 2 & 1 \\ 3 & -3 & 5 & 4 \\ 0 & 0 & 3 & -3 \\ 2 & -2 & 1 & 5 \end{bmatrix}$

In Exercises 19–26, a linear transformation T is given. (a) Find a basis for the range of T. (b) If the null space of T is nonzero, find a basis for the null space of T.

19. $T\left(\begin{bmatrix} x_1 \\ x_2 \\ x_3 \end{bmatrix} \right) = \begin{bmatrix} x_1 + 2x_2 + x_3 \\ 2x_1 + 3x_2 + 3x_3 \\ x_1 + 2x_2 + 4x_3 \end{bmatrix}$

20. $T\left(\begin{bmatrix} x_1 \\ x_2 \\ x_3 \end{bmatrix} \right) = \begin{bmatrix} x_1 + 2x_2 - x_3 \\ x_1 + x_2 \\ x_2 - x_3 \end{bmatrix}$

21. $T\left(\begin{bmatrix} x_1 \\ x_2 \\ x_3 \\ x_4 \end{bmatrix} \right) = \begin{bmatrix} x_1 - 2x_2 + x_3 + x_4 \\ 2x_1 - 5x_2 + x_3 + 3x_4 \\ x_1 - 3x_2 + 2x_4 \end{bmatrix}$

22. $T\left(\begin{bmatrix} x_1 \\ x_2 \\ x_3 \\ x_4 \end{bmatrix} \right) = \begin{bmatrix} x_1 + 2x_3 + x_4 \\ x_1 + 3x_3 + 2x_4 \\ -x_1 + x_3 \end{bmatrix}$

23. $T\left(\begin{bmatrix} x_1 \\ x_2 \\ x_3 \\ x_4 \end{bmatrix}\right) = \begin{bmatrix} x_1 + x_2 + 2x_3 - x_4 \\ 2x_1 + x_2 + x_3 \\ 0 \\ 3x_1 + x_2 + x_4 \end{bmatrix}$

24. $T\left(\begin{bmatrix} x_1 \\ x_2 \\ x_3 \\ x_4 \end{bmatrix}\right) = \begin{bmatrix} -2x_1 - x_2 + x_4 \\ 0 \\ x_1 + 2x_2 + 3x_3 + 4x_4 \\ 2x_1 + 3x_2 + 4x_3 + 5x_4 \end{bmatrix}$

25. $T\left(\begin{bmatrix} x_1 \\ x_2 \\ x_3 \\ x_4 \\ x_5 \end{bmatrix}\right) = \begin{bmatrix} x_1 + 2x_2 + 3x_3 + 4x_5 \\ 3x_1 + x_2 - x_3 - 3x_5 \\ 7x_1 + 4x_2 + x_3 - 2x_5 \end{bmatrix}$

26. $T\left(\begin{bmatrix} x_1 \\ x_2 \\ x_3 \\ x_4 \\ x_5 \end{bmatrix}\right) = \begin{bmatrix} -x_1 + x_2 + 4x_3 + 6x_4 + 9x_5 \\ x_1 + x_2 + 2x_3 + 4x_4 + 3x_5 \\ 3x_1 + x_2 + 2x_4 - 3x_5 \\ x_1 + 2x_2 + 5x_3 + 9x_4 + 9x_5 \end{bmatrix}$

27. Show that $\left\{ \begin{bmatrix} 1 \\ -3 \\ 2 \\ 2 \end{bmatrix}, \begin{bmatrix} 2 \\ -2 \\ 0 \\ 9 \end{bmatrix}, \begin{bmatrix} 1 \\ -6 \\ 5 \\ 2 \end{bmatrix} \right\}$ is a basis for the subspace in Exercise 13.

28. Show that $\left\{ \begin{bmatrix} -2 \\ 1 \\ 4 \\ -8 \end{bmatrix}, \begin{bmatrix} -2 \\ 5 \\ 7 \\ 1 \end{bmatrix}, \begin{bmatrix} -1 \\ 1 \\ 5 \\ -9 \end{bmatrix} \right\}$ is a basis for the subspace in Exercise 14.

29. Show that $\left\{ \begin{bmatrix} 0 \\ 1 \\ 1 \\ 1 \end{bmatrix}, \begin{bmatrix} 2 \\ 2 \\ 1 \\ 1 \end{bmatrix} \right\}$ is a basis for the null space of the matrix in Exercise 15.

30. Show that $\left\{ \begin{bmatrix} 0 \\ 3 \\ 1 \\ 1 \end{bmatrix}, \begin{bmatrix} -1 \\ 2 \\ 1 \\ 1 \end{bmatrix} \right\}$ is a basis for the null space of the matrix in Exercise 18.

31. What is the dimension of Span $\{\mathbf{v}\}$, where $\mathbf{v} \neq \mathbf{0}$? Justify your answer.

32. What is the dimension of the subspace

$$\left\{ \begin{bmatrix} v_1 \\ v_2 \\ \vdots \\ v_n \end{bmatrix} \in \mathcal{R}^n \colon v_1 = 0 \right\}? \text{ Justify your answer.}$$

33. What is the dimension of the subspace

$$\left\{ \begin{bmatrix} v_1 \\ v_2 \\ \vdots \\ v_n \end{bmatrix} \in \mathcal{R}^n \colon v_1 = 0 \text{ and } v_2 = 0 \right\}? \text{ Justify your answer.}$$

34. Find the dimension of the subspace

$$\left\{ \begin{bmatrix} v_1 \\ v_2 \\ \vdots \\ v_n \end{bmatrix} \in \mathcal{R}^n \colon v_1 + v_2 + \cdots + v_n = 0 \right\}. \text{ Justify your answer.}$$

35. Let $\mathcal{A} = \{\mathbf{u}_1, \mathbf{u}_2, \ldots, \mathbf{u}_k\}$ be a basis for a k-dimensional subspace V of $\mathcal{R}^n$. For any nonzero scalars $c_1, c_2, \ldots, c_k$, prove that $\mathcal{B} = \{c_1\mathbf{u}_1, c_2\mathbf{u}_2, \ldots, c_k\mathbf{u}_k\}$ is also a basis for V.

36. Let $\mathcal{A} = \{\mathbf{u}_1, \mathbf{u}_2, \ldots, \mathbf{u}_k\}$ be a basis for a k-dimensional subspace V of $\mathcal{R}^n$. Prove that $\mathcal{B} = \{\mathbf{u}_1, \mathbf{u}_1 + \mathbf{u}_2, \mathbf{u}_1 + \mathbf{u}_3, \ldots, \mathbf{u}_1 + \mathbf{u}_k\}$ is also a basis for V.

37. Let $\mathcal{A} = \{\mathbf{u}_1, \mathbf{u}_2, \ldots, \mathbf{u}_k\}$ be a basis for a k-dimensional subspace V of $\mathcal{R}^n$. Prove that $\{\mathbf{v}, \mathbf{u}_2, \mathbf{u}_3, \ldots, \mathbf{u}_k\}$ is also a basis for V, where $\mathbf{v} = \mathbf{u}_1 + \mathbf{u}_2 + \cdots + \mathbf{u}_k$.

38. Let $\mathcal{A} = \{\mathbf{u}_1, \mathbf{u}_2, \ldots, \mathbf{u}_k\}$ be a basis for a k-dimensional subspace V of $\mathcal{R}^n$. Prove that $\mathcal{B} = \{\mathbf{u}_1 + \mathbf{u}_2 + \cdots + \mathbf{u}_k, \mathbf{u}_2 + \mathbf{u}_3 + \cdots + \mathbf{u}_k, \ldots, \mathbf{u}_{k-1} + \mathbf{u}_k, \mathbf{u}_k\}$ is also a basis for V.

39. Let $T \colon \mathcal{R}^n \to \mathcal{R}^m$ be a linear transformation and $\{\mathbf{u}_1, \mathbf{u}_2, \ldots, \mathbf{u}_n\}$ be a basis for $\mathcal{R}^n$.
　(a) Prove that $\mathcal{S} = \{T(\mathbf{u}_1), T(\mathbf{u}_2), \ldots, T(\mathbf{u}_n)\}$ is a spanning set for the range of T.
　(b) Give an example to show that $\mathcal{S}$ need not be a basis for the range of T.

40. Let $T \colon \mathcal{R}^n \to \mathcal{R}^m$ be a one-to-one linear transformation and V be a subspace of $\mathcal{R}^n$. Recall from Exercise 59 of Section 4.1 that $W = \{T(\mathbf{u}) \colon \mathbf{u} \in V\}$ is a subspace of $\mathcal{R}^m$.
　(a) Prove that if $\{\mathbf{u}_1, \mathbf{u}_2, \ldots, \mathbf{u}_k\}$ is a basis for V, $\{T(\mathbf{u}_1), T(\mathbf{u}_2), \ldots, T(\mathbf{u}_k)\}$ is a basis for W.
　(b) Prove that $\dim V = \dim W$.

41. Does the zero subspace have a basis? Justify your answer.

42. Let V be a subspace of $\mathcal{R}^n$. By Theorem 4.4(a), a linearly independent subset $\mathcal{L} = \{\mathbf{u}_1, \mathbf{u}_2, \ldots, \mathbf{u}_m\}$ of V is contained in a basis for V. Show that if $\mathcal{S} = \{\mathbf{b}_1, \mathbf{b}_2, \ldots, \mathbf{b}_k\}$ is any spanning set for V, then the pivot columns of the matrix $[\mathbf{u}_1 \ \mathbf{u}_2 \ \cdots \ \mathbf{u}_m \ \mathbf{b}_1 \ \mathbf{b}_2 \ \cdots \ \mathbf{b}_k]$ form a basis for V that contains $\mathcal{L}$.

In Exercises 43–46, use the procedure described in Exercise 42 to find a basis for the subspace V that contains the given linearly independent subset $\mathcal{L}$ of V.

43. $\mathcal{L} = \left\{ \begin{bmatrix} 2 \\ 3 \\ 0 \end{bmatrix} \right\}, V = \mathcal{R}^3$

44. $\mathcal{L} = \left\{ \begin{bmatrix} -1 \\ -1 \\ 6 \\ -7 \end{bmatrix}, \begin{bmatrix} 5 \\ -9 \\ -2 \\ -1 \end{bmatrix} \right\}$,

$V = \text{Span} \left\{ \begin{bmatrix} 1 \\ -2 \\ 0 \\ 1 \end{bmatrix}, \begin{bmatrix} 1 \\ -1 \\ -2 \\ 3 \end{bmatrix}, \begin{bmatrix} 0 \\ 1 \\ -2 \\ 10 \end{bmatrix} \right\}$

45. $\mathcal{L} = \left\{ \begin{bmatrix} 0 \\ 2 \\ 1 \\ 0 \end{bmatrix} \right\}$, $V = \text{Null} \begin{bmatrix} 1 & -1 & 2 & 1 \\ 2 & -2 & 4 & 2 \\ -3 & 3 & -6 & -3 \end{bmatrix}$

46. $\mathcal{L} = \left\{ \begin{bmatrix} 0 \\ 0 \\ 1 \\ 0 \end{bmatrix} \right\}$, $V = \text{Col} \begin{bmatrix} 1 & -1 & -3 & 1 \\ -1 & 1 & 3 & 2 \\ -3 & 1 & -1 & -1 \\ 2 & -2 & -6 & 1 \end{bmatrix}$

47. Let $V = \left\{ \begin{bmatrix} v_1 \\ v_2 \\ v_3 \end{bmatrix} \in \mathcal{R}^3 \colon v_1 - v_2 + v_3 = 0 \right\}$

and $\mathcal{S} = \left\{ \begin{bmatrix} 1 \\ -1 \\ 2 \end{bmatrix}, \begin{bmatrix} 2 \\ -1 \\ 3 \end{bmatrix}, \begin{bmatrix} 2 \\ 1 \\ 2 \end{bmatrix} \right\}$.

(a) Show that $\mathcal{S}$ is linearly independent.

(b) Show that

$$(-9v_1 + 6v_2) \begin{bmatrix} 1 \\ -1 \\ 2 \end{bmatrix} + (7v_1 - 5v_2) \begin{bmatrix} 2 \\ -1 \\ 3 \end{bmatrix}$$

$$+ (-2v_1 + 2v_2) \begin{bmatrix} 2 \\ 1 \\ 2 \end{bmatrix} = \begin{bmatrix} v_1 \\ v_2 \\ v_3 \end{bmatrix}$$

for every vector $\begin{bmatrix} v_1 \\ v_2 \\ v_3 \end{bmatrix}$ in V.

(c) Determine if $\mathcal{S}$ is a basis for V. Justify your answer.

48. Let $V = \left\{ \begin{bmatrix} v_1 \\ v_2 \\ v_3 \\ v_4 \end{bmatrix} \in \mathcal{R}^4 \colon 3v_1 - v_3 = 0 \text{ and } v_4 = 0 \right\}$

and $\mathcal{S} = \left\{ \begin{bmatrix} 1 \\ 3 \\ 1 \\ 2 \end{bmatrix}, \begin{bmatrix} 2 \\ 5 \\ 3 \\ 3 \end{bmatrix}, \begin{bmatrix} 1 \\ -1 \\ 3 \\ 0 \end{bmatrix} \right\}$.

(a) Show that $\mathcal{S}$ is linearly independent.

(b) Show that

$$(9v_1 - 1.5v_2 - 3.5v_3) \begin{bmatrix} 1 \\ 3 \\ 1 \\ 2 \end{bmatrix} + (-5v_1 + v_2 + 2v_3) \begin{bmatrix} 2 \\ 5 \\ 3 \\ 3 \end{bmatrix}$$

$$+ (2v_1 - 0.5v_2 - 0.5v_3) \begin{bmatrix} 1 \\ -1 \\ 3 \\ 0 \end{bmatrix} = \begin{bmatrix} v_1 \\ v_2 \\ v_3 \\ v_4 \end{bmatrix}$$

for every vector $\begin{bmatrix} v_1 \\ v_2 \\ v_3 \\ v_4 \end{bmatrix}$ in V.

(c) Determine if $\mathcal{S}$ is a basis for V. Justify your answer.

In Exercises 49–52, use a calculator with matrix capabilities or computer software such as MATLAB to solve the problem.

49. Let

$$A = \begin{bmatrix} 0.1 & 0.2 & 0.34 & 0.5 & -0.09 \\ 0.7 & 0.9 & 1.23 & -0.5 & -1.98 \\ -0.5 & 0.5 & 1.75 & -0.5 & -2.50 \end{bmatrix}.$$

(a) Find a basis for the column space of A.

(b) Find a basis for the null space of A.

50. Show that

$$\left\{ \begin{bmatrix} 29.0 \\ -57.1 \\ 16.0 \\ 4.9 \\ -7.0 \end{bmatrix}, \begin{bmatrix} -26.6 \\ 53.8 \\ -7.0 \\ -9.1 \\ 13.0 \end{bmatrix} \right\}$$

is a basis for the null space of the matrix A in Exercise 49.

51. Show that

$$\left\{ \begin{bmatrix} 1.1 \\ -7.8 \\ -9.0 \end{bmatrix}, \begin{bmatrix} -2.7 \\ 7.6 \\ 4.0 \end{bmatrix}, \begin{bmatrix} 2.5 \\ -4.5 \\ -6.5 \end{bmatrix} \right\}$$

is a basis for the column space of the matrix A in Exercise 49.

52. Let

$$A = \begin{bmatrix} -0.1 & -0.21 & 0.2 & 0.58 & 0.4 & 0.61 \\ 0.3 & 0.63 & -0.1 & -0.59 & -0.5 & -0.81 \\ 1.2 & 2.52 & 0.6 & -0.06 & 0.6 & 0.12 \\ -0.6 & -1.26 & 0.2 & 1.18 & -0.2 & 0.30 \end{bmatrix}.$$

(a) Compute the rank of A, the dimension of Col A, and the dimension of Row A.

(b) Use the result of (a) to make a conjecture about the relationships among the rank of A, the dimension of Col A, and the dimension of Row A for an arbitrary matrix A.

(c) Test your conjecture using randomly generated 4×7 and 6×3 matrices.

Solutions to the Practice Problems

1. The reduced row echelon form of the given matrix A is

$$\begin{bmatrix} 1 & -2 & 0 & 3 \\ 0 & 0 & 1 & 2 \\ 0 & 0 & 0 & 0 \end{bmatrix}.$$

Hence a basis for the column space of A is

$$\left\{ \begin{bmatrix} -1 \\ 2 \\ 1 \end{bmatrix}, \begin{bmatrix} 1 \\ -3 \\ 0 \end{bmatrix} \right\},$$

the set consisting of the pivot columns of A. Thus the dimension of Col A is 2.

Solving the homogeneous system of linear equations having the reduced row echelon form of A as its coefficient matrix, we obtain the parametric form of the general solution to $A\mathbf{x} = \mathbf{0}$, which is

$$x_2 \begin{bmatrix} 2 \\ 1 \\ 0 \\ 0 \end{bmatrix} + x_4 \begin{bmatrix} -3 \\ 0 \\ -2 \\ 1 \end{bmatrix}.$$

The set of vectors in this representation,

$$\left\{ \begin{bmatrix} 2 \\ 1 \\ 0 \\ 0 \end{bmatrix}, \begin{bmatrix} -3 \\ 0 \\ -2 \\ 1 \end{bmatrix} \right\},$$

is a basis for the null space of A. Thus the dimension of Null A is also 2.

2. Let A be the matrix in problem 1. Since each of the vectors in the given set $\mathcal{B}$ is a solution to $A\mathbf{x} = \mathbf{0}$, $\mathcal{B}$ is a subset of the null space of A. Moreover, $\mathcal{B}$ is linearly independent since neither vector in $\mathcal{B}$ is a multiple of the other. Because problem 1 shows that Null A has dimension 2, it follows from Theorem 4.5 that $\mathcal{B}$ is a basis for Null A.

3. Since the given set contains 2 vectors and $\mathcal{R}^3$ has dimension 3, this set cannot be a basis for $\mathcal{R}^3$.

4.3 The Dimensions of Subspaces Associated with a Matrix

In Section 4.2, we defined the dimension of a nonzero subspace V of $\mathcal{R}^n$ to be the unique number of vectors in every basis for V. We now define the dimension of the zero subspace $\{\mathbf{0}\}$ to be 0. We denote this by writing dim $\{\mathbf{0}\} = 0$.

In this section, we investigate the dimensions of several important subspaces, including those defined in Section 4.1.

The first example illustrates an important general result.

Example 1   For the matrix

$$A = \begin{bmatrix} 1 & 2 & -1 & 2 & 1 & 2 \\ -1 & -2 & 1 & 2 & 3 & 6 \\ 2 & 4 & -3 & 2 & 0 & 3 \\ -3 & -6 & 2 & 0 & 3 & 9 \end{bmatrix} \qquad rref$$

in Example 1 of Section 4.2, we saw that the set of pivot columns of A,

$$\mathcal{B} = \left\{ \begin{bmatrix} 1 \\ -1 \\ 2 \\ -3 \end{bmatrix}, \begin{bmatrix} -1 \\ 1 \\ -3 \\ 2 \end{bmatrix}, \begin{bmatrix} 2 \\ 2 \\ 2 \\ 0 \end{bmatrix} \right\},$$

is a basis for Col A. Hence the dimension of Col A is 3.

As in Example 1, the pivot columns of any matrix form a basis for its column space. Hence the dimension of the column space of a matrix equals the number of pivot columns in the matrix. However, the number of pivot columns in a matrix is its rank.

> The dimension of the column space of a matrix equals the rank of the matrix.

The dimensions of the other subspaces associated with a matrix are also determined by the rank of the matrix. For instance, when finding the general solution to a homogeneous system $A\mathbf{x} = \mathbf{0}$, we have seen that the number of free variables that appear is the nullity of A. As in Example 2 of Section 4.2, each free variable in the parametric form of the general solution is multiplied by a vector in a basis for the solution set. Hence the dimension of the solution set of $A\mathbf{x} = \mathbf{0}$ is the nullity of A.

> The dimension of the null space of a matrix equals the nullity of the matrix.

It is useful to be able to find the dimension of a subspace without actually determining a basis for the subspace. In Example 3 of Section 4.2, for instance, we showed that

$$
\mathcal{B} = \left\{ \begin{bmatrix} 1 \\ -1 \\ 1 \\ 0 \end{bmatrix}, \begin{bmatrix} 1 \\ 0 \\ 1 \\ -1 \end{bmatrix}, \begin{bmatrix} 0 \\ 1 \\ 1 \\ -1 \end{bmatrix} \right\}
$$

is a basis for

$$
V = \left\{ \begin{bmatrix} v_1 \\ v_2 \\ v_3 \\ v_4 \end{bmatrix} \in \mathcal{R}^4 \colon v_1 + v_2 + v_4 = 0 \right\}.
$$

Our approach required that we know the dimension of V, and so we solved the equation $v_1 + v_2 + v_4 = 0$ to obtain a basis for V. There is an easier method, however, because $V = \text{Null} \begin{bmatrix} 1 & 1 & 0 & 1 \end{bmatrix}$. Since $\begin{bmatrix} 1 & 1 & 0 & 1 \end{bmatrix}$ is in reduced row echelon form, we see that its rank is 1 and hence its nullity is $4 - 1 = 3$. Thus $\dim V = 3$.

Example 2   Show that

$$
\mathcal{B} = \left\{ \begin{bmatrix} -2 \\ 1 \\ 1 \\ 2 \\ 1 \end{bmatrix}, \begin{bmatrix} 3 \\ -6 \\ -2 \\ -2 \\ -1 \end{bmatrix} \right\}
$$

is a basis for the null space of

$$
A = \begin{bmatrix} 3 & 1 & -2 & 1 & 5 \\ 1 & 0 & 1 & 0 & 1 \\ -5 & -2 & 5 & -5 & -3 \\ -2 & -1 & 3 & 2 & -10 \end{bmatrix}.
$$

Solution   Because each vector in $\mathcal{B}$ is a solution to $A\mathbf{x} = \mathbf{0}$, the set $\mathcal{B}$ is contained in Null A. Moreover, neither vector in $\mathcal{B}$ is a multiple of the other, and so $\mathcal{B}$ is linearly independent. Since the reduced row echelon form of A is

$$\begin{bmatrix} 1 & 0 & 1 & 0 & 1 \\ 0 & 1 & -5 & 0 & 4 \\ 0 & 0 & 0 & 1 & -2 \\ 0 & 0 & 0 & 0 & 0 \end{bmatrix},$$

the rank of A is 3 and its nullity is $5 - 3 = 2$. Hence Null A has dimension 2, and so $\mathcal{B}$ is a basis for Null A by Theorem 4.5.   ○

We now know how the dimensions of the column space and the null space of a matrix are related to the rank of the matrix. So it is natural to turn our attention to the other subspace associated with a matrix, its row space. Because the row space of a matrix A is defined to be the subspace spanned by its rows, we are able to obtain a basis for Row A using the technique in the proof of Theorem 4.3. This amounts to finding a basis for the column space of A^T. Hence this approach shows us that the dimension of Row A is the rank of A^T. In order to express the dimension of Row A in terms of the rank of A, we need another method for finding a basis for Row A.

It is important to note that, unlike the column space of a matrix, the row space is unaffected by elementary row operations. Consider the matrix A and its reduced row echelon form R given by

$$A = \begin{bmatrix} 1 & 1 \\ 2 & 2 \end{bmatrix} \quad \text{and} \quad R = \begin{bmatrix} 1 & 1 \\ 0 & 0 \end{bmatrix}.$$

Notice that

$$\text{Row } A = \text{Row } R = \text{Span}\{[1 \ \ 1]\},$$

but

$$\text{Col } A = \text{Span}\left\{ \begin{bmatrix} 1 \\ 2 \end{bmatrix} \right\} \neq \text{Col } R = \text{Span}\left\{ \begin{bmatrix} 1 \\ 0 \end{bmatrix} \right\}.$$

It follows from the next theorem that the row space of a matrix and the row space of its reduced row echelon form are always equal.

Theorem 4.6

The nonzero rows of the reduced row echelon form of a matrix form a basis for the row space of the matrix.

Proof   Let R be the reduced row echelon form of a matrix A. Since R is in reduced row echelon form, the leading entry of each nonzero row of R is the only nonzero entry in its column. Thus no nonzero row of R is a linear combination of other rows; hence the nonzero rows of R are linearly independent, and clearly they are also a spanning set for Row R. Therefore the nonzero rows of R are a basis for Row R.

To complete the proof, we need only show that Row $A = $ Row R. Because R is obtained from A by elementary row operations, each row of R is a linear combination of the rows of A. Thus Row R is contained in Row A by Theorem 1.6(b). Since elementary row operations are reversible, each row of A must also be a linear combination of the rows of R. Therefore Row A is also contained in Row R. It follows that Row $A = $ Row R, completing the proof.   ❑

Example 3   Recall that for the matrix

$$A = \begin{bmatrix} 3 & 1 & -2 & 1 & 5 \\ 1 & 0 & 1 & 0 & 1 \\ -5 & -2 & 5 & -5 & -3 \\ -2 & -1 & 3 & 2 & -10 \end{bmatrix}$$

in Example 2, the reduced row echelon form is

$$\begin{bmatrix} 1 & 0 & 1 & 0 & 1 \\ 0 & 1 & -5 & 0 & 4 \\ 0 & 0 & 0 & 1 & -2 \\ 0 & 0 & 0 & 0 & 0 \end{bmatrix}.$$

Hence a basis for Row A is

$$\{[1 \quad 0 \quad 1 \quad 0 \quad 1], [0 \quad 1 \quad -5 \quad 0 \quad 4], [0 \quad 0 \quad 0 \quad 1 \quad -2]\}.$$

Thus the dimension of Row A is 3, which is the rank of A.   ○

As Example 3 illustrates, Theorem 4.6 yields the following important fact.

> The dimension of the row space of a matrix equals its rank.

We have noted that the row and column spaces of a matrix are rarely equal. (For instance, the row space of the matrix A in Example 3 is a subspace of $\mathcal{R}^5$, whereas its column space is a subspace of $\mathcal{R}^4$.) Nevertheless, the results of this section show that their dimensions are always the same. It follows that $\dim(\text{Row } A) = \dim(\text{Col } A) = \dim(\text{Row } A^T)$, and so we have the following result.

> The rank of any matrix equals the rank of its transpose.

The results in this section are easily extended from matrices to linear transformations. Recall from Section 4.1 that the null space of a linear transformation $T: \mathcal{R}^n \to \mathcal{R}^m$ is equal to that of its standard matrix A, and the range of T is equal to the column space of A. Hence the dimension of the null space of T is the nullity of A, and the dimension of the range of T is the rank of A. It follows that *the sum of the dimensions of the null space and range of T is the dimension of the domain of T* (see Exercise 35).

Subspaces Contained within Subspaces

Suppose that both V and W are subspaces of $\mathcal{R}^n$ such that V is contained in W. Because V is contained in W, it is natural to expect that the dimension of V would be less than or equal to the dimension of W. This expectation is indeed correct, as our next result shows.

Theorem 4.7
If V and W are subspaces of $\mathcal{R}^n$ such that V is contained in W, then $\dim V \leq \dim W$. *Moreover, if V and W also have the same dimension, then $V = W$.*

Proof It is easy to verify this theorem if V is the zero subspace. So assume that V is a nonzero subspace, and let $\mathcal{B}$ be a basis for V. By the extension principle, $\mathcal{B}$ is contained in a basis for W, and so $\dim V \leq \dim W$.

Suppose also that both V and W have dimension k. Then $\mathcal{B}$ is a linearly independent subset of W that contains k vectors, and so $\mathcal{B}$ is a basis for W by Theorem 4.5. Therefore $V = \operatorname{Span} \mathcal{B} = W$.   ∎

Theorem 4.7 enables us to characterize the subspaces of $\mathcal{R}^n$ by dimension. For example, this theorem shows that a subspace of $\mathcal{R}^3$ must have dimension 0, 1, 2, or 3. First, a subspace of dimension 0 must be the zero subspace. Second, a subspace of dimension 1 must have a basis $\{\mathbf{u}\}$ consisting of a single nonzero vector, and so the subspace must consist of all vectors that are multiples of $\mathbf{u}$. As noted in Example 5 in Section 4.1, such a set can be depicted as a line through the origin of $\mathcal{R}^3$. Third, a subspace of dimension 2 must have a basis $\{\mathbf{u}, \mathbf{v}\}$ consisting of two vectors, neither of which is a multiple of the other. In this case the subspace consists of all vectors in $\mathcal{R}^3$ having the form $a\mathbf{u} + b\mathbf{v}$ for some scalars a and b. As in Example 2 of Section 1.6, such a set can be depicted as a plane through the origin of $\mathcal{R}^3$. Finally, a subspace of $\mathcal{R}^3$ having dimension 3 must be $\mathcal{R}^3$ itself, by Theorem 4.7.

We conclude this section with a table summarizing some of the facts from Sections 4.1 and 4.3 concerning important subspaces associated with an $m \times n$ matrix A. (This table also applies to a linear transformation $T: \mathcal{R}^n \to \mathcal{R}^m$ by taking A to be the standard matrix of T.) Again, we remind the reader that the three subspaces mentioned here are usually all distinct.

The Dimensions of the Subspaces Associated with an $m \times n$ Matrix A

Subspace	Containing space	Dimension
Null A	$\mathcal{R}^n$	nullity $A = n - \operatorname{rank} A$
Row A	$\mathcal{R}^n$	rank A
Col A	$\mathcal{R}^m$	rank A

Practice Problems

1. Find the dimensions of the column space, null space, and row space of
$$\begin{bmatrix} 1 & 2 & 3 & -2 & -1 \\ 0 & 0 & 1 & 3 & 2 \\ 2 & 4 & 7 & 0 & 1 \\ 3 & 6 & 11 & 1 & 2 \end{bmatrix}.$$

2. Show that $\left\{ \begin{bmatrix} 0 \\ -2 \\ 1 \\ -1 \\ 1 \end{bmatrix}, \begin{bmatrix} 2 \\ 1 \\ -1 \\ 1 \\ -1 \end{bmatrix} \right\}$ is a basis for the null space of the matrix in problem 1.

Exercises

1. Determine if the following statements are true or false.

   (a) If V and W are subspaces of $\mathcal{R}^n$ having the same dimension, then $V = W$.

   (b) If V is a subspace of $\mathcal{R}^n$ having dimension n, then $V = \mathcal{R}^n$.

   (c) If V is a subspace of $\mathcal{R}^n$ having dimension 0, then $V = \{\mathbf{0}\}$.

   (d) The dimension of the null space of a matrix equals the rank of the matrix.

   (e) The dimension of the column space of a matrix equals the nullity of the matrix.

   (f) The dimension of the row space of a matrix equals the rank of the matrix.

   (g) The row space of any matrix equals the row space of its reduced row echelon form.

(h) The column space of any matrix equals the column space of its reduced row echelon form.

(i) The null space of any matrix equals the null space of its reduced row echelon form.

In Exercises 2–4, determine the dimension of the given subspace.

2. $\left\{ \begin{bmatrix} s \\ 0 \\ 2s \end{bmatrix} \in \mathcal{R}^3 \colon s \in \mathcal{R} \right\}$

3. $\left\{ \begin{bmatrix} -3s + 4t \\ s - 2t \\ 2s \end{bmatrix} \in \mathcal{R}^3 \colon s, t \in \mathcal{R} \right\}$

4. $\left\{ \begin{bmatrix} s + 2t \\ 0 \\ 3t \end{bmatrix} \in \mathcal{R}^3 \colon s, t \in \mathcal{R} \right\}$

In Exercises 5–8, the reduced row echelon form of a matrix A is given. Determine the dimension of (a) Col A, (b) Null A, (c) Row A, and (d) Null A^T.

5. $\begin{bmatrix} 1 & -3 & 0 & 2 \\ 0 & 0 & 1 & -4 \\ 0 & 0 & 0 & 0 \end{bmatrix}$

6. $\begin{bmatrix} 1 & 0 & -2 & 0 \\ 0 & 1 & 5 & 0 \\ 0 & 0 & 0 & 1 \end{bmatrix}$

7. $\begin{bmatrix} 1 & -1 & 0 & 2 & 0 \\ 0 & 0 & 1 & 6 & 0 \\ 0 & 0 & 0 & 0 & 1 \end{bmatrix}$

8. $\begin{bmatrix} 1 & 0 & 0 & -4 & 2 \\ 0 & 1 & 0 & 2 & -1 \\ 0 & 0 & 1 & -3 & 1 \\ 0 & 0 & 0 & 0 & 0 \end{bmatrix}$

In Exercises 9–16, a matrix A is given. Determine the dimension of (a) Col A, (b) Null A, (c) Row A, and (d) Null A^T.

9. $[2 \quad -8 \quad -4 \quad 6]$

10. $\begin{bmatrix} 1 & 2 & -3 \\ 0 & -1 & -1 \\ 1 & 4 & -1 \end{bmatrix}$

11. $\begin{bmatrix} 1 & -1 & 2 \\ 2 & -3 & 1 \end{bmatrix}$

12. $\begin{bmatrix} 1 & -2 & 3 \\ -3 & 6 & -9 \end{bmatrix}$

13. $\begin{bmatrix} 1 & 1 & 2 & 1 \\ -1 & -2 & 2 & -2 \\ 2 & 3 & 0 & 3 \end{bmatrix}$

14. $\begin{bmatrix} -1 & 2 & 1 & -1 & -2 \\ 2 & -4 & 1 & 5 & 7 \\ 2 & -4 & -3 & 1 & 3 \end{bmatrix}$

15. $\begin{bmatrix} 1 & 1 & 1 \\ 1 & -1 & 5 \\ 2 & 1 & 4 \\ 0 & 2 & -4 \end{bmatrix}$

16. $\begin{bmatrix} 0 & -1 & 1 \\ 1 & 2 & -3 \\ 3 & 1 & -2 \\ -1 & 0 & 4 \end{bmatrix}$

In Exercises 17–24, determine the dimension of the (a) range and (b) null space of the given linear transformation T. Use this information to determine if T is one-to-one or onto.

17. $T\left(\begin{bmatrix} x_1 \\ x_2 \end{bmatrix} \right) = \begin{bmatrix} x_1 + x_2 \\ 2x_1 + x_2 \end{bmatrix}$

18. $T\left(\begin{bmatrix} x_1 \\ x_2 \end{bmatrix} \right) = \begin{bmatrix} x_1 - 3x_2 \\ -3x_1 + 9x_2 \end{bmatrix}$

19. $T\left(\begin{bmatrix} x_1 \\ x_2 \\ x_3 \end{bmatrix} \right) = \begin{bmatrix} -x_1 + 2x_2 + x_3 \\ x_1 - 2x_2 - x_3 \end{bmatrix}$

20. $T\left(\begin{bmatrix} x_1 \\ x_2 \\ x_3 \end{bmatrix} \right) = \begin{bmatrix} -x_1 + x_2 + 2x_3 \\ x_1 - 3x_3 \end{bmatrix}$

21. $T\left(\begin{bmatrix} x_1 \\ x_2 \end{bmatrix} \right) = \begin{bmatrix} x_1 \\ 2x_1 + x_2 \\ -x_2 \end{bmatrix}$

22. $T\left(\begin{bmatrix} x_1 \\ x_2 \end{bmatrix} \right) = \begin{bmatrix} x_1 - x_2 \\ -2x_1 + 3x_2 \\ x_1 \end{bmatrix}$

23. $T\left(\begin{bmatrix} x_1 \\ x_2 \\ x_3 \end{bmatrix} \right) = \begin{bmatrix} -x_1 - x_2 + x_3 \\ x_1 + 2x_2 + x_3 \end{bmatrix}$

24. $T\left(\begin{bmatrix} x_1 \\ x_2 \\ x_3 \end{bmatrix} \right) = \begin{bmatrix} -x_1 - x_2 + x_3 \\ x_1 + 2x_2 + x_3 \\ x_1 + x_2 \end{bmatrix}$

In Exercises 25–32, use the results of this section to show that B is a basis for the given subspace V.

25. $B = \{e_1, e_2\}, V = \left\{ \begin{bmatrix} 2s - t \\ s + 3t \end{bmatrix} \in \mathcal{R}^2 \colon s, t \in \mathcal{R} \right\}$

26. $B = \{e_1, e_2\}, V = \left\{ \begin{bmatrix} -2t \\ 5s + 3t \end{bmatrix} \in \mathcal{R}^2 \colon s, t \in \mathcal{R} \right\}$

27. $B = \left\{ \begin{bmatrix} 3 \\ 1 \\ 0 \end{bmatrix}, \begin{bmatrix} 2 \\ 1 \\ -1 \end{bmatrix} \right\}, V = \left\{ \begin{bmatrix} 4t \\ s + t \\ -3s + t \end{bmatrix} \in \mathcal{R}^3 \colon s, t \in \mathcal{R} \right\}$

28. $B = \left\{ \begin{bmatrix} 0 \\ 1 \\ 4 \end{bmatrix}, \begin{bmatrix} 4 \\ -7 \\ 0 \end{bmatrix} \right\}, V = \left\{ \begin{bmatrix} -s + t \\ 2s - t \\ s + 3t \end{bmatrix} \in \mathcal{R}^3 \colon s, t \in \mathcal{R} \right\}$

29. $B = \left\{ \begin{bmatrix} 1 \\ 0 \\ 0 \\ 0 \end{bmatrix}, \begin{bmatrix} 1 \\ 0 \\ -1 \\ -1 \end{bmatrix}, \begin{bmatrix} 1 \\ 0 \\ 1 \\ 2 \end{bmatrix} \right\},$

$V = \left\{ \begin{bmatrix} -r + 3s \\ 0 \\ s - t \\ r - 2t \end{bmatrix} \in \mathcal{R}^4 \colon r, s, t \in \mathcal{R} \right\}$

30. $\mathcal{B} = \left\{ \begin{bmatrix} 0 \\ 1 \\ 0 \\ 2 \end{bmatrix}, \begin{bmatrix} 1 \\ -1 \\ 0 \\ 0 \end{bmatrix}, \begin{bmatrix} 1 \\ -1 \\ 0 \\ 1 \end{bmatrix} \right\}$,

$V = \left\{ \begin{bmatrix} -r + s \\ 4s - 3t \\ 0 \\ 3r - t \end{bmatrix} \in \mathcal{R}^4 : r, s, t \in \mathcal{R} \right\}$

31. $\mathcal{B} = \left\{ \begin{bmatrix} 4 \\ 1 \\ 1 \\ -1 \end{bmatrix}, \begin{bmatrix} 0 \\ 9 \\ -3 \\ -8 \end{bmatrix}, \begin{bmatrix} 3 \\ 0 \\ 15 \\ 4 \end{bmatrix} \right\}$,

$V = \left\{ \begin{bmatrix} r - s + 3t \\ 2r - t \\ -r + 3s + 2t \\ -2r + s + t \end{bmatrix} \in \mathcal{R}^4 : r, s, t \in \mathcal{R} \right\}$

32. $\mathcal{B} = \left\{ \begin{bmatrix} 1 \\ 0 \\ 5 \\ -4 \end{bmatrix}, \begin{bmatrix} -4 \\ 2 \\ -5 \\ 5 \end{bmatrix}, \begin{bmatrix} 1 \\ 9 \\ 8 \\ -7 \end{bmatrix} \right\}$,

$V = \left\{ \begin{bmatrix} 2s - 5t \\ 3r + s - 2t \\ r - 4s + 3t \\ -r + 2s \end{bmatrix} \in \mathcal{R}^4 : r, s, t \in \mathcal{R} \right\}$

33. (a) Find bases for the row space and null space of the matrix in Exercise 13.
(b) Show that the union of the two bases in (a) is a basis for $\mathcal{R}^4$.

34. (a) Find bases for the row space and null space of the matrix in Exercise 15.
(b) Show that the union of the two bases in (a) is a basis for $\mathcal{R}^3$.

35. Let $T: \mathcal{R}^n \to \mathcal{R}^m$ be a linear transformation. Prove the *dimension theorem:* The sum of the dimensions of the null space and range of T is n.

36. Is there a 3×3 matrix whose null space and column space are equal? Justify your answer.

37. Prove that, for any $m \times n$ matrix A and any $n \times p$ matrix B, the column space of AB is contained in the column space of A.

38. Prove that, for any $m \times n$ matrix A and any $n \times p$ matrix B, the null space of B is contained in the null space of AB.

39. Use Exercise 37 to prove that, for any $m \times n$ matrix A and any $n \times p$ matrix B, rank $AB \leq$ rank A.

40. Use Exercise 39 to prove that, for any $m \times n$ matrix A and any $n \times n$ matrix B, nullity $A \leq$ nullity AB.

41. Use Exercise 39 to prove that, for any $m \times n$ matrix A and any $n \times p$ matrix B, rank $AB \leq$ rank B. *Hint:* The ranks of AB and $(AB)^T$ are equal.

42. Use Exercise 41 to prove that, for any $m \times n$ matrix A and any $n \times p$ matrix B, nullity $B \leq$ nullity AB.

43. Find two-dimensional subspaces V and W of $\mathcal{R}^5$ such that $V \cap W = \{\mathbf{0}\}$.

44. Prove that if V and W are three-dimensional subspaces of $\mathcal{R}^5$, then $V \cap W \neq \{\mathbf{0}\}$.

45. Let V be a k-dimensional subspace of $\mathcal{R}^n$ having a basis $\{\mathbf{u}_1, \mathbf{u}_2, \ldots, \mathbf{u}_k\}$. Define a function $T: \mathcal{R}^k \to \mathcal{R}^n$ by

$$T\left(\begin{bmatrix} x_1 \\ x_2 \\ \vdots \\ x_k \end{bmatrix} \right) = x_1 \mathbf{u}_1 + x_2 \mathbf{u}_2 + \cdots + x_k \mathbf{u}_k.$$

(a) Prove that T is a linear transformation.
(b) Prove that T is one-to-one.
(c) Prove that the range of T is V.

46. Let V be a subspace of $\mathcal{R}^n$ and $\mathbf{u}$ be a vector in $\mathcal{R}^n$ that is not in V. Define

$$W = \{\mathbf{v} + c\mathbf{u}: \mathbf{v} \text{ is in } V \text{ and } c \text{ is a scalar}\}.$$

(a) Prove that W is a subspace of $\mathcal{R}^n$.
(b) Determine the dimension of W. Justify your answer.

47. (a) Show that for any vector $\mathbf{u}$ in $\mathcal{R}^n$, $\mathbf{u}^T \mathbf{u} = 0$ if and only if $\mathbf{u} = \mathbf{0}$.
(b) Prove that for any matrix A, if $\mathbf{u} \in$ Row A and $\mathbf{v} \in$ Null A, then $\mathbf{u}^T \mathbf{v} = 0$. *Hint:* The row space of A equals the column space of A^T.
(c) Show that, for any matrix A, if $\mathbf{u}$ belongs to both Row A and Null A, then $\mathbf{u} = \mathbf{0}$.

48. Show that, for any $m \times n$ matrix A, the union of a basis for Row A and a basis for Null A is a basis for $\mathcal{R}^n$. *Hint:* Use Exercise 47(c).

49. Let

$$A = \begin{bmatrix} 1 & 0 & -1 & -2 \\ -1 & 1 & 2 & 1 \\ 1 & 3 & 2 & -5 \\ -1 & 6 & 7 & -4 \end{bmatrix}.$$

(a) Find a 4×4 matrix B with rank 2 such that $AB = O$.
(b) Prove that if C is a 4×4 matrix such that $AC = O$, then rank $C \leq 2$.

In Exercises 50–52, use a calculator with matrix capabilities or computer software such as MATLAB *to solve the problem.*

50. Let

$$\mathcal{B} = \left\{ \begin{bmatrix} -1 \\ 1 \\ 1 \\ 1 \\ 0 \end{bmatrix}, \begin{bmatrix} 0 \\ 1 \\ -1 \\ 1 \\ 1 \end{bmatrix}, \begin{bmatrix} 1 \\ 0 \\ 3 \\ -1 \\ -2 \end{bmatrix} \right\}.$$

Show that $\mathcal{B}$ is linearly independent and hence is a basis for a subspace W of $\mathcal{R}^5$.

51. Let

$$\mathcal{A}_1 = \left\{ \begin{bmatrix} 1.0 \\ 2.0 \\ 1.0 \\ 1.0 \\ 0.1 \end{bmatrix}, \begin{bmatrix} -1.0 \\ 3.0 \\ 4.0 \\ 2.0 \\ -0.6 \end{bmatrix}, \begin{bmatrix} 2.0 \\ -1.0 \\ -1.0 \\ -1.4 \\ -0.3 \end{bmatrix} \right\}$$

and $$\mathcal{A}_2 = \left\{ \begin{bmatrix} 2 \\ 1 \\ 0 \\ 0 \\ 0 \end{bmatrix}, \begin{bmatrix} -3 \\ 0 \\ 1 \\ 1 \\ 0 \end{bmatrix}, \begin{bmatrix} 1 \\ 0 \\ -2 \\ 0 \\ 1 \end{bmatrix} \right\}.$$

(a) Is $\mathcal{A}_1$ a basis for the subspace W in Exercise 50? Justify your answer.

(b) Is $\mathcal{A}_2$ a basis for the subspace W in Exercise 50? Justify your answer.

(c) Let B, A_1, and A_2 be the matrices whose columns are the vectors in $\mathcal{B}$, $\mathcal{A}_1$, and $\mathcal{A}_2$, respectively. Compute the reduced row echelon forms of B, A_1, A_2, B^T, A_1^T, and A_2^T.

52. Let $\mathcal{B}$ be a basis for a subspace W of $\mathcal{R}^n$ and B be the matrix whose columns are the vectors in $\mathcal{B}$. Suppose that $\mathcal{A}$ is a set of vectors in $\mathcal{R}^n$ and A is the matrix whose columns are the vectors in $\mathcal{A}$. Use the results of Exercise 51 to devise a test for $\mathcal{A}$ to be a basis for W. (The test should involve the matrices A and B.) Then prove that your test is valid.

Solutions to the Practice Problems

1. The reduced row echelon form of the given matrix A is

$$\begin{bmatrix} 1 & 2 & 0 & 0 & 4 \\ 0 & 0 & 1 & 0 & -1 \\ 0 & 0 & 0 & 1 & 1 \\ 0 & 0 & 0 & 0 & 0 \end{bmatrix},$$

and so its rank is 3. The dimensions of both the column space and the row space of A equal the rank of A, and so these dimensions are 3. The dimension of the null space of A is the nullity of A, which is

$$5 - \text{rank } A = 5 - 3 = 2.$$

2. Clearly

$$\begin{bmatrix} 1 & 2 & 3 & -2 & -1 \\ 0 & 0 & 1 & 3 & 2 \\ 2 & 4 & 7 & 0 & 1 \\ 3 & 6 & 11 & 1 & 2 \end{bmatrix} \begin{bmatrix} 0 \\ -2 \\ 1 \\ -1 \\ 1 \end{bmatrix} = \begin{bmatrix} 0 \\ 0 \\ 0 \\ 0 \end{bmatrix}$$

and

$$\begin{bmatrix} 1 & 2 & 3 & -2 & -1 \\ 0 & 0 & 1 & 3 & 2 \\ 2 & 4 & 7 & 0 & 1 \\ 3 & 6 & 11 & 1 & 2 \end{bmatrix} \begin{bmatrix} 2 \\ 1 \\ -1 \\ 1 \\ -1 \end{bmatrix} = \begin{bmatrix} 0 \\ 0 \\ 0 \\ 0 \end{bmatrix}.$$

Thus both

$$\begin{bmatrix} 0 \\ -2 \\ 1 \\ -1 \\ 1 \end{bmatrix} \quad \text{and} \quad \begin{bmatrix} 2 \\ 1 \\ -1 \\ 1 \\ -1 \end{bmatrix}$$

belong to Null A. Moreover, neither of these vectors is a multiple of the other, and so they are linearly independent. Since Null A has dimension 2, it follows from Theorem 4.5 that

$$\left\{ \begin{bmatrix} 0 \\ -2 \\ 1 \\ -1 \\ 1 \end{bmatrix}, \begin{bmatrix} 2 \\ 1 \\ -1 \\ 1 \\ -1 \end{bmatrix} \right\}$$

is a basis for Null A.

4.4 Coordinate Systems

Figure 4.3 shows an ellipse where the axes of symmetry are the lines $y = x$ and $y = -x$ and the lengths of the semimajor and semiminor axes are 3 and 2, respectively. Consider a new x', y'-coordinate system in which the x'-axis is the line $y = x$ and the y'-axis is the line $y = -x$. Because the ellipse is in standard position with respect to this coordinate system, the equation of this ellipse in the x', y'-coordinate system is

$$\frac{(x')^2}{3^2} + \frac{(y')^2}{2^2} = 1.$$

But what is the equation of the ellipse in the usual x, y-coordinate system?

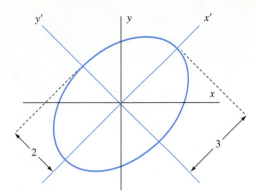

FIGURE 4.3

To answer this question requires a careful examination of what is meant by a coordinate system. When we say that $\mathbf{v} = \begin{bmatrix} 8 \\ 4 \end{bmatrix}$, we mean that $\mathbf{v} = 8\mathbf{e}_1 + 4\mathbf{e}_2$. That is, the components of $\mathbf{v}$ are the coefficients used to represent $\mathbf{v}$ as a linear combination of the standard vectors. See Figure 4.4.

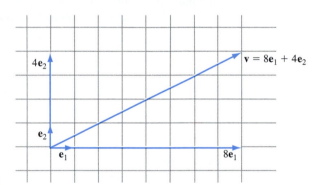

FIGURE 4.4

Any basis for a subspace provides a means for establishing a coordinate system on the subspace. For the ellipse in Figure 4.3, the standard basis is not the most useful one. A more useful basis is $\{\mathbf{b}_1, \mathbf{b}_2\}$, where $\mathbf{b}_1 = \begin{bmatrix} 1 \\ 1 \end{bmatrix}$ is a vector that lies along the x'-axis and $\mathbf{b}_2 = \begin{bmatrix} -1 \\ 1 \end{bmatrix}$ is a vector that lies along the y'-axis. (Note that since $\mathbf{b}_1$ and $\mathbf{b}_2$ are not multiples of one another, $\mathcal{B} = \{\mathbf{b}_1, \mathbf{b}_2\}$ is a linearly independent subset of two vectors from $\mathcal{R}^2$. Hence, by Theorem 4.5, $\{\mathbf{b}_1, \mathbf{b}_2\}$ is a basis for $\mathcal{R}^2$.) For this basis, the vector $\mathbf{v} = \begin{bmatrix} 8 \\ 4 \end{bmatrix}$ can be written $\mathbf{v} = 6\mathbf{b}_1 + (-2)\mathbf{b}_2$. Thus the coordinates of $\mathbf{v}$ relative to $\mathcal{B}$ are 6 and -2, as shown in Figure 4.5. In this case, the coordinates of $\mathbf{v}$ are different from those in the usual coordinate system for $\mathcal{R}^2$.

There is an obvious advantage to working with the standard basis for $\mathcal{R}^2$: the coordinates of any vector are easily determined. In fact, the coordinates of the vector

$$\mathbf{u} = \begin{bmatrix} a \\ b \end{bmatrix}$$

are simply a and b. On the other hand, the coordinates of a vector induced by the basis $\mathcal{B}$ are not as easily found. (They can be determined, however, by writing $\mathbf{u}$ as a linear combination of $\mathbf{b}_1$ and $\mathbf{b}_2$.)

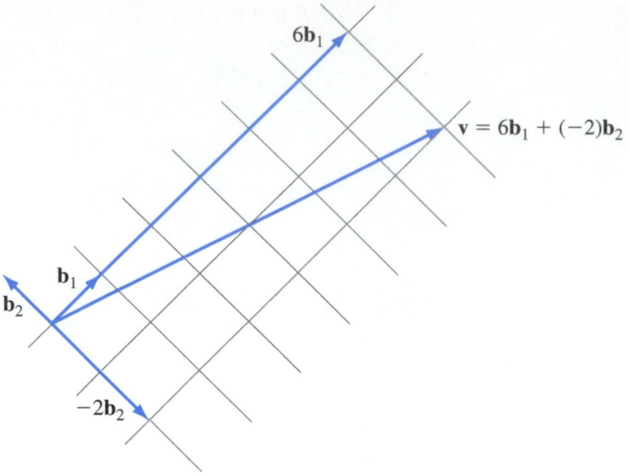

FIGURE 4.5

In order to identify the vector **v** with the coefficients 6 and -2 in the linear combination $\mathbf{v} = 6\mathbf{b}_1 + (-2)\mathbf{b}_2$, we need to know that **v** cannot be represented as another linear combination of $\mathbf{b}_1$ and $\mathbf{b}_2$. Consider, for example, the set

$$\left\{ \begin{bmatrix} 1 \\ 1 \end{bmatrix}, \begin{bmatrix} -1 \\ 1 \end{bmatrix}, \begin{bmatrix} 1 \\ 0 \end{bmatrix} \right\}.$$

We can represent **v** not only as

$$6 \begin{bmatrix} 1 \\ 1 \end{bmatrix} + (-2) \begin{bmatrix} -1 \\ 1 \end{bmatrix} + 0 \begin{bmatrix} 1 \\ 0 \end{bmatrix},$$

but also as

$$0 \begin{bmatrix} 1 \\ 1 \end{bmatrix} + 4 \begin{bmatrix} -1 \\ 1 \end{bmatrix} + 12 \begin{bmatrix} 1 \\ 0 \end{bmatrix}.$$

The following theorem assures us that this problem does not arise if the set of vectors forms a basis.

Theorem 4.8

Let $\mathcal{B} = \{\mathbf{u}_1, \mathbf{u}_2, \ldots, \mathbf{u}_k\}$ be a basis for a subspace V of $\mathcal{R}^n$. Any vector **v** in V can be uniquely represented as a linear combination of the vectors in $\mathcal{B}$; that is, there are unique scalars $a_1, a_2, \ldots, a_k$ such that $\mathbf{v} = a_1\mathbf{u}_1 + a_2\mathbf{u}_2 + \cdots + a_k\mathbf{u}_k$.

Proof Let $\mathbf{v} \in V$. Since $\mathcal{B}$ is a spanning set for V, every vector in V is a linear combination of $\mathbf{u}_1, \mathbf{u}_2, \ldots, \mathbf{u}_k$. Hence there exist scalars $a_1, a_2, \ldots, a_k$ such that $\mathbf{v} = a_1\mathbf{u}_1 + a_2\mathbf{u}_2 + \cdots + a_k\mathbf{u}_k$.

Assume also that $\mathbf{v} = b_1\mathbf{u}_1 + b_2\mathbf{u}_2 + \cdots + b_k\mathbf{u}_k$, where $b_1, b_2, \ldots, b_k$ are scalars. To show that the coefficients in the linear combination are unique, we prove that each b_i equals the corresponding a_i. Now

$$0 = \mathbf{v} - \mathbf{v}$$
$$= (a_1\mathbf{u}_1 + a_2\mathbf{u}_2 + \cdots + a_k\mathbf{u}_k) - (b_1\mathbf{u}_1 + b_2\mathbf{u}_2 + \cdots + b_k\mathbf{u}_k)$$
$$= (a_1 - b_1)\mathbf{u}_1 + (a_2 - b_2)\mathbf{u}_2 + \cdots + (a_k - b_k)\mathbf{u}_k.$$

Because $\mathcal{B}$ is linearly independent, this equation implies that $a_1 - b_1 = 0$, $a_2 - b_2 = 0, \ldots, a_k - b_k = 0$. Thus $a_1 = b_1, a_2 = b_2, \ldots, a_k = b_k$, which proves that the representation of **v** as a linear combination of the vectors in $\mathcal{B}$ is unique. ❑

The conclusion of Theorem 4.8 is of great practical value. A nonzero subspace V of $\mathcal{R}^n$ contains infinitely many vectors. But V has a finite basis $\mathcal{B}$, and every vector in V can be uniquely expressed as a linear combination of the vectors in $\mathcal{B}$. Thus we are able to study the infinitely many vectors in V by working with linear combinations of the finite number of vectors in $\mathcal{B}$.

Coordinate Vectors

We have seen that each vector in $\mathcal{R}^n$ can be uniquely represented as a linear combination of the vectors in any basis for $\mathcal{R}^n$. This representation can be used to introduce a coordinate system into $\mathcal{R}^n$ in the following manner.

Definition.   Let $\mathcal{B} = \{\mathbf{b}_1, \mathbf{b}_2, \ldots, \mathbf{b}_n\}$ be a basis[1] for $\mathcal{R}^n$. For each $\mathbf{v}$ in $\mathcal{R}^n$, there are unique scalars $c_1, c_2, \ldots, c_n$ such that $\mathbf{v} = c_1\mathbf{b}_1 + c_2\mathbf{b}_2 + \cdots + c_n\mathbf{b}_n$. The vector

$$\begin{bmatrix} c_1 \\ c_2 \\ \vdots \\ c_n \end{bmatrix} \in \mathcal{R}^n$$

is called the **coordinate vector** of $\mathbf{v}$ relative to $\mathcal{B}$ or the $\mathcal{B}$-**coordinate vector** of $\mathbf{v}$. We denote the $\mathcal{B}$-coordinate vector of $\mathbf{v}$ by $[\mathbf{v}]_\mathcal{B}$.

Example 1   Let

$$\mathcal{B} = \left\{ \begin{bmatrix} 1 \\ 1 \\ 1 \end{bmatrix}, \begin{bmatrix} 1 \\ -1 \\ 1 \end{bmatrix}, \begin{bmatrix} 1 \\ 2 \\ 2 \end{bmatrix} \right\}.$$

(Since $\mathcal{B}$ is a linearly independent set of three vectors in $\mathcal{R}^3$, $\mathcal{B}$ is a basis for $\mathcal{R}^3$.) Calculate $\mathbf{u}$ if

$$[\mathbf{u}]_\mathcal{B} = \begin{bmatrix} 3 \\ 6 \\ -2 \end{bmatrix}.$$

Solution   Since the components of $[\mathbf{u}]_\mathcal{B}$ are the coefficients that express $\mathbf{u}$ as a linear combination of the vectors in $\mathcal{B}$, we have

$$\mathbf{u} = 3 \begin{bmatrix} 1 \\ 1 \\ 1 \end{bmatrix} + 6 \begin{bmatrix} 1 \\ -1 \\ 1 \end{bmatrix} + (-2) \begin{bmatrix} 1 \\ 2 \\ 2 \end{bmatrix} = \begin{bmatrix} 7 \\ -7 \\ 5 \end{bmatrix}.$$

Example 2   For

$$\mathbf{v} = \begin{bmatrix} 1 \\ -4 \\ 4 \end{bmatrix} \quad \text{and} \quad \mathcal{B} = \left\{ \begin{bmatrix} 1 \\ 1 \\ 1 \end{bmatrix}, \begin{bmatrix} 1 \\ -1 \\ 1 \end{bmatrix}, \begin{bmatrix} 1 \\ 2 \\ 2 \end{bmatrix} \right\},$$

determine $[\mathbf{v}]_\mathcal{B}$.

---

[1] For the definition of a coordinate vector to be unambiguous, we must assume that the vectors in $\mathcal{B}$ are listed in the specific sequence $\mathbf{b}_1, \mathbf{b}_2, \ldots, \mathbf{b}_n$. When working with coordinate vectors, we always make this assumption. If we wish to emphasize this ordering, we refer to $\mathcal{B}$ as an *ordered basis*.

Solution   To find the $\mathcal{B}$-coordinate vector of $\mathbf{v}$, we must write $\mathbf{v}$ as a linear combination of the vectors in $\mathcal{B}$. As we learned in Chapter 1, this requires us to find scalars c_1, c_2, and c_3 such that

$$c_1 \begin{bmatrix} 1 \\ 1 \\ 1 \end{bmatrix} + c_2 \begin{bmatrix} 1 \\ -1 \\ 1 \end{bmatrix} + c_3 \begin{bmatrix} 1 \\ 2 \\ 2 \end{bmatrix} = \begin{bmatrix} 1 \\ -4 \\ 4 \end{bmatrix}.$$

From this equation, we obtain a system of linear equations with augmented matrix

$$\begin{bmatrix} 1 & 1 & 1 & 1 \\ 1 & -1 & 2 & -4 \\ 1 & 1 & 2 & 4 \end{bmatrix}.$$

Since the reduced row echelon form of this matrix is

$$\begin{bmatrix} 1 & 0 & 0 & -6 \\ 0 & 1 & 0 & 4 \\ 0 & 0 & 1 & 3 \end{bmatrix},$$

we see that the desired scalars are $c_1 = -6$, $c_2 = 4$, $c_3 = 3$. Thus

$$[\mathbf{v}]_{\mathcal{B}} = \begin{bmatrix} -6 \\ 4 \\ 3 \end{bmatrix}$$

is the $\mathcal{B}$-coordinate vector of $\mathbf{v}$. ○

Coordinate vectors relative to the standard basis $\mathcal{E}$ for $\mathcal{R}^n$ are easily computed. Because every vector

$$\mathbf{v} = \begin{bmatrix} v_1 \\ v_2 \\ \vdots \\ v_n \end{bmatrix}$$

in $\mathcal{R}^n$ can be written as $\mathbf{v} = v_1 \mathbf{e}_1 + v_2 \mathbf{e}_2 + \cdots + v_n \mathbf{e}_n$, we see that $[\mathbf{v}]_{\mathcal{E}} = \mathbf{v}$.

It is also easy to compute the $\mathcal{B}$-coordinate vector of any member of the basis $\mathcal{B}$. For if $\mathcal{B} = \{\mathbf{b}_1, \mathbf{b}_2, \ldots, \mathbf{b}_n\}$ is a basis for $\mathcal{R}^n$, we have

$$\mathbf{b}_i = 0\mathbf{b}_1 + 0\mathbf{b}_2 + \cdots + 0\mathbf{b}_{i-1} + 1\mathbf{b}_i + 0\mathbf{b}_{i+1} + \cdots + 0\mathbf{b}_n,$$

and so $[\mathbf{b}_i]_{\mathcal{B}} = \mathbf{e}_i$.

In general, the following result provides a simple method for calculating coordinate vectors relative to an arbitrary basis for $\mathcal{R}^n$.

Theorem 4.9
Let $\mathcal{B}$ be a basis for $\mathcal{R}^n$ and B be the matrix whose columns are the vectors in $\mathcal{B}$. Then B is invertible and, for every vector $\mathbf{v}$ in $\mathcal{R}^n$, $B[\mathbf{v}]_{\mathcal{B}} = \mathbf{v}$, or equivalently, $[\mathbf{v}]_{\mathcal{B}} = B^{-1}\mathbf{v}$.

Proof Let $\mathcal{B} = \{\mathbf{b}_1, \mathbf{b}_2, \ldots, \mathbf{b}_n\}$ be a basis for $\mathcal{R}^n$ and $\mathbf{v}$ be a vector in $\mathcal{R}^n$. If

$$[\mathbf{v}]_{\mathcal{B}} = \begin{bmatrix} c_1 \\ c_2 \\ \vdots \\ c_n \end{bmatrix},$$

then

$$\mathbf{v} = c_1\mathbf{b}_1 + c_2\mathbf{b}_2 + \cdots + c_n\mathbf{b}_n$$

$$= [\mathbf{b}_1 \ \ \mathbf{b}_2 \ \ \cdots \ \ \mathbf{b}_n] \begin{bmatrix} c_1 \\ c_2 \\ \vdots \\ c_n \end{bmatrix}$$

$$= B[\mathbf{v}]_{\mathcal{B}},$$

where $B = [\mathbf{b}_1 \ \ \mathbf{b}_2 \ \ \cdots \ \ \mathbf{b}_n]$. Since $\mathcal{B}$ is a basis, the columns of B are linearly independent. Hence B is invertible by Theorem 2.6, and so $B[\mathbf{v}]_{\mathcal{B}} = \mathbf{v}$ is equivalent to $[\mathbf{v}]_{\mathcal{B}} = B^{-1}\mathbf{v}$. $\square$

As an alternative to the calculation in Example 1, we can compute the vector $\mathbf{u}$ using Theorem 4.9. The result is

$$\mathbf{u} = B[\mathbf{u}]_{\mathcal{B}} = \begin{bmatrix} 1 & 1 & 1 \\ 1 & -1 & 2 \\ 1 & 1 & 2 \end{bmatrix} \begin{bmatrix} 3 \\ 6 \\ -2 \end{bmatrix} = \begin{bmatrix} 7 \\ -7 \\ 5 \end{bmatrix},$$

in agreement with Example 1.

Theorem 4.9 can also be used to compute coordinate vectors, as the next example demonstrates.

Example 3   Let

$$\mathcal{B} = \left\{ \begin{bmatrix} 1 \\ 1 \\ 1 \end{bmatrix}, \begin{bmatrix} 1 \\ -1 \\ 1 \end{bmatrix}, \begin{bmatrix} 1 \\ 2 \\ 2 \end{bmatrix} \right\} \qquad \text{and} \qquad \mathbf{v} = \begin{bmatrix} 1 \\ -4 \\ 4 \end{bmatrix}$$

as in Example 2, and let B be the matrix whose columns are the vectors in $\mathcal{B}$. Then

$$[\mathbf{v}]_{\mathcal{B}} = B^{-1}\mathbf{v} = \begin{bmatrix} -6 \\ 4 \\ 3 \end{bmatrix}.$$

Of course, this result agrees with that in Example 2. $\bigcirc$

Changing Coordinates

To obtain an equation for the ellipse in Figure 4.3, we must be able to switch between the x', y'-coordinate system and the usual coordinate system. This requires that we be able to convert coordinate vectors relative to an arbitrary basis $\mathcal{B}$ into coordinate vectors relative to the standard basis or vice versa. In other words, we must know the relationship between $[\mathbf{v}]_{\mathcal{B}}$ and $\mathbf{v}$ for an arbitrary vector $\mathbf{v}$ in $\mathcal{R}^n$. According to Theorem 4.9, this relationship is $[\mathbf{v}]_{\mathcal{B}} = B^{-1}\mathbf{v}$, where B is the matrix whose columns are the vectors in $\mathcal{B}$.

Consider the basis $\mathcal{B} = \{\mathbf{b}_1, \mathbf{b}_2\}$ obtained by rotating the vectors in the standard basis through $45°$. (For reasons that will become clear, the vectors $\mathbf{b}_1$ and $\mathbf{b}_2$ should be chosen to have length 1.) The components of these vectors can be computed using the 45°-rotation matrix as in Section 1.2:

$$\mathbf{b}_1 = A_{45°}\mathbf{e}_1 \quad \text{and} \quad \mathbf{b}_2 = A_{45°}\mathbf{e}_2.$$

In order to write the x', y'-equation

$$\frac{(x')^2}{3^2} + \frac{(y')^2}{2^2} = 1$$

as an equation in the usual x, y-coordinate system, we must use the relationship between a vector and its $\mathcal{B}$-coordinates. Let

$$\mathbf{v} = \begin{bmatrix} x \\ y \end{bmatrix}, \qquad [\mathbf{v}]_{\mathcal{B}} = \begin{bmatrix} x' \\ y' \end{bmatrix},$$

and

$$B = [\mathbf{b}_1 \quad \mathbf{b}_2] = [A_{45°}\mathbf{e}_1 \quad A_{45°}\mathbf{e}_2] = A_{45°}[\mathbf{e}_1 \quad \mathbf{e}_2] = A_{45°}I_2 = A_{45°}.$$

Since B is a rotation matrix, we have $B^{-1} = B^T$ (Exercise 2.3 in Section 2.3). Hence $[\mathbf{v}]_{\mathcal{B}} = B^{-1}\mathbf{v} = B^T\mathbf{v}$, and so

$$\begin{bmatrix} x' \\ y' \end{bmatrix} = B^T \begin{bmatrix} x \\ y \end{bmatrix} = \begin{bmatrix} \frac{\sqrt{2}}{2} & \frac{\sqrt{2}}{2} \\ -\frac{\sqrt{2}}{2} & \frac{\sqrt{2}}{2} \end{bmatrix} \begin{bmatrix} x \\ y \end{bmatrix} = \begin{bmatrix} \frac{\sqrt{2}}{2}x + \frac{\sqrt{2}}{2}y \\ -\frac{\sqrt{2}}{2}x + \frac{\sqrt{2}}{2}y \end{bmatrix}.$$

Therefore, substituting

$$x' = \frac{\sqrt{2}}{2}x + \frac{\sqrt{2}}{2}y$$

$$y' = -\frac{\sqrt{2}}{2}x + \frac{\sqrt{2}}{2}y$$

converts an equation in the x', y'-coordinate system into one in the x, y-coordinate system. So the equation of the given ellipse in the standard coordinate system is

$$\frac{\left(\frac{\sqrt{2}}{2}x + \frac{\sqrt{2}}{2}y\right)^2}{3^2} + \frac{\left(-\frac{\sqrt{2}}{2}x + \frac{\sqrt{2}}{2}y\right)^2}{2^2} = 1,$$

which simplifies to the form

$$13x^2 - 10xy + 13y^2 = 72.$$

Example 4   Write the equation $-\sqrt{3}x^2 + 2xy + \sqrt{3}y^2 = 12$ in terms of the x', y'-coordinate system, where the x'-axis and the y'-axis are obtained by rotating the usual x-axis and y-axis through the angle 30°.

Solution   Again a change of coordinates is required. This time, however, we must change from the x, y-coordinate system to the x', y'-coordinate system. Consider the basis $\mathcal{B} = \{\mathbf{b}_1, \mathbf{b}_2\}$ obtained by rotating the vectors in the standard basis through 30°. As above, let

$$\mathbf{v} = \begin{bmatrix} x \\ y \end{bmatrix}, \qquad [\mathbf{v}]_{\mathcal{B}} = \begin{bmatrix} x' \\ y' \end{bmatrix}, \qquad \text{and} \qquad B = [\mathbf{b}_1 \quad \mathbf{b}_2] = A_{30°}.$$

Since $\mathbf{v} = B[\mathbf{v}]_B$, we have

$$\begin{bmatrix} x \\ y \end{bmatrix} = B \begin{bmatrix} x' \\ y' \end{bmatrix} = \begin{bmatrix} \dfrac{\sqrt{3}}{2} & -\dfrac{1}{2} \\ \dfrac{1}{2} & \dfrac{\sqrt{3}}{2} \end{bmatrix} \begin{bmatrix} x' \\ y' \end{bmatrix} = \begin{bmatrix} \dfrac{\sqrt{3}}{2}x' - \dfrac{1}{2}y' \\ \dfrac{1}{2}x' + \dfrac{\sqrt{3}}{2}y' \end{bmatrix}.$$

Hence the equations relating the two coordinate systems are

$$x = \frac{\sqrt{3}}{2}x' - \frac{1}{2}y'$$

$$y = \frac{1}{2}x' + \frac{\sqrt{3}}{2}y'.$$

Making these substitutions transforms the equation $-\sqrt{3}x^2 + 2xy + \sqrt{3}y^2 = 12$ into $4x'y' = 12$, that is, $x'y' = 3$. From this form, we see that the graph of the given equation is a hyperbola (shown in Figure 4.6).

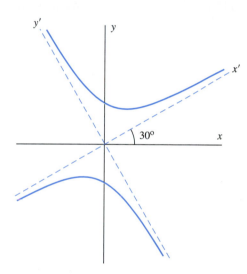

FIGURE 4.6

Practice Problems

1. Verify that $B = \left\{ \begin{bmatrix} 1 \\ 1 \\ 0 \end{bmatrix}, \begin{bmatrix} 1 \\ 1 \\ 1 \end{bmatrix}, \begin{bmatrix} 3 \\ 2 \\ 1 \end{bmatrix} \right\}$ is a basis for $\mathcal{R}^3$.

2. Find $\mathbf{u}$ if $[\mathbf{u}]_B = \begin{bmatrix} 5 \\ -2 \\ -1 \end{bmatrix}$ and B is as in problem 1.

3. Find $[\mathbf{v}]_B$ if $\mathbf{v} = \begin{bmatrix} -2 \\ 1 \\ 3 \end{bmatrix}$ and B is as in problem 1.

Exercises

1. Determine if the following statements are true or false.

   (a) If S is a spanning set for a subspace V of $\mathcal{R}^n$, then every vector in V can be uniquely represented as a linear combination of the vectors in S.

   (b) The entries of the B-coordinate vector of $\mathbf{u}$ are the coefficients that express $\mathbf{u}$ as a linear combination of the vectors in B.

   (c) If B is the standard basis for $\mathcal{R}^n$, then $[\mathbf{v}]_B = \mathbf{v}$ for all $\mathbf{v}$ in $\mathcal{R}^n$.

(d) The $\mathcal{B}$-coordinate vector of a vector in $\mathcal{B}$ is a standard vector.

(e) If $\mathcal{B}$ is a basis for $\mathcal{R}^n$ and B is the matrix whose columns are the vectors in $\mathcal{B}$, then $B^{-1}\mathbf{v} = [\mathbf{v}]_{\mathcal{B}}$ for all $\mathbf{v}$ in $\mathcal{R}^n$.

In Exercises 2–6, the $\mathcal{B}$-coordinate vector of a vector $\mathbf{v}$ is given. Find $\mathbf{v}$ if

$$B = \left\{ \begin{bmatrix} 1 \\ -1 \end{bmatrix}, \begin{bmatrix} -1 \\ 2 \end{bmatrix} \right\}.$$

2. $\begin{bmatrix} -3 \\ 2 \end{bmatrix}$
3. $\begin{bmatrix} 1 \\ 6 \end{bmatrix}$
4. $\begin{bmatrix} -1 \\ 4 \end{bmatrix}$
5. $\begin{bmatrix} 2 \\ 5 \end{bmatrix}$
6. $\begin{bmatrix} 5 \\ 2 \end{bmatrix}$

In Exercises 7–10, the $\mathcal{B}$-coordinate vector of a vector $\mathbf{v}$ is given. Find $\mathbf{v}$ if

$$B = \left\{ \begin{bmatrix} 0 \\ 1 \\ 1 \end{bmatrix}, \begin{bmatrix} -1 \\ 0 \\ 1 \end{bmatrix}, \begin{bmatrix} 1 \\ 1 \\ 1 \end{bmatrix} \right\}.$$

7. $\begin{bmatrix} 2 \\ -1 \\ 3 \end{bmatrix}$
8. $\begin{bmatrix} 3 \\ 1 \\ -4 \end{bmatrix}$
9. $\begin{bmatrix} -1 \\ 5 \\ -2 \end{bmatrix}$
10. $\begin{bmatrix} 3 \\ -4 \\ 2 \end{bmatrix}$

11. (a) Prove that $B = \left\{ \begin{bmatrix} 1 \\ 3 \end{bmatrix}, \begin{bmatrix} -2 \\ 1 \end{bmatrix} \right\}$ is a basis for $\mathcal{R}^2$.

(b) Find $[\mathbf{v}]_{\mathcal{B}}$ if $\mathbf{v} = 5\begin{bmatrix} 1 \\ 3 \end{bmatrix} - 3\begin{bmatrix} -2 \\ 1 \end{bmatrix}$.

12. (a) Prove that $B = \left\{ \begin{bmatrix} 3 \\ -2 \end{bmatrix}, \begin{bmatrix} -1 \\ 2 \end{bmatrix} \right\}$ is a basis for $\mathcal{R}^2$.

(b) Find $[\mathbf{v}]_{\mathcal{B}}$ if $\mathbf{v} = -2\begin{bmatrix} 3 \\ -2 \end{bmatrix} + 4\begin{bmatrix} -1 \\ 2 \end{bmatrix}$.

13. (a) Prove that $B = \left\{ \begin{bmatrix} -1 \\ 0 \\ 1 \end{bmatrix}, \begin{bmatrix} 2 \\ 1 \\ -1 \end{bmatrix}, \begin{bmatrix} 1 \\ -3 \\ 2 \end{bmatrix} \right\}$ is a basis for $\mathcal{R}^3$.

(b) Find $[\mathbf{v}]_{\mathcal{B}}$ if $\mathbf{v} = 3\begin{bmatrix} -1 \\ 0 \\ 1 \end{bmatrix} - \begin{bmatrix} 1 \\ -3 \\ 2 \end{bmatrix}$.

14. (a) Prove that $B = \left\{ \begin{bmatrix} 1 \\ -1 \\ 1 \end{bmatrix}, \begin{bmatrix} -1 \\ 1 \\ 1 \end{bmatrix}, \begin{bmatrix} 1 \\ 1 \\ 1 \end{bmatrix} \right\}$ is a basis for $\mathcal{R}^3$.

(b) Find $[\mathbf{v}]_{\mathcal{B}}$ if $\mathbf{v} = \begin{bmatrix} -1 \\ 1 \\ 1 \end{bmatrix} - 4\begin{bmatrix} 1 \\ 1 \\ 1 \end{bmatrix}$.

In Exercises 15–18, a vector is given. Find its $\mathcal{B}$-coordinate vector for the basis $\mathcal{B}$ in Exercises 2–6.

15. $\begin{bmatrix} -4 \\ 3 \end{bmatrix}$
16. $\begin{bmatrix} -1 \\ 2 \end{bmatrix}$
17. $\begin{bmatrix} 5 \\ -3 \end{bmatrix}$
18. $\begin{bmatrix} 3 \\ 2 \end{bmatrix}$

In Exercises 19–22, a vector is given. Find its $\mathcal{B}$-coordinate vector for the basis $\mathcal{B}$ in Exercises 7–10.

19. $\begin{bmatrix} 4 \\ 3 \\ 2 \end{bmatrix}$
20. $\begin{bmatrix} -2 \\ 6 \\ 3 \end{bmatrix}$
21. $\begin{bmatrix} 1 \\ -3 \\ -2 \end{bmatrix}$
22. $\begin{bmatrix} -1 \\ 5 \\ 2 \end{bmatrix}$

23. Find the unique representation of $\mathbf{u} = \begin{bmatrix} a \\ b \end{bmatrix}$ as a linear combination of

$$\mathbf{b}_1 = \begin{bmatrix} 3 \\ -1 \end{bmatrix} \quad \text{and} \quad \mathbf{b}_2 = \begin{bmatrix} -2 \\ 1 \end{bmatrix}.$$

24. Find the unique representation of $\mathbf{u} = \begin{bmatrix} a \\ b \\ c \end{bmatrix}$ as a linear combination of

$$\mathbf{b}_1 = \begin{bmatrix} 1 \\ -1 \\ 1 \end{bmatrix}, \quad \mathbf{b}_2 = \begin{bmatrix} -1 \\ 2 \\ 1 \end{bmatrix}, \quad \text{and} \quad \mathbf{b}_3 = \begin{bmatrix} 1 \\ 0 \\ 2 \end{bmatrix}.$$

25. Let $B = \{\mathbf{b}_1, \mathbf{b}_2\}$, where $\mathbf{b}_1 = \begin{bmatrix} 1 \\ 2 \end{bmatrix}$ and $\mathbf{b}_2 = \begin{bmatrix} 2 \\ 3 \end{bmatrix}$.

(a) Show that B is a basis for $\mathcal{R}^2$.
(b) Determine the matrix $A = [[\mathbf{e}_1]_{\mathcal{B}} \ [\mathbf{e}_2]_{\mathcal{B}}]$.
(c) What is the relationship between A and $B = [\mathbf{b}_1 \ \mathbf{b}_2]$?

26. Let $B = \{\mathbf{b}_1, \mathbf{b}_2\}$, where $\mathbf{b}_1 = \begin{bmatrix} 1 \\ 3 \end{bmatrix}$ and $\mathbf{b}_2 = \begin{bmatrix} -1 \\ -2 \end{bmatrix}$.

(a) Show that B is a basis for $\mathcal{R}^2$.
(b) Determine the matrix $A = [[\mathbf{e}_1]_{\mathcal{B}} \ [\mathbf{e}_2]_{\mathcal{B}}]$.
(c) What is the relationship between A and $B = [\mathbf{b}_1 \ \mathbf{b}_2]$?

27. Let $B = \{\mathbf{b}_1, \mathbf{b}_2, \mathbf{b}_3\}$, where $\mathbf{b}_1 = \begin{bmatrix} 2 \\ 1 \\ -1 \end{bmatrix}$, $\mathbf{b}_2 = \begin{bmatrix} -1 \\ -1 \\ 1 \end{bmatrix}$, and $\mathbf{b}_3 = \begin{bmatrix} -1 \\ -2 \\ 1 \end{bmatrix}$.

(a) Show that B is a basis for $\mathcal{R}^3$.
(b) Determine the matrix $A = [[\mathbf{e}_1]_{\mathcal{B}} \ [\mathbf{e}_2]_{\mathcal{B}} \ [\mathbf{e}_3]_{\mathcal{B}}]$.
(c) What is the relationship between the matrices A and $B = [\mathbf{b}_1 \ \mathbf{b}_2 \ \mathbf{b}_3]$?

28. Let $B = \{\mathbf{b}_1, \mathbf{b}_2, \mathbf{b}_3\}$, where $\mathbf{b}_1 = \begin{bmatrix} 1 \\ -2 \\ 1 \end{bmatrix}$, $\mathbf{b}_2 = \begin{bmatrix} -1 \\ 3 \\ 0 \end{bmatrix}$, and $\mathbf{b}_3 = \begin{bmatrix} 0 \\ 2 \\ 1 \end{bmatrix}$.

(a) Show that B is a basis for $\mathcal{R}^3$.
(b) Determine the matrix $A = [[\mathbf{e}_1]_{\mathcal{B}} \ [\mathbf{e}_2]_{\mathcal{B}} \ [\mathbf{e}_3]_{\mathcal{B}}]$.
(c) What is the relationship between the matrices A and $B = [\mathbf{b}_1 \ \mathbf{b}_2 \ \mathbf{b}_3]$?

In Exercises 29–32, an angle θ is given. Let $\mathbf{v} = \begin{bmatrix} x \\ y \end{bmatrix}$ and $[\mathbf{v}]_B = \begin{bmatrix} x' \\ y' \end{bmatrix}$, where B is the basis for $\mathcal{R}^2$ obtained by rotating $\mathbf{e}_1$ and $\mathbf{e}_2$ through the angle θ. Write equations expressing x' and y' in terms of x and y.

29. $\theta = 30°$ **30.** $\theta = 60°$ **31.** $\theta = 135°$ **32.** $\theta = 330°$

In Exercises 33–36, a basis B for $\mathcal{R}^3$ is given. If $\mathbf{v} = \begin{bmatrix} x \\ y \\ z \end{bmatrix}$ and $[\mathbf{v}]_B = \begin{bmatrix} x' \\ y' \\ z' \end{bmatrix}$, write equations expressing x', y', and z' in terms of x, y, and z.

33. $\left\{ \begin{bmatrix} 1 \\ 0 \\ 1 \end{bmatrix}, \begin{bmatrix} 1 \\ 1 \\ 0 \end{bmatrix}, \begin{bmatrix} 0 \\ -2 \\ 1 \end{bmatrix} \right\}$ **34.** $\left\{ \begin{bmatrix} -1 \\ -1 \\ 1 \end{bmatrix}, \begin{bmatrix} -2 \\ -2 \\ 1 \end{bmatrix}, \begin{bmatrix} 1 \\ 0 \\ 1 \end{bmatrix} \right\}$

35. $\left\{ \begin{bmatrix} 0 \\ 1 \\ 2 \end{bmatrix}, \begin{bmatrix} -1 \\ 0 \\ 1 \end{bmatrix}, \begin{bmatrix} -2 \\ -1 \\ 1 \end{bmatrix} \right\}$ **36.** $\left\{ \begin{bmatrix} -1 \\ 2 \\ 0 \end{bmatrix}, \begin{bmatrix} 1 \\ 2 \\ 1 \end{bmatrix}, \begin{bmatrix} 1 \\ 1 \\ 1 \end{bmatrix} \right\}$

In Exercises 37–40, an angle θ is given. Let $\mathbf{v} = \begin{bmatrix} x \\ y \end{bmatrix}$ and $[\mathbf{v}]_B = \begin{bmatrix} x' \\ y' \end{bmatrix}$, where B is the basis for $\mathcal{R}^2$ obtained by rotating $\mathbf{e}_1$ and $\mathbf{e}_2$ through the angle θ. Write equations expressing x and y in terms of x' and y'.

37. $\theta = 60°$ **38.** $\theta = 45°$ **39.** $\theta = 135°$ **40.** $\theta = 330°$

In Exercises 41–44, a basis B for $\mathcal{R}^3$ is given. If $\mathbf{v} = \begin{bmatrix} x \\ y \\ z \end{bmatrix}$ and $[\mathbf{v}]_B = \begin{bmatrix} x' \\ y' \\ z' \end{bmatrix}$, write equations expressing x, y, and z in terms of x', y', and z'.

41. $\left\{ \begin{bmatrix} 1 \\ 3 \\ 0 \end{bmatrix}, \begin{bmatrix} -1 \\ 1 \\ 1 \end{bmatrix}, \begin{bmatrix} 0 \\ -1 \\ 1 \end{bmatrix} \right\}$ **42.** $\left\{ \begin{bmatrix} 2 \\ -1 \\ 1 \end{bmatrix}, \begin{bmatrix} 0 \\ -1 \\ 1 \end{bmatrix}, \begin{bmatrix} 1 \\ -1 \\ 2 \end{bmatrix} \right\}$

43. $\left\{ \begin{bmatrix} 1 \\ -1 \\ 1 \end{bmatrix}, \begin{bmatrix} -1 \\ 3 \\ 2 \end{bmatrix}, \begin{bmatrix} -1 \\ 1 \\ 1 \end{bmatrix} \right\}$ **44.** $\left\{ \begin{bmatrix} -1 \\ 1 \\ 1 \end{bmatrix}, \begin{bmatrix} -1 \\ 2 \\ 2 \end{bmatrix}, \begin{bmatrix} 1 \\ -1 \\ 1 \end{bmatrix} \right\}$

In Exercises 45–48, an equation of a conic section is given in the x', y'-coordinate system. Determine the equation of the conic section in the usual x, y-coordinate system if the x'-axis and the y'-axis are obtained by rotating the usual x-axis and y-axis through the given angle θ.

45. $\dfrac{(x')^2}{4^2} + \dfrac{(y')^2}{5^2} = 1, \theta = 60°$

46. $\dfrac{(x')^2}{2^2} - \dfrac{(y')^2}{3^2} = 1, \theta = 45°$

47. $\dfrac{(x')^2}{3^2} - \dfrac{(y')^2}{5^2} = 1, \theta = 135°$

48. $\dfrac{(x')^2}{6} + \dfrac{(y')^2}{4} = 1, \theta = 150°$

In Exercises 49–52, an equation of a conic section is given in the x, y-coordinate system. Determine the equation of the conic section in the x', y'-coordinate system if the x'-axis and the y'-axis are obtained by rotating the usual x-axis and y-axis through the given angle θ.

49. $-3x^2 + 14xy - 3y^2 = 20, \theta = 45°$

50. $6x^2 - 2\sqrt{3}xy + 4y^2 = 21, \theta = 60°$

51. $15x^2 - 2\sqrt{3}xy + 13y^2 = 48, \theta = 150°$

52. $x^2 - 6xy + y^2 = 12, \theta = 135°$

53. Let $\mathcal{A} = \{\mathbf{u}_1, \mathbf{u}_2, \ldots, \mathbf{u}_n\}$ be a basis for $\mathcal{R}^n$ and $c_1, c_2, \ldots, c_n$ be nonzero scalars. Recall from Exercise 35 of Section 4.2 that $\mathcal{B} = \{c_1\mathbf{u}_1, c_2\mathbf{u}_2, \ldots, c_n\mathbf{u}_n\}$ is also a basis for $\mathcal{R}^n$. If $\mathbf{v}$ is a vector in $\mathcal{R}^n$ and

$$[\mathbf{v}]_{\mathcal{A}} = \begin{bmatrix} a_1 \\ a_2 \\ \vdots \\ a_n \end{bmatrix},$$

compute $[\mathbf{v}]_{\mathcal{B}}$.

54. Let $\mathcal{A} = \{\mathbf{u}_1, \mathbf{u}_2, \ldots, \mathbf{u}_n\}$ be a basis for $\mathcal{R}^n$. Recall that Exercise 36 of Section 4.2 shows that $\mathcal{B} = \{\mathbf{u}_1, \mathbf{u}_1 + \mathbf{u}_2, \ldots, \mathbf{u}_1 + \mathbf{u}_n\}$ is also a basis for $\mathcal{R}^n$. If $\mathbf{v}$ is a vector in $\mathcal{R}^n$ and

$$[\mathbf{v}]_{\mathcal{A}} = \begin{bmatrix} a_1 \\ a_2 \\ \vdots \\ a_n \end{bmatrix},$$

compute $[\mathbf{v}]_{\mathcal{B}}$.

55. Let $\mathcal{A} = \{\mathbf{u}_1, \mathbf{u}_2, \ldots, \mathbf{u}_n\}$ be a basis for $\mathcal{R}^n$. Recall that Exercise 37 of Section 4.2 shows that $\mathcal{B} = \{\mathbf{u}_1 + \mathbf{u}_2 + \cdots + \mathbf{u}_n, \mathbf{u}_2, \mathbf{u}_3, \ldots, \mathbf{u}_n\}$ is also a basis for $\mathcal{R}^n$. If $\mathbf{v}$ is a vector in $\mathcal{R}^n$ and

$$[\mathbf{v}]_{\mathcal{A}} = \begin{bmatrix} a_1 \\ a_2 \\ \vdots \\ a_n \end{bmatrix},$$

compute $[\mathbf{v}]_{\mathcal{B}}$.

56. Let $\mathcal{A} = \{\mathbf{u}_1, \mathbf{u}_2, \ldots, \mathbf{u}_n\}$ be a basis for $\mathcal{R}^n$. Recall that Exercise 38 of Section 4.2 shows that

$$\mathcal{B} = \{\mathbf{u}_1 + \mathbf{u}_2 + \cdots + \mathbf{u}_n, \mathbf{u}_2 + \mathbf{u}_3 + \cdots + \mathbf{u}_n,$$

$$\ldots, \mathbf{u}_{n-1} + \mathbf{u}_n, \mathbf{u}_n\}$$

is also a basis for $\mathcal{R}^n$. If $\mathbf{v}$ is a vector in $\mathcal{R}^n$ and

$$[\mathbf{v}]_{\mathcal{A}} = \begin{bmatrix} a_1 \\ a_2 \\ \vdots \\ a_n \end{bmatrix},$$

compute $[\mathbf{v}]_{\mathcal{B}}$.

57. Let $\mathcal{A}$ and $\mathcal{B}$ be two bases for $\mathcal{R}^n$. Suppose that $[\mathbf{v}]_{\mathcal{A}} = [\mathbf{v}]_{\mathcal{B}}$ for some nonzero vector $\mathbf{v}$ in $\mathcal{R}^n$. Must $\mathcal{A} = \mathcal{B}$? Justify your answer.

58. Let $\mathcal{A}$ and $\mathcal{B}$ be two bases for $\mathcal{R}^n$. Suppose that $[\mathbf{v}]_{\mathcal{A}} = [\mathbf{v}]_{\mathcal{B}}$ for every vector $\mathbf{v}$ in $\mathcal{R}^n$. Must $\mathcal{A} = \mathcal{B}$? Justify your answer.

59. Prove that if $\mathcal{S}$ is linearly dependent, then every vector in the span of $\mathcal{S}$ can be written as a linear combination of the vectors in $\mathcal{S}$ in more than one way.

60. Let $\mathcal{A}$ and $\mathcal{B}$ be two bases for $\mathcal{R}^n$. Express $[\mathbf{v}]_{\mathcal{A}}$ in terms of $[\mathbf{v}]_{\mathcal{B}}$.

61. (a) Let $\mathcal{B}$ be a basis for $\mathcal{R}^n$. Prove that the function $T: \mathcal{R}^n \rightarrow \mathcal{R}^n$ defined for all $\mathbf{v}$ in $\mathcal{R}^n$ by $T(\mathbf{v}) = [\mathbf{v}]_{\mathcal{B}}$ is a linear transformation.
    (b) Prove that T is one-to-one and onto.

62. What is the standard matrix of the linear transformation T in Exercise 61?

63. Let V be a subspace of $\mathcal{R}^n$ and $\mathcal{B} = \{\mathbf{u}_1, \mathbf{u}_2, \ldots, \mathbf{u}_k\}$ be a subset of V. Prove that if every vector $\mathbf{v}$ in V can be uniquely represented as a linear combination of the vectors in $\mathcal{B}$ (that is, if there are unique scalars $a_1, a_2, \ldots, a_k$ such that $\mathbf{v} = a_1\mathbf{u}_1 + a_2\mathbf{u}_2 + \cdots + a_k\mathbf{u}_k$), then $\mathcal{B}$ is a basis for V. (This is the converse of Theorem 4.8.)

64. Let $V = \left\{ \begin{bmatrix} v_1 \\ v_2 \\ v_3 \end{bmatrix} \in \mathcal{R}^3: -2v_1 + v_2 + v_3 = 0 \right\}$ and

$$\mathcal{S} = \left\{ \begin{bmatrix} -1 \\ 1 \\ 1 \end{bmatrix}, \begin{bmatrix} 1 \\ 0 \\ 1 \end{bmatrix}, \begin{bmatrix} 1 \\ -2 \\ -2 \end{bmatrix} \right\}.$$

(a) Show that $\mathcal{S}$ is linearly independent.
(b) Show that

$$(2v_1 - 5v_2) \begin{bmatrix} -1 \\ 1 \\ 1 \end{bmatrix} + (2v_1 - 2v_2) \begin{bmatrix} 1 \\ 0 \\ 1 \end{bmatrix}$$

$$+ (v_1 - 3v_2) \begin{bmatrix} 1 \\ -2 \\ -2 \end{bmatrix} = \begin{bmatrix} v_1 \\ v_2 \\ v_3 \end{bmatrix}$$

for every vector $\begin{bmatrix} v_1 \\ v_2 \\ v_3 \end{bmatrix}$ in $\mathcal{S}$.

(c) Is $\mathcal{S}$ a basis for V? Justify your answer.

65. Let $\mathcal{B}$ be a basis for $\mathcal{R}^n$ and $\{\mathbf{u}_1, \mathbf{u}_2, \ldots, \mathbf{u}_k\}$ be a subset of $\mathcal{R}^n$. Prove that $\{\mathbf{u}_1, \mathbf{u}_2, \ldots, \mathbf{u}_k\}$ is linearly independent if and only if $\{[\mathbf{u}_1]_{\mathcal{B}}, [\mathbf{u}_2]_{\mathcal{B}}, \ldots, [\mathbf{u}_k]_{\mathcal{B}}\}$ is linearly independent.

66. Let $\mathcal{B}$ be a basis for $\mathcal{R}^n$, $\{\mathbf{u}_1, \mathbf{u}_2, \ldots, \mathbf{u}_k\}$ be a subset of $\mathcal{R}^n$, and $\mathbf{v}$ be a vector in $\mathcal{R}^n$. Prove that $\mathbf{v}$ is a linear combination of $\{\mathbf{u}_1, \mathbf{u}_2, \ldots, \mathbf{u}_k\}$ if and only if $[\mathbf{v}]_{\mathcal{B}}$ is a linear combination of $\{[\mathbf{u}_1]_{\mathcal{B}}, [\mathbf{u}_2]_{\mathcal{B}}, \ldots, [\mathbf{u}_k]_{\mathcal{B}}\}$.

In Exercises 67–70, use a calculator with matrix capabilities or computer software such as MATLAB to solve the problem.

67. Let

$$\mathcal{B} = \left\{ \begin{bmatrix} 0 \\ 25 \\ -21 \\ 23 \\ 12 \end{bmatrix}, \begin{bmatrix} 14 \\ 73 \\ -66 \\ 64 \\ 42 \end{bmatrix}, \begin{bmatrix} -6 \\ -56 \\ 47 \\ -50 \\ -29 \end{bmatrix}, \begin{bmatrix} -14 \\ -68 \\ 60 \\ -59 \\ -39 \end{bmatrix}, \begin{bmatrix} -12 \\ -118 \\ 102 \\ -106 \\ -62 \end{bmatrix} \right\}$$

and $\quad \mathbf{v} = \begin{bmatrix} -2 \\ 3 \\ 1 \\ 2 \\ -1 \end{bmatrix}$.

(a) Show that $\mathcal{B}$ is a basis for $\mathcal{R}^5$.
(b) Find $[\mathbf{v}]_{\mathcal{B}}$.

68. For the basis $\mathcal{B}$ in Exercise 67, find a vector $\mathbf{u}$ in $\mathcal{R}^5$ such that $\mathbf{u} = [\mathbf{u}]_{\mathcal{B}}$.

69. For the basis $\mathcal{B}$ in Exercise 67, find a vector $\mathbf{v}$ in $\mathcal{R}^5$ such that $[\mathbf{v}]_{\mathcal{B}} = .5\mathbf{v}$.

70. Let $\mathcal{B}$ and $\mathbf{v}$ be as in Exercise 67, and let

$$\mathbf{u}_1 = \begin{bmatrix} 1 \\ 0 \\ -1 \\ 1 \\ 0 \end{bmatrix}, \quad \mathbf{u}_2 = \begin{bmatrix} 2 \\ -1 \\ 0 \\ 1 \\ 1 \end{bmatrix}, \quad \text{and} \quad \mathbf{u}_3 = \begin{bmatrix} 1 \\ 0 \\ -6 \\ 0 \\ -2 \end{bmatrix}.$$

(a) Show that $\mathbf{v}$ is a linear combination of $\mathbf{u}_1, \mathbf{u}_2$, and $\mathbf{u}_3$.
(b) Show that $[\mathbf{v}]_{\mathcal{B}}$ is a linear combination of $[\mathbf{u}_1]_{\mathcal{B}}$, $[\mathbf{u}_2]_{\mathcal{B}}$, and $[\mathbf{u}_3]_{\mathcal{B}}$.
(c) Make a conjecture that generalizes the results of (a) and (b).

Solutions to the Practice Problems

1. Let B be the matrix whose columns are the vectors in $\mathcal{B}$. Since the reduced row echelon form of B is I_3, the columns of B are linearly independent. Thus $\mathcal{B}$ is a linearly independent set of 3 vectors from $\mathcal{R}^3$, and hence it is a basis for $\mathcal{R}^3$.

2. By Theorem 4.9, we have

$$\mathbf{u} = B[\mathbf{u}]_{\mathcal{B}} = \begin{bmatrix} 1 & 1 & 3 \\ 1 & 1 & 2 \\ 0 & 1 & 1 \end{bmatrix} \begin{bmatrix} 5 \\ -2 \\ -1 \end{bmatrix} = \begin{bmatrix} 0 \\ 1 \\ -3 \end{bmatrix}.$$

3. By Theorem 4.9, we also have

$$[\mathbf{v}]_{\mathcal{B}} = B^{-1}\mathbf{v} = \begin{bmatrix} 1 & 1 & 3 \\ 1 & 1 & 2 \\ 0 & 1 & 1 \end{bmatrix}^{-1} \begin{bmatrix} -2 \\ 1 \\ 3 \end{bmatrix} = \begin{bmatrix} 1 \\ 6 \\ -3 \end{bmatrix}.$$

4.5   Matrix Representations of Linear Operators

The function that reflects a point in the Euclidean plane $\mathcal{R}^2$ about a line $\mathcal{L}$ through the origin arises frequently in geometry. Consider, for example, the reflection about the line with equation $y = \frac{1}{2}x$ (see Figure 4.7). What is the rule for this function?

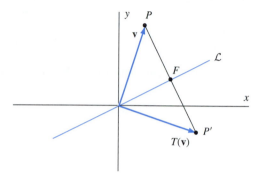

FIGURE 4.7

In general, a **reflection** about a line $\mathcal{L}$ through the origin of $\mathcal{R}^2$ is a function $T\colon \mathcal{R}^2 \to \mathcal{R}^2$ defined in the following manner. Let $\mathbf{v}$ be a vector in $\mathcal{R}^2$ with endpoint P. Construct a line from P perpendicular to $\mathcal{L}$, and let F denote the point of intersection of this perpendicular line with $\mathcal{L}$. Extend segment $\overline{PF}$ through F to a point P' such that $\overline{PF} = \overline{FP'}$. The vector with endpoint P' is $T(\mathbf{v})$ (see Figure 4.7).

For this function, as with most of the functions that we will encounter from now on, the domain and codomain are equal. A linear transformation where the domain and codomain equal $\mathcal{R}^n$ is called a **linear operator on** $\mathcal{R}^n$.

It can be shown that reflections are linear operators. This is clear for the special case that $\mathcal{L}$ is the x-axis, in which case the reflection T' about $\mathcal{L}$ is given by

$$T'\left(\begin{bmatrix} x_1 \\ x_2 \end{bmatrix}\right) = \begin{bmatrix} x_1 \\ -x_2 \end{bmatrix}.$$

For this reflection, $T'(\mathbf{e}_1) = \mathbf{e}_1$ and $T'(\mathbf{e}_2) = -\mathbf{e}_2$, because $\mathbf{e}_1$ is in the direction of the line of reflection, and $\mathbf{e}_2$ is in the direction perpendicular to the line of reflection. Therefore the standard matrix of T' is given by

$$[T'(\mathbf{e}_1) \quad T'(\mathbf{e}_2)] = [\mathbf{e}_1 \quad -\mathbf{e}_2] = \begin{bmatrix} 1 & 0 \\ 0 & -1 \end{bmatrix}.$$

More generally, suppose that $T\colon \mathcal{R}^2 \to \mathcal{R}^2$ is a reflection about a line $\mathcal{L}$. Then we can select nonzero vectors $\mathbf{b}_1$ and $\mathbf{b}_2$ so that $\mathbf{b}_1$ is in the direction of $\mathcal{L}$ and $\mathbf{b}_2$ is in a direction perpendicular to $\mathcal{L}$. It then follows that $T(\mathbf{b}_1) = \mathbf{b}_1$ and $T(\mathbf{b}_2) = -\mathbf{b}_2$ (see Figure 4.8).

Observe that $\mathcal{B} = \{\mathbf{b}_1, \mathbf{b}_2\}$ is a basis for $\mathcal{R}^2$ because $\mathcal{B}$ is a linearly independent subset of $\mathcal{R}^2$ consisting of two vectors. Moreover, we can describe

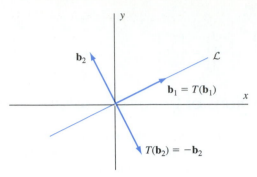

FIGURE 4.8

the action of the reflection T on $\mathcal{B}$. In particular, since

$$T(\mathbf{b}_1) = 1\mathbf{b}_1 + 0\mathbf{b}_2 \qquad \text{and} \qquad T(\mathbf{b}_2) = 0\mathbf{b}_1 + (-1)\mathbf{b}_2,$$

the coordinate vectors of $T(\mathbf{b}_1)$ and $T(\mathbf{b}_2)$ relative to $\mathcal{B}$ are given by

$$[T(\mathbf{b}_1)]_{\mathcal{B}} = \begin{bmatrix} 1 \\ 0 \end{bmatrix} \qquad \text{and} \qquad [T(\mathbf{b}_2)]_{\mathcal{B}} = \begin{bmatrix} 0 \\ -1 \end{bmatrix}.$$

We can collect these columns to form a matrix

$$[\,[T(\mathbf{b}_1)]_{\mathcal{B}} \quad [T(\mathbf{b}_2)]_{\mathcal{B}}\,] = \begin{bmatrix} 1 & 0 \\ 0 & -1 \end{bmatrix}$$

that captures the essence of the behavior of T in relation to the basis $\mathcal{B}$. In the case of the previously described reflection T' about the x-axis, $\mathcal{B}$ is the standard basis for $\mathcal{R}^2$, and the matrix $[\,[T'(\mathbf{b}_1)]_{\mathcal{B}} \quad [T'(\mathbf{b}_2)]_{\mathcal{B}}\,]$ is the standard matrix of T'. This motivates the following definition.

Definition. Let T be a linear operator on $\mathcal{R}^n$ and $\mathcal{B} = \{\mathbf{b}_1, \mathbf{b}_2, \ldots, \mathbf{b}_n\}$ be a basis for $\mathcal{R}^n$. The matrix

$$[\,[T(\mathbf{b}_1)]_{\mathcal{B}} \quad [T(\mathbf{b}_2)]_{\mathcal{B}} \quad \cdots \quad [T(\mathbf{b}_n)]_{\mathcal{B}}\,]$$

is called the **matrix representation of T with respect to $\mathcal{B}$** or the **$\mathcal{B}$-matrix of T**. It is denoted $[T]_{\mathcal{B}}$.

For a linear operator T on $\mathcal{R}^n$, if $\mathcal{B} = \mathcal{E}$ is the standard basis for $\mathcal{R}^n$, then

$$[T]_{\mathcal{E}} = [\,[T(\mathbf{b}_1)]_{\mathcal{E}} \quad [T(\mathbf{b}_2)]_{\mathcal{E}} \quad \cdots \quad [T(\mathbf{b}_n)]_{\mathcal{E}}\,] = [T(\mathbf{e}_1) \quad T(\mathbf{e}_2) \quad \cdots \quad T(\mathbf{e}_n)],$$

which is the standard matrix of T. Thus this definition extends the notion of a standard matrix to the context of an arbitrary basis for $\mathcal{R}^n$.

For the reflection T about the line $\mathcal{L}$ and the basis $\mathcal{B} = \{\mathbf{b}_1, \mathbf{b}_2\}$ given earlier in this section, the matrix representation of T with respect to $\mathcal{B}$ is given by

$$[T]_{\mathcal{B}} = [\,[T(\mathbf{b}_1)]_{\mathcal{B}} \quad [T(\mathbf{b}_2)]_{\mathcal{B}}\,] = \begin{bmatrix} 1 & 0 \\ 0 & -1 \end{bmatrix}.$$

Example 1   Let T be the linear operator on $\mathcal{R}^3$ defined by

$$T\left(\begin{bmatrix} x_1 \\ x_2 \\ x_3 \end{bmatrix}\right) = \begin{bmatrix} 3x_1 + x_3 \\ x_1 + x_2 \\ -x_1 - x_2 + 3x_3 \end{bmatrix}.$$

Calculate the matrix representation of T with respect to the basis $\mathcal{B} = \{\mathbf{b}_1, \mathbf{b}_2, \mathbf{b}_3\}$, where

$$\mathbf{b}_1 = \begin{bmatrix} 1 \\ 1 \\ 1 \end{bmatrix}, \qquad \mathbf{b}_2 = \begin{bmatrix} 1 \\ 2 \\ 3 \end{bmatrix}, \qquad \text{and} \qquad \mathbf{b}_3 = \begin{bmatrix} 2 \\ 1 \\ 1 \end{bmatrix}.$$

Solution   Applying T to each of the vectors in $\mathcal{B}$, we obtain

$$T(\mathbf{b}_1) = \begin{bmatrix} 4 \\ 2 \\ 1 \end{bmatrix}, \qquad T(\mathbf{b}_2) = \begin{bmatrix} 6 \\ 3 \\ 6 \end{bmatrix}, \qquad \text{and} \qquad T(\mathbf{b}_3) = \begin{bmatrix} 7 \\ 3 \\ 0 \end{bmatrix}.$$

We must now compute the *coordinate vectors* of these images with respect to $\mathcal{B}$. Let $B = [\mathbf{b}_1 \ \ \mathbf{b}_2 \ \ \mathbf{b}_3]$. Then

$$[T(\mathbf{b}_1)]_{\mathcal{B}} = B^{-1}T(\mathbf{b}_1) = \begin{bmatrix} 3 \\ -1 \\ 1 \end{bmatrix}, \qquad [T(\mathbf{b}_2)]_{\mathcal{B}} = B^{-1}T(\mathbf{b}_2) = \begin{bmatrix} -9 \\ 3 \\ 6 \end{bmatrix},$$

and

$$[T(\mathbf{b}_3)]_{\mathcal{B}} = B^{-1}T(\mathbf{b}_3) = \begin{bmatrix} 8 \\ -3 \\ 1 \end{bmatrix},$$

from which it follows that the $\mathcal{B}$-matrix of T is

$$[T]_{\mathcal{B}} = \begin{bmatrix} 3 & -9 & 8 \\ -1 & 3 & -3 \\ 1 & 6 & 1 \end{bmatrix}.$$

As in Section 4.4, it is natural to ask how the matrix representation of T with respect to $\mathcal{B}$ is related to the standard matrix of T (which is the matrix representation of T with respect to the standard basis for $\mathcal{R}^n$). The answer is provided by the following theorem.

Theorem 4.10

Let T be a linear operator on $\mathcal{R}^n$, $\mathcal{B}$ a basis for $\mathcal{R}^n$, B the matrix whose columns are the vectors in $\mathcal{B}$, and A the standard matrix of T. Then $[T]_{\mathcal{B}} = B^{-1}AB$, or equivalently, $A = B[T]_{\mathcal{B}}B^{-1}$.

Proof   Let $\mathcal{B} = \{\mathbf{b}_1, \mathbf{b}_2, \ldots, \mathbf{b}_n\}$ and $B = [\mathbf{b}_1 \ \ \mathbf{b}_2 \ \ \cdots \ \ \mathbf{b}_n]$. Recall that $T(\mathbf{u}) = A\mathbf{u}$ and $[\mathbf{v}]_{\mathcal{B}} = B^{-1}\mathbf{v}$ for all $\mathbf{u}$ and $\mathbf{v}$ in $\mathcal{R}^n$. Hence

$$\begin{aligned} [T]_{\mathcal{B}} &= [\,[T(\mathbf{b}_1)]_{\mathcal{B}} \quad [T(\mathbf{b}_2)]_{\mathcal{B}} \quad \cdots \quad [T(\mathbf{b}_n)]_{\mathcal{B}}\,] \\ &= [\,[A\mathbf{b}_1]_{\mathcal{B}} \quad [A\mathbf{b}_2]_{\mathcal{B}} \quad \cdots \quad [A\mathbf{b}_n]_{\mathcal{B}}\,] \\ &= [B^{-1}(A\mathbf{b}_1) \quad B^{-1}(A\mathbf{b}_2) \quad \cdots \quad B^{-1}(A\mathbf{b}_n)] \\ &= [(B^{-1}A)\mathbf{b}_1 \quad (B^{-1}A)\mathbf{b}_2 \quad \cdots \quad (B^{-1}A)\mathbf{b}_n] \\ &= B^{-1}A[\mathbf{b}_1 \quad \mathbf{b}_2 \quad \cdots \quad \mathbf{b}_n] \\ &= B^{-1}AB. \end{aligned}$$

Thus $[T]_B = B^{-1}AB$, which is equivalent to

$$B[T]_B B^{-1} = B(B^{-1}AB)B^{-1} = A. \qquad \square$$

If two square matrices A and B are such that $B = P^{-1}AP$ for some invertible matrix P, then A is said to be **similar** to B. It is easily seen that A is similar to B if and only if B is similar to A (see Exercise 48 of Section 2.4). Hence we will usually describe this situation by saying that A *and* B *are similar.*

Theorem 4.10 shows that the B-matrix representation of a linear operator on $\mathcal{R}^n$ is similar to its standard matrix. Theorem 4.10 not only gives us the relationship between $[T]_B$ and the standard matrix of T, it also provides a practical method for computing one of these matrices from the other. The following examples illustrate these computations.

Example 2   Calculate $[T]_B$ using Theorem 4.10 if T and B are the linear operator and the basis given in Example 1:

$$T\left(\begin{bmatrix} x_1 \\ x_2 \\ x_3 \end{bmatrix}\right) = \begin{bmatrix} 3x_1 + x_3 \\ x_1 + x_2 \\ -x_1 - x_2 + 3x_3 \end{bmatrix} \qquad \text{and} \qquad B = \left\{ \begin{bmatrix} 1 \\ 1 \\ 1 \end{bmatrix}, \begin{bmatrix} 1 \\ 2 \\ 3 \end{bmatrix}, \begin{bmatrix} 2 \\ 1 \\ 1 \end{bmatrix} \right\}.$$

Solution   The standard matrix of T is

$$A = [T(\mathbf{e}_1) \ \ T(\mathbf{e}_2) \ \ T(\mathbf{e}_3)] = \begin{bmatrix} 3 & 0 & 1 \\ 1 & 1 & 0 \\ -1 & -1 & 3 \end{bmatrix}.$$

Taking B to be the matrix whose columns are the vectors in B, we have

$$[T]_B = B^{-1}AB = \begin{bmatrix} 3 & -9 & 8 \\ -1 & 3 & -3 \\ 1 & 6 & 1 \end{bmatrix}.$$

Note that our answer agrees with that in Example 1.   ○

Example 3   Let T be a linear operator on $\mathcal{R}^3$ such that

$$T\left(\begin{bmatrix} 1 \\ 1 \\ 0 \end{bmatrix}\right) = \begin{bmatrix} 1 \\ 2 \\ -1 \end{bmatrix}, \quad T\left(\begin{bmatrix} 1 \\ 0 \\ 1 \end{bmatrix}\right) = \begin{bmatrix} 3 \\ -1 \\ 1 \end{bmatrix}, \quad \text{and} \quad T\left(\begin{bmatrix} 0 \\ 1 \\ 1 \end{bmatrix}\right) = \begin{bmatrix} 2 \\ 0 \\ 1 \end{bmatrix}.$$

Find the standard matrix of T.

Solution   Let A denote the standard matrix of T. To facilitate the discussion, let

$$\mathbf{b}_1 = \begin{bmatrix} 1 \\ 1 \\ 0 \end{bmatrix}, \qquad \mathbf{b}_2 = \begin{bmatrix} 1 \\ 0 \\ 1 \end{bmatrix}, \qquad \text{and} \qquad \mathbf{b}_3 = \begin{bmatrix} 0 \\ 1 \\ 1 \end{bmatrix},$$

and let

$$\mathbf{c}_1 = \begin{bmatrix} 1 \\ 2 \\ -1 \end{bmatrix}, \qquad \mathbf{c}_2 = \begin{bmatrix} 3 \\ -1 \\ 1 \end{bmatrix}, \qquad \text{and} \qquad \mathbf{c}_3 = \begin{bmatrix} 2 \\ 0 \\ 1 \end{bmatrix}.$$

First observe that $\mathcal{B} = \{\mathbf{b}_1, \mathbf{b}_2, \mathbf{b}_3\}$ is linearly independent and hence is a basis for $\mathcal{R}^3$. Let $B = [\mathbf{b}_1 \ \ \mathbf{b}_2 \ \ \mathbf{b}_3]$. Then

$$
\begin{aligned}
[T]_\mathcal{B} &= [\ [T(\mathbf{b}_1)]_\mathcal{B} \quad [T(\mathbf{b}_2)]_\mathcal{B} \quad [T(\mathbf{b}_1)]_\mathcal{B} \ ] \\
&= \left[B^{-1}T(\mathbf{b}_1) \quad B^{-1}T(\mathbf{b}_2) \quad B^{-1}T(\mathbf{b}_3) \right] \\
&= \left[B^{-1}\mathbf{c}_1 \quad B^{-1}\mathbf{c}_2 \quad B^{-1}\mathbf{c}_3 \right] \\
&= B^{-1}[\mathbf{c}_1 \quad \mathbf{c}_2 \quad \mathbf{c}_3].
\end{aligned}
$$

Thus

$$
\begin{aligned}
A &= B[T]_\mathcal{B}B^{-1} \\
&= BB^{-1}[\mathbf{c}_1 \quad \mathbf{c}_2 \quad \mathbf{c}_3]B^{-1} \\
&= [\mathbf{c}_1 \quad \mathbf{c}_2 \quad \mathbf{c}_3]B^{-1} \\
&= \begin{bmatrix} 1 & 3 & 2 \\ 2 & -1 & 0 \\ -1 & 1 & 1 \end{bmatrix} \begin{bmatrix} .5 & .5 & -.5 \\ .5 & -.5 & .5 \\ -.5 & .5 & .5 \end{bmatrix} \\
&= \begin{bmatrix} 1 & 0 & 2 \\ .5 & 1.5 & -1.5 \\ -.5 & -.5 & 1.5 \end{bmatrix}.
\end{aligned}
$$

Example 3 suggests that a linear operator is uniquely determined by its action on a basis because we are able to determine the standard matrix for the operator in this example solely from this information. This is indeed the case, as Exercise 52 shows.

Calling $[T]_\mathcal{B}$ a *matrix representation* of T suggests that this matrix in some way describes the action of T. Recall that if A is the standard matrix of a linear operator T on $\mathcal{R}^n$, then $T(\mathbf{v}) = A\mathbf{v}$ for all vectors $\mathbf{v}$ in $\mathcal{R}^n$. Since $[\mathbf{v}]_\mathcal{E} = \mathbf{v}$ for every $\mathbf{v}$ in $\mathcal{R}^n$, we see that $[T]_\mathcal{E}[v]_\mathcal{E} = A\mathbf{v} = T(\mathbf{v}) = [T(\mathbf{v})]_\mathcal{E}$. This result explains how the standard matrix of T is related to the image of a vector under T. The following theorem shows that an analogous result is true for $[T]_\mathcal{B}$ and so justifies our name for $[T]_\mathcal{B}$.

Theorem 4.11

Let T be a linear operator on $\mathcal{R}^n$ and $\mathcal{B}$ be a basis for $\mathcal{R}^n$. The $\mathcal{B}$-matrix of T is the unique $n \times n$ matrix such that $[T(\mathbf{v})]_\mathcal{B} = [T]_\mathcal{B}[\mathbf{v}]_\mathcal{B}$ for all $\mathbf{v}$ in $\mathcal{R}^n$.

Proof Let A be the standard matrix of T and B be the matrix whose columns are the vectors in $\mathcal{B}$. Then $[T]_\mathcal{B} = B^{-1}AB$ by Theorem 4.10. Since $B[\mathbf{v}]_\mathcal{B} = \mathbf{v}$, $B^{-1}\mathbf{v} = [\mathbf{v}]_\mathcal{B}$, and $T(\mathbf{v}) = A\mathbf{v}$ for all $\mathbf{v}$ in $\mathcal{R}^n$, we have

$$
\begin{aligned}
[T(\mathbf{v})]_\mathcal{B} &= B^{-1}(T(\mathbf{v})) = B^{-1}(A\mathbf{v}) = (B^{-1}A)\mathbf{v} \\
&= (B^{-1}A)(B[\mathbf{v}]_\mathcal{B}) = (B^{-1}AB)[\mathbf{v}]_\mathcal{B} = [T]_\mathcal{B}[\mathbf{v}]_\mathcal{B}.
\end{aligned}
$$

In order to show uniqueness, suppose that P is another $n \times n$ matrix such that $[T(\mathbf{v})]_\mathcal{B} = P[\mathbf{v}]_\mathcal{B}$ for all $\mathbf{v}$ in $\mathcal{R}^n$. Then $P[\mathbf{v}]_\mathcal{B} = [T]_\mathcal{B}[\mathbf{v}]_\mathcal{B}$ for all $\mathbf{v}$ in $\mathcal{R}^n$. Let $\mathcal{B} = \{\mathbf{b}_1, \mathbf{b}_2, \ldots, \mathbf{b}_n\}$. Since $[\mathbf{b}_j]_\mathcal{B} = \mathbf{e}_j$ for each j, we have

$$
P\mathbf{e}_j = P[\mathbf{b}_j]_\mathcal{B} = [T]_\mathcal{B}[\mathbf{b}_j]_\mathcal{B} = [T]_\mathcal{B}\mathbf{e}_j
$$

for every j. But $P\mathbf{e}_j$ is the jth column of P and $[T]_\mathcal{B}\mathbf{e}_j$ is the jth column of $[T]_\mathcal{B}$, and so the corresponding columns of P and $[T]_\mathcal{B}$ are equal. Thus $P = [T]_\mathcal{B}$. ☐

To conclude this section, we apply Theorem 4.10 to find an explicit description of the reflection T of $\mathcal{R}^2$ about the line $\mathcal{L}$ with equation $y = \frac{1}{2}x$. In the earlier discussion of this problem, we have seen that nonzero vectors $\mathbf{b}_1$ and $\mathbf{b}_2$ in $\mathcal{R}^2$ must be selected so that $\mathbf{b}_1$ lies on $\mathcal{L}$ and $\mathbf{b}_2$ is perpendicular to $\mathcal{L}$ (see Figure 4.8). One selection that works is

$$\mathbf{b}_1 = \begin{bmatrix} 2 \\ 1 \end{bmatrix} \quad \text{and} \quad \mathbf{b}_2 = \begin{bmatrix} -1 \\ 2 \end{bmatrix},$$

because $\mathbf{b}_1$ lies on $\mathcal{L}$, which has slope $\frac{1}{2}$, and $\mathbf{b}_2$ lies on the line perpendicular to $\mathcal{L}$, which has slope -2. Let $\mathcal{B} = \{\mathbf{b}_1, \mathbf{b}_2\}$, $B = [\mathbf{b}_1 \quad \mathbf{b}_2]$, and A be the standard matrix of T. Recall that

$$[T]_{\mathcal{B}} = [\,[T(\mathbf{b}_1)]_{\mathcal{B}} \quad T(\mathbf{b}_2)]_{\mathcal{B}}\,] = \begin{bmatrix} 1 & 0 \\ 0 & -1 \end{bmatrix}.$$

Then by Theorem 4.10

$$A = B[T]_{\mathcal{B}}B^{-1} = \begin{bmatrix} .6 & .8 \\ .8 & -.6 \end{bmatrix}.$$

It follows that the reflection of $\mathcal{R}^2$ about the line with equation $y = \frac{1}{2}x$ is given by

$$T\left(\begin{bmatrix} x_1 \\ x_2 \end{bmatrix} \right) = A \begin{bmatrix} x_1 \\ x_2 \end{bmatrix} = \begin{bmatrix} .6 & .8 \\ .8 & -.6 \end{bmatrix} \begin{bmatrix} x_1 \\ x_2 \end{bmatrix} = \begin{bmatrix} .6x_1 + .8x_2 \\ .8x_1 - .6x_2 \end{bmatrix}.$$

Practice Problems

1. Find the $\mathcal{B}$-matrix representation of $T: \mathcal{R}^3 \rightarrow \mathcal{R}^3$ if

$$T\left(\begin{bmatrix} x_1 \\ x_2 \\ x_3 \end{bmatrix} \right) = \begin{bmatrix} -x_1 + 2x_3 \\ x_1 + x_2 \\ -x_2 + x_3 \end{bmatrix}$$

and

$$\mathcal{B} = \left\{ \begin{bmatrix} 1 \\ 1 \\ 0 \end{bmatrix}, \begin{bmatrix} 1 \\ 1 \\ 1 \end{bmatrix}, \begin{bmatrix} 3 \\ 2 \\ 1 \end{bmatrix} \right\}.$$

2. If $U: \mathcal{R}^3 \rightarrow \mathcal{R}^3$ is a linear transformation such that

$$[U]_{\mathcal{B}} = \begin{bmatrix} 3 & 0 & 0 \\ 0 & 2 & 0 \\ 0 & 0 & 2 \end{bmatrix}$$

for the basis $\mathcal{B}$ in problem 1, determine an explicit definition of U.

Exercises

1. Determine if the following statements are true or false.

   (a) The matrix representation of a linear operator on $\mathcal{R}^n$ with respect to a basis for $\mathcal{R}^n$ is an $n \times n$ matrix.

   (b) If T is a linear operator on $\mathcal{R}^n$ and $\mathcal{B} = \{\mathbf{b}_1, \mathbf{b}_2, \ldots, \mathbf{b}_n\}$ is a basis for $\mathcal{R}^n$, then the matrix representation of T with respect to $\mathcal{B}$ is

   $$[T(\mathbf{b}_1) \quad T(\mathbf{b}_2) \quad \cdots \quad T(\mathbf{b}_n)].$$

   (c) If $\mathcal{E}$ is the standard basis for $\mathcal{R}^n$, then $[T]_{\mathcal{E}}$ is the standard matrix of T.

   (d) If T is a linear operator on $\mathcal{R}^n$, B is a basis for $\mathcal{R}^n$, B is the matrix whose columns are the vectors in $\mathcal{B}$, and A is the standard matrix of T, then $[T]_{\mathcal{B}} = BAB^{-1}$.

   (e) If T is a reflection of $\mathcal{R}^2$ about a line, then there is a basis $\mathcal{B}$ for $\mathcal{R}^2$ such that $[T]_{\mathcal{B}} = \begin{bmatrix} 1 & 0 \\ 0 & -1 \end{bmatrix}$.

   (f) If T is a linear operator on $\mathcal{R}^n$ and $\mathcal{B}$ is a basis for $\mathcal{R}^n$, then $[T]_{\mathcal{B}}$ is the unique $n \times n$ matrix such that $[T]_{\mathcal{B}}[\mathbf{v}]_{\mathcal{B}} = [T(\mathbf{v})]_{\mathcal{B}}$ for all $\mathbf{v}$ in $\mathcal{R}^n$.

2. Let $\mathcal{B} = \{\mathbf{b}_1, \mathbf{b}_2\}$ be a basis for $\mathcal{R}^2$ and T be a linear operator on $\mathcal{R}^2$ such that $T(\mathbf{b}_1) = 2\mathbf{b}_1 - 5\mathbf{b}_2$ and $T(\mathbf{b}_2) = -\mathbf{b}_1 + 3\mathbf{b}_2$. Determine $[T]_{\mathcal{B}}$.

3. Let $\mathcal{B} = \{\mathbf{b}_1, \mathbf{b}_2\}$ be a basis for $\mathcal{R}^2$ and T be a linear operator on $\mathcal{R}^2$ such that $T(\mathbf{b}_1) = \mathbf{b}_1 + 4\mathbf{b}_2$ and $T(\mathbf{b}_2) = -3\mathbf{b}_1$. Determine $[T]_{\mathcal{B}}$.

4. Let $\mathcal{B} = \{\mathbf{b}_1, \mathbf{b}_2, \mathbf{b}_3\}$ be a basis for $\mathcal{R}^3$ and T be a linear operator on $\mathcal{R}^3$ such that $T(\mathbf{b}_1) = \mathbf{b}_1 - 2\mathbf{b}_2 + 3\mathbf{b}_3$,

$T(\mathbf{b}_2) = 6\mathbf{b}_2 - \mathbf{b}_3$, and $T(\mathbf{b}_3) = 5\mathbf{b}_1 + 2\mathbf{b}_2 - 4\mathbf{b}_3$. Determine $[T]_\mathcal{B}$.

5. Let $\mathcal{B} = \{\mathbf{b}_1, \mathbf{b}_2, \mathbf{b}_3\}$ be a basis for $\mathcal{R}^3$ and T be a linear operator on $\mathcal{R}^3$ such that $T(\mathbf{b}_1) = -5\mathbf{b}_2 + 4\mathbf{b}_3$, $T(\mathbf{b}_2) = 2\mathbf{b}_1 - 7\mathbf{b}_3$, and $T(\mathbf{b}_3) = 3\mathbf{b}_1 + \mathbf{b}_3$. Determine $[T]_\mathcal{B}$.

6. Let $\mathcal{B} = \{\mathbf{b}_1, \mathbf{b}_2, \mathbf{b}_3, \mathbf{b}_4\}$ be a basis for $\mathcal{R}^4$ and T be a linear operator on $\mathcal{R}^4$ such that $T(\mathbf{b}_1) = \mathbf{b}_1 - \mathbf{b}_2 + \mathbf{b}_3 - \mathbf{b}_4$, $T(\mathbf{b}_2) = 2\mathbf{b}_2 - \mathbf{b}_4$, $T(\mathbf{b}_3) = -3\mathbf{b}_1 + 5\mathbf{b}_3$, and $T(\mathbf{b}_4) = 4\mathbf{b}_2 - \mathbf{b}_3 + 3\mathbf{b}_4$. Determine $[T]_\mathcal{B}$.

In Exercises 7–10, determine (a) $[T]_\mathcal{B}$, (b) the standard matrix of T, and (c) an explicit formula for $T(\mathbf{x})$ from the given information.

7. $\mathcal{B} = \left\{ \begin{bmatrix} 1 \\ 1 \end{bmatrix}, \begin{bmatrix} 1 \\ 2 \end{bmatrix} \right\}$, $T\left(\begin{bmatrix} 1 \\ 1 \end{bmatrix} \right) = \begin{bmatrix} 1 \\ 2 \end{bmatrix}$,

$T\left(\begin{bmatrix} 1 \\ 2 \end{bmatrix} \right) = 3 \begin{bmatrix} 1 \\ 1 \end{bmatrix}$

8. $\mathcal{B} = \left\{ \begin{bmatrix} 1 \\ 3 \end{bmatrix}, \begin{bmatrix} 1 \\ 0 \end{bmatrix} \right\}$, $T\left(\begin{bmatrix} 1 \\ 3 \end{bmatrix} \right) = \begin{bmatrix} 1 \\ 3 \end{bmatrix} - 2 \begin{bmatrix} 1 \\ 0 \end{bmatrix}$,

$T\left(\begin{bmatrix} 1 \\ 0 \end{bmatrix} \right) = 2 \begin{bmatrix} 1 \\ 3 \end{bmatrix} - \begin{bmatrix} 1 \\ 0 \end{bmatrix}$

9. $\mathcal{B} = \left\{ \begin{bmatrix} 1 \\ 0 \\ 1 \end{bmatrix}, \begin{bmatrix} 0 \\ 1 \\ 0 \end{bmatrix}, \begin{bmatrix} 1 \\ 1 \\ 0 \end{bmatrix} \right\}$, $T\left(\begin{bmatrix} 1 \\ 0 \\ 1 \end{bmatrix} \right) = - \begin{bmatrix} 0 \\ 1 \\ 0 \end{bmatrix}$,

$T\left(\begin{bmatrix} 0 \\ 1 \\ 0 \end{bmatrix} \right) = 2 \begin{bmatrix} 1 \\ 1 \\ 0 \end{bmatrix}$, $T\left(\begin{bmatrix} 1 \\ 1 \\ 0 \end{bmatrix} \right) = \begin{bmatrix} 1 \\ 0 \\ 1 \end{bmatrix} + 2 \begin{bmatrix} 0 \\ 1 \\ 0 \end{bmatrix}$

10. $\mathcal{B} = \left\{ \begin{bmatrix} 1 \\ 1 \\ -1 \end{bmatrix}, \begin{bmatrix} 0 \\ 1 \\ 1 \end{bmatrix}, \begin{bmatrix} 1 \\ 2 \\ 3 \end{bmatrix} \right\}$, $T\left(\begin{bmatrix} 1 \\ 1 \\ -1 \end{bmatrix} \right) =$

$\begin{bmatrix} 0 \\ 1 \\ 1 \end{bmatrix} + 2 \begin{bmatrix} 1 \\ 2 \\ 3 \end{bmatrix}$, $T\left(\begin{bmatrix} 0 \\ 1 \\ 1 \end{bmatrix} \right) = 4 \begin{bmatrix} 1 \\ 1 \\ -1 \end{bmatrix} - \begin{bmatrix} 1 \\ 2 \\ 3 \end{bmatrix}$,

$T\left(\begin{bmatrix} 1 \\ 2 \\ 3 \end{bmatrix} \right) = - \begin{bmatrix} 1 \\ 1 \\ -1 \end{bmatrix} + 3 \begin{bmatrix} 0 \\ 1 \\ 1 \end{bmatrix} + 2 \begin{bmatrix} 1 \\ 2 \\ 3 \end{bmatrix}$

In Exercises 11–20, determine $[T]_\mathcal{B}$ for the given linear operator T and basis $\mathcal{B}$.

11. $T\left(\begin{bmatrix} x_1 \\ x_2 \end{bmatrix} \right) = \begin{bmatrix} 2x_1 + x_2 \\ x_1 - x_2 \end{bmatrix}$ and $\mathcal{B} = \left\{ \begin{bmatrix} 2 \\ 1 \end{bmatrix}, \begin{bmatrix} 1 \\ 0 \end{bmatrix} \right\}$

12. $T\left(\begin{bmatrix} x_1 \\ x_2 \end{bmatrix} \right) = \begin{bmatrix} x_1 - x_2 \\ x_2 \end{bmatrix}$ and $\mathcal{B} = \left\{ \begin{bmatrix} 2 \\ 3 \end{bmatrix}, \begin{bmatrix} 1 \\ 1 \end{bmatrix} \right\}$

13. $T\left(\begin{bmatrix} x_1 \\ x_2 \end{bmatrix} \right) = \begin{bmatrix} x_1 + 2x_2 \\ x_1 + x_2 \end{bmatrix}$ and $\mathcal{B} = \left\{ \begin{bmatrix} 1 \\ 1 \end{bmatrix}, \begin{bmatrix} 2 \\ 1 \end{bmatrix} \right\}$

14. $T\left(\begin{bmatrix} x_1 \\ x_2 \end{bmatrix} \right) = \begin{bmatrix} -2x_1 + x_2 \\ x_1 + 3x_2 \end{bmatrix}$ and $\mathcal{B} = \left\{ \begin{bmatrix} 1 \\ 3 \end{bmatrix}, \begin{bmatrix} 2 \\ 5 \end{bmatrix} \right\}$

15. $T\left(\begin{bmatrix} x_1 \\ x_2 \\ x_3 \end{bmatrix} \right) = \begin{bmatrix} x_1 + x_2 \\ x_2 - 2x_3 \\ 2x_1 - x_2 + 3x_3 \end{bmatrix}$ and

$\mathcal{B} = \left\{ \begin{bmatrix} 1 \\ 1 \\ 1 \end{bmatrix}, \begin{bmatrix} 2 \\ 3 \\ 2 \end{bmatrix}, \begin{bmatrix} 1 \\ 2 \\ 2 \end{bmatrix} \right\}$

16. $T\left(\begin{bmatrix} x_1 \\ x_2 \\ x_3 \end{bmatrix} \right) = \begin{bmatrix} x_1 + x_3 \\ x_2 - x_3 \\ 2x_1 - x_2 \end{bmatrix}$ and

$\mathcal{B} = \left\{ \begin{bmatrix} 0 \\ -1 \\ 1 \end{bmatrix}, \begin{bmatrix} 1 \\ 0 \\ -1 \end{bmatrix}, \begin{bmatrix} 1 \\ 1 \\ -1 \end{bmatrix} \right\}$

17. $T\left(\begin{bmatrix} x_1 \\ x_2 \\ x_3 \end{bmatrix} \right) = \begin{bmatrix} 4x_2 \\ x_1 + 2x_3 \\ -2x_2 + 3x_3 \end{bmatrix}$ and

$\mathcal{B} = \left\{ \begin{bmatrix} 1 \\ 0 \\ 1 \end{bmatrix}, \begin{bmatrix} 1 \\ -2 \\ 0 \end{bmatrix}, \begin{bmatrix} -1 \\ 3 \\ 1 \end{bmatrix} \right\}$

18. $T\left(\begin{bmatrix} x_1 \\ x_2 \\ x_3 \end{bmatrix} \right) = \begin{bmatrix} x_1 - 2x_2 + 4x_3 \\ 3x_1 \\ -3x_2 + 2x_3 \end{bmatrix}$ and

$\mathcal{B} = \left\{ \begin{bmatrix} 1 \\ -2 \\ 1 \end{bmatrix}, \begin{bmatrix} 0 \\ -1 \\ 1 \end{bmatrix}, \begin{bmatrix} 1 \\ -5 \\ 3 \end{bmatrix} \right\}$

19. $T\left(\begin{bmatrix} x_1 \\ x_2 \\ x_3 \\ x_4 \end{bmatrix} \right) = \begin{bmatrix} x_1 + x_2 \\ x_2 - x_3 \\ x_1 + 2x_4 \\ x_2 - x_3 + 3x_4 \end{bmatrix}$ and

$\mathcal{B} = \left\{ \begin{bmatrix} 1 \\ -1 \\ 2 \\ 3 \end{bmatrix}, \begin{bmatrix} 1 \\ -2 \\ 1 \\ 4 \end{bmatrix}, \begin{bmatrix} 1 \\ -2 \\ 0 \\ 3 \end{bmatrix}, \begin{bmatrix} 0 \\ 1 \\ 1 \\ -2 \end{bmatrix} \right\}$

20. $T\left(\begin{bmatrix} x_1 \\ x_2 \\ x_3 \\ x_4 \end{bmatrix} \right) = \begin{bmatrix} x_1 - x_2 + x_3 + 2x_4 \\ 2x_1 - 3x_4 \\ x_1 + x_2 + x_3 \\ -3x_3 + x_4 \end{bmatrix}$ and

$\mathcal{B} = \left\{ \begin{bmatrix} 1 \\ 1 \\ 1 \\ 2 \end{bmatrix}, \begin{bmatrix} 2 \\ 3 \\ 3 \\ 3 \end{bmatrix}, \begin{bmatrix} 1 \\ 3 \\ 4 \\ 1 \end{bmatrix}, \begin{bmatrix} 4 \\ 5 \\ 8 \\ 8 \end{bmatrix} \right\}$

In Exercises 21–24, determine the standard matrix of the linear operator T, using the given basis $\mathcal{B}$ and the matrix representation of T with respect to $\mathcal{B}$.

21. $[T]_B = \begin{bmatrix} 1 & 4 \\ -3 & 5 \end{bmatrix}$ and $B = \left\{ \begin{bmatrix} -1 \\ 0 \end{bmatrix}, \begin{bmatrix} 3 \\ 1 \end{bmatrix} \right\}$

22. $[T]_B = \begin{bmatrix} 2 & 0 \\ 1 & -1 \end{bmatrix}$ and $B = \left\{ \begin{bmatrix} 1 \\ -2 \end{bmatrix}, \begin{bmatrix} -2 \\ 3 \end{bmatrix} \right\}$

23. $[T]_B = \begin{bmatrix} 1 & 0 & -3 \\ -2 & 1 & 2 \\ -1 & 1 & 1 \end{bmatrix}$ and

$B = \left\{ \begin{bmatrix} -2 \\ -1 \\ 1 \end{bmatrix}, \begin{bmatrix} -1 \\ -2 \\ 1 \end{bmatrix}, \begin{bmatrix} -1 \\ -1 \\ 1 \end{bmatrix} \right\}$

24. $[T]_B = \begin{bmatrix} -1 & 1 & -2 \\ 0 & 2 & 1 \\ 1 & 2 & 0 \end{bmatrix}$ and

$B = \left\{ \begin{bmatrix} 1 \\ 0 \\ 1 \end{bmatrix}, \begin{bmatrix} 1 \\ -2 \\ 0 \end{bmatrix}, \begin{bmatrix} -1 \\ 3 \\ 1 \end{bmatrix} \right\}$

25. Find $T(3\mathbf{b}_1 - 2\mathbf{b}_2)$ for the operator T and basis B in Exercise 3.

26. Find $T(-\mathbf{b}_1 + 4\mathbf{b}_2)$ for the operator T and basis B in Exercise 2.

27. Find $T(\mathbf{b}_1 - 3\mathbf{b}_2)$ for the operator T and basis B in Exercise 5.

28. Find $T(\mathbf{b}_2 - 2\mathbf{b}_3)$ for the operator T and basis B in Exercise 4.

29. Let I be the identity operator on $\mathcal{R}^n$, and let B be any basis for $\mathcal{R}^n$. Determine the matrix representation of I with respect to B.

30. Let T be the zero linear operator on $\mathcal{R}^n$, and let B be any basis for $\mathcal{R}^n$. Determine the matrix representation of T with respect to B.

31. Find an explicit description of the reflection T of $\mathcal{R}^2$ about the line with equation $y = \frac{1}{3}x$.

32. Find an explicit description of the reflection T of $\mathcal{R}^2$ about the line with equation $y = 2x$.

33. Find an explicit description of the reflection T of $\mathcal{R}^2$ about the line with equation $y = -2x$.

34. Find an explicit description of the reflection T of $\mathcal{R}^2$ about the line with equation $y = mx$.

The **orthogonal projection** of $\mathcal{R}^2$ onto line $\mathcal{L}$ is a function $U: \mathcal{R}^2 \to \mathcal{R}^2$ defined in the following manner. Let $\mathbf{v}$ be a vector in $\mathcal{R}^2$ with endpoint P. Construct a line from P perpendicular to $\mathcal{L}$, and let F denote the point of intersection of this perpendicular line with $\mathcal{L}$. The vector with endpoint F is $U(\mathbf{v})$ (see Figure 4.9).

35. Find $U\left(\begin{bmatrix} x_1 \\ x_2 \end{bmatrix} \right)$, where U is the orthogonal projection of $\mathcal{R}^2$ onto the line with equation $y = x$. *Hint:* First find

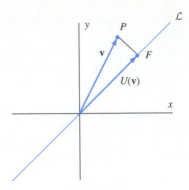

FIGURE 4.9

$[U]_B$, where $B = \left\{ \begin{bmatrix} 1 \\ 1 \end{bmatrix}, \begin{bmatrix} -1 \\ 1 \end{bmatrix} \right\}$.

36. Find $U\left(\begin{bmatrix} x_1 \\ x_2 \end{bmatrix} \right)$, where U is the orthogonal projection of $\mathcal{R}^2$ onto the line with equation $y = -\frac{1}{2}x$.

37. Find $U\left(\begin{bmatrix} x_1 \\ x_2 \end{bmatrix} \right)$, where U is the orthogonal projection of $\mathcal{R}^2$ onto the line with equation $y = -3x$.

38. Find $U\left(\begin{bmatrix} x_1 \\ x_2 \end{bmatrix} \right)$, where U is the orthogonal projection of $\mathcal{R}^2$ onto the line with equation $y = mx$.

Let W be a plane through the origin of $\mathcal{R}^3$, and let $\mathbf{v}$ be a vector in $\mathcal{R}^3$ with endpoint P. Construct a line from P perpendicular to W, and let F denote the point of intersection of this perpendicular line with W. Denote the vector with endpoint F as $U(\mathbf{v})$. Now extend the perpendicular from P to F an equal distance to a point P' on the other side of W, and denote the vector with endpoint P' as $T(\mathbf{v})$. The functions U and T defined in these ways can be shown to be linear operators on $\mathcal{R}^3$. We call U the **orthogonal projection** of $\mathcal{R}^3$ onto W and T the **reflection** of $\mathcal{R}^3$ about W (see Figure 4.10).

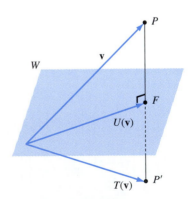

FIGURE 4.10

39. Let T be the reflection of $\mathcal{R}^3$ about the plane W in $\mathcal{R}^3$ with equation $x + 2y - 3z = 0$, and let $\mathcal{B} = \left\{ \begin{bmatrix} -2 \\ 1 \\ 0 \end{bmatrix}, \begin{bmatrix} 3 \\ 0 \\ 1 \end{bmatrix}, \begin{bmatrix} 1 \\ 2 \\ -3 \end{bmatrix} \right\}$. Note that the first two vectors in $\mathcal{B}$ lie in W, and the third vector is perpendicular to W.

  (a) Show that $\mathcal{B}$ is a basis for $\mathcal{R}^3$.
  (b) Find $[T]_{\mathcal{B}}$.
  (c) Find the standard matrix of T.
  (d) Determine an explicit formula for $T\left(\begin{bmatrix} x_1 \\ x_2 \\ x_3 \end{bmatrix} \right)$.

40. Let W be the plane in $\mathcal{R}^3$ with equation $2x - y + z = 0$ and T be the reflection of $\mathcal{R}^3$ about W. Determine an explicit formula for $T\left(\begin{bmatrix} x_1 \\ x_2 \\ x_3 \end{bmatrix} \right)$.

41. Let W and $\mathcal{B}$ be as in Exercise 39, and let U be the orthogonal projection of $\mathcal{R}^3$ onto W.

  (a) Find $[U]_{\mathcal{B}}$.
  (b) Find the standard matrix of U.
  (c) Determine an explicit formula for $U\left(\begin{bmatrix} x_1 \\ x_2 \\ x_3 \end{bmatrix} \right)$.

42. Let W be the plane in $\mathcal{R}^3$ with equation $x + y - 2z = 0$ and U be the orthogonal projection of $\mathcal{R}^3$ onto W. Determine an explicit formula for $U\left(\begin{bmatrix} x_1 \\ x_2 \\ x_3 \end{bmatrix} \right)$.

43. Let $\mathcal{B}$ be a basis for $\mathcal{R}^n$ and T be a linear operator on $\mathcal{R}^n$. Prove that T is invertible if and only if $[T]_{\mathcal{B}}$ is invertible.

44. Let $\mathcal{B}$ be a basis for $\mathcal{R}^n$ and T and U be linear operators on $\mathcal{R}^n$. Prove that $[UT]_{\mathcal{B}} = [U]_{\mathcal{B}}[T]_{\mathcal{B}}$.

45. Let $\mathcal{B}$ be a basis for $\mathcal{R}^n$ and T be a linear operator on $\mathcal{R}^n$. Prove that the dimension of the range of T equals the rank of $[T]_{\mathcal{B}}$.

46. Let $\mathcal{B}$ be a basis for $\mathcal{R}^n$ and T be a linear operator on $\mathcal{R}^n$. Prove that the dimension of the null space of T equals the nullity of $[T]_{\mathcal{B}}$.

47. Let $\mathcal{B}$ be a basis for $\mathcal{R}^n$ and T and U be linear operators on $\mathcal{R}^n$. Prove that $[T + U]_{\mathcal{B}} = [T]_{\mathcal{B}} + [U]_{\mathcal{B}}$. (See page 159 for the definition of $T + U$.)

48. Let $\mathcal{B}$ be a basis for $\mathcal{R}^n$ and T be a linear operator on $\mathcal{R}^n$. Prove that $[cT]_{\mathcal{B}} = c[T]_{\mathcal{B}}$ for any scalar c. (See page 159 for the definition of cT.)

49. Let T be a linear operator on $\mathcal{R}^n$, and let $\mathcal{A}$ and $\mathcal{B}$ be two bases for $\mathcal{R}^n$. Prove that $[T]_{\mathcal{A}}$ and $[T]_{\mathcal{B}}$ are similar.

50. Let A and B be similar matrices. Find bases $\mathcal{A}$ and $\mathcal{B}$ for $\mathcal{R}^n$ such that $[T_A]_{\mathcal{A}} = A$ and $[T_A]_{\mathcal{B}} = B$. (This proves that similar matrices are matrix representations of the same linear operator.)

51. Show that if A is the standard matrix of a reflection of $\mathcal{R}^2$ about a line, then $\det A = -1$.

52. Let $\mathcal{B} = \{\mathbf{b}_1, \mathbf{b}_2, \ldots, \mathbf{b}_n\}$ be a basis for $\mathcal{R}^n$, and let $\mathbf{c}_1, \mathbf{c}_2, \ldots, \mathbf{c}_n$ be (not necessarily distinct) vectors in $\mathcal{R}^n$. Prove that there exists a unique linear operator T on $\mathcal{R}^n$ such that $T(\mathbf{b}_j) = \mathbf{c}_j$ for each j.

In Exercises 53–57, use a calculator with matrix capabilities or computer software such as MATLAB *to solve the problem.*

53. Let T and U be the linear operators on $\mathcal{R}^4$ defined by

$$T\left(\begin{bmatrix} x_1 \\ x_2 \\ x_3 \\ x_4 \end{bmatrix} \right) = \begin{bmatrix} x_1 - 2x_2 \\ x_3 \\ -x_1 + 3x_3 \\ 2x_2 - x_4 \end{bmatrix}$$

and

$$U\left(\begin{bmatrix} x_1 \\ x_2 \\ x_3 \\ x_4 \end{bmatrix} \right) = \begin{bmatrix} x_2 - x_3 + 2x_4 \\ -2x_1 + 3x_4 \\ 2x_2 - x_3 \\ 3x_1 + x_4 \end{bmatrix},$$

and let $\mathcal{B} = \{\mathbf{b}_1, \mathbf{b}_2, \mathbf{b}_3, \mathbf{b}_4\}$, where

$$\mathbf{b}_1 = \begin{bmatrix} 0 \\ 1 \\ 1 \\ 1 \end{bmatrix}, \quad \mathbf{b}_2 = \begin{bmatrix} 0 \\ 1 \\ 2 \\ -1 \end{bmatrix}, \quad \mathbf{b}_3 = \begin{bmatrix} 1 \\ 1 \\ -1 \\ 0 \end{bmatrix}, \text{ and } \mathbf{b}_4 = \begin{bmatrix} 1 \\ 0 \\ -2 \\ -2 \end{bmatrix}.$$

  (a) Compute $[T]_{\mathcal{B}}$, $[U]_{\mathcal{B}}$, and $[UT]_{\mathcal{B}}$.
  (b) Determine a relationship among $[T]_{\mathcal{B}}$, $[U]_{\mathcal{B}}$, and $[UT]_{\mathcal{B}}$.

54. Let T and U be linear operators on $\mathcal{R}^n$ and $\mathcal{B}$ be a basis for $\mathcal{R}^n$. Use the result of Exercise 53(b) to conjecture a relationship among $[T]_{\mathcal{B}}$, $[U]_{\mathcal{B}}$, and $[UT]_{\mathcal{B}}$, and then prove that your conjecture is true.

55. Let $\mathcal{B}$ and $\mathbf{b}_1, \mathbf{b}_2, \mathbf{b}_3, \mathbf{b}_4$ be as in Exercise 53.

  (a) Compute $[T]_{\mathcal{B}}$, where T is the linear operator on $\mathcal{R}^4$ such that $T(\mathbf{b}_1) = \mathbf{b}_2, T(\mathbf{b}_2) = \mathbf{b}_3, T(\mathbf{b}_3) = \mathbf{b}_4$, and $T(\mathbf{b}_4) = \mathbf{b}_1$.
  (b) Determine an explicit formula for $T(\mathbf{x})$, where $\mathbf{x}$ is an arbitrary vector in $\mathcal{R}^4$.

56. Let $\mathcal{B} = \{\mathbf{b}_1, \mathbf{b}_2, \mathbf{b}_3, \mathbf{b}_4\}$ be as in Exercise 53, and let T be the linear operator on $\mathcal{R}^4$ defined by

$$T\left(\begin{bmatrix} x_1 \\ x_2 \\ x_3 \\ x_4 \end{bmatrix} \right) = \begin{bmatrix} x_1 + 2x_2 - 3x_3 - 2x_4 \\ -x_1 - 2x_2 + 4x_3 + 6x_4 \\ 2x_1 + 3x_2 - 5x_3 - 4x_4 \\ -x_1 + x_2 - x_3 - x_4 \end{bmatrix}.$$

  (a) Determine an explicit formula for $T^{-1}(\mathbf{x})$, where $\mathbf{x}$ is an arbitrary vector in $\mathcal{R}^4$.
  (b) Compute $[T]_{\mathcal{B}}$ and $[T^{-1}]_{\mathcal{B}}$.
  (c) Determine a relationship between $[T]_{\mathcal{B}}$ and $[T^{-1}]_{\mathcal{B}}$.

57. Let T be an invertible linear operator on $\mathcal{R}^n$ and $\mathcal{B}$ be a basis for $\mathcal{R}^n$. Use the results of Exercise 56(c) to conjecture a relationship between $[T]_{\mathcal{B}}$ and $[T^{-1}]_{\mathcal{B}}$, and then prove that your conjecture is true.

Solutions to the Practice Problems

1. The standard matrix of T is

$$A = \begin{bmatrix} -1 & 0 & 2 \\ 1 & 1 & 0 \\ 0 & -1 & 1 \end{bmatrix}.$$

Let

$$B = \begin{bmatrix} 1 & 1 & 3 \\ 1 & 1 & 2 \\ 0 & 1 & 1 \end{bmatrix}.$$

be the matrix whose columns are the vectors in $\mathcal{B}$. Then the $\mathcal{B}$-matrix representation of T is

$$B^{-1}AB = \begin{bmatrix} 6 & 3 & 12 \\ 2 & 1 & 5 \\ -3 & -1 & -6 \end{bmatrix}.$$

2. The standard matrix of U is

$$A = B[U]_{\mathcal{B}}B^{-1} = \begin{bmatrix} -2 & 5 & -1 \\ -3 & 6 & -1 \\ -1 & 1 & 2 \end{bmatrix}.$$

Hence

$$U\left(\begin{bmatrix} x_1 \\ x_2 \\ x_3 \end{bmatrix}\right) = A\begin{bmatrix} x_1 \\ x_2 \\ x_3 \end{bmatrix} = \begin{bmatrix} -2x_1 + 5x_2 - x_3 \\ -3x_1 + 6x_2 - x_3 \\ -x_1 + x_2 + 2x_3 \end{bmatrix}.$$

Chapter 4 Review Exercises

1. Determine if the following statements are true or false.

(a) If $\mathbf{u}_1, \mathbf{u}_2, \ldots, \mathbf{u}_k$ are vectors in a subspace V of $\mathcal{R}^n$, then every linear combination of $\mathbf{u}_1, \mathbf{u}_2, \ldots, \mathbf{u}_k$ belongs to V.

(b) The span of a finite nonempty subset of $\mathcal{R}^n$ is a subspace of $\mathcal{R}^n$.

(c) The null space of an $m \times n$ matrix is contained in $\mathcal{R}^m$.

(d) The column space of an $m \times n$ matrix is contained in $\mathcal{R}^n$.

(e) The row space of an $m \times n$ matrix is contained in $\mathcal{R}^m$.

(f) The range of every linear transformation is a subspace.

(g) The null space of every linear transformation equals the null space of its standard matrix.

(h) The range of every linear transformation equals the row space of its standard matrix.

(i) Every nonzero subspace of $\mathcal{R}^n$ has a unique basis.

(j) It is possible for different bases for a particular subspace to contain different numbers of vectors.

(k) Every finite spanning set for a nonzero subspace contains a basis for the subspace.

(l) The pivot columns of every matrix form a basis for its column space.

(m) The vectors in the parametric representation of the general solution to $A\mathbf{x} = \mathbf{0}$ form a basis for the null space of A.

(n) No subspace of $\mathcal{R}^n$ has dimension greater than n.

(o) There is only one subspace of $\mathcal{R}^n$ having dimension n.

(p) There is only one subspace of $\mathcal{R}^n$ having dimension 0.

(q) The dimension of the null space of a matrix equals the rank of the matrix.

(r) The dimension of the column space of a matrix equals the rank of the matrix.

(s) The dimension of the row space of a matrix equals the nullity of the matrix.

(t) The column space of any matrix equals the column space of its reduced row echelon form.

(u) The null space of any matrix equals the null space of its reduced row echelon form.

(v) If $\mathcal{B}$ is a basis for $\mathcal{R}^n$ and B is the matrix whose columns are the vectors in $\mathcal{B}$, then $B^{-1}\mathbf{v} = [\mathbf{v}]_{\mathcal{B}}$ for all $\mathbf{v}$ in $\mathcal{R}^n$.

(w) If T is a linear operator on $\mathcal{R}^n$, $\mathcal{B}$ is a basis for $\mathcal{R}^n$, B is the matrix whose columns are the vectors in $\mathcal{B}$, and A is the standard matrix of T, then $[T]_{\mathcal{B}} = BAB^{-1}$.

(x) If T is a linear operator on $\mathcal{R}^n$ and $\mathcal{B}$ is a basis for $\mathcal{R}^n$, then $[T]_{\mathcal{B}}$ is the unique $n \times n$ matrix such that $[T]_{\mathcal{B}}[\mathbf{v}]_{\mathcal{B}} = [T(\mathbf{v})]_{\mathcal{B}}$ for all $\mathbf{v}$ in $\mathcal{R}^n$.

2. Determine if each of the following phrases is a misuse of terminology. If so, explain what is wrong with each one.

(a) a basis for a matrix

(b) the rank of a subspace

(c) the dimension of a square matrix

(d) the dimension of the zero subspace

(e) the dimension of a basis for a subspace

(f) the column space of a linear transformation

(g) the dimension of a linear transformation

(h) the coordinate vector of a linear operator

3. Let V be a subspace of $\mathcal{R}^n$ with dimension k. Say as much as possible about the number of vectors in each of the following.

(a) a spanning set for V

(b) a linearly independent subset of V

4. Let A be the standard matrix of a linear transformation $T: \mathcal{R}^5 \to \mathcal{R}^7$. If the range of T has dimension 2, determine the dimension of each of the following subspaces.

(a) Col A (b) Null A (c) Row A

(d) Null A^T (e) the null space of T

In Exercises 5 and 6, determine if the given set is a subspace of $\mathcal{R}^4$. Justify your answer.

5. $\left\{ \begin{bmatrix} u_1 \\ u_2 \\ u_3 \\ u_4 \end{bmatrix} \text{ in } \mathcal{R}^4 \colon u_1^2 = u_3^3, \ u_2 = 0, \text{ and } u_4 = 0 \right\}$

6. $\left\{ \begin{bmatrix} u_1 \\ u_2 \\ u_3 \\ u_4 \end{bmatrix} \text{ in } \mathcal{R}^4 \colon u_2 = 5u_3, \ u_1 = 0, \text{ and } u_4 = 0 \right\}$

In Exercises 7 and 8, find bases for (a) the null space if it is nonzero, (b) the column space, and (c) the row space of the given matrix.

7. $\begin{bmatrix} 1 & 2 & -1 \\ -1 & -1 & -1 \\ 2 & 1 & 4 \\ 1 & 4 & -5 \end{bmatrix}$ 8. $\begin{bmatrix} -1 & 1 & 2 & 2 & 1 \\ 2 & -2 & -1 & -3 & 2 \\ 1 & -1 & 1 & 1 & 2 \\ 1 & -1 & 4 & 8 & 3 \end{bmatrix}$

In Exercises 9 and 10, a linear transformation T is given. (a) Find a basis for the range of T. (b) If the null space of T is nonzero, find a basis for the null space of T.

9. $T \colon \mathcal{R}^3 \to \mathcal{R}^4$ defined by

$$T \left(\begin{bmatrix} x_1 \\ x_2 \\ x_3 \end{bmatrix} \right) = \begin{bmatrix} x_2 - 2x_3 \\ -x_1 + 3x_2 + x_3 \\ x_1 - 4x_2 + x_3 \\ 2x_1 - x_2 + 3x_3 \end{bmatrix}$$

10. $T \colon \mathcal{R}^4 \to \mathcal{R}^2$ defined by

$$T \left(\begin{bmatrix} x_1 \\ x_2 \\ x_3 \\ x_4 \end{bmatrix} \right) = \begin{bmatrix} x_1 - 2x_2 + x_3 - 3x_4 \\ -2x_1 + 3x_2 - 3x_3 + 2x_4 \end{bmatrix}$$

11. Prove that $\left\{ \begin{bmatrix} -1 \\ 2 \\ 2 \\ -1 \end{bmatrix}, \begin{bmatrix} 1 \\ 5 \\ 3 \\ -2 \end{bmatrix} \right\}$ is a basis for the null space

of the linear transformation in Exercise 10.

12. Prove that $\left\{ \begin{bmatrix} 1 \\ 0 \\ -1 \\ -5 \end{bmatrix}, \begin{bmatrix} 1 \\ -7 \\ -4 \\ -3 \end{bmatrix}, \begin{bmatrix} 1 \\ -5 \\ -1 \\ 5 \end{bmatrix} \right\}$ is a basis for the

column space of the matrix in Exercise 8.

13. Let $\mathcal{B} = \left\{ \begin{bmatrix} 0 \\ -1 \\ 1 \end{bmatrix}, \begin{bmatrix} 1 \\ 0 \\ -1 \end{bmatrix}, \begin{bmatrix} -1 \\ -1 \\ 1 \end{bmatrix} \right\}$.

(a) Prove that $\mathcal{B}$ is a basis for $\mathcal{R}^3$.

(b) Find $\mathbf{v}$ if $[\mathbf{v}]_\mathcal{B} = \begin{bmatrix} 4 \\ -3 \\ -2 \end{bmatrix}$.

(c) Find $[\mathbf{w}]_\mathcal{B}$ if $\mathbf{w} = \begin{bmatrix} -2 \\ 5 \\ 3 \end{bmatrix}$.

14. Let $\mathcal{B} = \{\mathbf{b}_1, \mathbf{b}_2, \mathbf{b}_3\}$ be a basis for $\mathcal{R}^3$ and T be the linear operator on $\mathcal{R}^3$ such that

$$T(\mathbf{b}_1) = -2\mathbf{b}_2 + \mathbf{b}_3, \qquad T(\mathbf{b}_2) = 4\mathbf{b}_1 - 3\mathbf{b}_3,$$

$$\text{and} \qquad T(\mathbf{b}_3) = 5\mathbf{b}_1 - 4\mathbf{b}_2 + 2\mathbf{b}_3.$$

(a) Determine $[T]_\mathcal{B}$.

(b) Determine $T(\mathbf{v})$ if $\mathbf{v} = 3\mathbf{b}_1 - \mathbf{b}_2 - 2\mathbf{b}_3$.

15. Determine (a) $[T]_\mathcal{B}$, (b) the standard matrix of T, and (c) an explicit formula for $T(\mathbf{x})$ from the given information about the linear operator T on $\mathcal{R}^2$:

$$\mathcal{B} = \left\{ \begin{bmatrix} 1 \\ -2 \end{bmatrix}, \begin{bmatrix} -2 \\ 3 \end{bmatrix} \right\}, \qquad T \left(\begin{bmatrix} 1 \\ -2 \end{bmatrix} \right) = \begin{bmatrix} 3 \\ 4 \end{bmatrix},$$

$$\text{and} \qquad T \left(\begin{bmatrix} -2 \\ 3 \end{bmatrix} \right) = \begin{bmatrix} -1 \\ 1 \end{bmatrix}.$$

16. Let T be the linear operator on $\mathcal{R}^2$ and $\mathcal{B}$ be the basis for $\mathcal{R}^2$ defined by

$$T \left(\begin{bmatrix} x_1 \\ x_2 \end{bmatrix} \right) = \begin{bmatrix} 2x_1 - x_2 \\ x_1 - 2x_2 \end{bmatrix} \quad \text{and} \quad \mathcal{B} = \left\{ \begin{bmatrix} 1 \\ 2 \end{bmatrix}, \begin{bmatrix} 3 \\ 7 \end{bmatrix} \right\}.$$

Determine $[T]_\mathcal{B}$.

17. Determine an explicit description of $T(\mathbf{x})$ using the given basis $\mathcal{B}$ and the matrix representation of T with respect to $\mathcal{B}$, where

$$[T]_\mathcal{B} = \begin{bmatrix} 1 & 2 & -1 \\ -1 & 3 & 2 \\ 2 & 1 & 2 \end{bmatrix} \quad \text{and}$$

$$\mathcal{B} = \left\{ \begin{bmatrix} 2 \\ 1 \\ 1 \end{bmatrix}, \begin{bmatrix} 1 \\ 2 \\ 1 \end{bmatrix}, \begin{bmatrix} 1 \\ 1 \\ 1 \end{bmatrix} \right\}.$$

18. Let T be the linear operator on $\mathcal{R}^3$ such that

$$T \left(\begin{bmatrix} 1 \\ 0 \\ 1 \end{bmatrix} \right) = \begin{bmatrix} 2 \\ 1 \\ -2 \end{bmatrix}, \qquad T \left(\begin{bmatrix} 0 \\ -1 \\ 1 \end{bmatrix} \right) = \begin{bmatrix} 1 \\ 3 \\ -1 \end{bmatrix},$$

$$\text{and} \qquad T \left(\begin{bmatrix} -1 \\ 1 \\ -1 \end{bmatrix} \right) = \begin{bmatrix} -2 \\ 1 \\ 3 \end{bmatrix}.$$

Find an explicit description of $T(\mathbf{x})$.

In Exercises 19 and 20, an equation of a conic section is given in the x', y'-coordinate system. Determine the equation of the conic section in the usual x, y-coordinate system if the x'-axis and the y'-axis are obtained by rotating the usual x-axis and y-axis through the given angle θ.

19. $\dfrac{(x')^2}{2^2} + \dfrac{(y')^2}{3^2} = 1, \theta = 120°$

20. $-\sqrt{3}(x')^2 + 2x'y' + \sqrt{3}(y')^2 = 12, \theta = 330°$

In Exercises 21 and 22, an equation of a conic section is given in the x, y-coordinate system. Determine the equation of the conic section in the x', y'-coordinate system if the x'-axis and the y'-axis are obtained by rotating the usual x-axis and y-axis through the given angle θ.

21. $29x^2 - 42xy + 29y^2 = 200, \theta = 315°$

22. $-39x^2 - 50\sqrt{3}xy + 11y^2 = 576, \theta = 210°$

23. Find an explicit description of the reflection T of $\mathcal{R}^2$ about the line with equation $y = -\frac{3}{2}x$.

24. Find an explicit description of the orthogonal projection U of $\mathcal{R}^2$ on the line with equation $y = -\frac{3}{2}x$.

25. Let V be a subspace of $\mathcal{R}^n$ with dimension k, and let S be a subset of V. What can be said about m, the number of vectors in S, under the following conditions?

(a) S is linearly independent
(b) S is linearly dependent
(c) S is a spanning set for V

26. Let V and W be subspaces of $\mathcal{R}^n$. Prove that $V \cup W$ is a subspace of $\mathcal{R}^n$ if and only if V is contained in W or W is contained in V.

27. Let $\mathcal{B}$ be a basis for $\mathcal{R}^n$ and T be an invertible linear operator on $\mathcal{R}^n$. Prove that $[T^{-1}]_\mathcal{B} = ([T]_\mathcal{B})^{-1}$.

*Let V and W be subsets of $\mathcal{R}^n$. We define the **sum** of V and W, denoted $V + W$, to be the set*

$$\{\mathbf{u} \text{ in } \mathcal{R}^n \colon \mathbf{u} = \mathbf{v} + \mathbf{w} \text{ for some } \mathbf{v} \text{ in } V \text{ and some } \mathbf{w} \text{ in } W\}.$$

In Exercises 28–30, use the preceding definition.

28. Prove that if V and W are subspaces of $\mathcal{R}^n$, then $V + W$ is also a subspace of $\mathcal{R}^n$.

29. Let

$$V = \left\{ \begin{bmatrix} v_1 \\ v_2 \\ v_3 \end{bmatrix} \text{ in } \mathcal{R}^3 \colon v_1 + v_2 = 0 \text{ and } 2v_1 - v_3 = 0 \right\}$$

and

$$W = \left\{ \begin{bmatrix} w_1 \\ w_2 \\ w_3 \end{bmatrix} \text{ in } \mathcal{R}^3 \colon w_1 - 2w_3 = 0 \text{ and } w_2 + w_3 = 0 \right\}.$$

Find a basis for $V + W$.

30. Let S_1 and S_2 be subsets of $\mathcal{R}^n$, and let $S = S_1 \cup S_2$. Prove that if $V = \text{Span } S_1$ and $W = \text{Span } S_2$, then $\text{Span } S = V + W$.

CHAPTER 5

Eigenvalues, Eigenvectors, and Diagonalization

In many applications, it is important to understand how vectors in $\mathcal{R}^n$ are transformed when they are multiplied by a square matrix. In this chapter we will see that in many circumstances the problem can be reformulated so that the original matrix can be replaced by a diagonal matrix, which simplifies the problem. We have already seen an example of this in Section 4.5, where we learned how to describe a reflection in the plane in terms of a diagonal matrix. One important class of problems in which this approach is also useful involves sequences of matrix–vector products of the form $A\mathbf{p}$, $A^2\mathbf{p}$, $A^3\mathbf{p}$, ... For example, such a sequence arises in the study of long-term population trends considered in Example 6 of Section 2.1.

The investigation of matrices that can be replaced by diagonal matrices, the *diagonalizable* matrices, is the central theme of this chapter. We must begin this investigation with the introduction of special scalars and vectors, called *eigenvalues* and *eigenvectors*, which provide us with the tools necessary to describe diagonalizable matrices.

5.1 Eigenvalues and Eigenvectors

In Section 4.5, we discussed the reflection T of $\mathcal{R}^2$ about the line with equation $y = \frac{1}{2}x$. Recall that the vectors

$$\mathbf{b}_1 = \begin{bmatrix} 2 \\ 1 \end{bmatrix} \quad \text{and} \quad \mathbf{b}_2 = \begin{bmatrix} -1 \\ 2 \end{bmatrix}$$

played an essential role in determining the rule for T. The key to this computation is that $T(\mathbf{b}_1)$ is a multiple of $\mathbf{b}_1$ and $T(\mathbf{b}_2)$ is a multiple of $\mathbf{b}_2$ (see Figure 5.1). Nonzero vectors that are mapped to a multiple of themselves play an important role in understanding the behavior of linear operators and square matrices.

251

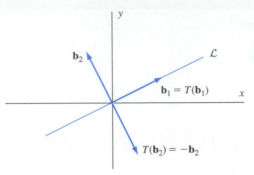

FIGURE 5.1

Definitions. Let T be a linear operator on $\mathcal{R}^n$. A nonzero vector $\mathbf{v}$ in $\mathcal{R}^n$ is called an **eigenvector** of T if $T(\mathbf{v})$ is a multiple of $\mathbf{v}$, that is, $T(\mathbf{v}) = \lambda\mathbf{v}$ for some scalar λ. The scalar[1] λ is called the **eigenvalue** of T that corresponds to $\mathbf{v}$.

For the reflection T of $\mathcal{R}^2$ about line $\mathcal{L}$ and the vectors $\mathbf{b}_1$ and $\mathbf{b}_2$, where $\mathbf{b}_1$ is in the direction of $\mathcal{L}$ and $\mathbf{b}_2$ is in a direction perpendicular to $\mathcal{L}$, we have

$$T(\mathbf{b}_1) = \mathbf{b}_1 = 1\mathbf{b}_1 \qquad \text{and} \qquad T(\mathbf{b}_2) = -\mathbf{b}_2 = (-1)\mathbf{b}_2.$$

Therefore $\mathbf{b}_1$ is an eigenvector of T with corresponding eigenvalue 1, and $\mathbf{b}_2$ is an eigenvector of T with corresponding eigenvalue -1. No nonzero vectors other than multiples of $\mathbf{b}_1$ and $\mathbf{b}_2$ have this property.

Since square matrices are intimately associated with linear operators, the concepts of eigenvector and eigenvalue can also be defined for square matrices.

Definitions. Let A be an $n \times n$ matrix. A nonzero vector $\mathbf{v}$ in $\mathcal{R}^n$ is called an **eigenvector** of A if $A\mathbf{v} = \lambda\mathbf{v}$ for some scalar[2] λ. The scalar λ is called the **eigenvalue** of A that corresponds to $\mathbf{v}$.

For example, let $A = \begin{bmatrix} .6 & .8 \\ .8 & -.6 \end{bmatrix}$. In Section 4.5, we showed that A is the standard matrix of the reflection of $\mathcal{R}^2$ about the line $\mathcal{L}$ with the equation $y = \frac{1}{2}x$. Consider the vectors

$$\mathbf{u}_1 = \begin{bmatrix} -5 \\ 5 \end{bmatrix}, \qquad \mathbf{u}_2 = \begin{bmatrix} 7 \\ 6 \end{bmatrix}, \qquad \mathbf{b}_1 = \begin{bmatrix} 2 \\ 1 \end{bmatrix}, \qquad \text{and} \qquad \mathbf{b}_2 = \begin{bmatrix} 1 \\ -2 \end{bmatrix}.$$

It is a simple matter to verify directly that

$$A\mathbf{u}_1 = \begin{bmatrix} 1 \\ -7 \end{bmatrix}, \qquad A\mathbf{u}_2 = \begin{bmatrix} 9 \\ 2 \end{bmatrix}, \qquad A\mathbf{b}_1 = \begin{bmatrix} 2 \\ 1 \end{bmatrix} = 1\mathbf{b}_1,$$

and

$$A\mathbf{b}_2 = \begin{bmatrix} -1 \\ 2 \end{bmatrix} = (-1)\mathbf{b}_2.$$

---

[1] In this book we are concerned primarily with scalars that are real numbers. Therefore, unless an explicit statement is made to the contrary, the term *eigenvalue* should be interpreted as meaning *real eigenvalue*. There are situations, however, where it is useful to allow eigenvalues to be complex numbers. When complex eigenvalues are permitted, the definition of an eigenvector must be changed to allow vectors in $\mathcal{C}^n$, the set of n-tuples having complex number components.

[2] See footnote 1.

Therefore neither $\mathbf{u}_1$ nor $\mathbf{u}_2$ is an eigenvector of A. However, $\mathbf{b}_1$ is an eigenvector of A with corresponding eigenvalue 1, and $\mathbf{b}_2$ is an eigenvector of A with corresponding eigenvalue -1. Because A is the standard matrix of the reflection T of $\mathcal{R}^2$ about the line $\mathcal{L}$, we can verify that $\mathbf{b}_1$ is an eigenvector of A without calculating the matrix product $A\mathbf{b}_1$. Indeed, since $\mathbf{b}_1$ is a vector in the direction of $\mathcal{L}$, we have $A\mathbf{b}_1 = T(\mathbf{b}_1) = \mathbf{b}_1$. Similarly, we can also verify that $\mathbf{b}_2$ is an eigenvector of A from the fact that $\mathbf{b}_2$ is an eigenvector of T.

In general, if T is a linear operator on $\mathcal{R}^n$, then T is a matrix transformation. Let A be the standard matrix of T. Since the equation $T(\mathbf{v}) = \lambda\mathbf{v}$ can be rewritten as $A\mathbf{v} = \lambda\mathbf{v}$, we can determine the eigenvalues and eigenvectors of T from A.

> The eigenvectors and corresponding eigenvalues of a linear operator are the same as those of its standard matrix.

In view of the relationship between a linear operator and its standard matrix, eigenvalues and eigenvectors of linear operators can be studied simultaneously with those of matrices. Example 1 shows how to verify that a given vector $\mathbf{v}$ is an eigenvector of a matrix A.

Example 1   For

$$\mathbf{v} = \begin{bmatrix} 1 \\ -1 \\ 1 \end{bmatrix} \quad \text{and} \quad A = \begin{bmatrix} 5 & 2 & 1 \\ -2 & 1 & -1 \\ 2 & 2 & 4 \end{bmatrix},$$

show that $\mathbf{v}$ is an eigenvector of A.

Solution   Because $\mathbf{v}$ is nonzero, to verify that $\mathbf{v}$ is an eigenvector of A, we need only show that $A\mathbf{v}$ is a multiple of $\mathbf{v}$. Since

$$A\mathbf{v} = \begin{bmatrix} 5 & 2 & 1 \\ -2 & 1 & -1 \\ 2 & 2 & 4 \end{bmatrix} \begin{bmatrix} 1 \\ -1 \\ 1 \end{bmatrix} = \begin{bmatrix} 4 \\ -4 \\ 4 \end{bmatrix} = 4\begin{bmatrix} 1 \\ -1 \\ 1 \end{bmatrix} = 4\mathbf{v},$$

we see that $\mathbf{v}$ is an eigenvector of A with corresponding eigenvalue 4. Note that $7\mathbf{v}$ and $(-3)\mathbf{v}$ are other eigenvectors with corresponding eigenvalue 4. In fact, for any nonzero scalar c, the vector $c\mathbf{v}$ is an eigenvector of A with corresponding eigenvalue 4 because $A(c\mathbf{v}) = c(A\mathbf{v}) = c(4\mathbf{v}) = 4(c\mathbf{v})$. ○

Example 1 illustrates that if $\mathbf{v}$ is an eigenvector of a matrix A, then the scalar λ such that $A\mathbf{v} = \lambda\mathbf{v}$ is unique. On the other hand, if λ is an eigenvalue of A, then there are infinitely many eigenvectors of A that correspond to λ.

The process of finding the eigenvectors of an $n \times n$ matrix that correspond to a particular eigenvalue is also straightforward. Note that $\mathbf{v}$ is an eigenvector of A corresponding to eigenvalue λ if and only if $\mathbf{v}$ is a nonzero vector such that

$$A\mathbf{v} = \lambda\mathbf{v}$$

$$A\mathbf{v} - \lambda\mathbf{v} = \mathbf{0}$$

$$A\mathbf{v} - \lambda I_n\mathbf{v} = \mathbf{0}$$

$$(A - \lambda I_n)\mathbf{v} = \mathbf{0}.$$

Thus **v** is a nonzero solution to the system of linear equations $(A - \lambda I_n)\mathbf{x} = \mathbf{0}$.

> Let A be an $n \times n$ matrix with eigenvalue λ. The eigenvectors of A corresponding to λ are the nonzero solutions of $(A - \lambda I_n)\mathbf{x} = \mathbf{0}$.

In this context, the set of solutions to $(A - \lambda I_n)\mathbf{x} = \mathbf{0}$ is called the **eigenspace of A corresponding to eigenvalue λ**. This is just the null space of $A - \lambda I_n$, and hence it is a subspace of $\mathcal{R}^n$. Note that the eigenspace of A corresponding to λ consists of the zero vector and all the eigenvectors corresponding to λ.

Similarly, if λ is an eigenvalue of a linear operator T on $\mathcal{R}^n$, the set of vectors **v** in $\mathcal{R}^n$ such that $T(\mathbf{v}) = \lambda\mathbf{v}$ is called the **eigenspace** of T corresponding to λ. In Section 5.4, we will see that under certain conditions the bases for the various eigenspaces of a linear operator on $\mathcal{R}^n$ can be combined to form a basis for $\mathcal{R}^n$. This basis enables us to find a very simple matrix representation of the operator.

Example 2 Show that 3 and -2 are eigenvalues of the linear operator T on $\mathcal{R}^2$ defined by

$$T\left(\begin{bmatrix} x_1 \\ x_2 \end{bmatrix}\right) = \begin{bmatrix} -2x_2 \\ -3x_1 + x_2 \end{bmatrix},$$

and find bases for the corresponding eigenspaces.

Solution The standard matrix of T is

$$A = \begin{bmatrix} 0 & -2 \\ -3 & 1 \end{bmatrix}.$$

To show that 3 is an eigenvalue of T, we show that 3 is an eigenvalue of A. Thus we must find a nonzero vector **u** such that $A\mathbf{u} = 3\mathbf{u}$. In other words, we must show that the solution set of the system of equations $(A - 3I_2)\mathbf{x} = \mathbf{0}$, which is the null space of $A - 3I_2$, contains nonzero vectors. Because the reduced row echelon form of $A - 3I_2$ is

$$\begin{bmatrix} 1 & \frac{2}{3} \\ 0 & 0 \end{bmatrix},$$

a matrix of nullity

$$2 - \text{rank}\,(A - 3I_2) = 2 - 1 = 1,$$

nonzero solutions exist. Furthermore, we see that the eigenvectors of A corresponding to eigenvalue 3 have the form

$$\begin{bmatrix} x_1 \\ x_2 \end{bmatrix} = \begin{bmatrix} -\frac{2}{3}x_2 \\ x_2 \end{bmatrix} = x_2 \begin{bmatrix} -\frac{2}{3} \\ 1 \end{bmatrix}$$

for $x_2 \neq 0$. It follows that

$$\left\{ \begin{bmatrix} -\frac{2}{3} \\ 1 \end{bmatrix} \right\}$$

is a basis for the eigenspace of A corresponding to eigenvalue 3. Notice that by taking $x_2 = 3$ above, we obtain another basis for this eigenspace (one consisting of a vector with integer components), namely

$$\left\{ \begin{bmatrix} -2 \\ 3 \end{bmatrix} \right\}.$$

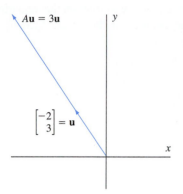

$Au = 3u$

$\begin{bmatrix} -2 \\ 3 \end{bmatrix} = u$

FIGURE 5.2

In a similar fashion, we must show that there is a nonzero vector $\mathbf{v}$ such that $A\mathbf{v} = (-2)\mathbf{v}$. In this case, we must show that the system of equations $(A + 2I_2)\mathbf{x} = \mathbf{0}$ has nonzero solutions. Since the reduced row echelon form of $A + 2I_2$ is

$$\begin{bmatrix} 1 & -1 \\ 0 & 0 \end{bmatrix},$$

a matrix of nullity 1, such solutions exist. From the parametric form of the general solution to this system, we see that the eigenvectors corresponding to eigenvalue -2 have the form

$$\begin{bmatrix} x_1 \\ x_2 \end{bmatrix} = \begin{bmatrix} x_2 \\ x_2 \end{bmatrix} = x_2 \begin{bmatrix} 1 \\ 1 \end{bmatrix}$$

for $x_2 \neq 0$. Thus a basis for the eigenspace of A corresponding to eigenvalue -2 is

$$\left\{ \begin{bmatrix} 1 \\ 1 \end{bmatrix} \right\}$$

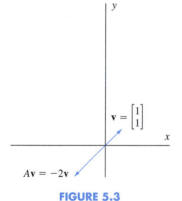

$\mathbf{v} = \begin{bmatrix} 1 \\ 1 \end{bmatrix}$

$A\mathbf{v} = -2\mathbf{v}$

FIGURE 5.3

(see Figures 5.2 and 5.3).

Since 3 and -2 are eigenvalues of A, they are also eigenvalues of T. Moreover, the eigenspaces of T have the same bases as those of A, namely,

$$\left\{ \begin{bmatrix} -2 \\ 3 \end{bmatrix} \right\} \quad \text{and} \quad \left\{ \begin{bmatrix} 1 \\ 1 \end{bmatrix} \right\}.$$

The two eigenspaces in Example 2 have dimension 1. This need not always be the case, as our next example shows.

Example 3   Show that 3 is an eigenvalue of

$$B = \begin{bmatrix} 3 & 0 & 0 \\ 0 & 1 & 2 \\ 0 & 2 & 1 \end{bmatrix},$$

and find a basis for the corresponding eigenspace.

Solution   As in Example 2, we must show that the null space of $B - 3I_3$ contains nonzero vectors. The reduced row echelon form of $B - 3I_3$ is

$$\begin{bmatrix} 0 & 1 & -1 \\ 0 & 0 & 0 \\ 0 & 0 & 0 \end{bmatrix}.$$

Hence the vectors in the eigenspace of B corresponding to eigenvalue 3 satisfy $x_2 - x_3 = 0$, and so the general solution to $(B - 3I_3)\mathbf{x} = \mathbf{0}$ is

$$x_1 \quad \text{free}$$
$$x_2 = x_3$$
$$x_2 \quad \text{free}.$$

(Notice that the variable x_1, which is not a basic variable in the equation $x_2 - x_3 = 0$, is a free variable.) Thus the vectors in the eigenspace of B corresponding to eigenvalue 3 have the form

$$\begin{bmatrix} x_1 \\ x_2 \\ x_3 \end{bmatrix} = \begin{bmatrix} x_1 \\ x_3 \\ x_3 \end{bmatrix} = x_1 \begin{bmatrix} 1 \\ 0 \\ 0 \end{bmatrix} + x_3 \begin{bmatrix} 0 \\ 1 \\ 1 \end{bmatrix}.$$

Therefore

$$\left\{ \begin{bmatrix} 1 \\ 0 \\ 0 \end{bmatrix}, \begin{bmatrix} 0 \\ 1 \\ 1 \end{bmatrix} \right\}$$

is a basis for the eigenspace of B corresponding to eigenvalue 3.   ○

It should be noted that not all square matrices and linear operators on $\mathcal{R}^n$ have real eigenvalues. (Such matrices and operators have no eigenvectors with components that are real numbers either.) Consider, for example, the linear operator T on $\mathcal{R}^2$ that rotates a vector by 90°. If this operator had a real eigenvalue λ, then there would be a nonzero vector $\mathbf{v}$ in $\mathcal{R}^2$ such that $T(\mathbf{v}) = \lambda\mathbf{v}$. But for any nonzero vector $\mathbf{v}$, the vector $T(\mathbf{v})$ obtained by rotating $\mathbf{v}$ through 90° is not a multiple of $\mathbf{v}$ (see Figure 5.4). Hence $\mathbf{v}$ cannot be an eigenvector of T, and so T has no real eigenvalues. Note that this argument also shows that the standard matrix of T, which is the 90°-rotation matrix

$$\begin{bmatrix} 0 & -1 \\ 1 & 0 \end{bmatrix},$$

has no real eigenvalues.

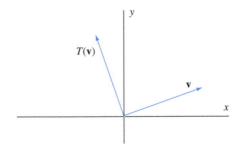

FIGURE 5.4

Practice Problems

1. Show that

$$\mathbf{u} = \begin{bmatrix} -2 \\ 1 \\ 2 \end{bmatrix} \quad \text{and} \quad \mathbf{v} = \begin{bmatrix} 1 \\ -3 \\ 4 \end{bmatrix}$$

are eigenvectors of

$$A = \begin{bmatrix} 5 & 2 & 1 \\ -2 & 1 & -1 \\ 2 & 2 & 4 \end{bmatrix}.$$

To what eigenvalues do $\mathbf{u}$ and $\mathbf{v}$ correspond?

2. Show that 1 is an eigenvalue of the linear operator on $\mathcal{R}^3$ defined by

$$T\left(\begin{bmatrix} x_1 \\ x_2 \\ x_3 \end{bmatrix}\right) = \begin{bmatrix} x_1 + 2x_2 \\ -x_1 - x_2 + x_3 \\ x_2 + x_3 \end{bmatrix},$$

and find a basis for the corresponding eigenspace.

Exercises

1. Determine if the following statements are true or false.

   (a) If $A\mathbf{v} = \lambda\mathbf{v}$ for some vector $\mathbf{v}$, then λ is an eigenvalue of matrix A.

   (b) If $A\mathbf{v} = \lambda\mathbf{v}$ for some vector $\mathbf{v}$, then $\mathbf{v}$ is an eigenvector of matrix A.

   (c) A scalar λ is an eigenvalue of an $n \times n$ matrix A if and only if the equation $(A - \lambda I_n)\mathbf{x} = \mathbf{0}$ has a nonzero solution.

   (d) If $\mathbf{v}$ is an eigenvector of a matrix, then there is a unique eigenvalue of the matrix that corresponds to $\mathbf{v}$.

   (e) If λ is an eigenvalue of a linear operator, then there are infinitely many eigenvectors of the operator that correspond to λ.

   (f) The eigenspace of an $n \times n$ matrix A corresponding to eigenvalue λ is the column space of $A - \lambda I_n$.

   (g) The eigenvalues of a linear operator on $\mathcal{R}^n$ are the same as those of its standard matrix.

   (h) The eigenspaces of a linear operator on $\mathcal{R}^n$ are the same as those of its standard matrix.

   (i) Every linear operator on $\mathcal{R}^n$ has real eigenvalues.

In Exercises 2–12, a matrix and a vector are given. Show that the vector is an eigenvector of the matrix and determine the corresponding eigenvalue.

2. $\begin{bmatrix} 12 & -14 \\ 7 & -9 \end{bmatrix}$, $\begin{bmatrix} 1 \\ 1 \end{bmatrix}$

3. $\begin{bmatrix} -5 & -4 \\ 8 & 7 \end{bmatrix}$, $\begin{bmatrix} 1 \\ -2 \end{bmatrix}$

4. $\begin{bmatrix} 15 & 24 \\ -4 & -5 \end{bmatrix}$, $\begin{bmatrix} -2 \\ 1 \end{bmatrix}$

5. $\begin{bmatrix} 19 & -7 \\ 42 & -16 \end{bmatrix}$, $\begin{bmatrix} 1 \\ 3 \end{bmatrix}$

6. $\begin{bmatrix} -9 & -8 & 5 \\ 7 & 6 & -5 \\ -6 & -6 & 4 \end{bmatrix}$, $\begin{bmatrix} 3 \\ -2 \\ 1 \end{bmatrix}$

7. $\begin{bmatrix} 4 & 6 & -5 \\ 9 & 7 & -11 \\ 8 & 8 & -11 \end{bmatrix}$, $\begin{bmatrix} -1 \\ 2 \\ 1 \end{bmatrix}$

8. $\begin{bmatrix} -3 & 14 & 10 \\ -2 & 5 & 2 \\ 2 & -10 & -7 \end{bmatrix}$, $\begin{bmatrix} -3 \\ -1 \\ 2 \end{bmatrix}$

9. $\begin{bmatrix} 2 & -6 & 6 \\ 1 & 9 & -6 \\ -2 & 16 & -13 \end{bmatrix}$, $\begin{bmatrix} -1 \\ 1 \\ 2 \end{bmatrix}$

10. $\begin{bmatrix} -5 & -1 & 2 \\ 2 & -1 & -2 \\ -7 & -2 & 2 \end{bmatrix}$, $\begin{bmatrix} 1 \\ -2 \\ 1 \end{bmatrix}$

11. $\begin{bmatrix} 5 & 6 & 12 \\ 3 & 2 & 6 \\ -3 & -3 & -7 \end{bmatrix}$, $\begin{bmatrix} -2 \\ -1 \\ 1 \end{bmatrix}$

12. $\begin{bmatrix} 6 & 5 & 15 \\ 5 & 6 & 15 \\ -5 & -5 & -14 \end{bmatrix}$, $\begin{bmatrix} -1 \\ -1 \\ 1 \end{bmatrix}$

In Exercises 13–24, a matrix and a scalar λ are given. Show that λ is an eigenvalue of the matrix and determine a basis for its eigenspace.

13. $\begin{bmatrix} 10 & 7 \\ -14 & -11 \end{bmatrix}$, $\lambda = 3$

14. $\begin{bmatrix} -11 & 14 \\ -7 & 10 \end{bmatrix}$, $\lambda = -4$

15. $\begin{bmatrix} 11 & 18 \\ -3 & -4 \end{bmatrix}$, $\lambda = 5$

16. $\begin{bmatrix} -11 & 5 \\ -30 & 14 \end{bmatrix}$, $\lambda = -1$

17. $\begin{bmatrix} -2 & -5 & 2 \\ 4 & 7 & -2 \\ -3 & -3 & 5 \end{bmatrix}$, $\lambda = 3$

18. $\begin{bmatrix} 6 & 9 & -10 \\ 6 & 3 & -4 \\ 7 & 7 & -9 \end{bmatrix}$, $\lambda = 5$

19. $\begin{bmatrix} -3 & 12 & 6 \\ -3 & 6 & 0 \\ 3 & -9 & -3 \end{bmatrix}$, $\lambda = 0$

20. $\begin{bmatrix} 3 & -2 & 2 \\ -4 & 1 & -2 \\ -5 & 1 & -2 \end{bmatrix}$, $\lambda = 2$

21. $\begin{bmatrix} -13 & -4 & 8 \\ 24 & 7 & -16 \\ -12 & -4 & 7 \end{bmatrix}$, $\lambda = -1$

22. $\begin{bmatrix} -2 & -2 & -4 \\ -1 & -1 & -2 \\ 1 & 1 & 2 \end{bmatrix}$, $\lambda = 0$

23. $\begin{bmatrix} 4 & -3 & -3 \\ -3 & 4 & 3 \\ 3 & -3 & -2 \end{bmatrix}$, $\lambda = 1$

24. $\begin{bmatrix} 5 & 3 & 9 \\ 3 & 5 & 9 \\ -3 & -3 & -7 \end{bmatrix}$, $\lambda = 2$

In Exercises 25–30, a linear operator and a vector are given. Show that the vector is an eigenvector of the operator and determine the corresponding eigenvalue.

25. $T\left(\begin{bmatrix} x_1 \\ x_2 \end{bmatrix}\right) = \begin{bmatrix} -3x_1 - 6x_2 \\ 12x_1 + 14x_2 \end{bmatrix}$, $\begin{bmatrix} -2 \\ 3 \end{bmatrix}$

26. $T\left(\begin{bmatrix} x_1 \\ x_2 \end{bmatrix}\right) = \begin{bmatrix} 8x_1 - 2x_2 \\ 6x_1 + x_2 \end{bmatrix}$, $\begin{bmatrix} 1 \\ 2 \end{bmatrix}$

27. $T\left(\begin{bmatrix} x_1 \\ x_2 \\ x_3 \end{bmatrix}\right) = \begin{bmatrix} -8x_1 + 9x_2 - 3x_3 \\ -5x_1 + 6x_2 - 3x_3 \\ -x_1 + x_2 - 2x_3 \end{bmatrix}$, $\begin{bmatrix} 3 \\ 2 \\ 1 \end{bmatrix}$

28. $T\left(\begin{bmatrix} x_1 \\ x_2 \\ x_3 \end{bmatrix}\right) = \begin{bmatrix} -2x_1 - x_2 - 3x_3 \\ -3x_1 - 4x_2 - 9x_3 \\ x_1 + x_2 + 2x_3 \end{bmatrix}$, $\begin{bmatrix} -1 \\ -3 \\ 1 \end{bmatrix}$

29. $T\left(\begin{bmatrix} x_1 \\ x_2 \\ x_3 \end{bmatrix}\right) = \begin{bmatrix} 6x_1 + x_2 - 2x_3 \\ -6x_1 + x_2 + 6x_3 \\ -2x_1 - x_2 + 6x_3 \end{bmatrix}$, $\begin{bmatrix} -1 \\ 3 \\ 1 \end{bmatrix}$

30. $T\left(\begin{bmatrix} x_1 \\ x_2 \\ x_3 \end{bmatrix}\right) = \begin{bmatrix} 4x_1 + 9x_2 + 8x_3 \\ -2x_1 - x_2 - 2x_3 \\ 2x_1 - 3x_2 - 2x_3 \end{bmatrix}$, $\begin{bmatrix} -1 \\ 0 \\ 1 \end{bmatrix}$

In Exercises 31–36, a linear operator and a scalar λ are given. Show that λ is an eigenvalue of the operator and determine a basis for its eigenspace.

31. $T\left(\begin{bmatrix} x_1 \\ x_2 \end{bmatrix}\right) = \begin{bmatrix} x_1 - 2x_2 \\ 6x_1 - 6x_2 \end{bmatrix}$, $\lambda = -2$

32. $T\left(\begin{bmatrix} x_1 \\ x_2 \end{bmatrix}\right) = \begin{bmatrix} 4x_1 + 6x_2 \\ -12x_1 - 13x_2 \end{bmatrix}$, $\lambda = -5$

33. $T\left(\begin{bmatrix} x_1 \\ x_2 \\ x_3 \end{bmatrix}\right) = \begin{bmatrix} x_1 - x_2 - 3x_3 \\ -3x_1 - x_2 - 9x_3 \\ x_1 + x_2 + 5x_3 \end{bmatrix}$, $\lambda = 2$

34. $T\left(\begin{bmatrix} x_1 \\ x_2 \\ x_3 \end{bmatrix}\right) = \begin{bmatrix} 4x_1 - 2x_2 - 5x_3 \\ 3x_1 - x_2 - 5x_3 \\ 4x_1 - 4x_2 - 3x_3 \end{bmatrix}$, $\lambda = -3$

35. $T\left(\begin{bmatrix} x_1 \\ x_2 \\ x_3 \end{bmatrix}\right) = \begin{bmatrix} x_1 + 4x_2 + 5x_3 \\ 2x_1 + 6x_2 + 2x_3 \\ -2x_1 - 10x_2 - 6x_3 \end{bmatrix}$, $\lambda = 3$

36. $T\left(\begin{bmatrix} x_1 \\ x_2 \\ x_3 \end{bmatrix}\right) = \begin{bmatrix} 5x_1 + 2x_2 - 4x_3 \\ -12x_1 - 5x_2 + 12x_3 \\ -4x_1 - 2x_2 + 5x_3 \end{bmatrix}$, $\lambda = 1$

37. What are the eigenvalues of the identity linear operator on $\mathcal{R}^n$? Justify your answer. Describe each eigenspace.

38. What are the eigenvalues of the zero linear operator on $\mathcal{R}^n$? Justify your answer. Describe each eigenspace.

39. Prove that if $\mathbf{v}$ is an eigenvector of an $n \times n$ matrix A, then for any nonzero scalar c, $c\mathbf{v}$ is also an eigenvector of A.

40. Prove that if $\mathbf{v}$ is an eigenvector of an $n \times n$ matrix A, then there is a unique scalar λ such that $A\mathbf{v} = \lambda\mathbf{v}$.

41. Suppose that 0 is an eigenvalue of a matrix A. Give another name for the eigenspace of A corresponding to 0.

42. Prove that an $n \times n$ matrix is invertible if and only if 0 is not an eigenvalue.

43. Prove that if λ is an eigenvalue of an invertible matrix A, then $\lambda \neq 0$ and $1/\lambda$ is an eigenvalue of A^{-1}.

44. An $n \times n$ matrix A is called *nilpotent* if, for some positive integer k, $A^k = O$, where O is the $n \times n$ zero matrix. Prove that 0 is the only eigenvalue of a nilpotent matrix.

45. Prove that if λ is an eigenvalue of a matrix A, then λ^2 is an eigenvalue of A^2.

46. State and prove a generalization of Exercise 45.

47. Determine necessary and sufficient conditions on a vector $\mathbf{v}$ such that the span of $\{A\mathbf{v}\}$ equals the span of $\{\mathbf{v}\}$.

48. Suppose that A is a square matrix in which the sum of the entries of each row equals the same scalar r. Show that r is an eigenvalue of A by finding an eigenvector of A corresponding to r.

49. Let $\mathbf{v}_1$ and $\mathbf{v}_2$ be eigenvectors of a linear operator T on $\mathcal{R}^n$, and let λ_1 and λ_2, respectively, be the corresponding eigenvalues. Prove that if $\lambda_1 \neq \lambda_2$, then $\{\mathbf{v}_1, \mathbf{v}_2\}$ is linearly independent.

50. State and prove a generalization of Exercise 49 for k eigenvectors.

51. Let T be a linear operator on $\mathcal{R}^2$ with an eigenspace of dimension 2. Prove that $T = \lambda I$ for some scalar λ.

In Exercises 52–58, use a calculator with matrix capabilities or computer software such as MATLAB to solve the problem.

52. Let
$$A = \begin{bmatrix} -1.9 & 14.4 & -8.4 & 34.8 \\ 1.6 & -2.7 & 3.2 & -1.6 \\ 1.2 & -8.0 & 4.7 & -18.2 \\ 1.6 & -1.6 & 3.2 & -2.7 \end{bmatrix}.$$

Show that
$$\mathbf{v}_1 = \begin{bmatrix} -9 \\ 1 \\ 5 \\ 1 \end{bmatrix}, \quad \mathbf{v}_2 = \begin{bmatrix} -2 \\ 0 \\ 1 \\ 0 \end{bmatrix}, \quad \mathbf{v}_3 = \begin{bmatrix} -3 \\ 1 \\ 2 \\ 0 \end{bmatrix},$$

and
$$\mathbf{v}_4 = \begin{bmatrix} -3 \\ -5 \\ 0 \\ 2 \end{bmatrix}$$

are eigenvectors of A. What are the eigenvalues corresponding to each of these eigenvectors?

53. Are the eigenvalues of A (determined in Exercise 52) also eigenvalues of $3A$? If so, find an eigenvector corresponding to each eigenvalue.

54. Are $\mathbf{v}_1, \mathbf{v}_2, \mathbf{v}_3$, and $\mathbf{v}_4$ in Exercise 52 also eigenvectors of $3A$? If so, what eigenvalue corresponds to each of these eigenvectors?

55. (a) Based on the results of Exercises 52–54, make a conjecture about the relationship between the eigenvalues of an $n \times n$ matrix B and those of cB, where c is a scalar.
    (b) Based on the results of Exercises 52–54, make a conjecture about the relationship between the eigenvectors of an $n \times n$ matrix B and those of cB, where c is a scalar.
    (c) Justify the conjectures made in (a) and (b).

56. Are $\mathbf{v}_1, \mathbf{v}_2, \mathbf{v}_3$, and $\mathbf{v}_4$ in Exercise 52 also eigenvectors of A^T? If so, what eigenvalue corresponds to each of these eigenvectors?

57. Are the eigenvalues of A (determined in Exercise 52) also eigenvalues of A^T? If so, find an eigenvector corresponding to each eigenvalue.

58. Based on the results of Exercises 56 and 57, make a conjecture about any possible relationship between the eigenvalues or eigenvectors of an $n \times n$ matrix B and those of B^T.

Solutions to the Practice Problems

1. Since

$$A\mathbf{u} = \begin{bmatrix} 5 & 2 & 1 \\ -2 & 1 & -1 \\ 2 & 2 & 4 \end{bmatrix} \begin{bmatrix} -2 \\ 1 \\ 2 \end{bmatrix} = \begin{bmatrix} -6 \\ 3 \\ 6 \end{bmatrix} = 3 \begin{bmatrix} -2 \\ 1 \\ 2 \end{bmatrix} = 3\mathbf{u},$$

$\mathbf{u}$ is an eigenvector of A corresponding to eigenvalue 3. Likewise,

$$A\mathbf{v} = \begin{bmatrix} 5 & 2 & 1 \\ -2 & 1 & -1 \\ 2 & 2 & 4 \end{bmatrix} \begin{bmatrix} 1 \\ -3 \\ 4 \end{bmatrix} = \begin{bmatrix} 3 \\ -9 \\ 12 \end{bmatrix} = 3 \begin{bmatrix} 1 \\ -3 \\ 4 \end{bmatrix} = 3\mathbf{v},$$

so that $\mathbf{v}$ is also an eigenvector of A corresponding to eigenvalue 3.

2. The standard matrix of T is

$$A = \begin{bmatrix} 1 & 2 & 0 \\ -1 & -1 & 1 \\ 0 & 1 & 1 \end{bmatrix},$$

and the row echelon form of $A - I_3$ is

$$\begin{bmatrix} 1 & 0 & -1 \\ 0 & 1 & 0 \\ 0 & 0 & 0 \end{bmatrix}.$$

Hence the parametric form of the general solution to $(A - I_3)\mathbf{x} = \mathbf{0}$ is

$$\begin{bmatrix} x_1 \\ x_2 \\ x_3 \end{bmatrix} = \begin{bmatrix} x_3 \\ 0 \\ x_3 \end{bmatrix} = x_3 \begin{bmatrix} 1 \\ 0 \\ 1 \end{bmatrix},$$

and so

$$\left\{ \begin{bmatrix} 1 \\ 0 \\ 1 \end{bmatrix} \right\}$$

is a basis for the eigenspace of T corresponding to eigenvalue 1.

5.2   The Characteristic Polynomial

In Section 5.1, we learned how to find the eigenvalue corresponding to a given eigenvector and the eigenvectors for a given eigenvalue. But ordinarily we know neither the eigenvalues nor the eigenvectors of a matrix. Suppose, for instance, that we want to find the eigenvalues and the eigenspaces of an $n \times n$ matrix A. If λ is an eigenvalue of A, there must be a nonzero vector $\mathbf{v}$ in $\mathcal{R}^n$ such that $A\mathbf{v} = \lambda\mathbf{v}$. Thus $\mathbf{v}$ is a nonzero solution to $(A - \lambda I_n)\mathbf{x} = \mathbf{0}$. But in order for the homogeneous system of linear equations $(A - \lambda I_n)\mathbf{x} = \mathbf{0}$ to have nonzero solutions, the rank of $A - \lambda I_n$ must be less than n (Theorem 2.6). Hence the $n \times n$ matrix $A - \lambda I_n$ is not invertible, and so its determinant must be 0.

> The eigenvalues of a square matrix A are the values of t that satisfy
> $$\det(A - tI_n) = 0.$$

The equation $\det(A - tI_n) = 0$ is called the **characteristic equation** of A, and $\det(A - tI_n)$ is called the **characteristic polynomial** of A. Thus the eigenvalues of matrix A are the roots of the characteristic polynomial of A.

Example 1   Determine the eigenvalues of

$$A = \begin{bmatrix} -4 & -3 \\ 3 & 6 \end{bmatrix},$$

and then find a basis for each eigenspace.

Solution   We begin by forming the matrix

$$A - tI_2 = \begin{bmatrix} -4-t & -3 \\ 3 & 6-t \end{bmatrix}.$$

The characteristic polynomial of A is the determinant of this matrix, which is

$$\det(A - tI_2) = (-4-t)(6-t) - (-3) \cdot 3$$
$$= (-24 - 2t + t^2) + 9$$
$$= t^2 - 2t - 15$$
$$= (t+3)(t-5).$$

Therefore the roots of the characteristic polynomial are -3 and 5; so these are the eigenvalues of A.

As in Section 5.1, we solve $(A + 3I_2)\mathbf{x} = \mathbf{0}$ and $(A - 5I_2)\mathbf{x} = \mathbf{0}$ to find bases for the eigenspaces. Since the reduced row echelon form of $A + 3I_2$ is

$$\begin{bmatrix} 1 & 3 \\ 0 & 0 \end{bmatrix},$$

the parametric form of the general solution to $(A + 3I_2)\mathbf{x} = \mathbf{0}$ is

$$\begin{bmatrix} x_1 \\ x_2 \end{bmatrix} = \begin{bmatrix} -3x_2 \\ x_2 \end{bmatrix} = x_2 \begin{bmatrix} -3 \\ 1 \end{bmatrix}.$$

Hence

$$\left\{ \begin{bmatrix} -3 \\ 1 \end{bmatrix} \right\}$$

is a basis for the eigenspace of A corresponding to the eigenvalue -3.

In a similar manner, the reduced row echelon form of $A - 5I_2$, which is

$$\begin{bmatrix} 1 & \frac{1}{3} \\ 0 & 0 \end{bmatrix},$$

produces the basis

$$\left\{ \begin{bmatrix} -1 \\ 3 \end{bmatrix} \right\}$$

for the eigenspace of A corresponding to the eigenvalue 5.   ○

For the 2×2 matrix A in Example 1, the characteristic polynomial is $t^2 - 2t - 15 = (t+3)(t-5)$, a polynomial of degree 2. In general, *the characteristic polynomial of an $n \times n$ matrix can be shown to be a polynomial of degree n.*

Note also that the reduced row echelon form of the matrix A in Example 1 is I_2, which has a characteristic polynomial of $(t-1)^2$. Thus the characteristic polynomial of a matrix is not usually equal to the characteristic polynomial of its reduced row echelon form. In general, therefore, the eigenvalues of a matrix and of its reduced row echelon form are not the same. It is also easy to see that the eigenvectors of a matrix and of its reduced row echelon form are usually not the same. Consequently, there is no way to apply elementary row operations to a matrix in hopes of finding its eigenvalues or eigenvectors.

Computing the characteristic polynomial of a 2×2 matrix is straight-forward, as Example 1 shows. On the other hand, calculating the character-istic polynomial of a larger matrix by hand can be quite tedious. Although a programmable calculator or computer software can be used to determine the characteristic polynomial of matrices that are not too large, there is no way to find the precise roots of the characteristic polynomial of an arbitrary matrix that is larger than 4×4. Hence for large matrices, some sort of approximation method is required to calculate the eigenvalues. Because of the difficulty in computing the characteristic polynomial, in this book we usually use only 2×2 and 3×3 matrices in our examples and exercises.

Recall that two matrices A and B are similar if there exists an invertible matrix P such that $B = P^{-1}AP$. By Theorem 3.4, we have

$$\det (B - tI_n) = \det (P^{-1}AP - P^{-1}(tI_n)P)$$

$$= \det (P^{-1}(A - tI_n)P)$$

$$= (\det P^{-1})[\det (A - tI_n)](\det P)$$

$$= \left(\frac{1}{\det P}\right)[\det (A - tI_n)](\det P)$$

$$= \det (A - tI_n).$$

It follows that the characteristic polynomial of A is the same as that of B. Thus we have proved the following result.

> Similar $n \times n$ matrices have the same characteristic polynomial.

Recall that the problem of finding eigenvalues and eigenvectors of a linear operator can be turned into the corresponding problem for its standard matrix. In the context of a linear operator T, the characteristic equation of the standard matrix of T is called the **characteristic equation** of T, and the characteristic polynomial of the standard matrix of T is called the **characteristic polynomial** of T. Thus *the characteristic polynomial of a linear operator on $\mathcal{R}^n$ is a polynomial of degree n whose roots are the eigenvalues of T.*

Example 2   We have noted on page 256 that the linear operator T on $\mathcal{R}^2$ that rotates a vector by $90°$ has no real eigenvalues. Equivalently, the $90°$-rotation matrix has no real eigenvalues. In fact, the characteristic polynomial of T, which is also the characteristic polynomial of the $90°$-rotation matrix, is given by

$$\det \left(\begin{bmatrix} 0 & -1 \\ 1 & 0 \end{bmatrix} - tI_2\right) = \det \begin{bmatrix} -t & -1 \\ 1 & -t \end{bmatrix} = t^2 + 1,$$

which has no real roots. This confirms the observation made in Section 5.1 that T, and hence the $90°$-rotation matrix, has no real eigenvalues.   ○

The Multiplicity of an Eigenvalue

Consider the matrix

$$A = \begin{bmatrix} -1 & 0 & 0 \\ 0 & 1 & 2 \\ 0 & 2 & 1 \end{bmatrix}.$$

Using the cofactor expansion along the first row, we see that

$$\det(A - tI_3) = \det \begin{bmatrix} -1-t & 0 & 0 \\ 0 & 1-t & 2 \\ 0 & 2 & 1-t \end{bmatrix}$$

$$= (-1-t) \cdot \det \begin{bmatrix} 1-t & 2 \\ 2 & 1-t \end{bmatrix}$$

$$= (-1-t)[(1-t)^2 - 4]$$

$$= (-1-t)(t^2 - 2t - 3)$$

$$= -(t+1)(t+1)(t-3)$$

$$= -(t+1)^2(t-3).$$

Hence the eigenvalues of A are -1 and 3. A similar calculation shows that the characteristic polynomial of

$$B = \begin{bmatrix} 3 & 0 & 0 \\ 0 & 1 & 2 \\ 0 & 2 & 1 \end{bmatrix}$$

is $-(t+1)(t-3)^2$. Therefore the eigenvalues of B are also -1 and 3. But the status of the eigenvalues -1 and 3 is different in A and B.

If λ is an eigenvalue of an $n \times n$ matrix M, then the largest positive integer k such that $(t - \lambda)^k$ is a factor of the characteristic polynomial of M is called the **multiplicity**[3] of λ. Thus if

$$\det(M - tI_n) = (t-5)^2(t+6)(t-7)^3(t-8)^4,$$

then the eigenvalues of M are 5 (with multiplicity 2), -6 (with multiplicity 1), 7 (with multiplicity 3), and 8 (with multiplicity 4).

For the preceding matrices A and B, the eigenvalues -1 and 3 have different multiplicities. For A, the multiplicity of -1 is 2 and the multiplicity of 3 is 1, whereas for B the multiplicity of -1 is 1 and the multiplicity of 3 is 2. It is instructive to investigate the eigenspaces of A and B corresponding to the same eigenvalue, say 3. Since the reduced row echelon form of $A - 3I_3$ is

$$\begin{bmatrix} 1 & 0 & 0 \\ 0 & 1 & -1 \\ 0 & 0 & 0 \end{bmatrix},$$

the parametric form of the general solution of $(A - 3I_3)\mathbf{x} = \mathbf{0}$ is

$$\begin{bmatrix} x_1 \\ x_2 \\ x_3 \end{bmatrix} = \begin{bmatrix} 0 \\ x_3 \\ x_3 \end{bmatrix} = x_3 \begin{bmatrix} 0 \\ 1 \\ 1 \end{bmatrix}.$$

Hence

$$\left\{ \begin{bmatrix} 0 \\ 1 \\ 1 \end{bmatrix} \right\}$$

---

[3]Some authors use the term *algebraic multiplicity*. In this case, the dimension of the eigenspace corresponding to λ is usually called the *geometric multiplicity* of λ.

is a basis for the eigenspace of A corresponding to eigenvalue 3. Therefore this eigenspace has dimension 1. On the other hand, in Example 3 of Section 5.1 we saw that

$$\left\{ \begin{bmatrix} 1 \\ 0 \\ 0 \end{bmatrix}, \begin{bmatrix} 0 \\ 1 \\ 1 \end{bmatrix} \right\}$$

is a basis for the eigenspace of B corresponding to eigenvalue 3. Therefore this eigenspace has dimension 2.

For these matrices A and B, the dimension of the eigenspace corresponding to eigenvalue 3 equals the multiplicity of the eigenvalue. This need not always be the case, but there is a connection between the dimension of an eigenspace and the multiplicity of the corresponding eigenvalue. It is described in our next theorem.

Theorem 5.1

Let λ be an eigenvalue of a matrix A. The dimension of the eigenspace of A corresponding to λ is less than or equal to the multiplicity of λ.

Proof Let A be an $n \times n$ matrix, and let k denote the dimension of the eigenspace of A corresponding to the eigenvalue λ. By Theorem 4.4(a), a basis $\{\mathbf{v}_1, \mathbf{v}_2, \ldots, \mathbf{v}_k\}$ for this eigenspace can be extended to a basis $\mathcal{B} = \{\mathbf{v}_1, \mathbf{v}_2, \ldots, \mathbf{v}_n\}$ for $\mathcal{R}^n$. Consider the linear operator T_A on $\mathcal{R}^n$, the matrix transformation induced by A. Recall that column j of the $\mathcal{B}$-matrix representation of T_A is $[T_A(\mathbf{v}_j)]_\mathcal{B}$. In this case,

$$[T_A(\mathbf{v}_j)]_\mathcal{B} = [A\mathbf{v}_j]_\mathcal{B} = [\lambda\mathbf{v}_j]_\mathcal{B} = \lambda[\mathbf{v}_j]_\mathcal{B} = \lambda\mathbf{e}_j$$

for $1 \leq j \leq k$, because $\mathbf{v}_1, \mathbf{v}_2, \ldots, \mathbf{v}_k$ are eigenvectors of A corresponding to eigenvalue λ. Thus the $\mathcal{B}$-matrix representation of this operator has the block form

$$M = \begin{bmatrix} \lambda I_k & C \\ O & D \end{bmatrix},$$

where O is the $(n-k) \times k$ zero matrix.

We claim that the characteristic polynomial of A is the same as that of M. Since A is the standard matrix of T_A, Theorem 4.10 shows that A and M are similar. Thus by the discussion on page 261, the characteristic polynomial of A is the same as that of M. But the characteristic polynomial of M is

$$\det(M - tI_n) = \det \begin{bmatrix} \lambda I_k - tI_k & C \\ O & D - tI_{n-k} \end{bmatrix}$$

$$= \det \begin{bmatrix} (\lambda - t)I_k & C \\ O & D - tI_{n-k} \end{bmatrix}$$

$$= [\det(\lambda - t)I_k] \cdot [\det(D - tI_{n-k})]$$

$$= (\lambda - t)^k [\det(D - tI_{n-k})]$$

by Example 3 of Section 3.2. It follows that λ is an eigenvalue of A having multiplicity k or more. ∎

Example 3

Determine the eigenvalues, their multiplicities, and a basis for each eigenspace of the linear operator on $\mathcal{R}^3$ defined by

$$T\left(\begin{bmatrix} x_1 \\ x_2 \\ x_3 \end{bmatrix}\right) = \begin{bmatrix} -x_1 \\ 2x_1 - x_2 - x_3 \\ -x_3 \end{bmatrix}.$$

Solution

The standard matrix of T is

$$A = \begin{bmatrix} -1 & 0 & 0 \\ 2 & -1 & -1 \\ 0 & 0 & -1 \end{bmatrix}.$$

Hence the characteristic polynomial of T is

$$\det(A - tI_3) = \det \begin{bmatrix} -1-t & 0 & 0 \\ 2 & -1-t & -1 \\ 0 & 0 & -1-t \end{bmatrix}$$

$$= (-1-t) \cdot \det \begin{bmatrix} -1-t & -1 \\ 0 & -1-t \end{bmatrix}$$

$$= (-1-t)^3$$

$$= -(t+1)^3.$$

Therefore the only eigenvalue of T is -1, and it is of multiplicity 3. The eigenspace of T corresponding to -1 is the solution set to $(A + I_3)\mathbf{x} = \mathbf{0}$. Since the reduced row echelon form of $A + I_3$ is

$$\begin{bmatrix} 1 & 0 & -.5 \\ 0 & 0 & 0 \\ 0 & 0 & 0 \end{bmatrix},$$

we see that

$$\left\{ \begin{bmatrix} 0 \\ 1 \\ 0 \end{bmatrix}, \begin{bmatrix} 1 \\ 0 \\ 2 \end{bmatrix} \right\}$$

is a basis for the eigenspace of T corresponding to -1. Note that this eigenspace is 2-dimensional and the multiplicity of -1 is 3, in accordance with Theorem 5.1.   ○

Complex Eigenvalues*

We have seen in Example 2 that not all $n \times n$ matrices or linear operators on $\mathcal{R}^n$ have real eigenvalues and eigenvectors. The characteristic polynomial of such a matrix must have no real roots. However, it is a consequence of the fundamental theorem of algebra (see the appendix on complex numbers) that every $n \times n$ matrix has complex eigenvalues. In fact, the fundamental theorem of algebra implies that the characteristic polynomial of every $n \times n$ matrix can

---

*The remainder of this section is used only in the description of harmonic motion (an optional part of Section 5.5).

be written in the form

$$c(t - \lambda_1)(t - \lambda_2) \cdots (t - \lambda_n)$$

for some complex numbers $c, \lambda_1, \lambda_2, \ldots, \lambda_n$. Thus, if we count each eigenvalue as often as its multiplicity, every $n \times n$ matrix has exactly n complex eigenvalues. However, some or all of these may not be real numbers.

There are applications (in such disciplines as physics and electrical engineering) where complex eigenvalues provide useful information about real-world problems. For the most part, the mathematical theory is no different for complex numbers than for real numbers. In the complex case, we allow complex entries in matrices and vectors. The set of all $n \times 1$ matrices with complex entries, denoted by $\mathcal{C}^n$, replaces the usual space of vectors $\mathcal{R}^n$, and the set $\mathcal{C}$ of complex numbers replaces $\mathcal{R}$ as the set of scalars.

Example 4 illustrates the calculations required to find eigenvalues and eigenvectors involving complex numbers. However, with the exception of an application discussed in Section 5.5, in this book we restrict our attention to real eigenvalues and eigenvectors having real components.

Example 4   Determine the complex eigenvalues and a basis for each eigenspace of

$$A = \begin{bmatrix} 1 & -10 \\ 2 & 5 \end{bmatrix}.$$

Solution   The characteristic polynomial of A is

$$\det (A - t I_2) = \begin{bmatrix} 1 - t & -10 \\ 2 & 5 - t \end{bmatrix} = (1 - t)(5 - t) + 20 = t^2 - 6t + 25.$$

Applying the quadratic formula, we find that the roots of the characteristic polynomial of A are

$$t = \frac{6 \pm \sqrt{(-6)^2 - 4(1)(25)}}{2} = \frac{6 \pm \sqrt{-64}}{2} = \frac{6 \pm 8i}{2} = 3 \pm 4i.$$

Hence the eigenvalues of A are $3 + 4i$ and $3 - 4i$. As with real eigenvalues, we find the eigenvectors in $\mathcal{C}^2$ corresponding to $3 + 4i$ by solving $(A - (3 + 4i)I_2)\mathbf{x} = \mathbf{0}$. The reduced row echelon form of $A - (3 + 4i)I_2$ is

$$\begin{bmatrix} 1 & 1 - 2i \\ 0 & 0 \end{bmatrix}.$$

Thus the vectors in the eigenspace corresponding to $3 + 4i$ have the form

$$\begin{bmatrix} x_1 \\ x_2 \end{bmatrix} = \begin{bmatrix} (-1 + 2i)x_2 \\ x_2 \end{bmatrix} = x_2 \begin{bmatrix} -1 + 2i \\ 1 \end{bmatrix},$$

and so a basis for the eigenspace corresponding to $3 + 4i$ is

$$\left\{ \begin{bmatrix} -1 + 2i \\ 1 \end{bmatrix} \right\}.$$

Similarly, the reduced row echelon form of $A - (3 - 4i)I_2$ is

$$\begin{bmatrix} 1 & 1 + 2i \\ 0 & 0 \end{bmatrix}.$$

Hence the vectors in the eigenspace corresponding to $3 - 4i$ have the form

$$\begin{bmatrix} x_1 \\ x_2 \end{bmatrix} = \begin{bmatrix} (-1 - 2i)x_2 \\ x_2 \end{bmatrix} = x_2 \begin{bmatrix} -1 - 2i \\ 1 \end{bmatrix},$$

and so a basis for the eigenspace corresponding to $3 - 4i$ is

$$\left\{ \begin{bmatrix} -1 - 2i \\ 1 \end{bmatrix} \right\}.$$

When the entries of A are real numbers, the characteristic polynomial of A has real coefficients. Under these conditions, if some nonreal number is a root of the characteristic polynomial of A, then its complex conjugate can also be shown to be a root. Thus *the nonreal eigenvalues of a real matrix occur in complex conjugate pairs.* Moreover, if $\mathbf{v}$ is an eigenvector of A corresponding to a nonreal eigenvalue λ, then the complex conjugate of $\mathbf{v}$ (the vector whose components are the complex conjugates of the components of $\mathbf{v}$) can be shown to be an eigenvector of A corresponding to the complex conjugate of λ. Note this relationship in Example 4.

Practice Problems

1. If $-(t-3)(t+5)^2(t-8)^4$ is the characteristic polynomial of a matrix, determine the eigenvalues of the matrix and their multiplicities.

2. Determine the eigenvalues of

$$A = \begin{bmatrix} 1 & -1 & -1 \\ 4 & -3 & -5 \\ 0 & 0 & 2 \end{bmatrix},$$

their multiplicities, and a basis for each eigenspace.

3. Determine the complex eigenvalues and a basis for each eigenspace of the 90°-rotation matrix

$$A = \begin{bmatrix} 0 & -1 \\ 1 & 0 \end{bmatrix}.$$

Exercises

1. Determine if the following statements are true or false.

   (a) If two $n \times n$ matrices have the same characteristic polynomial, then they have the same eigenvectors.
   (b) If two $n \times n$ matrices have the same characteristic polynomial, then they have the same eigenvalues.
   (c) The characteristic polynomial of an $n \times n$ matrix is a polynomial of degree n.
   (d) The eigenvalues of a matrix are equal to those of its reduced row echelon form.
   (e) The eigenvectors of a matrix are equal to those of its reduced row echelon form.
   (f) An $n \times n$ matrix has n distinct eigenvalues.
   (g) Every $n \times n$ matrix has an eigenvector in $\mathcal{R}^n$.
   (h) Every square matrix has a complex eigenvalue.
   (i) The characteristic polynomial of an $n \times n$ matrix can be written $c(t - \lambda_1)(t - \lambda_2) \cdots (t - \lambda_n)$ for some real numbers $c, \lambda_1, \lambda_2, \ldots, \lambda_n$.
   (j) The characteristic polynomial of an $n \times n$ matrix can be written $c(t - \lambda_1)(t - \lambda_2) \cdots (t - \lambda_n)$ for some complex numbers $c, \lambda_1, \lambda_2, \ldots, \lambda_n$.
   (k) If $(t - 4)^2$ divides the characteristic polynomial of A, then 4 is an eigenvalue of A with multiplicity 2.
   (l) The multiplicity of an eigenvalue equals the dimension of the corresponding eigenspace.

2. What is the coefficient of t^n in the characteristic polynomial of an $n \times n$ matrix?

3. Suppose that A is a 4×4 matrix with no nonreal eigenvalues and exactly two real eigenvalues, 5 and -9. Let W_1 and W_2 be the eigenspaces of A corresponding to 5 and -9, respectively. Write all the possible characteristic polynomials of A that are consistent with the following information.

   (a) dim $W_1 = 3$
   (b) dim $W_2 = 1$
   (c) dim $W_1 = 2$

4. Suppose that A is a 5×5 matrix with no nonreal eigenvalues and exactly three real eigenvalues, 4, 6, and 7. Let W_1, W_2, and W_3 be the eigenspaces corresponding to 4, 6, and 7, respectively. Write all the possible characteristic

polynomials of A that are consistent with the following information.

(a) dim $W_2 = 3$
(b) dim $W_1 = 2$
(c) dim $W_1 = 1$ and dim $W_2 = 2$
(d) dim $W_2 = 2$ and dim $W_3 = 2$

In Exercises 5–16, a matrix and its characteristic polynomial are given. Find the eigenvalues of the matrix and determine a basis for each eigenspace.

5. $\begin{bmatrix} 3 & -3 \\ 2 & 8 \end{bmatrix}$, $(t-5)(t-6)$

6. $\begin{bmatrix} -7 & 1 \\ -6 & -2 \end{bmatrix}$, $(t+4)(t+5)$

7. $\begin{bmatrix} -10 & 6 \\ -15 & 9 \end{bmatrix}$, $t(t+1)$

8. $\begin{bmatrix} -9 & -7 \\ 14 & 12 \end{bmatrix}$, $(t+2)(t-5)$

9. $\begin{bmatrix} 6 & -5 & -4 \\ 5 & -3 & -5 \\ 4 & -5 & -2 \end{bmatrix}$, $-(t+3)(t-2)^2$

10. $\begin{bmatrix} -2 & -6 & -6 \\ -3 & 2 & -2 \\ 3 & 2 & 6 \end{bmatrix}$, $-(t+2)(t-4)^2$

11. $\begin{bmatrix} 6 & -4 & -4 \\ -8 & 2 & 4 \\ 8 & -4 & -6 \end{bmatrix}$, $-(t-6)(t+2)^2$

12. $\begin{bmatrix} -5 & 6 & 1 \\ -1 & 2 & 1 \\ -8 & 6 & 4 \end{bmatrix}$, $-(t+4)(t-2)(t-3)$

13. $\begin{bmatrix} 0 & 2 & 1 \\ 1 & -1 & -1 \\ 4 & 4 & -3 \end{bmatrix}$, $-(t+3)(t+2)(t-1)$

14. $\begin{bmatrix} 3 & 2 & 2 \\ -2 & -1 & -2 \\ 2 & 2 & 3 \end{bmatrix}$, $-(t-3)(t-1)^2$

15. $\begin{bmatrix} -1 & 4 & -4 & -4 \\ 5 & -2 & 1 & 6 \\ 0 & 0 & -1 & 0 \\ 5 & -5 & 5 & 9 \end{bmatrix}$, $(t-3)(t-4)(t+1)^2$

16. $\begin{bmatrix} 1 & 6 & -6 & -6 \\ 6 & 7 & -6 & -12 \\ 3 & 3 & -2 & -6 \\ 3 & 9 & -9 & -11 \end{bmatrix}$, $(t+5)(t+2)(t-1)^2$

In Exercises 17–28, find the eigenvalues of the given matrix and determine a basis for each eigenspace.

17. $\begin{bmatrix} 6 & -5 \\ 10 & -9 \end{bmatrix}$

18. $\begin{bmatrix} 8 & 2 \\ -12 & -2 \end{bmatrix}$

19. $\begin{bmatrix} -3 & -4 \\ 12 & 11 \end{bmatrix}$

20. $\begin{bmatrix} 3 & 2 \\ -10 & -6 \end{bmatrix}$

21. $\begin{bmatrix} -7 & 5 & 4 \\ 0 & -3 & 0 \\ -8 & 9 & 5 \end{bmatrix}$

22. $\begin{bmatrix} -3 & -12 & 0 \\ 0 & 3 & 0 \\ -4 & -8 & 1 \end{bmatrix}$

23. $\begin{bmatrix} -7 & 6 & 6 \\ 0 & -1 & 0 \\ -12 & 12 & 11 \end{bmatrix}$

24. $\begin{bmatrix} 3 & 0 & 0 \\ 9 & 3 & 10 \\ -5 & 0 & -2 \end{bmatrix}$

25. $\begin{bmatrix} -4 & 0 & 2 \\ 2 & 4 & -8 \\ 2 & 0 & -4 \end{bmatrix}$

26. $\begin{bmatrix} 3 & 0 & 7 \\ 7 & -4 & 7 \\ 0 & 0 & -4 \end{bmatrix}$

27. $\begin{bmatrix} -2 & -1 & -1 & 4 \\ 0 & 2 & 0 & 0 \\ 1 & -2 & 0 & -1 \\ 0 & 0 & 0 & 2 \end{bmatrix}$

28. $\begin{bmatrix} 1 & 0 & 0 & 0 \\ 9 & -2 & -3 & 3 \\ -6 & 0 & 1 & -3 \\ -6 & 0 & 0 & -2 \end{bmatrix}$

In Exercises 29–34, a linear operator and its characteristic polynomial are given. Find the eigenvalues of the operator and determine a basis for each eigenspace.

29. $T\left(\begin{bmatrix} x_1 \\ x_2 \end{bmatrix} \right) = \begin{bmatrix} -x_1 + 6x_2 \\ -8x_1 + 13x_2 \end{bmatrix}$, $(t-5)(t-7)$

30. $T\left(\begin{bmatrix} x_1 \\ x_2 \end{bmatrix} \right) = \begin{bmatrix} -x_1 + 2x_2 \\ -4x_1 - 7x_2 \end{bmatrix}$, $(t+5)(t+3)$

31. $T\left(\begin{bmatrix} x_1 \\ x_2 \\ x_3 \end{bmatrix} \right) = \begin{bmatrix} -2x_2 + 4x_3 \\ -3x_1 + x_2 + 3x_3 \\ -x_1 + x_2 + 5x_3 \end{bmatrix}$, $-(t+2)(t-4)^2$

32. $T\left(\begin{bmatrix} x_1 \\ x_2 \\ x_3 \end{bmatrix} \right) = \begin{bmatrix} -8x_1 - 5x_2 - 7x_3 \\ 6x_2 + 3x_2 + 7x_3 \\ 8x_1 + 8x_2 - 9x_3 \end{bmatrix}$, $-(t+3)(t+2)(t+9)$

33. $T\left(\begin{bmatrix} x_1 \\ x_2 \\ x_3 \end{bmatrix} \right) = \begin{bmatrix} 3x_1 + 2x_2 - 2x_3 \\ 2x_1 + 6x_2 - 4x_3 \\ 3x_1 + 6x_2 - 4x_3 \end{bmatrix}$, $-(t-1)(t-2)^2$

34. $T\left(\begin{bmatrix} x_1 \\ x_2 \\ x_3 \end{bmatrix} \right) = \begin{bmatrix} 3x_1 + 4x_2 - 4x_3 \\ 8x_1 + 7x_2 - 8x_3 \\ 8x_1 + 8x_2 - 9x_3 \end{bmatrix}$, $-(t+1)^2(t-3)$

In Exercises 35–40, find the eigenvalues of the given linear operator and determine a basis for each eigenspace.

35. $T\left(\begin{bmatrix} x_1 \\ x_2 \end{bmatrix}\right) = \begin{bmatrix} -4x_2 + x_2 \\ -2x_1 - x_2 \end{bmatrix}$

36. $T\left(\begin{bmatrix} x_1 \\ x_2 \end{bmatrix}\right) = \begin{bmatrix} 6x_1 - x_2 \\ 6x_1 + x_2 \end{bmatrix}$

37. $T\left(\begin{bmatrix} x_1 \\ x_2 \\ x_3 \end{bmatrix}\right) = \begin{bmatrix} 7x_1 - 10x_2 \\ 5x_1 - 8x_2 \\ -x_1 + x_2 + 2x_3 \end{bmatrix}$

38. $T\left(\begin{bmatrix} x_1 \\ x_2 \\ x_3 \end{bmatrix}\right) = \begin{bmatrix} -6x_1 - 5x_2 + 5x_3 \\ -x_2 \\ -10x_1 - 10x_2 + 9x_3 \end{bmatrix}$

39. $T\left(\begin{bmatrix} x_1 \\ x_2 \\ x_3 \end{bmatrix}\right) = \begin{bmatrix} -3x_1 \\ -8x_1 + x_2 \\ -12x_1 + x_3 \end{bmatrix}$

40. $T\left(\begin{bmatrix} x_1 \\ x_2 \\ x_3 \end{bmatrix}\right) = \begin{bmatrix} -4x_1 + 6x_2 \\ 2x_2 \\ -5x_1 + 5x_2 + x_3 \end{bmatrix}$

41. Show that $\begin{bmatrix} 6 & -7 \\ 4 & -3 \end{bmatrix}$ has no real eigenvalues.

42. Show that $\begin{bmatrix} 4 & -5 \\ 3 & -2 \end{bmatrix}$ has no real eigenvalues.

43. Let A be an $n \times n$ matrix, and suppose that, for a particular scalar c, the reduced row echelon form of $A - cI_n$ is I_n. What can be said about c?

44. If $f(t)$ is the characteristic polynomial of a square matrix A, what is $f(0)$?

45. Show that if A is an upper triangular or a lower triangular $n \times n$ matrix, then λ is an eigenvalue of A with multiplicity k if and only if λ appears exactly k times on the diagonal of A.

46. Show that the rotation matrix A_θ has no real eigenvalues if $0° < \theta < 180°$.

47. (a) Determine a basis for each eigenspace of
$$A = \begin{bmatrix} 3 & 2 \\ -1 & 0 \end{bmatrix}.$$
(b) Determine a basis for each eigenspace of $(-3)A$.
(c) Determine a basis for each eigenspace of $5A$.
(d) Establish a relationship between the eigenvectors of any square matrix B and those of cB for any scalar c.
(e) Establish a relationship between the eigenvalues of any square matrix B and those of cB for any scalar c.

48. (a) Determine a basis for each eigenspace of
$$A = \begin{bmatrix} 5 & -2 \\ 1 & 8 \end{bmatrix}.$$
(b) Determine a basis for each eigenspace of $A + 4I_2$.
(c) Determine a basis for each eigenspace of $A - 6I_2$.
(d) Establish a relationship between the eigenvectors of any $n \times n$ matrix B and those of $B + cI_n$ for any

scalar c.
(e) Establish a relationship between the eigenvalues of any $n \times n$ matrix B and those of $B + cI_n$ for any scalar c.

49. (a) Determine the characteristic polynomial of A^T, where A is the matrix in Exercise 48.
(b) Establish a relationship between the characteristic polynomial of any square matrix B and that of B^T.
(c) What does (b) imply about the relationship between the eigenvalues of a square matrix B and those of B^T?
(d) Is there a relationship between the eigenvectors of a square matrix B and those of B^T?

50. (a) Prove that similar matrices have the same eigenvalues.
(b) Must similar matrices have the same eigenvectors? Justify your answer.

51. Let A be a symmetric 2×2 matrix. Prove that A has real eigenvalues.

52. (a) The characteristic polynomial of $A = \begin{bmatrix} a & b \\ c & d \end{bmatrix}$ has the form $t^2 + rt + s$ for some scalars r and s. Determine r and s in terms of a, b, c, and d.
(b) Show that $A^2 + rA + sI_2 = O$, the 2×2 zero matrix. (A similar result is true for any square matrix (see Exercise 58 of Section 5.3). It is called the *Cayley–Hamilton theorem*.)

In Exercises 53–57, use a calculator with matrix capabilities or computer software such as MATLAB *to solve the problem.*

53. Compute the characteristic polynomial of
$$\begin{bmatrix} 1 & \frac{1}{2} & \frac{1}{3} \\ \frac{1}{2} & \frac{1}{3} & \frac{1}{4} \\ \frac{1}{3} & \frac{1}{4} & \frac{1}{5} \end{bmatrix}.$$
(This matrix is called the 3×3 *Hilbert matrix*. Computations with Hilbert matrices are subject to significant roundoff errors.)

54. Compute the characteristic polynomial of
$$\begin{bmatrix} 0 & 0 & 0 & -17 \\ 1 & 0 & 0 & -18 \\ 0 & 1 & 0 & -19 \\ 0 & 0 & 1 & -20 \end{bmatrix}.$$

55. Use the result of Exercise 54 to find a 4×4 matrix whose characteristic polynomial is $t^4 - 11t^3 + 23t^2 + 7t - 5$.

56. Let A be a randomly generated 4×4 matrix.
(a) Compute the characteristic polynomials of A and A^T.
(b) Formulate a conjecture about the characteristic polynomials of B and B^T, where B is an arbitrary $n \times n$ matrix. Test your conjecture using an arbitrary 5×5 matrix.
(c) Prove that your conjecture in (d) is valid.

57. Let

$$A = \begin{bmatrix} 6.5 & -3.5 \\ 7.0 & -4.0 \end{bmatrix}.$$

(a) Find the eigenvalues of A and an eigenvector corresponding to each eigenvalue.
(b) Show that A is invertible, and then find the eigenvalues of A^{-1} and an eigenvector corresponding to each eigenvalue.
(c) Use the results of (a) and (b) to formulate a conjecture about the relationship between the eigenvectors and eigenvalues of an invertible $n \times n$ matrix and those of its inverse.
(d) Test your conjecture in (c) on the invertible matrix

$$\begin{bmatrix} 3 & -2 & 2 \\ -4 & 8 & -10 \\ -5 & 2 & -4 \end{bmatrix}.$$

(e) Prove that your conjecture in (c) is valid.

Solutions to the Practice Problems

1. The eigenvalues of the matrix are the roots of its characteristic polynomial, which is $-(t-3)(t+5)^2(t-8)^4$. Thus the eigenvalues of the matrix are 3, -5, and 8. The multiplicity of eigenvalue λ is the number of factors of $t - \lambda$ that appear in the characteristic polynomial. Hence 3 is an eigenvalue of multiplicity 1, -5 is an eigenvalue of multiplicity 2, and 8 is an eigenvalue of multiplicity 4.

2. Form the matrix

$$B = A - tI_3 = \begin{bmatrix} 1-t & -1 & -1 \\ 4 & -3-t & -5 \\ 0 & 0 & 2-t \end{bmatrix}.$$

To evaluate the determinant of B, we use the cofactor expansion along the third row. Then

$$\det B = (-1)^{3+1}b_{31} \cdot \det B_{31} + (-1)^{3+2}b_{32} \cdot \det B_{32}$$
$$+ (-1)^{3+3}b_{33} \cdot \det B_{33}$$
$$= 0 + 0 + (-1)^6(2-t) \cdot \det \begin{bmatrix} 1-t & -1 \\ 4 & -3-t \end{bmatrix}$$
$$= (2-t)[(1-t)(-3-t) + 4]$$
$$= (2-t)[(t^2 + 2t - 3) + 4]$$
$$= (2-t)(t^2 + 2t + 1)$$
$$= -(t-2)(t+1)^2.$$

Hence the eigenvalues of A are -1 (with multiplicity 2) and 2 (with multiplicity 1). Because the reduced row echelon form of $A + I_3$ is

$$\begin{bmatrix} 1 & -.5 & 0 \\ 0 & 0 & 1 \\ 0 & 0 & 0 \end{bmatrix},$$

we see that the parametric form of the general solution to $(A + I_3)\mathbf{x} = \mathbf{0}$ is

$$\begin{bmatrix} x_1 \\ x_2 \\ x_3 \end{bmatrix} = \begin{bmatrix} .5x_2 \\ x_2 \\ 0 \end{bmatrix} = x_2 \begin{bmatrix} .5 \\ 1 \\ 0 \end{bmatrix}.$$

Taking $x_2 = 2$, we obtain the basis

$$\left\{ \begin{bmatrix} 1 \\ 2 \\ 0 \end{bmatrix} \right\}$$

for the eigenspace of A corresponding to eigenvalue -1. Also,

$$\begin{bmatrix} 1 & 0 & 0 \\ 0 & 1 & 1 \\ 0 & 0 & 0 \end{bmatrix}$$

is the reduced row echelon form of $A - 2I_3$. Therefore a basis for the eigenspace of A corresponding to eigenvalue 2 is

$$\left\{ \begin{bmatrix} 0 \\ -1 \\ 1 \end{bmatrix} \right\}.$$

3. The characteristic polynomial of A is

$$\det(A - tI_2) = \det \begin{bmatrix} -t & -1 \\ 1 & -t \end{bmatrix} = t^2 + 1 = (t+i)(t-i).$$

Hence A has eigenvalues of $-i$ and i. Since the reduced row echelon form of $A + iI_2$ is

$$\begin{bmatrix} 1 & i \\ 0 & 0 \end{bmatrix},$$

a basis for the eigenspace corresponding to eigenvalue $-i$ is

$$\left\{ \begin{bmatrix} -i \\ 1 \end{bmatrix} \right\}.$$

Furthermore, the reduced row echelon form of $A - iI_2$ is

$$\begin{bmatrix} 1 & -i \\ 0 & 0 \end{bmatrix},$$

and so a basis for the eigenspace corresponding to eigenvalue i is

$$\left\{ \begin{bmatrix} i \\ 1 \end{bmatrix} \right\}.$$

5.3 Diagonalization of Matrices

In Example 6 of Section 2.1, we considered a metropolitan area in which the current population (in thousands) of the city and suburbs are given by

$$\begin{matrix} \text{City} \\ \text{Suburbs} \end{matrix} \begin{bmatrix} 500 \\ 700 \end{bmatrix} = \mathbf{p},$$

and the population shifts between the city and suburbs are described by the matrix

$$\begin{matrix} & & \text{From} \\ & & \text{City} \quad \text{Suburbs} \\ \text{To} & \begin{matrix} \text{City} \\ \text{Suburbs} \end{matrix} & \begin{bmatrix} .85 & .03 \\ .15 & .97 \end{bmatrix} = A. \end{matrix}$$

We saw in this example that the population of the city and suburbs after m years is given by the matrix–vector product $A^m\mathbf{p}$.

In this section, we discuss a technique for computing $A^m\mathbf{p}$. Note that when m is a large positive integer, a direct computation of $A^m\mathbf{p}$ involves considerable work. This calculation would be quite easy, however, if A were a diagonal matrix such as

$$D = \begin{bmatrix} .82 & 0 \\ 0 & 1 \end{bmatrix}.$$

For in this case, the powers of D are diagonal matrices and hence can be easily determined by the method described in Section 2.1. In fact,

$$D^m = \begin{bmatrix} (.82)^m & 0 \\ 0 & 1^m \end{bmatrix} = \begin{bmatrix} (.82)^m & 0 \\ 0 & 1 \end{bmatrix}.$$

Although $A \neq D$, it can be checked that $A = PDP^{-1}$, where

$$P = \begin{bmatrix} -1 & 1 \\ 1 & 5 \end{bmatrix}.$$

This relationship enables us to compute the powers of A in terms of those of D. For example,

$$A^3 = (PDP^{-1})(PDP^{-1})(PDP^{-1}) = PD^3P^{-1}.$$

In a similar manner, it can be shown that

$$
\begin{aligned}
A^m &= PD^mP^{-1} \\
&= \begin{bmatrix} -1 & 1 \\ 1 & 5 \end{bmatrix} \begin{bmatrix} (.82)^m & 0 \\ 0 & 1 \end{bmatrix} \begin{bmatrix} -1 & 1 \\ 1 & 5 \end{bmatrix}^{-1} \\
&= \begin{bmatrix} -1 & 1 \\ 1 & 5 \end{bmatrix} \begin{bmatrix} (.82)^m & 0 \\ 0 & 1 \end{bmatrix} \begin{bmatrix} -\frac{5}{6} & \frac{1}{6} \\ \frac{1}{6} & \frac{1}{6} \end{bmatrix} \\
&= \frac{1}{6} \begin{bmatrix} 1 + 5(.82)^m & 1 - (.82)^m \\ 5 - 5(.82)^m & 5 + (.82)^m \end{bmatrix}.
\end{aligned}
$$

Hence

$$A^m \mathbf{p} = \frac{1}{6} \begin{bmatrix} 1 + 5(.82)^m & 1 - (.82)^m \\ 5 - 5(.82)^m & 5 + (.82)^m \end{bmatrix} \begin{bmatrix} 500 \\ 700 \end{bmatrix}$$

$$= \frac{1}{6} \begin{bmatrix} 1200 + 1800(.82)^m \\ 6000 - 1800(.82)^m \end{bmatrix}$$

$$= \begin{bmatrix} 200 + 300(.82)^m \\ 1000 - 300(.82)^m \end{bmatrix}.$$

Because $\lim_{m \to \infty} (.82)^m = 0$, we see that the limit of $A^m \mathbf{p}$ is

$$\begin{bmatrix} 200 \\ 1000 \end{bmatrix}.$$

Hence after many years, the population of the metropolitan area will consist of about 200 thousand city dwellers and 1 million suburbanites.

In the computation above, note that calculating $PD^m P^{-1}$ requires only two matrix multiplications instead of the $m - 1$ multiplications needed to compute A^m directly. This simplification of the calculation is possible because A can be written in the form PDP^{-1} for some diagonal matrix D and some invertible matrix P.

Definition. An $n \times n$ matrix A is called **diagonalizable** if $A = PDP^{-1}$ for some diagonal $n \times n$ matrix D and some invertible $n \times n$ matrix P.

Thus the matrix

$$A = \begin{bmatrix} .85 & .03 \\ .15 & .97 \end{bmatrix}$$

in Example 6 of Section 2.1 is diagonalizable because $A = PDP^{-1}$ for the matrices

$$P = \begin{bmatrix} -1 & 1 \\ 1 & 5 \end{bmatrix} \quad \text{and} \quad D = \begin{bmatrix} .82 & 0 \\ 0 & 1 \end{bmatrix}.$$

It is clear that every diagonal matrix is diagonalizable (see Exercise 49). However, not every matrix is diagonalizable, as the following example shows.

Example 1   Show that the matrix

$$A = \begin{bmatrix} 0 & 1 \\ 0 & 0 \end{bmatrix}$$

is not diagonalizable.

Solution   Suppose to the contrary that $A = PDP^{-1}$, where P is an invertible 2×2 matrix and D is a diagonal 2×2 matrix. A direct calculation shows that $A^2 = O$, where O is the 2×2 zero matrix. But because $A = PDP^{-1}$, we also have $A^2 = PD^2 P^{-1}$. Thus $D^2 = P^{-1}A^2 P = P^{-1}OP = O$. Since D is a diagonal matrix, it follows that $D = O$. Hence $A = PDP^{-1} = POP^{-1} = O$, a contradiction.

We now study the conditions under which a matrix A is diagonalizable. From this investigation comes a procedure for finding an invertible matrix P and a diagonal matrix D such that $A = PDP^{-1}$.

Suppose that A is a diagonalizable $n \times n$ matrix, $P = [\mathbf{p}_1 \quad \mathbf{p}_2 \quad \cdots \quad \mathbf{p}_n]$ is an invertible $n \times n$ matrix, and D is a diagonal $n \times n$ matrix such that $A = PDP^{-1}$. For P to be invertible, its rank must be n. Thus its columns form a spanning set, and hence a basis, for $\mathcal{R}^n$. Rewrite the preceding equation as $AP = PD$, and note that

$$AP = [A\mathbf{p}_1 \quad A\mathbf{p}_2 \quad \cdots \quad A\mathbf{p}_n].$$

Since D is a diagonal matrix, its columns are multiples of standard vectors. Specifically, we have

$$D = [d_1\mathbf{e}_1 \quad d_2\mathbf{e}_2 \quad \cdots \quad d_n\mathbf{e}_n],$$

where d_j is the (j, j)-entry of D. Thus we have

$$\begin{aligned} PD &= P[d_1\mathbf{e}_1 \quad d_2\mathbf{e}_2 \quad \cdots \quad d_n\mathbf{e}_n] \\ &= [P(d_1\mathbf{e}_1) \quad P(d_2\mathbf{e}_2) \quad \cdots \quad P(d_n\mathbf{e}_n)] \\ &= [d_1(P\mathbf{e}_1) \quad d_2(P\mathbf{e}_2) \quad \cdots \quad d_n(P\mathbf{e}_n)] \\ &= [d_1\mathbf{p}_1 \quad d_2\mathbf{p}_2 \quad \cdots \quad d_n\mathbf{p}_n]. \end{aligned}$$

Equating the jth columns of AP and PD, we obtain $A\mathbf{p}_j = d_j\mathbf{p}_j$ for $j = 1, 2, \ldots, n$. Thus we see that each column of P is an eigenvector of A. Moreover, the eigenvalue corresponding to this eigenvector is d_j, the jth diagonal entry of D. We have now established half of the following result.

Theorem 5.2

(a) An $n \times n$ matrix A is diagonalizable if and only if there is a basis for $\mathcal{R}^n$ consisting of eigenvectors of A.

(b) If P is an invertible $n \times n$ matrix and D is a diagonal $n \times n$ matrix, then $A = PDP^{-1}$ if and only if the columns of P are a basis for $\mathcal{R}^n$ consisting of eigenvectors of A and the diagonal entries of D are the eigenvalues corresponding to the respective columns of P.

Proof The preceding discussion proves half of both (a) and (b). To establish the remainder, suppose that there is a basis $\{\mathbf{p}_1, \mathbf{p}_2, \ldots, \mathbf{p}_n\}$ for $\mathcal{R}^n$ consisting of eigenvectors of A. Let $P = [\mathbf{p}_1 \quad \mathbf{p}_2 \quad \cdots \quad \mathbf{p}_n]$ and

$$D = \begin{bmatrix} \lambda_1 & 0 & \cdots & 0 \\ 0 & \lambda_2 & \cdots & 0 \\ \vdots & \vdots & & \vdots \\ 0 & 0 & \cdots & \lambda_n \end{bmatrix},$$

where λ_j is the eigenvalue of A corresponding to $\mathbf{p}_j$. Then the columns of P are linearly independent, and hence P is invertible by Theorem 2.6. In addition, we have

$$\begin{aligned} PD &= P[\lambda_1\mathbf{e}_1 \quad \lambda_2\mathbf{e}_2 \quad \cdots \quad \lambda_n\mathbf{e}_n] \\ &= [P(\lambda_1\mathbf{e}_1) \quad P(\lambda_2\mathbf{e}_2) \quad \cdots \quad P(\lambda_n\mathbf{e}_n)] \\ &= [\lambda_1(P\mathbf{e}_1) \quad \lambda_2(P\mathbf{e}_2) \quad \cdots \quad \lambda_n(P\mathbf{e}_n)] \\ &= [\lambda_1\mathbf{p}_1 \quad \lambda_2\mathbf{p}_2 \quad \cdots \quad \lambda_n\mathbf{p}_n] \end{aligned}$$

$$= [A\mathbf{p}_1 \quad A\mathbf{p}_2 \quad \cdots \quad A\mathbf{p}_n]$$

$$= A[\mathbf{p}_1 \quad \mathbf{p}_2 \quad \cdots \quad \mathbf{p}_n]$$

$$= AP.$$

Hence $PDP^{-1} = A$. ❑

Theorem 5.2 shows us how to diagonalize a matrix such as

$$A = \begin{bmatrix} .85 & .03 \\ .15 & .97 \end{bmatrix}$$

in the population example discussed above. The characteristic polynomial of A is

$$\det(A - tI_2) = \det \begin{bmatrix} .85 - t & .03 \\ .15 & .97 - t \end{bmatrix}$$

$$= (.85 - t)(.97 - t) - .03(.15)$$

$$= t^2 - 1.82t + .82$$

$$= (t - .82)(t - 1),$$

and so the eigenvalues of A are .82 and 1. Since the reduced row echelon form of $A - .82I_2$ is

$$\begin{bmatrix} 1 & 1 \\ 0 & 0 \end{bmatrix},$$

we see that

$$\mathcal{B}_1 = \left\{ \begin{bmatrix} -1 \\ 1 \end{bmatrix} \right\}$$

is a basis for the eigenspace of A corresponding to .82. Likewise,

$$\begin{bmatrix} 1 & -.2 \\ 0 & 0 \end{bmatrix}$$

is the reduced row echelon form of $A - I_2$, and so

$$\mathcal{B}_2 = \left\{ \begin{bmatrix} 1 \\ 5 \end{bmatrix} \right\}$$

is a basis for the eigenspace of A corresponding to 1. Note that the set

$$\mathcal{B}_1 \cup \mathcal{B}_2 = \left\{ \begin{bmatrix} -1 \\ 1 \end{bmatrix}, \begin{bmatrix} 1 \\ 5 \end{bmatrix} \right\}$$

is linearly independent, since neither of its vectors is a multiple of the other. Therefore it is a basis for $\mathcal{R}^2$, and it consists of eigenvectors of A. Thus Theorem 5.2 guarantees that A is diagonalizable. Notice that the columns of the matrix

$$P = \begin{bmatrix} -1 & 1 \\ 1 & 5 \end{bmatrix}$$

used to diagonalize A are the vectors in this basis, and the diagonal matrix

$$D = \begin{bmatrix} .82 & 0 \\ 0 & 1 \end{bmatrix}$$

has as its diagonal entries the eigenvalues of A corresponding to the respective columns of P.

The matrices P and D such that $PDP^{-1} = A$ are not unique. For example, taking

$$P = \begin{bmatrix} 2 & -3 \\ 10 & 3 \end{bmatrix} \quad \text{and} \quad D = \begin{bmatrix} 1 & 0 \\ 0 & .82 \end{bmatrix}$$

also gives $PDP^{-1} = A$, because these matrices satisfy the hypotheses of Theorem 5.2. Note, however, that although the matrix D in Theorem 5.2 is not unique, any two such matrices differ only in the sequence in which the eigenvalues of A are listed along the diagonal of D.

Example 2   Show that the matrix

$$A = \begin{bmatrix} -4 & -3 \\ 3 & 6 \end{bmatrix}$$

is diagonalizable, and find an invertible matrix P and a diagonal matrix D such that $A = PDP^{-1}$.

Solution   In Example 1 of Section 5.2, we saw that -3 and 5 are the eigenvalues of A and that

$$\mathcal{B}_1 = \left\{ \begin{bmatrix} -3 \\ 1 \end{bmatrix} \right\} \quad \text{and} \quad \mathcal{B}_2 = \left\{ \begin{bmatrix} -1 \\ 3 \end{bmatrix} \right\}$$

are bases for the corresponding eigenspaces. Then $\mathcal{B}_1 \cup \mathcal{B}_2$ is a linearly independent set, and hence it is a basis for $\mathcal{R}^2$ consisting of eigenvectors of A. Thus A is diagonalizable by Theorem 5.2. Furthermore, this theorem also guarantees that if

$$P = \begin{bmatrix} -3 & -1 \\ 1 & 3 \end{bmatrix} \quad \text{and} \quad D = \begin{bmatrix} -3 & 0 \\ 0 & 5 \end{bmatrix},$$

then $PDP^{-1} = A$, as is easily checked.   ○

When using Theorem 5.2 to show that a matrix is diagonalizable, we normally need to know that the union of bases for distinct eigenspaces is linearly independent (as is true for $\mathcal{B}_1 \cup \mathcal{B}_2$ in Example 2). Actually, this is always the case, as the following result shows.[4]

Theorem 5.3
Let $\mathcal{L}_1, \mathcal{L}_2, \ldots, \mathcal{L}_k$ be linearly independent subsets of different eigenspaces of a matrix A. Then $\mathcal{L}_1 \cup \mathcal{L}_2 \cup \cdots \cup \mathcal{L}_k$ is linearly independent.

A useful special case of Theorem 5.3 is the fact that a set of eigenvectors corresponding to distinct eigenvalues is linearly independent.

---
[4]The proof of Theorem 5.3 is given at the end of this section.

When Is a Matrix Diagonalizable?

We seek a method that can be used to determine whether or not a given matrix is diagonalizable. For this purpose, we examine two properties that distinguish diagonalizable matrices from nondiagonalizable matrices.

If A is a diagonalizable $n \times n$ matrix, there is a diagonal matrix D and an invertible matrix P such that $A = PDP^{-1}$. We have seen that the diagonal entries of D are the eigenvalues of A, and the number of occurrences of any eigenvalue on the diagonal of D is the multiplicity of that eigenvalue. Let

$$D = \begin{bmatrix} \lambda_1 & 0 & \cdots & 0 \\ 0 & \lambda_2 & \cdots & 0 \\ \vdots & \vdots & & \vdots \\ 0 & 0 & \cdots & \lambda_n \end{bmatrix}.$$

Then by Example 2 of Section 5.2,

$$\det (A - t I_n) = \det (D - t I_n)$$

$$= \det \begin{bmatrix} \lambda_1 - t & 0 & \cdots & 0 \\ 0 & \lambda_2 - t & \cdots & 0 \\ \vdots & \vdots & & \vdots \\ 0 & 0 & \cdots & \lambda_n - t \end{bmatrix}$$

$$= (-1)^n (t - \lambda_1)(t - \lambda_2) \cdots (t - \lambda_n).$$

Thus the characteristic polynomial of A is

$$(-1)^n (t - \lambda_1)(t - \lambda_2) \cdots (t - \lambda_n).$$

In other words, *the characteristic polynomial of a diagonalizable matrix factors into a product of linear factors* (that is, polynomials of the form $t - \lambda_i$). Moreover, the numbers λ_i that appear in this factorization are the eigenvalues of A, and the number of occurrences of the factor $t - \lambda_i$ is the multiplicity of λ_i.

For example, consider the matrix

$$B = \begin{bmatrix} 5 & 0 & 0 \\ 0 & 0 & -2 \\ 0 & 2 & 0 \end{bmatrix}$$

with characteristic polynomial

$$\det (B - I_3) = \det \begin{bmatrix} 5 - t & 0 & 0 \\ 0 & -t & -2 \\ 0 & 2 & -t \end{bmatrix} = -(t - 5)(t^2 + 4)^2.$$

Because $t^2 + 4$ cannot be factored into linear polynomials (with real coefficients), the characteristic polynomial of B does not factor into a product of linear factors. Therefore B is not diagonalizable.

Note, however, that even when its characteristic polynomial can be factored into linear polynomials, a matrix might not be diagonalizable. For instance, the matrix

$$A = \begin{bmatrix} 0 & 1 \\ 0 & 0 \end{bmatrix}$$

in Example 1 is not diagonalizable even though its characteristic polynomial is

$$\det \begin{bmatrix} 0 & 1 \\ 0 & 0 \end{bmatrix} = t^2 = (t - 0)(t - 0),$$

a product of linear factors.

Another example of this phenomenon is the matrix

$$B = \begin{bmatrix} 2 & 0 & 0 \\ 0 & 4 & 0 \\ 1 & -1 & 2 \end{bmatrix}.$$

The characteristic polynomial of B,

$$\det (B - I_3) = (2 - t)(4 - t)(2 - t) = -(t - 4)(t - 2)^2,$$

factors into a product of linear polynomials. However, B is not diagonalizable because there are simply not enough linearly independent eigenvectors to constitute a basis for $\mathcal{R}^3$. For suppose that $\mathcal{B}$ is a linearly independent subset of $\mathcal{R}^3$ consisting of eigenvectors of B. Observe that the reduced row echelon form of $B - 2I_3$ is

$$\begin{bmatrix} 1 & 0 & 0 \\ 0 & 1 & 0 \\ 0 & 0 & 0 \end{bmatrix},$$

a matrix with nullity 1. It follows that the eigenspace of B corresponding to eigenvalue 2, which is the null space of $B - 2I_3$, has dimension 1. Thus every basis for the eigenspace of B corresponding to eigenvalue 2 contains exactly 1 vector. So $\mathcal{B}$ could contain at most one eigenvector corresponding to eigenvalue 2. By Theorem 5.1, the eigenspace of B corresponding to eigenvalue 4 has dimension 1 because this eigenvalue has multiplicity 1. Hence $\mathcal{B}$ must contain at most one eigenvector corresponding to eigenvalue 4. Since the only eigenvalues of B are 2 and 4, $\mathcal{B}$ cannot contain more than two vectors. So $\mathcal{B}$ is not a basis for $\mathcal{R}^3$, and hence B is not diagonalizable.

More generally, consider any $n \times n$ matrix A whose characteristic polynomial factors into a product of linear polynomials. Suppose first that the sum of the dimensions of the eigenspaces of A is less than n. Then, as for the matrix B above, A is not diagonalizable because it is impossible to obtain a set of n linearly independent eigenvectors. Now suppose that the sum of the dimensions of the eigenspaces of A is greater than or equal to n. In this case, the union of bases for each eigenspace is a set of at least n eigenvectors that is linearly independent by Theorem 5.3. It follows that this set consists of exactly n vectors (Theorem 1.9) and is a basis for $\mathcal{R}^n$ (Theorem 4.5). So A is diagonalizable. Thus *an $n \times n$ matrix A whose characteristic polynomial factors into a product of linear polynomials is diagonalizable if and only if the sum of the dimensions of the eigenspaces of A is equal to n. Furthermore, if A is diagonalizable, then a basis for $\mathcal{R}^n$ consisting of eigenvectors of A can be obtained by taking the union of bases for each eigenspace.* Finally, we examine the condition above in the light of Theorem 5.1. Again let A be an $n \times n$ matrix whose characteristic polynomial factors into a product of factors of the form $t - \lambda_i$. Since each λ_i is an eigenvalue of A, the sum of the multiplicities of the eigenvalues of A is equal to n. By Theorem 5.1, the dimension of the eigenspace of A corresponding to λ_i is less than or equal to the multiplicity of λ_i. Therefore, *the sum of the dimensions of the eigenspaces of A is equal* to n *if and only if, for each eigenvalue* λ *of A, the dimension of the eigenspace corresponding to* λ *is equal to the multiplicity of* λ.

Combining the preceding italicized statements, we obtain the following method for checking if a matrix is diagonalizable.

Test for a Diagonalizable Matrix

An $n \times n$ matrix A is diagonalizable if and only if both of the following conditions are met.

(1) The characteristic polynomial of A factors into a product of linear factors.[5]

(2) For each eigenvalue λ of A, the multiplicity of λ equals the dimension of the corresponding eigenspace, which is $n - \text{rank}\,(A - \lambda I_n)$.

Note that, by Theorem 5.1, condition 2 above is always satisfied for an eigenvalue of multiplicity 1. Hence *condition 2 can fail only for an eigenvalue of multiplicity greater than 1.* Therefore, to check if the previous matrix

$$B = \begin{bmatrix} 2 & 0 & 0 \\ 0 & 4 & 0 \\ 1 & -1 & 2 \end{bmatrix}$$

is diagonalizable, it is necessary to check condition 2 only for the eigenvalue 2 (the one with multiplicity greater than 1). In particular, we note the following important special case.

An $n \times n$ matrix having n distinct eigenvalues is diagonalizable.

Of course, the $n \times n$ identity matrix is diagonalizable but has only one eigenvalue. So the converse of this statement is false.

The discussion preceding the test for a matrix to be diagonalizable also yields a procedure for diagonalizing a matrix.

Algorithm for Matrix Diagonalization

Let A be a diagonalizable $n \times n$ matrix. The union of bases for each eigenspace of A is a basis for $\mathcal{R}^n$ consisting of eigenvectors of A. Therefore, if P is the matrix whose columns are the vectors in this union, then $A = PDP^{-1}$, where D is a diagonal matrix whose diagonal entries are eigenvalues of A corresponding to the respective columns of P.

Example 3 Determine if the matrix

$$B = \begin{bmatrix} 3 & 0 & 0 \\ 0 & 1 & 2 \\ 0 & 2 & 1 \end{bmatrix}$$

is diagonalizable. If so, find an invertible matrix P and a diagonal matrix D such that $B = PDP^{-1}$.

---

[5]It follows from the fundamental theorem of algebra that condition 1 is always satisfied if complex eigenvalues are allowed.

Solution   In Section 5.2, we saw that B has eigenvalues of -1 (with multiplicity 1) and 3 (with multiplicity 2). Because the reduced row echelon form of $B - 3I_3$ is

$$\begin{bmatrix} 0 & 1 & -1 \\ 0 & 0 & 0 \\ 0 & 0 & 0 \end{bmatrix},$$

a matrix of rank 1, the dimension of the eigenspace of B corresponding to 3 is $3 - 1 = 2$. Hence the dimension of this eigenspace equals the multiplicity of the eigenvalue 3. Since 3 is the only eigenvalue of B with multiplicity greater than 1, it follows that B is diagonalizable.

To find an invertible matrix P and a diagonal matrix D such that $B = PDP^{-1}$, we must find a basis for each eigenspace of A. From the preceding reduced row echelon form of $B - 3I_3$, we see that

$$\mathcal{B}_1 = \left\{ \begin{bmatrix} 1 \\ 0 \\ 0 \end{bmatrix}, \begin{bmatrix} 0 \\ 1 \\ 1 \end{bmatrix} \right\}$$

is a basis for the eigenspace of B corresponding to 3. Also, the reduced row echelon form of $B + I_3$ is

$$\begin{bmatrix} 1 & 0 & 0 \\ 0 & 1 & 1 \\ 0 & 0 & 0 \end{bmatrix}.$$

Hence

$$\mathcal{B}_2 = \left\{ \begin{bmatrix} 0 \\ -1 \\ 1 \end{bmatrix} \right\}$$

is a basis for the eigenspace of B corresponding to -1. Therefore, by Theorem 5.3,

$$\mathcal{B}_1 \cup \mathcal{B}_2 = \left\{ \begin{bmatrix} 1 \\ 0 \\ 0 \end{bmatrix}, \begin{bmatrix} 0 \\ 1 \\ 1 \end{bmatrix}, \begin{bmatrix} 0 \\ -1 \\ 1 \end{bmatrix} \right\}$$

is linearly independent. Thus $\mathcal{B}_1 \cup \mathcal{B}_2$ is a basis for $\mathcal{R}^3$ consisting of eigenvectors of B, and so B is diagonalizable by Theorem 5.2. Note that for

$$P = \begin{bmatrix} 1 & 0 & 0 \\ 0 & 1 & -1 \\ 0 & 1 & 1 \end{bmatrix} \quad \text{and} \quad D = \begin{bmatrix} 3 & 0 & 0 \\ 0 & 3 & 0 \\ 0 & 0 & -1 \end{bmatrix},$$

we have

$$PDP^{-1} = \begin{bmatrix} 1 & 0 & 0 \\ 0 & 1 & -1 \\ 0 & 1 & 1 \end{bmatrix} \begin{bmatrix} 3 & 0 & 0 \\ 0 & 3 & 0 \\ 0 & 0 & -1 \end{bmatrix} \begin{bmatrix} 1 & 0 & 0 \\ 0 & 1 & -1 \\ 0 & 1 & 1 \end{bmatrix}^{-1}$$

$$= \begin{bmatrix} 3 & 0 & 0 \\ 0 & 1 & 2 \\ 0 & 2 & 1 \end{bmatrix} = B,$$

as guaranteed by the algorithm for diagonalization.

Example 4   Determine if the matrix

$$A = \begin{bmatrix} -1 & 0 & 0 \\ -4 & -2 & 5 \\ -4 & -5 & 8 \end{bmatrix}$$

is diagonalizable. If so, find an invertible matrix P and a diagonal matrix D such that $A = PDP^{-1}$.

Solution   The characteristic polynomial of A is

$$\det(A - tI_3) = -(t+1)(t-3)^2,$$

and so A has eigenvalues of -1 (with multiplicity 1) and 3 (with multiplicity 2). The reduced row echelon form of $A - 3I_3$ is

$$\begin{bmatrix} 1 & 0 & 0 \\ 0 & 1 & -1 \\ 0 & 0 & 0 \end{bmatrix},$$

a matrix of rank 2. Since $3 - \text{rank}(A - 3I_3) = 3 - 2 = 1$, which does not equal the multiplicity of the eigenvalue 3, the second condition in the test for a matrix to be diagonalizable fails. Therefore A is not diagonalizable.   ○

In Section 5.4, we will consider what it means for a linear operator to be diagonalizable. First, however, we conclude this section by proving Theorem 5.3.

Theorem 5.3
Let $\mathcal{L}_1, \mathcal{L}_2, \ldots, \mathcal{L}_k$ be linearly independent subsets of different eigenspaces of a matrix A. Then $\mathcal{L}_1 \cup \mathcal{L}_2 \cup \cdots \cup \mathcal{L}_k$ is linearly independent.

Proof   We prove the contrapositive: If $\mathcal{L}_1, \mathcal{L}_2, \ldots, \mathcal{L}_k$ are finite subsets of different eigenspaces and $\mathcal{L}_1 \cup \mathcal{L}_2 \cup \cdots \cup \mathcal{L}_k$ is linearly dependent, then some $\mathcal{L}_i$ is linearly dependent. If $\mathcal{L}_1$ is linearly dependent, then we are done; so suppose that $\mathcal{L}_1$ is linearly independent. Let m denote the smallest positive integer such that the set $\mathcal{L}_1 \cup \mathcal{L}_2 \cup \cdots \cup \mathcal{L}_m$ is linearly dependent. Note that since $\mathcal{L}_1$ is linearly independent, we have $m \geq 2$.

Let $\mathcal{L}_1 \cup \mathcal{L}_2 \cup \cdots \cup \mathcal{L}_{m-1} = \{\mathbf{u}_1, \mathbf{u}_2, \ldots, \mathbf{u}_p\}$ and $\mathcal{L}_m = \{\mathbf{v}_1, \mathbf{v}_2, \ldots, \mathbf{v}_q\}$. Since $\mathcal{L}_1 \cup \mathcal{L}_2 \cup \cdots \cup \mathcal{L}_m$ is linearly dependent, there are scalars $a_1, a_2, \ldots, a_p$ and $b_1, b_2, \ldots, b_q$, not all zero, such that

$$a_1\mathbf{u}_1 + a_2\mathbf{u}_2 + \cdots + a_p\mathbf{u}_p + b_1\mathbf{v}_1 + b_2\mathbf{v}_2 + \cdots + b_q\mathbf{v}_q = \mathbf{0}. \qquad (1)$$

Let $\mathbf{w} = b_1\mathbf{v}_1 + b_2\mathbf{v}_2 + \cdots + b_q\mathbf{v}_q$. Then (1) can be rewritten as

$$a_1\mathbf{u}_1 + a_2\mathbf{u}_2 + \cdots + a_p\mathbf{u}_p = -\mathbf{w}. \qquad (2)$$

Since $\mathbf{w}$ is a linear combination of vectors from $\mathcal{L}_m$, $\mathbf{w}$ is an element of some eigenspace of A, say the eigenspace corresponding to eigenvalue λ_0. Moreover, each $\mathbf{u}_i$ is an element of some eigenspace of A; so there exists an eigenvalue λ_i of A such that $A\mathbf{u}_i = \lambda_i\mathbf{u}_i$. Because $\mathcal{L}_1, \mathcal{L}_2, \ldots, \mathcal{L}_k$ are subsets of different eigenspaces, we have $\lambda_i \neq \lambda_0$ for $i = 1, 2, \ldots, p$.

If we multiply $\mathbf{u}_i$ by the matrix $A - \lambda_0 I_n$, we obtain

$$(A - \lambda_0 I_n)\mathbf{u}_i = A\mathbf{u}_i - \lambda_0 I_n \mathbf{u}_i = \lambda_i\mathbf{u}_i - \lambda_0\mathbf{u}_i = (\lambda_i - \lambda_0)\mathbf{u}_i.$$

On the other hand, if we multiply $\mathbf{w}$ by the matrix $A - \lambda_0 I_n$, we obtain

$$(A - \lambda_0 I_n)\mathbf{w} = A\mathbf{w} - \lambda_0 I_n\mathbf{w} = \lambda_0\mathbf{w} - \lambda_0\mathbf{w} = \mathbf{0}.$$

Thus if we multiply both sides of (2) by $A - \lambda_0 I_n$ and use Theorem 1.2, we obtain

$$(A - \lambda_0 I_n)(a_1\mathbf{u}_1 + a_2\mathbf{u}_2 + \cdots + a_p\mathbf{u}_p) = (A - \lambda_0 I_n)(-\mathbf{w})$$

$$a_1(A - \lambda_0 I_n)\mathbf{u}_1 + a_2(A - \lambda_0 I_n)\mathbf{u}_2 + \cdots + a_p(A - \lambda_0 I_n)\mathbf{u}_p$$
$$= -(A - \lambda_0 I_n)(\mathbf{w})$$

$$a_1(\lambda_1 - \lambda_0)\mathbf{u}_1 + a_2(\lambda_2 - \lambda_0)\mathbf{u}_2 + \cdots + a_p(\lambda_p - \lambda_0)\mathbf{u}_p = \mathbf{0}.$$

Because $\mathcal{L}_1 \cup \mathcal{L}_2 \cup \cdots \cup \mathcal{L}_{m-1} = \{\mathbf{u}_1, \mathbf{u}_2, \ldots, \mathbf{u}_p\}$ is linearly independent, we have

$$a_1(\lambda_1 - \lambda_0) = a_2(\lambda_2 - \lambda_0) = \cdots = a_p(\lambda_p - \lambda_0) = 0.$$

But since $\lambda_i \neq \lambda_0$ for each i, it follows that $a_1 = a_2 = \cdots = a_p = 0$. Thus (2) reduces to $\mathbf{0} = -\mathbf{w}$. Hence $\mathbf{w} = \mathbf{0}$, that is,

$$b_1\mathbf{v}_1 + b_2\mathbf{v}_2 + \cdots + b_q\mathbf{v}_q = \mathbf{0}. \tag{3}$$

Because $a_1, a_2, \ldots, a_p, b_1, b_2, \ldots, b_q$ are not all zero but each $a_i = 0$, it follows from (3) that $\mathcal{L}_m$ is linearly dependent. ∎

Practice Problems

1. The characteristic polynomial of the matrix $\begin{bmatrix} -4 & -6 & 0 \\ 3 & 5 & 0 \\ 3 & 3 & 2 \end{bmatrix}$ is $-(t+1)(t-2)^2$. Determine if the matrix is diagonalizable. If so, find an invertible matrix P and a diagonal matrix D such that $A = PDP^{-1}$.

2. The characteristic polynomial of the matrix $\begin{bmatrix} 2 & 2 & 1 \\ 0 & 0 & 3 \\ 0 & -1 & 0 \end{bmatrix}$ is $-(t-2)(t^2+3)$. Determine if the matrix is diagonalizable. If so, find an invertible matrix P and a diagonal matrix D such that $A = PDP^{-1}$.

3. The characteristic polynomial of the matrix $\begin{bmatrix} 5 & 5 & -6 \\ 0 & -1 & 0 \\ 3 & 2 & -4 \end{bmatrix}$ is $-(t-2)(t+1)^2$. Determine if the matrix is diagonalizable. If so, find an invertible matrix P and a diagonal matrix D such that $A = PDP^{-1}$.

Exercises

1. Determine if the following statements are true or false.

   (a) Every $n \times n$ matrix is diagonalizable.
   (b) An $n \times n$ matrix A is diagonalizable if and only if there is a basis for $\mathcal{R}^n$ consisting of eigenvectors of A.
   (c) If P is an invertible $n \times n$ matrix and D is a diagonal $n \times n$ matrix such that $A = PDP^{-1}$, then the columns of P are a basis for $\mathcal{R}^n$ consisting of eigenvectors of A.
   (d) If P is an invertible $n \times n$ matrix and D is a diagonal $n \times n$ matrix such that $A = PDP^{-1}$, then the eigenvalues of A are the diagonal entries of D.
   (e) If A is a diagonalizable $n \times n$ matrix, then there exists a unique diagonal matrix D such that $A = PDP^{-1}$.
   (f) If an $n \times n$ matrix has n distinct eigenvectors, then it is diagonalizable.
   (g) Every diagonalizable $n \times n$ matrix has n distinct eigenvalues.
   (h) If $\mathcal{B}_1, \mathcal{B}_2, \ldots, \mathcal{B}_k$ are bases for distinct eigenspaces of an $n \times n$ matrix A, then $\mathcal{B}_1 \cup \mathcal{B}_2 \cup \cdots \cup \mathcal{B}_k$ is linearly independent.
   (i) If the sum of the multiplicities of the eigenvalues of an $n \times n$ matrix A equals n, then A is diagonalizable.
   (j) If, for each eigenvalue λ of A, the multiplicity of λ equals the dimension of the corresponding eigenspace, then A is diagonalizable.

(k) If A is a diagonalizable 6×6 matrix having two distinct eigenvalues with multiplicities 2 and 4, then the corresponding eigenspaces of A must be two-dimensional and four-dimensional.

(l) If λ is an eigenvalue of A, the dimension of the eigenspace corresponding to λ is the rank of $A - \lambda I_n$.

2. A 4×4 matrix has eigenvalues of $-3, -1, 2$, and 5. Is the matrix diagonalizable? Justify your answer.

3. A 4×4 matrix has eigenvalues of $-3, -1, 2$. Eigenvalue -1 has multiplicity 2.

   (a) Under what conditions is the matrix diagonalizable? Justify your answer.
   (b) Under what conditions is it not diagonalizable? Justify your answer.

4. A 5×5 matrix has eigenvalues of -4 (with multiplicity 3) and 6 (with multiplicity 2). The eigenspace corresponding to eigenvalue 6 has dimension 2.

   (a) Under what conditions is the matrix diagonalizable? Justify your answer.
   (b) Under what conditions is it not diagonalizable? Justify your answer.

5. A 5×5 matrix has eigenvalues of -3 (with multiplicity 4) and 7 (with multiplicity 1).

   (a) Under what conditions is the matrix diagonalizable? Justify your answer.
   (b) Under what conditions is it not diagonalizable? Justify your answer.

6. Let A be a 4×4 matrix with exactly the eigenvalues 2 and 7, and corresponding eigenspaces W_1 and W_2. For each of the following parts, information is given. Either write the characteristic polynomial of A, or state why there is insufficient information to determine the characteristic polynomial.

   (a) $\dim W_1 = 3$
   (b) $\dim W_2 = 2$
   (c) A is diagonalizable and $\dim W_2 = 2$

7. Let A be a 5×5 matrix with exactly the eigenvalues 4, 5, and 8, and corresponding eigenspaces W_1, W_2, and W_3. For each of the following parts, information is given. Either write the characteristic polynomial of A, or state why there is insufficient information to determine the characteristic polynomial.

   (a) $\dim W_1 = 2$ and $\dim W_3 = 2$
   (b) A is diagonalizable and $\dim W_2 = 2$
   (c) A is diagonalizable, $\dim W_1 = 1$, and $\dim W_2 = 2$

8. Let $A = \begin{bmatrix} 1 & -2 \\ 1 & -2 \end{bmatrix}$ and $B = \begin{bmatrix} 2 & 0 \\ 1 & 0 \end{bmatrix}$.

   (a) Show that AB and BA have the same eigenvalues.
   (b) Is AB diagonalizable? Justify your answer.
   (c) Is BA diagonalizable? Justify your answer.

In Exercises 9–20, a matrix A and its characteristic polynomial are given. Find, if possible, an invertible matrix P

and a diagonal matrix D such that $A = PDP^{-1}$. Otherwise, explain why A is not diagonalizable.

9. $\begin{bmatrix} 7 & 6 \\ -1 & 2 \end{bmatrix}$ $(t-4)(t-5)$

10. $\begin{bmatrix} -2 & 7 \\ -1 & 2 \end{bmatrix}$ $t^2 + 3$

11. $\begin{bmatrix} 8 & 9 \\ -4 & -4 \end{bmatrix}$ $(t-2)^2$

12. $\begin{bmatrix} 9 & 15 \\ -6 & -10 \end{bmatrix}$ $t(t+1)$

13. $\begin{bmatrix} 3 & 2 & -2 \\ -8 & 0 & -5 \\ -8 & -2 & -3 \end{bmatrix}$ $-(t+5)(t-2)(t-3)$

14. $\begin{bmatrix} -9 & 8 & -8 \\ -4 & 3 & -4 \\ 2 & -2 & 1 \end{bmatrix}$ $-(t+3)(t+1)^2$

15. $\begin{bmatrix} 3 & -5 & 6 \\ 1 & 3 & -6 \\ 0 & 3 & -5 \end{bmatrix}$ $-(t-1)(t^2+2)$

16. $\begin{bmatrix} -2 & 6 & 3 \\ -2 & -8 & -2 \\ 4 & 6 & -1 \end{bmatrix}$ $-(t+5)(t+4)(t+2)$

17. $\begin{bmatrix} 1 & -2 & 2 \\ 8 & 11 & -8 \\ 4 & 4 & -1 \end{bmatrix}$ $-(t-5)(t-3)^2$

18. $\begin{bmatrix} 5 & 1 & 2 \\ 1 & 4 & 1 \\ -3 & -2 & 0 \end{bmatrix}$ $-(t-3)^3$

19. $\begin{bmatrix} -1 & 0 & 0 & 0 \\ 0 & -1 & 0 & 0 \\ 5 & 5 & 4 & -5 \\ 0 & 0 & 0 & -1 \end{bmatrix}$ $(t+1)^3(t-4)$

20. $\begin{bmatrix} -8 & 0 & -10 & 0 \\ -5 & 2 & -5 & 0 \\ 5 & 0 & 7 & 0 \\ -5 & 0 & -5 & 2 \end{bmatrix}$ $(t+3)(t-2)^3$

In Exercises 21–28, a matrix A is given. Find, if possible, an invertible matrix P and a diagonal matrix D such that $A = PDP^{-1}$. Otherwise, explain why A is not diagonalizable.

21. $\begin{bmatrix} 16 & -9 \\ 25 & -14 \end{bmatrix}$

22. $\begin{bmatrix} -1 & 2 \\ 3 & 4 \end{bmatrix}$

23. $\begin{bmatrix} 6 & 6 \\ -2 & -1 \end{bmatrix}$

24. $\begin{bmatrix} 1 & 5 \\ -1 & -1 \end{bmatrix}$

25. $\begin{bmatrix} -1 & 2 & -1 \\ 0 & -3 & 1 \\ 0 & 0 & 2 \end{bmatrix}$

26. $\begin{bmatrix} -3 & 0 & -5 \\ 0 & 2 & 0 \\ 2 & 0 & 3 \end{bmatrix}$

27. $\begin{bmatrix} 0 & 0 & 0 \\ 1 & 1 & 0 \\ 0 & -1 & 0 \end{bmatrix}$

28. $\begin{bmatrix} 2 & 0 & -1 \\ 1 & 3 & -1 \\ 2 & 0 & 5 \end{bmatrix}$

In Exercises 29–34, an $n \times n$ matrix A, a basis for $\mathcal{R}^n$ consisting of eigenvectors of A, and the corresponding eigenvalues are given. Calculate A^k for an arbitrary positive integer k.

29. $\begin{bmatrix} 2 & 2 \\ -1 & 5 \end{bmatrix}; \left\{ \begin{bmatrix} 1 \\ 1 \end{bmatrix}, \begin{bmatrix} 2 \\ 1 \end{bmatrix} \right\}; 4, 3$

30. $\begin{bmatrix} -4 & 1 \\ -2 & -1 \end{bmatrix}; \left\{ \begin{bmatrix} 1 \\ 2 \end{bmatrix}, \begin{bmatrix} 1 \\ 1 \end{bmatrix} \right\}; -2, -3$

31. $\begin{bmatrix} 5 & 6 \\ -1 & 0 \end{bmatrix}; \left\{ \begin{bmatrix} 2 \\ -1 \end{bmatrix}, \begin{bmatrix} -3 \\ 1 \end{bmatrix} \right\}; 2, 3$

32. $\begin{bmatrix} 7 & 5 \\ -10 & -8 \end{bmatrix}; \left\{ \begin{bmatrix} -1 \\ 2 \end{bmatrix}, \begin{bmatrix} -1 \\ 1 \end{bmatrix} \right\}; -3, 2$

33. $\begin{bmatrix} -3 & -8 & 0 \\ 4 & 9 & 0 \\ 0 & 0 & 5 \end{bmatrix}; \left\{ \begin{bmatrix} -1 \\ 1 \\ 0 \end{bmatrix}, \begin{bmatrix} 0 \\ 0 \\ 1 \end{bmatrix}, \begin{bmatrix} -2 \\ 1 \\ 0 \end{bmatrix} \right\}; 5, 5, 1$

34. $\begin{bmatrix} -1 & 0 & 2 \\ 0 & 2 & 0 \\ -4 & 0 & 5 \end{bmatrix}; \left\{ \begin{bmatrix} 1 \\ 0 \\ 1 \end{bmatrix}, \begin{bmatrix} 0 \\ 1 \\ 0 \end{bmatrix}, \begin{bmatrix} 1 \\ 0 \\ 2 \end{bmatrix} \right\}; 1, 2, 3$

In Exercises 35–44, a matrix and its characteristic polynomial are given. Determine all values of the scalar c for which the given matrix is not diagonalizable.

35. $\begin{bmatrix} 1 & 0 & -1 \\ -2 & c & -2 \\ 2 & 0 & 4 \end{bmatrix}$

$-(t-c)(t-2)(t-3)$

36. $\begin{bmatrix} -7 & -1 & 2 \\ 0 & c & 0 \\ -10 & 3 & 3 \end{bmatrix}$

$-(t-c)(t+3)(t+2)$

37. $\begin{bmatrix} c & 0 & 0 \\ -1 & 1 & 4 \\ 3 & -2 & -1 \end{bmatrix}$

$-(t-c)(t^2+7)$

38. $\begin{bmatrix} 0 & 0 & -2 \\ -4 & c & -4 \\ 4 & 0 & 6 \end{bmatrix}$

$-(t-c)(t-2)(t-4)$

39. $\begin{bmatrix} 1 & -1 & 0 \\ 6 & 6 & 0 \\ 0 & 0 & c \end{bmatrix}$

$-(t-c)(t-3)(t-4)$

40. $\begin{bmatrix} 2 & -4 & -1 \\ 3 & -2 & 1 \\ 0 & 0 & c \end{bmatrix}$

$-(t-c)(t^2+8)$

41. $\begin{bmatrix} -3 & 0 & -2 \\ -6 & c & -2 \\ 1 & 0 & 0 \end{bmatrix}$

$-(t-c)(t+2)(t+1)$

42. $\begin{bmatrix} 3 & 0 & 0 \\ 0 & c & 0 \\ 1 & 0 & -2 \end{bmatrix}$

$-(t-c)(t+2)(t-3)$

43. $\begin{bmatrix} c & -9 & -3 & -15 \\ 0 & -7 & 0 & -6 \\ 0 & 7 & 2 & 13 \\ 0 & 4 & 0 & 3 \end{bmatrix}$

$(t-c)(t+3)(t+1)(t-2)$

44. $\begin{bmatrix} c & 6 & 2 & 10 \\ 0 & -12 & 0 & -15 \\ 0 & -11 & 1 & -15 \\ 0 & 10 & 0 & 13 \end{bmatrix}$

$(t-c)(t+2)(t-1)(t-3)$

45. Find a 2×2 matrix having eigenvalues of -3 and 5, with corresponding eigenvectors of $\begin{bmatrix} 1 \\ 1 \end{bmatrix}$ and $\begin{bmatrix} 1 \\ 3 \end{bmatrix}$.

46. Find a 3×3 matrix having eigenvalues of $3, 2,$ and 2 with corresponding eigenvectors $\begin{bmatrix} 2 \\ 1 \\ 1 \end{bmatrix}, \begin{bmatrix} 1 \\ 0 \\ 1 \end{bmatrix},$ and $\begin{bmatrix} 1 \\ 1 \\ 1 \end{bmatrix}$.

47. Give an example of diagonalizable $n \times n$ matrices A and

B such that $A + B$ is not diagonalizable.

48. Give an example of diagonalizable $n \times n$ matrices A and B such that AB is not diagonalizable.

49. Show that every diagonal $n \times n$ matrix is diagonalizable.

50. (a) Let A be an $n \times n$ matrix having a single eigenvalue c. Show that if A is diagonalizable, then $A = cI_n$.

   (b) Use (a) to explain why $\begin{bmatrix} 2 & 1 \\ 0 & 2 \end{bmatrix}$ is not diagonalizable.

51. If A is a diagonalizable $n \times n$ matrix, prove that A^T is diagonalizable.

52. If A is an invertible matrix that is diagonalizable, prove that A^{-1} is diagonalizable.

53. If A is a diagonalizable $n \times n$ matrix, prove that A^2 is diagonalizable.

54. If A is a diagonalizable $n \times n$ matrix, prove that A^k is diagonalizable for any positive integer k.

55. Suppose that A and B are similar $n \times n$ matrices such that $B = PAP^{-1}$ for some invertible matrix P.

   (a) Show that A is diagonalizable if and only if B is diagonalizable.

   (b) How are the eigenvalues of A related to the eigenvalues of B? Justify your answer.

   (c) How are the eigenvectors of A related to the eigenvectors of B? Justify your answer.

56. A matrix B is called a *cube root* of a matrix A if $B^3 = A$. Prove that every diagonalizable matrix has a cube root.

57. Prove that if a nilpotent matrix is diagonalizable, then it must be the zero matrix. *Hint:* Use Exercise 44 of Section 5.1.

58. Let A be a diagonalizable $n \times n$ matrix. Prove that if $f(t) = a_n t^n + a_{n-1} t^{n-1} + \cdots + a_1 t + a_0$ is the characteristic polynomial of A, then $f(A) = O$, where $f(A) = a_n A^n + a_{n-1} A^{n-1} + \cdots + a_1 A + a_0 I_n$. (This result is called the *Cayley–Hamilton theorem*.[6]) *Hint:* If $A = PDP^{-1}$, show that $f(A) = Pf(D)P^{-1}$.

59. The **trace** of a square matrix is the sum of its diagonal entries.

   (a) Prove that if A is a diagonalizable matrix, then the trace of A equals the sum of the eigenvalues of A. *Hint:* For all $n \times n$ matrices A and B, show that the

---

[6]The Cayley–Hamilton theorem first appeared in 1858. Arthur Cayley (1821–1895) was an English mathematician who contributed greatly to the development of both algebra and geometry. He was one of the first to study matrices, and this work contributed to the development of quantum mechanics. The Irish mathematician William Rowan Hamilton (1805–1865) is perhaps best known for his use of algebra in optics. His 1833 paper first gave a formal structure to ordered pairs of real numbers that yielded the system of complex numbers, and led to his later development of *quaternions*.

trace of AB equals the trace of BA.

(b) Let A be a diagonalizable $n \times n$ matrix with $p(t) = (-1)^n (t-\lambda_1)(t-\lambda_2) \cdots (t-\lambda_n)$ as its characteristic polynomial. Prove that the coefficient of t^{n-1} in $p(t)$ is $(-1)^{n-1}$ times the trace of A.

(c) For A as in (b), what is the constant term of the characteristic polynomial of A?

In Exercises 60–64, use a calculator with matrix capabilities or computer software such as MATLAB to solve the problem. For each of the matrices A in Exercises 60–63, find, if possible, an invertible matrix P and a diagonal matrix D such that $A = PDP^{-1}$. If no such matrices exist, explain why.

60.
$$\begin{bmatrix} 2 & 1 & 1 & 1 \\ 1 & 2 & 1 & 1 \\ -2 & 2 & 2 & 3 \\ 0 & 2 & 1 & 2 \end{bmatrix}$$

61.
$$\begin{bmatrix} -4 & -5 & -7 & -4 \\ -1 & -6 & -4 & -3 \\ 1 & 1 & 1 & 1 \\ 1 & 7 & 5 & 3 \end{bmatrix}$$

62.
$$\begin{bmatrix} 7 & 6 & 24 & -2 & 14 \\ 6 & 5 & 18 & 0 & 12 \\ -8 & -6 & -25 & 2 & -14 \\ -12 & -8 & -36 & 3 & -20 \\ 6 & 4 & 18 & -2 & 9 \end{bmatrix}$$

63.
$$\begin{bmatrix} 4 & 13 & -5 & -29 & -17 \\ -3 & -11 & 0 & 32 & 24 \\ 0 & -3 & 7 & 3 & -3 \\ -2 & -5 & -5 & 18 & 17 \\ 1 & 2 & 5 & -10 & -11 \end{bmatrix}$$

64. Let
$$A = \begin{bmatrix} 1.00 & 4.0 & c \\ 0.16 & 0.0 & 0 \\ 0.00 & -0.5 & 0 \end{bmatrix} \quad \text{and} \quad \mathbf{u} = \begin{bmatrix} 1 \\ 3 \\ -5 \end{bmatrix}.$$

(a) If $c = 8.1$, what appears to happen to the vector $A^m \mathbf{u}$ as m increases?

(b) What are the eigenvalues of A when $c = 8.1$?

(c) If $c = 8.0$, what appears to happen to the vector $A^m \mathbf{u}$ as m increases?

(d) What are the eigenvalues of A when $c = 8.0$?

(e) If $c = 7.9$, what appears to happen to the vector $A^m \mathbf{u}$ as m increases?

(f) What are the eigenvalues of A when $c = 7.9$?

(g) Let B be an $n \times n$ matrix having n distinct eigenvalues, all of which have absolute value less than 1. Let $\mathbf{u}$ be any vector in $\mathcal{R}^n$. Based on the answers to (a) through (f), make a conjecture about the behavior of $B^m \mathbf{u}$ as m increases. Then prove that your conjecture is valid.

Solutions to the Practice Problems

1. The given matrix A has two eigenvalues, -1 (with multiplicity 1) and 2 (with multiplicity 2). Thus the first condition in the test for a diagonalizable matrix is satisfied. Checking the eigenvalue 2 (the eigenvalue with multiplicity greater than 1), we see that the reduced row echelon form of $A - 2I_3$ is

$$\begin{bmatrix} 1 & 1 & 0 \\ 0 & 0 & 0 \\ 0 & 0 & 0 \end{bmatrix}.$$

Since this matrix has nullity $3 - 1 = 2$, the eigenspace of A corresponding to eigenvalue 2 has dimension 2. Thus the second condition in the test for a diagonalizable matrix is satisfied for every eigenvalue with multiplicity greater than 1, and so A is diagonalizable.

From the reduced row echelon form of $A - 2I_3$, we see that the parametric form of the general solution to $(A - 2I_3)\mathbf{x} = \mathbf{0}$ is

$$\begin{bmatrix} x_1 \\ x_2 \\ x_3 \end{bmatrix} = \begin{bmatrix} -x_2 \\ x_2 \\ x_3 \end{bmatrix} = x_2 \begin{bmatrix} -1 \\ 1 \\ 0 \end{bmatrix} + x_3 \begin{bmatrix} 0 \\ 0 \\ 1 \end{bmatrix}.$$

Hence

$$\left\{ \begin{bmatrix} -1 \\ 1 \\ 0 \end{bmatrix}, \begin{bmatrix} 0 \\ 0 \\ 1 \end{bmatrix} \right\}$$

is a basis for the eigenspace of A corresponding to eigenvalue 2. Likewise, from the reduced row echelon form of

$A + I_3$, which is

$$\begin{bmatrix} 1 & 0 & 2 \\ 0 & 1 & -1 \\ 0 & 0 & 0 \end{bmatrix},$$

we see that

$$\left\{ \begin{bmatrix} -2 \\ 1 \\ 1 \end{bmatrix} \right\}$$

is a basis for the eigenspace of A corresponding to eigenvalue -1.

An invertible matrix P such that $A = PDP^{-1}$ is

$$P = \begin{bmatrix} -1 & 0 & -2 \\ 1 & 0 & 1 \\ 0 & 1 & 1 \end{bmatrix},$$

the matrix whose columns are the vectors in the eigenspace bases. The corresponding diagonal matrix D is

$$D = \begin{bmatrix} 2 & 0 & 0 \\ 0 & 2 & 0 \\ 0 & 0 & -1 \end{bmatrix},$$

the one whose diagonal entries are the eigenvalues that correspond to the respective columns of P.

2. Since the characteristic polynomial of this matrix does not factor into linear factors, the first condition in the test

for a diagonalizable matrix is not met. Hence the matrix is not diagonalizable.

3. The given matrix A has two eigenvalues, 2 (with multiplicity 1) and -1 (with multiplicity 2). So the first condition in the test for a matrix to be diagonalizable is met. Checking the eigenvalue -1 (the eigenvalue with multiplicity greater than 1), we see that the reduced row

echelon form of $A + I_3$ is

$$\begin{bmatrix} 1 & 0 & -1 \\ 0 & 1 & 0 \\ 0 & 0 & 0 \end{bmatrix}.$$

Since the rank of this matrix is 2, the second condition in the test for a matrix to be diagonalizable is not met. Hence A is not diagonalizable.

5.4   Diagonalization of Linear Operators

In Section 5.3, we defined a diagonalizable matrix and saw that an $n \times n$ matrix is diagonalizable if and only if there is a basis for $\mathcal{R}^n$ consisting of eigenvectors of the matrix (Theorem 5.2). We now define a linear operator on $\mathcal{R}^n$ to be **diagonalizable** if there is a basis for $\mathcal{R}^n$ consisting of eigenvectors of the operator.

Since the eigenvalues and eigenvectors of a linear operator are the same as those of its standard matrix, the procedure for finding a basis of eigenvectors for a linear operator is the same as that for a matrix. Moreover, a basis of eigenvectors of either the operator or its standard matrix is also a basis of eigenvectors of the other. Therefore *a linear operator is diagonalizable if and only if its standard matrix is diagonalizable.* So the test for a diagonalizable matrix can be translated into the language of linear operators as follows.

Test for a Diagonalizable Linear Operator

A linear operator T on $\mathcal{R}^n$ with standard matrix A is diagonalizable if and only if both of the following conditions are met.

1. The characteristic polynomial of T factors into a product of linear factors.

2. The multiplicity of each eigenvalue of T equals the dimension of the corresponding eigenspace, which is $n - \text{rank}\,(A - \lambda I_n)$.

Example 1   Find, if possible, a basis for $\mathcal{R}^3$ consisting of eigenvectors of the linear operator $T: \mathcal{R}^3 \to \mathcal{R}^3$ defined by

$$T\left(\begin{bmatrix} x_1 \\ x_2 \\ x_3 \end{bmatrix}\right) = \begin{bmatrix} 8x_1 + 9x_2 \\ -6x_1 - 7x_2 \\ 3x_1 + 3x_2 - x_3 \end{bmatrix}.$$

Solution   The standard matrix of T is

$$A = \begin{bmatrix} 8 & 9 & 0 \\ -6 & -7 & 0 \\ 3 & 3 & -1 \end{bmatrix}.$$

Since T is diagonalizable if and only if A is, we must determine the eigenvalues and eigenspaces of A. The characteristic polynomial of A is $-(t+1)^2(t-2)$. Thus the first condition in the test for a diagonalizable linear operator is met, and the eigenvalues of A are -1 (with multiplicity 2) and 2 (with multiplicity 1).

The reduced row echelon form of $A + I_3$ is

$$\begin{bmatrix} 1 & 1 & 0 \\ 0 & 0 & 0 \\ 0 & 0 & 0 \end{bmatrix},$$

and so the eigenspace corresponding to eigenvalue -1 has

$$\mathcal{B}_1 = \left\{ \begin{bmatrix} -1 \\ 1 \\ 0 \end{bmatrix}, \begin{bmatrix} 0 \\ 0 \\ 1 \end{bmatrix} \right\}$$

as a basis. Since -1 is the only eigenvalue of multiplicity greater than 1 and the dimension of its eigenspace equals its multiplicity, A (and hence T) is diagonalizable. To obtain a basis of eigenvectors, we need the reduced row echelon form of $A - 2I_3$, which is

$$\begin{bmatrix} 1 & 0 & -3 \\ 0 & 1 & 2 \\ 0 & 0 & 0 \end{bmatrix}.$$

It follows that

$$\mathcal{B}_2 = \left\{ \begin{bmatrix} 3 \\ -2 \\ 1 \end{bmatrix} \right\}$$

is a basis for the eigenspace of A corresponding to eigenvalue 2. By Theorem 5.3,

$$\mathcal{B}_1 \cup \mathcal{B}_2 = \left\{ \begin{bmatrix} -1 \\ 1 \\ 0 \end{bmatrix}, \begin{bmatrix} 0 \\ 0 \\ 1 \end{bmatrix}, \begin{bmatrix} 3 \\ -2 \\ 1 \end{bmatrix} \right\}$$

is a basis for $\mathcal{R}^3$ consisting of eigenvectors of A. This set is also a basis for $\mathcal{R}^3$ consisting of eigenvectors of T. ○

Example 2   Find, if possible, a basis for $\mathcal{R}^3$ consisting of eigenvectors of the linear operator $T\colon \mathcal{R}^3 \to \mathcal{R}^3$ defined by

$$T\left(\begin{bmatrix} x_1 \\ x_2 \\ x_3 \end{bmatrix} \right) = \begin{bmatrix} -x_1 + x_2 + 2x_3 \\ x_1 - x_2 \\ 0 \end{bmatrix}.$$

Solution   The standard matrix of T is

$$A = \begin{bmatrix} -1 & 1 & 2 \\ 1 & -1 & 0 \\ 0 & 0 & 0 \end{bmatrix}.$$

We must determine if A is diagonalizable. The characteristic polynomial of A is $-t^2(t + 2)$, and so the first condition in the test for a diagonalizable linear operator is met. In this case, A has eigenvalues of 0 (with multiplicity 2) and -2 (with multiplicity 1). As noted in Section 5.3, A is diagonalizable if and only if the eigenspace corresponding to the eigenvalue of multiplicity 2 has dimension 2. Therefore we examine the eigenspace corresponding to eigenvalue 0. Because the reduced row echelon form of $A - 0I_3 = A$ is

$$\begin{bmatrix} 1 & -1 & 0 \\ 0 & 0 & 1 \\ 0 & 0 & 0 \end{bmatrix},$$

we see that the rank of $A - 0I_3$ is 2. Thus the eigenspace corresponding to eigenvalue 0 has dimension 1, and so A, and hence T, is not diagonalizable. ○

Suppose that T is a linear operator on $\mathcal{R}^n$ for which there is a basis $\mathcal{B} = \{\mathbf{v}_1, \mathbf{v}_2, \ldots, \mathbf{v}_n\}$ consisting of eigenvectors of T. Then for each i, we have $T(\mathbf{v}_i) = \lambda_i \mathbf{v}_i$, where λ_i is the eigenvalue corresponding to $\mathbf{v}_i$. Therefore $[T(\mathbf{v}_i)]_\mathcal{B} = \lambda_i \mathbf{e}_i$ for each i, and so

$$[T]_\mathcal{B} = [[T(\mathbf{v}_1)]_\mathcal{B} \quad [T(\mathbf{v}_2)]_\mathcal{B} \quad \cdots \quad [T(\mathbf{v}_n)]_\mathcal{B}] = [\lambda_1 \mathbf{e}_1 \quad \lambda_1 \mathbf{e}_2 \quad \cdots \quad \lambda_n \mathbf{e}_n]$$

is a diagonal matrix. The converse of this result is also true, which explains the use of the term *diagonalizable* for such a linear operator.

> A linear operator T on $\mathcal{R}^n$ is diagonalizable if and only if there is a basis $\mathcal{B}$ for $\mathcal{R}^n$ such that $[T]_\mathcal{B}$, the $\mathcal{B}$-matrix of T, is a diagonal matrix. Such a basis $\mathcal{B}$ must consist of eigenvectors of T.

Recall from Theorem 4.10 that the $\mathcal{B}$-matrix of T is given by $[T]_\mathcal{B} = B^{-1}AB$, where B is the matrix whose columns are the vectors in $\mathcal{B}$ and A is the standard matrix of T. Thus if we take

$$\mathcal{B} = \left\{ \begin{bmatrix} -1 \\ 1 \\ 0 \end{bmatrix}, \begin{bmatrix} 0 \\ 0 \\ 1 \end{bmatrix}, \begin{bmatrix} 3 \\ -2 \\ 1 \end{bmatrix} \right\}$$

in Example 1, we have

$$[T]_\mathcal{B} = B^{-1}AB = \begin{bmatrix} -1 & 0 & 0 \\ 0 & -1 & 0 \\ 0 & 0 & 2 \end{bmatrix},$$

which is the diagonal matrix whose diagonal entries are the eigenvalues corresponding to the respective columns of B.

An Orthogonal Projection Operator

In Chapter 6, we will encounter a particular type of linear operator on $\mathcal{R}^n$ that arises in many applications. We conclude this section by obtaining an explicit formula for an operator of this type. Consider the set

$$W = \left\{ \begin{bmatrix} x_1 \\ x_2 \\ x_3 \end{bmatrix} \in \mathcal{R}^3 : x_1 - x_2 + x_3 = 0 \right\},$$

which is a plane in $\mathcal{R}^3$ containing the origin. Thus W is also a two-dimensional subspace of $\mathcal{R}^3$. Consider the mapping $U: \mathcal{R}^3 \to \mathcal{R}^3$ defined as follows: For each vector $\mathbf{u}$ in $\mathcal{R}^3$, $U(\mathbf{u})$ is the vector in W whose endpoint is the point of intersection of W and the line through $\mathbf{u}$ perpendicular to W (see Figure 5.5). The mapping U is called the *orthogonal projection operator onto W*. It can be shown that U is linear (see Exercise 45).

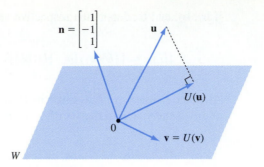

FIGURE 5.5

By applying what we have learned about matrix representations of linear operators, we can obtain the standard matrix of U, which leads to an explicit formula for U. For this purpose, we use a familiar fact about the equation of a plane with equation $ax + by + cz = d$: The vector

$$\begin{bmatrix} a \\ b \\ c \end{bmatrix},$$

whose components are the coefficients of the equation of the plane, is normal (perpendicular) to the plane. Thus for the plane W, the vector

$$\mathbf{n} = \begin{bmatrix} 1 \\ -1 \\ 1 \end{bmatrix}$$

is normal to W (see Figure 5.5). It follows from this that $U(\mathbf{n}) = \mathbf{0}$. Furthermore, for any vector $\mathbf{v}$ in W, the line through $\mathbf{v}$ perpendicular to W intersects W at $\mathbf{v}$ itself, and hence $U(\mathbf{v}) = \mathbf{v}$. Thus $\mathbf{n}$ is an eigenvector of U corresponding to eigenvalue 0, and every nonzero vector in W is an eigenvector of U corresponding to eigenvalue 1. These facts enable us to select a basis for $\mathcal{R}^3$ consisting of eigenvectors of U. The vectors in W have the form

$$\begin{bmatrix} x_1 \\ x_2 \\ x_3 \end{bmatrix} = \begin{bmatrix} x_2 - x_3 \\ x_2 \\ x_3 \end{bmatrix} = x_2 \begin{bmatrix} 1 \\ 1 \\ 0 \end{bmatrix} + x_3 \begin{bmatrix} -1 \\ 0 \\ 1 \end{bmatrix},$$

and so

$$\mathcal{B}_1 = \left\{ \begin{bmatrix} 1 \\ 1 \\ 0 \end{bmatrix}, \begin{bmatrix} -1 \\ 0 \\ 1 \end{bmatrix} \right\}$$

is a basis for W. Since $\mathbf{n}$ is not in W, we can adjoin it to $\mathcal{B}_1$ to obtain the linearly independent subset

$$\mathcal{B} = \left\{ \begin{bmatrix} 1 \\ 1 \\ 0 \end{bmatrix}, \begin{bmatrix} -1 \\ 0 \\ 1 \end{bmatrix}, \begin{bmatrix} 1 \\ -1 \\ 1 \end{bmatrix} \right\}$$

of $\mathcal{R}^3$. Because $\mathcal{B}$ also contains 3 vectors, it must be a basis for $\mathcal{R}^3$. Therefore, if $\mathbf{b}_1$, $\mathbf{b}_2$, and $\mathbf{b}_3$ denote the respective vectors in $\mathcal{B}$, then

$$[U]_{\mathcal{B}} = [\,[U(\mathbf{b}_1)]_{\mathcal{B}} \quad [U(\mathbf{b}_2)]_{\mathcal{B}} \quad [U(\mathbf{b}_3)]_{\mathcal{B}}\,] = \begin{bmatrix} 1 & 0 & 0 \\ 0 & 1 & 0 \\ 0 & 0 & 0 \end{bmatrix}.$$

Also, the $\mathcal{B}$-matrix of U is given by $[U]_{\mathcal{B}} = B^{-1}AB$, where $B = [\mathbf{b}_1 \quad \mathbf{b}_2 \quad \mathbf{b}_3]$ and A is the standard matrix of U. Hence the standard matrix of U is

$$A = B[U]_{\mathcal{B}}B^{-1} = \begin{bmatrix} \frac{2}{3} & \frac{1}{3} & -\frac{1}{3} \\ \frac{1}{3} & \frac{2}{3} & \frac{1}{3} \\ -\frac{1}{3} & \frac{1}{3} & \frac{2}{3} \end{bmatrix}.$$

Thus

$$U\left(\begin{bmatrix} x_1 \\ x_2 \\ x_3 \end{bmatrix} \right) = A \begin{bmatrix} x_1 \\ x_2 \\ x_3 \end{bmatrix} = \begin{bmatrix} \frac{2}{3}x_1 + \frac{1}{3}x_2 - \frac{1}{3}x_3 \\ \frac{1}{3}x_1 + \frac{2}{3}x_2 + \frac{1}{3}x_3 \\ -\frac{1}{3}x_1 + \frac{1}{3}x_2 + \frac{2}{3}x_3 \end{bmatrix}$$

is an explicit formula for U.

Practice Problems

1. The characteristic polynomial of the linear operator

$$T\left(\begin{bmatrix} x_1 \\ x_2 \\ x_3 \end{bmatrix} \right) = \begin{bmatrix} x_1 + 2x_2 + x_3 \\ 2x_2 \\ -x_1 + 2x_2 + 3x_3 \end{bmatrix}$$

is $-(t-2)^3$. Determine if this linear operator is diagonalizable. If so, find a basis for $\mathcal{R}^3$ consisting of eigenvectors of T.

2. Determine if the linear operator

$$T\left(\begin{bmatrix} x_1 \\ x_2 \end{bmatrix} \right) = \begin{bmatrix} -7x_1 - 10x_2 \\ 3x_1 + 4x_2 \end{bmatrix}$$

is diagonalizable. If so, find a basis for $\mathcal{R}^2$ consisting of eigenvectors of T.

Exercises

1. Determine if the following statements are true or false.

   (a) If a linear operator on $\mathcal{R}^n$ is diagonalizable, then its standard matrix is a diagonal matrix.

   (b) For every linear operator on $\mathcal{R}^n$, there is a basis $\mathcal{B}$ for $\mathcal{R}^n$ such that $[T]_{\mathcal{B}}$ is a diagonal matrix.

   (c) A linear operator on $\mathcal{R}^n$ is diagonalizable if and only if its standard matrix is diagonalizable.

   (d) If T is a diagonalizable linear operator on $\mathcal{R}^n$, there is a unique basis $\mathcal{B}$ such that $[T]_{\mathcal{B}}$ is a diagonal matrix.

   (e) If T is a diagonalizable linear operator on $\mathcal{R}^n$, there is a unique diagonal matrix D such that $[T]_{\mathcal{B}} = D$.

   (f) Let W be a two-dimensional subspace of $\mathcal{R}^3$. The orthogonal projection operator onto W is one-to-one.

   (g) Let W be a two-dimensional subspace of $\mathcal{R}^3$. The orthogonal projection operator onto W is onto.

In Exercises 2–6, a linear operator T on $\mathcal{R}^3$ and a basis $\mathcal{B}$ for $\mathcal{R}^3$ are given. Compute $[T]_{\mathcal{B}}$, and determine if $\mathcal{B}$ is a basis for $\mathcal{R}^3$ consisting of eigenvectors of T.

2. $T\left(\begin{bmatrix} x_1 \\ x_2 \\ x_3 \end{bmatrix} \right) = \begin{bmatrix} 7x_1 + 5x_2 + 4x_3 \\ -4x_1 - 2x_2 - 2x_3 \\ -8x_1 - 7x_2 - 5x_3 \end{bmatrix}$,

$$\mathcal{B} = \left\{ \begin{bmatrix} -1 \\ 0 \\ 2 \end{bmatrix}, \begin{bmatrix} -1 \\ -1 \\ 3 \end{bmatrix}, \begin{bmatrix} 1 \\ -2 \\ 1 \end{bmatrix} \right\}$$

3. $T\left(\begin{bmatrix} x_1 \\ x_2 \\ x_3 \end{bmatrix} \right) = \begin{bmatrix} -4x_1 + 2x_2 - 2x_3 \\ -7x_1 - 3x_2 - 7x_3 \\ 7x_1 + x_2 + 5x_3 \end{bmatrix}$,

$$\mathcal{B} = \left\{ \begin{bmatrix} 0 \\ 1 \\ -1 \end{bmatrix}, \begin{bmatrix} -1 \\ 0 \\ 1 \end{bmatrix}, \begin{bmatrix} -1 \\ -1 \\ 1 \end{bmatrix} \right\}$$

4. $T\left(\begin{bmatrix} x_1 \\ x_2 \\ x_3 \end{bmatrix}\right) = \begin{bmatrix} -5x_1 - 2x_2 \\ 5x_1 - 6x_3 \\ 4x_1 + 4x_2 + 7x_3 \end{bmatrix}$,

$\mathcal{B} = \left\{ \begin{bmatrix} 1 \\ -4 \\ 2 \end{bmatrix}, \begin{bmatrix} 0 \\ -1 \\ 1 \end{bmatrix}, \begin{bmatrix} 2 \\ -4 \\ 1 \end{bmatrix} \right\}$

5. $T\left(\begin{bmatrix} x_1 \\ x_2 \\ x_3 \end{bmatrix}\right) = \begin{bmatrix} -3x_1 + 5x_2 - 5x_3 \\ 2x_1 - 3x_2 + 2x_3 \\ 2x_1 - 5x_2 + 4x_3 \end{bmatrix}$,

$\mathcal{B} = \left\{ \begin{bmatrix} -1 \\ 0 \\ 1 \end{bmatrix}, \begin{bmatrix} 0 \\ 1 \\ 1 \end{bmatrix}, \begin{bmatrix} -1 \\ 1 \\ 1 \end{bmatrix} \right\}$

6. $T\left(\begin{bmatrix} x_1 \\ x_2 \\ x_3 \end{bmatrix}\right) = \begin{bmatrix} -x_1 + x_2 + 3x_3 \\ 2x_1 + 6x_3 \\ -x_1 - x_2 - 5x_3 \end{bmatrix}$,

$\mathcal{B} = \left\{ \begin{bmatrix} -2 \\ -1 \\ 1 \end{bmatrix}, \begin{bmatrix} -1 \\ -2 \\ 1 \end{bmatrix}, \begin{bmatrix} -1 \\ -3 \\ 1 \end{bmatrix} \right\}$

In Exercises 7–18, a linear operator T on $\mathcal{R}^n$ and its characteristic polynomial are given. Find, if possible, a basis for $\mathcal{R}^n$ consisting of eigenvectors of T. If no such basis exists, explain why.

7. $T\left(\begin{bmatrix} x_1 \\ x_2 \end{bmatrix}\right) = \begin{bmatrix} 7x_1 - 6x_2 \\ 9x_1 - 7x_2 \end{bmatrix}$

$t^2 + 5$

8. $T\left(\begin{bmatrix} x_1 \\ x_2 \end{bmatrix}\right) = \begin{bmatrix} x_1 + x_2 \\ -9x_1 - 5x_2 \end{bmatrix}$

$(t + 2)^2$

9. $T\left(\begin{bmatrix} x_1 \\ x_2 \end{bmatrix}\right) = \begin{bmatrix} 7x_1 - 5x_2 \\ 10x_1 - 8x_2 \end{bmatrix}$

$(t + 3)(t - 2)$

10. $T\left(\begin{bmatrix} x_1 \\ x_2 \end{bmatrix}\right) = \begin{bmatrix} -7x_1 - 4x_2 \\ 8x_1 + 5x_2 \end{bmatrix}$

$(t + 3)(t - 1)$

11. $T\left(\begin{bmatrix} x_1 \\ x_2 \\ x_3 \end{bmatrix}\right) = \begin{bmatrix} -5x_1 \\ 7x_1 + 2x_2 \\ -7x_1 + x_2 + 3x_3 \end{bmatrix}$

$-(t + 5)(t - 2)(t - 3)$

12. $T\left(\begin{bmatrix} x_1 \\ x_2 \\ x_3 \end{bmatrix}\right) = \begin{bmatrix} -3x_1 \\ 4x_1 + x_2 \\ x_3 \end{bmatrix}$

$-(t + 3)(t - 1)^2$

13. $T\left(\begin{bmatrix} x_1 \\ x_2 \\ x_3 \end{bmatrix}\right) = \begin{bmatrix} -x_1 - x_2 \\ -x_2 \\ x_1 + x_2 \end{bmatrix}$

$-t(t + 1)^2$

14. $T\left(\begin{bmatrix} x_1 \\ x_2 \\ x_3 \end{bmatrix}\right) = \begin{bmatrix} 3x_1 + 2x_2 \\ x_2 \\ 4x_1 - 3x_2 \end{bmatrix}$

$-t(t - 1)(t - 3)$

15. $T\left(\begin{bmatrix} x_1 \\ x_2 \\ x_3 \end{bmatrix}\right) = \begin{bmatrix} 6x_1 - 9x_2 + 9x_3 \\ -3x_2 + 7x_3 \\ 4x_3 \end{bmatrix}$

$-(t + 3)(t - 4)(t - 6)$

16. $T\left(\begin{bmatrix} x_1 \\ x_2 \\ x_3 \end{bmatrix}\right) = \begin{bmatrix} -x_1 \\ -x_2 \\ x_1 - 2x_2 - x_3 \end{bmatrix}$

$-(t + 1)^3$

17. $T\left(\begin{bmatrix} x_1 \\ x_2 \\ x_3 \\ x_4 \end{bmatrix}\right) = \begin{bmatrix} -7x_1 - 4x_2 + 4x_3 - 4x_4 \\ x_2 \\ -8x_1 - 4x_2 + 5x_3 - 4x_4 \\ x_4 \end{bmatrix}$

$(t + 3)(t - 1)^3$

18. $T\left(\begin{bmatrix} x_1 \\ x_2 \\ x_3 \\ x_4 \end{bmatrix}\right) = \begin{bmatrix} 3x_1 - 5x_3 \\ 3x_2 - 5x_3 \\ -2x_3 \\ 5x_3 + 3x_4 \end{bmatrix}$

$(t + 2)(t - 3)^3$

In Exercises 19–26, a linear operator T on $\mathcal{R}^n$ is given. Find, if possible, a basis $\mathcal{B}$ for $\mathcal{R}^n$ such that $[T]_\mathcal{B}$ is a diagonal matrix. If no such basis exists, explain why.

19. $T\left(\begin{bmatrix} x_1 \\ x_2 \end{bmatrix}\right) = \begin{bmatrix} x_1 - x_2 \\ 3x_1 - x_2 \end{bmatrix}$

20. $T\left(\begin{bmatrix} x_1 \\ x_2 \end{bmatrix}\right) = \begin{bmatrix} -x_1 + 3x_2 \\ -4x_1 + 6x_2 \end{bmatrix}$

21. $T\left(\begin{bmatrix} x_1 \\ x_2 \end{bmatrix}\right) = \begin{bmatrix} -2x_1 + 3x_2 \\ 4x_1 - 3x_2 \end{bmatrix}$

22. $T\left(\begin{bmatrix} x_1 \\ x_2 \end{bmatrix}\right) = \begin{bmatrix} 11x_1 - 9x_2 \\ 16x_1 - 13x_2 \end{bmatrix}$

23. $T\left(\begin{bmatrix} x_1 \\ x_2 \\ x_3 \end{bmatrix}\right) = \begin{bmatrix} -x_1 \\ 3x_1 - x_2 + 3x_3 \\ 3x_1 + 2x_3 \end{bmatrix}$

24. $T\left(\begin{bmatrix} x_1 \\ x_2 \\ x_3 \end{bmatrix}\right) = \begin{bmatrix} 4x_1 - 5x_2 \\ -x_2 \\ -x_3 \end{bmatrix}$

25. $T\left(\begin{bmatrix} x_1 \\ x_2 \\ x_3 \end{bmatrix}\right) = \begin{bmatrix} x_1 \\ -x_1 + x_2 - x_3 \\ x_3 \end{bmatrix}$

26. $T\left(\begin{bmatrix} x_1 \\ x_2 \\ x_3 \end{bmatrix}\right) = \begin{bmatrix} 3x_1 - x_2 - 3x_3 \\ 3x_2 - 4x_3 \\ -x_3 \end{bmatrix}$

In Exercises 27–36, a linear operator and its characteristic polynomial are given. Determine all the values of the scalar c for which the given linear operator on $\mathcal{R}^3$ is not diagonalizable.

27. $T\left(\begin{bmatrix} x_1 \\ x_2 \\ x_3 \end{bmatrix}\right) = \begin{bmatrix} 12x_1 + 10x_3 \\ -5x_1 + cx_2 - 5x_3 \\ -5x_1 - 3x_3 \end{bmatrix}$
$-(t - c)(t - 2)(t - 7)$

28. $T\left(\begin{bmatrix} x_1 \\ x_2 \\ x_3 \end{bmatrix}\right) = \begin{bmatrix} x_1 + 2x_2 - x_3 \\ cx_2 \\ 6x_1 - x_2 + 6x_3 \end{bmatrix}$
$-(t - c)(t - 3)(t - 4)$

29. $T\left(\begin{bmatrix} x_1 \\ x_2 \\ x_3 \end{bmatrix}\right) = \begin{bmatrix} cx_1 \\ -x_1 - 3x_2 - x_3 \\ -8x_1 + x_2 - 5x_3 \end{bmatrix}$
$-(t - c)(t + 4)^2$

30. $T\left(\begin{bmatrix} x_1 \\ x_2 \\ x_3 \end{bmatrix}\right) = \begin{bmatrix} -4x_1 + x_2 \\ -4x_2 \\ cx_3 \end{bmatrix}$
$-(t - c)(t + 4)^2$

31. $T\left(\begin{bmatrix} x_1 \\ x_2 \\ x_3 \end{bmatrix}\right) = \begin{bmatrix} cx_1 \\ 2x_1 - 3x_2 + 2x_3 \\ -3x_1 - x_3 \end{bmatrix}$
$-(t - c)(t + 3)(t + 1)$

32. $T\left(\begin{bmatrix} x_1 \\ x_2 \\ x_3 \end{bmatrix}\right) = \begin{bmatrix} -4x_1 - 2x_2 \\ cx_2 \\ 4x_1 + 4x_2 - 2x_3 \end{bmatrix}$
$-(t - c)(t + 4)(t + 2)$

33. $T\left(\begin{bmatrix} x_1 \\ x_2 \\ x_3 \end{bmatrix}\right) = \begin{bmatrix} -5x_1 + 9x_2 + 3x_3 \\ cx_2 \\ -9x_1 + 13x_2 + 5x_3 \end{bmatrix}$
$-(t - c)(t^2 + 2)$

34. $T\left(\begin{bmatrix} x_1 \\ x_2 \\ x_3 \end{bmatrix}\right) = \begin{bmatrix} cx_1 \\ 10x_2 - 2x_3 \\ 6x_2 + 3x_3 \end{bmatrix}$
$-(t - c)(t - 6)(t - 7)$

35. $T\left(\begin{bmatrix} x_1 \\ x_2 \\ x_3 \end{bmatrix}\right) = \begin{bmatrix} -7x_1 + 2x_2 \\ -10x_1 + 2x_2 \\ cx_3 \end{bmatrix}$
$-(t - c)(t + 3)(t + 2)$

36. $T\left(\begin{bmatrix} x_1 \\ x_2 \\ x_3 \end{bmatrix}\right) = \begin{bmatrix} 3x_1 + 7x_3 \\ x_1 + cx_2 + 2x_3 \\ -2x_1 - 3x_3 \end{bmatrix}$
$-(t - c)(t^2 + 5)$

In Exercises 37–42, the equation of a plane W through the origin of $\mathcal{R}^3$ is given. Determine an explicit formula for the orthogonal projection of $\mathcal{R}^3$ onto W.

37. $x + y + z = 0$
38. $2x + y + z = 0$
39. $x + 2y - z = 0$
40. $x + z = 0$
41. $2x + 2y + z = 0$
42. $x + 2y + z = 0$

In Exercises 43–48, the equation of a plane W through the origin of $\mathcal{R}^3$ is given. Determine an explicit formula for the reflection of $\mathcal{R}^3$ about W. (See the Exercises of Section 4.5 for the definition of the reflection of $\mathcal{R}^3$ about W.)

43. $x + y + z = 0$
44. $2x + y + z = 0$
45. $x + 2y - z = 0$
46. $x + z = 0$
47. $2x + 2y + z = 0$
48. $x + 2y + z = 0$

49. Let $\{\mathbf{u}, \mathbf{v}, \mathbf{w}\}$ be a basis for $\mathcal{R}^3$, and let $T\colon \mathcal{R}^3 \to \mathcal{R}^3$ be the function defined by

$$T(a\mathbf{u} + b\mathbf{v} + c\mathbf{w}) = a\mathbf{u} + b\mathbf{v}$$

for all scalars a, b, and c. Prove that T is a linear transformation.

50. Let $\{\mathbf{u}, \mathbf{v}, \mathbf{w}\}$ be a basis for $\mathcal{R}^3$, and let $T\colon \mathcal{R}^3 \to \mathcal{R}^3$ be the function defined by

$$T(a\mathbf{u} + b\mathbf{v} + c\mathbf{w}) = a\mathbf{u} + b\mathbf{v} - c\mathbf{w}$$

for all scalars a, b, and c. Prove that T is a linear transformation.

51. Let T be a linear operator on $\mathcal{R}^n$ and $\mathcal{B}$ be a basis for $\mathcal{R}^n$ such that $[T]_\mathcal{B}$ is a diagonal matrix. Prove that $\mathcal{B}$ must consist of eigenvectors of T.

52. If T and U are diagonalizable linear operators on $\mathcal{R}^n$, must $T + U$ be a diagonalizable linear operator on $\mathcal{R}^n$? Justify your answer.

53. If T is a diagonalizable linear operator on $\mathcal{R}^n$, must cT be a diagonalizable linear operator on $\mathcal{R}^n$ for any scalar c? Justify your answer.

54. If T and U are diagonalizable linear operators on $\mathcal{R}^n$, must TU be a diagonalizable linear operator on $\mathcal{R}^n$? Justify your answer.

55. Let T be a linear operator on $\mathcal{R}^n$, and suppose that $\mathbf{v}_1, \mathbf{v}_2, \ldots, \mathbf{v}_k$ are eigenvectors of T corresponding to distinct nonzero eigenvalues. Prove that the set $\{T(\mathbf{v}_1), T(\mathbf{v}_2), \ldots, T(\mathbf{v}_k)\}$ is linearly independent.

56. Let T and U be diagonalizable linear operators on $\mathcal{R}^n$. Prove that if there is a basis for $\mathcal{R}^n$ consisting of eigenvectors of both T and U, then $TU = UT$.

57. Let T and U be linear operators on $\mathcal{R}^n$. If $T^2 = U$ (where $T^2 = TT$), then T is called a *square root* of U. Show that if U is diagonalizable and has only nonnegative eigenvalues, then U has a square root.

58. Let T be a linear operator on $\mathcal{R}^n$ and $\mathcal{B}_1, \mathcal{B}_2, \ldots, \mathcal{B}_k$ be bases for all the distinct eigenspaces of T. Prove that T is diagonalizable if and only if $\mathcal{B}_1 \cup \mathcal{B}_2 \cup \cdots \cup \mathcal{B}_k$ is a spanning set for $\mathcal{R}^n$.

In Exercises 59 and 60, use a calculator with matrix capabilities or computer software such as MATLAB *to find a basis for* $\mathcal{R}^5$ *consisting of eigenvectors of the linear operator T, or explain why no such basis exists.*

59. T is the linear operator on $\mathcal{R}^5$ defined by

$$T\begin{bmatrix} x_1 \\ x_2 \\ x_3 \\ x_4 \\ x_5 \end{bmatrix} = \begin{bmatrix} -11x_1 - 9x_2 + 13x_3 + 18x_4 - 9x_5 \\ 6x_1 + 5x_2 - 6x_3 - 8x_4 + 4x_5 \\ 6x_1 + 3x_2 - 4x_3 - 6x_4 + 3x_5 \\ -2x_1 + 2x_3 + 3x_4 - 2x_5 \\ 14x_1 + 12x_2 - 14x_3 - 20x_4 + 9x_5 \end{bmatrix}$$

60. T is the linear operator on $\mathcal{R}^5$ defined by

$$T\begin{bmatrix} x_1 \\ x_2 \\ x_3 \\ x_4 \\ x_5 \end{bmatrix} = \begin{bmatrix} -2x_1 - 4x_2 - 9x_3 - 5x_4 - 16x_5 \\ x_1 + 4x_2 + 6x_3 + 5x_4 + 12x_5 \\ 4x_1 + 10x_2 + 20x_3 + 14x_4 + 37x_5 \\ 3x_1 + 2x_2 + 3x_3 + 2x_4 + 4x_5 \\ -4x_1 - 6x_2 - 12x_3 - 8x_4 - 21x_5 \end{bmatrix}$$

Solutions to the Practice Problems

1. Since the characteristic polynomial of T is $-(t-2)^3$, the only eigenvalue of T is 2 (with multiplicity 3). The standard matrix of T is

$$A = \begin{bmatrix} 1 & 2 & 1 \\ 0 & 2 & 0 \\ -1 & 2 & 3 \end{bmatrix},$$

and the reduced row echelon form of $A - 2I_3$ is

$$\begin{bmatrix} 1 & -2 & -1 \\ 0 & 0 & 0 \\ 0 & 0 & 0 \end{bmatrix},$$

a matrix of rank 1. Since $3 - 1 \neq 3$, the second condition in the test for a diagonalizable linear operator is not met, and so T is not diagonalizable.

2. The standard matrix of T is

$$A = \begin{bmatrix} -7 & -10 \\ 3 & 4 \end{bmatrix},$$

and its characteristic polynomial is $(t+1)(t+2)$. So T has two eigenvalues (-2 and -1), each of multiplicity 1. Thus the test for a diagonalizable linear operator guarantees that T is diagonalizable.

To find a basis of eigenvectors of T, we find bases for each of its eigenspaces. Since the reduced row echelon

form of $A + 2I_2$ is

$$\begin{bmatrix} 1 & 2 \\ 0 & 0 \end{bmatrix},$$

we see that

$$\left\{ \begin{bmatrix} -2 \\ 1 \end{bmatrix} \right\}$$

is a basis for the eigenspace of T corresponding to -2. Also, the reduced row echelon form of $A + I_2$ is

$$\begin{bmatrix} 1 & \frac{5}{3} \\ 0 & 0 \end{bmatrix}.$$

Hence

$$\left\{ \begin{bmatrix} -5 \\ 3 \end{bmatrix} \right\}$$

is a basis for the eigenspace of T corresponding to -1. Therefore, by Theorem 5.3,

$$\left\{ \begin{bmatrix} -2 \\ 1 \end{bmatrix}, \begin{bmatrix} -5 \\ 3 \end{bmatrix} \right\}$$

is a basis for $\mathcal{R}^2$ consisting of eigenvectors of T.

5.5*   Applications of Eigenvalues

In this section, we discuss three applications involving eigenvalues. The first, to Markov chains, requires knowledge of Section 5.1 only; the second, to systems of differential equations, requires knowledge of Sections 5.1–5.3. The third application, to difference equations, also requires knowledge of Sections 5.1–5.3. The three applications are independent of one another, and any may be omitted without loss of continuity.

Markov Chains

Markov chains have been used to analyze situations as diverse as land use in Toronto, Canada [3], economic development in New Zealand [6], and the game

*This section can be omitted without loss of continuity.

of Monopoly [1, 2]. This concept is named after the Russian mathematician Andrei Markov (1856–1922), who developed the fundamentals of the theory at the beginning of the twentieth century.

A **Markov chain** is a process that consists of a finite number of **states** and known probabilities p_{ij}, where p_{ij} represents the probability of moving from state j to state i. Note that this probability depends only on the present state j and the future state i. The movement of population between the city and suburbs described in Example 6 of Section 2.1 is an example of a Markov chain with two states (living in the city and living in the suburbs), where p_{ij} represents the probability of moving from one location to another during the coming year. Other possible examples include political affiliation (Democrat, Republican, or independent), where p_{ij} represents the probability of a son having affiliation i if his father has affiliation j; cholesterol level (high, normal, and low), where p_{ij} represents the probability of moving from one level to another in a fixed amount of time; or market share of competing products, where p_{ij} represents the probability that a consumer switches brands in a certain amount of time.

Consider a Markov chain with n states, where the probability of moving from state j to state i during a certain period of time is p_{ij} for $1 \leq i, j \leq n$. The $n \times n$ matrix A with (i, j)-entry equal to p_{ij} is called the **transition matrix** of this Markov chain. It is a stochastic matrix, that is, a matrix with nonnegative entries whose column sums are all 1. A Markov chain often has the property that it is possible to move from any state to any other during some period of time. In such a case, the transition matrix of the Markov chain is said to be **regular**. It can be shown that the transition matrix of a Markov chain is regular if and only if some power of it contains no zero entries. Thus if

$$A = \begin{bmatrix} .5 & 0 & .3 \\ 0 & .4 & .7 \\ .5 & .6 & 0 \end{bmatrix},$$

then A is regular because

$$A^2 = \begin{bmatrix} .40 & .18 & .15 \\ .35 & .58 & .28 \\ .25 & .24 & .57 \end{bmatrix}$$

has no zero entries. On the other hand,

$$B = \begin{bmatrix} .5 & 0 & .3 \\ .0 & 1 & .7 \\ .5 & 0 & 0 \end{bmatrix},$$

is not a regular transition matrix because, for every positive integer k, B^k contains at least one zero entry, for example, the $(1, 2)$-entry.

Suppose that A is the transition matrix of a Markov chain with n states. Recall that a probability vector is a vector with nonnegative components that sum to 1. If $\mathbf{p} \in \mathcal{R}^n$ is a probability vector whose components represent the probability of being in each state of the Markov chain at some time, then the entries of the probability vector $A^m\mathbf{p}$ give the probabilities of being in each state after m time periods.

The behavior of the vectors $A^m\mathbf{p}$, which we saw in Section 5.3, is often of interest in the study of a Markov chain. When A is a regular transition matrix, the behavior of these vectors can be easily described. A proof of the following theorem can be found in [4: 288–89].

Theorem 5.4

If A is a regular $n \times n$ transition matrix and $\mathbf{p} \in \mathcal{R}^n$ is a probability vector, then

(a) 1 is an eigenvalue of A;

(b) there is a unique probability vector $\mathbf{v}$ of A that is also an eigenvector corresponding to eigenvalue 1; and

(c) the vectors $A^m \mathbf{p}$ approach $\mathbf{v}$ for $m = 1, 2, 3, \ldots$.

A probability vector $\mathbf{v}$ such that $A\mathbf{v} = \mathbf{v}$ is called a **steady-state vector**. Such a vector is a probability vector that is also an eigenvector of A corresponding to eigenvalue 1. Theorem 5.4 asserts that a regular Markov chain has a unique steady-state vector and, moreover, the vectors $A^m \mathbf{p}$ approach $\mathbf{v}$ for $m = 1, 2, 3, \ldots$ no matter what the original probability vector $\mathbf{p}$.

To illustrate these ideas, suppose that Amy jogs or rides her bicycle every day for exercise. If she jogs today, then tomorrow she will flip a fair coin and jog if it lands heads or ride her bicycle if it lands tails. If she rides her bicycle today, then she will always jog tomorrow. This situation can be modeled by a Markov chain with two states (jog and ride) and transition matrix

$$
\begin{array}{cc}
 & \text{Today} \\
 & \text{Jog \quad Ride} \\
\text{Tomorrow} \quad \begin{array}{c} \text{Jog} \\ \text{Ride} \end{array} & \begin{bmatrix} .5 & 1 \\ .5 & 0 \end{bmatrix} = A.
\end{array}
$$

For example, the $(1, 1)$-entry of A is .5 because if Amy jogs today, there is a .5 probability that she will jog tomorrow.

Suppose that Amy jogs on Monday. If we take $\mathbf{p} = \begin{bmatrix} 1 \\ 0 \end{bmatrix}$, then

$$
A\mathbf{p} = \begin{bmatrix} .5 & 1 \\ .5 & 0 \end{bmatrix} \begin{bmatrix} 1 \\ 0 \end{bmatrix} = \begin{bmatrix} .5 \\ .5 \end{bmatrix}
$$

and

$$
A^2\mathbf{p} = A(A\mathbf{p}) = \begin{bmatrix} .5 & 1 \\ .5 & 0 \end{bmatrix} \begin{bmatrix} .5 \\ .5 \end{bmatrix} = \begin{bmatrix} .75 \\ .25 \end{bmatrix}.
$$

Therefore Amy will jog with probability .5 or ride her bicycle with probability .5 on Tuesday, and will jog with probability .75 or ride her bicycle with probability .25 on Wednesday.

Since $A^2 = \begin{bmatrix} .75 & .5 \\ .25 & .5 \end{bmatrix}$ has no zero entries, A is a regular transition matrix. Thus A has a unique steady-state vector $\mathbf{v}$, and the vectors $A^m \mathbf{p}$ (for $m = 1, 2, 3, \ldots$) approach $\mathbf{v}$ by Theorem 5.4. The steady-state vector is a solution to $A\mathbf{v} = \mathbf{v}$, that is, to $(A - I_2)\mathbf{v} = \mathbf{0}$. Since the reduced row echelon form of $A - I_2$ is

$$
\begin{bmatrix} 1 & -2 \\ 0 & 0 \end{bmatrix},
$$

we see that the solutions to $(A - I_2)\mathbf{v} = \mathbf{0}$ have the form

$$
v_1 = 2v_2
$$

$$
v_2 \quad \text{free}.
$$

In order that $\mathbf{v} = \begin{bmatrix} v_1 \\ v_2 \end{bmatrix}$ be a probability vector, we must have $v_1 + v_2 = 1$.

Hence

$$2v_2 + \ v_2 = 1$$
$$3v_2 = 1$$
$$v_2 = \tfrac{1}{3}.$$

Thus the unique steady-state vector for A is $\mathbf{v} = \begin{bmatrix} \frac{2}{3} \\ \frac{1}{3} \end{bmatrix}$. Hence, over the long run, Amy jogs $\frac{2}{3}$ of the time and rides her bicycle $\frac{1}{3}$ of the time.

Systems of Differential Equations

The decay of radioactive material and the unrestricted growth of bacteria and other organisms are examples of processes in which the quantity of a substance changes at every instant of time in proportion to the amount present. If $y = f(t)$ represents the amount of such a substance present at time t and k represents the constant of proportionality, then this type of growth is described by the differential equation $f'(t) = kf(t)$ or

$$y' = ky. \tag{4}$$

In calculus, it is shown that the **general solution** to (4) is

$$y = ae^{kt},$$

where a is an arbitrary constant. That is, if we substitute ae^{kt} for y (and its derivative ake^{kt} for y') in (4), we obtain an identity. To find the value of a in the general solution, we need an **initial condition**. For instance, we need to know how much of the substance is present at a particular time, say $t = 0$. If 3 units of the substance are present initially, then $y(0) = 3$. Therefore

$$3 = y(0) = ae^{k(0)} = a \cdot 1 = a,$$

and the **particular solution** to (4) is $y = 3e^{kt}$.

Now suppose that we have a system of three differential equations

$$y_1' = 3y_1$$
$$y_2' = 4y_2$$
$$y_3' = 5y_3.$$

This system is just as easy to solve as (4) because each of the three equations can be solved independently. Its general solution is

$$y_1 = ae^{3t}$$
$$y_2 = be^{4t}$$
$$y_3 = ce^{5t}.$$

Moreover, if there are initial conditions $y_1(0) = 10$, $y_2(0) = 12$, and $y_3(0) = 15$, then the particular solution is given by

$$y_1 = 10e^{3t}$$
$$y_2 = 12e^{4t}$$
$$y_3 = 15e^{5t}.$$

As for systems of linear equations, the system of differential equations above can be represented by the matrix equation

$$\begin{bmatrix} y_1' \\ y_2' \\ y_3' \end{bmatrix} = \begin{bmatrix} 3 & 0 & 0 \\ 0 & 4 & 0 \\ 0 & 0 & 5 \end{bmatrix} \begin{bmatrix} y_1 \\ y_2 \\ y_3 \end{bmatrix}.$$

Letting

$$\mathbf{y} = \begin{bmatrix} y_1 \\ y_2 \\ y_3 \end{bmatrix}, \qquad \mathbf{y}' = \begin{bmatrix} y_1' \\ y_2' \\ y_3' \end{bmatrix}, \qquad \text{and} \qquad D = \begin{bmatrix} 3 & 0 & 0 \\ 0 & 4 & 0 \\ 0 & 0 & 5 \end{bmatrix},$$

we can represent the system as the matrix equation

$$\mathbf{y}' = D\mathbf{y}$$

with initial condition

$$\mathbf{y}(0) = \begin{bmatrix} 10 \\ 12 \\ 15 \end{bmatrix}.$$

More generally, the system of linear differential equations

$$y_1' = a_{11}y_1 + a_{12}y_2 + \cdots + a_{1n}y_n$$
$$y_2' = a_{21}y_1 + a_{22}y_2 + \cdots + a_{2n}y_n$$
$$\vdots$$
$$y_n' = a_{n1}y_1 + a_{n2}y_2 + \cdots + a_{nn}y_n$$

can be written as

$$\mathbf{y}' = A\mathbf{y}, \tag{5}$$

where A is an $n \times n$ matrix. Such a system could describe the numbers of animals of three species that are dependent upon one another, so that the growth rate of each species depends on the present number of animals of all three species. This type of system arises in the context of *predator–prey models*. For example, y_1 and y_2 might represent the numbers of rabbits and foxes in an ecosystem (see Exercise 43) or the numbers of food fish and sharks.

To obtain a solution to (5), we make an appropriate *change of variable*. Define $\mathbf{z} = P^{-1}\mathbf{y}$ (or equivalently, $\mathbf{y} = P\mathbf{z}$), where P is an invertible matrix. It is not hard to prove that $\mathbf{y}' = P\mathbf{z}'$ (see Exercise 42). Therefore, substituting $P\mathbf{z}$ for $\mathbf{y}$ and $P\mathbf{z}'$ for $\mathbf{y}'$ in (5) yields

$$P\mathbf{z}' = AP\mathbf{z}$$

or

$$\mathbf{z}' = P^{-1}AP\mathbf{z}.$$

Thus if there is an invertible matrix P such that $P^{-1}AP$ is a diagonal matrix D, then we obtain the system $\mathbf{z}' = D\mathbf{z}$, which is of the same simple form as the one solved above. Moreover, the solution to (5) can be obtained easily from $\mathbf{z}$ because $\mathbf{y} = P\mathbf{z}$.

Our knowledge of diagonalization enables us to find such a P if A is diagonalizable. We merely choose P to be a matrix whose columns form a basis for $\mathcal{R}^n$ consisting of eigenvectors of A. Of course, the diagonal entries of D are the eigenvalues of A. This method for solving a system of differential equations with a diagonalizable coefficient matrix is summarized below.

Solution to $y' = Ay$ When A is Diagonalizable

1. Find the eigenvalues of A and a basis for each eigenspace.

2. Let P be a matrix whose columns consist of basis vectors for each eigenspace of A, and let D be the diagonal matrix whose diagonal entries are the eigenvalues of A corresponding to the respective columns of P.

3. Solve the system $\mathbf{z}' = D\mathbf{z}$.

4. The solution to the original system is $\mathbf{y} = P\mathbf{z}$.

Consider, for example, the system

$$y_1' = 4y_1 + \ y_2$$
$$y_2' = 3y_1 + 2y_2.$$

The matrix form of this system is $\mathbf{y}' = A\mathbf{y}$, where

$$\mathbf{y} = \begin{bmatrix} y_1 \\ y_2 \end{bmatrix} \quad \text{and} \quad A = \begin{bmatrix} 4 & 1 \\ 3 & 2 \end{bmatrix}.$$

Using the techniques of Section 5.3, we see that A is diagonalizable because it has distinct eigenvalues of 1 and 5. Moreover,

$$\left\{ \begin{bmatrix} -1 \\ 3 \end{bmatrix} \right\} \quad \text{and} \quad \left\{ \begin{bmatrix} 1 \\ 1 \end{bmatrix} \right\}$$

are bases for the corresponding eigenspaces of A. Hence we take

$$P = \begin{bmatrix} -1 & 1 \\ 3 & 1 \end{bmatrix} \quad \text{and} \quad D = \begin{bmatrix} 1 & 0 \\ 0 & 5 \end{bmatrix}.$$

Now solve the system $\mathbf{z}' = D\mathbf{z}$, which is

$$z_1' = \ z_1$$
$$z_2' = 5z_2.$$

The solution to this system is

$$\mathbf{z} = \begin{bmatrix} ae^t \\ be^{5t} \end{bmatrix}.$$

Thus the general solution to the original system is

$$\mathbf{y} = P\mathbf{z} = \begin{bmatrix} -1 & 1 \\ 3 & 1 \end{bmatrix} \begin{bmatrix} ae^t \\ be^{5t} \end{bmatrix} = \begin{bmatrix} -ae^t + be^{5t} \\ 3ae^t + be^{5t} \end{bmatrix},$$

or

$$y_1 = -ae^t + be^{5t}$$
$$y_2 = \ 3ae^t + be^{5t}.$$

Note that it is not necessary to compute P^{-1}.

If, additionally, we are given the initial conditions $y_1(0) = 120$ and $y_2(0) = 40$, then the particular solution to the system can be found. To do so, we must solve the system of linear equations

$$120 = y_1(0) = -ae^0 + be^{5(0)} = -a + b$$
$$40 = y_2(0) = 3ae^0 + be^{5(0)} = 3a + b$$

for a and b. Since the solution to this system of linear equations is $a = -20$ and $b = 100$, the particular solution to the original system of differential equations is

$$y_1 = \quad 20e^t + 100e^{5t}$$
$$y_2 = -60e^t + 100e^{5t}.$$

It should be noted that, with only a slight modification, the procedure we have presented for solving $\mathbf{y}' = A\mathbf{y}$ can also be used to solve a nonhomogeneous system $\mathbf{y}' = A\mathbf{y} + \mathbf{b}$, where $\mathbf{b} \neq \mathbf{0}$.

However, this procedure for solving $\mathbf{y}' = A\mathbf{y}$ cannot be used if A is not diagonalizable. In such a case, a similar technique can be developed utilizing the *Jordan canonical form* of A. Because a discussion of the Jordan canonical form is beyond the scope of this book, such a procedure is not pursued here. An interested reader should see [4: 472].

Under some circumstances, the system of differential equations (5) can be used to solve a higher-order differential equation. We illustrate this technique by solving the third-order differential equation

$$y''' - 6y'' + 5y' + 12y = 0.$$

By making the substitutions $y_1 = y$, $y_2 = y'$, and $y_3 = y''$, we obtain the system

$$y_1' = \qquad\qquad y_2$$
$$y_2' = \qquad\qquad\qquad y_3$$
$$y_3' = -12y_1 - 5y_2 + 6y_3.$$

The matrix form of this system is

$$\begin{bmatrix} y_1' \\ y_2' \\ y_3' \end{bmatrix} = \begin{bmatrix} 0 & 1 & 0 \\ 0 & 0 & 1 \\ -12 & -5 & 6 \end{bmatrix} \begin{bmatrix} y_1 \\ y_2 \\ y_3 \end{bmatrix}.$$

The characteristic equation of the 3×3 matrix

$$A = \begin{bmatrix} 0 & 1 & 0 \\ 0 & 0 & 1 \\ -12 & -5 & 6 \end{bmatrix}$$

is $t^3 - 6t^2 + 5t + 12 = 0$. This equation resembles the original differential equation $y''' - 6y'' + 5y' + 12y = 0$. This resemblance is no accident (see Exercise 44). Thus A has distinct eigenvalues of -1, 3, and 4, and so must be diagonalizable. Using the method previously described, we can solve for y_1, y_2, and y_3. Of course, in this case we are interested only in $y_1 = y$. The general solution to the given third-order equation can be shown to be

$$y = ae^{-t} + be^{3t} + ce^{4t}.$$

Harmonic Motion

Many real-world problems can be modeled by a differential equation that can be solved by the preceding method. Consider, for instance, a body of weight w that is suspended from a spring (see Figure 5.6). Suppose that the body is moved from its resting position and set in motion. Let $y(t)$ denote the distance of the body from its resting point at time t, where positive distances are measured

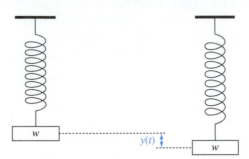

w $y(t)$ w **FIGURE 5.6**

downward. If k is the spring constant, g is the acceleration due to gravity (32 feet per second per second), and $-by'(t)$ is the *damping force*,[7] then the motion of the body satisfies the differential equation

$$\frac{w}{g}y''(t) + by'(t) + ky(t) = 0.$$

For example, if the body weighs 8 pounds, the spring constant is 2.125 pounds per foot, and $b = 0.75$, then the differential equation above reduces to the form

$$y'' + 3y' + 8.5y = 0.$$

By making the substitutions $y_1 = y$ and $y_2 = y'$, we obtain the system

$$y_1' = \qquad\qquad y_2$$
$$y_2' = -8.5y_1 - 3y_2,$$

or, in matrix form,

$$\begin{bmatrix} y_1' \\ y_2' \end{bmatrix} = \begin{bmatrix} 0 & 1 \\ -8.5 & -3 \end{bmatrix} \begin{bmatrix} y_1 \\ y_2 \end{bmatrix}.$$

The characteristic polynomial of the preceding matrix is

$$t^2 + 3t + 8.5,$$

which has nonreal roots of $-1.5 + 2.5i$ and $-1.5 - 2.5i$.

The general solution to the differential equation can be shown to be

$$y = ae^{(-1.5+2.5i)t} + be^{(-1.5-2.5i)t}.$$

Using Euler's formula (see the Appendix), we obtain

$$y = ae^{-1.5t}(\cos 2.5t + i \sin 2.5t) + be^{-1.5t}(\cos 2.5t - i \sin 2.5t),$$

which can be written as

$$y = ce^{-1.5t} \cos 2.5t + de^{-1.5t} \sin 2.5t.$$

---

[7]The damping force is a frictional force that reflects the viscosity of the medium in which the motion occurs. It is proportional to the velocity but acts in the opposite direction.

The constants c and d can be determined from initial conditions, such as the initial displacement of the body from its resting position, $y(0)$, and its initial velocity, $y'(0)$. From this solution, it can be shown that the body oscillates with amplitudes that decrease to zero.

Difference Equations

To introduce difference equations, we begin with a counting problem. This problem is typical of the type that occurs in the study of combinatorial analysis, which has gained considerable attention in recent years because of its applications to computer science and operations research.

Suppose that we have a large number of blocksof three colors: yellow, red, and green. Each yellow block fills one space, and each red or green block fills two spaces. How many different ways are there of arranging the blocks in a line so that they fill n spaces? Denote the answer to this question by r_n, and let Y, R, and G represent a yellow, red, and green block, respectively.

For convenience, we allow for the possibility that $n = 0$ and adopt the useful convention that in this case there is one way to arrange the blocks to fill zero spaces, namely, the "empty" way. So $r_0 = 1$. The following table lists the possible arrangements for this problem in the cases $n = 0, 1, 2$, and 3.

n	Arrangements	r_n
0	empty arrangement	1
1	Y	1
2	YY, R, G	3
3	YYY, YR, YG, RY, GY	5

Now suppose we have to fill n spaces. Consider the three possible cases.

Case 1. The last block is yellow.
In this case, there are $n - 1$ spaces left to fill. The number of ways of doing this is r_{n-1}.

Case 2. The last block is red.
In this case, there are $n - 2$ spaces left to fill. The number of ways of doing this is r_{n-2}.

Case 3. The last block is green.
This is similar to case 2, and so the total number of ways is r_{n-2}.

Putting these cases together, we have

$$r_n = r_{n-1} + 2r_{n-2}. \tag{6}$$

Notice that this equation agrees with the table for $n = 2$, in which case,

$$r_2 = r_1 + 2r_0 = 3 + 2 \cdot 1 = 5.$$

With this formula, we can easily compute the number of arrangements for $n = 4$. In this case

$$r_4 = r_3 + 2r_2 = 5 + 2 \cdot 3 = 11.$$

Equation (6) is an example of a **difference equation** or **recurrence relation**. Difference equations are analogous to differential equations except that the independent variable is treated as discrete in a difference equation and as continuous in a differential equation.

But how do we find a formula expressing r_n as a function of n? One way is to rewrite (6) as a matrix equation. First, write

$$r_{n+1} = r_n + 2r_{n-1}$$

$$r_n = r_n.$$

(The first equation is formed by replacing n by $n + 1$ in (6).) The system can now be written in the matrix form:

$$\begin{bmatrix} r_{n+1} \\ r_n \end{bmatrix} = \begin{bmatrix} 1 & 2 \\ 1 & 0 \end{bmatrix} \begin{bmatrix} r_n \\ r_{n-1} \end{bmatrix},$$

or $\mathbf{s}_n = A\mathbf{s}_{n-1}$, where

$$\mathbf{s}_n = \begin{bmatrix} r_{n+1} \\ r_n \end{bmatrix} \qquad \text{and} \qquad A = \begin{bmatrix} 1 & 2 \\ 1 & 0 \end{bmatrix}.$$

Furthermore, from the solutions for $n = 0$ and $n = 1$, we have that

$$\mathbf{s}_0 = \begin{bmatrix} r_1 \\ r_0 \end{bmatrix} = \begin{bmatrix} 1 \\ 1 \end{bmatrix}.$$

Thus

$$\mathbf{s}_n = A\mathbf{s}_{n-1} = A^2\mathbf{s}_{n-2} = \cdots = A^n\mathbf{s}_0 = A^n \begin{bmatrix} 1 \\ 1 \end{bmatrix}.$$

To compute $\mathbf{s}_n$, we must compute powers of a matrix, a problem we considered in Section 5.3 for the case that the matrix is diagonalizable. Using the methods developed earlier in this chapter, we find matrices

$$P = \begin{bmatrix} 2 & 1 \\ 1 & -1 \end{bmatrix} \qquad \text{and} \qquad D = \begin{bmatrix} 2 & 0 \\ 0 & -1 \end{bmatrix}$$

such that $A = PDP^{-1}$. By the same reasoning as in Section 5.3, we have $A^n = PD^nP^{-1}$. Thus

$$\mathbf{s}_n = PD^nP^{-1}\mathbf{s}_0,$$

or

$$\begin{bmatrix} r_{n+1} \\ r_n \end{bmatrix} = \begin{bmatrix} 2 & 1 \\ 1 & -1 \end{bmatrix} \begin{bmatrix} 2^n & 0 \\ 0 & (-1)^n \end{bmatrix} \begin{bmatrix} \frac{1}{3} & \frac{1}{3} \\ \frac{1}{3} & -\frac{2}{3} \end{bmatrix} \begin{bmatrix} 1 \\ 1 \end{bmatrix}$$

$$= \frac{1}{3} \begin{bmatrix} 2^{n+2} + (-1)^{n-1} \\ 2^{n+1} + (-1)^n \end{bmatrix}.$$

Therefore

$$r_n = \frac{2^{n+1} + (-1)^n}{3}.$$

As a check, observe that this formula gives $r_0 = 1$, $r_1 = 1$, $r_2 = 3$, and $r_3 = 5$, which agrees with the table. It is now easy with a calculator to compute r_n for larger values of n. For example, $r_{10} = 683$, $r_{20} = 669,051$, and r_{32} is almost 3 billion! Needless to say, a complete listing of the various ways that these three kinds of blocks can be arranged in 32 spaces would be impractical.

In general, a **kth-order homogeneous difference equation** (or **recurrence relation**) is an equation of the form

$$r_n = a_{n-1}r_{n-1} + a_{n-2}r_{n-2} + \cdots + a_{n-k}r_{n-k}, \tag{7}$$

where the a_i are scalars, n and k are positive integers such that $n > k$, and $a_{n-k} \neq 0$.

Equation (6) enables us to compute successive values of r_n if we know two consecutive values. In the block problem, we found that $r_0 = 1$ and $r_1 = 1$ by enumerating the possibilities for filling 0 spaces and 1 space. Such a set of consecutive values of r_n is called a set of **initial conditions**. More generally, in (7), we need to know k consecutive values of r_n to have a set of initial conditions. Thus the number of consecutive values required equals the order of the difference equation.

As in the example above, we can represent the kth-order equation (7) by a matrix equation of the form $\mathbf{s}_n = A\mathbf{s}_{n-1}$, where A is a $k \times k$ matrix and $\mathbf{s}_n$ is a vector in $\mathcal{R}^k$ (see Exercise 54). It can be shown (see Exercise 55) that if A has k distinct eigenvalues, $\lambda_1, \lambda_2, \ldots, \lambda_k$, then the general solution has the form

$$r_n = b_1\lambda_1^n + b_2\lambda_2^n + \cdots + b_k\lambda_k^n. \tag{8}$$

The b_i are determined by the initial conditions, which are given by the components of the vector $\mathbf{s}_0$. Furthermore, the λ_i, which are the distinct eigenvalues of A, can also be obtained as solutions to the equation

$$\lambda^k = a_{n-1}\lambda^{k-1} + a_{n-2}\lambda^{k-2} + \cdots + a_{n-k+1}\lambda + a_{n-k} \tag{9}$$

(see Exercise 56). The case in which the eigenvalues are not distinct is not discussed in this text.

Equations (8) and (9) offer us an alternative method for finding r_n *without* computing the eigenvectors of A. We illustrate this method with another example.

It is known that rabbits reproduce at a very rapid rate. For the sake of simplicity, we assume that a pair of rabbits does not produce any offspring during the first month of their lives, but that they produce exactly one pair (male and female) each month thereafter. Suppose that initially we have one male–female pair of newborn rabbits, and that no rabbits die. How many rabbits will there be after n months?

Let r_n denote the number of pairs of rabbits after n months. Let's try to answer this question for $n = 0, 1, 2$, and 3. After zero months, we have only the initial pair. Similarly, after 1 month, we still have only the initial pair. So $r_1 = r_0 = 1$. After 2 months, we have the initial pair and their offspring, that is, $r_2 = 2$. After 3 months, we have what we had before and, in addition, the offspring of the pair that we had over a month ago; that is, $r_3 = r_2 + r_1 = 2 + 1 = 3$. In general, after n months, we have the rabbits we had last month and the offspring of those rabbits that are over 1 month old. Thus

$$r_n = r_{n-1} + r_{n-2}, \tag{10}$$

a second-order difference equation. The numbers generated by (10) are 1, 1, 2, 3, 5, 8, 13, 21, 34, ... Each number is the sum of the preceding two numbers. A sequence with this property is called a **Fibonacci sequence**. It occurs naturally in a variety of settings, including the number of spirals of various plants.

Now we use (8) and (9) to obtain a formula for r_n. By (10) and (9), we have

$$\lambda^2 = \lambda + 1,$$

which has solutions $(1 \pm \sqrt{5})/2$. Thus, by (8), there are scalars b_1 and b_2 such that

$$r_n = b_1 \left(\frac{1}{2} + \frac{\sqrt{5}}{2} \right)^n + b_2 \left(\frac{1}{2} - \frac{\sqrt{5}}{2} \right)^n.$$

To find b_1 and b_2, we use the initial conditions

$$1 = r_0 = \qquad (1)b_1 + \qquad (1)b_2$$
$$1 = r_1 = \left(\tfrac{1}{2} + \tfrac{\sqrt{5}}{2} \right) b_1 + \left(\tfrac{1}{2} - \tfrac{\sqrt{5}}{2} \right) b_2.$$

This system has the solution

$$b_1 = \frac{1}{\sqrt{5}} \left(\frac{1}{2} + \frac{\sqrt{5}}{2} \right) \qquad \text{and} \qquad b_2 = -\frac{1}{\sqrt{5}} \left(\frac{1}{2} - \frac{\sqrt{5}}{2} \right).$$

Thus, in general,

$$r_n = \frac{1}{\sqrt{5}} \left(\frac{1}{2} + \frac{\sqrt{5}}{2} \right)^{n+1} - \frac{1}{\sqrt{5}} \left(\frac{1}{2} - \frac{\sqrt{5}}{2} \right)^{n+1}.$$

This complicated formula should surprise most readers because r_n is a positive integer for every value of n. To find the fiftieth Fibonacci number, we compute r_{50}, which is over 20 billion!

Our final example, which can also by solved using differential equations, involves an application to heat loss.

The water in a hot tub loses heat to the surrounding air so that the difference between the temperature of the water and the temperature of the surrounding air is reduced by 5% each minute. The temperature of the water is 120°F now and the temperature of the surrounding air is a constant 70°F. Let r_n denote the temperature difference at the end of n minutes. Then

$$r_n = .95 r_{n-1} \text{ for each } n \qquad \text{and} \qquad r_0 = 120 - 70 = 50°\text{F}.$$

By (8) and (9), $r_n = b\lambda^n$ and $\lambda = 0.95$, and hence $r_n = b(.95)^n$. Furthermore, $50 = r_0 = b(.95)^0 = b$, and thus $r_n = 50(.95)^n$. For example, at the end of 10 minutes, $r_n = 50(.95)^{10} \approx 30°\text{F}$, and so the water temperature is approximately $70 + 30 = 100°\text{F}$.

In Exercises 57–59, we show how to find a formula for the solution to the first-order nonhomogeneous equation $r_n = a r_{n-1} + c$, where a and c are scalars. This equation occurs frequently in financial applications, such as annuities (see, for example, Exercise 60).

Practice Problems

1. A car survey has found that 80% of those who were driving a car five years ago are now driving a car, 10% are now driving a minivan, and 10% are now driving a sport utility vehicle. Of those who were driving a minivan five years ago, 20% are now driving a car, 70% are now driving a minivan, and 10% are now driving a sport utility vehicle. Finally, of those who were driving a sport utility vehicle five years ago, 10% are now driving a car, 30% are now driving a minivan, and 60% are now driving a sport utility vehicle.

(a) Determine the transition matrix for this Markov chain.

(b) Suppose that 70% of those questioned were driving cars five years ago, 20% were driving minivans, and 10% were driving sport utility vehicles. Estimate the percentage of these persons driving each type of vehicle now.

(c) Under the conditions in (b), estimate the percentage of these persons who will be driving each type of vehicle five years from now.

(d) Determine the percentage of these persons driving each type of vehicle in the long run, assuming that the present trend continues indefinitely.

2. Find the general solution to the following system of differential equations:

$$y_1' = -5y_1 - 4y_2$$
$$y_2' = 8y_1 - 7y_2.$$

3. Find the particular solution to the system of differential equations in problem 5 that satisfies the initial conditions $y_1(0) = 1$ and $y_2(0) = 4$.

4. In the back room of a book store there are a large number of copies of three books: a novel by Nabokov, a novel by Updike, and a calculus book. The novels are each one inch thick, and the calculus book is two inches thick. Find r_n, the number of ways of arranging these copies in a stack n inches high.

Exercises

1. Determine if the following statements are true or false.

(a) The row sums of the transition matrix of a Markov chain are all 1.

(b) If the transition matrix of a Markov chain contains zero entries, then it is not regular.

(c) If A is the transition matrix of a Markov chain and $\mathbf{p}$ is any probability vector, then $A\mathbf{p}$ is a probability vector.

(d) If A is the transition matrix of a Markov chain and $\mathbf{p}$ is any probability vector, then the vectors $A^m\mathbf{p}$ approach a probability vector for $m = 1, 2, 3, \ldots$.

(e) If A is the transition matrix of a regular Markov chain, then for $m = 1, 2, 3, \ldots$ the vectors $A^m\mathbf{p}$ approach the same probability vector for every probability vector $\mathbf{p}$.

(f) The general solution to $y' = ky$ is $y = ke^t$.

(g) If $P^{-1}AP$ is a diagonal matrix D, then the change of variable $\mathbf{z} = P\mathbf{y}$ transforms the matrix equation $\mathbf{y}' = A\mathbf{y}$ into $\mathbf{z}' = D\mathbf{z}$.

(h) A differential equation $a_3 y''' + a_2 y'' + a_1 y' + a_0 y = 0$, where $a_n, \ldots, a_1, a_0$ are scalars, can be written as a system of linear differential equations.

In Exercises 2–5, determine whether the given transition matrix is regular.

2. $\begin{bmatrix} 0 & .5 \\ 1 & .5 \end{bmatrix}$

3. $\begin{bmatrix} 0.25 & 0 \\ 0.75 & 1 \end{bmatrix}$

4. $\begin{bmatrix} .9 & .5 & .4 \\ 0 & .5 & 0 \\ .1 & 0 & .6 \end{bmatrix}$

5. $\begin{bmatrix} .5 & 0 & .7 \\ .5 & 0 & .3 \\ 0 & 1 & 0 \end{bmatrix}$

In Exercises 6–11, a regular transition matrix is given. Determine its steady-state vector.

6. $\begin{bmatrix} .6 & .1 \\ .4 & .9 \end{bmatrix}$

7. $\begin{bmatrix} .9 & .3 \\ .1 & .7 \end{bmatrix}$

8. $\begin{bmatrix} .7 & .1 & .6 \\ 0 & .9 & 0 \\ .3 & 0 & .4 \end{bmatrix}$

9. $\begin{bmatrix} .5 & .1 & .2 \\ .2 & .6 & .1 \\ .3 & .3 & .7 \end{bmatrix}$

10. $\begin{bmatrix} .7 & 0 & .2 \\ 0 & .4 & .8 \\ .3 & .6 & 0 \end{bmatrix}$

11. $\begin{bmatrix} .8 & 0 & .1 \\ 0 & .4 & .9 \\ .2 & .6 & 0 \end{bmatrix}$

12. Suppose that the probability that the child of a college-educated parent also becomes college-educated is 0.75, and that the probability that the child of a non–college-educated parent becomes college-educated is 0.35.

(a) Assuming that the information above describes a Markov chain, write a transition matrix for this situation.

(b) If 30% of parents are college-educated, what (approximate) proportion of the population will be college-educated in one, two, and three generations?

(c) Without any knowledge of the present proportion of college-educated parents, determine the eventual proportion of college-educated people.

13. When Alison goes to her favorite ice cream store, she buys either a root beer float or a chocolate sundae. If she bought a float on her last visit, there is a .25 probability that she will buy a float on her next visit. If she bought a sundae on her last visit, there is a .5 probability that she will buy a float on her next visit.

(a) Assuming that the information above describes a Markov chain, write a transition matrix for this situation.

(b) If Alison bought a sundae on her next-to-last visit, what is the probability that she will buy a float on her next visit?

(c) Over the long run, on what proportion of Alison's trips does she buy a sundae?

14. Suppose that a particular region with a fixed population is divided into three areas: city, suburbs, and country. The probability that a person living in the city moves to the suburbs (in one year) is 0.10 and moves to the country is 0.50. The probability that a person living in the suburbs moves to the city is 0.20 and moves to the country is 0.10. The probability that a person living in the country moves

to the city is 0.20 and moves to the suburbs is 0.20. Suppose initially that 50% of the people live in the city, 30% live in the suburbs, and 20% live in the country.
(a) Determine the transition matrix for the three states.
(b) Determine the percentage of people living in each area after 1, 2, and 3 years.
(c) Use a calculator with matrix capabilities or computer software such as MATLAB to find the percentage of people living in each area after 5 and 8 years.
(d) Determine the eventual percentages of people in each area.

15. A supermarket sells three brands of baking powder. Of those who bought brand A last, 70% will buy brand A the next time they buy baking powder, 10% will buy brand B, and 20% will buy brand C. Of those who bought brand B last, 10% will buy brand A the next time they buy baking powder, 60% will buy brand B, and 30% will buy brand C. Of those who bought brand C last, 10% will buy each of brands A and B the next time they buy baking powder and 80% will buy brand C.

(a) Assuming that the information describes a Markov chain, write a transition matrix for this situation.
(b) If a customer last bought brand B, what is the probability that his or her next purchase of baking powder is brand B?
(c) If a customer last bought brand A, what is the probability that his or her second future purchase of baking powder is brand C?
(d) Over the long run, what proportion of the supermarket's baking powder sales are accounted for by each brand?

16. A company leases rental cars at three Chicago offices (located at Midway Airport, O'Hare Field, and the Loop). Its records show that 60% of the cars rented at Midway are returned there and 20% are returned to each of the other locations. Also, 80% of the cars rented at O'Hare are returned there and 10% are returned to each of the other locations. Finally, 70% of the cars rented at the Loop are returned there, 10% are returned to Midway, and 20% are returned to O'Hare.

(a) Assuming that the information above describes a Markov chain, write a transition matrix for this situation.
(b) If a car is rented at O'Hare, what is the probability that it will be returned to the Loop?
(c) If a car is rented at Midway, what is the probability that it will be returned to the Loop after its second rental?
(d) Over the long run, if all of the cars are returned, what proportion of the company's fleet will be located at each office?

In Exercise 17, use a calculator with matrix capabilities or computer software such as MATLAB.

17. In [5], Gabriel and Neumann found that a Markov chain could be used to describe the occurrence of rainfall at Tel

Aviv during the rainy seasons from 1923–24 to 1949–50. A day was classified as *wet* if at least 0.1 mm of precipitation fell at a certain location in Tel Aviv during the period from 8 a.m. one day to 8 a.m. the next day; otherwise the day was classified as *dry*. The data for November follow.

	Current day wet	Current day dry
Next day wet	117	80
Next day dry	78	535

(a) Assuming that the information above describes a Markov chain, write a transition matrix for this situation.
(b) If a November day was dry, what is the probability that the following day will be dry?
(c) If a Tuesday in November was dry, what is the probability that the following Thursday will be dry?
(d) If a Wednesday in November was wet, what is the probability that the following Saturday will be dry?
(e) Over the long run, what is the probability that a November day in Tel Aviv is wet?

18. Give an example of a 3×3 regular transition matrix A such that A, A^2, and A^3 all contain zero entries.

19. Suppose that the transition matrix of a Markov chain is

$$A = \begin{bmatrix} .90 & .1 & .3 \\ .05 & .8 & .3 \\ .05 & .1 & .4 \end{bmatrix}.$$

(a) What is the probability that an object in the first state will next move to each of the other states?
(b) What is the probability that an object in the second state will next move to each of the other states?
(c) What is the probability that an object in the third state will next move to each of the other states?
(d) Use your answers to (a), (b), and (c) to predict the steady-state vector for A.
(e) Verify your prediction in (d).

20. Use ideas from Exercise 19 to construct two regular 3×3 stochastic matrices having

$$\begin{bmatrix} .4 \\ .2 \\ .4 \end{bmatrix}$$

as their steady-state vector.

21. Let A be an $n \times n$ stochastic matrix, and let **u** be the vector in $\mathcal{R}^n$ with all components equal to 1.
(a) Compute $A^T \mathbf{u}$.
(b) What does (a) imply about the eigenvalues of A^T?
(c) Prove that $\det (A - I_n) = 0$.
(d) What does (c) imply about the eigenvalues of A?

22. Let A be the 2×2 stochastic matrix $\begin{bmatrix} a & 1-b \\ 1-a & b \end{bmatrix}$.

(a) Determine the eigenvalues of A.
(b) Determine a basis for each eigenspace of A.
(c) Under what conditions is A diagonalizable?

23. Prove that if A is a stochastic matrix and **p** is a probability vector, then $A\mathbf{p}$ is a probability vector.

24. Prove that if A and B are $n \times n$ stochastic matrices, then AB is a stochastic matrix.

25. Let A be an $n \times n$ stochastic matrix.

(a) Let $\mathbf{v}$ be any vector in $\mathcal{R}^n$ and k be an index such that $|v_j| \leq |v_k|$ for each j. Prove that the absolute value of every component of $A^T \mathbf{v}$ is less than or equal to $|v_k|$.

(b) Use (a) to show that if $\mathbf{v}$ is an eigenvector of A^T that corresponds to eigenvalue λ, then $|\lambda| \cdot |v_k| \leq |v_k|$.

(c) Deduce that if λ is an eigenvalue of A, then $|\lambda| \leq 1$.

In Exercises 26–31, find the general solution to the given system of differential equations.

26. $y_1' = y_1 + 2y_2$
$y_2' = -y_1 + 4y_2$

27. $y_1' = 3y_1 + 2y_2$
$y_2' = 3y_1 - 2y_2$

28. $y_1' = -5y_1 + 6y_2$
$y_2' = -15y_1 + 14y_2$

29. $y_1' = 2y_1 + 4y_2$
$y_2' = -6y_1 - 8y_2$

30. $y_1' = y_1 + 2y_2 - y_3$
$y_2' = y_1 + y_3$
$y_3' = 4y_1 - 4y_2 + 5y_3$

31. $y_1' = 2y_1$
$y_2' = 3y_1 + 2y_2 + 3y_3$
$y_3' = -3y_1 - y_3$

In Exercises 32–37, find the particular solution to the given system of differential equations that satisfies the given initial conditions.

32. $y_1' = 2y_1 + 2y_2$
$y_2' = -y_1 + 5y_2$
$y_1(0) = 7, \ y_2(0) = 5$

33. $y_1' = y_1 + y_2$
$y_2' = 4y_1 + y_2$
$y_1(0) = 15, \ y_2(0) = -10$

34. $y_1' = -5y_1 - 8y_2$
$y_2' = 4y_1 + 7y_2$
$y_1(0) = 1, \ y_2(0) = -3$

35. $y_1' = 8y_1 + 2y_2$
$y_2' = -4y_1 + 2y_2$
$y_1(0) = 2, \ y_2(0) = 1$

36. $y_1' = y_1 + 2y_3$
$y_2' = 2y_1 + 3y_2 - 2y_3$
$y_3' = 3y_3$
$y_1(0) = -1, \ y_2(0) = 1, \ y_3(0) = 2$

37. $y_1' = 6y_1 - 5y_2 - 7y_3$
$y_2' = y_1 - y_3$
$y_3' = 3y_1 - 3y_2 - 4y_3$
$y_1(0) = 0, \ y_2(0) = 2, \ y_3(0) = 1$

38. Convert the third-order differential equation
$$y''' - 2y'' - 8y' = 0$$
into a system of differential equations, and then find its general solution.

39. Convert the third-order differential equation
$$y''' - 2y'' - y' + 2y = 0$$
into a system of differential equations, and then find the particular solution such that $y(0) = 2$, $y'(0) = -3$, and $y''(0) = 5$.

40. Find the general solution to the differential equation that describes the harmonic motion of a weight of 4 pounds that is attached to a spring, where the spring constant is 1.5 pounds per foot and the damping force constant is $b = 0.5$.

41. Find the general solution to the differential equation that describes the harmonic motion of a weight of 10 pounds that is attached to a spring, where the spring constant is 1.25 pounds per foot and the damping force constant is $b = 0.625$.

42. Let $\mathbf{z}$ be a 3×1 column vector of differentiable functions, and let P be any 3×3 matrix. Prove that if $\mathbf{y} = P\mathbf{z}$, then $\mathbf{y}' = P\mathbf{z}'$.

43. Let y_1 denote the number of rabbits in a certain area at time t and y_2 denote the number of foxes in this area at time t. Suppose that at time 0 there are 900 rabbits and 300 foxes in this area, and assume that the system of differential equations
$$y_1' = 2y_1 - 4y_2$$
$$y_2' = y_1 - 3y_2$$
expresses the rate at which the number of animals of each type changes.

(a) Find the particular solution to this system.

(b) Approximately how many of each species will be present at times $t = 1, 2,$ and 3? For each of these times, compute the ratio of foxes to rabbits.

(c) Approximately what is the eventual ratio of foxes to rabbits in this area? Does this number depend on the initial numbers of rabbits and foxes in the area?

44. Show that the characteristic polynomial of
$$\begin{bmatrix} 0 & 1 & 0 \\ 0 & 0 & 1 \\ -c & -b & -a \end{bmatrix}$$
is $-t^3 - at^2 - bt - c$.

45. Prove that if λ_1, λ_2, and λ_3 are the distinct roots of $t^3 + at^2 + bt + c$, then $y = pe^{\lambda_1 t} + qe^{\lambda_2 t} + re^{\lambda_3 t}$ is the general solution to $y''' + ay'' + by' + cy = 0$. *Hint:* Express the differential equation as a system of differential equations $\mathbf{y}' = A\mathbf{y}$, and show that

$$\left\{ \begin{bmatrix} 1 \\ \lambda_1 \\ \lambda_1^2 \end{bmatrix}, \begin{bmatrix} 1 \\ \lambda_2 \\ \lambda_2^2 \end{bmatrix}, \begin{bmatrix} 1 \\ \lambda_3 \\ \lambda_3^2 \end{bmatrix} \right\}$$

is a basis for $\mathcal{R}^3$ consisting of eigenvectors of A.

In Exercises 46–50, use either of the two methods developed in this section to find a formula for r_n. Then use your result to find r_6.

46. $r_n = 2r_{n-1}; r_0 = 5$

47. $r_n = r_{n-1} + 2r_{n-2}; r_0 = 7$ and $r_1 = 2$

48. $r_n = 3r_{n-1} - 2r_{n-2}; r_0 = 1$ and $r_1 = 3$

49. $r_n = 3r_{n-1} + 4r_{n-2}; r_0 = 1$ and $r_1 = 1$

50. $r_n = 2r_{n-1} + r_{n-2} - 2r_{n-3}; r_0 = 3, r_1 = 1,$ and $r_2 = 3$

51. Suppose that we have a large number of blocks. The blocks are of five colors: red, yellow, green, orange, and blue. Each of the red and yellow blocks weighs one ounce, and each of the green, orange, and blue blocks weighs two ounces. Let r_n be the number of ways the blocks can be arranged in a stack that weighs n ounces.

   (a) Determine $r_0, r_1, r_2,$ and r_3 by listing the possibilities.
   (b) Write the difference equation involving r_n.
   (c) Use (b) to find a formula for r_n.
   (d) Use your answer to (c) to check your answers in (a).

52. Suppose that a bank pays interest of 8% compounded annually on savings. Use the appropriate difference equation to determine how much money would be in a savings account after n years if initially there was $1000 in the account. What is the value of the account after 5 years, 10 years, and 15 years?

53. Write the third-order difference equation

$$r_n = 4r_{n-1} - 2r_{n-2} + 5r_{n-3}$$

   in matrix notation, $\mathbf{s}_n = A\mathbf{s}_{n-1}$, as we did in this section.

54. Justify the matrix form of (7) given in this section: $\mathbf{s}_n = A\mathbf{s}_{n-1}$, where

$$\mathbf{s}_n = \begin{bmatrix} r_{n+k-1} \\ r_{n+k-2} \\ \vdots \\ r_{n+1} \\ r_n \end{bmatrix}$$

and

$$A = \begin{bmatrix} a_{n-1} & a_{n-2} & a_{n-3} & \cdots & a_{n-k+1} & a_{n-k} \\ 1 & 0 & 0 & \cdots & 0 & 0 \\ 0 & 1 & 0 & \cdots & 0 & 0 \\ \vdots & \vdots & \vdots & & \vdots & \vdots \\ 0 & 0 & 0 & \cdots & 1 & 0 \end{bmatrix}.$$

55. Consider a kth-order difference equation of the form (7) with a set of k initial conditions, and let the matrix form

of the equation be $\mathbf{s}_n = A\mathbf{s}_{n-1}$. Suppose, furthermore, that A has k distinct eigenvalues $\lambda_1, \lambda_2, \ldots, \lambda_k$, and that $\mathbf{v}_1, \mathbf{v}_2, \ldots, \mathbf{v}_k$ are corresponding eigenvectors.

   (a) Prove that there exist scalars $t_1, t_2, \ldots, t_k$ such that

$$\mathbf{s}_0 = t_1\mathbf{v}_1 + t_2\mathbf{v}_2 + \cdots + t_k\mathbf{v}_k.$$

   (b) Prove that for any positive integer n,

$$\mathbf{s}_n = \lambda_1^n t_1\mathbf{v}_1 + \lambda_2^n t_2\mathbf{v}_2 + \cdots + \lambda_k^n t_k\mathbf{v}_k.$$

   (c) Derive (8) by comparing the last components of the vector equation in (b).

56. Prove that a scalar λ is an eigenvalue of the matrix A in Exercise 54 if and only if λ is a solution to (9). *Hint:* Let

$$\mathbf{w}_\lambda = \begin{bmatrix} \lambda^{k-1} \\ \lambda^{k-2} \\ \vdots \\ \lambda \\ 1 \end{bmatrix},$$

   and prove each of the following results.

   (i) If $\mathbf{w}_\lambda$ is an eigenvector of A, then λ is a solution to (9).
   (ii) If λ is a solution to (9), then $\mathbf{w}_\lambda$ is an eigenvector of A, and λ is the corresponding eigenvalue.

In Exercises 57–59, we examine the **nonhomogeneous** *first-order difference equation, which is of the form*

$$r_n = ar_{n-1} + c,$$

where a and c are constants. For the purpose of these exercises, we let

$$\mathbf{s}_n = \begin{bmatrix} 1 \\ r_n \end{bmatrix} \quad \text{and} \quad A = \begin{bmatrix} 1 & 0 \\ c & a \end{bmatrix}.$$

57. Prove that $\mathbf{s}_n = A^n\mathbf{s}_0$ for any positive integer n.

58. For this exercise, we assume that $a = 1$.

   (a) Prove that $A^n = \begin{bmatrix} 1 & 0 \\ nc & 1 \end{bmatrix}$ for any positive integer n.
   (b) Use (a) to derive the solution $r_n = r_0 + nc$.

59. For this exercise, we assume that $a \neq 1$.

   (a) Prove that 1 and a are eigenvalues of A.
   (b) Prove that there exist eigenvectors $\mathbf{v}_1$ and $\mathbf{v}_2$ of A corresponding to 1 and a, respectively, and scalars b_1 and b_2 such that

$$\mathbf{s}_0 = b_1\mathbf{v}_1 + b_2\mathbf{v}_2.$$

   (c) Use (b) to prove that there exist scalars b_1 and b_2 such that $r_n = cb_1 + b_2a^n$, and show that

$$b_1 = \frac{-1}{a-1} \quad \text{and} \quad b_2 = r_0 + \frac{c}{a-1}.$$

60. An investor opened a savings account on March 1 with an initial deposit of $5000. Each year thereafter, he

added $2000 to the account on March 1. If the account earns interest at the rate of 6% per year, find a formula for the value of this account after n years. *Hint:* Use Exercise 59(c).

In Exercises 61 and 62, use a calculator with matrix capabilities or computer software such as MATLAB *to solve the problem.*

61. Solve the system of differential equations

$$y_1' = 3.2y_1 + 4.1y_2 + 7.7y_3 + 3.7y_4$$
$$y_2' = -0.3y_1 + 1.2y_2 + 0.2y_3 + 0.5y_4$$
$$y_3' = -1.8y_1 - 1.8y_2 - 4.4y_3 - 1.8y_4$$
$$y_4' = 1.7y_1 - 0.7y_2 + 2.9y_3 + 0.4y_4$$

subject to the initial conditions $y_1(0) = 1$, $y_2(0) = -4$, $y_3(0) = 2$, $y_4(0) = 3$.

62. In [3], Bourne examined the changes in land use in Toronto, Canada during the years 1950–62. Land was classified in the following ten ways.

1. low-density residential   2. high-density residential
3. office   4. general commercial
5. automobile commercial   6. parking
7. warehousing   8. industry
9. transportation   10. vacant

The following transition matrix shows the changes in land use from 1952 to 1962.

		1	2	3	4	5	6	7	8	9	10
	1	.13	.02	.00	.02	.00	.08	.01	.01	.01	.25
	2	.34	.41	.07	.01	.00	.05	.03	.02	.18	.08
	3	.10	.05	.43	.09	.11	.14	.02	.02	.14	.03
	4	.04	.04	.05	.30	.07	.08	.12	.03	.04	.03
Use in 1962	5	.04	.00	.01	.09	.70	.12	.03	.03	.10	.05
	6	.22	.04	.28	.27	.06	.39	.11	.08	.39	.15
	7	.03	.00	.14	.05	.00	.04	.38	.18	.03	.22
	8	.02	.00	.00	.08	.01	.00	.21	.61	.03	.13
	9	.00	.00	.00	.01	.00	.01	.01	.00	.08	.00
	10	.08	.44	.02	.08	.05	.09	.08	.02	.00	.06

(Column group header: Use in 1952)

Assume that the trend in land-use changes from 1952 to 1962 continues indefinitely.

(a) Suppose that at some time the percentage of land use for each purpose is as follows: 10%, 20%, 25%, 0%, 0%, 5%, 15%, 10%, 10%, and 5%, respectively. What percentage of land will be used for each purpose two decades later?

(b) Show that the transition matrix is regular.

(c) After many decades, what percentage of land will be used for each purpose?

Solutions to the Practice Problems

1. (a) The Markov chain has three states, which correspond to the type of vehicle driven—car, van, or sport utility vehicle (suv). The transition matrix for this Markov chain is

$$\text{Now} \quad \begin{array}{c} \text{car} \\ \text{van} \\ \text{suv} \end{array} \begin{bmatrix} .8 & .2 & .1 \\ .1 & .7 & .3 \\ .1 & .1 & .6 \end{bmatrix} = A.$$

(Column header: Five years ago — car van suv)

(b) The probability vector that gives the probability of driving each type of vehicle five years ago is

$$\mathbf{p} = \begin{bmatrix} .70 \\ .20 \\ .10 \end{bmatrix}.$$

Hence the probability that someone in the survey is now driving each type of vehicle is given by

$$A\mathbf{p} = \begin{bmatrix} .8 & .2 & .1 \\ .1 & .7 & .3 \\ .1 & .1 & .6 \end{bmatrix} \begin{bmatrix} .70 \\ .20 \\ .10 \end{bmatrix} = \begin{bmatrix} .61 \\ .24 \\ .15 \end{bmatrix}.$$

Therefore 61% of those surveyed are now driving cars, 24% are now driving minivans, and 15% are now driving sport utility vehicles.

(c) Five years from now the probability that someone in the survey will be driving each type of vehicle is given by

$$A(A\mathbf{p}) = \begin{bmatrix} .8 & .2 & .1 \\ .1 & .7 & .3 \\ .1 & .1 & .6 \end{bmatrix} \begin{bmatrix} .61 \\ .24 \\ .15 \end{bmatrix} = \begin{bmatrix} .551 \\ .274 \\ .175 \end{bmatrix}.$$

So in five years, we estimate that 55.1% of those surveyed will drive cars, 27.4% minivans, and 17.5% sport utility vehicles.

(d) Note that A is a regular transition matrix, and so by Theorem 5.4 A has a steady-state vector $\mathbf{v}$. This vector is a solution to $A\mathbf{v} = \mathbf{v}$, that is, to the equation $(A - I_3)\mathbf{v} = \mathbf{0}$. Since the reduced row echelon form of A is

$$\begin{bmatrix} 1 & 0 & -2.25 \\ 0 & 1 & -1.75 \\ 0 & 0 & 0 \end{bmatrix},$$

we see that the solutions to $(A - I_3)\mathbf{v} = \mathbf{0}$ have the form

$$v_1 = 2.25v_3$$
$$v_2 = 1.75v_3$$
$$v_3 \quad \text{free.}$$

In order for $\mathbf{v} = \begin{bmatrix} v_1 \\ v_2 \\ v_3 \end{bmatrix}$ to be a probability vector, we must have $v_1 + v_2 + v_3 = 1$. Hence

$$2.25v_3 + 1.75v_3 + v_3 = 1$$
$$5v_3 = 1$$
$$v_3 = .2.$$

So $v_1 = .45, v_2 = .35$, and $v_3 = .2$. Thus, in the long run, we expect that 45% of those surveyed will drive cars, 35% minivans, and 20% sport utility vehicles.

2. The matrix form of the given system of differential equations is $\mathbf{y}' = A\mathbf{y}$, where

$$\mathbf{y} = \begin{bmatrix} y_1 \\ y_2 \end{bmatrix} \quad \text{and} \quad A = \begin{bmatrix} -5 & -4 \\ 8 & 7 \end{bmatrix}.$$

The characteristic polynomial of A is $(t + 1)(t - 3)$, and so A has eigenvalues of -1 and 3. Since each eigenvalue of A has multiplicity 1, A is diagonalizable. In the usual manner, we find that

$$\left\{ \begin{bmatrix} -1 \\ 1 \end{bmatrix} \right\} \quad \text{and} \quad \left\{ \begin{bmatrix} -1 \\ 2 \end{bmatrix} \right\}$$

are bases for the eigenspaces of A. Take

$$P = \begin{bmatrix} -1 & -1 \\ 1 & 2 \end{bmatrix} \quad \text{and} \quad D = \begin{bmatrix} -1 & 0 \\ 0 & 3 \end{bmatrix}.$$

Then the change of variable $\mathbf{y} = P\mathbf{z}$ transforms $\mathbf{y}' = A\mathbf{y}$ into $\mathbf{z}' = D\mathbf{z}$, which is

$$z_1' = -z_1$$
$$z_2' = 3z_2.$$

Hence

$$\mathbf{z} = \begin{bmatrix} z_1 \\ z_2 \end{bmatrix} = \begin{bmatrix} ae^{-t} \\ be^{3t} \end{bmatrix}.$$

Therefore the general solution to the given system of differential equations is

$$\mathbf{y} = P\mathbf{z} = \begin{bmatrix} -1 & -1 \\ 1 & 2 \end{bmatrix} \begin{bmatrix} ae^{-t} \\ be^{3t} \end{bmatrix} = \begin{bmatrix} -ae^{-t} - be^{3t} \\ ae^{-t} + 2be^{3t} \end{bmatrix},$$

that is,

$$y_1 = -ae^{-t} - be^{3t}$$
$$y_2 = ae^{-t} + 2be^{3t}.$$

3. To satisfy the initial conditions $y_1(0) = 1$ and $y_2(0) = 4$, the constants a and b must satisfy the system of linear equations

$$1 = y_1(0) = -ae^0 - be^{3(0)} = -a - b$$
$$4 = y_2(0) = ae^0 + 2be^{3(0)} = a + 2b.$$

It is easily checked that $a = -6$ and $b = 5$, and so the desired particular solution is

$$y_1 = 6e^{-t} - 5e^{3t}$$
$$y_2 = -6e^{-t} + 10e^{3t}.$$

4. As in the block example discussed earlier, $r_0 = 1$ because there is one "empty" stack. Furthermore, $r_1 = 2$ because each of the two novels is one inch high. Now suppose we have to pile a stack n inches high. There are three cases to consider.

Case 1. The bottom book is the novel by Nabokov.
Since this novel is one inch thick, there are r_{n-1} ways of stacking the rest of the books.

Case 2. The bottom book is the novel by Updike.
This is similar to case 1, and so there are r_{n-1} ways of stacking the rest of the books.

Case 3. The bottom book is a calculus book.
Since this book is two inches thick, there are r_{n-2} ways of stacking the rest of the books.

Combining these three cases, we have the second-order difference equation

$$r_n = 2r_{n-1} + r_{n-2}.$$

We use (9) to obtain

$$\lambda^2 = 2\lambda + 1,$$

which has solutions $1 \pm \sqrt{2}$. Thus, by (8), there are scalars b_1 and b_2 such that

$$r_n = b_1(1 + \sqrt{2})^n + b_2(1 - \sqrt{2})^n.$$

To find b_1 and b_2, we use the initial conditions

$$1 = r(0) = (1)b_1 + (1)b_2$$
$$2 = r(1) = (1 + \sqrt{2})b_1 + (1 - \sqrt{2})b_2.$$

This system has the solution

$$b_1 = \frac{\sqrt{2} + 1}{2\sqrt{2}} \quad \text{and} \quad b_2 = \frac{\sqrt{2} - 1}{2\sqrt{2}}.$$

Thus, in general,

$$r_n = \frac{1}{2\sqrt{2}} \left[(1 + \sqrt{2})^{n+1} - (1 - \sqrt{2})^{n+1} \right].$$

Chapter 5 Review Exercises

1. Determine if the following statements are true or false.

   (a) A scalar λ is an eigenvalue of an $n \times n$ matrix A if and only if $\det(A - \lambda I_n) = 0$.
   (b) If λ is an eigenvalue of a matrix, then there is a unique eigenvector of the matrix that corresponds to λ.
   (c) If $\mathbf{v}$ is an eigenvector of a matrix, then there is a unique eigenvalue of the matrix that corresponds to $\mathbf{v}$.
   (d) The eigenspace of an $n \times n$ matrix A corresponding to eigenvalue λ is the null space of $A - \lambda I_n$.

(e) The eigenvalues of a linear operator on $\mathcal{R}^n$ are the same as those of its standard matrix.

(f) The eigenspaces of a linear operator on $\mathcal{R}^n$ are the same as those of its standard matrix.

(g) Every linear operator on $\mathcal{R}^n$ has real eigenvalues.

(h) Every $n \times n$ matrix has n distinct eigenvalues.

(i) Every diagonalizable $n \times n$ matrix has n distinct eigenvalues.

(j) If two $n \times n$ matrices have the same characteristic polynomial, then they have the same eigenvectors.

(k) The multiplicity of an eigenvalue need not equal the dimension of the corresponding eigenspace.

(l) An $n \times n$ matrix A is diagonalizable if and only if there is a basis for $\mathcal{R}^n$ consisting of eigenvectors of A.

(m) If P is an invertible $n \times n$ matrix and D is a diagonal $n \times n$ matrix such that $A = P^{-1}DP$, then the columns of P are a basis for $\mathcal{R}^n$ consisting of eigenvectors of A.

(n) If P is an invertible $n \times n$ matrix and D is a diagonal $n \times n$ matrix such that $A = PDP^{-1}$, then the eigenvalues of A are the diagonal entries of D.

(o) If λ is an eigenvalue of an $n \times n$ matrix A, then the dimension of the eigenspace corresponding to λ is the nullity of $A - \lambda I_n$.

(p) A linear operator on $\mathcal{R}^n$ is diagonalizable if and only if its standard matrix is diagonalizable.

2. Show that $\begin{bmatrix} 1 & 2 \\ -3 & -2 \end{bmatrix}$ has no real eigenvalues.

In Exercises 3–6, determine the eigenvalues of the given matrix and a basis for each eigenspace.

3. $\begin{bmatrix} 5 & 6 \\ -2 & -2 \end{bmatrix}$

4. $\begin{bmatrix} 1 & -9 \\ 1 & -5 \end{bmatrix}$

5. $\begin{bmatrix} -2 & 0 & 2 \\ 1 & -1 & 0 \\ 0 & 0 & -2 \end{bmatrix}$

6. $\begin{bmatrix} -1 & 0 & 0 \\ 1 & 0 & 1 \\ -1 & -1 & -2 \end{bmatrix}$

In Exercises 6–9, a matrix A is given. Find, if possible, an invertible matrix P and a diagonal matrix D such that $A = PDP^{-1}$. If no such matrices exist, explain why.

7. $\begin{bmatrix} 1 & 2 \\ -3 & 8 \end{bmatrix}$

8. $\begin{bmatrix} -1 & 1 \\ -1 & -3 \end{bmatrix}$

9. $\begin{bmatrix} 1 & 0 & 0 \\ -2 & 0 & 1 \\ 2 & -1 & -2 \end{bmatrix}$

10. $\begin{bmatrix} -2 & 0 & 0 \\ -4 & 2 & 0 \\ 4 & -3 & -1 \end{bmatrix}$

In Exercises 11–14, a linear operator T on $\mathcal{R}^n$ is given. Find, if possible, a basis for $\mathcal{R}^n$ consisting of eigenvectors of T. If no such basis exists, explain why.

11. $T\left(\begin{bmatrix} x_1 \\ x_2 \end{bmatrix} \right) = \begin{bmatrix} 4x_1 + 2x_2 \\ -4x_1 - 5x_2 \end{bmatrix}$

12. $T\left(\begin{bmatrix} x_1 \\ x_2 \end{bmatrix} \right) = \begin{bmatrix} x_1 - 2x_2 \\ 4x_1 - x_2 \end{bmatrix}$

13. $T\left(\begin{bmatrix} x_1 \\ x_2 \\ x_3 \end{bmatrix} \right) = \begin{bmatrix} 2x_1 \\ 2x_2 \\ -3x_1 + 3x_2 - x_3 \end{bmatrix}$

14. $T\left(\begin{bmatrix} x_1 \\ x_2 \\ x_3 \end{bmatrix} \right) = \begin{bmatrix} x_1 \\ 3x_1 + x_2 - 3x_3 \\ 3x_1 - 2x_3 \end{bmatrix}$

In Exercises 15–18, a matrix and its characteristic polynomial are given. Determine all values of the scalar c for which the given matrix is not diagonalizable.

15. $\begin{bmatrix} 1 & 0 & 1 \\ 0 & c & 0 \\ -2 & 0 & 4 \end{bmatrix}$, $-(t-c)(t-2)(t-3)$

16. $\begin{bmatrix} 5 & 1 & -3 \\ 0 & c & 0 \\ 6 & 2 & -4 \end{bmatrix}$, $-(t-c)(t+1)(t-2)$

17. $\begin{bmatrix} c & -1 & 2 \\ 0 & -10 & -8 \\ 0 & 12 & 10 \end{bmatrix}$, $-(t-c)(t-2)(t+2)$

18. $\begin{bmatrix} 3 & 1 & 0 \\ -1 & 1 & 0 \\ 0 & 0 & c \end{bmatrix}$, $-(t-c)(t-2)^2$

In Exercises 19–20, find A^k for an arbitrary positive integer k.

19. $\begin{bmatrix} 5 & -6 \\ 3 & -4 \end{bmatrix}$

20. $\begin{bmatrix} 11 & 8 \\ -12 & -9 \end{bmatrix}$

21. Let T be the linear operator on $\mathcal{R}^3$ defined by

$$T\left(\begin{bmatrix} x_1 \\ x_2 \\ x_3 \end{bmatrix} \right) = \begin{bmatrix} -4x_1 - 3x_2 - 3x_3 \\ -x_2 \\ 6x_1 + 6x_2 + 5x_3 \end{bmatrix}.$$

Find a basis $\mathcal{B}$ such that $[T]_\mathcal{B}$ is a diagonal matrix.

22. Find a 3×3 matrix having eigenvalues of $-1, 2$, and 3 with corresponding eigenvectors $\begin{bmatrix} -1 \\ 1 \\ 1 \end{bmatrix}$, $\begin{bmatrix} -2 \\ 1 \\ 2 \end{bmatrix}$, and $\begin{bmatrix} -1 \\ 1 \\ 2 \end{bmatrix}$.

23. Prove that $\begin{bmatrix} a & 1 & 0 \\ 0 & a & 0 \\ 0 & 0 & b \end{bmatrix}$ is not diagonalizable for any scalars a and b.

24. Suppose that A is an $n \times n$ matrix having two distinct eigenvalues, λ_1 and λ_2, where λ_1 has multiplicity 1. State and prove a necessary and sufficient condition for A to be diagonalizable.

25. Prove that $I_n - A$ is invertible if and only if 1 is *not* an eigenvalue of A.

26. Two $n \times n$ matrices A and B are called *simultaneously diagonalizable* if there exists an invertible matrix P such that both $P^{-1}AP$ and $P^{-1}BP$ are diagonal matrices. Prove that if A and B are simultaneously diagonalizable, then $AB = BA$.

27. Let T be a linear operator on $\mathcal{R}^n$, $\mathcal{B}$ be a basis for $\mathcal{R}^n$, and A be the standard matrix of T. Prove that $[T]_\mathcal{B}$ and A have the same characteristic polynomial.

28. Let T be a linear operator on $\mathcal{R}^n$. A subspace W of $\mathcal{R}^n$ is called *T-invariant* if $T(\mathbf{w})$ is in W for each $\mathbf{w}$ in W. Prove that if V is an eigenspace of T, then V is T-invariant.

CHAPTER 6

Orthogonality

Thus far, we have focused our attention on two operations with vectors—namely, addition and scalar multiplication. In this chapter, we consider such geometric concepts as *length* and *perpendicularity* of vectors. By combining the geometry of vectors with matrices and linear transformations, we obtain powerful techniques for solving a wide variety of problems. For example, we apply these new tools to such areas as least-squares approximation, the graphing of conic sections, and computer graphics. The key to most of these solutions is the construction of a basis of perpendicular eigenvectors for a given matrix or linear transformation.

To accomplish this construction, we show how to convert any basis for a subspace of $\mathcal{R}^n$ into one in which all of the vectors are perpendicular to each other. Once this is done, we determine conditions that guarantee that there is a basis for $\mathcal{R}^n$ consisting of perpendicular eigenvectors of a matrix or a linear transformation. Surprisingly, for a matrix, a necessary and sufficient condition that such a basis exists is that the matrix is symmetric.

6.1   The Geometry of Vectors

In this section, we introduce the concepts of length and perpendicularity of vectors in $\mathcal{R}^n$. Many familiar geometric properties seen in earlier courses extend to this more general space. In particular, the Pythagorean theorem, which relates the squared lengths of sides of a right triangle, also holds in high-dimensional subspaces. To show that many of these results hold in $\mathcal{R}^n$, we define and develop the notion of *dot product*. The dot product is fundamental in the sense that, from it, we can define length and perpendicularity.

Perhaps the most basic concept of geometry is length. In Figure 6.1(a), an application of the Pythagorean theorem suggests that we define the *length* of the vector **u** to be $\sqrt{u_1^2 + u_2^2}$. This definition in $\mathcal{R}^2$ easily extends to $\mathcal{R}^n$.

Definitions.   Let **v** be any vector in $\mathcal{R}^n$. The **norm** **(length)** of **v**, denoted $\|\mathbf{v}\|$, is defined by

$$\|\mathbf{v}\| = \sqrt{v_1^2 + v_2^2 + \cdots + v_n^2}.$$

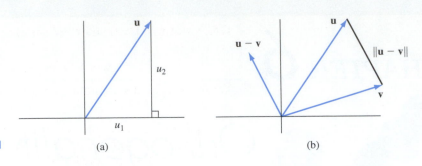

FIGURE 6.1   (a)   (b)

The **distance** between two vectors **u** and **v** in $\mathcal{R}^n$ is defined by $\|\mathbf{u} - \mathbf{v}\|$ (see Figure 6.1(b)).

Example 1   Find $\|\mathbf{u}\|$, $\|\mathbf{v}\|$, and the distance between **u** and **v** if

$$\mathbf{u} = \begin{bmatrix} 1 \\ 2 \\ 3 \end{bmatrix} \quad \text{and} \quad \mathbf{v} = \begin{bmatrix} 2 \\ -3 \\ 0 \end{bmatrix}.$$

Solution   By definition, $\|\mathbf{u}\| = \sqrt{1^2 + 2^2 + 3^2} = \sqrt{14}$, $\|\mathbf{v}\| = \sqrt{2^2 + (-3)^2 + 0^2} = \sqrt{13}$, and $\|\mathbf{u} - \mathbf{v}\| = \sqrt{(1-2)^2 + (2-(-3))^2 + (3-0)^2} = \sqrt{35}$.   ○

Just as we used the Pythagorean theorem in $\mathcal{R}^2$ to motivate the definition of the norm of a vector, we again use this theorem to examine what it means for two vectors **u** and **v** in $\mathcal{R}^2$ to be perpendicular. According to the Pythagorean theorem (see Figure 6.2), we see that **u** and **v** are perpendicular if and only if

$$\|\mathbf{v} - \mathbf{u}\|^2 = \|\mathbf{u}\|^2 + \|\mathbf{v}\|^2$$
$$(v_1 - u_1)^2 + (v_2 - u_2)^2 = u_1^2 + u_2^2 + v_1^2 + v_2^2$$
$$v_1^2 - 2u_1v_1 + u_1^2 + v_2^2 - 2u_2v_2 + u_2^2 = u_1^2 + u_2^2 + v_1^2 + v_2^2$$
$$-2u_1v_1 - 2u_2v_2 = 0$$
$$u_1v_1 + u_2v_2 = 0.$$

The expression $u_1v_1 + u_2v_2$ in the last equation is called the *dot product* of **u** and **v**, and is denoted by **u** · **v**. So **u** and **v** are perpendicular if and only if their dot product equals zero. It is this definition that we extend to vectors in $\mathcal{R}^n$.

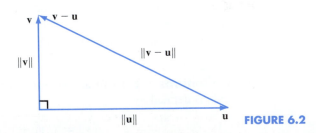

FIGURE 6.2

Definitions. Let **u** and **v** be vectors in $\mathcal{R}^n$. The **dot product** of **u** and **v** is defined by

$$\mathbf{u} \cdot \mathbf{v} = u_1 v_1 + u_2 v_2 + \cdots + u_n v_n.$$

We say that **u** and **v** are **orthogonal (perpendicular)** if $\mathbf{u} \cdot \mathbf{v} = 0$.

Notice that in $\mathcal{R}^n$ the dot product of two vectors is a scalar and that the dot product of **0** with every vector is zero. Hence **0** is orthogonal to every vector in $\mathcal{R}^n$. Also, as noted above, the property of being orthogonal in $\mathcal{R}^2$ is equivalent to the usual geometric definition of perpendicularity.

Example 2   Let

$$\mathbf{u} = \begin{bmatrix} 2 \\ -1 \\ 3 \end{bmatrix}, \qquad \mathbf{v} = \begin{bmatrix} 1 \\ 4 \\ -2 \end{bmatrix}, \qquad \text{and} \qquad \mathbf{w} = \begin{bmatrix} -8 \\ 3 \\ 2 \end{bmatrix}.$$

Determine which pairs of these vectors are orthogonal.

Solution   We need only check which pairs have dot products equal to zero.

$$\mathbf{u} \cdot \mathbf{v} = (2)(1) + (-1)(4) + (3)(-2) = -8$$

$$\mathbf{u} \cdot \mathbf{w} = (2)(-8) + (-1)(3) + (3)(2) = -13$$

$$\mathbf{v} \cdot \mathbf{w} = (1)(-8) + (4)(3) + (-2)(2) = 0$$

We see that **v** and **w** are the only orthogonal vectors.

It is useful to observe that the dot product of **u** and **v** can also be represented as the matrix product $\mathbf{u}^T \mathbf{v}$.

$$\mathbf{u}^T \mathbf{v} = \begin{bmatrix} u_1 & u_2 & \cdots & u_n \end{bmatrix} \begin{bmatrix} v_1 \\ v_2 \\ \vdots \\ v_n \end{bmatrix} = u_1 v_1 + u_2 v_2 + \cdots + u_n v_n = \mathbf{u} \cdot \mathbf{v}.$$

Notice that we have treated the 1×1 matrix $\mathbf{u}^T \mathbf{v}$ as a scalar by writing it as $u_1 v_1 + u_2 v_2 + \cdots + u_n v_n$ instead of $[u_1 v_1 + u_2 v_2 + \cdots + u_n v_n]$.

One useful consequence of identifying a dot product as a matrix product is that it enables us to move a matrix from one side of a dot product to the other. More precisely, if A is an $m \times n$ matrix, **u** is in $\mathcal{R}^n$, and **v** is in $\mathcal{R}^m$, then

$$A\mathbf{u} \cdot \mathbf{v} = \mathbf{u} \cdot A^T \mathbf{v}.$$

This follows because

$$A\mathbf{u} \cdot \mathbf{v} = (A\mathbf{u})^T \mathbf{v} = \mathbf{u}^T (A^T \mathbf{v}) = \mathbf{u} \cdot A^T \mathbf{v}.$$

Just as there are arithmetic properties of vector addition and scalar multiplication, there is an arithmetic for the dot product and norm.

Theorem 6.1

Let **u**, **v**, *and* **w** *be vectors in* $\mathcal{R}^n$ *and* c *be a scalar.*

(a) $\mathbf{u} \cdot \mathbf{u} = \|\mathbf{u}\|^2$.

(b) $\mathbf{u} \cdot \mathbf{u} = 0$ if and only if $\mathbf{u} = \mathbf{0}$.

(c) $\mathbf{u} \cdot \mathbf{v} = \mathbf{v} \cdot \mathbf{u}$.

(d) $\mathbf{u} \cdot (\mathbf{v} + \mathbf{w}) = \mathbf{u} \cdot \mathbf{v} + \mathbf{u} \cdot \mathbf{w}$.

(e) $(\mathbf{v} + \mathbf{w}) \cdot \mathbf{u} = \mathbf{v} \cdot \mathbf{u} + \mathbf{w} \cdot \mathbf{u}$.

(f) $(c\mathbf{u}) \cdot \mathbf{v} = c(\mathbf{u} \cdot \mathbf{v}) = \mathbf{u} \cdot (c\mathbf{v})$.

(g) $\|c\mathbf{u}\| = |c| \, \|\mathbf{u}\|$.

Proof We prove (d) and (g) and leave the rest as exercises.

(d) Using matrix properties, we have

$$\mathbf{u} \cdot (\mathbf{v} + \mathbf{w}) = \mathbf{u}^T (\mathbf{v} + \mathbf{w})$$

$$= \mathbf{u}^T \mathbf{v} + \mathbf{u}^T \mathbf{w}$$

$$= \mathbf{u} \cdot \mathbf{v} + \mathbf{u} \cdot \mathbf{w}.$$

(g) By (a) and (f), we have

$$\|c\mathbf{u}\|^2 = (c\mathbf{u}) \cdot (c\mathbf{u})$$

$$= c^2 \mathbf{u} \cdot \mathbf{u}$$

$$= c^2 \|\mathbf{u}\|^2.$$

By taking the square root of both sides and using $\sqrt{c^2} = |c|$, we are finished.

Because of Theorem 6.1(f), there is no ambiguity in writing $c\mathbf{u} \cdot \mathbf{v}$ for any of the three expressions in (f).

This theorem allows us to treat expressions with dot products and norms just as we would algebraic expressions. For example, compare the similarity of the algebraic result

$$(2x + 3y)^2 = 4x^2 + 12xy + 9y^2$$

with

$$\|2\mathbf{u} + 3\mathbf{v}\|^2 = 4\|\mathbf{u}\|^2 + 12\mathbf{u} \cdot \mathbf{v} + 9\|\mathbf{v}\|^2.$$

The proof of the preceding result relies heavily on Theorem 6.1:

$$\|2\mathbf{u} + 3\mathbf{v}\|^2 = (2\mathbf{u} + 3\mathbf{v}) \cdot (2\mathbf{u} + 3\mathbf{v}) \qquad \text{by (a)}$$

$$= (2\mathbf{u}) \cdot (2\mathbf{u} + 3\mathbf{v}) + (3\mathbf{v}) \cdot (2\mathbf{u} + 3\mathbf{v}) \qquad \text{by (e)}$$

$$= (2\mathbf{u}) \cdot (2\mathbf{u}) + (2\mathbf{u}) \cdot (3\mathbf{v}) + (3\mathbf{v}) \cdot (2\mathbf{u}) + (3\mathbf{v}) \cdot (3\mathbf{v}) \qquad \text{by (d)}$$

$$= 4(\mathbf{u} \cdot \mathbf{u}) + 6(\mathbf{u} \cdot \mathbf{v}) + 6(\mathbf{v} \cdot \mathbf{u}) + 9(\mathbf{v} \cdot \mathbf{v}) \qquad \text{by (f)}$$

$$= 4\|\mathbf{u}\|^2 + 6(\mathbf{u} \cdot \mathbf{v}) + 6(\mathbf{u} \cdot \mathbf{v}) + 9\|\mathbf{v}\|^2 \qquad \text{by (a) and (c)}$$

$$= 4\|\mathbf{u}\|^2 + 12(\mathbf{u} \cdot \mathbf{v}) + 9\|\mathbf{v}\|^2.$$

As noted, by (f) we can write the last expression as $4\|\mathbf{u}\|^2 + 12\mathbf{u} \cdot \mathbf{v} + 9\|\mathbf{v}\|^2$. From this point on, we will omit the steps above when computing with dot products and norms. *Caution*: Symbols such as $\mathbf{u}^2$ and $\mathbf{uv}$ are *not* defined.

It is easy to extend (d) and (e) to linear combinations, namely,

$$\mathbf{u} \cdot (c_1 \mathbf{v}_1 + c_2 \mathbf{v}_2 + \cdots + c_p \mathbf{v}_p) = c_1 \mathbf{u} \cdot \mathbf{v}_1 + c_2 \mathbf{u} \cdot \mathbf{v}_2 + \cdots + c_p \mathbf{u} \cdot \mathbf{v}_p$$

and

$$(c_1 \mathbf{v}_1 + c_2 \mathbf{v}_2 + \cdots + c_p \mathbf{v}_p) \cdot \mathbf{u} = c_1 \mathbf{v}_1 \cdot \mathbf{u} + c_2 \mathbf{v}_2 \cdot \mathbf{u} + \cdots + c_p \mathbf{v}_p \cdot \mathbf{u}.$$

As an application of these arithmetic properties, we show that the Pythagorean theorem holds in $\mathcal{R}^n$.

Theorem 6.2   (Pythagorean theorem in $\mathcal{R}^n$)

Let $\mathbf{u}$ *and* $\mathbf{v}$ *be vectors in* $\mathcal{R}^n$. *Then* $\mathbf{u}$ *and* $\mathbf{v}$ *are orthogonal if and only if*

$$\|\mathbf{u} + \mathbf{v}\|^2 = \|\mathbf{u}\|^2 + \|\mathbf{v}\|^2.$$

Proof Applying the arithmetic of dot products and norms to the vectors $\mathbf{u}$ and $\mathbf{v}$, we have

$$\|\mathbf{u} + \mathbf{v}\|^2 = \|\mathbf{u}\|^2 + 2\mathbf{u} \cdot \mathbf{v} + \|\mathbf{v}\|^2.$$

Because $\mathbf{u}$ and $\mathbf{v}$ are orthogonal if and only if $\mathbf{u} \cdot \mathbf{v} = 0$, the result follows immediately. $\square$

Recall that a *rhombus* is a parallelogram with all sides of equal length. We use Theorem 6.1 to prove the following result from geometry:

The diagonals of a parallelogram are orthogonal if and only if the parallelogram is a rhombus.

The diagonals of the rhombus are $\mathbf{u} + \mathbf{v}$ and $\mathbf{u} - \mathbf{v}$ (see Figure 6.3). Applying the arithmetic of dot products and norms, we obtain

$$(\mathbf{u} + \mathbf{v}) \cdot (\mathbf{u} - \mathbf{v}) = \|\mathbf{u}\|^2 - \|\mathbf{v}\|^2.$$

From this result, we see that the diagonals are orthogonal if and only if the preceding dot product is zero. This occurs if and only if $\|\mathbf{u}\|^2 = \|\mathbf{v}\|^2$, that is, if the sides have equal lengths.

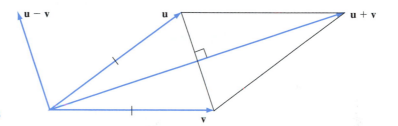

FIGURE 6.3

Orthogonal Projection of a Vector onto a Line

In Sections 2.6 and 5.4, we considered orthogonal projections of $\mathcal{R}^3$ onto a plane W. Now we use properties of the dot product and norm to obtain a formula for the orthogonal projection onto a line through the origin in $\mathcal{R}^2$. With this result in hand, we obtain a simple method for finding the distance of a point from a line. In Section 6.3, we extend this result to obtain the *distance* from a point to any subspace of $\mathcal{R}^n$. The problem of finding the distance from a point to a line or a plane is not only fundamental in geometry, but has wide applications in statistics (as we see later).

Suppose we have a line $\mathcal{L}$ through the origin and a point P in $\mathcal{R}^2$ not on $\mathcal{L}$. Let $\mathbf{v}$ be the vector with endpoint P. The vector $\mathbf{w}$ whose endpoint is the point of intersection of $\mathcal{L}$ and the line through P perpendicular to $\mathcal{L}$ (see Figure 6.4) is called the **orthogonal projection of v onto** $\mathcal{L}$. To find $\mathbf{w}$ in terms of $\mathbf{v}$ and $\mathcal{L}$, let $\mathbf{u}$ be *any* nonzero vector along $\mathcal{L}$, and let $\mathbf{z} = \mathbf{v} - \mathbf{w}$. Then $\mathbf{w} = c\mathbf{u}$ for some scalar c. Notice that $\mathbf{z}$ and $\mathbf{u}$ are orthogonal, that is,

$$0 = \mathbf{z} \cdot \mathbf{u} = (\mathbf{v} - \mathbf{w}) \cdot \mathbf{u} = (\mathbf{v} - c\mathbf{u}) \cdot \mathbf{u} = \mathbf{v} \cdot \mathbf{u} - c\mathbf{u} \cdot \mathbf{u} = \mathbf{v} \cdot \mathbf{u} - c\|\mathbf{u}\|^2.$$

So $c = \dfrac{\mathbf{v} \cdot \mathbf{u}}{\|\mathbf{u}\|^2}$, and thus $\mathbf{w} = \dfrac{\mathbf{v} \cdot \mathbf{u}}{\|\mathbf{u}\|^2}\mathbf{u}.$

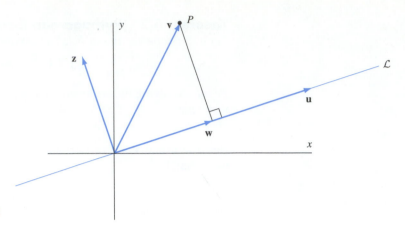

FIGURE 6.4

To find the distance from P to $\mathcal{L}$, we compute the distance from $\mathbf{v}$ to $\mathbf{w}$, the orthogonal projection of $\mathbf{v}$ onto $\mathcal{L}$. That is, we compute $\left\| \mathbf{v} - \dfrac{\mathbf{v} \cdot \mathbf{u}}{\|\mathbf{u}\|^2} \mathbf{u} \right\|$.

To illustrate these ideas, suppose that we want to find the distance from the point $(4, 1)$ to the line whose equation is $y = \frac{1}{2}x$. Following our preceeding derivation, we let

$$\mathbf{v} = \begin{bmatrix} 4 \\ 1 \end{bmatrix}, \qquad \mathbf{u} = \begin{bmatrix} 2 \\ 1 \end{bmatrix}, \qquad \text{and} \qquad \frac{\mathbf{v} \cdot \mathbf{u}}{\|\mathbf{u}\|^2} \mathbf{u} = \frac{9}{5} \begin{bmatrix} 2 \\ 1 \end{bmatrix}.$$

Then the desired distance is $\left\| \begin{bmatrix} 4 \\ 1 \end{bmatrix} - \dfrac{9}{5} \begin{bmatrix} 2 \\ 1 \end{bmatrix} \right\| = \dfrac{1}{5} \left\| \begin{bmatrix} 2 \\ -4 \end{bmatrix} \right\| = \dfrac{2}{5}\sqrt{5}.$

The Cauchy–Schwarz and Triangle Inequalities*

The reader is undoubtedly aware that, in every triangle, the length of any side is less than the sum of the lengths of the other two sides. This simple result may be stated in the language of norms of vectors. Using Figure 6.5, we see that this statement is equivalent to the *triangle inequality* in $\mathcal{R}^2$:

$$\|\mathbf{u} + \mathbf{v}\| \le \|\mathbf{u}\| + \|\mathbf{v}\|.$$

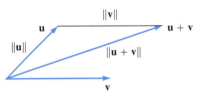

FIGURE 6.5

What we show now is that this inequality holds not just for vectors in $\mathcal{R}^2$, but also for vectors in $\mathcal{R}^n$. A preliminary result is the *Cauchy–Schwarz inequality*.

Theorem 6.3   (Cauchy–Schwarz inequality[1])

For any vectors $\mathbf{u}$ and $\mathbf{v}$ in $\mathcal{R}^n$, we have

$$|\mathbf{u} \cdot \mathbf{v}| \le \|\mathbf{u}\| \cdot \|\mathbf{v}\|.$$

Proof The inequality certainly holds if $\mathbf{u} = \mathbf{0}$. So assume that $\mathbf{u} \ne \mathbf{0}$. The same computation performed previously for the orthogonal projection in $\mathcal{R}^2$

---

*The remainder of this section may be omitted without loss of continuity. However, this material is frequently used in later courses.

[1]The Cauchy–Schwarz inequality was developed independently by the French mathematician Augustin-Louis Cauchy (1789–1857), the German Amandus Schwarz (1843–1921), and the Russian Viktor Yakovlevich Bunyakovsky (1804–99). The result first appeared in Cauchy's 1821 text for an analysis course at the École Polytechnique in Paris. It was later proved for functions by Bunyakovsky in 1859 and by Schwarz in 1884.

extends to $\mathcal{R}^n$ (see Exercise 41). Specifically, if we let $\mathbf{w} = \dfrac{\mathbf{v} \cdot \mathbf{u}}{\|\mathbf{u}\|^2}\mathbf{u}$ and $\mathbf{z} = \mathbf{v} - \mathbf{w}$, then $\mathbf{v} = \mathbf{w} + \mathbf{z}$ and $\mathbf{w} \cdot \mathbf{z} = 0$ (see Figure 6.4). By the Pythagorean theorem and Theorem 6.1(g), we have

$$\|\mathbf{v}\|^2 = \|\mathbf{w} + \mathbf{z}\|^2 = \|\mathbf{w}\|^2 + \|\mathbf{z}\|^2 \geq \|\mathbf{w}\|^2 = \left\| \frac{\mathbf{v} \cdot \mathbf{u}}{\|\mathbf{u}\|^2}\mathbf{u} \right\|^2$$

$$= \left| \frac{\mathbf{v} \cdot \mathbf{u}}{\|\mathbf{u}\|^2} \right|^2 \|\mathbf{u}\|^2 = \frac{|\mathbf{v} \cdot \mathbf{u}|^2}{\|\mathbf{u}\|^2}.$$

Multiplying the inequality above by $\|\mathbf{u}\|^2$ yields

$$\|\mathbf{u}\|^2\|\mathbf{v}\|^2 \geq |\mathbf{v} \cdot \mathbf{u}|^2,$$

and taking square roots establishes the desired result.   ❑

The case where equality is achieved is examined in the exercises. At the end of this section, we see an interesting application of the Cauchy–Schwarz inequality. Example 4 contains another consequence of the Cauchy–Schwarz inequality.

Example 3   Verify the Cauchy–Schwarz inequality for the vectors

$$\mathbf{u} = \begin{bmatrix} 2 \\ -3 \\ 4 \end{bmatrix} \quad \text{and} \quad \mathbf{v} = \begin{bmatrix} 1 \\ -2 \\ -5 \end{bmatrix}.$$

Solution   It is easy to show that $\mathbf{u} \cdot \mathbf{v} = -12$, $\|\mathbf{u}\| = \sqrt{29}$, and $\|\mathbf{v}\| = \sqrt{30}$. So

$$|\mathbf{u} \cdot \mathbf{v}|^2 = 144 \leq 870 = (29)(30) = \|\mathbf{u}\|^2 \cdot \|\mathbf{v}\|^2.$$

Taking square roots confirms the Cauchy–Schwarz inequality for these vectors.   ○

Example 4   For any real numbers $a_1, a_2, a_3, b_1, b_2,$ and b_3, show that

$$|a_1b_1 + a_2b_2 + a_3b_3| \leq \sqrt{a_1^2 + a_2^2 + a_3^2} \, \sqrt{b_1^2 + b_2^2 + b_3^2}.$$

Solution   Apply the Cauchy–Schwarz inequality to $\mathbf{u} = \begin{bmatrix} a_1 \\ a_2 \\ a_3 \end{bmatrix}$ and $\mathbf{v} = \begin{bmatrix} b_1 \\ b_2 \\ b_3 \end{bmatrix}$.   ○

Our next result is the promised generalization to $\mathcal{R}^n$ of the triangle inequality.

Theorem 6.4   (Triangle inequality)
For any vectors $\mathbf{u}$ and $\mathbf{v}$ in $\mathcal{R}^n$, we have

$$\|\mathbf{u} + \mathbf{v}\| \leq \|\mathbf{u}\| + \|\mathbf{v}\|.$$

Proof   Applying the Cauchy–Schwarz inequality, we obtain

$$\|\mathbf{u} + \mathbf{v}\|^2 = \|\mathbf{u}\|^2 + 2\mathbf{u} \cdot \mathbf{v} + \|\mathbf{v}\|^2 \leq \|\mathbf{u}\|^2 + 2\|\mathbf{u}\| \cdot \|\mathbf{v}\| + \|\mathbf{v}\|^2$$

$$= (\|\mathbf{u}\| + \|\mathbf{v}\|)^2.$$

Taking square roots of both sides yields the triangle inequality.   ❑

The case where equality is achieved is examined in the exercises.

Example 5   Verify the triangle inequality for the vectors $\mathbf{u}$ and $\mathbf{v}$ in Example 3.

Solution   Since $\mathbf{u} + \mathbf{v} = \begin{bmatrix} 3 \\ -5 \\ -1 \end{bmatrix}$, it follows that $\|\mathbf{u} + \mathbf{v}\| = \sqrt{35}$. Recalling that $\|\mathbf{u}\| = \sqrt{29}$ and $\|\mathbf{v}\| = \sqrt{30}$, the triangle inequality follows from the observation that

$$\|\mathbf{u} + \mathbf{v}\| = \sqrt{35} < \sqrt{36} = 6 < 5 + 5$$
$$= \sqrt{25} + \sqrt{25} < \sqrt{29} + \sqrt{30} = \|\mathbf{u}\| + \|\mathbf{v}\|.$$

An Application of the Cauchy–Schwarz Inequality

A private school for the training of computer programmers advertised that its average class size is 20. After receiving complaints of false advertising, an investigator for the Office of Consumer Affairs obtained a list of the 60 students enrolled in the school. He polled each student and obtained the student's class size. He added these numbers and divided the total by 60. The result was 27.6, a figure significantly higher than the advertised number 20. As a result, he initiated a complaint against the school. However, the complaint was withdrawn by his supervisor after doing some work of her own.

Using the same enrollment list, the supervisor polled all 60 students. She found that the students were divided among three classes. The first class had 25 students, the second class had 3 students, and the third class had 32 students. Notice that the sum of these three enrollments is 60. She then divided 60 by 3 to obtain a class average of 20, confirming the advertised class average.

To see why there was a difference between the results of these two computations, we apply some linear algebra to a more general situation. Suppose that we have a total of m students who are distributed among n classes consisting respectively of $v_1, v_2, \ldots, v_n$ students. Using this notation, we see that the average of the class sizes is given by

$$\bar{v} = \frac{1}{n}(v_1 + v_2 + \cdots + v_n) = \frac{m}{n}.$$

This is the method used by the supervisor.

Now consider the method used by the investigator. A student in the ith class responds that his or her class size is v_i. Because there are v_i students who give this response, the poll yields a sum of $v_i v_i = v_i^2$ that is contributed by the ith class. Because this is done for each class, the total sum of the responses is

$$v_1^2 + v_2^2 + \cdots + v_n^2.$$

Since m students are polled, this sum is divided by m to obtain the investigator's "average," say v^*, given by

$$v^* = \frac{1}{m}\left(v_1^2 + v_2^2 + \cdots + v_n^2\right).$$

To see the relationship between $\bar{v}$ and v^*, we define the vectors

$$\mathbf{u} = \frac{1}{n}\begin{bmatrix} 1 \\ 1 \\ \vdots \\ 1 \end{bmatrix} \qquad \text{and} \qquad \mathbf{v} = \begin{bmatrix} v_1 \\ v_2 \\ \vdots \\ v_n \end{bmatrix}.$$

Using the definition of dot product and norm in $\mathcal{R}^n$, we compute the following quantities.

$$\mathbf{u} \cdot \mathbf{v} = \frac{1}{n}(v_1 + v_2 + \cdots + v_n) = \bar{v},$$

$$\|\mathbf{u}\|^2 = \mathbf{u} \cdot \mathbf{u} = \frac{1}{n^2}n = \frac{1}{n},$$

and

$$\|\mathbf{v}\|^2 = v_1^2 + v_2^2 + \cdots + v_n^2.$$

We are now in the position to apply the Cauchy–Schwarz inequality. We have

$$\bar{v} = \frac{m}{n} = \frac{n}{m}\left(\frac{m}{n}\right)^2$$

$$= \frac{n}{m}\bar{v}^2$$

$$= \frac{n}{m}(\mathbf{u} \cdot \mathbf{v})^2$$

$$\leq \frac{n}{m}\|\mathbf{u}\|^2\|\mathbf{v}\|^2$$

$$= \frac{n}{m}\frac{1}{n}(v_1^2 + v_2^2 + \cdots + v_n^2)$$

$$= v^*.$$

Consequently, we always have that $\bar{v} \leq v^*$. Using Exercise 42, it can be shown that $\bar{v} = v^*$ if and only if all of the class sizes are equal.

Practice Problems

1. Let

$$\mathbf{u} = \begin{bmatrix} 1 \\ -2 \\ 2 \end{bmatrix} \quad \text{and} \quad \mathbf{v} = \begin{bmatrix} 6 \\ 2 \\ 3 \end{bmatrix}.$$

Compute $\|\mathbf{u}\|$, $\|\mathbf{v}\|$, $\mathbf{u} \cdot \mathbf{v}$, and the distance between $\mathbf{u}$ and $\mathbf{v}$.

2. Determine which pairs of the vectors

$$\mathbf{u} = \begin{bmatrix} 1 \\ -1 \\ 2 \end{bmatrix}, \quad \mathbf{v} = \begin{bmatrix} -2 \\ -5 \\ 3 \end{bmatrix}, \quad \text{and} \quad \mathbf{w} = \begin{bmatrix} -3 \\ 1 \\ 2 \end{bmatrix}$$

are orthogonal.

3. Find the orthogonal projection of $\mathbf{v}$ onto the line through the origin with direction $\mathbf{u}$, where $\mathbf{u}$ and $\mathbf{v}$ are as in problem 2.

Exercises

1. Determine if the following statements are true or false.

   (a) Vectors must be of the same size for their dot product to be defined.
   (b) The dot product of two vectors in $\mathcal{R}^n$ is a vector in $\mathcal{R}^n$.
   (c) The norm of a vector equals the dot product of the vector with itself.
   (d) The norm of a multiple of a vector is the same multiple of the norm of the vector.
   (e) The norm of a sum of vectors is the sum of the norms of the vectors.

   (f) The squared norm of a sum of orthogonal vectors is the sum of the squared norms of the vectors.
   (g) The orthogonal projection of a vector onto a line is a vector that lies along the line.

In Exercises 2–7, two vectors $\mathbf{u}$ and $\mathbf{v}$ are given. Compute the norms of the vectors and the distance d between them.

2. $\mathbf{u} = \begin{bmatrix} 1 \\ 2 \end{bmatrix}$ and $\mathbf{v} = \begin{bmatrix} 3 \\ 7 \end{bmatrix}$ 3. $\mathbf{u} = \begin{bmatrix} 1 \\ -1 \end{bmatrix}$ and $\mathbf{v} = \begin{bmatrix} 2 \\ 1 \end{bmatrix}$

4. $\mathbf{u} = \begin{bmatrix} 1 \\ 3 \\ 1 \end{bmatrix}$ and $\mathbf{v} = \begin{bmatrix} -1 \\ 4 \\ 2 \end{bmatrix}$ **5.** $\mathbf{u} = \begin{bmatrix} 1 \\ -1 \\ 3 \end{bmatrix}$ and $\mathbf{v} = \begin{bmatrix} 2 \\ 1 \\ 0 \end{bmatrix}$

6. $\mathbf{u} = \begin{bmatrix} 1 \\ 2 \\ 1 \\ -1 \end{bmatrix}$ and $\mathbf{v} = \begin{bmatrix} 2 \\ 3 \\ 2 \\ 0 \end{bmatrix}$ **7.** $\mathbf{u} = \begin{bmatrix} 1 \\ -1 \\ -2 \\ 1 \end{bmatrix}$ and $\mathbf{v} = \begin{bmatrix} 2 \\ 3 \\ 1 \\ 1 \end{bmatrix}$

In Exercises 8–13, two vectors are given. Compute the dot product of the vectors, and determine if the vectors are orthogonal.

8. $\mathbf{u} = \begin{bmatrix} 1 \\ 2 \end{bmatrix}$ and $\mathbf{v} = \begin{bmatrix} 3 \\ 7 \end{bmatrix}$ **9.** $\mathbf{u} = \begin{bmatrix} 1 \\ -1 \end{bmatrix}$ and $\mathbf{v} = \begin{bmatrix} 2 \\ 1 \end{bmatrix}$

10. $\mathbf{u} = \begin{bmatrix} 1 \\ 3 \\ 1 \end{bmatrix}$ and $\mathbf{v} = \begin{bmatrix} -1 \\ 4 \\ 2 \end{bmatrix}$ **11.** $\mathbf{u} = \begin{bmatrix} 1 \\ -2 \\ 3 \end{bmatrix}$ and $\mathbf{v} = \begin{bmatrix} 2 \\ 1 \\ 0 \end{bmatrix}$

12. $\mathbf{u} = \begin{bmatrix} 1 \\ 2 \\ -3 \\ -1 \end{bmatrix}$ and $\mathbf{v} = \begin{bmatrix} 2 \\ 3 \\ 2 \\ 0 \end{bmatrix}$ **13.** $\mathbf{u} = \begin{bmatrix} 1 \\ -1 \\ -2 \\ 1 \end{bmatrix}$ and $\mathbf{v} = \begin{bmatrix} 2 \\ 3 \\ 1 \\ 1 \end{bmatrix}$

In Exercises 14–17, two orthogonal vectors $\mathbf{u}$ and $\mathbf{v}$ are given. Compute the quantities $\|\mathbf{u}\|^2$, $\|\mathbf{v}\|^2$, and $\|\mathbf{u} + \mathbf{v}\|^2$. Use your results to illustrate the Pythagorean theorem.

14. $\mathbf{u} = \begin{bmatrix} 3 \\ 1 \end{bmatrix}$ and $\mathbf{v} = \begin{bmatrix} -1 \\ 3 \end{bmatrix}$ **15.** $\mathbf{u} = \begin{bmatrix} 2 \\ 3 \end{bmatrix}$ and $\mathbf{v} = \begin{bmatrix} 0 \\ 0 \end{bmatrix}$

16. $\mathbf{u} = \begin{bmatrix} 1 \\ -1 \\ 2 \end{bmatrix}$ and $\mathbf{v} = \begin{bmatrix} -2 \\ 0 \\ 1 \end{bmatrix}$ **17.** $\mathbf{u} = \begin{bmatrix} 1 \\ 2 \\ 3 \end{bmatrix}$ and $\mathbf{v} = \begin{bmatrix} -11 \\ 4 \\ 1 \end{bmatrix}$

In Exercises 18–21, two vectors $\mathbf{u}$ and $\mathbf{v}$ are given. Compute the quantities $\|\mathbf{u}\|$, $\|\mathbf{v}\|$, and $\|\mathbf{u} + \mathbf{v}\|$. Use your results to illustrate the triangle inequality.

18. $\begin{bmatrix} 2 \\ 1 \end{bmatrix}$ and $\begin{bmatrix} 3 \\ -2 \end{bmatrix}$ **19.** $\begin{bmatrix} 4 \\ 2 \end{bmatrix}$ and $\begin{bmatrix} 3 \\ -1 \end{bmatrix}$

20. $\begin{bmatrix} 2 \\ -3 \\ 1 \end{bmatrix}$ and $\begin{bmatrix} 1 \\ 1 \\ 2 \end{bmatrix}$ **21.** $\begin{bmatrix} 2 \\ -1 \\ 3 \end{bmatrix}$ and $\begin{bmatrix} 4 \\ 0 \\ 1 \end{bmatrix}$

In Exercises 22–25, two vectors $\mathbf{u}$ and $\mathbf{v}$ are given. Compute the quantities $\|\mathbf{u}\|$, $\|\mathbf{v}\|$, and $\mathbf{u} \cdot \mathbf{v}$. Use your results to illustrate the Cauchy–Schwarz inequality.

22. $\begin{bmatrix} 2 \\ 5 \end{bmatrix}$ and $\begin{bmatrix} 3 \\ 4 \end{bmatrix}$ **23.** $\begin{bmatrix} 4 \\ 1 \end{bmatrix}$ and $\begin{bmatrix} 0 \\ -2 \end{bmatrix}$

24. $\begin{bmatrix} 0 \\ 1 \\ 1 \end{bmatrix}$ and $\begin{bmatrix} -2 \\ 1 \\ 3 \end{bmatrix}$ **25.** $\begin{bmatrix} 4 \\ 2 \\ 1 \end{bmatrix}$ and $\begin{bmatrix} 2 \\ -1 \\ -1 \end{bmatrix}$

In Exercises 26–29, a vector $\mathbf{v}$ and a line $\mathcal{L}$ in $\mathcal{R}^2$ are given. Compute the orthogonal projection $\mathbf{w}$ of $\mathbf{v}$ onto $\mathcal{L}$, and use it to compute the distance d from the endpoint of $\mathbf{v}$ to $\mathcal{L}$.

26. $\mathbf{v} = \begin{bmatrix} 2 \\ 3 \end{bmatrix}$ and $y = 2x$ **27.** $\mathbf{v} = \begin{bmatrix} 3 \\ 4 \end{bmatrix}$ and $y = -x$

28. $\mathbf{v} = \begin{bmatrix} 3 \\ 4 \end{bmatrix}$ and $y = -2x$ **29.** $\mathbf{v} = \begin{bmatrix} 4 \\ 1 \end{bmatrix}$ and $y = 3x$

For Exercises 30–35, suppose that $\mathbf{u}$, $\mathbf{v}$, and $\mathbf{w}$ are vectors in $\mathcal{R}^n$ such that $\|\mathbf{u}\| = 2$, $\|\mathbf{v}\| = 3$, $\|\mathbf{w}\| = 5$, $\mathbf{u} \cdot \mathbf{v} = -1$, $\mathbf{u} \cdot \mathbf{w} = 1$, and $\mathbf{v} \cdot \mathbf{w} = -4$.

30. Compute $\|4\mathbf{w}\|$.   **31.** Compute $(\mathbf{u} + \mathbf{v}) \cdot \mathbf{w}$.

32. Compute $(\mathbf{u} + \mathbf{w}) \cdot \mathbf{v}$.   **33.** Compute $\|\mathbf{u} + \mathbf{v}\|^2$.

34. Compute $\|2\mathbf{u} + 3\mathbf{v}\|^2$.   **35.** Compute $\|\mathbf{v} - 4\mathbf{w}\|^2$.

36. Prove (a) of Theorem 6.1.

37. Prove (b) of Theorem 6.1.

38. Prove (c) of Theorem 6.1.

39. Prove (f) of Theorem 6.1.

40. Prove that if $\mathbf{v}$ is any nonzero vector in $\mathcal{R}^n$ and $\mathbf{u} = \dfrac{1}{\|\mathbf{v}\|}\mathbf{v}$, then $\|\mathbf{u}\| = 1$.

41. Prove that the vectors $\mathbf{w}$ and $\mathbf{z}$ defined in the proof of the Cauchy–Schwarz inequality are orthogonal.

42. State and prove a necessary and sufficient condition for equality in the Cauchy–Schwarz inequality.

43. State and prove a necessary and sufficient condition for equality in the triangle inequality.

44. Use the triangle inequality to prove that $|\,\|\mathbf{v}\| - \|\mathbf{w}\|\,| \le \|\mathbf{v} - \mathbf{w}\|$ for all vectors $\mathbf{v}$ and $\mathbf{w}$ in $\mathcal{R}^n$.

45. Prove $(\mathbf{u} + \mathbf{v}) \cdot \mathbf{w} = \mathbf{u} \cdot \mathbf{w} + \mathbf{v} \cdot \mathbf{w}$ for all vectors $\mathbf{u}$, $\mathbf{v}$, and $\mathbf{w}$ in $\mathcal{R}^n$.

46. Let $\mathbf{z}$ be a vector in $\mathcal{R}^n$. Let $W = \{\mathbf{u} \in \mathcal{R}^n : \mathbf{u} \cdot \mathbf{z} = 0\}$. Prove that W is a subspace of $\mathcal{R}^n$.

47. Let $\mathcal{S}$ be a subset of $\mathcal{R}^n$ and $W = \{\mathbf{u} \in \mathcal{R}^n : \mathbf{u} \cdot \mathbf{z} = 0$ for all $\mathbf{z}$ in $\mathcal{S}\}$. Prove that W is a subspace of $\mathcal{R}^n$.

48. Let W denote the set of all vectors that lie along the line $y = 2x$. Find a vector $\mathbf{z}$ in $\mathcal{R}^2$ such that W has the form $W = \{\mathbf{u} \in \mathcal{R}^2 : \mathbf{u} \cdot \mathbf{z} = 0\}$. Justify your answer.

49. Prove the *parallelogram law* for vectors in $\mathcal{R}^n$:

$$\|\mathbf{u} + \mathbf{v}\|^2 + \|\mathbf{u} - \mathbf{v}\|^2 = 2\|\mathbf{u}\|^2 + 2\|\mathbf{v}\|^2.$$

50. Prove that if $\mathbf{u}$ and $\mathbf{v}$ are orthogonal nonzero vectors in $\mathcal{R}^n$, then they are linearly independent.

51.[2] Let A be any $m \times n$ matrix.

(a) Prove that $A^T A$ and A have the same null space. *Hint:* Let $\mathbf{v}$ be a vector in $\mathcal{R}^n$ such that $A^T A\mathbf{v} = \mathbf{0}$. Observe that $A^T A\mathbf{v} \cdot \mathbf{v} = A\mathbf{v} \cdot A\mathbf{v} = 0$.

(b) Use (a) to prove that rank $A^T A =$ rank A.

52.[3] Let $\mathbf{u}$ and $\mathbf{v}$ be nonzero vectors in $\mathcal{R}^2$ or $\mathcal{R}^3$, and let θ be

---

[2]The results of this exercise are used in Section 6.6 (page 365).
[3]The result of this exercise is used in Section 6.7 (page 384).

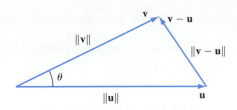

FIGURE 6.6

the angle between **u** and **v**. Then **u**, **v**, and **v** − **u** determine a triangle (see Figure 6.6). The relationship between the lengths of the sides of this triangle and θ is called the *law of cosines*. In the notation of Figure 6.6, it states that

$$\|\mathbf{v} - \mathbf{u}\|^2 = \|\mathbf{u}\|^2 + \|\mathbf{v}\|^2 - 2\|\mathbf{u}\|\,\|\mathbf{v}\|\cos\theta.$$

Use the law of cosines and Theorem 6.1 to derive the formula

$$\mathbf{u} \cdot \mathbf{v} = \|\mathbf{u}\|\,\|\mathbf{v}\|\cos\theta.$$

In Exercise 53, use a calculator with matrix capabilities or computer software such as MATLAB to solve the problem.

53. In every triangle, the length of any side is less than the sum of the lengths of the other two sides. When this observation is generalized to $\mathcal{R}^n$, we obtain the *triangle inequality* (Theorem 6.4), which states that

$$\|\mathbf{u} + \mathbf{v}\| \leq \|\mathbf{u}\| + \|\mathbf{v}\|$$

for any vectors **u** and **v** in $\mathcal{R}^n$. Let

$$\mathbf{u} = \begin{bmatrix} 1 \\ 2 \\ 3 \\ 4 \end{bmatrix}, \quad \mathbf{v} = \begin{bmatrix} -8 \\ -6 \\ 4 \\ 5 \end{bmatrix}, \quad \mathbf{v}_1 = \begin{bmatrix} 2.01 \\ 4.01 \\ 6.01 \\ 8.01 \end{bmatrix}, \quad \text{and} \quad \mathbf{v}_2 = \begin{bmatrix} 3.01 \\ 6.01 \\ 9.01 \\ 12.01 \end{bmatrix}.$$

(a) Verify the triangle inequality for **u** and **v**.
(b) Verify the triangle inequality for **u** and $\mathbf{v}_1$.
(c) Verify the triangle inequality for **u** and $\mathbf{v}_2$.
(d) From what you have observed in (b) and (c), make a conjecture about when equality occurs in the triangle inequality (see Exercise 43).
(e) Interpret your conjecture in (d) geometrically in $\mathcal{R}^2$.

Solutions to the Practice Problems

1.
$$\|\mathbf{u}\| = \sqrt{1^2 + (-2)^2 + 2^2} = 2,$$
$$\|\mathbf{v}\| = \sqrt{6^2 + 2^2 + 3^2} = 7,$$

and

$$\mathbf{u} \cdot \mathbf{v} = (1)(6) + (-2)(2) + (2)(3) = 8.$$

2. Taking dot products, we obtain

$$\mathbf{u} \cdot \mathbf{v} = (1)(-2) + (-1)(-5) + (2)(3) = 9$$
$$\mathbf{u} \cdot \mathbf{w} = (1)(-3) + (-1)(1) + (2)(2) = 0$$
$$\mathbf{v} \cdot \mathbf{w} = (-2)(-3) + (-5)(1) + (3)(2) = 7.$$

So **u** and **w** are orthogonal, but **u** and **v** are not orthogonal and **v** and **w** are not orthogonal.

3. Let **w** be the required orthogonal projection. Then

$$\mathbf{w} = \frac{\mathbf{v} \cdot \mathbf{u}}{\|\mathbf{u}\|^2} \mathbf{u} = \frac{(-2)(1) + (-5)(-1) + (3)(2)}{1^2 + (-1)^2 + 2^2} \begin{bmatrix} 1 \\ -1 \\ 2 \end{bmatrix}$$
$$= \frac{3}{2} \begin{bmatrix} 1 \\ -1 \\ 2 \end{bmatrix}.$$

6.2   Orthogonal Vectors

The property of orthogonality can be extended from two vectors to any set of vectors in a natural way, by requiring that distinct vectors in the set be orthogonal to each other. Bases with this property play a major role in this chapter.

Definition.   A subset of $\mathcal{R}^n$ is called an **orthogonal set** if every pair of distinct vectors in the set is orthogonal.

For example, the set

$$\mathcal{S} = \left\{ \begin{bmatrix} 1 \\ 2 \\ 3 \end{bmatrix}, \begin{bmatrix} 1 \\ 1 \\ -1 \end{bmatrix}, \begin{bmatrix} 5 \\ -4 \\ 1 \end{bmatrix} \right\}$$

is an orthogonal set because the dot product of every pair of distinct vectors in $\mathcal{S}$ is equal to zero. Also, the standard basis for $\mathcal{R}^n$ is an orthogonal set.

Notice that any set consisting of a single vector is an orthogonal set by definition because any two *distinct* elements in the set are orthogonal.

(This is automatically true because the set does not contain two distinct elements.)

Our first result asserts that in most circumstances orthogonal sets are linearly independent.

Theorem 6.5

Any orthogonal set of nonzero vectors is linearly independent.

Proof Let $\mathcal{S} = \{\mathbf{v}_1, \mathbf{v}_2, \ldots, \mathbf{v}_k\}$ be an orthogonal subset of $\mathcal{R}^n$ consisting of nonzero vectors, and let $c_1, c_2, \ldots, c_k$ be scalars such that

$$c_1\mathbf{v}_1 + c_2\mathbf{v}_2 + \cdots + c_k\mathbf{v}_k = \mathbf{0}.$$

Then for any $\mathbf{v}_i$ we have

$$
\begin{aligned}
0 &= \mathbf{0} \cdot \mathbf{v}_i \\
&= (c_1\mathbf{v}_1 + c_2\mathbf{v}_2 + \cdots + c_i\mathbf{v}_i + \cdots + c_k\mathbf{v}_k) \cdot \mathbf{v}_i \\
&= c_1\mathbf{v}_1 \cdot \mathbf{v}_i + c_2\mathbf{v}_2 \cdot \mathbf{v}_i + \cdots + c_i\mathbf{v}_i \cdot \mathbf{v}_i + \cdots + c_k\mathbf{v}_k \cdot \mathbf{v}_i \\
&= c_i(\mathbf{v}_i \cdot \mathbf{v}_i) \\
&= c_i\|\mathbf{v}_i\|^2.
\end{aligned}
$$

But $\|\mathbf{v}_i\|^2 \neq 0$ because $\mathbf{v}_i \neq \mathbf{0}$, and hence $c_i = 0$. We conclude that $\mathcal{S}$ is linearly independent. $\qquad\square$

An orthogonal set that is also a basis for a subspace of $\mathcal{R}^n$ is called an **orthogonal basis** for the subspace. For example, the standard basis for $\mathcal{R}^n$ is an orthogonal basis for $\mathcal{R}^n$.

Suppose that $\mathcal{S} = \{\mathbf{v}_1, \mathbf{v}_2, \ldots, \mathbf{v}_k\}$ is an orthogonal basis for a subspace V of $\mathcal{R}^n$. We can adapt the proof of Theorem 6.5 to obtain a simple method of representing any vector in V as a linear combination of the vectors of $\mathcal{S}$. This method uses dot products in contrast to the tedious task of solving systems of linear equations. Consider any vector $\mathbf{v}$ in V, and suppose that

$$\mathbf{v} = c_1\mathbf{v}_1 + c_2\mathbf{v}_2 + \cdots + c_k\mathbf{v}_k.$$

To obtain c_i, we observe that

$$
\begin{aligned}
\mathbf{v} \cdot \mathbf{v}_i &= (c_1\mathbf{v}_1 + c_2\mathbf{v}_2 + \cdots + c_i\mathbf{v}_i + \cdots + c_k\mathbf{v}_k) \cdot \mathbf{v}_i \\
&= c_1\mathbf{v}_1 \cdot \mathbf{v}_i + c_2\mathbf{v}_2 \cdot \mathbf{v}_i + \cdots + c_i\mathbf{v}_i \cdot \mathbf{v}_i + \cdots + c_k\mathbf{v}_k \cdot \mathbf{v}_i \\
&= c_i(\mathbf{v}_i \cdot \mathbf{v}_i) \\
&= c_i\|\mathbf{v}_i\|^2,
\end{aligned}
$$

and hence

$$c_i = \frac{\mathbf{v} \cdot \mathbf{v}_i}{\|\mathbf{v}_i\|^2}.$$

Summarizing, we have the following result.

Representation of a Vector in Terms of an Orthogonal Basis

Let $\mathcal{S} = \{\mathbf{v}_1, \mathbf{v}_2, \ldots, \mathbf{v}_k\}$ be an orthogonal basis for a subspace V of $\mathcal{R}^n$, and let $\mathbf{v}$ be a vector in V. Then

$$\mathbf{v} = \frac{\mathbf{v} \cdot \mathbf{v}_1}{\|\mathbf{v}_1\|^2}\mathbf{v}_1 + \frac{\mathbf{v} \cdot \mathbf{v}_2}{\|\mathbf{v}_2\|^2}\mathbf{v}_2 + \cdots + \frac{\mathbf{v} \cdot \mathbf{v}_k}{\|\mathbf{v}_k\|^2}\mathbf{v}_k.$$

Example 1   Let $S = \{\mathbf{v}_1, \mathbf{v}_2, \mathbf{v}_3\}$, where

$$\mathbf{v}_1 = \begin{bmatrix} 1 \\ 2 \\ 3 \end{bmatrix}, \qquad \mathbf{v}_2 = \begin{bmatrix} 1 \\ 1 \\ -1 \end{bmatrix}, \qquad \text{and} \qquad \mathbf{v}_3 = \begin{bmatrix} 5 \\ -4 \\ 1 \end{bmatrix}.$$

We have seen that S is an orthogonal subset of $\mathcal{R}^3$. Since the vectors in S are nonzero, Theorem 6.5 tells us that S is linearly independent. Therefore, by Theorem 4.5, S is a basis for $\mathcal{R}^3$. So S is an orthogonal basis for $\mathcal{R}^3$.

Let $\mathbf{v} = \begin{bmatrix} 3 \\ 2 \\ 1 \end{bmatrix}$. We now use the method previously described to obtain the coefficients that represent $\mathbf{v}$ as a linear combination of the vectors of S. Suppose that

$$\mathbf{v} = c_1 \mathbf{v}_1 + c_2 \mathbf{v}_2 + c_3 \mathbf{v}_3.$$

Then

$$c_1 = \frac{\mathbf{v} \cdot \mathbf{v}_1}{\|\mathbf{v}_1\|^2} = \frac{10}{14}, \quad c_2 = \frac{\mathbf{v} \cdot \mathbf{v}_2}{\|\mathbf{v}_2\|^2} = \frac{4}{3}, \quad \text{and} \quad c_3 = \frac{\mathbf{v} \cdot \mathbf{v}_3}{\|\mathbf{v}_3\|^2} = \frac{8}{42}.$$

The reader should verify that

$$\mathbf{v} = \frac{10}{14}\mathbf{v}_1 + \frac{4}{3}\mathbf{v}_2 + \frac{8}{42}\mathbf{v}_3.$$

Since we have seen the advantages of using orthogonal bases for subspaces, two questions naturally arise. Does every subspace have an orthogonal basis; and, if so, how can it be found? The next theorem provides us with a computational method, called the *Gram–Schmidt (orthogonalization) process*, for replacing a linearly independent set with an orthogonal set having the same span.[4] Thus this result gives us a method of converting any basis into an orthogonal basis. It follows that *every subspace has an orthogonal basis*. The Gram–Schmidt process is the extension of the procedure in Section 6.1 for finding the orthogonal projection of a vector onto a line. In the notation used there, linearly independent vectors $\mathbf{u}$ and $\mathbf{v}$ are given, and $\mathbf{w}$ is the orthogonal projection of $\mathbf{v}$ onto $\mathcal{L}$ (see Figure 6.7). Notice that the vectors $\mathbf{u}$ and $\mathbf{v} - \mathbf{w} = \dfrac{\mathbf{v} \cdot \mathbf{u}}{\|\mathbf{u}\|^2}\mathbf{u}$ are nonzero orthogonal vectors that span the same subspace as the vectors $\mathbf{u}$ and $\mathbf{v}$.

Theorem 6.6   (Gram–Schmidt process[5])

Let $S = \{\mathbf{u}_1, \mathbf{u}_2, \dots, \mathbf{u}_k\}$ be a linearly independent subset of $\mathcal{R}^n$. Let $S' = \{\mathbf{v}_1, \mathbf{v}_2, \dots, \mathbf{v}_k\}$ be the set obtained from S as follows:

$$\mathbf{v}_1 = \mathbf{u}_1,$$

and, if $\mathbf{v}_1, \mathbf{v}_2, \dots, \mathbf{v}_{i-1}$ have been defined, then $\mathbf{v}_i$ is defined by

$$\mathbf{v}_i = \mathbf{u}_i - \frac{\mathbf{u}_i \cdot \mathbf{v}_1}{\|\mathbf{v}_1\|^2}\mathbf{v}_1 - \frac{\mathbf{u}_i \cdot \mathbf{v}_2}{\|\mathbf{v}_2\|^2}\mathbf{v}_2 - \cdots - \frac{\mathbf{u}_i \cdot \mathbf{v}_{i-1}}{\|\mathbf{v}_{i-1}\|^2}\mathbf{v}_{i-1} \quad \text{for } 2 \leq i \leq k.$$

---

[4]A modification of this procedure, usually called the *modified Gram–Schmidt process,* is more computationally efficient.

[5]The Gram–Schmidt process first appeared in an 1833 paper by the Danish mathematician Jorgen P. Gram (1850–1916). A later paper by the German mathematician Erhard Schmidt (1876–1959) contained a detailed proof of the result.

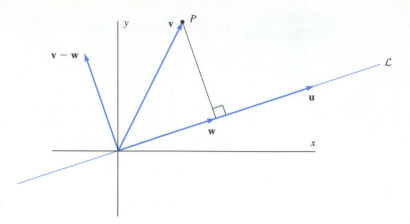

FIGURE 6.7

Then $\mathcal{S}'$ is an orthogonal set of nonzero vectors, and the span of $\mathcal{S}'$ equals the span of $\mathcal{S}$.

Proof First, observe that $\mathbf{v}_1 = \mathbf{u}_1 \neq \mathbf{0}$ because $\mathbf{u}_1$ lies in $\mathcal{S}$, which is linearly independent. Therefore $\{\mathbf{v}_1\}$ is an orthogonal set of nonzero vectors contained in the span of $\{\mathbf{u}_1\}$. Now suppose that for some i, $2 \leq i \leq k$, we have established that $\{\mathbf{v}_1, \mathbf{v}_2, \ldots, \mathbf{v}_{i-1}\}$ is an orthogonal set of nonzero vectors contained in the span of $\{\mathbf{u}_1, \mathbf{u}_2, \ldots, \mathbf{u}_{i-1}\}$. Since

$$\mathbf{v}_i = \mathbf{u}_i - \frac{\mathbf{u}_i \cdot \mathbf{v}_1}{\|\mathbf{v}_1\|^2}\mathbf{v}_1 - \frac{\mathbf{u}_i \cdot \mathbf{v}_2}{\|\mathbf{v}_2\|^2}\mathbf{v}_2 - \cdots - \frac{\mathbf{u}_i \cdot \mathbf{v}_{i-1}}{\|\mathbf{v}_{i-1}\|^2}\mathbf{v}_{i-1},$$

$\mathbf{v}_i$ is a linear combination of vectors contained in the span of $\{\mathbf{u}_1, \mathbf{u}_2, \ldots, \mathbf{u}_i\}$. Furthermore, $\mathbf{v}_i \neq \mathbf{0}$ because otherwise $\mathbf{u}_i$ would be contained in the span of $\{\mathbf{u}_1, \mathbf{u}_2, \ldots, \mathbf{u}_{i-1}\}$, which is not the case because $\{\mathbf{u}_1, \mathbf{u}_2, \ldots, \mathbf{u}_i\}$ is linearly independent. Next, observe that for any $j < i$

$$\mathbf{v}_i \cdot \mathbf{v}_j = \left(\mathbf{u}_i - \frac{\mathbf{u}_i \cdot \mathbf{v}_1}{\|\mathbf{v}_1\|^2}\mathbf{v}_1 - \frac{\mathbf{u}_i \cdot \mathbf{v}_2}{\|\mathbf{v}_2\|^2}\mathbf{v}_2 - \cdots - \frac{\mathbf{u}_i \cdot \mathbf{v}_j}{\|\mathbf{v}_j\|^2}\mathbf{v}_j - \cdots - \frac{\mathbf{u}_i \cdot \mathbf{v}_{i-1}}{\|\mathbf{v}_{i-1}\|^2}\mathbf{v}_{i-1}\right) \cdot \mathbf{v}_j$$

$$= \mathbf{u}_i \cdot \mathbf{v}_j - \frac{\mathbf{u}_i \cdot \mathbf{v}_1}{\|\mathbf{v}_1\|^2}\mathbf{v}_1 \cdot \mathbf{v}_j - \frac{\mathbf{u}_i \cdot \mathbf{v}_2}{\|\mathbf{v}_2\|^2}\mathbf{v}_2 \cdot \mathbf{v}_j$$

$$- \cdots - \frac{\mathbf{u}_i \cdot \mathbf{v}_j}{\|\mathbf{v}_j\|^2}\mathbf{v}_j \cdot \mathbf{v}_j - \cdots - \frac{\mathbf{u}_i \cdot \mathbf{v}_{i-1}}{\|\mathbf{v}_{i-1}\|^2}\mathbf{v}_{i-1} \cdot \mathbf{v}_j$$

$$= \mathbf{u}_i \cdot \mathbf{v}_j - \frac{\mathbf{u}_i \cdot \mathbf{v}_j}{\|\mathbf{v}_j\|^2}\mathbf{v}_j \cdot \mathbf{v}_j$$

$$= \mathbf{u}_i \cdot \mathbf{v}_j - \mathbf{u}_i \cdot \mathbf{v}_j$$

$$= 0.$$

It follows that $\{\mathbf{v}_1, \mathbf{v}_2, \ldots, \mathbf{v}_i\}$ is an orthogonal set of nonzero vectors contained in the span of $\{\mathbf{u}_1, \mathbf{u}_2, \ldots, \mathbf{u}_i\}$. Applying this procedure $k - 1$ times starting at $i = 2$, we obtain an orthogonal set $\mathcal{S}' = \{\mathbf{v}_1, \mathbf{v}_2, \ldots, \mathbf{v}_k\}$ of nonzero vectors contained in the span of $\mathcal{S} = \{\mathbf{u}_1, \mathbf{u}_2, \ldots, \mathbf{u}_k\}$.

It follows that Span $\mathcal{S}'$ is a k-dimensional subspace of Span $\mathcal{S}$, which also has dimension k. Hence by Theorem 4.7, Span $\mathcal{S}' =$ Span $\mathcal{S}$. $\quad\square$

Example 2   Let W be the span of $\mathcal{S} = \{\mathbf{u}_1, \mathbf{u}_2, \mathbf{u}_3\}$, where

$$\mathbf{u}_1 = \begin{bmatrix} 1 \\ 1 \\ 1 \\ 1 \end{bmatrix}, \qquad \mathbf{u}_2 = \begin{bmatrix} 2 \\ 1 \\ 0 \\ 1 \end{bmatrix}, \qquad \text{and} \qquad \mathbf{u}_3 = \begin{bmatrix} 1 \\ 1 \\ 2 \\ 1 \end{bmatrix},$$

is a linearly independent subset of $\mathcal{R}^4$. Apply the Gram-Schmidt process to $\mathcal{S}$ to obtain an orthogonal basis $\mathcal{S}'$ for W.

Solution   Let

$$\mathbf{v}_1 = \mathbf{u}_1 = \begin{bmatrix} 1 \\ 1 \\ 1 \\ 1 \end{bmatrix},$$

$$\mathbf{v}_2 = \mathbf{u}_2 - \frac{\mathbf{u}_2 \cdot \mathbf{v}_1}{\|\mathbf{v}_1\|^2}\mathbf{v}_1 = \begin{bmatrix} 2 \\ 1 \\ 0 \\ 1 \end{bmatrix} - \frac{4}{4}\begin{bmatrix} 1 \\ 1 \\ 1 \\ 1 \end{bmatrix} = \begin{bmatrix} 1 \\ 0 \\ -1 \\ 0 \end{bmatrix},$$

and

$$\mathbf{v}_3 = \mathbf{u}_3 - \frac{\mathbf{u}_3 \cdot \mathbf{v}_1}{\|\mathbf{v}_1\|^2}\mathbf{v}_1 - \frac{\mathbf{u}_3 \cdot \mathbf{v}_2}{\|\mathbf{v}_2\|^2}\mathbf{v}_2 = \begin{bmatrix} 1 \\ 1 \\ 2 \\ 1 \end{bmatrix} - \frac{5}{4}\begin{bmatrix} 1 \\ 1 \\ 1 \\ 1 \end{bmatrix} - \frac{(-1)}{2}\begin{bmatrix} 1 \\ 0 \\ -1 \\ 0 \end{bmatrix} = \frac{1}{4}\begin{bmatrix} 1 \\ -1 \\ 1 \\ -1 \end{bmatrix}.$$

Then $\mathcal{S}' = \{\mathbf{v}_1, \mathbf{v}_2, \mathbf{v}_3\}$ is an orthogonal basis for W. $\bigcirc$

Replacing a vector in an orthogonal set by a scalar multiple of the vector results in a new set that is also orthogonal. If the scalar is not zero and the orthogonal set consists of nonzero vectors, then the new set is linearly independent and spans the same subspace as the original set. So multiplying vectors in an orthogonal basis by nonzero scalars produces a new orthogonal basis for the same subspace (see Exercise 29). For example, consider the orthogonal basis $\mathcal{S}' = \{\mathbf{v}_1, \mathbf{v}_2, \mathbf{v}_3\}$ produced in Example 2. In order to eliminate the fractional components in $\mathbf{v}_3$, we may replace $\mathbf{v}_3$ with the vector $4\mathbf{v}_3$ to obtain another orthogonal basis $\mathcal{S}'' = \{\mathbf{v}_1, \mathbf{v}_2, 4\mathbf{v}_3\}$ for W.

For theoretical purposes (to simplify statements and proofs of theorems), it is often desirable to select an orthogonal basis in which each vector has norm equal to one. A vector that has norm equal to one is called a **unit vector**. Note that any nonzero vector $\mathbf{v}$ can be transformed into a unit vector by multiplying it by the scalar $\dfrac{1}{\|\mathbf{v}\|}$. For if $\mathbf{u} = \left(\dfrac{1}{\|\mathbf{v}\|}\right)\mathbf{v}$, then

$$\|\mathbf{u}\| = \left\|\left(\frac{1}{\|\mathbf{v}\|}\right)\mathbf{v}\right\| = \left|\frac{1}{\|\mathbf{v}\|}\right| \cdot \|\mathbf{v}\| = \frac{1}{\|\mathbf{v}\|}\|\mathbf{v}\| = 1.$$

Thus we can transform any orthogonal basis into an orthogonal basis of unit vectors by simply multiplying each vector in the basis by the reciprocal of its norm.

An orthogonal basis consisting of unit vectors is called an **orthonormal basis**. For example, the standard basis is an orthonormal basis for $\mathcal{R}^n$. It is a consequence of the Gram–Schmidt orthogonalization process that *every subspace has an orthonormal basis*.

Example 3   Replace the orthogonal basis $S' = \{\mathbf{v}_1, \mathbf{v}_2, \mathbf{v}_3\}$ for W obtained in Example 2 by an orthonormal basis B.

Solution   In view of the preceding comments, $B = \left\{ \dfrac{1}{\|\mathbf{v}_1\|}\mathbf{v}_1, \dfrac{1}{\|\mathbf{v}_2\|}\mathbf{v}_2, \dfrac{1}{\|\mathbf{v}_3\|}\mathbf{v}_3 \right\}$ is an orthonormal basis for W. Notice that

$$\|\mathbf{v}_1\| = 2, \qquad \|\mathbf{v}_2\| = \sqrt{2}, \qquad \text{and} \qquad \|\mathbf{v}_3\| = \frac{1}{2}.$$

Therefore the desired orthonormal basis is $B = \{\mathbf{w}_1, \mathbf{w}_2, \mathbf{w}_3\}$, where

$$\mathbf{w}_1 = \frac{1}{2}\mathbf{v}_1 = \frac{1}{2}\begin{bmatrix} 1 \\ 1 \\ 1 \\ 1 \end{bmatrix}, \qquad \mathbf{w}_2 = \frac{1}{\sqrt{2}}\mathbf{v}_2 = \frac{1}{\sqrt{2}}\begin{bmatrix} 1 \\ 0 \\ -1 \\ 0 \end{bmatrix},$$

and

$$\mathbf{w}_3 = 2\mathbf{v}_3 = \frac{1}{2}\begin{bmatrix} 1 \\ -1 \\ 1 \\ -1 \end{bmatrix}.$$

One useful property of orthonormal bases is that the representation of a vector as a linear combination of vectors in an orthogonal basis is simplified if the basis is orthonormal. For suppose that $S = \{\mathbf{w}_1, \mathbf{w}_2, \ldots, \mathbf{w}_k\}$ is an orthonormal basis for a subspace W of $\mathcal{R}^n$. If $\mathbf{v} = c_1\mathbf{w}_1 + c_2\mathbf{w}_2 + \cdots + c_k\mathbf{w}_k$, then

$$c_i = \frac{\mathbf{v} \cdot \mathbf{w}_i}{\|\mathbf{w}_i\|^2} = \mathbf{v} \cdot \mathbf{w}_i$$

for each i. Combining these two equations, we obtain the following result.

Representation of a Vector in Terms of an Orthonormal Basis

Let $S = \{\mathbf{w}_1, \mathbf{w}_2, \ldots, \mathbf{w}_k\}$ be an orthonormal basis for a subspace W of $\mathcal{R}^n$. Then for any vector $\mathbf{v}$ in W,

$$\mathbf{v} = (\mathbf{v} \cdot \mathbf{w}_1)\mathbf{w}_1 + (\mathbf{v} \cdot \mathbf{w}_2)\mathbf{w}_2 + \cdots + (\mathbf{v} \cdot \mathbf{w}_k)\mathbf{w}_k.$$

Example 4   The vector

$$\mathbf{v} = \begin{bmatrix} 2 \\ 3 \\ 5 \\ 3 \end{bmatrix}$$

is an element of the subspace W in Example 3. Represent $\mathbf{v}$ as a linear combination of the vectors in B, the orthonormal basis for W obtained in Example 3.

Solution   To represent $\mathbf{v}$ as a linear combination of the vectors in the orthonormal basis $B = \{\mathbf{w}_1, \mathbf{w}_2, \mathbf{w}_3\}$, we compute

$$\mathbf{v} \cdot \mathbf{w}_1 = \frac{13}{2}, \qquad \mathbf{v} \cdot \mathbf{w}_2 = \frac{-3}{\sqrt{2}}, \qquad \text{and} \qquad \mathbf{v} \cdot \mathbf{w}_3 = \frac{1}{2}.$$

Thus

$$\mathbf{v} = (\mathbf{v} \cdot \mathbf{w}_1)\mathbf{w}_1 + (\mathbf{v} \cdot \mathbf{w}_2)\mathbf{w}_2 + (\mathbf{v} \cdot \mathbf{w}_3)\mathbf{w}_3$$

$$= \frac{13}{2}\mathbf{w}_1 + \left(\frac{-3}{\sqrt{2}}\right)\mathbf{w}_2 + \frac{1}{2}\mathbf{w}_3.$$

Orthogonal Complements

We have seen the advantage of using orthogonal vectors. We now extend this idea to sets. Consider, for example, the set

$$S_1 = \left\{ \begin{bmatrix} -1 \\ -2 \end{bmatrix}, \begin{bmatrix} 2 \\ 4 \end{bmatrix} \right\}.$$

What vectors are orthogonal to each of the vectors in S_1?

Note that the vectors in S_1 lie along the line with equation $y = 2x$. Hence the vectors orthogonal to the vectors in S_1 lie along the line with equation $y = -\frac{1}{2}x$ (see Figure 6.8). The set S_2 of vectors that lie along the line $y = -\frac{1}{2}x$ is called the *orthogonal complement* of S_1. More generally, we have the following definition.

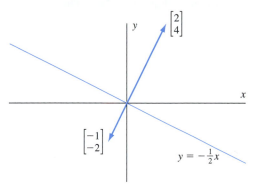

FIGURE 6.8

Definition.   The **orthogonal complement** of a nonempty subset S of $\mathcal{R}^n$, denoted $S^\perp$ (read "S perp"), is the set of all vectors in $\mathcal{R}^n$ that are orthogonal to every vector in S. That is,

$$S^\perp = \{\mathbf{v} \in \mathcal{R}^n \colon \mathbf{v} \cdot \mathbf{u} = 0 \text{ for every } \mathbf{u} \text{ in } S\}.$$

For example, if $S = \mathcal{R}^n$, then $S^\perp = \{\mathbf{0}\}$, and conversely. Note also that if S is any nonempty subset of $\mathcal{R}^n$, then $\mathbf{0}$ is in $S^\perp$ because $\mathbf{0}$ is orthogonal to every vector in S. Moreover, if $\mathbf{v}$ and $\mathbf{w}$ are in $S^\perp$, then, for every vector $\mathbf{u}$ in S,

$$(\mathbf{v} + \mathbf{w}) \cdot \mathbf{u} = \mathbf{v} \cdot \mathbf{u} + \mathbf{w} \cdot \mathbf{u} = 0 + 0 = 0,$$

and therefore, $\mathbf{v} + \mathbf{w}$ is in $S^\perp$. So $S^\perp$ is closed under vector addition. A similar argument can be used to show that $S^\perp$ is closed under scalar multiplication. Therefore $S^\perp$ is a subspace of $\mathcal{R}^n$. Hence we have the following result.

> The orthogonal complement of any nonempty subset of $\mathcal{R}^n$ is a subspace of $\mathcal{R}^n$.

For the sets $\mathcal{S}_1$ and $\mathcal{S}_2$ above, the span of $\mathcal{S}_1$ can be visualized as the line with equation $y = 2x$. Note that $\mathcal{S}_2 = \mathcal{S}_1^{\perp} = (\mathrm{Span}\,\mathcal{S}_1)^{\perp}$. More generally, *for any nonempty subset $\mathcal{S}$ of $\mathcal{R}^n$, $(\mathrm{Span}\,\mathcal{S})^{\perp} = \mathcal{S}^{\perp}$ (see Exercise 31).* One consequence of this is that the orthogonal complement of a basis for a subspace of $\mathcal{R}^n$ is the same as the orthogonal complement of the subspace.

Example 5   Let W denote the xy-plane viewed as a subspace of $\mathcal{R}^3$, that is,

$$W = \left\{ \begin{bmatrix} v_1 \\ v_2 \\ 0 \end{bmatrix} : v_1 \text{ and } v_2 \text{ are real numbers} \right\}.$$

Then a vector $\mathbf{v} = \begin{bmatrix} v_1 \\ v_2 \\ v_3 \end{bmatrix}$ lies in $W^{\perp}$ if and only if $v_1 = v_2 = 0$. For suppose $v_1 = v_2 = 0$. Then clearly $\mathbf{v} \cdot \mathbf{u} = 0$ for all $\mathbf{u}$ in W, and hence $\mathbf{v}$ is in $W^{\perp}$. Now suppose that $\mathbf{v}$ is in $W^{\perp}$. Then

$$v_1 = \mathbf{e}_1 \cdot \mathbf{v} = 0$$

because $\mathbf{e}_1$ is in W. Similarly, $v_2 = 0$ because $\mathbf{e}_2$ is in W. Thus

$$W^{\perp} = \left\{ \begin{bmatrix} 0 \\ 0 \\ v_3 \end{bmatrix} : v_3 \text{ is a real number} \right\},$$

and so $W^{\perp}$ can be identified with the z-axis.

The next example shows how orthogonal complements arise in the study of systems of linear equations.

Example 6   Find a basis for the orthogonal complement of $W = \mathrm{Span}\,\{\mathbf{u}_1, \mathbf{u}_2\}$, where

$$\mathbf{u}_1 = \begin{bmatrix} 1 \\ 1 \\ -1 \\ 4 \end{bmatrix} \quad \text{and} \quad \mathbf{u}_2 = \begin{bmatrix} 1 \\ -1 \\ 1 \\ 2 \end{bmatrix}.$$

Solution   A vector $\mathbf{v} = \begin{bmatrix} x_1 \\ x_2 \\ x_3 \\ x_4 \end{bmatrix}$ lies in $W^{\perp}$ if and only if $\mathbf{u}_1 \cdot \mathbf{v} = 0$ and $\mathbf{u}_2 \cdot \mathbf{v} = 0$.

Notice that these two equations can be written as the homogeneous system of linear equations

$$\begin{aligned} x_1 + x_2 - x_3 + 4x_4 &= 0 \\ x_1 - x_2 + x_3 + 2x_4 &= 0. \end{aligned} \tag{1}$$

From the reduced row echelon form of the augmented matrix, we see that the parametric representation of the general solution to 1 is

$$\begin{bmatrix} x_1 \\ x_2 \\ x_3 \\ x_4 \end{bmatrix} = \begin{bmatrix} -3x_4 \\ x_3 - x_4 \\ x_3 \\ x_4 \end{bmatrix} = x_3 \begin{bmatrix} 0 \\ 1 \\ 1 \\ 0 \end{bmatrix} + x_4 \begin{bmatrix} -3 \\ -1 \\ 0 \\ 1 \end{bmatrix}.$$

Thus

$$\mathcal{B} = \left\{ \begin{bmatrix} 0 \\ 1 \\ 1 \\ 0 \end{bmatrix}, \begin{bmatrix} -3 \\ -1 \\ 0 \\ 1 \end{bmatrix} \right\}$$

is a basis for $W^\perp$.

Let A denote the coefficient matrix of 1. Notice that the vectors $\mathbf{u}_1$ and $\mathbf{u}_2$ in that example are the rows of A, and hence W is the row space of A. Furthermore the set of solutions to 1 is the null space of A. It follows that

$$W^\perp = (\text{Row } A)^\perp = \text{Null } A.$$

This observation is valid for any matrix.

> For any matrix A,
>
> $$(\text{Row } A)^\perp = \text{Null } A.$$

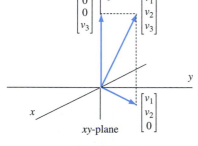

FIGURE 6.9

In the notation of Example 5, any vector $\begin{bmatrix} v_1 \\ v_2 \\ v_3 \end{bmatrix}$ in $\mathcal{R}^3$ can be written as the sum of the vector $\begin{bmatrix} v_1 \\ v_2 \\ 0 \end{bmatrix}$ in W and the vector $\begin{bmatrix} 0 \\ 0 \\ v_3 \end{bmatrix}$ in $W^\perp$. So in some sense, a vector in $\mathcal{R}^3$ is subdivided into two pieces, one residing in W and the other in $W^\perp$ (see Figure 6.9). This is true in general.

Theorem 6.7
Let W be a subspace of $\mathcal{R}^n$. Then, for any vector $\mathbf{v}$ in $\mathcal{R}^n$, there exists a unique vector $\mathbf{w}$ in W and a unique vector $\mathbf{z}$ in $W^\perp$ such that $\mathbf{v} = \mathbf{w} + \mathbf{z}$.

Proof Consider any vector $\mathbf{v}$ in $\mathcal{R}^n$. Choose $\mathcal{B} = \{\mathbf{v}_1, \mathbf{v}_2, \ldots, \mathbf{v}_k\}$ to be an orthonormal basis for W, and let

$$\mathbf{w} = (\mathbf{v} \cdot \mathbf{v}_1)\mathbf{v}_1 + (\mathbf{v} \cdot \mathbf{v}_2)\mathbf{v}_2 + \cdots + (\mathbf{v} \cdot \mathbf{v}_k)\mathbf{v}_k. \tag{2}$$

Then $\mathbf{w}$ is in W because it is a linear combination of the basis vectors for W. (Notice that (2) resembles the equation given in the representation of a vector in terms of an orthonormal basis. Indeed $\mathbf{v} = \mathbf{w}$ if and only if $\mathbf{v}$ is in W.)

Let $\mathbf{z} = \mathbf{v} - \mathbf{w}$. Then clearly $\mathbf{v} = \mathbf{w} + \mathbf{z}$. We show that $\mathbf{z}$ is in $W^\perp$. By Exercise 31, it suffices to show that $\mathbf{z}$ is orthogonal to every vector in $\mathcal{B}$. For any $\mathbf{v}_i$ in $\mathcal{B}$, $\mathbf{w} \cdot \mathbf{v}_i = \mathbf{v} \cdot \mathbf{v}_i$ by 2, and therefore

$$\mathbf{z} \cdot \mathbf{v}_i = (\mathbf{v} - \mathbf{w}) \cdot \mathbf{v}_i = \mathbf{v} \cdot \mathbf{v}_i - \mathbf{w} \cdot \mathbf{v}_i = \mathbf{v} \cdot \mathbf{v}_i - \mathbf{v} \cdot \mathbf{v}_i = 0.$$

Hence $\mathbf{z}$ is in $W^\perp$.

Next, we show that this representation is unique. Suppose that $\mathbf{v} = \mathbf{w}' + \mathbf{z}'$, where $\mathbf{w}'$ is in W and $\mathbf{z}'$ is in $W^\perp$. Then $\mathbf{w} + \mathbf{z} = \mathbf{w}' + \mathbf{z}'$, and hence $\mathbf{w} - \mathbf{w}' = \mathbf{z}' - \mathbf{z}$. But $\mathbf{w} - \mathbf{w}'$ is in W and $\mathbf{z}' - \mathbf{z}$ is in $W^\perp$. Thus $\mathbf{w} - \mathbf{w}'$ lies in both W and $W^\perp$. This means that $\mathbf{w} - \mathbf{w}'$ is orthogonal to itself. But by Theorem 6.1(b), $\mathbf{0}$ is the only vector with this property. Hence $\mathbf{w} - \mathbf{w}' = \mathbf{0}$, and therefore, $\mathbf{w} = \mathbf{w}'$. Similarly, $\mathbf{z} = \mathbf{z}'$, and we conclude that the representation is unique. ∎

Suppose that $\mathcal{B}_1$ is a basis for a subspace W of $\mathcal{R}^n$ and that $\mathcal{B}_2$ is a basis for $W^\perp$. It follows from Theorem 6.7 that $\mathcal{B}_1 \cup \mathcal{B}_2$ is a basis for $\mathcal{R}^n$ (see Exercise 32). One simple consequence of this observation is the following useful result.

> For any subspace W of $\mathcal{R}^n$,
>
> $$\dim W + \dim W^\perp = n.$$

The Orthogonal Projection of a Vector onto a Subspace

The proof of Theorem 6.7 gives us a computational method for representing a given vector as the sum of a vector in a subspace and a vector in the orthogonal complement of the subspace. The following example illustrates this method.

Example 7 Let W be the two-dimensional subspace of $\mathcal{R}^3$ defined by

$$x_1 - x_2 + 2x_3 = 0,$$

and let $\mathbf{v} = \begin{bmatrix} 1 \\ 3 \\ 4 \end{bmatrix}$. Find the vectors $\mathbf{w}$ in W and $\mathbf{z}$ in $W^\perp$ such that $\mathbf{v} = \mathbf{w} + \mathbf{z}$.

Solution First observe that

$$\mathcal{B} = \{\mathbf{w}_1, \mathbf{w}_2\} = \left\{ \frac{1}{\sqrt{2}} \begin{bmatrix} 1 \\ 1 \\ 0 \end{bmatrix}, \frac{1}{\sqrt{3}} \begin{bmatrix} -1 \\ 1 \\ 1 \end{bmatrix} \right\}$$

is an orthonormal basis for W. (An orthonormal basis such as $\mathcal{B}$ can be obtained by applying the Gram–Schmidt process to an ordinary basis for W.) We use $\mathcal{B}$ to find $\mathbf{w}$, as in the proof of Theorem 6.7. By (2), we have

$$\mathbf{w} = (\mathbf{v} \cdot \mathbf{w}_1)\mathbf{w}_1 + (\mathbf{v} \cdot \mathbf{w}_2)\mathbf{w}_2$$

$$= \frac{4}{\sqrt{2}}\mathbf{w}_1 + \frac{6}{\sqrt{3}}\mathbf{w}_2$$

$$= 2 \begin{bmatrix} 1 \\ 1 \\ 0 \end{bmatrix} + 2 \begin{bmatrix} -1 \\ 1 \\ 1 \end{bmatrix}$$

$$= \begin{bmatrix} 0 \\ 4 \\ 2 \end{bmatrix}.$$

Therefore

$$\mathbf{z} = \mathbf{v} - \mathbf{w} = \begin{bmatrix} 1 \\ 3 \\ 4 \end{bmatrix} - \begin{bmatrix} 0 \\ 4 \\ 2 \end{bmatrix} = \begin{bmatrix} 1 \\ -1 \\ 2 \end{bmatrix}.$$

Note that $\mathbf{z}$ is orthogonal to $\mathbf{w}_1$ and $\mathbf{w}_2$, confirming that $\mathbf{z}$ is in $W^{\perp}$.

The vector $\mathbf{w}$ in Example 7 can be viewed as the result of dropping a perpendicular from the endpoint of $\mathbf{v}$ to the subspace W, which, in this case, is a plane. In this context, the vector $\mathbf{z} = \mathbf{v} - \mathbf{w}$ connects the endpoint of $\mathbf{w}$ to the endpoint of $\mathbf{v}$ and is perpendicular (orthogonal) to W (see Figure 6.10). In general, consider any subspace W of $\mathcal{R}^n$ and any vector $\mathbf{v}$ in $\mathcal{R}^n$. The unique vector $\mathbf{w}$ in W such that $\mathbf{v} = \mathbf{w} + \mathbf{z}$, where $\mathbf{z}$ is in $W^{\perp}$, is called the **orthogonal projection of $\mathbf{v}$ onto** W. Thus the orthogonal projection of $\mathbf{v}$ onto W is the unique vector $\mathbf{w}$ in W such that $\mathbf{v} - \mathbf{w}$ is in $W^{\perp}$. For example, in the context of Example 7, $\mathbf{w} = \begin{bmatrix} 0 \\ 4 \\ 2 \end{bmatrix}$ is the orthogonal projection of $\mathbf{v}$ onto W.

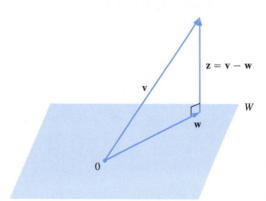

FIGURE 6.10

$\mathbf{w}$ is the orthogonal projection of $\mathbf{v}$ onto W.

Figure 6.10 suggests that the orthogonal projection of a vector $\mathbf{v}$ in $\mathcal{R}^n$ onto a subspace W of $\mathcal{R}^n$ is the vector in W that is nearest to $\mathbf{v}$. As we now show, this is indeed the case. For let $\mathbf{w}$ be the orthogonal projection of $\mathbf{v}$ onto W, and $\mathbf{w}'$ be any vector in W. Since $\mathbf{v} - \mathbf{w}$ is in $W^{\perp}$, it is orthogonal to $\mathbf{w} - \mathbf{w}'$, which lies in W. Thus, by the Pythagorean theorem in $\mathcal{R}^n$ (Theorem 6.2),

$$\| \mathbf{v} - \mathbf{w}' \|^2 = \| (\mathbf{v} - \mathbf{w}) + (\mathbf{w} - \mathbf{w}') \|^2 = \| \mathbf{v} - \mathbf{w} \|^2 + \| \mathbf{w} - \mathbf{w}' \|^2 \geq \| \mathbf{v} - \mathbf{w} \|^2.$$

Moreover, the final inequality is a strict inequality if $\mathbf{w} \neq \mathbf{w}'$. Therefore we have established the following result, which is often useful in applications.

Closest Vector Property

Let W be a subspace of $\mathcal{R}^n$ and $\mathbf{v}$ be a vector in $\mathcal{R}^n$. Among all vectors in W, the vector closest to $\mathbf{v}$ is the orthogonal projection of $\mathbf{v}$ onto W.

We define the **distance from a vector $\mathbf{v}$ in $\mathcal{R}^n$ to a subspace W of $\mathcal{R}^n$** to be the distance between $\mathbf{v}$ and the orthogonal projection of $\mathbf{v}$ onto W. So the distance between $\mathbf{v}$ and W is the minimum distance between $\mathbf{v}$ and every vector in W.

In the context of Example 7, the distance between $\mathbf{v}$ and W is

$$\|\mathbf{v} - \mathbf{w}\| = \|\mathbf{z}\| = \left\| \begin{bmatrix} 1 \\ -1 \\ 2 \end{bmatrix} \right\| = \sqrt{6}.$$

Practice Problems

1. Find an orthonormal basis $\mathcal{B}$ for the column space of

$$A = \begin{bmatrix} 1 & -1 & 1 \\ -1 & 3 & 6 \\ -1 & -3 & 3 \\ 1 & 5 & -4 \end{bmatrix}.$$

2. The vector

$$\mathbf{v} = \begin{bmatrix} 1 \\ 4 \\ 7 \\ -10 \end{bmatrix}$$

is in the column space of the matrix A in problem 1. Write $\mathbf{v}$ as a linear combination of the vectors in the basis $\mathcal{B}$ obtained in problem 1.

3. Let

$$W = \text{Span} \left\{ \begin{bmatrix} 1 \\ 1 \\ -1 \\ 1 \end{bmatrix}, \begin{bmatrix} 3 \\ 2 \\ -1 \\ 0 \end{bmatrix} \right\} \quad \text{and} \quad \mathbf{v} = \begin{bmatrix} 0 \\ 7 \\ 4 \\ 7 \end{bmatrix}.$$

(a) Find a basis for $W^{\perp}$.
(b) Find vectors $\mathbf{w}$ in W and $\mathbf{z}$ in $W^{\perp}$ such that $\mathbf{w} + \mathbf{z} = \mathbf{v}$.
(c) Find the orthogonal projection of $\mathbf{v}$ onto W.
(d) Find the distance from $\mathbf{v}$ to W.

Exercises

1. Determine if the following statements are true or false.

   (a) Any orthogonal subset of $\mathcal{R}^n$ is linearly independent.
   (b) Every subspace of $\mathcal{R}^n$ has an orthogonal basis.
   (c) For any nonempty subset S of $\mathcal{R}^n$, $(S^{\perp})^{\perp} = S$.
   (d) If F and G are subsets of $\mathcal{R}^n$ and $F^{\perp} = G^{\perp}$, then $F = G$.
   (e) Any subset of $\mathcal{R}^n$ consisting of a single vector is an orthogonal set.
   (f) If S is an orthogonal set of n nonzero vectors in $\mathcal{R}^n$, then S is a basis for $\mathcal{R}^n$.

In Exercises 2–6, determine if the given set is orthogonal.

2. $\left\{ \begin{bmatrix} 1 \\ 1 \end{bmatrix}, \begin{bmatrix} 1 \\ -1 \end{bmatrix} \right\}$

3. $\left\{ \begin{bmatrix} 1 \\ 2 \\ 1 \end{bmatrix}, \begin{bmatrix} 1 \\ -1 \\ 1 \end{bmatrix}, \begin{bmatrix} 2 \\ -1 \\ 0 \end{bmatrix} \right\}$

4. $\left\{ \begin{bmatrix} 1 \\ 0 \\ 1 \end{bmatrix}, \begin{bmatrix} -1 \\ 0 \\ 1 \end{bmatrix}, \begin{bmatrix} 0 \\ -1 \\ 0 \end{bmatrix} \right\}$

5. $\left\{ \begin{bmatrix} 1 \\ 2 \\ 3 \\ -3 \end{bmatrix}, \begin{bmatrix} 1 \\ 1 \\ -1 \\ 0 \end{bmatrix}, \begin{bmatrix} 3 \\ -3 \\ 0 \\ -1 \end{bmatrix} \right\}$

6. $\left\{ \begin{bmatrix} 2 \\ 1 \\ -1 \\ 1 \end{bmatrix}, \begin{bmatrix} 1 \\ 1 \\ 3 \\ 0 \end{bmatrix}, \begin{bmatrix} 1 \\ -1 \\ 0 \\ 1 \end{bmatrix} \right\}$

In Exercises 7–12, (a) apply the Gram–Schmidt process to replace the given linearly independent set S by an orthogonal set of nonzero vectors with the same span, and (b) obtain an orthonormal set with the same span as S.

7. $\left\{ \begin{bmatrix} 1 \\ 1 \\ 1 \end{bmatrix}, \begin{bmatrix} 5 \\ -1 \\ 2 \end{bmatrix} \right\}$

8. $\left\{ \begin{bmatrix} 1 \\ -2 \\ 1 \end{bmatrix}, \begin{bmatrix} 1 \\ -1 \\ 0 \end{bmatrix} \right\}$

9. $\left\{ \begin{bmatrix} 0 \\ 1 \\ 1 \\ 1 \end{bmatrix}, \begin{bmatrix} 1 \\ 0 \\ 1 \\ 1 \end{bmatrix}, \begin{bmatrix} 1 \\ 1 \\ 0 \\ 1 \end{bmatrix} \right\}$

10. $\left\{ \begin{bmatrix} 1 \\ -1 \\ 0 \\ 2 \end{bmatrix}, \begin{bmatrix} 1 \\ 1 \\ 1 \\ 3 \end{bmatrix}, \begin{bmatrix} 3 \\ 1 \\ 1 \\ 5 \end{bmatrix} \right\}$

11. $\left\{ \begin{bmatrix} 1 \\ 0 \\ -1 \\ 1 \end{bmatrix}, \begin{bmatrix} 2 \\ 1 \\ -1 \\ 0 \end{bmatrix}, \begin{bmatrix} 2 \\ -1 \\ -1 \\ 3 \end{bmatrix} \right\}$

12. $\left\{ \begin{bmatrix} 1 \\ -1 \\ 0 \\ 1 \\ 1 \end{bmatrix}, \begin{bmatrix} 2 \\ -1 \\ 0 \\ 3 \\ 2 \end{bmatrix}, \begin{bmatrix} 1 \\ -1 \\ 1 \\ 1 \\ 1 \end{bmatrix}, \begin{bmatrix} 3 \\ 1 \\ 1 \\ 1 \\ 1 \end{bmatrix} \right\}$

In Exercises 13–16, an orthogonal set S and a vector $\mathbf{v}$ in Span S are given. Use dot products (not systems of linear equations) to represent $\mathbf{v}$ as a linear combination of the vectors in S.

13. $S = \left\{ \begin{bmatrix} 2 \\ 1 \end{bmatrix}, \begin{bmatrix} -1 \\ 2 \end{bmatrix} \right\}, \quad \mathbf{v} = \begin{bmatrix} 1 \\ 8 \end{bmatrix}$

14. $S = \left\{ \begin{bmatrix} 1 \\ 1 \\ 1 \end{bmatrix}, \begin{bmatrix} 1 \\ 2 \\ -3 \end{bmatrix} \right\}, \quad \mathbf{v} = \begin{bmatrix} 2 \\ 1 \\ 6 \end{bmatrix}$

15. $S = \left\{ \begin{bmatrix} 1 \\ 0 \\ 1 \end{bmatrix}, \begin{bmatrix} 1 \\ 2 \\ -1 \end{bmatrix}, \begin{bmatrix} 1 \\ -1 \\ -1 \end{bmatrix} \right\}, \quad \mathbf{v} = \begin{bmatrix} 3 \\ 1 \\ 2 \end{bmatrix}$

16. $S = \left\{ \begin{bmatrix} 1 \\ 1 \\ 1 \\ 1 \end{bmatrix}, \begin{bmatrix} 1 \\ -1 \\ 1 \\ -1 \end{bmatrix}, \begin{bmatrix} 1 \\ -1 \\ -1 \\ 1 \end{bmatrix}, \begin{bmatrix} 1 \\ 1 \\ -1 \\ -1 \end{bmatrix} \right\}, \quad \mathbf{v} = \begin{bmatrix} 2 \\ 1 \\ -1 \\ 2 \end{bmatrix}$

In Exercises 17–20, find a basis for $S^{\perp}$.

17. $S = \left\{ \begin{bmatrix} 1 \\ -1 \\ 2 \end{bmatrix} \right\}$ **18.** $S = \left\{ \begin{bmatrix} 1 \\ 1 \\ 1 \end{bmatrix}, \begin{bmatrix} 1 \\ -1 \\ -1 \end{bmatrix} \right\}$

19. $S = \left\{ \begin{bmatrix} 1 \\ -2 \\ 1 \\ 1 \end{bmatrix}, \begin{bmatrix} 1 \\ -1 \\ 3 \\ 2 \end{bmatrix} \right\}$

20. $S = \left\{ \begin{bmatrix} 1 \\ -1 \\ 1 \\ 1 \end{bmatrix}, \begin{bmatrix} 1 \\ 1 \\ -1 \\ 1 \end{bmatrix}, \begin{bmatrix} 1 \\ 1 \\ 1 \\ -1 \end{bmatrix} \right\}$

In Exercises 21–23, a vector $\mathbf{v}$ in $\mathcal{R}^n$ and an orthonormal basis S for a subspace W of $\mathcal{R}^n$ are given.

(a) Use S to obtain the unique vectors $\mathbf{w}$ in W and $\mathbf{z}$ in $W^{\perp}$ so that $\mathbf{v} = \mathbf{w} + \mathbf{z}$.

(b) Find the orthogonal projection of $\mathbf{v}$ onto W.

(c) Find the distance from $\mathbf{v}$ to W.

21. $\mathbf{v} = \begin{bmatrix} 1 \\ 3 \end{bmatrix}, \quad S = \left\{ \frac{1}{\sqrt{2}} \begin{bmatrix} 1 \\ -1 \end{bmatrix} \right\}$

22. $\mathbf{v} = \begin{bmatrix} 2 \\ 3 \\ -1 \end{bmatrix}, \quad S = \left\{ \frac{1}{\sqrt{2}} \begin{bmatrix} 1 \\ 1 \\ 0 \end{bmatrix}, \frac{1}{\sqrt{3}} \begin{bmatrix} 1 \\ -1 \\ 1 \end{bmatrix} \right\}$

23. $\mathbf{v} = \begin{bmatrix} 1 \\ 4 \\ -1 \end{bmatrix}, \quad S = \left\{ \frac{1}{\sqrt{6}} \begin{bmatrix} -1 \\ 2 \\ 1 \end{bmatrix}, \frac{1}{\sqrt{3}} \begin{bmatrix} 1 \\ 1 \\ -1 \end{bmatrix} \right\}$

24. $\mathbf{v} = \begin{bmatrix} 2 \\ 4 \\ 1 \\ 3 \end{bmatrix}, \quad S = \left\{ \frac{1}{\sqrt{3}} \begin{bmatrix} 1 \\ 0 \\ 1 \\ 1 \end{bmatrix}, \frac{1}{\sqrt{3}} \begin{bmatrix} 1 \\ 1 \\ 0 \\ -1 \end{bmatrix}, \frac{1}{\sqrt{3}} \begin{bmatrix} -1 \\ 1 \\ 1 \\ 0 \end{bmatrix} \right\}$

In Exercises 25–28, a vector $\mathbf{v}$ in $\mathcal{R}^n$ and a subspace W of $\mathcal{R}^n$ are given. Combine what you have learned in this section to obtain the unique vectors $\mathbf{w}$ in W and $\mathbf{z}$ in $W^{\perp}$ such that $\mathbf{v} = \mathbf{w} + \mathbf{z}$. Use your results to find the orthogonal projection of $\mathbf{v}$ onto W and the distance from $\mathbf{v}$ to W.

25. $\mathbf{v} = \begin{bmatrix} -10 \\ 5 \end{bmatrix}, \quad W = \text{Span} \left\{ \begin{bmatrix} -3 \\ 4 \end{bmatrix} \right\}$

26. $\mathbf{v} = \begin{bmatrix} 1 \\ 3 \\ 7 \end{bmatrix}$; W is the solution set to the equation

$$x_1 - 2x_2 + 3x_3 = 0$$

27. $\mathbf{v} = \begin{bmatrix} 1 \\ 2 \\ -1 \end{bmatrix}$; W is the solution set to the system of equations

$$x_1 + x_2 - x_3 = 0$$
$$x_1 - x_2 + 3x_3 = 0$$

28. $\mathbf{v} = \begin{bmatrix} 7 \\ 4 \\ 1 \\ 2 \end{bmatrix}, \quad W = \text{Span} \left\{ \begin{bmatrix} 1 \\ 2 \\ 1 \\ -1 \end{bmatrix}, \begin{bmatrix} 1 \\ 3 \\ 2 \\ 2 \end{bmatrix} \right\}$

29. Let $\{\mathbf{v}_1, \mathbf{v}_2, \ldots, \mathbf{v}_k\}$ be an orthogonal subset of $\mathcal{R}^n$. Prove that for any scalars $c_1, c_2, \ldots, c_k$, the set $\{c_1\mathbf{v}_1, c_2\mathbf{v}_2, \ldots, c_k\mathbf{v}_k\}$ is also orthogonal.

30. Suppose that S is a nonempty *orthogonal* subset of $\mathcal{R}^n$ consisting of nonzero vectors, and suppose that S' is obtained by applying the Gram–Schmidt process to S. Prove that $S' = S$.

31. Let S be a nonempty finite subset of $\mathcal{R}^n$, and suppose that $W = \text{Span } S$. Prove that $W^{\perp} = S^{\perp}$.

32. Let W be a subspace of $\mathcal{R}^n$, and let $\mathcal{B}_1$ and $\mathcal{B}_2$ be bases for W and $W^{\perp}$, respectively. Apply Theorem 6.7 to prove the following.

(a) $\mathcal{B}_1 \cup \mathcal{B}_2$ is a basis for $\mathcal{R}^n$.

(b) $\dim W + \dim W^{\perp} = n$.

33.[6] Suppose that $\{\mathbf{v}_1, \mathbf{v}_2, \ldots, \mathbf{v}_n\}$ is an orthogonal basis for $\mathcal{R}^n$. Let $W = \text{Span } \{\mathbf{v}_1, \mathbf{v}_2, \ldots, \mathbf{v}_k\}$ for some k, where $1 \le k < n$. Prove that $\{\mathbf{v}_{k+1}, \mathbf{v}_{k+2}, \ldots, \mathbf{v}_n\}$ is a basis for $W^{\perp}$.

34. Let $\{\mathbf{w}_1, \mathbf{w}_2, \ldots, \mathbf{w}_n\}$ be an orthonormal basis for $\mathcal{R}^n$. Prove that the following statements are true for any vectors $\mathbf{u}$ and $\mathbf{v}$ in $\mathcal{R}^n$.

(a) $\mathbf{u} + \mathbf{v} = (\mathbf{u} \cdot \mathbf{w}_1 + \mathbf{v} \cdot \mathbf{w}_1)\mathbf{w}_1 + (\mathbf{u} \cdot \mathbf{w}_2 + \mathbf{v} \cdot \mathbf{w}_2)\mathbf{w}_2 + \cdots + (\mathbf{u} \cdot \mathbf{w}_n + \mathbf{v} \cdot \mathbf{w}_n)\mathbf{w}_n$.

(b) $\mathbf{u} \cdot \mathbf{v} = (\mathbf{u} \cdot \mathbf{w}_1)(\mathbf{v} \cdot \mathbf{w}_1) + (\mathbf{u} \cdot \mathbf{w}_2)(\mathbf{v} \cdot \mathbf{w}_2) + \cdots + (\mathbf{u} \cdot \mathbf{w}_n)(\mathbf{v} \cdot \mathbf{w}_n)$.

(This result is known as *Parseval's identity*.)

(c) $\|\mathbf{u}\|^2 = (\mathbf{u} \cdot \mathbf{w}_1)^2 + (\mathbf{u} \cdot \mathbf{w}_2)^2 + \cdots + (\mathbf{u} \cdot \mathbf{w}_n)^2$.

35. Prove that for any subspace W of $\mathcal{R}^n$, $(W^{\perp})^{\perp} = W$.

36. Prove that for any matrix A, $(\text{Row } A)^{\perp} = \text{Null } A$.

---

[6]The results of this exercise are used in Section 6.6 (page 372).

37. Prove that if V and W are subspaces of $\mathcal{R}^n$ such that V is contained in W, then $W^\perp$ is contained in $V^\perp$.

38. Prove that for any nonempty finite subset S of $\mathcal{R}^n$, $(S^\perp)^\perp = \operatorname{Span} S$.

39. Use the fact that $(\operatorname{Row} A)^\perp = \operatorname{Null} A$ for any matrix A to give another proof that $\dim W + \dim W^\perp = n$ for any subspace W of $\mathcal{R}^n$. (*Hint:* Let A be a $k \times n$ matrix whose rows constitute a basis for W.)

40. Let $S = \{\mathbf{v}_1, \mathbf{v}_2, \ldots, \mathbf{v}_k\}$ be an orthonormal subset of $\mathcal{R}^n$, and let $\mathbf{u}$ be a vector in $\mathcal{R}^n$. Prove the following.

    (a) $(\mathbf{u} \cdot \mathbf{v}_1)^2 + (\mathbf{u} \cdot \mathbf{v}_2)^2 + \cdots + (\mathbf{u} \cdot \mathbf{v}_k)^2 \le \|\mathbf{u}\|^2$.
    (b) The inequality in (a) is an equality if and only if $\mathbf{u}$ lies in $\operatorname{Span} S$.

41.[7] Suppose that $\{\mathbf{v}_1, \mathbf{v}_2, \ldots, \mathbf{v}_k\}$ is an orthonormal subset of $\mathcal{R}^n$. Combine Theorem 4.4 on page 212 with the Gram–Schmidt process to prove that this set can be extended to an orthonormal basis $\{\mathbf{v}_1, \mathbf{v}_2, \ldots, \mathbf{v}_k, \ldots, \mathbf{v}_n\}$ for $\mathcal{R}^n$.

---

[7]The results of this exercise are used in Section 6.6 (page 365).

In Exercises 42 and 43, use a calculator with matrix capabilities or computer software such as MATLAB to solve the problem.

42. Let $W = \operatorname{Span} S$, where

$$S = \left\{ \begin{bmatrix} 0 \\ -3 \\ 9 \\ 0 \\ -4 \end{bmatrix}, \begin{bmatrix} -8 \\ 9 \\ -8 \\ 0 \\ 2 \end{bmatrix}, \begin{bmatrix} -4 \\ 8 \\ 1 \\ -1 \\ 8 \end{bmatrix}, \begin{bmatrix} -9 \\ 5 \\ 5 \\ 6 \\ -7 \end{bmatrix} \right\} \quad \text{and} \quad \mathbf{v} = \begin{bmatrix} -9 \\ 4 \\ 7 \\ 2 \\ 4 \end{bmatrix}.$$

    (a) Find an orthonormal basis for W.
    (b) Use your answer to (a) to find the orthogonal projection of $\mathbf{v}$ onto W.
    (c) Use your answer to (b) to find the distance from $\mathbf{v}$ to W.

43. Use your answer to (a) of Exercise 42 to create a 5×4 matrix Q whose columns form an orthonormal set of vectors.

    (a) Verify that $Q^T Q = I_4$.
    (b) Prove, in general, that if A is an $m \times n$ matrix whose columns form an orthonormal set, then $A^T A = I_n$.

Solutions to the Practice Problems

1. We apply the Gram–Schmidt process to $\{\mathbf{a}_1, \mathbf{a}_2, \mathbf{a}_3\}$, the columns of A, to replace the set of columns by the orthogonal set $\{\mathbf{v}_1, \mathbf{v}_2, \mathbf{v}_3\}$, where

$$\mathbf{v}_1 = \mathbf{a}_1 = \begin{bmatrix} 1 \\ -1 \\ -1 \\ 1 \end{bmatrix},$$

$$\mathbf{v}_2 = \mathbf{a}_2 - \frac{\mathbf{a}_2 \cdot \mathbf{v}_1}{\|\mathbf{v}_1\|^2} \mathbf{v}_1 = \begin{bmatrix} -1 \\ 3 \\ -3 \\ 5 \end{bmatrix} - \frac{4}{4} \begin{bmatrix} 1 \\ -1 \\ -1 \\ 1 \end{bmatrix} = \begin{bmatrix} -2 \\ 4 \\ -2 \\ 4 \end{bmatrix},$$

and

$$\mathbf{v}_3 = \mathbf{a}_3 - \frac{\mathbf{a}_3 \cdot \mathbf{v}_1}{\|\mathbf{v}_1\|^2} \mathbf{v}_1 - \frac{\mathbf{a}_3 \cdot \mathbf{v}_2}{\|\mathbf{v}_2\|^2} \mathbf{v}_2$$

$$= \begin{bmatrix} 1 \\ 6 \\ 3 \\ -4 \end{bmatrix} - \frac{-12}{4} \begin{bmatrix} 1 \\ -1 \\ -1 \\ 1 \end{bmatrix} - 0 \begin{bmatrix} -2 \\ 4 \\ -2 \\ 4 \end{bmatrix} = \begin{bmatrix} 4 \\ 3 \\ 0 \\ -1 \end{bmatrix}.$$

Thus

$$\mathcal{B} = \left\{ \frac{1}{\|\mathbf{v}_1\|} \mathbf{v}_1, \frac{1}{\|\mathbf{v}_2\|} \mathbf{v}_2, \frac{1}{\|\mathbf{v}_3\|} \mathbf{v}_3 \right\}$$

$$= \left\{ \frac{1}{2} \begin{bmatrix} 1 \\ -1 \\ -1 \\ 1 \end{bmatrix}, \frac{1}{\sqrt{10}} \begin{bmatrix} -1 \\ 2 \\ -1 \\ 2 \end{bmatrix}, \frac{1}{\sqrt{26}} \begin{bmatrix} 4 \\ 3 \\ 0 \\ -1 \end{bmatrix} \right\}.$$

2. For each i, let $\mathbf{w}_i = \dfrac{1}{\|\mathbf{v}_i\|} \mathbf{v}_i$, where $\mathbf{v}_i$ is as in the solution to problem 1. Then

$$\mathbf{v} = (\mathbf{v} \cdot \mathbf{w}_1)\mathbf{w}_1 + (\mathbf{v} \cdot \mathbf{w}_2)\mathbf{w}_2 + (\mathbf{v} \cdot \mathbf{w}_3)\mathbf{w}_3$$

$$= (-10)\mathbf{w}_1 + (-2\sqrt{10})\mathbf{w}_2 + \sqrt{26}\mathbf{w}_3.$$

3. (a) A vector $\mathbf{u} = \begin{bmatrix} x_1 \\ x_2 \\ x_3 \\ x_4 \end{bmatrix}$ is in $W^\perp$ if and only if it is a solution to the homogeneous system of linear equations

$$x_1 + x_2 - x_3 + x_4 = 0$$
$$3x_1 + 2x_2 - x_3 \qquad = 0.$$

The parametric form of the general solution to this system is

$$\begin{bmatrix} x_1 \\ x_2 \\ x_3 \\ x_4 \end{bmatrix} = x_3 \begin{bmatrix} -1 \\ 2 \\ 1 \\ 0 \end{bmatrix} + x_4 \begin{bmatrix} 2 \\ -3 \\ 0 \\ 1 \end{bmatrix}.$$

Thus

$$\left\{ \begin{bmatrix} -1 \\ 2 \\ 1 \\ 0 \end{bmatrix}, \begin{bmatrix} 2 \\ -3 \\ 0 \\ 1 \end{bmatrix} \right\}$$

is a basis for $W^\perp$.

(b) Apply the methods used in problem 1 to obtain an orthonormal basis $\{\mathbf{w}_1, \mathbf{w}_2\}$ for W, where

$$\mathbf{w}_1 = \frac{1}{2}\begin{bmatrix} 1 \\ 1 \\ -1 \\ 1 \end{bmatrix} \quad \text{and} \quad \mathbf{w}_2 = \frac{1}{2\sqrt{5}}\begin{bmatrix} 3 \\ 1 \\ 1 \\ -3 \end{bmatrix}.$$

Thus

$$\mathbf{w} = (\mathbf{v} \cdot \mathbf{w}_1)\mathbf{w}_1 + (\mathbf{v} \cdot \mathbf{w}_2)\mathbf{w}_2$$

$$= (5) \cdot \frac{1}{2}\begin{bmatrix} 1 \\ 1 \\ -1 \\ 1 \end{bmatrix} + (-\sqrt{5}) \cdot \frac{1}{2\sqrt{5}}\begin{bmatrix} 3 \\ 1 \\ 1 \\ -3 \end{bmatrix}$$

$$= \begin{bmatrix} 1 \\ 2 \\ -3 \\ 4 \end{bmatrix}.$$

Finally,

$$\mathbf{z} = \mathbf{v} - \mathbf{w} = \begin{bmatrix} 0 \\ 7 \\ 4 \\ 7 \end{bmatrix} - \begin{bmatrix} 1 \\ 2 \\ -3 \\ 4 \end{bmatrix} = \begin{bmatrix} -1 \\ 5 \\ 7 \\ 3 \end{bmatrix}.$$

(c) $\mathbf{w} = \begin{bmatrix} 1 \\ 2 \\ -3 \\ 4 \end{bmatrix}.$

(d) $\|\mathbf{z}\| = \left\| \begin{bmatrix} -1 \\ 5 \\ 7 \\ 3 \end{bmatrix} \right\| = \sqrt{84}.$

6.3   Least-Squares Approximation and Orthogonal Projection Matrices

In almost all areas of empirical research, there is an interest in finding simple mathematical relationships between variables. In economics, the variables might be the gross domestic product, the unemployment rate, and the annual deficit. In the life sciences, the variables of interest might be the incidence of smoking and heart disease. In sociology, it might be birth order and frequency of juvenile delinquency.

Many relationships in science are *deterministic,* that is, information about one variable completely determines the value of another variable. For example, the relationship between force f and acceleration a of an object of mass m is given by the equation $f = ma$ (Newton's second law). Another example is the height of a freely falling object and the time that it has been falling. On the other hand, the relationship between the height and the weight of an individual is not deterministic. There are many people with the same height but different weights. Yet there exist charts in hospitals which give the recommended weight for a given height. Relationships that are not deterministic are often called *probabilistic* or *stochastic.*

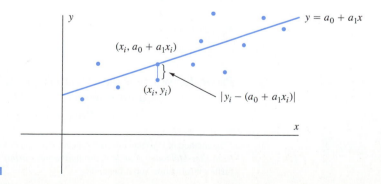

FIGURE 6.11

We can apply what we know about orthogonal projection matrices to obtain probabilistic relationships between variables. We begin with a given set of data $(x_1, y_1), (x_2, y_2), \ldots, (x_n, y_n)$ obtained by empirical measurement. For example, we might have a randomly selected sample of n people, where x_i represents the number of years of education and y_i represents the annual income of the ith person. The data are plotted as in Figure 6.11. Notice that there is an approximately linear (straight line) relationship between x and y. To obtain this relationship, we would like to find the line $y = a_0 + a_1 x$ that *best fits* the data. The usual criterion that statisticians use for defining the line of best fit is that the sum of the squared vertical distances of the data from the line is smaller than from any other line. From Figure 6.11, we see that we must find a_0 and a_1 so that the quantity

$$E = [y_1 - (a_0 + a_1 x_1)]^2 + [y_2 - (a_0 + a_1 x_2)]^2 + \cdots + [y_n - (a_0 + a_1 x_n)]^2 \tag{3}$$

is minimized. This technique is called the **method of least squares**,[8] E is called the **error sum of squares**, and the line for which E is minimized is called the **least-squares line**.

To find the least-squares line, we consider

$$\mathbf{v}_1 = \begin{bmatrix} 1 \\ 1 \\ \vdots \\ 1 \end{bmatrix}, \quad \mathbf{v}_2 = \begin{bmatrix} x_1 \\ x_2 \\ \vdots \\ x_n \end{bmatrix}, \quad \mathbf{y} = \begin{bmatrix} y_1 \\ y_2 \\ \vdots \\ y_n \end{bmatrix}, \quad \text{and} \quad C = [\mathbf{v}_1 \quad \mathbf{v}_2].$$

With this notation, (3) can be rewritten in the notation of vectors as

$$E = \|\mathbf{y} - (a_0 \mathbf{v}_1 + a_1 \mathbf{v}_2)\|^2 \tag{4}$$

(see Exercise 43). Notice that $\sqrt{E} = \|\mathbf{y} - (a_0\mathbf{v}_1 + a_1\mathbf{v}_2)\|$ is the distance from $\mathbf{y}$ to the vector $a_0\mathbf{v}_1 + a_1\mathbf{v}_2$, which lies in $W = \text{Span}\{\mathbf{v}_1, \mathbf{v}_2\}$. So to minimize E, we need only choose the vector in W that is nearest to $\mathbf{y}$. But from the closest vector property, this vector is the orthogonal projection of $\mathbf{y}$ onto W. Thus we want $a_0\mathbf{v}_1 + a_1\mathbf{v}_2 = C \begin{bmatrix} a_0 \\ a_1 \end{bmatrix}$ to be the orthogonal projection of $\mathbf{y}$ onto W.

For any reasonable set of data, the x_i's are not all equal, and hence $\mathbf{v}_1$ and $\mathbf{v}_2$ are not multiples of one another. Thus the vectors $\mathbf{v}_1$ and $\mathbf{v}_2$ are linearly independent, and so $\mathcal{B} = \{\mathbf{v}_1, \mathbf{v}_2\}$ is a basis for W. In this setting, the following theorem provides an explicit formula for the orthogonal projection of a vector onto W.

Theorem 6.8

Let C be an $n \times k$ matrix whose columns form a basis for a subspace W of $\mathcal{R}^n$. For any vector $\mathbf{v}$ in $\mathcal{R}^n$, the orthogonal projection of $\mathbf{v}$ onto W is

$$C(C^T C)^{-1} C^T \mathbf{v}.$$

Proof Let $\mathbf{w}$ denote the orthogonal projection of $\mathbf{v}$ onto W. Since $W = \text{Col } C$, we have $\mathbf{w} = C\mathbf{u}$ for some $\mathbf{u}$ in $\mathcal{R}^k$. Then $\mathbf{v} - \mathbf{w}$ is in

$$W^\perp = (\text{Col } C)^\perp = (\text{Row } C^T)^\perp = \text{Null } C^T.$$

---

[8]The method of least squares first appeared in a paper by Adrien Marie Legendre (1752–1833) entitled *Nouvelles Méthodes pour le détermination des orbites des comètes*. See the footnote on p. 435 for further information about Legendre.

Hence

$$0 = C^T(\mathbf{v} - \mathbf{w}) = C^T\mathbf{v} - C^T\mathbf{w} = C^T\mathbf{v} - C^T C\mathbf{u}.$$

Thus

$$C^T C\mathbf{u} = C^T\mathbf{v}. \tag{5}$$

We claim that the $k \times k$ matrix $C^T C$ is invertible. Suppose that $C^T C\mathbf{b} = \mathbf{0}$. Recall from Section 6.1 that the dot product $(C\mathbf{b}) \cdot (C\mathbf{b})$ can be written as the matrix product $(C\mathbf{b})^T C\mathbf{b}$. Then

$$\|C\mathbf{b}\|^2 = (C\mathbf{b}) \cdot (C\mathbf{b}) = (C\mathbf{b})^T C\mathbf{b} = \mathbf{b}^T C^T C\mathbf{b} = \mathbf{b}^T\mathbf{0} = 0,$$

and so $C\mathbf{b} = \mathbf{0}$. Since the columns of C are linearly independent, it follows that $\mathbf{b} = \mathbf{0}$. Thus $\mathbf{0}$ is the only solution to $C^T C\mathbf{x} = \mathbf{0}$, and hence $C^T C$ is invertible by Theorem 2.6.

It now follows from (5) that $\mathbf{u} = (C^T C)^{-1} C^T\mathbf{v}$, and so the orthogonal projection of $\mathbf{v}$ onto W is

$$\mathbf{w} = C\mathbf{u} = C(C^T C)^{-1} C^T\mathbf{v}. \qquad \square$$

It follows from (5) that the minimum value of E in (4) occurs when

$$C^T C \begin{bmatrix} a_0 \\ a_1 \end{bmatrix} = C^T\mathbf{v}.$$

This matrix equation $C^T C\mathbf{x} = C^T\mathbf{v}$ corresponds to a system of linear equations called the **normal equations**. Thus the line of best fit occurs when $\begin{bmatrix} a_0 \\ a_1 \end{bmatrix}$ is the solution to the normal equations. Since the proof of Theorem 6.8 shows that the matrix $C^T C$ is invertible, we see that the least-squares line has the equation $y = a_0 + a_1 x$, where

$$\begin{bmatrix} a_0 \\ a_1 \end{bmatrix} = (C^T C)^{-1} C^T\mathbf{y}.$$

| **Example 1** | In the manufacture of refrigerators, it is necessary to finish connecting rods. If the weight of the finished rod is above a certain amount, the rod must be discarded. As the finishing process is expensive, it would be of considerable value to the manufacturer to be able to estimate the relationship between the finished weight and the initial rough weight. In this way, those rods whose rough weights are too high could be discarded. From past experience, the manufacturer knows that this relationship is approximately linear.

From a sample of five rods, we let x_i be the rough weight and y_i the finished weight of the ith rod. The data are given in the following table.

Rough weight x_i (lb)	Finished weight y_i (lb)
2.60	2.00
2.72	2.10
2.75	2.10
2.67	2.03
2.68	2.04

From this information, we let

$$C = \begin{bmatrix} 1 & 2.60 \\ 1 & 2.72 \\ 1 & 2.75 \\ 1 & 2.67 \\ 1 & 2.68 \end{bmatrix} \quad \text{and} \quad \mathbf{y} = \begin{bmatrix} 2.00 \\ 2.10 \\ 2.10 \\ 2.03 \\ 2.04 \end{bmatrix}.$$

Then

$$C^T C = \begin{bmatrix} 5.0000 & 13.4200 \\ 13.4200 & 36.0322 \end{bmatrix},$$

and

$$\begin{bmatrix} a_0 \\ a_1 \end{bmatrix} = (C^T C)^{-1} C^T \mathbf{y} \approx \begin{bmatrix} 0.056 \\ 0.745 \end{bmatrix}.$$

Thus, the approximate relationship between the finished weight y and the rough weight x is given by the equation of the least-squares line

$$y = 0.056 + 0.745x.$$

For example, if the rough weight of a rod is 2.65 pounds, then the finished weight is approximately

$$0.056 + 0.745(2.65) \approx 2.030 \text{ pounds.} \quad \bigcirc$$

The method we have developed can be extended to find the best fit by a quadratic polynomial $y = a_0 + a_1 x + a_2 x^2$ to data points $(x_1, y_1), (x_2, y_2), \ldots,$ (x_n, y_n). The only modification to the linear case is that the new error sum of squares is

$$E = \left[y_1 - \left(a_0 + a_1 x_1 + a_2 x_1^2 \right) \right]^2 + \cdots + \left[y_n - \left(a_0 + a_1 x_n + a_2 x_n^2 \right) \right]^2.$$

In this case, let

$$\mathbf{v}_1 = \begin{bmatrix} 1 \\ 1 \\ \vdots \\ 1 \end{bmatrix}, \quad \mathbf{v}_2 = \begin{bmatrix} x_1 \\ x_2 \\ \vdots \\ x_n \end{bmatrix}, \quad \mathbf{v}_3 = \begin{bmatrix} x_1^2 \\ x_2^2 \\ \vdots \\ x_n^2 \end{bmatrix}, \quad \text{and} \quad \mathbf{y} = \begin{bmatrix} y_1 \\ y_2 \\ \vdots \\ y_n \end{bmatrix}.$$

If $\mathbf{v}_1, \mathbf{v}_2,$ and $\mathbf{v}_3$ are linearly independent (which in practice is always the case), then they form a basis for a three-dimensional subspace W of $\mathcal{R}^n$. Adapting the method in the linear case to the quadratic case, we let C be the $n \times 3$ matrix $C = [\mathbf{v}_1 \quad \mathbf{v}_2 \quad \mathbf{v}_3]$. It follows that

$$\begin{bmatrix} a_0 \\ a_1 \\ a_2 \end{bmatrix} = (C^T C)^{-1} C^T \mathbf{y}.$$

Example 2   It is known from physics that if a ball is thrown upward at a velocity of v_0 feet per second from a building of height s_0 feet, then the height of the ball after t seconds is given by $s = s_0 + v_0 t + \frac{1}{2} g t^2$, where g represents the acceleration due to gravity. To provide an empirical estimate of g, a ball is thrown upward from a building 100 feet high at a velocity of 30 feet per second. The height of the ball is observed at the times given in the following table.

Time (seconds)	Height (feet)
0	100
1	118
2	92
3	48
3.5	7

For these data, we let

$$C = \begin{bmatrix} 1 & 0 & 0 \\ 1 & 1 & 1 \\ 1 & 2 & 4 \\ 1 & 3 & 9 \\ 1 & 3.5 & 12.25 \end{bmatrix} \quad \text{and} \quad \mathbf{y} = \begin{bmatrix} 100 \\ 118 \\ 92 \\ 48 \\ 7 \end{bmatrix}.$$

Thus the quadratic polynomial $y = a_0 + a_1 x + a_2 x^2$ of best fit satisfies

$$\begin{bmatrix} s_0 \\ v_0 \\ \frac{1}{2}g \end{bmatrix} = \begin{bmatrix} a_0 \\ a_1 \\ a_2 \end{bmatrix} = (C^T C)^{-1} C^T \mathbf{y} \approx \begin{bmatrix} 101.00 \\ 29.77 \\ -16.11 \end{bmatrix}.$$

This yields the approximate relationship

$$s = 101.00 + 29.77t - 16.11t^2$$

(see Figure 6.12). Setting $\frac{1}{2}g = -16.11$, we obtain -32.22 feet per second per second as the estimate for g. ○

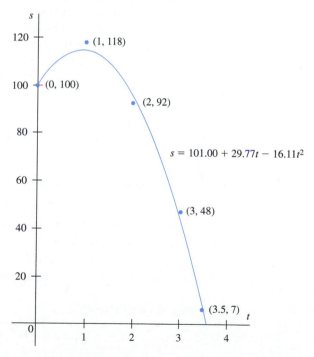

FIGURE 6.12

It should be pointed out that the same method may be extended to find the best-fitting polynomial of any desired maximum degree, provided that the data set is sufficiently large. Furthermore, by using the appropriate change of variable, many more complicated relationships may be estimated by the same type of matrix computations.

Orthogonal Projection Matrices

If C is an $n \times k$ matrix whose columns form a basis for a subspace W of $\mathcal{R}^n$, then Theorem 6.8 shows that the orthogonal projection of a vector $\mathbf{v}$ in $\mathcal{R}^n$ onto W is

$$C(C^T C)^{-1} C^T \mathbf{v}.$$

We call $P_W = C(C^T C)^{-1} C^T$ the **orthogonal projection matrix** for W. It is a surprising consequence of Theorem 6.8 that P_W is independent of the choice of a basis for W, and hence it is completely determined by W.

The Orthogonal Projection Matrix for a Subspace

Let W be a nonzero subspace of $\mathcal{R}^n$, and let $\{\mathbf{v}_1, \mathbf{v}_2, \cdots, \mathbf{v}_k\}$ be a basis for W. If $C = [\mathbf{v}_1 \ \mathbf{v}_2 \ \cdots \ \mathbf{v}_k]$, then $C^T C$ is invertible, and the orthogonal projection matrix P_W for W is given by

$$P_W = C(C^T C)^{-1} C^T. \tag{6}$$

In the special case that $\{\mathbf{v}_1, \mathbf{v}_2, \ldots, \mathbf{v}_k\}$ is an orthonormal basis for W, we have $C^T C = I_k$. Thus $P_W = CC^T$ (see Exercise 35).

The orthogonal projection matrix P_W generalizes the ideas about orthogonal projections presented in Sections 6.1 and 4.5 to an arbitrary subspace W. Although (6) provides a useful theoretical representation of P_W, it often proves to be unreliable for numerical computations in which $\det(C^T C)$ is very small. For our purposes, however, this method of computing P_W is adequate.

For example, in Section 6.1 we determined the orthogonal projection of $\mathbf{v} = \begin{bmatrix} 4 \\ 1 \end{bmatrix}$ onto the line with equation $y = \frac{1}{2}x$ by taking the vector $\mathbf{u} = \begin{bmatrix} 2 \\ 1 \end{bmatrix}$ lying on the line and computing

$$\mathbf{w} = \frac{\mathbf{v} \cdot \mathbf{u}}{\|\mathbf{u}\|^2}\mathbf{u} = \frac{9}{5}\begin{bmatrix} 2 \\ 1 \end{bmatrix}.$$

Since $\left\{ \begin{bmatrix} 2 \\ 1 \end{bmatrix} \right\}$ is a basis for the subspace W that consists of the vectors on the line with equation $y = \frac{1}{2}x$, we can apply 6 by taking

$$C = \begin{bmatrix} 2 \\ 1 \end{bmatrix}.$$

Then the orthogonal projection of $\mathbf{v}$ onto W is

$$P_W \mathbf{v} = C(C^T C)^{-1} C^T \mathbf{v} = \begin{bmatrix} 2 \\ 1 \end{bmatrix} [5]^{-1} [2 \ \ 1] \begin{bmatrix} 4 \\ 1 \end{bmatrix} = \frac{1}{5}\begin{bmatrix} 2 \\ 1 \end{bmatrix} [2 \ \ 1] \begin{bmatrix} 4 \\ 1 \end{bmatrix}$$

$$= \frac{1}{5}\begin{bmatrix} 4 & 2 \\ 2 & 1 \end{bmatrix} \begin{bmatrix} 4 \\ 1 \end{bmatrix} = \frac{1}{5}\begin{bmatrix} 18 \\ 9 \end{bmatrix} = \frac{9}{5}\begin{bmatrix} 2 \\ 1 \end{bmatrix}.$$

As another example, recall that in Section 5.4 we found an explicit description of the orthogonal projection operator U onto the subspace W of $\mathcal{R}^3$ defined by the equation $x_1 - x_2 + x_3 = 0$. In the earlier discussion of this problem, we found that

$$\left\{ \begin{bmatrix} 1 \\ 1 \\ 0 \end{bmatrix}, \begin{bmatrix} -1 \\ 0 \\ 1 \end{bmatrix} \right\}$$

is a basis for W. Hence by taking

$$C = \begin{bmatrix} 1 & -1 \\ 1 & 0 \\ 0 & 1 \end{bmatrix},$$

we can calculate P_W. The result is

$$P_W = C(C^T C)^{-1} C^T$$

$$= \begin{bmatrix} 1 & -1 \\ 1 & 0 \\ 0 & 1 \end{bmatrix} \begin{bmatrix} 2 & -1 \\ -1 & 2 \end{bmatrix}^{-1} \begin{bmatrix} 1 & 1 & 0 \\ -1 & 0 & 1 \end{bmatrix}$$

$$= \begin{bmatrix} \frac{2}{3} & \frac{1}{3} & -\frac{1}{3} \\ \frac{1}{3} & \frac{2}{3} & \frac{1}{3} \\ -\frac{1}{3} & \frac{1}{3} & \frac{2}{3} \end{bmatrix}.$$

Therefore

$$U\left(\begin{bmatrix} x_1 \\ x_2 \\ x_3 \end{bmatrix} \right) = P_W \begin{bmatrix} x_1 \\ x_2 \\ x_3 \end{bmatrix} = \begin{bmatrix} \frac{2}{3}x_1 + \frac{1}{3}x_2 - \frac{1}{3}x_3 \\ \frac{1}{3}x_1 + \frac{2}{3}x_2 + \frac{1}{3}x_3 \\ -\frac{1}{3}x_1 + \frac{1}{3}x_2 + \frac{2}{3}x_3 \end{bmatrix},$$

in agreement with our previous calculation.

We conclude this section with an example in which P_W is calculated for a particular subspace W.

Example 3   Let

$$\mathcal{B} = \left\{ \begin{bmatrix} 1 \\ 0 \\ 1 \\ 0 \end{bmatrix}, \begin{bmatrix} 1 \\ -1 \\ 1 \\ 0 \end{bmatrix} \right\} \quad \text{and} \quad \mathbf{v} = \begin{bmatrix} 3 \\ -2 \\ -1 \\ 4 \end{bmatrix}.$$

Find the orthogonal projection matrix for $W = \text{Span}\,\mathcal{B}$ and the orthogonal projection of $\mathbf{v}$ onto W.

Solution   Note that $\mathcal{B}$ is a basis for W, and let C be the matrix whose columns are the vectors in $\mathcal{B}$. Then

$$C^T C = \begin{bmatrix} 1 & 0 & 1 & 0 \\ 1 & -1 & 1 & 0 \end{bmatrix} \begin{bmatrix} 1 & 1 \\ 0 & -1 \\ 1 & 1 \\ 0 & 0 \end{bmatrix} = \begin{bmatrix} 2 & 2 \\ 2 & 3 \end{bmatrix}.$$

Therefore

$$P_W = C(C^T C)^{-1} C^T = \begin{bmatrix} 1 & 1 \\ 0 & -1 \\ 1 & 1 \\ 0 & 0 \end{bmatrix} \frac{1}{2} \begin{bmatrix} 3 & -2 \\ -2 & 2 \end{bmatrix} \begin{bmatrix} 1 & 0 & 1 & 0 \\ 1 & -1 & 1 & 0 \end{bmatrix}$$

$$= \begin{bmatrix} .5 & 0 & .5 & 0 \\ 0 & 1 & 0 & 0 \\ .5 & 0 & .5 & 0 \\ 0 & 0 & 0 & 0 \end{bmatrix}.$$

Moreover, the orthogonal projection of $\mathbf{v}$ onto W is

$$P_W \mathbf{v} = \begin{bmatrix} .5 & 0 & .5 & 0 \\ 0 & 1 & 0 & 0 \\ .5 & 0 & .5 & 0 \\ 0 & 0 & 0 & 0 \end{bmatrix} \begin{bmatrix} 3 \\ -2 \\ -1 \\ 4 \end{bmatrix} = \begin{bmatrix} 1 \\ -2 \\ 1 \\ 0 \end{bmatrix}.$$

Practice Problems

1. Find the orthogonal projection matrix P_W for the subspace W of $\mathcal{R}^4$, where

$$W = \text{Span} \left\{ \begin{bmatrix} 1 \\ -1 \\ 2 \\ -1 \end{bmatrix}, \begin{bmatrix} 1 \\ 1 \\ -3 \\ 2 \end{bmatrix} \right\}.$$

2. For the subspace W in problem 1, find the vector in W that is closest to

$$\mathbf{v} = \begin{bmatrix} 4 \\ 0 \\ -3 \\ 8 \end{bmatrix}.$$

3. Find the equation of the least-squares line for the data $(1, 62)$, $(3, 54)$, $(4, 50)$, $(5, 48)$, and $(7, 40)$.

Exercises

1. Determine if the following statements are true or false.

   (a) If $\mathbf{v}$ is in $\mathcal{R}^n$ and W is a subspace of $\mathcal{R}^n$, then $P\mathbf{v}$ is the vector in W that is closest to $\mathbf{v}$.

   (b) Every orthogonal projection matrix is invertible.

   (c) If W is a subspace of $\mathcal{R}^n$, then, for any vector $\mathbf{v}$ in $\mathcal{R}^n$, the vector $\mathbf{v} - P_W \mathbf{v}$ is orthogonal to every vector in W.

   (d) For a given set of data plotted in the xy-plane, the least-squares line is the unique line in the plane that minimizes the sum of the vertical distances from the data points to the line.

   (e) If $\begin{bmatrix} a_0 \\ a_1 \end{bmatrix}$ is a solution to the normal equations for the data, then $y = a_0 + a_1 x$ is the equation of the least-squares line.

In Exercises 2–6, a basis for a subspace W is given. Find P_W.

2. $\left\{ \begin{bmatrix} 1 \\ 2 \end{bmatrix} \right\}$ 3. $\left\{ \begin{bmatrix} 1 \\ -1 \\ 1 \\ 0 \end{bmatrix} \right\}$ 4. $\left\{ \begin{bmatrix} 1 \\ -1 \\ 1 \end{bmatrix}, \begin{bmatrix} -2 \\ 1 \\ -1 \end{bmatrix} \right\}$

5. $\left\{ \begin{bmatrix} 1 \\ 1 \\ 0 \\ 1 \end{bmatrix}, \begin{bmatrix} 0 \\ 1 \\ 0 \\ 1 \end{bmatrix} \right\}$ 6. $\left\{ \begin{bmatrix} 1 \\ 0 \\ 1 \\ 1 \end{bmatrix}, \begin{bmatrix} 1 \\ -1 \\ 1 \\ 1 \end{bmatrix}, \begin{bmatrix} 1 \\ 1 \\ 0 \\ 0 \end{bmatrix} \right\}$

In Exercises 7–11, find P_W for the given subspace W.

7. W is the solution set to

$$x_1 + 2x_2 - x_3 = 0.$$

8. W is the solution set to

$$x_1 + 2x_2 - 3x_3 = 0$$
$$x_1 + x_2 - 3x_3 = 0.$$

9. W is the solution set to

$$x_1 + x_2 - x_3 = 0$$
$$x_1 + 2x_2 + 3x_3 = 0.$$

10. W is the solution set to

$$x_1 + x_2 - x_3 - x_4 = 0.$$

11. W is the solution set to

$$x_1 + x_2 - x_3 + x_4 = 0$$
$$x_1 - x_2 + 3x_3 + x_4 = 0.$$

In Exercises 12–16, find the vector in the given subspace W of $\mathcal{R}^n$ that is closest to the given vector $\mathbf{v}$ in $\mathcal{R}^n$. Then find the distance from $\mathbf{v}$ to W.

12. $W = \text{Span} \left\{ \begin{bmatrix} 1 \\ -1 \end{bmatrix} \right\}$, $\mathbf{v} = \begin{bmatrix} 1 \\ 5 \end{bmatrix}$

13. $W = \text{Span}\left\{\begin{bmatrix} 1 \\ 2 \\ 1 \end{bmatrix}\right\}$, $\mathbf{v} = \begin{bmatrix} -1 \\ 1 \\ 3 \end{bmatrix}$

14. $W = \text{Span}\left\{\begin{bmatrix} 1 \\ -1 \\ 1 \end{bmatrix}, \begin{bmatrix} 1 \\ 1 \\ 1 \end{bmatrix}\right\}$, $\mathbf{v} = \begin{bmatrix} 2 \\ -1 \\ 4 \end{bmatrix}$

15. $W = \text{Span}\left\{\begin{bmatrix} 1 \\ 0 \\ 1 \\ -1 \end{bmatrix}, \begin{bmatrix} 1 \\ 2 \\ 0 \\ 1 \end{bmatrix}\right\}$, $\mathbf{v} = \begin{bmatrix} 1 \\ 1 \\ 1 \\ 1 \end{bmatrix}$

16. $W = \text{Span}\left\{\begin{bmatrix} 1 \\ 0 \\ 1 \\ 1 \end{bmatrix}, \begin{bmatrix} 1 \\ 0 \\ 0 \\ 1 \end{bmatrix}, \begin{bmatrix} 1 \\ -1 \\ 1 \\ 1 \end{bmatrix}\right\}$, $\mathbf{v} = \begin{bmatrix} 2 \\ 1 \\ -1 \\ 4 \end{bmatrix}$

In Exercises 17–22, find the equation of the least-squares line for the given data.

17. (1, 14), (3, 17), (5, 19), (7, 20)

18. (1, 30), (2, 27), (4, 21), (7, 14)

19. (1, 5), (2, 6), (3, 8), (4, 10), (5, 11)

20. (1, 2), (2, 4), (3, 7), (4, 8), (5, 10)

21. (1, 40), (3, 36), (7, 23), (8, 21), (10, 13)

22. (1, 19), (2, 17), (3, 16), (4, 14), (5, 12)

23. Suppose that a force y is applied to one end of a spring that has its other end fixed, thus stretching it to a length x. In physics, *Hooke's law* states that (within certain limits) there is a linear relation between x and y. That is, there are constants a and b such that $y = a + bx$. The coefficient b is called the *spring constant*. Use the following data to estimate the spring constant.

Length x (in.)	Force y (lb.)
3.5	1.0
4.0	2.2
4.5	2.8
5.0	4.3

In Exercises 24 and 25, use a calculator with matrix capabilities or computer software such as MATLAB.

24. A space vehicle is launched from a space platform near a space station. The vehicle moves in a direction away from the station at a constant acceleration, so that t seconds after launch, its distance y from the station (in meters) is given by the formula $y = a + bt + ct^2$. (Here a is the distance from the station to the platform at the time of launch, b is the speed of the platform relative to the space station, and $2c$ is the acceleration of the vehicle.) Use the method of least squares to obtain the best quadratic fit to the data below.

t	5	10	15	20	25	30
y	140	290	560	910	1400	2000

25. Use the method of least squares to find the best cubic fit for the points: $(-2, -4)$, $(-1, 1)$, $(0, 1)$, $(2, 10)$, and $(3, 26)$.

26.[9] Let W be a subspace of $\mathcal{R}^n$. Prove that $(P_W)^2 = P_W$.

27.[10] Let W be a subspace of $\mathcal{R}^n$. Prove that $(P_W)^T = P_W$.

28. Let W be a subspace of $\mathcal{R}^n$. Prove that, for any $\mathbf{v}$ in $\mathcal{R}^n$, $P_W\mathbf{v} = \mathbf{v}$ if and only if $\mathbf{v}$ is in W.

29. Let W be a subspace of $\mathcal{R}^n$. Prove that, for any $\mathbf{v}$ in $\mathcal{R}^n$, $P_W\mathbf{v} = \mathbf{0}$ if and only if $\mathbf{v}$ is in $W^\perp$.

30. Let W be a subspace of $\mathcal{R}^n$. Prove that a vector $\mathbf{v}$ in $\mathcal{R}^n$ is an eigenvector of P_W if and only if $\mathbf{v}$ is an eigenvector of $P_{W^\perp}$.

31. Let W be a subspace of $\mathcal{R}^n$. Prove that $(P_W\mathbf{u}) \cdot \mathbf{v} = \mathbf{u} \cdot (P_W\mathbf{v})$ for every $\mathbf{u}$ and $\mathbf{v}$ in $\mathcal{R}^n$.

32. Let W be a subspace of $\mathcal{R}^n$. Prove that $P_W P_{W^\perp} = P_{W^\perp} P_W = O$.

33. Let W be a subspace of $\mathcal{R}^n$. Prove that $P_W + P_{W^\perp} = I_n$.

34. Let V and W be subspaces of $\mathcal{R}^n$ such that for any $\mathbf{v} \in V$ and $\mathbf{w} \in W$, the vectors $\mathbf{v}$ and $\mathbf{w}$ are orthogonal. Prove that $P_V + P_W$ is an orthogonal projection matrix. Describe the subspace Z of $\mathcal{R}^n$ such that $P_Z = P_V + P_W$.

35. Suppose that $\mathcal{B} = \{\mathbf{v}_1, \mathbf{v}_2, \ldots, \mathbf{v}_k\}$ is an orthonormal basis for a subspace W of $\mathcal{R}^n$. Let C be the $n \times k$ matrix whose columns are the vectors in $\mathcal{B}$. Prove that $C^T C = I_k$.

36. Show that for any vector $\mathbf{v}$ in $\mathcal{R}^n$, the orthogonal projection of $\mathbf{v}$ on $\{\mathbf{0}\}$ is $\mathbf{0}$.

37.[11] Let W be a subspace of $\mathcal{R}^n$ having dimension k, where $0 < k < n$.
   (a) Prove that 1 and 0 are the only eigenvalues of P_W.
   (b) Prove that W and $W^\perp$ are eigenspaces of P_W corresponding to the eigenvalues 1 and 0, respectively.
   (c) Let $\mathcal{B}_1$ and $\mathcal{B}_2$ be bases for W and $W^\perp$, respectively. Recall from Exercise 32 of Section 6.2 that $\mathcal{B} = \mathcal{B}_1 \cup \mathcal{B}_2$ is a basis for $\mathcal{R}^n$. Prove that if B is the matrix whose columns are the vectors in $\mathcal{B}$, and if D is the diagonal $n \times n$ matrix whose first k diagonal entries are 1s and whose other entries are 0s, then $P_W = BDB^{-1}$.

38. Let $V = \text{Row } A$, where

$$A = \begin{bmatrix} -1 & 1 & 0 & -1 \\ 0 & 1 & -2 & 1 \\ -3 & 1 & 4 & -5 \\ 1 & 1 & -4 & 3 \end{bmatrix}.$$

Use the method described in Exercise 37 to find P_V. *Hint:* Obtain a basis for $V^\perp$ as in Example 6 of Section 6.2.

---

[9] The results of this exercise are used in Section 6.5 (page 360).
[10] The results of this exercise are used in Section 6.5 (page 360).
[11] The results of this exercise are used in Section 6.6 (page 372).

39. Let $W = \text{Null } A$, where A is the matrix in Exercise 38. Use the method described in Exercise 37 to find P_W. *Hint:* Obtain a basis for $W^{\perp}$ as in Example 6 of Section 6.2.

40. Let V and W be as in Exercises 38 and 39. Compute $P_V + P_W$. What accounts for this occurrence?

41. Let $W = \text{Col } A$, where

$$A = \begin{bmatrix} 1 & 0 & 5 & -3 \\ 0 & 1 & 2 & 4 \\ -1 & -2 & -9 & -5 \\ 1 & 1 & 7 & 1 \end{bmatrix}.$$

Use the method described in Exercise 37 to find P_W.

42. Suppose that A is an $n \times n$ matrix such that $A^2 = A^T = A$. Prove that A is the orthogonal projection matrix P_W, where $W = \text{Col } A = \{A\mathbf{v}: \mathbf{v} \in \mathcal{R}^n\}$. *Hint:* Show that for any $\mathbf{v}$ in $\mathcal{R}^n$, $\mathbf{v} = A\mathbf{v} + (I_n - A)\mathbf{v}$, $A\mathbf{v} \in W$, and $(I_n - A)\mathbf{v} \in W^{\perp}$.

43. Let E be the error sum of squares for the data $(x_1, y_1), (x_2, y_2), \ldots, (x_n, y_n)$, as defined in 4. Prove that $E = \|\mathbf{y} - (a_0 \mathbf{v}_1 + a_1 \mathbf{v}_2)\|^2$, where

$$\mathbf{v}_1 = \begin{bmatrix} 1 \\ 1 \\ \vdots \\ 1 \end{bmatrix}, \quad \mathbf{v}_2 = \begin{bmatrix} x_1 \\ x_2 \\ \vdots \\ x_n \end{bmatrix}, \quad \text{and} \quad \mathbf{y} = \begin{bmatrix} y_1 \\ y_2 \\ \vdots \\ y_n \end{bmatrix}.$$

In Exercises 44 and 45, use a calculator with matrix capabilities or computer software such as MATLAB *to solve the problem.*

44. The following table gives the approximate values of the function $y = 10 \sin x$ over the interval $[0, 2\pi]$. We will use the method of least squares to approximate this function by linear and cubic polynomials.

x	$y = 10 \sin x$
0.00000	0.00000
0.62832	5.87786
1.25664	9.51057
1.88496	9.51055
2.51328	5.87781
3.14160	−0.00007
3.76992	−5.87792
4.39824	−9.51060
5.02656	−9.51053
5.65488	−5.87775
6.28320	0.00014

(a) Use the method of least squares to find the equation of the least-squares line for the data in the table.

(b) Compute the error sum of squares associated with (a).

(c) Graph $y = 10 \sin x$ and the least-squares line obtained in (a) using the same set of axes.

(d) Use the method of least squares to produce the best cubic fit for the data.

(e) Compute the error sum of squares associated with (d).

(f) Graph $y = 10 \sin x$ and the cubic polynomial obtained in (d) using the same set of axes.

45. Let

$$\mathbf{v} = \begin{bmatrix} 6 \\ -4 \\ -2 \\ 1 \\ -1 \end{bmatrix} \quad \text{and} \quad W = \text{Span}\left\{ \begin{bmatrix} -9 \\ 5 \\ 5 \\ 6 \\ -7 \end{bmatrix}, \begin{bmatrix} -9 \\ 4 \\ 7 \\ 2 \\ 4 \end{bmatrix}, \begin{bmatrix} 4 \\ 9 \\ 7 \\ -5 \\ -4 \end{bmatrix} \right\}.$$

Compute the orthogonal projection matrix P_W, and use it to find the distance from $\mathbf{v}$ to W.

Solutions to the Practice Problems

1. Let

$$C = \begin{bmatrix} 1 & 1 \\ -1 & 1 \\ 2 & -3 \\ -1 & 2 \end{bmatrix}.$$

Then

$$P_W = C(C^T C)^{-1} C^T = \frac{1}{41} \begin{bmatrix} 38 & -8 & 1 & 7 \\ -8 & 6 & -11 & 5 \\ 1 & -11 & 27 & -16 \\ 7 & 5 & -16 & 11 \end{bmatrix}.$$

2. Let P_W be the orthogonal projection matrix obtained in problem 1. Then the vector in W that is closest to $\mathbf{v}$ is given by

$$P_W \mathbf{v} = \frac{1}{41} \begin{bmatrix} 38 & -8 & 1 & 7 \\ -8 & 6 & -11 & 5 \\ 1 & -11 & 27 & -16 \\ 7 & 5 & -16 & 11 \end{bmatrix} \begin{bmatrix} 4 \\ 0 \\ -3 \\ 8 \end{bmatrix} = \begin{bmatrix} 5 \\ 1 \\ -5 \\ 4 \end{bmatrix}.$$

3. Let

$$C = \begin{bmatrix} 1 & 1 \\ 1 & 3 \\ 1 & 4 \\ 1 & 5 \\ 1 & 7 \end{bmatrix} \quad \text{and} \quad \mathbf{y} = \begin{bmatrix} 62 \\ 54 \\ 50 \\ 48 \\ 40 \end{bmatrix}.$$

Then $y = a_0 + a_1 x$, where

$$\begin{bmatrix} a_0 \\ a_1 \end{bmatrix} = (C^T C)^{-1} C^T \mathbf{y}$$

$$= \begin{bmatrix} 65.2 \\ -3.6 \end{bmatrix}.$$

Hence the equation of the least-squares line is

$$y = 65.2 - 3.6x.$$

6.4   Orthogonal Matrices and Operators

In Chapter 2, we studied the functions from $\mathcal{R}^n$ to $\mathcal{R}^n$ that preserve the operations of vector addition and scalar multiplication. Now that we have introduced the concept of the norm of a vector, it is natural to ask which linear operators on $\mathcal{R}^n$ also preserve norms; that is, which operators T satisfy $\|T(\mathbf{u})\| = \|\mathbf{u}\|$ for every vector $\mathbf{u}$ in $\mathcal{R}^n$. These linear operators and their standard matrices are extremely useful in numerical calculations because they do not magnify any roundoff or experimental error. Because such operators on $\mathcal{R}^2$ also preserve the angle between nonzero vectors (see Exercise 40), it follows that they also preserve many familiar properties from geometry.

It is clear that an arbitrary operator on $\mathcal{R}^n$ does not have this property. For if an operator U on $\mathcal{R}^n$ has an eigenvalue λ other than ± 1 with corresponding eigenvector $\mathbf{v}$, then $\|U(\mathbf{v})\| = \|\lambda \mathbf{v}\| = |\lambda| \cdot \|\mathbf{v}\| \neq \|\mathbf{v}\|$. There are, however, familiar operators that do have this property, as our first example shows.

Example 1 Let T be the linear operator on $\mathcal{R}^2$ that rotates a vector through an angle θ. Clearly $T(\mathbf{v})$ has the same length as $\mathbf{v}$ for every $\mathbf{v}$ in $\mathcal{R}^2$, and therefore $\|T(\mathbf{v})\| = \|\mathbf{v}\|$ for every $\mathbf{v}$ in $\mathcal{R}^2$.

A linear operator that rotates every vector in $\mathcal{R}^2$ through a fixed angle is called a **rotation operator** or simply a **rotation**. Clearly a linear operator on $\mathcal{R}^2$ is a rotation if and only if its standard matrix is a rotation matrix.

Because of the connection between linear operators and their standard matrices, we can study linear operators on $\mathcal{R}^n$ that preserve norms by studying the $n \times n$ matrices Q such that $\|Q\mathbf{u}\| = \|\mathbf{u}\|$ for every $\mathbf{u}$ in $\mathcal{R}^n$. Consider an arbitrary column $\mathbf{q}_j$ of such a matrix. Since

$$\|\mathbf{q}_j\| = \|Q\mathbf{e}_j\| = \|\mathbf{e}_j\| = 1, \tag{7}$$

the norm of every column of Q is 1. Moreover, for any two distinct columns of Q, say $\mathbf{q}_i$ and $\mathbf{q}_j$, we have

$$\|\mathbf{q}_i + \mathbf{q}_j\|^2 = \|Q\mathbf{e}_i + Q\mathbf{e}_j\|^2 = \|Q(\mathbf{e}_i + \mathbf{e}_j)\|^2 = \|\mathbf{e}_i + \mathbf{e}_j\|^2$$

$$= 2 = \|\mathbf{q}_i\|^2 + \|\mathbf{q}_j\|^2. \tag{8}$$

Hence $\mathbf{q}_i$ and $\mathbf{q}_j$ are orthogonal by Theorem 6.2. It follows that the columns of Q form an orthonormal set of distinct vectors, and so constitue an orthonormal basis for $\mathcal{R}^n$. This motivates the following definitions.

Definitions. An $n \times n$ matrix Q is called an **orthogonal matrix** (or simply **orthogonal**) if the columns of Q form an orthonormal basis for $\mathcal{R}^n$. A linear operator T on $\mathcal{R}^n$ is called an **orthogonal operator** (or simply **orthogonal**) if its standard matrix is an orthogonal matrix.

To verify that an $n \times n$ matrix Q is orthogonal, it suffices to show that the columns of Q are distinct and form an orthonormal set.

Example 2 Consider the θ-rotation matrix

$$A_\theta = \begin{bmatrix} \cos \theta & -\sin \theta \\ \sin \theta & \cos \theta \end{bmatrix}.$$

Since

$$\begin{bmatrix} \cos\theta \\ \sin\theta \end{bmatrix} \cdot \begin{bmatrix} -\sin\theta \\ \cos\theta \end{bmatrix} = (\cos\theta)(-\sin\theta) + (\sin\theta)(\cos\theta) = 0,$$

$$\begin{bmatrix} \cos\theta \\ \sin\theta \end{bmatrix} \cdot \begin{bmatrix} \cos\theta \\ \sin\theta \end{bmatrix} = \cos^2\theta + \sin^2\theta = 1,$$

and

$$\begin{bmatrix} -\sin\theta \\ \cos\theta \end{bmatrix} \cdot \begin{bmatrix} -\sin\theta \\ \cos\theta \end{bmatrix} = \sin^2\theta + \cos^2\theta = 1,$$

A_θ is an orthogonal matrix because its columns form an orthonormal set of two distinct vectors in $\mathcal{R}^2$.   ○

The following theorem lists several conditions that are equivalent to a matrix being orthogonal.

Theorem 6.9
Let Q be an $n \times n$ matrix. Then the following conditions are equivalent.

    (a) Q *is orthogonal.*
    (b) Q *is invertible, and* $Q^T = Q^{-1}$.
    (c) $Q\mathbf{u} \cdot Q\mathbf{v} = \mathbf{u} \cdot \mathbf{v}$ *for any* $\mathbf{u}$ *and* $\mathbf{v}$ *in* $\mathcal{R}^n$. (Q *preserves dot products.*)
    (d) $\|Q\mathbf{u}\| = \|\mathbf{u}\|$ *for any* $\mathbf{u}$ *in* $\mathcal{R}^n$. (Q *preserves norms.*)

Proof We show that (a) $\Rightarrow$ (b) $\Rightarrow$ (c) $\Rightarrow$ (d) $\Rightarrow$ (a) to establish the equivalence of these conditions.

    (a) *implies* (b) Suppose that Q is orthogonal. Since the columns of Q form an orthonormal basis for $\mathcal{R}^n$, Q has rank n, and hence is invertible. Next observe that the (i, j)-entry of $Q^T Q$ is the dot product of the ith row of Q^T and $\mathbf{q}_j$. But the ith row of Q^T equals $\mathbf{q}_i$ and hence the (i, j)-entry of $Q^T Q$ equals $\mathbf{q}_i \cdot \mathbf{q}_j$. Since $\mathbf{q}_i \cdot \mathbf{q}_j = 1$ if $i = j$ and $\mathbf{q}_i \cdot \mathbf{q}_j = 0$ if $i \neq j$, we see that $Q^T Q = I_n$. Therefore $Q^T = Q^{-1}$ by Theorem 2.6.

    (b) *implies* (c) Assume (b). Then for any $\mathbf{u}$ and $\mathbf{v}$ in $\mathcal{R}^n$,

$$Q\mathbf{u} \cdot Q\mathbf{v} = \mathbf{u} \cdot Q^T Q\mathbf{v} = \mathbf{u} \cdot Q^{-1} Q\mathbf{v} = \mathbf{u} \cdot \mathbf{v}.$$

    (c) *implies* (d) Assume (c). Then for any $\mathbf{u}$ in $\mathcal{R}^n$

$$\|Q\mathbf{u}\| = \sqrt{Q\mathbf{u} \cdot Q\mathbf{u}} = \sqrt{\mathbf{u} \cdot \mathbf{u}} = \|\mathbf{u}\|.$$

The proof that (d) implies (a) follows from 7 and 8   ❑

    One immediate consequence of Theorem 6.9 is that an $n \times n$ matrix Q is orthogonal if and only if $QQ^T = I_n$, and hence Q *is orthogonal if and only if the rows of Q form an orthonormal basis for* $\mathcal{R}^n$ (see Exercise 22).
    The following general result lists some important properties of orthogonal matrices.

Theorem 6.10
Let P and Q be $n \times n$ orthogonal matrices. Then the following statements are true.

    (a) $\det Q = \pm 1$.
    (b) PQ *is an orthogonal matrix.*
    (c) Q^{-1} *is an orthogonal matrix.*

Proof (a) Since Q is an orthogonal matrix, $QQ^T = I_n$, and hence by Theorem 6.9(b),

$$1 = \det I_n = \det QQ^T = (\det Q)(\det Q^T) = (\det Q)(\det Q) = (\det Q)^2.$$

Therefore $\det Q = \pm 1$.

(b) Because P and Q are orthogonal, they are invertible, and hence PQ is invertible. Therefore by Theorem 6.9(b)

$$(PQ)^T = Q^T P^T = Q^{-1} P^{-1} = (PQ)^{-1},$$

and hence PQ is an orthogonal matrix by Theorem 6.9(b).

(c) Since $(Q^{-1})^T = (Q^T)^{-1} = (Q^{-1})^{-1}$, we see that Q^{-1} is orthogonal by Theorem 6.9(b). ∎

Since a linear operator is orthogonal if and only if its standard matrix is orthogonal, parts of Theorems 6.9 and 6.10 can be restated in terms of orthogonal operators. We state these results below, leaving the proofs as exercises.

If T is a linear operator on $\mathcal{R}^n$, then the following statements are equivalent.

(a) T is an orthogonal operator.

(b) $T(\mathbf{u}) \cdot T(\mathbf{v}) = \mathbf{u} \cdot \mathbf{v}$ for all $\mathbf{u}$ and $\mathbf{v}$ in $\mathcal{R}^n$. (T preserves dot products.)

(c) $\|T(\mathbf{u})\| = \|\mathbf{u}\|$ for all $\|\mathbf{u}\|$ in $\mathcal{R}^n$. (T preserves norms.)

If T and U are orthogonal operators on $\mathcal{R}^n$, then TU and T^{-1} are orthogonal operators on $\mathcal{R}^n$.

In Examples 1 and 2, we saw that rotations of the plane are orthogonal operators. As an application of Theorem 6.10, we now prove that reflections of the plane, as defined in Section 4.5, are also orthogonal operators.

Example 3

Let $\mathcal{L}$ be a line in $\mathcal{R}^2$ that contains $\mathbf{0}$, and let $T: \mathcal{R}^2 \to \mathcal{R}^2$ be the reflection about $\mathcal{L}$. Show that T is an orthogonal operator.

Solution

Let $\mathbf{b}_1$ be a unit vector in the direction of $\mathcal{L}$, and let $\mathbf{b}_2$ be a unit vector in a direction perpendicular to $\mathcal{L}$ (see Figure 6.13). Then $\mathcal{B} = \{\mathbf{b}_1, \mathbf{b}_2\}$ is an orthonormal basis for $\mathcal{R}^2$, and hence $P = [\mathbf{b}_1 \quad \mathbf{b}_2]$ is an orthogonal matrix.

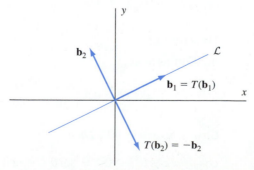

FIGURE 6.13

Furthermore, $T(\mathbf{b}_1) = \mathbf{b}_1$, and $T(\mathbf{b}_2) = -\mathbf{b}_2$, and hence $\mathbf{b}_1$ and $\mathbf{b}_2$ are eigenvectors of T with corresponding eigenvalues 1 and -1, respectively. Then $[T]_\mathcal{B} = PQP^{-1}$ by Theorem 4.10, where Q is the standard matrix of T. Since $\mathcal{B}$ is a basis for $\mathcal{R}^2$ consisting of eigenvectors of T, $[T]_\mathcal{B}$ is the diagonal matrix $D = \begin{bmatrix} 1 & 0 \\ 0 & -1 \end{bmatrix}$. Notice that D is an orthogonal matrix, and hence $Q = PDP^{-1}$ is a product of orthogonal matrices. So by Theorem 6.10, Q is an orthogonal matrix. Therefore T is an orthogonal operator.   ○

Orthogonal Operators on $\mathcal{R}^2$

By Examples 2 and 3, we see that rotations and reflections are orthogonal operators on $\mathcal{R}^2$. We now show that these are the only orthogonal operators on $\mathcal{R}^2$. Furthermore, we see how to use determinants to distinguish between these two kinds of operators.

Theorem 6.11
Let T be an orthogonal linear operator on $\mathcal{R}^2$ with standard matrix Q. Then the following statements are true.

(a) *If $\det Q = 1$, then T is a rotation.*
(b) *If $\det Q = -1$, then T is a reflection.*

Proof Suppose that $Q = \begin{bmatrix} a & c \\ b & d \end{bmatrix}$. Since the columns of Q are unit vectors, $a^2 + b^2 = 1$ and $c^2 + d^2 = 1$. Thus there exist angles θ and μ such that $a = \cos\theta$, $b = \sin\theta$, $c = \cos\mu$, and $d = \sin\mu$. Since the two columns of Q are orthogonal, θ and μ can be chosen so that they differ by 90°, that is, $\mu = \theta \pm 90°$ (see Figure 6.14). We consider each case separately.

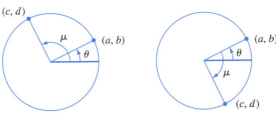

Case 1: $\mu = \theta + 90°$ Case 2: $\mu = \theta - 90°$ **FIGURE 6.14**

Case 1. $\mu = \theta + 90°$. *Then*

$$\cos\mu = \cos(\theta + 90°) = -\sin\theta \quad \text{and} \quad \sin\mu = \sin(\theta + 90°) = \cos\theta.$$

Therefore $Q = \begin{bmatrix} \cos\theta & -\sin\theta \\ \sin\theta & \cos\theta \end{bmatrix}$, which we recognize as the rotation matrix A_θ. Furthermore,

$$\det Q = \det \begin{bmatrix} \cos\theta & -\sin\theta \\ \sin\theta & \cos\theta \end{bmatrix} = \cos^2\theta + \sin^2\theta = 1.$$

Case 2. $\mu = \theta - 90°$. *Then*

$$\cos\mu = \cos(\theta - 90°) = \sin\theta \quad \text{and} \quad \sin\mu = \sin(\theta - 90°) = -\cos\theta.$$

Therefore $Q = \begin{bmatrix} \cos\theta & \sin\theta \\ \sin\theta & -\cos\theta \end{bmatrix}$, and so the characteristic polynomial of Q is

$$\det(Q - tI_2) = \det \begin{bmatrix} \cos\theta - t & \sin\theta \\ \sin\theta & -\cos\theta - t \end{bmatrix}$$

$$= (\cos\theta - t)(-\cos\theta - t) - \sin^2\theta$$

$$= t^2 - \cos^2\theta - \sin^2\theta$$

$$= t^2 - 1$$

$$= (t+1)(t-1).$$

It follows that Q, and hence T, has the eigenvalues 1 and -1. Let $\mathbf{b}_1$ and $\mathbf{b}_2$ be eigenvectors corresponding to the eigenvalues 1 and -1, respectively. Then $T(\mathbf{b}_1) = \mathbf{b}_1$ and $T(\mathbf{b}_2) = -\mathbf{b}_2$. Moreover

$$\mathbf{b}_1 \cdot \mathbf{b}_2 = T(\mathbf{b}_1) \cdot T(\mathbf{b}_2) = \mathbf{b}_1 \cdot (-\mathbf{b}_2) = -\mathbf{b}_1 \cdot \mathbf{b}_2.$$

Therefore, $2\mathbf{b}_1 \cdot \mathbf{b}_2 = 0$, and hence $\mathbf{b}_1 \cdot \mathbf{b}_2 = 0$. So $\mathbf{b}_1$ and $\mathbf{b}_2$ are orthogonal. Now let $\mathcal{L}$ be the line containing $\mathbf{0}$ in the direction of $\mathbf{b}_1$. Then $\mathbf{b}_2$ is a nonzero vector in a direction perpendicular to $\mathcal{L}$. It follows that T is a reflection about $\mathcal{L}$. Furthermore,

$$\det Q = \det \begin{bmatrix} \cos\theta & \sin\theta \\ \sin\theta & -\cos\theta \end{bmatrix} = -\cos^2\theta - \sin^2\theta = -1.$$

To summarize, we have shown that under case 1, T is a rotation and $\det Q = 1$, and that under case 2, T is a reflection and $\det Q = -1$. Since these are the only two cases, the result is established. ❑

Example 4   For the matrix $Q = \begin{bmatrix} 0.6 & 0.8 \\ 0.8 & -0.6 \end{bmatrix}$, verify that Q is the standard matrix of a reflection, and find the equation of the line of reflection $\mathcal{L}$.

Solution   First observe that

$$QQ^T = \begin{bmatrix} 0.6 & 0.8 \\ 0.8 & -0.6 \end{bmatrix} \begin{bmatrix} 0.6 & 0.8 \\ 0.8 & -0.6 \end{bmatrix} = I_2,$$

and hence Q is invertible and $Q^T = Q^{-1}$. Next observe that

$$\det Q = -0.6^2 - 0.8^2 = -1,$$

and hence Q is the standard matrix of a reflection by Theorem 6.11. To determine the equation of $\mathcal{L}$, we first find an eigenvector of Q corresponding to the eigenvalue 1. Such a vector is a nonzero solution to the homogeneous system of equations

$$(Q - I_2)\mathbf{x} = \mathbf{0},$$

that is,

$$-0.4x_1 + 0.8x_2 = 0$$
$$0.8x_1 - 1.6x_2 = 0.$$

The vector $\mathbf{b} = \begin{bmatrix} 2 \\ 1 \end{bmatrix}$ is such a solution. Notice that $\mathbf{b}$ lies on the line with equation $y = 0.5x$, which is, therefore, the equation of $\mathcal{L}$. ○

Example 5   For the matrix $Q = \begin{bmatrix} -0.6 & 0.8 \\ -0.8 & -0.6 \end{bmatrix}$, verify that Q is the standard matrix of a rotation, and find θ, the angle of rotation.

Solution   Observe that $QQ^T = I_2$ and $\det Q = 1$, and hence Q is an orthogonal matrix and is the standard matrix of a rotation by Theorem 6.11. Thus Q is a rotation matrix, and hence

$$Q = \begin{bmatrix} -0.6 & 0.8 \\ -0.8 & -0.6 \end{bmatrix} = \begin{bmatrix} \cos\theta & -\sin\theta \\ \sin\theta & \cos\theta \end{bmatrix} = A_\theta,$$

where θ is the angle of rotation. Equating corresponding entries in the first column we see that

$$\cos\theta = -0.6 \quad \text{and} \quad \sin\theta = -0.8.$$

It follows that θ is in the third quadrant and

$$\theta = 180° + \cos^{-1}(0.6) \approx 233.2°.$$

We have seen that the composition of two rotations on $\mathcal{R}^2$ is also a rotation. But what about the composition of two reflections, or the composition of a reflection and a rotation? The next theorem, which is an easy consequence of Theorem 6.11, answers these questions.

Theorem 6.12
Let T and U be orthogonal operators on $\mathcal{R}^2$. Then the following statements are true.
    (a) *If both T and U are reflections, then TU is a rotation.*
    (b) *If one of T or U is a reflection and the other is a rotation, then TU is a reflection.*

Proof Let P and Q be the standard matrices of T and U, respectively. Then PQ is the standard matrix of TU. Furthermore, TU is an orthogonal operator since both T and U are orthogonal operators.
    (a) Since both T and U are reflections, $\det P = \det Q = -1$ by Theorem 6.11. Hence

$$\det PQ = (\det P)(\det Q) = (-1)(-1) = 1,$$

and therefore PQ is a rotation by Theorem 6.11.
    (b) The proof of (b) is similar and is left as an exercise. ❑

Rigid Motions*

A function $F: \mathcal{R}^n \to \mathcal{R}^n$ is called a **rigid motion** if

$$\|F(\mathbf{u}) - F(\mathbf{v})\| = \|\mathbf{u} - \mathbf{v}\|$$

for all $\mathbf{u}$ and $\mathbf{v}$ in $\mathcal{R}^n$. In geometric terms, a rigid motion preserves distances between vectors and their images.

---

*This subsection can be omitted without loss of continuity.

Any orthogonal operator is a rigid motion because if T is an orthogonal operator on $\mathcal{R}^n$, then for any $\mathbf{u}$ and $\mathbf{v}$ in $\mathcal{R}^n$

$$\|T(\mathbf{u}) - T(\mathbf{v})\| = \|T(\mathbf{u} - \mathbf{v})\| = \|\mathbf{u} - \mathbf{v}\|.$$

Furthermore, *any rigid motion that is also linear is an orthogonal operator* because if F is a linear rigid motion, then $F(\mathbf{0}) = \mathbf{0}$, and hence, for any vector $\mathbf{v}$ in $\mathcal{R}^n$,

$$\|F(\mathbf{v})\| = \|F(\mathbf{v}) - \mathbf{0}\| = \|F(\mathbf{v}) - F(\mathbf{0})\| = \|\mathbf{v} - \mathbf{0}\| = \|\mathbf{v}\|.$$

Therefore F is an orthogonal operator by Theorem 6.9(d).

There is one kind of rigid motion that is not usually linear, namely, a *translation*. For any $\mathbf{b}$ in $\mathcal{R}^n$, the function $F_{\mathbf{b}}\colon \mathcal{R}^n \to \mathcal{R}^n$ defined by $F_{\mathbf{b}}(\mathbf{v}) = \mathbf{v} + \mathbf{b}$ is called the **translation by b**. If $\mathbf{b} \neq \mathbf{0}$, then F is not linear because $F_{\mathbf{b}}(\mathbf{0}) = \mathbf{b} \neq \mathbf{0}$. However, $F_{\mathbf{b}}$ is a rigid motion because for any $\mathbf{u}$ and $\mathbf{v}$ in $\mathcal{R}^n$

$$\|F_{\mathbf{b}}(\mathbf{u}) - F_{\mathbf{b}}(\mathbf{v})\| = \|(\mathbf{u} + \mathbf{b}) - (\mathbf{v} + \mathbf{b})\| = \|\mathbf{u} - \mathbf{v}\|.$$

We can use function composition to combine rigid motions to create new ones because *the composition of two rigid motions on $\mathcal{R}^n$ is a rigid motion on $\mathcal{R}^n$* (see Exercise 32). It follows, for example, that if $F_{\mathbf{b}}$ is a translation and T is an orthogonal operator on $\mathcal{R}^n$, then the composition $F_{\mathbf{b}}T$ is a rigid motion. It is remarkable that the converse of this is also true, that is, any rigid motion on $\mathcal{R}^n$ can be represented as the composition of an orthogonal operator followed by a translation. To establish this result, we prove the following theorem.

Theorem 6.13

Let $T\colon \mathcal{R}^n \to \mathcal{R}^n$ be a rigid motion such that $T(\mathbf{0}) = \mathbf{0}$. Then the following statements are true.

(a) $\|T(\mathbf{u})\| = \|\mathbf{u}\|$ for every $\mathbf{u}$ in $\mathcal{R}^n$.
(b) $T(\mathbf{u}) \cdot T(\mathbf{v}) = \mathbf{u} \cdot \mathbf{v}$ for all $\mathbf{u}$ and $\mathbf{v}$ in $\mathcal{R}^n$.
(c) T is linear.
(d) T is an orthogonal operator.

Proof The proof of (a) is left as an exercise.

(b) Let $\mathbf{u}$ and $\mathbf{v}$ be in $\mathcal{R}^n$. Observe that

$$\|T(\mathbf{u}) - T(\mathbf{v})\|^2 = \|T(\mathbf{u})\|^2 - 2T(\mathbf{u}) \cdot T(\mathbf{v}) + \|T(\mathbf{v})\|^2,$$

and

$$\|\mathbf{u} - \mathbf{v}\|^2 = \|\mathbf{u}\|^2 - 2\mathbf{u} \cdot \mathbf{v} + \|\mathbf{v}\|^2.$$

Since T is a rigid motion, $\|T(\mathbf{u}) - T(\mathbf{v})\|^2 = \|\mathbf{u} - \mathbf{v}\|^2$. Hence (b) follows from the two preceding equations and (a).

(c) Let $\mathbf{u}$ and $\mathbf{v}$ be in $\mathcal{R}^n$. Then, by (a) and (b), we have

$$\|T(\mathbf{u} + \mathbf{v}) - T(\mathbf{u}) - T(\mathbf{v})\|^2$$

$$= [T(\mathbf{u} + \mathbf{v}) - T(\mathbf{u}) - T(\mathbf{v})] \cdot [T(\mathbf{u} + \mathbf{v}) - T(\mathbf{u}) - T(\mathbf{v})]$$

$$= \|T(\mathbf{u} + \mathbf{v})\|^2 + \|T(\mathbf{u})\|^2 + \|T(\mathbf{v})\|^2 - 2T(\mathbf{u} + \mathbf{v}) \cdot T(\mathbf{u})$$
$$\quad -2T(\mathbf{u} + \mathbf{v}) \cdot T(\mathbf{v}) + 2T(\mathbf{u}) \cdot T(\mathbf{v})$$

$$= \|\mathbf{u} + \mathbf{v}\|^2 + \|\mathbf{u}\|^2 + \|\mathbf{v}\|^2 - 2(\mathbf{u} + \mathbf{v}) \cdot \mathbf{u}$$
$$\quad -2(\mathbf{u} + \mathbf{v}) \cdot \mathbf{v} + 2\mathbf{u} \cdot \mathbf{v}.$$

We leave it to the reader to show that the last expression equals 0. Therefore $T(\mathbf{u}+\mathbf{v})-T(\mathbf{u})-T(\mathbf{v}) = \mathbf{0}$, and so $T(\mathbf{u}+\mathbf{v}) = T(\mathbf{u})+T(\mathbf{v})$. Thus T preserves vector addition. Similarly (see Exercise 34), T preserves scalar multiplication, and hence T is linear.

(d) This now follows from (c) and (a). ☐

Consider any rigid motion F on $\mathcal{R}^n$, and let $T\colon \mathcal{R}^n \to \mathcal{R}^n$ be defined by

$$T(\mathbf{v}) = F(\mathbf{v}) - F(\mathbf{0}).$$

Then T is a rigid motion and $T(\mathbf{0}) = F(\mathbf{0}) - F(\mathbf{0}) = \mathbf{0}$. Therefore T is an orthogonal operator by Theorem 6.13. Furthermore

$$F(\mathbf{v}) = T(\mathbf{v}) + F(\mathbf{0})$$

for any $\mathbf{v}$ in $\mathcal{R}^n$. Thus, setting $\mathbf{b} = F(\mathbf{0})$, we obtain

$$F(\mathbf{v}) = F_{\mathbf{b}}T(\mathbf{v})$$

for any $\mathbf{v}$ in $\mathcal{R}^n$, and hence F is the composition $F = F_{\mathbf{b}}T$. Combining this observation with Theorem 6.11 yields the following result.

> Any rigid motion on $\mathcal{R}^n$ is the composition of an orthogonal operator followed by a translation. Hence any rigid motion on $\mathcal{R}^2$ is the composition of a rotation or a reflection followed by a translation.

Practice Problems

1. Determine if each of the following matrices is orthogonal.

(a) $\begin{bmatrix} .7 & -.3 \\ .3 & .7 \end{bmatrix}$

(b) $\begin{bmatrix} .3\sqrt{2} & -.8 & .3\sqrt{2} \\ .4\sqrt{2} & .6 & .4\sqrt{2} \\ .5\sqrt{2} & 0 & -.5\sqrt{2} \end{bmatrix}$

2. Show that each of the following functions $T\colon \mathcal{R}^2 \to \mathcal{R}^2$ is an orthogonal operator on $\mathcal{R}^2$. Then determine if it is a rotation or a reflection. If it is a rotation, give the angle of rotation; if it is a reflection, give the line of reflection.

(a) $T\left(\begin{bmatrix} x_1 \\ x_2 \end{bmatrix}\right) = \dfrac{1}{13}\begin{bmatrix} 5x_1 - 12x_2 \\ 12x_1 + 5x_2 \end{bmatrix}$

(b) $T\left(\begin{bmatrix} x_1 \\ x_2 \end{bmatrix}\right) = \dfrac{1}{61}\begin{bmatrix} -60x_1 + 11x_2 \\ 11x_1 + 60x_2 \end{bmatrix}$

Exercises

1. Determine if the following statements are true or false.

(a) The rows of an $n \times n$ orthogonal matrix form an orthonormal basis for $\mathcal{R}^n$.

(b) If $T\colon \mathcal{R}^n \to \mathcal{R}^n$ is a function with the property that $\|T(\mathbf{u}) - T(\mathbf{v})\| = \|\mathbf{u} - \mathbf{v}\|$ for all vectors $\mathbf{u}$ and $\mathbf{v}$ in $\mathcal{R}^n$, then T is an orthogonal operator.

(c) Every linear operator preserves dot products.

(d) If a linear operator preserves dot products, then it preserves norms.

(e) If P is an orthogonal matrix, then P^T is an orthogonal matrix.

(f) If P and Q are $n \times n$ orthogonal matrices, then PQ^T is an orthogonal matrix.

(g) If P and Q are $n \times n$ orthogonal matrices, then $P+Q$ is an orthogonal matrix.

(h) If P is an $n \times n$ matrix such that $\det P = \pm 1$, then P is an orthogonal matrix.

In Exercises 2–8, determine if the given matrix is orthogonal.

2. $\begin{bmatrix} 1 & 1 \\ 1 & -1 \end{bmatrix}$

3. $\begin{bmatrix} 0.6 & 0.4 \\ 0.4 & -0.6 \end{bmatrix}$

4. I_5

5. $\begin{bmatrix} 0 & 1 & 0 \\ 0 & 0 & 1 \\ 1 & 0 & 0 \end{bmatrix}$

6. $\dfrac{1}{\sqrt{3}}\begin{bmatrix} 1 & 1 & 1 \\ 1 & -1 & 1 \\ 1 & 0 & -2 \end{bmatrix}$

7. $\dfrac{1}{\sqrt{2}}\begin{bmatrix} 1 & 1 \\ 0 & 0 \\ 1 & -1 \end{bmatrix}$ 8. $\begin{bmatrix} \frac{2}{3} & \frac{\sqrt{2}}{2} & \frac{\sqrt{2}}{6} \\ \frac{2}{3} & -\frac{\sqrt{2}}{2} & \frac{\sqrt{2}}{6} \\ \frac{1}{3} & 0 & \frac{-2\sqrt{2}}{3} \end{bmatrix}$

In Exercises 9–16, determine if the given orthogonal matrix is the standard matrix of a rotation or a reflection. If the operator is a rotation, determine the angle of rotation. If the operator is a reflection, determine the equation of the line of reflection.

9. $\dfrac{1}{\sqrt{2}}\begin{bmatrix} 1 & 1 \\ 1 & -1 \end{bmatrix}$ 10. $\dfrac{1}{\sqrt{2}}\begin{bmatrix} 1 & -1 \\ 1 & 1 \end{bmatrix}$ 11. $\dfrac{1}{2}\begin{bmatrix} \sqrt{3} & -1 \\ 1 & \sqrt{3} \end{bmatrix}$

12. $\dfrac{1}{2}\begin{bmatrix} -\sqrt{3} & 1 \\ 1 & \sqrt{3} \end{bmatrix}$ 13. $\dfrac{1}{13}\begin{bmatrix} 5 & 12 \\ 12 & -5 \end{bmatrix}$ 14. $\begin{bmatrix} 0 & 1 \\ 1 & 0 \end{bmatrix}$

15. $\begin{bmatrix} 0 & 1 \\ -1 & 0 \end{bmatrix}$ 16. $\dfrac{1}{2}\begin{bmatrix} -1 & \sqrt{3} \\ \sqrt{3} & 1 \end{bmatrix}$

17. Let $0° < \theta < 180°$ be a fixed angle, and suppose that T is the linear operator on $\mathcal{R}^3$ such that

$$T(\mathbf{e}_1) = \cos\theta\mathbf{e}_1 + \sin\theta\mathbf{e}_2$$

$$T(\mathbf{e}_2) = -\sin\theta\mathbf{e}_1 + \cos\theta\mathbf{e}_2$$

$$T(\mathbf{e}_3) = \mathbf{e}_3.$$

(a) Prove that T is an orthogonal operator.
(b) Find the eigenvalues of T and a basis for each eigenspace.
(c) Give a geometric description of T.

18. Let T be the linear operator on $\mathcal{R}^3$ defined by
$$T\left(\begin{bmatrix} x_1 \\ x_2 \\ x_3 \end{bmatrix}\right) = \begin{bmatrix} -x_1 \\ x_2 \\ x_3 \end{bmatrix}.$$ Prove that T is an orthogonal operator.

19. Let Q be a 2×2 orthogonal matrix such that $Q \neq I_2$ and $Q \neq -I_2$. Prove that Q is diagonalizable if and only if Q is a reflection.

20. Let W be a subspace of $\mathcal{R}^n$. Let T be the linear operator on $\mathcal{R}^n$ defined by $T(\mathbf{v}) = \mathbf{w} - \mathbf{z}$, where $\mathbf{v} = \mathbf{w} + \mathbf{z}$, $\mathbf{w}$ is in W, and $\mathbf{z}$ is in $W^\perp$ (see Theorem 6.7).

(a) Prove that T is an orthogonal operator.
(b) Let $U: \mathcal{R}^n \rightarrow \mathcal{R}^n$ be the function defined by $U(\mathbf{v}) = (1/2)(\mathbf{v} + T(\mathbf{v}))$ for $\mathbf{v} \in \mathcal{R}^n$. Prove that the standard matrix of U is P_W, the orthogonal projection matrix for W.

21. Let $\{\mathbf{v}, \mathbf{w}\}$ be an orthonormal basis for $\mathcal{R}^2$, and let $T: \mathcal{R}^2 \rightarrow \mathcal{R}^2$ be the function defined by

$$T(\mathbf{u}) = (\mathbf{u} \cdot \mathbf{v}\cos\theta + \mathbf{u} \cdot \mathbf{w}\sin\theta)\mathbf{v}$$

$$+ (-\mathbf{u} \cdot \mathbf{v}\sin\theta + \mathbf{u} \cdot \mathbf{w}\cos\theta)\mathbf{w}.$$

Prove that T is an orthogonal operator.

22. Let Q be an $n \times n$ matrix. Prove that Q is an orthogonal

matrix if and only if the rows of Q form an orthonormal basis for $\mathcal{R}^n$. *Hint:* Interpret the (i, j)-entry of QQ^T as the dot product of the ith and jth rows of Q.

23. Use Theorem 6.10 to prove that if T and U are orthogonal operators on $\mathcal{R}^n$, then both TU and T^{-1} are orthogonal operators.

24. Prove Theorem 6.12(b).

25.[12] Prove that if Q is an orthogonal matrix and λ is a (real) eigenvalue of Q, then $\lambda = \pm 1$.

26. Let U be a reflection and T a rotation of $\mathcal{R}^2$. Prove the following equalities.

(a) $U^2 = I$, where I is the identity transformation on $\mathcal{R}^2$, and so $U^{-1} = U$.
(b) $TUT = U$. *Hint:* Consider TU.
(c) $UTU = T^{-1}$.

27. Let T be an orthogonal operator on $\mathcal{R}^2$.

(a) Prove that if T is a rotation, then T^{-1} is also a rotation. How is the angle of rotation of T^{-1} related to the angle of rotation of T?
(b) Prove that if T is a reflection, then T^{-1} is also a reflection. How is the line of reflection of T^{-1} related to the line of reflection of T?

28. Let U be a reflection of $\mathcal{R}^2$, and let T be the linear operator on $\mathcal{R}^2$ that rotates a vector by an angle θ. By Theorem 6.12, TU is a reflection. If U reflects about the line $\mathcal{L}$, we can describe the line about which TU reflects in terms of $\mathcal{L}$ and θ. To do so, let S be the linear operator on $\mathcal{R}^2$ that rotates a vector by the angle $\theta/2$, and let $\mathbf{b}$ be a nonzero vector in the direction of $\mathcal{L}$, so that $\mathbf{b}$ is an eigenvector of U corresponding to eigenvalue 1.

(a) Prove that $S(\mathbf{b})$ is an eigenvector of TU corresponding to eigenvalue 1. *Hint:* Show that $TS^{-1} = S$, and use Exercise 26.
(b) Prove that if $\mathcal{L}'$ is obtained by rotating $\mathcal{L}$ by the angle $\theta/2$, then TU is the reflection about $\mathcal{L}'$.

29. Let W be a one-dimensional subspace of $\mathcal{R}^2$. Regard W as a line containing the origin. Let P_W be the orthogonal projection matrix on W, and let $Q_W = 2P_W - I_2$. Prove the following results.

(a) $Q_W^T = Q_W$.
(b) $Q_W^2 = I_2$.
(c) Q_W is an orthogonal matrix.
(d) $Q_W\mathbf{w} = \mathbf{w}$ for all $\mathbf{w}$ in W.
(e) $Q_W\mathbf{v} = -\mathbf{v}$ for all $\mathbf{v}$ in $W^\perp$.
(f) Q_W is the standard matrix of the reflection of $\mathcal{R}^2$ about W.

30. Let T be a linear operator on $\mathcal{R}^n$, and suppose that $\{\mathbf{v}_1, \mathbf{v}_2, \ldots, \mathbf{v}_n\}$ is an orthonormal basis for $\mathcal{R}^n$. Prove that T is an orthogonal operator if and only if the set $\{T(\mathbf{v}_1), T(\mathbf{v}_2), \ldots, T(\mathbf{v}_n)\}$ is also an orthonormal basis for $\mathcal{R}^n$.

---

[12]The results of this exercise are used in Section 6.7 (page 390).

31. Suppose that $\{v_1, v_2, \ldots, v_n\}$ and $\{w_1, w_2, \ldots, w_n\}$ are orthonormal bases for $\mathcal{R}^n$. Prove that there exists a unique orthogonal operator T on $\mathcal{R}^n$ such that $T(v_i) = w_i$ for $1 \le i \le n$. (This is the converse of Exercise 30.)

32. Prove that the composition of two rigid motions on $\mathcal{R}^n$ is a rigid motion.

33. Prove Theorem 6.13(a).

34. Complete the proof of Theorem 6.13(c) by showing that T preserves scalar multiplication.

35. Let $F: \mathcal{R}^n \to \mathcal{R}^n$ be a rigid motion. By the final result of this section, there exists an $n \times n$ orthogonal matrix Q and a vector b in $\mathcal{R}^n$ such that

$$F(v) = Qv + b$$

for all v in $\mathcal{R}^n$. Prove that Q and b are unique.

36. Suppose that F and G are rigid motions on $\mathcal{R}^n$. By Exercise 35, there exist unique orthogonal matrices P and Q and unique vectors a and b such that

$$F(v) = Qv + b \qquad \text{and} \qquad G(v) = Pv + a$$

for all v in $\mathcal{R}^n$. By Exercise 32, the composition of F and G is a rigid motion, and hence by Exercise 35 there exist a unique orthogonal matrix R and a unique vector c such that $F(G(v)) = Rv + c$ for all v in $\mathcal{R}^n$. Find R and c in terms of P, Q, a, and b.

In Exercises 37–38, a rigid motion $F: \mathcal{R}^2 \to \mathcal{R}^2$ is given. Use the given information to find the orthogonal matrix Q and the vector b such that $F(v) = Qv + b$ for all for all v in $\mathcal{R}^2$.

37. $F\left(\begin{bmatrix} 1 \\ 0 \end{bmatrix}\right) = \begin{bmatrix} 2 \\ 4 \end{bmatrix}$, $F\left(\begin{bmatrix} 0 \\ 1 \end{bmatrix}\right) = \begin{bmatrix} 1 \\ 3 \end{bmatrix}$, and

$F\left(\begin{bmatrix} 1 \\ 1 \end{bmatrix}\right) = \begin{bmatrix} 2 \\ 3 \end{bmatrix}$

38. $F\left(\begin{bmatrix} 2 \\ 1 \end{bmatrix}\right) = \begin{bmatrix} 1 \\ 2 \end{bmatrix}$, $F\left(\begin{bmatrix} 1 \\ 3 \end{bmatrix}\right) = \begin{bmatrix} 2 \\ 0 \end{bmatrix}$, and

$F\left(\begin{bmatrix} 7 \\ 1 \end{bmatrix}\right) = \begin{bmatrix} 4 \\ 6 \end{bmatrix}$

39. Let $T: \mathcal{R}^n \to \mathcal{R}^n$ be a function such that $T(u) \cdot T(v) = u \cdot v$ for all u and v in $\mathcal{R}^n$. Prove that T is linear, and hence is an orthogonal operator. *Hint:* Apply Theorem 6.13.

40. Use Exercise 52 of Section 6.1 to prove that if T is an orthogonal operator on $\mathcal{R}^2$, then T preserves the angle

between any two nonzero vectors. That is, for any nonzero vectors u and v in $\mathcal{R}^2$, the angle between $T(u)$ and $T(v)$ equals the angle between u and v.

41. Let E_n be the $n \times n$ matrix all of whose entries are 1s. Let $A_n = I_n - \frac{2}{n} E_n$.
    (a) Determine A_n for $n = 2, 3, 6$.
    (b) Compute the matrix $A_n^T A_n$ for $n = 2, 3, 6$, and use Theorem 6.9(b) to conclude that A_n is an orthogonal matrix.
    (c) Prove that A_n is symmetric for all n.
    (d) Prove that A_n is an orthogonal matrix for all n. *Hint:* First prove that $E_n^2 = n E_n$.

42. In $\mathcal{R}^2$, let $\mathcal{L}$ be the line through the origin that makes an angle θ with the positive half of the x-axis, and let U be the reflection of $\mathcal{R}^2$ about $\mathcal{L}$. Prove that the standard matrix of U is

$$\begin{bmatrix} \cos 2\theta & \sin 2\theta \\ \sin 2\theta & -\cos 2\theta \end{bmatrix}.$$

43. In $\mathcal{R}^2$, let $\mathcal{L}$ be the line through the origin with slope m, and let U be the reflection of $\mathcal{R}^2$ about $\mathcal{L}$. Prove that the standard matrix of U is

$$\frac{1}{1 + m^2} \begin{bmatrix} 1 - m^2 & 2m \\ 2m & m^2 - 1 \end{bmatrix}.$$

In Exercises 44–47, use a calculator with matrix capabilities or computer software such as MATLAB to solve the problem.

44. Find the standard matrix of the reflection about the line in $\mathcal{R}^2$ that contains the origin and the point with coordinates $(2.43, -1.31)$.

45. Find the standard matrix of the reflection about the line in $\mathcal{R}^2$ that contains the origin and the point with coordinates $(3.27, 1.14)$.

46. According to Theorem 6.12, the composition of two reflections is a rotation. Find the angle, to the nearest degree, that a vector is rotated if it is reflected about the line with equation $y = 3.21x$ and the result is then reflected about the line with equation $y = 1.54x$.

47. According to Theorem 6.12, the composition of two reflections is a rotation. Find the angle, to the nearest degree, that a vector is rotated if it is reflected about the line with equation $y = 1.23x$ and the result is then reflected about the line with equation $y = -0.24x$.

Solutions to the Practice Problems

1. (a) The product of this matrix and its transpose is

$$\begin{bmatrix} .7 & -.3 \\ .3 & .7 \end{bmatrix} \begin{bmatrix} .7 & -.3 \\ .3 & .7 \end{bmatrix}^T = \begin{bmatrix} .7 & -.3 \\ .3 & .7 \end{bmatrix} \begin{bmatrix} .7 & .3 \\ -.3 & .7 \end{bmatrix}$$

$$= \begin{bmatrix} .58 & 0 \\ 0 & .58 \end{bmatrix} \ne I_2,$$

and hence the matrix is not orthogonal.

(b) The product of this matrix and its transpose is

$$\begin{bmatrix} .3\sqrt{2} & -.8 & .3\sqrt{2} \\ .4\sqrt{2} & .6 & .4\sqrt{2} \\ .5\sqrt{2} & 0 & -.5\sqrt{2} \end{bmatrix} \begin{bmatrix} .3\sqrt{2} & .4\sqrt{2} & .5\sqrt{2} \\ -.8 & .6 & 0 \\ .3\sqrt{2} & .4\sqrt{2} & -.5\sqrt{2} \end{bmatrix}$$

$$= I_3,$$

and hence this matrix is orthogonal.

2. (a) The standard matrix of T is

$$\begin{bmatrix} \frac{5}{13} & -\frac{12}{13} \\ \frac{12}{13} & \frac{5}{13} \end{bmatrix},$$

which has determinant equal to 1. Thus T is a rotation, and its standard matrix is a rotation matrix A_θ, where θ is the angle of rotation. Hence

$$\begin{bmatrix} \frac{5}{13} & -\frac{12}{13} \\ \frac{12}{13} & \frac{5}{13} \end{bmatrix} = \begin{bmatrix} \cos\theta & -\sin\theta \\ \sin\theta & \cos\theta \end{bmatrix}.$$

Comparing the corresponding entries of the first column, we have that

$$\cos\theta = \frac{5}{13} \quad \text{and} \quad \sin\theta = \frac{12}{13}.$$

Thus θ can be chosen as the angle in the first quadrant with

$$\theta = \cos^{-1}\left(\frac{5}{13}\right) \approx 67.4°.$$

(b) The standard matrix of T is

$$Q = \frac{1}{61}\begin{bmatrix} -60 & 11 \\ 11 & 60 \end{bmatrix},$$

which has determinant equal to -1. Hence T is a reflection. To determine the line of reflection, we first find an eigenvector of Q corresponding to the eigenvalue 1. One such eigenvector is $\mathbf{b} = \begin{bmatrix} 1 \\ 11 \end{bmatrix}$, which lies on the line with equation $y = 11x$. This is the line of reflection.

6.5   Symmetric Matrices

We have seen in Sections 5.3 and 5.4 that diagonalizable matrices and operators have important properties that allow us to solve difficult computational problems. For example, in the case of an $n \times n$ diagonalizable matrix A, the existence of an invertible matrix P and a diagonal matrix D such that $A = PDP^{-1}$ allows us to compute powers of A very easily. Recall that the columns of P form a basis for $\mathcal{R}^n$ consisting of eigenvectors of A, and the diagonal entries of D are the corresponding eigenvalues.

In this chapter, we have observed that orthonormal bases have particularly useful properties. Thus it is natural to examine the case where there is an *orthonormal basis* for $\mathcal{R}^n$ consisting of eigenvectors of A. Note that we saw an example of such a matrix in Example 3 of Section 6.4, namely, the standard matrix of a reflection about a line in $\mathcal{R}^2$. Suppose that A is a matrix for which there is an orthonormal basis for $\mathcal{R}^n$ consisting of eigenvectors of A. Then A can be expressed in the form $A = PDP^{-1}$, where P is an orthogonal matrix. By Theorem 6.9, we have that $P^T = P^{-1}$, and so

$$A^T = (PDP^{-1})^T = (PDP^T)^T = (P^T)^T D^T P^T = PDP^T = PDP^{-1} = A.$$

We see that the condition $A^T = A$ follows necessarily. Recall that in Section 2.1 we called a matrix with this property *symmetric*.

Therefore the preceding calculation shows that if there is an orthonormal basis consisting of eigenvectors of a matrix, then the matrix must be symmetric. What is truly remarkable is that the converse of this result is also true. (Its proof may be found in [4: 362].) Thus we have the following characterization of an $n \times n$ matrix such that there is an orthonormal basis for $\mathcal{R}^n$ consisting of eigenvectors of the matrix.

Theorem 6.14
An $n \times n$ matrix A is symmetric if and only if there is an orthonormal basis for $\mathcal{R}^n$ consisting of eigenvectors of A, in which case there exists an orthogonal matrix P and a diagonal matrix D such that $P^T AP = D$.

The following result about the eigenvectors of a symmetric matrix is useful in obtaining an orthonormal basis of eigenvectors.

Theorem 6.15

*If **u** and **v** are eigenvectors of a symmetric matrix that correspond to distinct eigenvalues, then **u** and **v** are orthogonal.*

Proof Let A be a symmetric matrix. Suppose that **u** and **v** are eigenvectors of A associated with distinct eigenvalues λ and μ, respectively. Then

$$A\mathbf{u} \cdot \mathbf{v} = \lambda \mathbf{u} \cdot \mathbf{v} = \lambda(\mathbf{u} \cdot \mathbf{v}).$$

Also,

$$A\mathbf{u} \cdot \mathbf{v} = \mathbf{u} \cdot A^T \mathbf{v} = \mathbf{u} \cdot A\mathbf{v} = \mathbf{u} \cdot \mu\mathbf{v} = \mu(\mathbf{u} \cdot \mathbf{v}).$$

So $\lambda(\mathbf{u} \cdot \mathbf{v}) = \mu(\mathbf{u} \cdot \mathbf{v})$. Because λ and μ are distinct, we have $\mathbf{u} \cdot \mathbf{v} = 0$, that is, **u** and **v** are orthogonal. ∎

With Theorem 6.15 in place, we are now able to formulate a strategy for finding an orthonormal basis of eigenvectors for a given symmetric matrix A. Suppose that the distinct eigenvalues of A are $\lambda_1, \lambda_2, \ldots, \lambda_k$ with corresponding eigenspaces $E_1, E_2, \ldots, E_k$. For each i, we can find an orthonormal basis $\mathcal{B}_i$ for E_i. Let $\mathcal{B} = \mathcal{B}_1 \cup \mathcal{B}_2 \cup \cdots \cup \mathcal{B}_k$. By the algorithm for matrix diagonalization on page 277, $\mathcal{B}$ is a basis for $\mathcal{R}^n$. However, because of Theorem 6.15, we know that for $i \neq j$ each vector in $\mathcal{B}_i$ is orthogonal to each vector in $\mathcal{B}_j$. Therefore $\mathcal{B}$ is an orthonormal basis of eigenvectors of A.

Example 1 For the matrix $A = \begin{bmatrix} 3 & -4 \\ -4 & -3 \end{bmatrix}$, find an orthogonal matrix P such that $P^T A P$ is a diagonal matrix.

Solution Because A is symmetric, we know that such a matrix P exists. We need to find an orthonormal basis for $\mathcal{R}^2$ consisting of eigenvectors of A. Using our past methods, we find eigenvalues 5 and -5 with corresponding eigenvectors $\begin{bmatrix} -2 \\ 1 \end{bmatrix}$ and $\begin{bmatrix} 1 \\ 2 \end{bmatrix}$. Notice that these two vectors are orthogonal, as predicted by Theorem 6.15. By multiplying each of these vectors by the reciprocal of its norm, we obtain the orthonormal basis $\left\{ \dfrac{1}{\sqrt{5}} \begin{bmatrix} -2 \\ 1 \end{bmatrix}, \dfrac{1}{\sqrt{5}} \begin{bmatrix} 1 \\ 2 \end{bmatrix} \right\}$ for $\mathcal{R}^2$. So a desired orthogonal matrix P and diagonal matrix D are

$$P = \begin{bmatrix} \frac{-2}{\sqrt{5}} & \frac{1}{\sqrt{5}} \\ \frac{1}{\sqrt{5}} & \frac{2}{\sqrt{5}} \end{bmatrix} = \frac{1}{\sqrt{5}} \begin{bmatrix} -2 & 1 \\ 1 & 2 \end{bmatrix} \quad \text{and} \quad D = \begin{bmatrix} 5 & 0 \\ 0 & -5 \end{bmatrix}.$$

Example 2 For the matrix

$$A = \begin{bmatrix} 4 & 2 & 2 \\ 2 & 4 & 2 \\ 2 & 2 & 4 \end{bmatrix},$$

find an orthogonal matrix P such that $P^T A P$ is a diagonal matrix D.

Solution As in Example 1, we know that because A is symmetric, such a matrix P exists. We compute the characteristic polynomial of A to be $-(t-2)^2(t-8)$.

It can be shown that the vectors

$$\begin{bmatrix} -1 \\ 1 \\ 0 \end{bmatrix} \quad \text{and} \quad \begin{bmatrix} -1 \\ 0 \\ 1 \end{bmatrix}$$

form a basis for the eigenspace corresponding to the eigenvalue 2. Because these vectors are not orthogonal, we apply the Gram–Schmidt process to these two vectors and obtain the orthogonal vectors

$$\begin{bmatrix} -1 \\ 1 \\ 0 \end{bmatrix} \quad \text{and} \quad -\frac{1}{2}\begin{bmatrix} 1 \\ 1 \\ -2 \end{bmatrix}.$$

These two vectors form an orthogonal basis for the eigenspace corresponding to eigenvalue 2. Furthermore, we can choose any eigenvector corresponding to eigenvalue 8, for example $\begin{bmatrix} 1 \\ 1 \\ 1 \end{bmatrix}$, because by Theorem 6.15 it must be orthogonal to the two preceding vectors. So the set

$$\left\{ \begin{bmatrix} \frac{-1}{\sqrt{2}} \\ \frac{1}{\sqrt{2}} \\ 0 \end{bmatrix}, \begin{bmatrix} \frac{1}{\sqrt{6}} \\ \frac{1}{\sqrt{6}} \\ \frac{-2}{\sqrt{6}} \end{bmatrix}, \begin{bmatrix} \frac{1}{\sqrt{3}} \\ \frac{1}{\sqrt{3}} \\ \frac{1}{\sqrt{3}} \end{bmatrix} \right\}$$

is an orthonormal basis of eigenvectors of A. Consequently, one possible choice of the orthogonal matrix P and diagonal matrix D is

$$P = \begin{bmatrix} \frac{-1}{\sqrt{2}} & \frac{1}{\sqrt{6}} & \frac{1}{\sqrt{3}} \\ \frac{1}{\sqrt{2}} & \frac{1}{\sqrt{6}} & \frac{1}{\sqrt{3}} \\ 0 & \frac{-2}{\sqrt{6}} & \frac{1}{\sqrt{3}} \end{bmatrix} \quad \text{and} \quad D = \begin{bmatrix} 2 & 0 & 0 \\ 0 & 2 & 0 \\ 0 & 0 & 8 \end{bmatrix}.$$

Quadratic Forms

Historically, the conic sections have played an important role in physics. For example, ellipses describe the motion of the planets, hyperbolas are used in the manufacture of telescopes, and parabolas describe the paths of projectiles. In the plane, the equations of all the conic sections (the circle, ellipse, parabola, and hyperbola) can be obtained from

$$ax^2 + 2bxy + cy^2 + dx + ey + f = 0 \tag{9}$$

by making various choices for the coefficients.[13] For example, $a = c = 1$, $b = d = e = 0$, and $f = -9$ yields the equation

$$x^2 + y^2 = 9,$$

which represents a circle with radius 3 and center at the origin. If we change d to 8 and complete the square, we obtain

$$(x + 4)^2 + y^2 = 25,$$

which represents a circle with radius 5 and center at the point $(-4, 0)$.

---

[13]The coefficient $2b$ is used for computational purposes.

In the equation of a conic section whose major axis is not parallel to either of the coordinate axes, the coefficient of the xy-term must be nonzero. In this case, the conic section is difficult to graph without a change of coordinates. We have encountered computations with different coordinate systems in Section 4.4. Now we use our knowledge of orthogonal and symmetric matrices to discover the appropriate change of coordinates to make the graphing easier and to gain a better understanding of the underlying geometry.

We begin by considering the **associated quadratic form** of (9), namely,

$$ax^2 + 2bcy + cy^2.$$

If we let

$$A = \begin{bmatrix} a & b \\ b & c \end{bmatrix} \qquad \text{and} \qquad \mathbf{v} = \begin{bmatrix} x \\ y \end{bmatrix},$$

then the associated quadratic form can be written as $\mathbf{v}^T A \mathbf{v}$. For example, the form $3x^2 + 4xy + 6y^2$ can be written as

$$\begin{bmatrix} x & y \end{bmatrix} \begin{bmatrix} 3 & 2 \\ 2 & 6 \end{bmatrix} \begin{bmatrix} x \\ y \end{bmatrix}.$$

Because A is symmetric, we may apply Theorem 6.14 to obtain an orthonormal basis of eigenvectors of A. It is these eigenvectors that determine our new coordinate system. If we let P be the matrix whose columns are these eigenvectors and D be the diagonal matrix whose diagonal entries are the corresponding eigenvalues λ_1 and λ_2 of A, then $P^T A P = D$. It follows from Theorem 4.9 that the coordinate vector $\mathbf{v}' = \begin{bmatrix} x' \\ y' \end{bmatrix}$ of $\mathbf{v}$ relative to this basis satisfies $P\mathbf{v}' = \mathbf{v}$. So

$$
\begin{aligned}
ax^2 + 2bxy + cy^2 &= \mathbf{v}^T A \mathbf{v} \\
&= (P\mathbf{v}')^T A (P\mathbf{v}') \\
&= (\mathbf{v}')^T P^T A P \mathbf{v}' \\
&= (\mathbf{v}')^T D \mathbf{v}' \\
&= \lambda_1 (x')^2 + \lambda_2 (y')^2.
\end{aligned}
$$

We see that by using the variables x' and y', we may rewrite the associated quadratic form with no $x'y'$-term.

To see how this works in practice, consider the equation

$$2x^2 - 4xy + 5y^2 - 36 = 0.$$

The associated quadratic form is $2x^2 - 4xy + 5y^2$. We let

$$A = \begin{bmatrix} 2 & -2 \\ -2 & 5 \end{bmatrix}.$$

It can be shown that the eigenvalues of A are 1 and 6 with corresponding eigenvectors

$$\frac{1}{\sqrt{5}} \begin{bmatrix} 2 \\ 1 \end{bmatrix} \qquad \text{and} \qquad \frac{1}{\sqrt{5}} \begin{bmatrix} -1 \\ 2 \end{bmatrix},$$

thus we let

$$P = \begin{bmatrix} \frac{2}{\sqrt{5}} & -\frac{1}{\sqrt{5}} \\ \frac{1}{\sqrt{5}} & \frac{2}{\sqrt{5}} \end{bmatrix}.$$

Using $\mathbf{v} = P\mathbf{v}'$, we arrive at the following equations that relate the two coordinate systems:

$$x = \frac{2}{\sqrt{5}}x' - \frac{1}{\sqrt{5}}y'$$

$$y = \frac{1}{\sqrt{5}}x' + \frac{2}{\sqrt{5}}y'.$$

Appealing to these equations, we have

$$x^2 - 4xy + 5y^2 = (x')^2 + 6(y')^2.$$

Thus the original equation becomes

$$(x')^2 + 6(y')^2 - 36 = 0,$$

or

$$\frac{(x')^2}{36} + \frac{(y')^2}{6} = 1.$$

From this form, we see that the equation represents an ellipse, but now the coordinates are relative to the basis of eigenvectors of A. The major and minor axes are in the directions of these eigenvectors (see Figure 6.15). Because the x'-axis is in the direction of the vector $\frac{1}{\sqrt{5}}\begin{bmatrix} 2 \\ 1 \end{bmatrix}$, we know that $\theta = \tan^{-1}\left(\frac{1}{2}\right) \approx 26.6°$. It is easy to see that the matrix P is the rotation matrix A_θ. Thus the change of coordinates $\mathbf{v}' = P^T\mathbf{v}$ can be achieved by a rotation of the x- and y-axes.

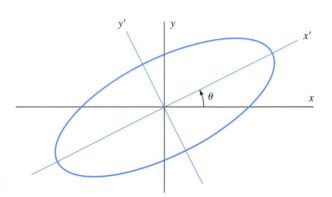

FIGURE 6.15

There is another possibility for P, however. Suppose we consider the eigenvalue 6 first and choose the eigenvector $\frac{1}{\sqrt{5}}\begin{bmatrix} 1 \\ -2 \end{bmatrix}$ instead of $\frac{1}{\sqrt{5}}\begin{bmatrix} -1 \\ 2 \end{bmatrix}$. Then we obtain the matrix

$$\begin{bmatrix} \frac{1}{\sqrt{5}} & \frac{2}{\sqrt{5}} \\ -\frac{2}{\sqrt{5}} & \frac{1}{\sqrt{5}} \end{bmatrix}.$$

This matrix represents a rotation through an angle $\theta = \tan^{-1}(-2) \approx -63.4°$. This choice produces the same ellipse as the one in Figure 6.15 but interchanges the names of the x'- and y'-axes.

However we choose the orthonormal basis of eigenvectors for A, it is clear that the matrix P is always an orthogonal matrix, and so by Theorem 6.10 we have $\det P = \pm 1$. If $\det P = 1$, then P is a rotation matrix by Theorem 6.11.

If det $P = -1$, we can interchange the columns of P, that is, select the same eigenvectors in reverse order, and now det $P = 1$. So we can always choose an orthonormal basis of eigenvectors so that the new set of axes is a rotation of the x- and y-axes.

Spectral Decomposition of a Matrix

It is interesting to note that every symmetric matrix can be decomposed into a sum of very simple matrices. With this decomposition in hand, it is an easy task to prove deep results about matrices.

Consider an $n \times n$ symmetric matrix A and an orthonormal basis $\{\mathbf{u}_1, \mathbf{u}_2, \ldots, \mathbf{u}_n\}$ for $\mathcal{R}^n$ consisting of eigenvectors of A. Let $\lambda_1, \lambda_2, \ldots, \lambda_n$ be the corresponding eigenvalues. Suppose that $P = [\mathbf{u}_1 \quad \mathbf{u}_2 \quad \cdots \quad \mathbf{u}_n]$, and let D denote the $n \times n$ diagonal matrix with diagonal entries $\lambda_1, \lambda_2, \ldots, \lambda_n$, respectively. Then

$$A = PDP^T$$

$$= P[\lambda_1 \mathbf{e}_1 \quad \lambda_2 \mathbf{e}_2 \quad \cdots \quad \lambda_n \mathbf{e}_n] \begin{bmatrix} \mathbf{u}_1^T \\ \mathbf{u}_2^T \\ \vdots \\ \mathbf{u}_2^T \end{bmatrix}$$

$$= [P(\lambda_1 \mathbf{e}_1) \quad P(\lambda_2 \mathbf{e}_2) \quad \cdots \quad P(\lambda_n \mathbf{e}_n)] \begin{bmatrix} \mathbf{u}_1^T \\ \mathbf{u}_2^T \\ \vdots \\ \mathbf{u}_2^T \end{bmatrix}$$

$$= [\lambda_1 P \mathbf{e}_1 \quad \lambda_2 P \mathbf{e}_2 \quad \cdots \quad \lambda_n P \mathbf{e}_n] \begin{bmatrix} \mathbf{u}_1^T \\ \mathbf{u}_2^T \\ \vdots \\ \mathbf{u}_2^T \end{bmatrix}$$

$$= [\lambda_1 \mathbf{u}_1 \quad \lambda_2 \mathbf{u}_2 \quad \cdots \quad \lambda_n \mathbf{u}_n] \begin{bmatrix} \mathbf{u}_1^T \\ \mathbf{u}_2^T \\ \vdots \\ \mathbf{u}_2^T \end{bmatrix}$$

$$= \lambda_1 \mathbf{u}_1 \mathbf{u}_1^T + \lambda_2 \mathbf{u}_2 \mathbf{u}_2^T + \cdots + \lambda_n \mathbf{u}_n \mathbf{u}_n^T.$$

Recall from Section 2.1 that the matrix product $P_i = \mathbf{u}_i \mathbf{u}_i^T$ is a matrix of rank 1. So we have expressed A as a sum of n matrices of rank 1. It can be shown (see Exercise 23) that P_i is the orthogonal projection matrix for the subspace spanned by the eigenvector $\mathbf{u}_i$. The representation

$$A = \lambda_1 P_1 + \lambda_2 P_2 + \cdots + \lambda_n P_n$$

is called a **spectral decomposition** of A.

By Exercises 27 and 26 of Section 6.3, we have that each P_i is symmetric and satisfies $P_i^2 = P_i$. Consequently, a number of other properties follow easily. They are given in the next theorem. The proofs of parts (b), (c), and (d) are left as exercises (see Exercises 24–26).

Theorem 6.16

Let A be an n × n symmetric matrix, and let $\{\mathbf{u}_1, \mathbf{u}_2, \ldots, \mathbf{u}_n\}$ be an orthonormal basis for $\mathcal{R}^n$ consisting of eigenvectors of A with corresponding eigenvalues $\lambda_1, \lambda_2, \ldots, \lambda_n$. Then there exist symmetric matrices $P_1, P_2, \ldots, P_n$ such that the following results hold.

(a) $A = \lambda_1 P_1 + \lambda_2 P_2 + \cdots + \lambda_n P_n$.
(b) rank $P_i = 1$ for all i.
(c) $P_i P_i = P_i$ for all i, and $P_i P_j = O$ if $i \neq j$.
(d) $P_i \mathbf{u}_i = \mathbf{u}_i$ for all i, and $P_i \mathbf{u}_j = \mathbf{0}$ if $i \neq j$.

Example 3   Find a spectral decomposition of the matrix $A = \begin{bmatrix} 3 & -4 \\ -4 & -3 \end{bmatrix}$ in Example 1.

Solution   Using the results of Example 1, we let $\mathbf{u}_1 = \dfrac{1}{\sqrt{5}}\begin{bmatrix} -2 \\ 1 \end{bmatrix}$, $\mathbf{u}_2 = \dfrac{1}{\sqrt{5}}\begin{bmatrix} 1 \\ 2 \end{bmatrix}$, $\lambda_1 = 5$, and $\lambda_2 = -5$. So

$$P_1 = \mathbf{u}_1\mathbf{u}_1^T = \begin{bmatrix} \frac{4}{5} & -\frac{2}{5} \\ -\frac{2}{5} & \frac{1}{5} \end{bmatrix} \qquad \text{and} \qquad P_2 = \mathbf{u}_2\mathbf{u}_2^T = \begin{bmatrix} \frac{1}{5} & \frac{2}{5} \\ \frac{2}{5} & \frac{4}{5} \end{bmatrix}.$$

Therefore

$$A = \lambda_1 P_1 + \lambda_2 P_2 = 5\begin{bmatrix} \frac{4}{5} & -\frac{2}{5} \\ -\frac{2}{5} & \frac{1}{5} \end{bmatrix} + (-5)\begin{bmatrix} \frac{1}{5} & \frac{2}{5} \\ \frac{2}{5} & \frac{4}{5} \end{bmatrix}.$$

A number of interesting consequences of the spectral decomposition are given in the exercises.

Practice Problems

1. (a) Find a symmetric matrix A so that the associated quadratic form of the equation

    $$34x^2 + 24xy + 41y^2 = 200$$

    may be written as $\mathbf{v}^T A\mathbf{v}$.
   (b) Find a rotation of the x- and y-axes to x'- and y'-axes that transforms the equation in (a) into one having no $x'y'$-term. Give the angle of rotation, the equations that relate x' and y' to x and y, and the transformed equation. Then identify the type of conic section.

2. Find a spectral decomposition of

$$A = \begin{bmatrix} 4 & 1 & -1 \\ 1 & 4 & -1 \\ -1 & -1 & 4 \end{bmatrix}.$$

Exercises

1. Determine if the following statements are true or false.

   (a) Every symmetric matrix is diagonalizable.
   (b) If P is a matrix whose columns are eigenvectors of a symmetric matrix, then P is orthogonal.
   (c) Eigenvectors of a matrix that correspond to distinct eigenvalues are orthogonal.
   (d) Distinct eigenvectors of a symmetric matrix are orthogonal.
   (e) By a suitable rotation of the xy-axes to $x'y'$-axes, the equation of any conic section can be written without an $x'y'$-term.
   (f) The associated quadratic form of an equation of any conic section can be written as $\mathbf{v}^T A\mathbf{v}$, where A is a 2×2 matrix and $\mathbf{v}$ is in $\mathcal{R}^2$.
   (g) Every symmetric matrix can be written as a sum of orthogonal projection matrices.
   (h) Every symmetric matrix can be written as a sum of multiples of orthogonal projection matrices.

In Exercises 2–12, answer the following parts for the given equation of a conic section.

(a) Find a symmetric matrix A so that the associated quadratic form may be written as $\mathbf{v}^T A\mathbf{v}$.

For parts (b)–(e), find a rotation of the x- and y-axes to x'- and y'-axes that transforms the given equation into one having no x'y'-term.

(b) Give the angle of rotation.

(c) Give the equations that relate x' and y' to x and y.

(d) Give the transformed equation.

(e) Identify the type of conic section.

2. $2x^2 + 2xy + 2y^2 - 1 = 0$

3. $x^2 - 12xy - 4y^2 = 40$

4. $3x^2 - 4xy + 3y^2 - 5 = 0$

5. $5x^2 + 4xy + 5y^2 - 9 = 0$

6. $11x^2 + 24xy + 4y^2 - 15 = 0$

7. $x^2 + 4xy + y^2 - 7 = 0$

8. $4x^2 + 6xy - 4y^2 = 180$

9. $2x^2 - 12xy - 7y^2 = 200$

10. $6x^2 + 5xy - 6y^2 = 26$

11. $x^2 + 2xy + y^2 + 8x + y = 0$

12. $52x^2 + 72xy + 73y^2 - 160x - 130y - 25 = 0$

In Exercises 13–20, a symmetric matrix A is given. Find an orthonormal basis of eigenvectors and their corresponding eigenvalues. Use this information to obtain a spectral decomposition of the matrix.

13. $\begin{bmatrix} 3 & 1 \\ 1 & 3 \end{bmatrix}$ **14.** $\begin{bmatrix} 7 & 6 \\ 6 & -2 \end{bmatrix}$ **15.** $\begin{bmatrix} 1 & 2 \\ 2 & 1 \end{bmatrix}$

16. $\begin{bmatrix} 1 & -1 \\ -1 & 1 \end{bmatrix}$ **17.** $\begin{bmatrix} 3 & 2 & 2 \\ 2 & 2 & 0 \\ 2 & 0 & 4 \end{bmatrix}$ **18.** $\begin{bmatrix} 0 & 2 & 2 \\ 2 & 0 & 2 \\ 2 & 2 & 0 \end{bmatrix}$

19. $\begin{bmatrix} -1 & 0 & 0 \\ 0 & 0 & 2 \\ 0 & 2 & 3 \end{bmatrix}$ **20.** $\begin{bmatrix} -2 & 0 & -36 \\ 0 & -3 & 0 \\ -36 & 0 & -23 \end{bmatrix}$

21. Show that a spectral decomposition is not unique by finding two different spectral decompositions for the matrix $2I_2$.

22. Show that a spectral decomposition is not unique by finding two different spectral decompositions for the matrix in Exercise 19.

23. Let $\mathbf{u}$ be a vector in $\mathcal{R}^n$ of norm 1, and let P be the $n \times n$ matrix $\mathbf{u}\mathbf{u}^T$. Prove that P is the orthogonal projection matrix for the subspace spanned by $\mathbf{u}$.

24. Prove Theorem 6.16(b).

25. Prove Theorem 6.16(c).

26. Prove Theorem 6.16(d).

In Exercises 27–34, let A be an $n \times n$ symmetric matrix with spectral decomposition $A = \lambda_1 P_1 + \lambda_2 P_2 + \cdots + \lambda_n P_n$. Assume that $\mu_1, \mu_2, \ldots, \mu_k$ are all the distinct eigenvalues of A and that Q_j denotes the sum of all the P_i that are associated with μ_j.

27. Prove that $A = \mu_1 Q_1 + \mu_1 Q_2 + \cdots + \mu_k Q_k$.

28. Prove that $Q_j Q_j = Q_j$ for all j, and $Q_i Q_j = O$ if $j \neq i$.

29. Prove that Q_j is symmetric for all j.

30. Prove that, for all j, Q_j is the orthogonal projection matrix for the eigenspace associated with μ_j.

31. Suppose that $\{\mathbf{w}_1, \mathbf{w}_2, \ldots, \mathbf{w}_s\}$ is an orthonormal basis for the eigenspace corresponding to μ_j. Represent Q_j as a sum of matrices each of rank 1.

32. Prove that the rank of Q_j equals the dimension of the eigenspace associated with μ_j.

33. Use a spectral decomposition of A to compute a spectral decomposition of A^s, where s is any positive integer greater than 1.

34. Use a spectral decomposition of A to find a spectral decomposition of an $n \times n$ matrix C such that $C^3 = A$.

Exercises 35–37 use the following definition: For a polynomial $g(t) = a_n t^n + a_{n-1} t^{n-1} + \cdots + a_1 t + a_0$ and an $n \times n$ matrix B, define the matrix g(B) by

$$g(B) = a_n B^n + a_{n-1} B^{n-1} + \cdots + a_1 B + a_0 I_n.$$

35. Use Exercises 27 and 28 to show that for any polynomial g,

$$g(A) = g(\mu_1) Q_1 + g(\mu_2) Q_2 + \cdots + g(\mu_k) Q_k,$$

where A, μ_i, and Q_i are as in Exercises 27–34.

36. Use Exercise 35 to prove a special case of the *Cayley–Hamilton theorem:* If f is the characteristic polynomial of a symmetric matrix A, then $f(A) = O$.

37. Let A, μ_i, and Q_i be as in Exercises 27–34. It can be shown that for any j, $1 \leq j \leq k$, there is a polynomial f_j such that $f_j(\mu_j) = 1$ and $f_j(\mu_i) = 0$ if $i \neq j$. Use this result along with Exercise 35 to show that $Q_j = f_j(A)$. So the Q_j are uniquely determined by the properties given in Exercises 27 and 28.

38. Use Exercise 37 to prove that an $n \times n$ matrix B commutes with A (that is, $AB = BA$) if and only if B commutes with each Q_j.

*An $n \times n$ matrix C is said to be **positive definite** if C is symmetric and $\mathbf{v}^T C \mathbf{v} > 0$ for every nonzero vector $\mathbf{v}$ in $\mathcal{R}^n$. We say C is **positive semidefinite** if C is symmetric and $\mathbf{v}^T C \mathbf{v} \geq 0$ for every vector $\mathbf{v}$ in $\mathcal{R}^n$.*

In Exercises 39–53, we assume the preceding definitions. The equation $\mathbf{v}^T A \mathbf{v} = \mathbf{v} \cdot A \mathbf{v}$ is often useful in solving these exercises.

39. Suppose that A is a symmetric matrix. Prove that A is positive definite if and only if all of its eigenvalues are positive.

40.[14] State and prove an analogous result to Exercise 39 if A is positive semidefinite.

41. Suppose that A is invertible and positive definite. Prove that A^{-1} is positive definite.

42. Suppose that A is positive definite and $c > 0$. Prove that cA is positive definite.

43. State and prove a result analogous to Exercise 42 if A is positive semidefinite.

44. Suppose that A and B are positive definite $n \times n$ matrices. Prove that $A + B$ is positive definite.

45. State and prove a result analogous to Exercise 44 if A and B are positive semidefinite.

46. Suppose that $A = QBQ^T$, where Q is an orthogonal matrix and B is positive definite. Prove that A is positive definite.

47. State and prove a result analogous to Exercise 46 if B is positive semidefinite.

48. Prove that if A is positive definite, then there exists a positive definite matrix B such that $B^2 = A$.

49. State and prove a result analogous to Exercise 48 if A is positive semidefinite.

50. Let A be an $n \times n$ symmetric matrix. Prove that A is

positive definite if and only if

$$\sum_{i,j} a_{ij} u_i u_j > 0$$

for all scalars $u_1, u_2, \ldots, u_n$, not all zero.

51. State and prove a result analogous to Exercise 50 if A is positive semidefinite.

52.[15] Prove that, for any matrix A, the matrices $A^T A$ and AA^T are positive semidefinite.

53. Prove that, for any invertible matrix A, the matrices $A^T A$ and AA^T are positive definite.

In Exercise 54, use a calculator with matrix capabilities or computer software such as MATLAB *to solve the problem.*

54. Let

$$A = \begin{bmatrix} 4 & 0 & 2 & 0 & 2 \\ 0 & 4 & 0 & 2 & 0 \\ 2 & 0 & 4 & 0 & 2 \\ 0 & 2 & 0 & 4 & 0 \\ 2 & 0 & 2 & 0 & 4 \end{bmatrix}.$$

(a) Verify that A is symmetric.
(b) Find the eigenvalues of A.
(c) Find an orthonormal basis of eigenvectors for $\mathcal{R}^5$.
(d) Use your answers to (b) and (c) to find a spectral decomposition of A.
(e) Compute A^6 by matrix multiplication.
(f) Use your answer to (d) to find a spectral decomposition of A^6.
(g) Use your answer to (f) to compute A^6.

---

[14] The results of this exercise are used in Section 6.6 (page 365).

[15] This results of this exercise are used in Section 6.6 (page 365).

Solutions to the Practice Problems

1. (a) The entries of A are obtained from the coefficients of the quadratic form: $a_{11} = 34$, $a_{22} = 41$, and $a_{12} = a_{21} = \frac{1}{2} \cdot 24 = 12$. Thus

$$A = \begin{bmatrix} 34 & 12 \\ 12 & 41 \end{bmatrix}.$$

(b) It can be shown that A has the eigenvalues 25 and 50 with corresponding unit eigenvectors $\begin{bmatrix} 0.8 \\ -0.6 \end{bmatrix}$ and $\begin{bmatrix} 0.6 \\ 0.8 \end{bmatrix}$, respectively. Let

$$P = \begin{bmatrix} 0.8 & 0.6 \\ -0.6 & 0.8 \end{bmatrix}, \quad D = \begin{bmatrix} 25 & 0 \\ 0 & 50 \end{bmatrix}, \quad \text{and} \quad \mathbf{v} = \begin{bmatrix} x \\ y \end{bmatrix}.$$

Then $\mathbf{v} = P\mathbf{v}'$, and hence

$$x = 0.8x' + 0.6y'$$
$$y = -0.6x' + 0.8y'.$$

Thus

$$34x^2 + 24xy + 41y^2 = 25(x')^2 + 50(y')^2 = 200,$$

and so the original equation becomes

$$\frac{(x')^2}{8} + \frac{(y')^2}{4} = 1,$$

which is the equation of an ellipse. Because the x'-axis is in the direction of $\begin{bmatrix} 0.8 \\ -0.6 \end{bmatrix}$, which is in the fourth quadrant, the angle of rotation θ of the coordinate system is

$$\theta = \sin^{-1}(-0.6) \approx -36.9°.$$

2. First observe (we omit the details) that $\lambda_1 = \lambda_2 = 3$ and $\lambda_3 = 6$ are the eigenvalues of A, with an orthonormal

basis of $\mathcal{R}^3$ consisting of corresponding eigenvectors

$$\{u_1, u_2, u_3\} = \left\{ \begin{bmatrix} \frac{1}{\sqrt{2}} \\ 0 \\ \frac{1}{\sqrt{2}} \end{bmatrix}, \begin{bmatrix} -\frac{1}{\sqrt{6}} \\ \frac{2}{\sqrt{6}} \\ \frac{1}{\sqrt{6}} \end{bmatrix}, \begin{bmatrix} \frac{1}{\sqrt{3}} \\ \frac{1}{\sqrt{3}} \\ -\frac{1}{\sqrt{3}} \end{bmatrix} \right\}.$$

Thus a spectral decomposition of A is given by

$$A = \lambda_1 u_1 u_1^T + \lambda_2 u_2 u_2^T + \lambda_3 u_3 u_3^T$$

$$= 3 \begin{bmatrix} \frac{1}{2} & 0 & \frac{1}{2} \\ 0 & 0 & 0 \\ \frac{1}{2} & 0 & \frac{1}{2} \end{bmatrix} + 3 \begin{bmatrix} \frac{1}{6} & -\frac{2}{6} & -\frac{1}{6} \\ -\frac{2}{6} & \frac{4}{6} & \frac{2}{6} \\ -\frac{1}{6} & \frac{2}{6} & \frac{1}{6} \end{bmatrix}$$

$$+ 6 \begin{bmatrix} \frac{1}{3} & \frac{1}{3} & -\frac{1}{3} \\ \frac{1}{3} & \frac{1}{3} & -\frac{1}{3} \\ -\frac{1}{3} & -\frac{1}{3} & \frac{1}{3} \end{bmatrix}.$$

6.6*   Singular Value Decomposition

The ideal situation under which to study a matrix is with an orthonormal basis of eigenvectors. The basis of eigenvectors gives us a complete insight into the way the matrix acts on vectors. If this basis is also an orthonormal set, then we have the added benefit of a set of mutually perpendicular coordinate axes that illuminate the geometric behavior of the matrix as it acts on vectors.

But alas, only the symmetric matrices enjoy all of these properties. And if the matrix is not square, then eigenvectors are not even available.

We now consider an approach in which we retain orthogonality but drop the requirement that the basis members are eigenvectors. At the same time, we want to maintain the flavor of eigenvalues and eigenvectors. One fruitful approach in this direction is to find two separate orthonormal bases ordered so that the product of the matrix and a vector in the first basis is a scalar multiple of the corresponding vector in the second basis. We show that such orthonormal bases and scalar multiples always exist, and that the nonzero scalars are unique provided that they are chosen to be positive, which is always possible. Since the two orthonormal bases are distinct, the number of vectors in one may be different from the number of vectors in the other. Thus the matrix need not be square. One fact that is useful in what follows is that if the rank of an $m \times n$ matrix is k, then $k \leq m$ and $k \leq n$.

With this in mind we state the principal theorem of this section.

Theorem 6.17

Let A be an $m \times n$ matrix of rank k. Then there exist orthonormal bases

$$\mathcal{B}_1 = \{v_1, v_2, \ldots, v_n\} \quad \text{for } \mathcal{R}^n$$

and

$$\mathcal{B}_2 = \{u_1, u_2, \ldots, u_m\} \quad \text{for } \mathcal{R}^m$$

and scalars

$$\sigma_1 \geq \sigma_2 \geq \cdots \geq \sigma_k > 0$$

such that

$$A v_i = \begin{cases} \sigma_i u_i & \text{if } 1 \leq i \leq k \\ 0 & \text{if } i > k \end{cases} \tag{10}$$

and

$$A^T u_i = \begin{cases} \sigma_i v_i & \text{if } 1 \leq i \leq k \\ 0 & \text{if } i > k. \end{cases} \tag{11}$$

---

*This section can be omitted without loss of continuity.

Proof By Exercise 52 of Section 6.5, $A^T A$ is an $n \times n$ positive semidefinite matrix, and hence there is an orthonormal basis $\mathcal{B}_1 = \{\mathbf{v}_1, \mathbf{v}_2, \dots, \mathbf{v}_n\}$ for $\mathcal{R}^n$ consisting of eigenvectors of $A^T A$ with corresponding eigenvalues λ_i that are nonnegative (Exercise 40 of Section 6.5). We order these eigenvalues and the vectors in $\mathcal{B}_1$ so that $\lambda_1 \geq \lambda_2 \geq \cdots \geq \lambda_n$. By Exercise 51 of Section 6.1, $A^T A$ has rank k, and hence the first k eigenvalues are positive and the last $n - k$ eigenvalues are zero. For each $i = 1, 2, \dots k$, let $\sigma_i = \sqrt{\lambda_i}$. Then $\sigma_1 \geq \sigma_2 \geq \cdots \geq \sigma_k > 0$.

Next, for each $i \leq k$, let $\mathbf{u}_i$ be the vector in $\mathcal{R}^m$ defined by $\mathbf{u}_i = \dfrac{1}{\sigma_i} A \mathbf{v}_i$. We show that $\{\mathbf{u}_1, \mathbf{u}_2, \dots, \mathbf{u}_k\}$ is an orthonormal subset of $\mathcal{R}^m$. Consider any $\mathbf{u}_i$ and $\mathbf{u}_j$. Then

$$
\begin{aligned}
\mathbf{u}_i \cdot \mathbf{u}_j &= \frac{1}{\sigma_i} A \mathbf{v}_i \cdot \frac{1}{\sigma_j} A \mathbf{v}_j \\[2mm]
&= \frac{1}{\sigma_i \sigma_j} A \mathbf{v}_i \cdot A \mathbf{v}_j \\[2mm]
&= \frac{1}{\sigma_i \sigma_j} \mathbf{v}_i \cdot A^T A \mathbf{v}_j \\[2mm]
&= \frac{1}{\sigma_i \sigma_j} \mathbf{v}_i \cdot \lambda_j \mathbf{v}_j \\[2mm]
&= \frac{\sigma_j^2}{\sigma_i \sigma_j} \mathbf{v}_i \cdot \mathbf{v}_j.
\end{aligned}
$$

Thus

$$
\mathbf{u}_i \cdot \mathbf{u}_j = \frac{\sigma_j}{\sigma_i} \mathbf{v}_i \cdot \mathbf{v}_j = \begin{cases} 0 & \text{if } i \neq j \\ 1 & \text{if } i = j, \end{cases}
$$

and it follows that $\{\mathbf{u}_1, \mathbf{u}_2, \dots, \mathbf{u}_k\}$ is an orthonormal set. By Exercise 41 of Section 6.2, this set extends to an orthonormal basis $\mathcal{B}_2 = \{\mathbf{u}_1, \mathbf{u}_2, \dots, \mathbf{u}_m\}$ for $\mathcal{R}^m$.

Our final task is to verify (11). Consider any $\mathbf{u}_i$ in $\mathcal{B}_2$. First suppose that $i \leq k$. Then

$$
\begin{aligned}
A^T \mathbf{u}_i &= A^T \left(\frac{1}{\sigma_i} A \mathbf{v}_i \right) \\[2mm]
&= \frac{1}{\sigma_i} A^T A \mathbf{v}_i \\[2mm]
&= \frac{1}{\sigma_i} \sigma_i^2 \mathbf{v}_i \\[2mm]
&= \sigma_i \mathbf{v}_i.
\end{aligned}
$$

Now suppose that $i > k$. We show that $A^T \mathbf{u}_i$ is orthogonal to every vector in $\mathcal{B}_1$. Since $\mathcal{B}_1$ is a basis for $\mathcal{R}^n$, it follows that $A^T \mathbf{u}_i = \mathbf{0}$. Consider any $\mathbf{v}_j$ in $\mathcal{B}_1$. First suppose that $j \leq k$. Then

$$
A^T \mathbf{u}_i \cdot \mathbf{v}_j = \mathbf{u}_i \cdot A \mathbf{v}_j = \mathbf{u}_i \cdot \sigma_j \mathbf{u}_j = \sigma_j \mathbf{u}_i \cdot \mathbf{u}_j = 0
$$

because $i \neq j$. Finally suppose that $j > k$. Then

$$A^T \mathbf{u}_i \cdot \mathbf{v}_j = \mathbf{u}_i \cdot A\mathbf{v}_j = \mathbf{u}_i \cdot \mathbf{0} = 0.$$

Thus $A^T \mathbf{u}_i$ is orthogonal to every vector in $\mathcal{B}_1$, and we conclude that $A^T \mathbf{u}_i = \mathbf{0}$.

□

In the notation of Theorem 6.17, the scalars $\sigma_1, \sigma_2, \ldots, \sigma_k$, which are the square roots of the nonzero eigenvalues of $A^T A$, are called the **singular values** of A. The vectors in $\mathcal{B}_1$ are called **right singular vectors** of A, and the vectors in $\mathcal{B}_2$ are called **left singular vectors** of A. The reason for the designations *left* and *right* will become clear later.

In the proof of Theorem 6.17, the vectors $\mathbf{v}_i$ are chosen to be eigenvectors of $A^T A$. It can be shown (see Exercise 35) that if $\{\mathbf{v}_1, \mathbf{v}_2, \ldots, \mathbf{v}_n\}$ and $\{\mathbf{u}_1, \mathbf{u}_2, \ldots, \mathbf{u}_m\}$ are any orthonormal bases for $\mathcal{R}^n$ and $\mathcal{R}^m$, respectively, that satisfy (10) and (11), then each $\mathbf{v}_i$ is an eigenvector of $A^T A$ corresponding to the eigenvalue σ_i^2 if $i \leq k$ and to the eigenvalue 0 if $i > k$. Furthermore, for $i = 1, 2, \ldots, k$, the vector $\mathbf{u}_i$ is an eigenvector of AA^T corresponding to eigenvalue σ_i^2 and, for $i > k$, the vector $\mathbf{u}_i$ is an eigenvector of AA^T corresponding to the eigenvalue 0.

Although the singular values of a matrix are unique, the left and right singular vectors are not unique. Of course, this is also the case for eigenvalues and eigenvectors of a matrix, even if the eigenvectors form an orthonormal basis.

Example 1   Find the singular values and orthonormal bases of right and left singular vectors for the matrix

$$A = \begin{bmatrix} 0 & 1 & 2 \\ 1 & 0 & 1 \end{bmatrix}.$$

Solution   The proof of Theorem 6.17 gives us the method for solving this problem. We first form the product

$$A^T A = \begin{bmatrix} 1 & 0 & 1 \\ 0 & 1 & 2 \\ 1 & 2 & 5 \end{bmatrix}.$$

Since A has rank 2, so does $A^T A$. In fact, it can be shown (we omit the details) that

$$\mathcal{B}_1 = \left\{ \frac{1}{\sqrt{30}} \begin{bmatrix} 1 \\ 2 \\ 5 \end{bmatrix}, \frac{1}{\sqrt{5}} \begin{bmatrix} 2 \\ -1 \\ 0 \end{bmatrix}, \frac{1}{\sqrt{6}} \begin{bmatrix} 1 \\ 2 \\ -1 \end{bmatrix} \right\}$$

is an orthonormal basis for $\mathcal{R}^3$ of eigenvectors of $A^T A$ with corresponding eigenvalues 6, 1, and 0. So $\sigma_1 = \sqrt{6}$, and $\sigma_2 = \sqrt{1} = 1$ are the singular values of A, and $\mathcal{B}_1$ is a set of right singular vectors. Let $\mathbf{v}_1$ and $\mathbf{v}_2$ denote the first two vectors in $\mathcal{B}_1$, which correspond to the singular values σ_1 and σ_2, respectively. Let

$$\mathbf{u}_1 = \frac{1}{\sigma_1} A\mathbf{v}_1 = \frac{1}{\sqrt{6}} \frac{1}{\sqrt{30}} \begin{bmatrix} 0 & 1 & 2 \\ 1 & 0 & 1 \end{bmatrix} \begin{bmatrix} 1 \\ 2 \\ 5 \end{bmatrix} = \frac{1}{6\sqrt{5}} \begin{bmatrix} 12 \\ 6 \end{bmatrix} = \frac{1}{\sqrt{5}} \begin{bmatrix} 2 \\ 1 \end{bmatrix},$$

and

$$\mathbf{u}_2 = \frac{1}{\sigma_2} A\mathbf{v}_2 = \frac{1}{\sqrt{5}} \begin{bmatrix} 0 & 1 & 2 \\ 1 & 0 & 1 \end{bmatrix} \begin{bmatrix} 2 \\ -1 \\ 0 \end{bmatrix} = \frac{1}{\sqrt{5}} \begin{bmatrix} -1 \\ 2 \end{bmatrix}.$$

Notice that $\mathcal{B}_2 = \{\mathbf{u}_1, \mathbf{u}_2\}$ is an orthonormal basis for $\mathcal{R}^2$, and hence we do not add vectors to this set. The vectors in $\mathcal{B}_2$ are left singular vectors of A.   ○

Example 2   Find the singular values and orthonormal bases of right and left singular vectors for the matrix

$$A = \begin{bmatrix} 1 & 3 & 2 & 1 \\ 3 & 1 & 2 & -1 \\ 1 & 1 & 1 & 0 \end{bmatrix}.$$

Solution   We first form the product

$$A^T A = \begin{bmatrix} 11 & 7 & 9 & -2 \\ 7 & 11 & 9 & 2 \\ 9 & 9 & 9 & 0 \\ -2 & 2 & 0 & 2 \end{bmatrix}.$$

It can be shown (we omit the details) that

$$\mathcal{B}_1 = \left\{ \frac{1}{\sqrt{3}} \begin{bmatrix} 1 \\ 1 \\ 1 \\ 0 \end{bmatrix}, \frac{1}{\sqrt{3}} \begin{bmatrix} 1 \\ -1 \\ 0 \\ -1 \end{bmatrix}, \frac{1}{\sqrt{3}} \begin{bmatrix} 1 \\ 0 \\ -1 \\ 1 \end{bmatrix}, \frac{1}{\sqrt{3}} \begin{bmatrix} 0 \\ -1 \\ 1 \\ 1 \end{bmatrix} \right\}$$

is an orthonormal basis for $\mathcal{R}^4$ of eigenvectors of $A^T A$ with corresponding eigenvalues 27, 6, 0, and 0, respectively. So $\sigma_1 = \sqrt{27}$ and $\sigma_2 = \sqrt{6}$ are the singular values of A, and $\mathcal{B}_1$ is a set of right singular vectors. Let $\mathbf{v}_1$ and $\mathbf{v}_2$ denote the first two vectors in $\mathcal{B}_1$, which correspond to the singular values σ_1 and σ_2, respectively. Let

$$\mathbf{u}_1 = \frac{1}{\sigma_1} A\mathbf{v}_1 = \frac{1}{\sqrt{27}} \frac{1}{\sqrt{3}} \begin{bmatrix} 1 & 3 & 2 & 1 \\ 3 & 1 & 2 & -1 \\ 1 & 1 & 1 & 0 \end{bmatrix} \begin{bmatrix} 1 \\ 1 \\ 1 \\ 0 \end{bmatrix} = \frac{1}{9} \begin{bmatrix} 6 \\ 6 \\ 3 \end{bmatrix} = \frac{1}{3} \begin{bmatrix} 2 \\ 2 \\ 1 \end{bmatrix},$$

and

$$\mathbf{u}_2 = \frac{1}{\sigma_2} A\mathbf{v}_2 = \frac{1}{\sqrt{6}} \frac{1}{\sqrt{3}} \begin{bmatrix} 1 & 3 & 2 & 1 \\ 3 & 1 & 2 & -1 \\ 1 & 1 & 1 & 0 \end{bmatrix} \begin{bmatrix} 1 \\ -1 \\ 0 \\ -1 \end{bmatrix} = \frac{1}{3\sqrt{2}} \begin{bmatrix} -3 \\ 3 \\ 0 \end{bmatrix}$$

$$= \frac{1}{\sqrt{2}} \begin{bmatrix} -1 \\ 1 \\ 0 \end{bmatrix}.$$

The vectors $\mathbf{u}_1$ and $\mathbf{u}_2$ can be used as the first two of three vectors in an orthonormal basis for $\mathcal{R}^3$ of left singular vectors of A. The only requirement for a third vector is that it be a unit vector orthogonal to both $\mathbf{u}_1$ and $\mathbf{u}_2$, since any orthonormal set of three vectors is a basis for $\mathcal{R}^3$. A nonzero vector is orthogonal to both $\mathbf{u}_1$ and $\mathbf{u}_2$ if and only if it is a nonzero solution to the system

$$2x_1 + 2x_2 + x_3 = 0$$
$$-x_1 + x_2 = 0.$$

For example, (we omit the details) the vector $\mathbf{c} = \begin{bmatrix} 1 \\ 1 \\ -4 \end{bmatrix}$ is a nonzero solution to this system. Therefore we let

$$\mathbf{u}_3 = \frac{1}{\|\mathbf{c}\|}\mathbf{c} = \frac{1}{\sqrt{18}}\begin{bmatrix} 1 \\ 1 \\ -4 \end{bmatrix} = \frac{1}{3\sqrt{2}}\begin{bmatrix} 1 \\ 1 \\ -4 \end{bmatrix}.$$

Thus $\mathcal{B}_2 = \{\mathbf{u}_1, \mathbf{u}_2, \mathbf{u}_3\}$ is an orthonormal basis for $\mathcal{R}^3$ of left singular vectors of A. ○

Consider a linear transformation $T \colon \mathcal{R}^n \to \mathcal{R}^m$. Since the image of a vector in $\mathcal{R}^n$ is a vector in $\mathcal{R}^m$, geometric objects in $\mathcal{R}^n$ that can be constructed from vectors are transformed by T into objects in $\mathcal{R}^m$. The singular values of the standard matrix A of T can be used to describe how the norms of vectors in $\mathcal{R}^n$ are affected by T. For example, a vector parallel to a left singular vector $\mathbf{v}_i$ of A has an image that is σ_i times its original norm. This is so because

$$\|T(\mathbf{v}_i)\| = \|A\mathbf{v}_i\| = \|\sigma_i \mathbf{u}_i\| = \sigma_i = \sigma_i \|\mathbf{v}_i\|,$$

where $\mathbf{u}_i$ is the right singular vector such that $A\mathbf{v}_i = \sigma_i \mathbf{u}_i$.

As a simple example of how singular values can be used to describe shape changes, we consider the image of the unit circle (the circle of radius 1 and center $\mathbf{0}$) under the matrix transformation T_A, where A is an invertible 2×2 matrix with distinct (nonzero) singular values.

| **Example 3** | Let S be the unit circle in $\mathcal{R}^2$, and let A be a 2×2 invertible matrix with the distinct singular values $\sigma_1 > \sigma_2 > 0$. Let $S' = T_A(S)$ be the image of S under the matrix transformation T_A. We describe S'. For this purpose, let $\mathcal{B}_1 = \{\mathbf{v}_1, \mathbf{v}_2\}$ and $\mathcal{B}_2 = \{\mathbf{u}_1, \mathbf{u}_2\}$ be orthonormal bases for $\mathcal{R}^2$ consisting of left and right singular vectors of A, respectively. For a vector $\mathbf{u}$ in $\mathcal{R}^2$, let |

$$\mathbf{u} = x_1'\mathbf{u}_1 + x_2'\mathbf{u}_2 \text{ for some scalars } x_1' \text{ and } x_2'. \text{ So } [\mathbf{u}]_{\mathcal{B}_2} = \begin{bmatrix} x_1' \\ x_2' \end{bmatrix}. \text{ We wish to}$$

characterize S' by means of an equation relating x_1' and x_2'.

For any vector $\mathbf{v} = x_1\mathbf{v}_1 + x_2\mathbf{v}_2$, the condition $\mathbf{u} = T_A(\mathbf{v})$ means that

$$x_1'\mathbf{u}_1 + x_2'\mathbf{u}_2 = T_A(x_1\mathbf{v}_1 + x_2\mathbf{v}_2) = x_1 A\mathbf{v}_1 + x_2 A\mathbf{v}_2 = x_1\sigma_1\mathbf{u}_1 + x_2\sigma_2\mathbf{u}_1,$$

and hence

$$x_1' = \sigma_1 x_1 \qquad \text{and} \qquad x_2' = \sigma_2 x_2.$$

Furthermore, $\mathbf{v}$ is in S if and only if $\|\mathbf{v}\|^2 = x_1^2 + x_2^2 = 1$. It follows that $\mathbf{u}$ is in S' if and only if $\mathbf{u} = T(\mathbf{v})$, where $\mathbf{v}$ is in S, that is,

$$\frac{(x_1')^2}{\sigma_1^2} + \frac{(x_2')^2}{\sigma_2^2} = x_1^2 + x_2^2 = 1.$$

This is the equation of an ellipse with the major and minor axes oriented along the lines generated by $\mathbf{u}_1$ and $\mathbf{u}_2$, respectively (see Figure 6.16). ○

The Singular Value Decomposition of a Matrix

Theorem 6.17 can be restated in the context of a single matrix equation that has many useful applications. Using the notation of Theorem 6.17, let A be an $m \times n$ matrix of rank k with singular values $\sigma_1 \geq \sigma_2 \geq \cdots \geq \sigma_k > 0$, let $\mathcal{B}_1 = \{\mathbf{v}_1, \mathbf{v}_2, \ldots, \mathbf{v}_n\}$ be an orthonormal basis for $\mathcal{R}^n$ consisting of right

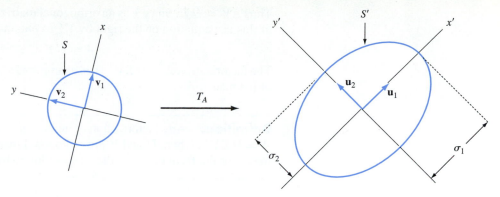

FIGURE 6.16

singular vectors of A, and let $\mathcal{B}_2 = \{\mathbf{u}_1, \mathbf{u}_2, \ldots, \mathbf{u}_m\}$ be an orthonormal basis for $\mathcal{R}^m$ consisting of left singular vectors of A. Let V and U be the $n \times n$ and $m \times m$ matrices, respectively, defined by

$$V = [\mathbf{v}_1 \quad \mathbf{v}_2 \quad \cdots \quad \mathbf{v}_n] \quad \text{and} \quad U = [\mathbf{u}_1 \quad \mathbf{u}_2 \quad \cdots \quad \mathbf{u}_m].$$

Notice that V and U are orthogonal matrices because their columns form orthonormal bases. Now let Σ be the $m \times n$ matrix whose (i, j)-entry, which we denote by s_{ij}, is given by

$$\begin{cases} s_{ii} = \sigma_i & \text{for } i = 1, 2, \ldots, k \\ s_{ij} = 0 & \text{otherwise.} \end{cases}$$

So

$$\Sigma = \left[\begin{array}{cccc|cccc} \sigma_1 & 0 & \cdots & 0 & 0 & 0 & \cdots & 0 \\ 0 & \sigma_2 & \ldots & 0 & 0 & 0 & \cdots & 0 \\ \vdots & \vdots & \ddots & \vdots & \vdots & \vdots & & \vdots \\ 0 & 0 & \cdots & \sigma_k & 0 & 0 & \cdots & 0 \\ \hline 0 & 0 & \cdots & 0 & 0 & 0 & \cdots & 0 \\ \vdots & \vdots & & \vdots & \vdots & \vdots & & \vdots \\ 0 & 0 & \cdots & 0 & 0 & 0 & \cdots & 0 \end{array} \right]. \tag{12}$$

Then, by (10),

$$AV = A[\mathbf{v}_1 \quad \mathbf{v}_2 \quad \cdots \quad \mathbf{v}_n]$$

$$= [A\mathbf{v}_1 \quad A\mathbf{v}_2 \quad \cdots \quad A\mathbf{v}_n]$$

$$= [\sigma_1\mathbf{u}_1 \quad \sigma_2\mathbf{u}_2 \quad \cdots \quad \sigma_k\mathbf{u}_k \quad \mathbf{0} \quad \cdots \quad \mathbf{0}]$$

$$= [\mathbf{u}_1 \quad \mathbf{u}_2 \quad \cdots \quad \mathbf{u}_m] \left[\begin{array}{cccc|cccc} \sigma_1 & 0 & \cdots & 0 & 0 & 0 & \cdots & 0 \\ 0 & \sigma_2 & \cdots & 0 & 0 & 0 & \cdots & 0 \\ \vdots & \vdots & \ddots & \vdots & \vdots & \vdots & & \vdots \\ 0 & 0 & \cdots & \sigma_k & 0 & 0 & \cdots & 0 \\ \hline 0 & 0 & \cdots & 0 & 0 & 0 & \ldots & 0 \\ \vdots & \vdots & & \vdots & \vdots & \vdots & & \vdots \\ 0 & 0 & \cdots & 0 & 0 & 0 & \cdots & 0 \end{array} \right]$$

$$= U\Sigma.$$

Thus $AV = U\Sigma$. Since V is an orthogonal matrix, we may multiply both sides of this last equation on the right by V^T to obtain

$$A = U\Sigma V^T.$$

The factorization $A = U\Sigma V^T$ is an example of a *singular value decomposition* of the matrix A.

Definition.   Any factorization of an $m \times n$ matrix A into a product $A = U\Sigma V^T$, where U and V are orthogonal matrices and Σ is an $m \times n$ matrix of the form (12), is called a **singular value decomposition** of A.

Thus we have the following result.

Theorem 6.18 (Singular value decomposition)

For any $m \times n$ matrix A of rank k, there exist $\sigma_1 \geq \sigma_2 \geq \cdots \geq \sigma_k > 0$, an $m \times m$ orthogonal matrix U, and an $n \times n$ orthogonal matrix V such that

$$A = U\Sigma V^T,$$

where Σ is the $m \times n$ matrix (12).

In the proof of Theorem 6.18, the nonzero entries of Σ are chosen to be the singular values of A, and the columns of U and V are chosen to be an orthonormal basis for $\mathcal{R}^m$ consisting of left singular vectors of A, and an orthonormal basis for $\mathcal{R}^n$ consisting of right singular vectors of A, respectively. So the designations of *left* and *right* come from the locations of U and V in the decomposition product.

It can be proved that if $A = U\Sigma V^T$ is any singular value decomposition of A, then the nonzero diagonal entries of Σ are the singular values of A, the columns of V constitute an orthonormal basis of right singular vectors of A, and the columns of U constitute an orthonormal basis of left singular vectors of A (see Exercise 37).

Example 4   Find a singular value decomposition of the matrix

$$A = \begin{bmatrix} 0 & 1 & 2 \\ 1 & 0 & 1 \end{bmatrix}$$

in Example 1.

Solution   We may use the results of Example 1 to obtain the decomposition. The columns of the matrix U are the vectors in $\mathcal{B}_2$, and the columns of V are the vectors in $\mathcal{B}_1$. Furthermore $\sigma_1 = \sqrt{6}$ and $\sigma_2 = 1$. Thus

$$A = U\Sigma V^T = \begin{bmatrix} \frac{2}{\sqrt{5}} & \frac{-1}{\sqrt{5}} \\ \frac{1}{\sqrt{5}} & \frac{2}{\sqrt{5}} \end{bmatrix} \begin{bmatrix} \sqrt{6} & 0 & 0 \\ 0 & 1 & 0 \end{bmatrix} \begin{bmatrix} \frac{1}{\sqrt{30}} & \frac{2}{\sqrt{5}} & \frac{1}{\sqrt{6}} \\ \frac{2}{\sqrt{30}} & \frac{-1}{\sqrt{5}} & \frac{2}{\sqrt{6}} \\ \frac{5}{\sqrt{30}} & 0 & \frac{-1}{\sqrt{6}} \end{bmatrix}^T .$$

Example 5   Find a singular value decomposition of the matrix

$$C = \begin{bmatrix} 0 & 1 \\ 1 & 0 \\ 2 & 1 \end{bmatrix}.$$

Solution Notice that C is the transpose of the matrix A in Examples 1 and 2. A comparison of (11) and (10) indicates that the singular values of $C = A^T$ are the same as the singular values of A, an orthonormal basis of the right singular vectors of A is also an orthonormal basis of left singular values of C, and an orthonormal basis of left singular vectors of A is also an orthonormal basis of right singular values of C. Thus in the singular value decomposition of C, the roles of the orthogonal matrices V and U in Example 2 are reversed. Let Σ' be the 3×2 matrix with $(1, 1)$-entry equal to $\sqrt{6}$, $(2, 2)$-entry equal to 1, and all other entries equal to 0. Then the singular value decomposition of C is

$$C = V\Sigma'U^T = \begin{bmatrix} \frac{1}{\sqrt{30}} & \frac{2}{\sqrt{5}} & \frac{1}{\sqrt{6}} \\ \frac{2}{\sqrt{30}} & \frac{-1}{\sqrt{5}} & \frac{2}{\sqrt{6}} \\ \frac{5}{\sqrt{30}} & 0 & \frac{-1}{\sqrt{6}} \end{bmatrix} \begin{bmatrix} \sqrt{6} & 0 \\ 0 & 1 \\ 0 & 0 \end{bmatrix} \begin{bmatrix} \frac{2}{\sqrt{5}} & \frac{-1}{\sqrt{5}} \\ \frac{1}{\sqrt{5}} & \frac{2}{\sqrt{5}} \end{bmatrix}^T.$$

We can also apply Exercise 38, which allows us to obtain the singular value decomposition of C directly by taking the transpose of the singular value decomposition of A. Since $A = U\Sigma V^T$, we have

$$C = A^T = (U\Sigma V^T)^T = (V^T)^T \Sigma^T U^T = V\Sigma^T U^T.$$

Notice that $\Sigma^T = \Sigma'$.

There are efficient and accurate methods for finding a singular value decomposition $U\Sigma V^T$ of an $m \times n$ matrix A, and many practical applications of linear algebra use this decomposition. Because the matrices U and V^T are orthogonal, multiplication by U and V^T does not change the norms of vectors or the angles between them. Thus any roundoff errors that arise in calculations involving A are due solely to the matrix Σ. For this reason, calculations involving A are most reliable if a singular value decomposition of A is used.

The Singular Value Decomposition and Systems of Linear Equations

Let A be an $m \times n$ matrix and $\mathbf{b}$ be in $\mathcal{R}^m$. We have seen that the system of linear equations $A\mathbf{x} = \mathbf{b}$ can be consistent or inconsistent.

In the case that the system is consistent, a vector $\mathbf{u}$ in $\mathcal{R}^n$ is a solution if and only if $\|A\mathbf{u} - \mathbf{b}\| = 0$.

In the case that the system is inconsistent, $\|A\mathbf{u} - \mathbf{b}\| > 0$ for every $\mathbf{u}$ in $\mathcal{R}^n$. However, it is often desirable to find a vector $\mathbf{z}$ in $\mathcal{R}^n$ that minimizes the distance between $A\mathbf{u}$ and $\mathbf{b}$, that is, such that

$$\|A\mathbf{z} - \mathbf{b}\| \le \|A\mathbf{u} - \mathbf{b}\| \qquad \text{for all } \mathbf{u} \text{ in } \mathcal{R}^n.$$

This problem, called the *least-squares problem*, was encountered in Section 6.3, where we obtained a method for computing the least-squares line to fit a given set of data.

So for either the consistent or the inconsistent case, we wish to choose a vector $\mathbf{z}$ that minimizes $\|A\mathbf{u} - \mathbf{b}\|$. One immediate approach to this problem is to make use of the fact that the set of all vectors of the form $A\mathbf{u}$ is a subspace of $\mathcal{R}^m$. Recall that this set is actually the column space, Col A, of A. For convenience, let $W = \text{Col } A$. Then a vector $A\mathbf{z}$ in W that minimizes the distances $\|A\mathbf{u} - \mathbf{b}\|$ is the one that is closest to $\mathbf{b}$. Recall by the closest vector property that the vector in W that is closest to $\mathbf{b}$ is $P_W\mathbf{b}$, the orthogonal projection of $\mathbf{b}$ onto W. Thus a vector $\mathbf{z}$ minimizes $\|A\mathbf{u} - \mathbf{b}\|$ if and only if

$$A\mathbf{z} = P_W\mathbf{b}. \tag{13}$$

One way to analyze this problem is to use the singular value decomposition of A. Suppose that A has rank k. Then $k \le m$ and $k \le n$. Let $A = U\Sigma V^T$ be a singular value decomposition of A, where the singular values of A are the first k diagonal entries of Σ, as in (12). By Exercise 37, the columns of V and U satisfy (10). Since V is invertible, the columns of V form a basis for $\mathcal{R}^n$, and hence by (10), the first k columns of U form a basis for W. Since U is an orthogonal matrix, the columns of U form an orthonormal basis of $\mathcal{R}^m$. Thus the first k columns of U form an orthonormal basis for W and, by Exercise 33 of Section 6.2, the last $m - k$ columns of U form an orthonormal basis for $W^\perp$. So by Exercise 37(c) of Section 6.3, we have that

$$P_W = UDU^T, \tag{14}$$

where D is the diagonal $m \times m$ matrix whose first k diagonal entries are 1s and whose other entries are 0s.

To solve the least-squares problem, we modify the $m \times n$ matrix Σ to obtain a new $n \times m$ matrix, $\Sigma^\dagger$, defined by

$$\Sigma^\dagger = \begin{bmatrix} \frac{1}{\sigma_1} & 0 & \cdots & 0 & 0 & 0 & \cdots & 0 \\ 0 & \frac{1}{\sigma_2} & \cdots & 0 & 0 & 0 & \cdots & 0 \\ \vdots & \vdots & \ddots & \vdots & \vdots & \vdots & & \vdots \\ 0 & 0 & \cdots & \frac{1}{\sigma_k} & 0 & 0 & \cdots & 0 \\ \hline 0 & 0 & \cdots & 0 & 0 & 0 & \cdots & 0 \\ \vdots & \vdots & & \vdots & \vdots & \vdots & & \vdots \\ 0 & 0 & \cdots & 0 & 0 & 0 & \cdots & 0 \end{bmatrix}. \tag{15}$$

Observe that $\Sigma\Sigma^\dagger = D$. With this new matrix, we set

$$\mathbf{z} = V\Sigma^\dagger U^T\mathbf{b}.$$

Then by (14)

$$A\mathbf{z} = (U\Sigma V^T)(V\Sigma^\dagger U^T\mathbf{b}) = U\Sigma\Sigma^\dagger U^T\mathbf{b} = UDU^T\mathbf{b} = P_W\mathbf{b},$$

and hence by (13), the vector $\mathbf{z}$ just defined, minimizes $\|A\mathbf{u} - \mathbf{b}\|$.

Besides the vector $\mathbf{z} = V\Sigma^\dagger U^T\mathbf{b}$, there may be other vectors in $\mathcal{R}^n$ that also minimize $\|A\mathbf{u} - \mathbf{b}\|$. However, we now show that among all such vectors, $\mathbf{z} = V\Sigma^\dagger U^T\mathbf{b}$ is the unique vector of least norm. Suppose that $\mathbf{y}$ is any vector in $\mathcal{R}^n$ different from $\mathbf{z}$ that also minimizes $\|A\mathbf{u} - \mathbf{b}\|$. Then $A\mathbf{y} = P_W\mathbf{b}$ by (13). Let $\mathbf{w} = \mathbf{y} - \mathbf{z}$. Then $\mathbf{w} \ne \mathbf{0}$, but

$$A\mathbf{w} = A\mathbf{y} - A\mathbf{z} = P_W\mathbf{b} - P_W\mathbf{b} = \mathbf{0}.$$

Substituting the singular value decomposition of A, we have

$$U\Sigma V^T\mathbf{w} = \mathbf{0}.$$

Since U is invertible, it follows that

$$\Sigma V^T\mathbf{w} = \mathbf{0}.$$

This last equation tells us that the first k components of $V^T\mathbf{w}$ are 0s. Furthermore, since $\mathbf{z} = V\Sigma^\dagger U^T\mathbf{b}$, we have that $V^T\mathbf{z} = \Sigma^\dagger U^T\mathbf{b}$, and hence the last

$n - k$ components of $V^T\mathbf{z}$ are 0s. It follows that $V^T\mathbf{w}$ and $V^T\mathbf{z}$ are orthogonal. Since V^T is an orthogonal matrix, it preserves dot products, and hence

$$\mathbf{z} \cdot \mathbf{w} = (V^T\mathbf{z}) \cdot (V^T\mathbf{w}) = 0.$$

Thus $\mathbf{z}$ and $\mathbf{w}$ are orthogonal. Since $\mathbf{y} = \mathbf{z} + \mathbf{w}$, we may apply the Pythagorean theorem to obtain that

$$\|\mathbf{y}\|^2 = \|\mathbf{z} + \mathbf{w}\|^2 = \|\mathbf{z}\|^2 + \|\mathbf{w}\|^2 > \|\mathbf{z}\|^2,$$

and hence $\|\mathbf{y}\| > \|\mathbf{z}\|$. It follows that $\mathbf{z} = V\Sigma^\dagger U^T\mathbf{b}$ is the vector of least norm that minimizes $\|A\mathbf{u} - \mathbf{b}\|$.

We summarize what we have learned in the following.

Let A be an $m \times n$ matrix with a singular value decomposition $A = U\Sigma V^T$, $\mathbf{b}$ be a vector in $\mathcal{R}^m$, and $\mathbf{z} = V\Sigma^\dagger U^T\mathbf{b}$, where $\Sigma^\dagger$ is defined as in (15). Then the following statements are true.

(a) If the system $A\mathbf{x} = \mathbf{b}$ is consistent, then $\mathbf{z}$ is the unique solution of least norm.

(b) If the system $A\mathbf{x} = \mathbf{b}$ is inconsistent, then $\mathbf{z}$ is the unique vector of least norm such that

$$\|A\mathbf{z} - \mathbf{b}\| \leq \|A\mathbf{u} - \mathbf{b}\|$$

for all $\mathbf{u}$ in $\mathcal{R}^n$.

Example 6   Use a singular value decomposition to find the solution of least norm to the system

$$x_2 + 2x_3 = 5$$
$$x_1 \quad + \quad x_3 = 1.$$

Solution   Let A denote the coefficient matrix of this system, and let $\mathbf{b} = \begin{bmatrix} 5 \\ 1 \end{bmatrix}$. A singular value decomposition of A was computed in Example 2, where we obtained

$$\begin{bmatrix} 0 & 1 & 2 \\ 1 & 0 & 1 \end{bmatrix} = A = U\Sigma V^T = \begin{bmatrix} \frac{2}{\sqrt{5}} & \frac{-1}{\sqrt{5}} \\ \frac{1}{\sqrt{5}} & \frac{2}{\sqrt{5}} \end{bmatrix} \begin{bmatrix} \sqrt{6} & 0 & 0 \\ 0 & 1 & 0 \end{bmatrix} \begin{bmatrix} \frac{1}{\sqrt{30}} & \frac{2}{\sqrt{5}} & \frac{1}{\sqrt{6}} \\ \frac{2}{\sqrt{30}} & \frac{-1}{\sqrt{5}} & \frac{2}{\sqrt{6}} \\ \frac{5}{\sqrt{30}} & 0 & \frac{-1}{\sqrt{6}} \end{bmatrix}^T.$$

Let $\mathbf{z}$ denote the solution of least norm to this system. Then

$$\mathbf{z} = V\Sigma^\dagger U^T\mathbf{b}$$

$$= \begin{bmatrix} \frac{1}{\sqrt{30}} & \frac{2}{\sqrt{5}} & \frac{1}{\sqrt{6}} \\ \frac{2}{\sqrt{30}} & \frac{-1}{\sqrt{5}} & \frac{2}{\sqrt{6}} \\ \frac{5}{\sqrt{30}} & 0 & \frac{-1}{\sqrt{6}} \end{bmatrix} \begin{bmatrix} \frac{1}{\sqrt{6}} & 0 \\ 0 & 1 \\ 0 & 0 \end{bmatrix} \begin{bmatrix} \frac{2}{\sqrt{5}} & \frac{-1}{\sqrt{5}} \\ \frac{1}{\sqrt{5}} & \frac{2}{\sqrt{5}} \end{bmatrix}^T \begin{bmatrix} 5 \\ 1 \end{bmatrix}$$

$$= \frac{1}{6} \begin{bmatrix} -5 \\ 8 \\ 11 \end{bmatrix}.$$

Example 7   Let

$$A = \begin{bmatrix} 1 & 1 & 2 \\ 1 & -1 & 3 \\ 1 & 3 & 1 \end{bmatrix} \quad \text{and} \quad \mathbf{b} = \begin{bmatrix} 1 \\ 4 \\ -1 \end{bmatrix}.$$

It is easy to show that the equation $A\mathbf{x} = \mathbf{b}$ has no solution. Find a vector $\mathbf{z}$ in $\mathcal{R}^3$ such that

$$\|A\mathbf{z} - \mathbf{b}\| \le \|A\mathbf{u} - \mathbf{b}\|$$

for all $\mathbf{u}$ in $\mathcal{R}^3$.

Solution   First observe (we omit the details) that

$$\mathcal{B}_1 = \left\{ \frac{1}{\sqrt{6}} \begin{bmatrix} 1 \\ 1 \\ 2 \end{bmatrix}, \frac{1}{\sqrt{5}} \begin{bmatrix} 0 \\ 2 \\ -1 \end{bmatrix}, \frac{1}{\sqrt{30}} \begin{bmatrix} -5 \\ 1 \\ 2 \end{bmatrix} \right\}$$

is an orthonormal basis of eigenvectors of $A^T A$ with corresponding eigenvalues 18, 10, and 0. So $\sigma_1 = \sqrt{18}$, and $\sigma_2 = \sqrt{10}$ are the singular values of A, and $\mathcal{B}_1$ is a set of right singular vectors of A. Let $\mathbf{v}_1$ and $\mathbf{v}_2$ denote the first two vectors in $\mathcal{B}_1$, which correspond to the singular values σ_1 and σ_2, respectively, and let

$$\mathbf{u}_1 = \frac{1}{\sigma_1} A\mathbf{v}_1 = \frac{1}{\sqrt{18}} \frac{1}{\sqrt{6}} \begin{bmatrix} 1 & 1 & 2 \\ 1 & -1 & 3 \\ 1 & 3 & 1 \end{bmatrix} \begin{bmatrix} 1 \\ 1 \\ 2 \end{bmatrix} = \frac{1}{\sqrt{3}} \begin{bmatrix} 1 \\ 1 \\ 1 \end{bmatrix}$$

and

$$\mathbf{u}_2 = \frac{1}{\sigma_2} A\mathbf{v}_2 = \frac{1}{\sqrt{10}} \frac{1}{\sqrt{5}} \begin{bmatrix} 1 & 1 & 2 \\ 1 & -1 & 3 \\ 1 & 3 & 1 \end{bmatrix} \begin{bmatrix} 0 \\ 2 \\ -1 \end{bmatrix} = \frac{1}{\sqrt{2}} \begin{bmatrix} 0 \\ -1 \\ 1 \end{bmatrix}.$$

As in Example 2, we are able to extend $\{\mathbf{u}_1, \mathbf{u}_2\}$ to an orthonormal basis for $\mathcal{R}^3$ by adjoining the vector $\mathbf{u}_3 = \dfrac{1}{\sqrt{6}} \begin{bmatrix} 2 \\ -1 \\ -1 \end{bmatrix}$. This produces a set of left singular vectors $\mathcal{B}_2 = \{\mathbf{u}_1, \mathbf{u}_2, \mathbf{u}_3\}$ of A. Next, set

$$U = [\mathbf{u}_1 \quad \mathbf{u}_2 \quad \mathbf{u}_3], \qquad V = [\mathbf{v}_1 \quad \mathbf{v}_2 \quad \mathbf{v}_3],$$

and

$$\Sigma = \begin{bmatrix} \sigma_1 & 0 & 0 \\ 0 & \sigma_2 & 0 \\ 0 & 0 & 0 \end{bmatrix} = \begin{bmatrix} \sqrt{18} & 0 & 0 \\ 0 & \sqrt{10} & 0 \\ 0 & 0 & 0 \end{bmatrix}.$$

Then $A = U\Sigma V^T$ is a singular value decomposition of A. Hence

$$\mathbf{z} = V\Sigma^\dagger U^T \mathbf{b}$$

$$= \begin{bmatrix} \frac{1}{\sqrt{6}} & 0 & \frac{-5}{\sqrt{30}} \\ \frac{1}{\sqrt{6}} & \frac{2}{\sqrt{5}} & \frac{1}{\sqrt{30}} \\ \frac{2}{\sqrt{6}} & \frac{-1}{\sqrt{5}} & \frac{2}{\sqrt{30}} \end{bmatrix} \begin{bmatrix} \frac{1}{\sqrt{18}} & 0 & 0 \\ 0 & \frac{1}{\sqrt{10}} & 0 \\ 0 & 0 & 0 \end{bmatrix} \begin{bmatrix} \frac{1}{\sqrt{3}} & 0 & \frac{2}{\sqrt{6}} \\ \frac{1}{\sqrt{3}} & \frac{-1}{\sqrt{2}} & \frac{-1}{\sqrt{6}} \\ \frac{1}{\sqrt{3}} & \frac{1}{\sqrt{2}} & \frac{-1}{\sqrt{6}} \end{bmatrix}^T \begin{bmatrix} 1 \\ 4 \\ -1 \end{bmatrix}$$

$$= \frac{1}{18} \begin{bmatrix} 4 \\ -14 \\ 17 \end{bmatrix}$$

is the vector of least norm that satisfies the condition $\|A\mathbf{z} - \mathbf{b}\| \leq \|A\mathbf{u} - \mathbf{b}\|$ for all $\mathbf{u}$ in $\mathcal{R}^3$.

In the discussion above, a singular value decomposition of the coefficient matrix $A = U\Sigma V^T$ of a system of linear equations $A\mathbf{x} = \mathbf{b}$ is used to find the solution of least norm or the vector of least norm that minimizes $A\mathbf{u} - \mathbf{b}$. The solution is the product of the matrix $V\Sigma^\dagger U^T$ and $\mathbf{b}$.

Although a singular value decomposition of a matrix $A = U\Sigma V^T$ is not unique, the matrix $V\Sigma^\dagger U^T$ is unique, that is, it is independent of the choice of a singular value decomposition of A. To see this, consider an $m \times n$ matrix A, and suppose that

$$A = U_1\Sigma V_1^T = U_2\Sigma V_2^T$$

are two singular value decompositions of A. Now consider any vector $\mathbf{b}$ in $\mathcal{R}^m$. Then we have seen that both $V_1\Sigma^\dagger U_1^T\mathbf{b}$ and $V_2\Sigma^\dagger U_2^T\mathbf{b}$ are unique vectors of least norm that minimize $\|A\mathbf{u} - \mathbf{b}\|$. So

$$V_1\Sigma^\dagger U_1^T\mathbf{b} = V_2\Sigma^\dagger U_2^T\mathbf{b}.$$

Since $\mathbf{b}$ is an arbitrarily chosen vector in $\mathcal{R}^n$, it follows that $V_1\Sigma^\dagger U_1^T = V_2\Sigma^\dagger U_2^T$.

For a given matrix $A = U\Sigma V^T$, the matrix $V\Sigma^\dagger U^T$ is called the **pseudoinverse** or **Moore–Penrose generalized inverse** of A and is denoted by $A^\dagger$. Note that the pseudoinverse of the matrix Σ in (12) is the matrix $\Sigma^\dagger$ in (15)—see Exercise 40. The terminology *pseudoinverse* is due to the fact that if A is invertible, then $A^\dagger = A^{-1}$, the ordinary inverse of A (see Exercise 41).

Example 8   Find the pseudoinverse of the matrix

$$A = \begin{bmatrix} 0 & 1 & 2 \\ 1 & 0 & 1 \end{bmatrix}$$

in Example 4.

Solution   From Example 4, the singular value decomposition of A is given by

$$A = U\Sigma V^T = \begin{bmatrix} \frac{2}{\sqrt{5}} & \frac{-1}{\sqrt{5}} \\ \frac{1}{\sqrt{5}} & \frac{2}{\sqrt{5}} \end{bmatrix} \begin{bmatrix} \sqrt{6} & 0 & 0 \\ 0 & 1 & 0 \end{bmatrix} \begin{bmatrix} \frac{1}{\sqrt{30}} & \frac{2}{\sqrt{5}} & \frac{1}{\sqrt{6}} \\ \frac{2}{\sqrt{30}} & \frac{-1}{\sqrt{5}} & \frac{2}{\sqrt{6}} \\ \frac{5}{\sqrt{30}} & 0 & \frac{-1}{\sqrt{6}} \end{bmatrix}^T.$$

Hence the pseudoinverse of A is

$$A^\dagger = V\Sigma^\dagger U^T = \begin{bmatrix} \frac{1}{\sqrt{30}} & \frac{2}{\sqrt{5}} & \frac{1}{\sqrt{6}} \\ \frac{2}{\sqrt{30}} & \frac{-1}{\sqrt{5}} & \frac{2}{\sqrt{6}} \\ \frac{5}{\sqrt{30}} & 0 & \frac{-1}{\sqrt{6}} \end{bmatrix} \begin{bmatrix} \frac{1}{\sqrt{6}} & 0 \\ 0 & 1 \\ 0 & 0 \end{bmatrix} \begin{bmatrix} \frac{2}{\sqrt{5}} & \frac{-1}{\sqrt{5}} \\ \frac{1}{\sqrt{5}} & \frac{2}{\sqrt{5}} \end{bmatrix}^T$$

$$= \begin{bmatrix} -\frac{1}{3} & \frac{5}{6} \\ \frac{1}{3} & -\frac{1}{3} \\ \frac{1}{3} & \frac{1}{6} \end{bmatrix}.$$

Practice Problems

1. Find the singular values and orthonormal bases of right and left singular vectors for

$$A = \begin{bmatrix} -2 & -20 & 8 \\ 14 & -10 & 19 \end{bmatrix}.$$

2. Find a singular value decomposition of the matrix A in problem 1.

3. Use your answer to problem 2 to find the solution of least norm to the system

$$-2x_1 - 20x_2 + 8x_3 = 5$$
$$14x_1 - 10x_2 + 19x_3 = -5.$$

4. Let

$$A = \begin{bmatrix} 1 & 1 & 2 \\ 1 & -1 & 3 \\ 1 & 3 & 1 \end{bmatrix} \quad \text{and} \quad b = \begin{bmatrix} 27 \\ 36 \\ -18 \end{bmatrix}.$$

Use the singular value decomposition in Example 7 to find a vector z in $\mathcal{R}^3$ such that $\|Az - b\| \le \|Au - b\|$ for all u in $\mathcal{R}^3$.

Exercises

1. Determine if the following statements are true or false.

   (a) If σ is a singular value of a matrix A, then σ is an eigenvalue of $A^T A$.

   (b) An orthonormal basis of right singular vectors of a matrix is also an orthonormal basis of left singular vectors of the transpose of the matrix.

   (c) If a matrix is square, then any orthonormal basis of left singular vectors is also an orthonormal basis of right singular vectors.

   (d) Every matrix has the same singular values as its transpose.

   (e) A matrix has a pseudoinverse if and only if it is not invertible.

In Exercises 2–10, find a singular value decomposition for the given matrix.

2. $\begin{bmatrix} 1 & 1 \\ 0 & 0 \end{bmatrix}$
3. $\begin{bmatrix} 1 \\ 2 \\ 2 \end{bmatrix}$
4. $\begin{bmatrix} 1 \\ 1 \\ -1 \\ 1 \end{bmatrix}$

5. $\begin{bmatrix} 1 & 1 \\ 1 & -1 \\ 1 & 2 \end{bmatrix}$
6. $\begin{bmatrix} 1 & 2 \\ 3 & -1 \\ 1 & 0 \\ 1 & 1 \end{bmatrix}$
7. $\begin{bmatrix} 1 & 1 & 1 \\ 1 & -1 & -1 \end{bmatrix}$

8. $\begin{bmatrix} 1 & 0 & 0 & 0 \\ 0 & 2 & 0 & 0 \end{bmatrix}$
9. $\begin{bmatrix} 1 & 1 & 2 \\ 2 & 0 & -1 \\ 1 & -1 & 0 \end{bmatrix}$

10. $\begin{bmatrix} 1 & -1 & 3 \\ 1 & -1 & -1 \\ 2 & 1 & -1 \end{bmatrix}$

In Exercises 11–14, find a singular value decomposition of the given matrix A. In each case, the characteristic polynomial of $A^T A$ is given.

11. $A = \begin{bmatrix} 3 & 5 & 4 & 1 \\ 4 & 0 & 2 & -2 \\ 0 & 0 & 0 & 0 \end{bmatrix}$
$t^2(t - 60)(t - 15)$

12. $A = \begin{bmatrix} 2 & -3 & 2 & -3 \\ 6 & 1 & 6 & 1 \\ 0 & 0 & 0 & 0 \end{bmatrix}$
$t^2(t - 80)(t - 20)$

13. $A = \begin{bmatrix} 3 & 0 & 1 & 3 \\ 0 & 3 & 1 & 0 \\ 0 & -3 & -1 & 0 \end{bmatrix}$
$t^2(t - 18)(t - 21)$

14. $A = \begin{bmatrix} -4 & 8 & 0 & -8 \\ 8 & -25 & 0 & 7 \\ 8 & -7 & 0 & 25 \end{bmatrix}$
$t^2(t - 324)(t - 1296)$

In Exercises 15 and 16, sketch the image of the unit circle under the matrix transformation T_A induced by the given matrix A.

15. $\begin{bmatrix} 2 & 1 \\ -2 & 1 \end{bmatrix}$
16. $\begin{bmatrix} 1 & 2 \\ 2 & 1 \end{bmatrix}$

In Exercises 17–20, find the unique solution of least norm to the given system of linear equations. In one case, you can use the results of Exercises 2–10.

17. $x_1 + x_2 = 2$
    $2x_1 + 2x_2 = 4$

18. $x_1 + x_3 = 3$
    $x_2 = 1$

19. $x_1 + x_2 + x_3 = 5$
    $x_1 - x_2 - x_3 = 1$

20. $x_1 + x_2 = 4$
    $x_1 - x_2 - x_3 = 1$

In Exercises 21–24, the systems are inconsistent. For each system $Ax = b$, find the unique vector z of least norm such that $\|Az - b\|$ is a minimum. In a few cases, you can use the results of Exercises 2–10.

21. $x_1 + x_2 = 3$
$x_1 - x_2 = 1$
$x_1 + 2x_2 = 2$

22. $x_1 + 2x_2 = 4$
$3x_1 - x_2 = 5$
$x_1 \quad = 1$
$x_1 + x_2 = 0$

23. $x_1 + x_2 - x_3 = 4$
$x_1 + x_2 + x_3 = 6$
$x_3 = 3$

24. $2x_1 + x_2 = \quad 1$
$x_1 - x_2 = -4$
$x_1 + 2x_2 = \quad 0$

In Exercises 25–30, find the pseudoinverse of the given matrix. You can use the results of Exercises 2–14.

25. $\begin{bmatrix} 1 \\ 2 \\ 2 \end{bmatrix}$

26. $\begin{bmatrix} 1 & 2 \\ 3 & -1 \\ 1 & 0 \\ 1 & 1 \end{bmatrix}$

27. $\begin{bmatrix} 1 & 1 \\ 1 & -1 \\ 1 & 2 \end{bmatrix}$

28. $\begin{bmatrix} 1 & 0 & 0 & 0 \\ 0 & 2 & 0 & 0 \end{bmatrix}$

29. $\begin{bmatrix} 1 & 1 & 1 \\ 1 & -1 & -1 \end{bmatrix}$

30. $\begin{bmatrix} 3 & 5 & 4 & 1 \\ 4 & 0 & 2 & -2 \\ 0 & 0 & 0 & 0 \end{bmatrix}$

In (14), a singular value decomposition of a matrix A is used to obtain the orthogonal projection onto the subspace Col A. *Use this method in Exercises 31–34 to compute the orthogonal projection matrix P_W onto the subspace W. You can use the results of Exercises 2–10.*

31. $W = \text{Span} \left\{ \begin{bmatrix} 1 \\ 2 \\ 2 \end{bmatrix} \right\}$

32. $W = \text{Span} \left\{ \begin{bmatrix} 1 \\ 1 \\ -1 \\ 1 \end{bmatrix} \right\}$

33. $W = \text{Span} \left\{ \begin{bmatrix} 1 \\ 1 \\ 1 \end{bmatrix}, \begin{bmatrix} 1 \\ -1 \\ 2 \end{bmatrix} \right\}$

34. $W = \text{Span} \left\{ \begin{bmatrix} 1 \\ 3 \\ 1 \\ 1 \end{bmatrix}, \begin{bmatrix} 2 \\ -1 \\ 0 \\ 1 \end{bmatrix} \right\}$

35. Suppose that A is an $m \times n$ matrix of rank k with singular values $\sigma_1 \geq \sigma_2 \geq \cdots \geq \sigma_k > 0$ and orthonormal bases of right and left singular vectors $\mathcal{B}_1 = \{\mathbf{v}_1, \mathbf{v}_2, \ldots, \mathbf{v}_n\}$ and $\mathcal{B}_2 = \{\mathbf{u}_1, \mathbf{u}_2, \ldots, \mathbf{u}_m\}$, respectively.

(a) Prove that $\mathcal{B}_1$ is a basis consisting of eigenvectors of $A^T A$ with corresponding eigenvalues $\sigma_1^2, \sigma_2^2, \ldots, \sigma_k^2, 0, \ldots, 0$.

(b) Prove that $\mathcal{B}_2$ is a basis consisting of eigenvectors of AA^T with corresponding eigenvalues $\sigma_1^2, \sigma_2^2, \ldots, \sigma_k^2, 0, \ldots, 0$.

(c) Prove that the singular values of a matrix are unique, but that the orthonormal bases of right and left singular vectors are not unique.

36. Let A be an $m \times n$ matrix of rank m with singular values

$$\sigma_1 \geq \sigma_2 \geq \cdots \geq \sigma_m > 0.$$

(a) Prove that $\sigma_m \|\mathbf{v}\| \leq \|A\mathbf{v}\| \leq \sigma_1 \|\mathbf{v}\|$ for every vector $\mathbf{v}$ in $\mathcal{R}^n$.

(b) Prove that there exist nonzero vectors $\mathbf{v}$ and $\mathbf{w}$ in $\mathcal{R}^n$ such that

$$\|A\mathbf{v}\| = \sigma_m \|\mathbf{v}\| \quad \text{and} \quad \|A\mathbf{w}\| = \sigma_1 \|\mathbf{w}\|.$$

37. Let A be an $m \times n$ matrix of rank k, and suppose that $A = U \Sigma V^T$ is a singular value decomposition of A.

(a) Prove that the nonzero diagonal entries of Σ are the singular values of A.

(b) Prove that the columns of V form an orthonormal basis of $\mathcal{R}^n$ of right singular vectors of A.

(c) Prove that the columns of U form an orthonormal basis of $\mathcal{R}^m$ of left singular vectors of A.

38. Prove that the transpose of a singular value decomposition of A is a singular value decomposition of A^T.

39. Prove that, for any matrix A, the matrices $A^T A$ and AA^T have the same nonzero eigenvalues.

40. Prove that the pseudoinverse of the matrix Σ in (12) is the matrix $\Sigma^\dagger$ in (15).

41. Prove that if A is an invertible matrix, then $A^\dagger = A^{-1}$.

42. Prove that, for any matrix A, $(A^T)^\dagger = (A^\dagger)^T$.

43. Prove that if A is a symmetric matrix, then the singular values of A are the absolute values of the nonzero eigenvalues of A.

44. Let A be an $n \times n$ symmetric matrix of rank k, with singular values $\sigma_1 \geq \sigma_2 \geq \cdots \geq \sigma_k > 0$, and let Σ be the $n \times n$ matrix in (12). Prove that A is positive semidefinite if and only if there is an $n \times n$ orthogonal matrix V such that $V \Sigma V^T$ is a singular value decomposition of A.

45. Let Q be an $n \times n$ orthogonal matrix.

(a) Determine the singular values of Q. Justify your answer.

(b) Describe a singular value decomposition of Q.

46. Let A be an $n \times n$ matrix of rank n. Prove that A is an orthogonal matrix if and only if 1 is the only singular value of A.

47. Let A be an $m \times n$ matrix with a singular value decomposition $A = U \Sigma V^T$. Suppose that P and Q are orthogonal matrices of sizes $m \times m$ and $n \times n$, respectively, such that $P \Sigma = \Sigma Q$.

(a) Prove that $(UP)\Sigma(VQ)^T$ is a singular value decomposition of A.

(b) Use (a) to find an example of a matrix with two distinct singular value decompositions.

(c) Prove the converse of (a): If $U_1 \Sigma V_1^T$ is any singular value decomposition of A, then there exist orthogonal matrices P and Q of sizes $m \times m$ and $n \times n$, respectively, such that $P \Sigma = \Sigma Q$, $U_1 = UP$, and $V_1 = VQ$.

48. Prove that if $A = U \Sigma V^T$ is a singular value decomposition of an $m \times n$ matrix of rank k, then $\Sigma \Sigma^\dagger$ is the

$m \times m$ diagonal matrix whose first k diagonal entries are ones and whose last $m - k$ diagonal entries are zeros.

49. Prove that, for any matrix A, $AA^{\dagger}$ is the orthogonal projection matrix for Col A.

50. Prove that, for any matrix A, $A^{\dagger}A$ is the orthogonal projection matrix for Row A.

In Exercises 51 and 52, use a calculator with matrix capabilities or computer software such as MATLAB *to find a singular value decomposition and the pseudoinverse of the given matrix A.*

51. $\begin{bmatrix} 1 & 2 & 1 & 3 \\ 2 & -1 & 1 & 4 \\ -1 & 0 & 1 & 2 \end{bmatrix}$ 52. $\begin{bmatrix} 2 & 0 & 1 & -1 \\ 1 & 3 & 1 & 2 \\ 1 & 1 & -1 & 1 \end{bmatrix}$

Solutions to the Practice Problems

1. First observe that

$$A^T A = \begin{bmatrix} 200 & -100 & 250 \\ -100 & 500 & -350 \\ 250 & -350 & 425 \end{bmatrix}.$$

Then

$$\mathcal{B}_1 = \{\mathbf{v}_1, \mathbf{v}_2, \mathbf{v}_3\} = \left\{ \frac{1}{3}\begin{bmatrix} 1 \\ -2 \\ 2 \end{bmatrix}, \frac{1}{3}\begin{bmatrix} 2 \\ 2 \\ 1 \end{bmatrix}, \frac{1}{3}\begin{bmatrix} 2 \\ -1 \\ -2 \end{bmatrix} \right\}$$

is an orthonormal basis for $\mathcal{R}^3$ consisting of eigenvectors of $A^T A$ with corresponding eigenvalues $\lambda_1 = 900$, $\lambda_2 = 225$, and $\lambda_3 = 0$. Thus the singular values of A are $\sigma_1 = \sqrt{\lambda_1} = 30$ and $\sigma_2 = \sqrt{\lambda_2} = 15$. Furthermore, $\mathcal{B}_1$ is an orthonormal basis of right singular vectors of A.

Next, set

$$\mathbf{u}_1 = \frac{1}{\sigma_1}A\mathbf{v}_1 = \frac{1}{30} \cdot \frac{1}{3}\begin{bmatrix} 54 \\ 72 \end{bmatrix} = \frac{1}{5}\begin{bmatrix} 3 \\ 4 \end{bmatrix}$$

and

$$\mathbf{u}_2 = \frac{1}{\sigma_2}A\mathbf{v}_2 = \frac{1}{15} \cdot \frac{1}{3}\begin{bmatrix} -36 \\ 27 \end{bmatrix} = \frac{1}{5}\begin{bmatrix} -4 \\ 3 \end{bmatrix}.$$

Then $\mathcal{B}_2 = \{\mathbf{u}_1, \mathbf{u}_2\}$ is an orthonormal basis of left singular vectors of A.

2. Let $\mathbf{v}_1, \mathbf{v}_2, \mathbf{v}_3, \mathbf{u}_1$, and $\mathbf{u}_2$ be as in the solution to problem 1. Define $V = [\mathbf{v}_1 \quad \mathbf{v}_2 \quad \mathbf{v}_3]$, $U = [\mathbf{u}_1 \quad \mathbf{u}_2]$, and

$$\Sigma = \begin{bmatrix} \sigma_1 & 0 & 0 \\ 0 & \sigma_2 & 0 \end{bmatrix} = \begin{bmatrix} 30 & 0 & 0 \\ 0 & 15 & 0 \end{bmatrix}.$$

Then

$$A = U\Sigma V^T = \begin{bmatrix} \frac{3}{5} & -\frac{4}{5} \\ \frac{4}{5} & \frac{3}{5} \end{bmatrix} \begin{bmatrix} 30 & 0 & 0 \\ 0 & 15 & 0 \end{bmatrix} \begin{bmatrix} \frac{1}{3} & \frac{2}{3} & \frac{2}{3} \\ -\frac{2}{3} & \frac{2}{3} & -\frac{1}{3} \\ \frac{2}{3} & \frac{1}{3} & -\frac{2}{3} \end{bmatrix}^T$$

is a singular value decomposition of A.

3. Let $\mathbf{b} = \begin{bmatrix} 5 \\ -5 \end{bmatrix}$. Then the solution of least norm is

$$\mathbf{z} = V\Sigma^{\dagger}U^T\mathbf{b}$$

$$= \begin{bmatrix} \frac{1}{3} & \frac{2}{3} & \frac{2}{3} \\ -\frac{2}{3} & \frac{2}{3} & -\frac{1}{3} \\ \frac{2}{3} & \frac{1}{3} & -\frac{2}{3} \end{bmatrix} \begin{bmatrix} \frac{1}{30} & 0 \\ 0 & \frac{1}{15} \\ 0 & 0 \end{bmatrix} \begin{bmatrix} \frac{3}{5} & -\frac{4}{5} \\ \frac{4}{5} & \frac{3}{5} \end{bmatrix}^T \begin{bmatrix} 5 \\ -5 \end{bmatrix}$$

$$= -\frac{1}{90}\begin{bmatrix} 29 \\ 26 \\ 16 \end{bmatrix}.$$

4. Using the matrices V, U, and Σ in Example 7, we see that the desired vector is

$$\mathbf{z} = V\Sigma^{\dagger}U^T\mathbf{b}$$

$$= \begin{bmatrix} \frac{1}{\sqrt{6}} & 0 & \frac{-5}{\sqrt{30}} \\ \frac{1}{\sqrt{6}} & \frac{2}{\sqrt{5}} & \frac{1}{\sqrt{30}} \\ \frac{2}{\sqrt{6}} & \frac{-1}{\sqrt{5}} & \frac{2}{\sqrt{30}} \end{bmatrix} \begin{bmatrix} \frac{1}{\sqrt{18}} & 0 & 0 \\ 0 & \frac{1}{\sqrt{10}} & 0 \\ 0 & 0 & 0 \end{bmatrix}$$

$$\times \begin{bmatrix} \frac{1}{\sqrt{3}} & 0 & \frac{2}{\sqrt{6}} \\ \frac{1}{\sqrt{3}} & \frac{-1}{\sqrt{2}} & \frac{-1}{\sqrt{6}} \\ \frac{1}{\sqrt{3}} & \frac{1}{\sqrt{2}} & \frac{-1}{\sqrt{6}} \end{bmatrix}^T \begin{bmatrix} 27 \\ 36 \\ -18 \end{bmatrix}$$

$$= \frac{1}{10}\begin{bmatrix} 25 \\ -83 \\ 104 \end{bmatrix}.$$

6.7* Rotations of $\mathcal{R}^3$ and Computer Graphics

In this section, we study rotations of $\mathcal{R}^3$ about a line that contains $\mathbf{0}$. These rotations can be described in terms of left multiplication by special 3×3 orthogonal matrices, just as rotations in the plane about the origin can be represented by left

---

*This section can be omitted without loss of continuity.

multiplication by 2×2 rotation matrices, which are orthogonal matrices. The most important lines about which rotations are performed are the x-, y-, and z-axes. We describe how to compute rotations about these axes and explain how they can be used for the graphical representation of three-dimensional objects. We then consider rotations about arbitrary lines containing $\mathbf{0}$.

We begin with a description of rotations about the z-axis. Fix an angle θ, and consider a vector $\begin{bmatrix} x \\ y \\ z \end{bmatrix}$ in $\mathcal{R}^3$. Let $\begin{bmatrix} x' \\ y' \end{bmatrix}$ be the result of rotating $\begin{bmatrix} x \\ y \end{bmatrix}$ by θ in the xy-plane. Then $\begin{bmatrix} x' \\ y' \\ z \end{bmatrix}$ is the rotation of $\begin{bmatrix} x \\ y \\ z \end{bmatrix}$ about the z-axis by θ (see Figure 6.17).

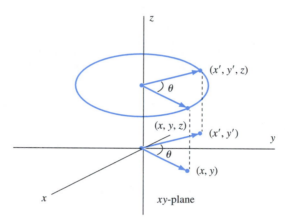

FIGURE 6.17

Rotation of $\mathcal{R}^3$ about the z-axis.

Hence we may use the rotation matrix A_θ introduced in Section 1.2 to obtain

$$\begin{bmatrix} x' \\ y' \end{bmatrix} = A_\theta \begin{bmatrix} x \\ y \end{bmatrix} = \begin{bmatrix} \cos\theta & -\sin\theta \\ \sin\theta & \cos\theta \end{bmatrix} \begin{bmatrix} x \\ y \end{bmatrix}.$$

It follows that

$$\begin{bmatrix} x' \\ y' \\ z \end{bmatrix} = \left[\begin{array}{cc|c} \cos\theta & -\sin\theta & 0 \\ \sin\theta & \cos\theta & 0 \\ \hline 0 & 0 & 1 \end{array} \right] \begin{bmatrix} x \\ y \\ z \end{bmatrix}.$$

Let

$$R_\theta = \begin{bmatrix} \cos\theta & -\sin\theta & 0 \\ \sin\theta & \cos\theta & 0 \\ 0 & 0 & 1 \end{bmatrix}.$$

Then $R_\theta \left(\begin{bmatrix} x \\ y \\ z \end{bmatrix} \right) = \begin{bmatrix} x' \\ y' \\ z \end{bmatrix}$ produces a rotation of $\begin{bmatrix} x \\ y \\ z \end{bmatrix}$ about the z-axis by the angle θ. The matrix R_θ is an example of a 3×3 *rotation matrix*. Using arguments similar to the one above, we can find the rotation matrices for rotating a vector by an angle θ about the other principal axes. Let P_θ and Q_θ be the matrices for rotations by an angle θ about the x-axis and the y-axis, respectively.

Then

$$P_\theta = \begin{bmatrix} 1 & 0 & 0 \\ 0 & \cos\theta & -\sin\theta \\ 0 & \sin\theta & \cos\theta \end{bmatrix} \quad \text{and} \quad Q_\theta = \begin{bmatrix} \cos\theta & 0 & \sin\theta \\ 0 & 1 & 0 \\ -\sin\theta & 0 & \cos\theta \end{bmatrix}.$$

In each case, the positive direction of the rotation is counterclockwise when viewed from a position along the positive direction of the axis of rotation. Note that all of these rotation matrices are orthogonal matrices.

We can combine rotations by taking products of rotation matrices. For example, if a vector $\mathbf{v}$ is rotated about the z-axis by an angle θ and the result is then rotated about the y-axis by an angle ϕ, the final position of the rotated vector is given by $Q_\phi(R_\theta \mathbf{v}) = (Q_\phi R_\theta)\mathbf{v}$. One should be careful to note the order in which the rotations are made, because $Q_\phi R_\theta$ is not necessarily equal to $R_\theta Q_\phi$. For example, let $\mathbf{v} = \mathbf{e}_1$, $\theta = 30°$, and $\phi = 45°$. Then

$$Q_\phi R_\theta \mathbf{v} = \begin{bmatrix} \frac{1}{\sqrt{2}} & 0 & \frac{1}{\sqrt{2}} \\ 0 & 1 & 0 \\ -\frac{1}{\sqrt{2}} & 0 & \frac{1}{\sqrt{2}} \end{bmatrix} \begin{bmatrix} \frac{\sqrt{3}}{2} & -\frac{1}{2} & 0 \\ \frac{1}{2} & \frac{\sqrt{3}}{2} & 0 \\ 0 & 0 & 1 \end{bmatrix} \begin{bmatrix} 1 \\ 0 \\ 0 \end{bmatrix}$$

$$= \begin{bmatrix} \frac{1}{\sqrt{2}} & 0 & \frac{1}{\sqrt{2}} \\ 0 & 1 & 0 \\ -\frac{1}{\sqrt{2}} & 0 & \frac{1}{\sqrt{2}} \end{bmatrix} \begin{bmatrix} \frac{\sqrt{3}}{2} \\ \frac{1}{2} \\ 0 \end{bmatrix}$$

$$= \begin{bmatrix} \frac{\sqrt{3}}{2\sqrt{2}} \\ \frac{1}{2} \\ -\frac{\sqrt{3}}{2\sqrt{2}} \end{bmatrix}.$$

On the other hand, a similar calculation shows that

$$R_\theta Q_\phi \mathbf{v} = \begin{bmatrix} \frac{\sqrt{3}}{2\sqrt{2}} \\ \frac{1}{2\sqrt{2}} \\ -\frac{1}{\sqrt{2}} \end{bmatrix},$$

which is not equal to $Q_\phi R_\theta \mathbf{v}$.

Rotation matrices are used in computer graphics to compute various orientations of the same three-dimensional shape. Although computers can store the information necessary to construct three-dimensional shapes, these shapes must be represented graphically on a two-dimensional surface such as the screen of a computer monitor or a sheet of paper. From a mathematical viewpoint, such a representation is projected on a plane. For example, the shape can be projected on the yz-plane by simply ignoring the first coordinates of the points that constitute the shape, and by plotting only the second and third coordinates. To get different views, the shape can be rotated in various ways before each projection is made. To illustrate the results of these procedures, a simple program that creates three-dimensional shapes consisting of points connected by lines was written for a computer. The coordinates of these points (vertices) and the information about which of these points are connected by lines (edges) are used as data in the program. The program plots the projection of the resulting shape on the yz-plane and represents the results as a printout. Before making such a plot, the computer rotates the shape about any one or a combination of the three coordinate axes by multiplying the vertices of the shape by the appropriate rotation matrix.

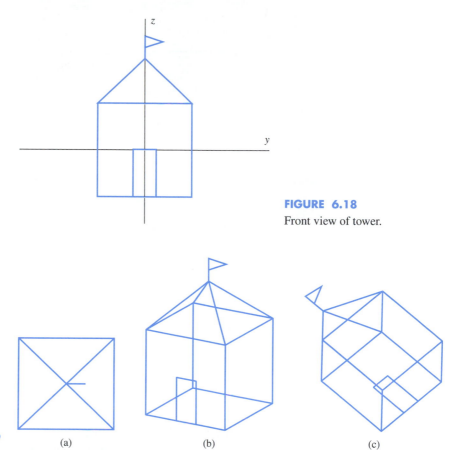

FIGURE 6.18
Front view of tower.

FIGURE 6.19

(a)   (b)   (c)

In Figure 6.18, we use a crude rendering of a tower, originally oriented as shown with the principal axes superimposed. What we see here is the projection of the figure on the yz-plane without rotations. Notice that the x-axis is not visible because it is perpendicular to the plane of this page. For each subsequent figure, the tower is rotated about one or two axes before being projected. In Figure 6.19(a), the tower is rotated $90°$ about the y-axis. In Figure 6.19(b), the tower is first rotated by $-30°$ about the z-axis, and then rotated by $20°$ about the y-axis. In Figure 6.19(c) the tower is first rotated by $45°$ about the x-axis and then by $30°$ about the y-axis.

Perspective

An object appears smaller when viewed from greater distances. This effect, called **perspective**, is apparent when the object is viewed directly, as well as when it is viewed from photographic images. In both cases, light reflected from the object converges to a point, called a **focal point**, and then diverges along lines to a plane. This correspondence is called a **perspective projection**. For example, the plane on which the image is projected could be film in a camera. Figure 6.20 illustrates this phenomenon. In this figure, the focal

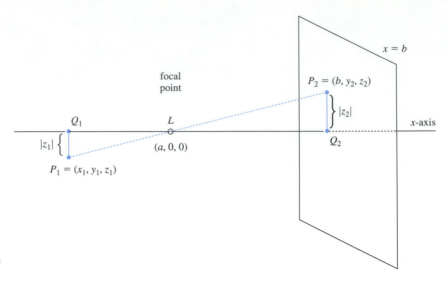

FIGURE 6.20

point L is situated on the x-axis at $(a, 0, 0)$, and the plane on which any image is projected is perpendicular to the x-axis at $x = b$. Notice that the image of any point lies on the opposite side of the x-axis. This effect causes the image of an object to be reversed.

We can use similar triangles to relate the location of a point with the location of its projected image. Consider an arbitrary point P_1 with coordinates (x_1, y_1, z_1). The point P_1 projects to the point $P_2 = (b, y_2, z_2)$ on the other side of the focal point (see Figure 6.20). Note that triangles $P_1 Q_1 L$ and $P_2 Q_2 L$ are similar, the length of $P_1 Q_1$ is $|z_1|$, and the length of $P_2 Q_2$ is $|z_2|$. It follows that

$$|z_2| = \frac{|z_1|(b - a)}{a - x_1}.$$

This equation tells us that the greater the distance that P_1 is from the focal point, the smaller $|z_2|$ is in comparison with $|z_1|$. That is, the larger the value of $a - x_1$, the smaller the size of the projected image. Since the image of P_1 is reversed, the signs of z_1 and z_2 are opposite, and hence

$$z_2 = \frac{-z_1(b - a)}{a - x_1}. \tag{16}$$

Similarly, we have that

$$y_2 = \frac{-y_1(b - a)}{a - x_1}. \tag{17}$$

The problem with (16) and (17) is that their use in the graphic projection of an actual object results in an image that is reversed. But if we simply replace y_2 by $-y_2$ and z_2 by $-z_2$, we invert the reversed image to obtain an image that is restored to its original orientation. Finally, we ignore the first coordinate, b, of the projected point, treating the plane $x = b$ as if it were the yz-plane. Thus we obtain a correspondence, called a *perspective projection*, that takes P_1, with coordinates (x_1, y_1, z_1), into P_2, with coordinates $\left(\dfrac{z_1(b - a)}{a - x_1}, \dfrac{y_1(b - a)}{a - x_1} \right)$. This correspondence enables us to create the illusion of perspective.

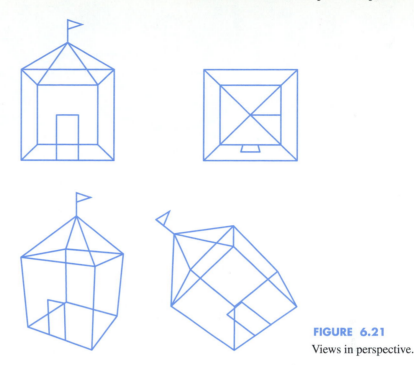

FIGURE 6.21
Views in perspective.

To illustrate the difference between a computer graphic with and without perspective, in Figure 6.21 we reproduce Figures 6.18 and 6.19 using perspective. For this purpose, we compute the graphic with the focus located on the x-axis at $x = a = 100$ and the projected plane located at $x = b = 180$.

Rotation Matrices

In addition to rotations of $\mathcal{R}^3$ about the principal axes, a rotation of $\mathcal{R}^3$ about any line L that contains $\mathbf{0}$ can be produced by left multiplication by the appropriate orthogonal matrix. Such a matrix is called a **rotation matrix**, and the line L is called the **axis of rotation**. Notice that an axis of rotation is a one-dimensional subspace of $\mathcal{R}^3$ and, conversely, any one-dimensional subspace of $\mathcal{R}^3$ is the axis of rotation for some rotation matrix.

In what follows, we discuss the problem of finding these more general rotation matrices. Let L be a one-dimensional subspace of $\mathcal{R}^3$, and let θ be an angle. We wish to find the rotation matrix P so that left multiplication by P causes a rotation of θ about the axis L. Before examining this problem, we must first decide what is meant by a rotation by the angle θ. We want to adopt the convention, as is done in the xy-plane, that the rotation is counterclockwise if $\theta > 0$ and is clockwise if $\theta < 0$. However, what is clockwise and what is counterclockwise literally depends on one's point of view. Suppose we could physically transport ourselves to a point $\mathbf{p}$ on L, where $\mathbf{p} \neq \mathbf{0}$. From this vantage point, we can view the two-dimensional subspace $L^{\perp}$, which is the plane perpendicular to L at $\mathbf{0}$, and observe the rotation of a vector $\mathbf{v}$ in $L^{\perp}$ in the counterclockwise direction to a vector $\mathbf{v}'$ in $L^{\perp}$. On the other hand, if we observe this same rotation from the opposite side of L, at the point $-\mathbf{p}$, then the same rotation is now seen to be in the clockwise direction (see Figure 6.22).

Thus the direction of a rotation is affected by which side of $\mathbf{0}$ contains the vantage point in L. A side is determined by the unit vector in L that points in

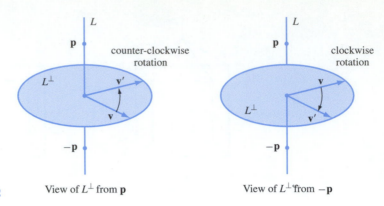

FIGURE 6.22   View of $L^\perp$ from $\mathbf{p}$ View of $L^\perp$ from $-\mathbf{p}$

its direction. Since there are two sides, and these can be identified with the two unit vectors in L, the direction of rotation can be described unambiguously by choosing among these two unit vectors. Such a choice is called an **orientation** of L.

Choose a unit vector $\mathbf{v}_3$ in L, which determines an orientation of L and hence the direction of a counterclockwise rotation on $L^\perp$. Having chosen the axis L, the orientation $\mathbf{v}_3$, and the angle θ, we are ready to find the rotation matrix P. Since L is a one-dimensional subspace of $\mathcal{R}^3$, its orthogonal complement, $L^\perp$, is a two-dimensional subspace. Select an orthonormal basis $\{\mathbf{v}_1, \mathbf{v}_2\}$ for $L^\perp$ such that $\mathbf{v}_2$ is the result of rotating $\mathbf{v}_1$ counterclockwise in $L^\perp$ by $90°$ with respect to the chosen orientation of L. Let $\mathcal{B} = \{\mathbf{v}_1, \mathbf{v}_2, \mathbf{v}_3\}$. Then $\mathcal{B}$ is an orthonormal basis for $\mathcal{R}^3$, and hence the rotation matrix P can be obtained by finding $P\mathbf{v}_1$, $P\mathbf{v}_2$, and $P\mathbf{v}_3$, and then applying what we know about matrix representations. Since $P\mathbf{v}_1$ makes an angle θ with $\mathbf{v}_1$, we may apply Exercise 52 of Section 6.1 to obtain

$$P\mathbf{v}_1 \cdot \mathbf{v}_1 = \|P\mathbf{v}_1\| \, \|\mathbf{v}_1\| \cos\theta = \cos\theta.$$

Because $\mathbf{v}_2$ is obtained by rotating $\mathbf{v}_1$ by the angle of $90°$ in the counterclockwise direction, if follows that the angle between $P\mathbf{v}_1$ and $\mathbf{v}_2$ is $90° - \theta$ if $\theta < 90°$ and is $\theta - 90°$ if $\theta > 90°$ (see Figure 6.23). In either case, $\cos(\theta - 90°) = \cos(90° - \theta) = \sin\theta$, and hence

$$P\mathbf{v}_1 \cdot \mathbf{v}_2 = \|P\mathbf{v}_1\| \, \|\mathbf{v}_2\| \cos(\pm(\theta - 90°)) = (1)(1)\sin\theta = \sin\theta.$$

Therefore

$$P\mathbf{v}_1 = (P\mathbf{v}_1 \cdot \mathbf{v}_1)\mathbf{v}_1 + (P\mathbf{v}_1 \cdot \mathbf{v}_2)\mathbf{v}_2 = (\cos\theta)\mathbf{v}_1 + (\sin\theta)\mathbf{v}_2. \qquad (18)$$

To obtain $P\mathbf{v}_2$, observe that $-\mathbf{v}_1$ can be obtained from $\mathbf{v}_2$ by a $90°$ counterclockwise rotation, and so we may apply the same arguments to the set $\{\mathbf{v}_2, -\mathbf{v}_1, \mathbf{v}_3\}$

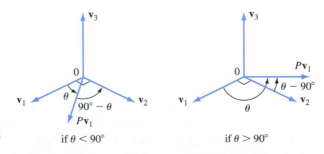

FIGURE 6.23   if $\theta < 90°$ if $\theta > 90°$

to obtain

$$Pv_2 = (\cos\theta)v_2 + (\sin\theta)(-v_1) = -(\sin\theta)(v_1) + (\cos\theta)v_2. \qquad (19)$$

Finally, since v_3 is in L and L remains unmoved by the rotation,

$$Pv_3 = v_3. \qquad (20)$$

We can now use (18), (19), and (20) to obtain the matrix representation of the matrix transformation T_P relative to $\mathcal{B}$:

$$[T_P]_\mathcal{B} = \begin{bmatrix} \cos\theta & -\sin\theta & 0 \\ \sin\theta & \cos\theta & 0 \\ 0 & 0 & 1 \end{bmatrix} = R_\theta. \qquad (21)$$

Let V be the 3×3 matrix $V = [v_1 \quad v_2 \quad v_3]$. Then V is an orthogonal matrix because its columns are the vectors in $\mathcal{B}$, an orthonormal basis for $\mathcal{R}^3$. Furthermore

$$V^{-1}PV = [T_P]_\mathcal{B}, \qquad (22)$$

since the columns of V are the vectors of $\mathcal{B}$. Combining (21) and (22), we obtain

$$P = VR_\theta V^{-1} = VR_\theta V^T. \qquad (23)$$

We summarize this fact about rotation matrices in the following.

> Any 3×3 rotation matrix has the form $VR_\theta V^T$ for some orthogonal matrix V and some angle θ.

Example 1   Find the rotation matrix P that rotates $\mathcal{R}^3$ by an angle of $30°$ about the axis L containing $\begin{bmatrix} 1 \\ 1 \\ 1 \end{bmatrix}$, with the orientation determined by the unit vector

$$v_3 = \frac{1}{\sqrt{3}} \begin{bmatrix} 1 \\ 1 \\ 1 \end{bmatrix}.$$

Solution   Our task is to find a pair of orthonormal vectors v_1 and v_2 in $L^\perp$ such that v_2 is the $90°$ counterclockwise rotation of v_1 with respect to the orientation determined by v_3. First, choose any nonzero vector w_1 orthogonal to v_3, for example, $w_1 = \begin{bmatrix} 1 \\ 0 \\ -1 \end{bmatrix}$. Now select a vector orthogonal to both v_3 and w_1.

Such a vector can be obtained by choosing a nontrivial solution to the system of linear equations

$$\sqrt{3}\,v_3 \cdot x = \begin{bmatrix} 1 \\ 1 \\ 1 \end{bmatrix} \cdot \begin{bmatrix} x_1 \\ x_2 \\ x_3 \end{bmatrix} = x_1 + x_2 + x_3 = 0$$

$$w_1 \cdot x = \begin{bmatrix} 1 \\ 0 \\ -1 \end{bmatrix} \cdot \begin{bmatrix} x_1 \\ x_2 \\ x_3 \end{bmatrix} = x_1 \qquad - x_3 = 0.$$

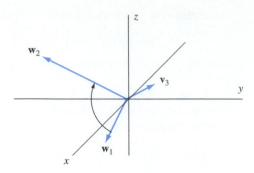

FIGURE 6.24

For example, $\mathbf{w}_2 = \begin{bmatrix} 1 \\ -2 \\ 1 \end{bmatrix}$ is such a solution. Then $\mathbf{w}_1$ and $\mathbf{w}_2$ are orthogonal vectors in $L^\perp$. We use these vectors to obtain $\mathbf{v}_1$ and $\mathbf{v}_2$, but we must first deal with a more difficult situation. For the orientation of L determined by $\mathbf{w}_3$, we must decide if the $90°$ rotation from $\mathbf{w}_1$ to $\mathbf{w}_2$ is clockwise or counterclockwise. From Figure 6.24 we see that the $90°$ rotation from $\mathbf{w}_1$ to $\mathbf{w}_2$ is clockwise. There are two ways to correct this situation. One possibility is to reverse the order of $\mathbf{w}_1$ and $\mathbf{w}_2$, since the $90°$ rotation from $\mathbf{w}_2$ to $\mathbf{w}_1$ is counterclockwise. Or, we can replace $\mathbf{w}_2$ by $-\mathbf{w}_2$ since the rotation from $\mathbf{w}_1$ to $-\mathbf{w}_2$ is also counterclockwise. Either way is acceptable. We arbitrarily select the first way. Finally we replace the $\mathbf{w}_i$ by unit vectors in the same direction. Thus we let

$$\mathbf{v}_1 = \frac{1}{\|\mathbf{w}_2\|}\mathbf{w}_2 = \frac{1}{\sqrt{6}}\begin{bmatrix} 1 \\ -2 \\ 1 \end{bmatrix} \qquad \text{and} \qquad \mathbf{v}_2 = \frac{1}{\|\mathbf{w}_1\|}\mathbf{w}_1 = \frac{1}{\sqrt{2}}\begin{bmatrix} 1 \\ 0 \\ -1 \end{bmatrix}.$$

Then $\mathcal{B} = \{\mathbf{v}_1, \mathbf{v}_2, \mathbf{v}_3\}$ is the required orthonormal basis. Let

$$V = [\mathbf{v}_1 \quad \mathbf{v}_2 \quad \mathbf{v}_3] = \begin{bmatrix} \frac{1}{\sqrt{6}} & \frac{1}{\sqrt{2}} & \frac{1}{\sqrt{3}} \\ \frac{-2}{\sqrt{6}} & 0 & \frac{1}{\sqrt{3}} \\ \frac{1}{\sqrt{6}} & \frac{-1}{\sqrt{2}} & \frac{1}{\sqrt{3}} \end{bmatrix}.$$

Then, by (23),

$$P = V R_{30°} V^T$$

$$= \begin{bmatrix} \frac{1}{\sqrt{6}} & \frac{1}{\sqrt{2}} & \frac{1}{\sqrt{3}} \\ \frac{-2}{\sqrt{6}} & 0 & \frac{1}{\sqrt{3}} \\ \frac{1}{\sqrt{6}} & \frac{-1}{\sqrt{2}} & \frac{1}{\sqrt{3}} \end{bmatrix} \begin{bmatrix} \cos 30° & -\sin 30° & 0 \\ \sin 30° & \cos 30° & 0 \\ 0 & 0 & 1 \end{bmatrix} \begin{bmatrix} \frac{1}{\sqrt{6}} & \frac{-2}{\sqrt{6}} & \frac{1}{\sqrt{6}} \\ \frac{1}{\sqrt{2}} & 0 & \frac{-1}{\sqrt{6}} \\ \frac{1}{\sqrt{3}} & \frac{1}{\sqrt{3}} & \frac{1}{\sqrt{3}} \end{bmatrix}$$

$$= \begin{bmatrix} \frac{1}{\sqrt{6}} & \frac{1}{\sqrt{2}} & \frac{1}{\sqrt{3}} \\ \frac{-2}{\sqrt{6}} & 0 & \frac{1}{\sqrt{3}} \\ \frac{1}{\sqrt{6}} & \frac{-1}{\sqrt{2}} & \frac{1}{\sqrt{3}} \end{bmatrix} \begin{bmatrix} \frac{\sqrt{3}}{2} & \frac{-1}{2} & 0 \\ \frac{1}{2} & \frac{\sqrt{3}}{2} & 0 \\ 0 & 0 & 1 \end{bmatrix} \begin{bmatrix} \frac{1}{\sqrt{6}} & \frac{-2}{\sqrt{6}} & \frac{1}{\sqrt{6}} \\ \frac{1}{\sqrt{2}} & 0 & \frac{-1}{\sqrt{6}} \\ \frac{1}{\sqrt{3}} & \frac{1}{\sqrt{3}} & \frac{1}{\sqrt{3}} \end{bmatrix}$$

$$= \frac{1}{3}\begin{bmatrix} 1+\sqrt{3} & 1-\sqrt{3} & 1 \\ 1 & 1+\sqrt{3} & 1-\sqrt{3} \\ 1-\sqrt{3} & 1 & 1+\sqrt{3} \end{bmatrix}.$$

We now make an important observation about the eigenvectors of an arbitrary 3×3 rotation matrix P with axis of rotation L. For any vector $\mathbf{v}$ in L, $P\mathbf{v} = \mathbf{v}$, and hence 1 is an eigenvalue of P and L is contained in the eigenspace of P corresponding to the eigenvalue 1. If the angle of rotation of P is $\theta = 0°$ (or, more generally, $\theta = 360n$ degrees for some integer n), then the matrix transformation T_P rotates every vector in $\mathcal{R}^3$ to itself, and hence $P = I_3$. In this case, the axis of rotation can be taken to be any one-dimensional subspace of $\mathcal{R}^3$. With the exception of this trivial rotation, P has a unique axis of rotation L. Furthermore, if $P \neq I_3$, then for any $\mathbf{v}$ not in L, we have $P\mathbf{v} \neq \mathbf{v}$ (see Exercise 20), and hence L is the eigenspace of P corresponding to eigenvalue 1. Therefore the eigenspace of P corresponding to eigenvalue 1 has dimension equal to 1 except for the trivial case that $P = I_3$, in which case the eigenspace has dimension equal to 3.

Next we make an important observation about the determinant of P. By (23), there is an orthogonal matrix V such that $P = V R_\theta V^{-1}$. By Exercise 21, $\det R_\theta = 1$, and hence

$$\det P = \det (V R_\theta V^{-1})$$

$$= (\det V)(\det R_\theta)(\det V^{-1})$$

$$= (\det V)(\det R_\theta)(\det V)^{-1}$$

$$= \det R_\theta$$

$$= 1.$$

The converse is also true. We state the full result here but defer the proof of the converse, which is more substantial, to the end of this section. Thus this determinant condition gives us a simple characterization of 3×3 rotation matrices. As an application, we give a simple proof that the transpose of a rotation matrix and the product of rotation matrices are rotation matrices.

Theorem 6.19

Let P and Q be 3×3 orthogonal matrices.

(a) P is a rotation matrix if and only if $\det P = 1$.
(b) If P is a rotation matrix, then P^T is a rotation matrix.
(c) If P and Q are rotation matrices, then PQ is a rotation matrix.

Proof The proof of (a) can be found on page 389.

(b) Suppose that P is a rotation matrix. Since P is an orthogonal matrix, P^T is an orthogonal matrix and $\det P^T = \det P = 1$ by (a). Thus P^T is a rotation matrix by (a).

(c) Suppose that P and Q are rotation matrices. Then $\det P = \det Q = 1$ by (a). Since P and Q are each orthogonal matrices, PQ is also an orthogonal matrix. Furthermore,

$$\det(PQ) = (\det P)(\det Q) = 1 \cdot 1 = 1.$$

Therefore PQ is a rotation matrix by (a).   ∎

There is something unsatisfactory about our solution to Example 1: We relied on a figure to make the judgment that a certain $90°$ rotation is counter-clockwise when viewed from a certain direction. In general, there is a problem with this approach because drawing and interpreting three-dimensional figures relies on the ability to visualize space, and this can fail us. However,

Theorem 6.19 gives us a way out of this predicament because it leads to a computational method for determining if a certain rotation is counterclockwise. The following result describes this method.

Theorem 6.20

Let $\{\mathbf{v}_1, \mathbf{v}_2, \mathbf{v}_3\}$ be an orthonormal basis for $\mathcal{R}^3$. The $90°$ rotation from $\mathbf{v}_1$ to $\mathbf{v}_2$ is counterclockwise as viewed from $\mathbf{v}_3$ if and only if $\det [\mathbf{v}_1 \quad \mathbf{v}_2 \quad \mathbf{v}_3] = 1$.

Proof First suppose that $\det [\mathbf{v}_1 \quad \mathbf{v}_2 \quad \mathbf{v}_3] = 1$. Since $V = [\mathbf{v}_1 \quad \mathbf{v}_2 \quad \mathbf{v}_3]$ is an orthogonal matrix, it is a rotation matrix by Theorem 6.19. Observe that the $90°$ rotation from $\mathbf{e}_1$ to $\mathbf{e}_2$ is counterclockwise as viewed from $\mathbf{e}_3$. Since V is a rotation, the relative positions of $\mathbf{e}_1, \mathbf{e}_2$ and $\mathbf{e}_3$ are the same as those of $V\mathbf{e}_1 = \mathbf{v}_1$, $V\mathbf{e}_2 = \mathbf{v}_2$, and $V\mathbf{e}_3 = \mathbf{v}_3$. Therefore the $90°$ rotation from $\mathbf{v}_1$ to $\mathbf{v}_2$ is counterclockwise as viewed from $\mathbf{v}_3$.

Now suppose that $\det [\mathbf{v}_1 \quad \mathbf{v}_2 \quad \mathbf{v}_3] \neq 1$. Since $[\mathbf{v}_1 \quad \mathbf{v}_2 \quad \mathbf{v}_3]$ is an orthogonal matrix, $\det [\mathbf{v}_1 \quad \mathbf{v}_2 \quad \mathbf{v}_3] = -1$ by Theorem 6.10. Therefore,

$$\det [\mathbf{v}_2 \quad \mathbf{v}_1 \quad \mathbf{v}_3] = -\det [\mathbf{v}_1 \quad \mathbf{v}_2 \quad \mathbf{v}_3] = (-1)(-1) = 1.$$

By what was discussed above, we can deduce that the $90°$ rotation from $\mathbf{v}_2$ to $\mathbf{v}_1$ as viewed from $\mathbf{v}_3$ is counterclockwise. It follows that the $90°$ rotation from $\mathbf{v}_1$ to $\mathbf{v}_2$ as viewed from $\mathbf{v}_3$ is not counterclockwise. ∎

Revisiting Example 1, we can apply Theorem 6.20 to verify that our choice of orthonormal vectors $\mathbf{v}_1, \mathbf{v}_2$, and $\mathbf{v}_3$ satisfies the requirement that the $90°$ rotation from $\mathbf{v}_1$ to $\mathbf{v}_2$ is counterclockwise as viewed from $\mathbf{v}_3$. In this case,

$$\det [\mathbf{v}_1 \quad \mathbf{v}_2 \quad \mathbf{v}_3] = \det \begin{bmatrix} \frac{1}{\sqrt{6}} & \frac{1}{\sqrt{2}} & \frac{1}{\sqrt{3}} \\ \frac{-2}{\sqrt{6}} & 0 & \frac{1}{\sqrt{3}} \\ \frac{1}{\sqrt{6}} & \frac{-1}{\sqrt{2}} & \frac{1}{\sqrt{3}} \end{bmatrix} = 1,$$

and hence the choice of orthonormal vectors $\mathbf{v}_1, \mathbf{v}_2$, and $\mathbf{v}_3$ made in Example 1 is acceptable.

Example 2

By Theorem 6.19(b), $P_\phi R_\theta$ is a rotation matrix for any angles ϕ and θ. Describe the axis of rotation for $P_\phi R_\theta$, where $\phi = 45°$ and $\theta = 30°$.

Solution

The axis of rotation for $P_\phi R_\theta$ is spanned by an eigenvector $\mathbf{v}$ of $P_\phi R_\theta$ corresponding to eigenvalue 1. So $P_\phi R_\theta \mathbf{v} = \mathbf{v}$, and hence $R_\theta \mathbf{v} = P_\phi^{-1} \mathbf{v} = P_\phi^T \mathbf{v}$. Therefore $(R_\theta - P_\phi^T)\mathbf{v} = \mathbf{0}$. Conversely, any nonzero solution to the equation $(R_\theta - P_\phi^T)\mathbf{x} = \mathbf{0}$ is an eigenvector of $P_\phi R_\theta$ corresponding to eigenvalue 1. Therefore we require a nonzero solution to the equation

$$\left(R_\theta - P_\phi^T\right) \begin{bmatrix} x_1 \\ x_2 \\ x_3 \end{bmatrix} = \frac{1}{2} \begin{bmatrix} \sqrt{3}-2 & -1 & 0 \\ 1 & \sqrt{3}-2 & -\sqrt{2} \\ 0 & \sqrt{2} & 2-\sqrt{2} \end{bmatrix} \begin{bmatrix} x_1 \\ x_2 \\ x_3 \end{bmatrix} = \begin{bmatrix} 0 \\ 0 \\ 0 \end{bmatrix}.$$

The vector

$$\mathbf{v} = \begin{bmatrix} 1-\sqrt{2} \\ (\sqrt{3}-2)(1-\sqrt{2}) \\ \sqrt{3}-2 \end{bmatrix}$$

is such a solution. Thus the axis of rotation for $P_\phi R_\theta$ is the subspace of $\mathcal{R}^3$ spanned by **v**.   ◯

Example 3   Find the angle of rotation induced by the rotation matrix $P_\phi R_\theta$ of Example 2.

Solution   Let α be the angle of rotation. Since no orientation of the axis of rotation is given, we assume that $\alpha > 0$. Choose a nonzero vector **w** in $L^\perp$, where L is the axis of rotation, and observe that α is the angle between $P_\phi R_\theta \mathbf{w}$ and **w**. Any nonzero vector **w** orthogonal to the vector **v** in Example 2 will suffice,

for example, $\mathbf{w} = \begin{bmatrix} \sqrt{3} - 2 \\ 0 \\ \sqrt{2} - 1 \end{bmatrix}$. Since $P_\phi^T = P_\phi^{-1}$ is an orthogonal matrix and

orthogonal matrices preserve dot products and norms, it follows that

$$\cos\alpha = \frac{(P_\phi R_\theta \mathbf{w}) \cdot \mathbf{w}}{\|P_\phi R_\theta \mathbf{w}\| \, \|\mathbf{w}\|} = \frac{(P_\phi^T P_\phi R_\theta \mathbf{w}) \cdot (P_\phi^T \mathbf{w})}{\|P_\phi R_\theta \mathbf{w}\| \, \|\mathbf{w}\|} = \frac{(R_\theta \mathbf{w}) \cdot (P_\phi^T \mathbf{w})}{\|\mathbf{w}\|^2}.$$

Thus

$$\cos\alpha = \frac{\dfrac{1}{4} \begin{bmatrix} 3 - 2\sqrt{3} \\ \sqrt{3} - 2 \\ 2\sqrt{2} - 2 \end{bmatrix} \cdot \begin{bmatrix} 2\sqrt{3} - 4 \\ 2 - \sqrt{2} \\ 2 - \sqrt{2} \end{bmatrix}}{10 - 4\sqrt{3} - 2\sqrt{2}}$$

$$= \frac{16\sqrt{3} - 36 - \sqrt{6} + 8\sqrt{2}}{4(10 - 4\sqrt{3} - 2\sqrt{2})}$$

$$\approx 0.59275.$$

Therefore

$$\alpha \approx \cos^{-1}(0.59275) \approx 53.65°.$$ ◯

Finally we complete the proof of Theorem 6.19 as promised.

Proof (Theorem 6.19(a))   Since the comments given immediately before the statement of Theorem 6.19 prove that the determinant of every 3×3 rotation matrix equals 1, only the converse need be proved.

Let P be a 3×3 orthogonal matrix such that $\det P = 1$. We first prove that 1 is an eigenvalue of P. We recall a few facts to prepare us for the calculation that follows. Since P is an orthogonal matrix, $P^{-1} = P^T$. Furthermore, $\det P^T = \det P$, and hence

$$\det A = \det P^T \det A = \det P^T A$$

for any 3×3 matrix A. Finally, $\det(-A) = -\det A$ for any 3×3 matrix A. Let $f(t)$ be the characteristic polynomial of P. Then

$$f(1) = \det(P - I_3)$$

$$= \det(P^T(P - I_3))$$

$$= \det(P^T P - P^T)$$

$$= \det(I_3 - P^T)$$

$$f(1) = \det (I_3 - P)^T$$
$$= \det (I_3 - P)$$
$$= \det (-(P - I_3))$$
$$= -\det (P - I_3)$$
$$= -f(1),$$

and hence $2f(1) = 0$. Thus $f(1) = 0$, and it follows that 1 is an eigenvalue of P.

Let L be the eigenspace of P corresponding to the eigenvalue 1. We now establish that for any $\mathbf{w}$ in $L^{\perp}$, $P\mathbf{w}$ is in $L^{\perp}$. To see this, consider any $\mathbf{v}$ in L. Then $P\mathbf{v} = \mathbf{v}$, and hence $P^T\mathbf{v} = P^{-1}\mathbf{v} = \mathbf{v}$. It now follows that

$$(P\mathbf{w})\cdot\mathbf{v} = (P\mathbf{w})^T\mathbf{v} = (\mathbf{w}^T P^T)\mathbf{v} = \mathbf{w}^T(P^T\mathbf{v}) = \mathbf{w}^T\mathbf{v} = \mathbf{w}\cdot\mathbf{v} = 0.$$

Thus $P\mathbf{w}$ is in $L^{\perp}$.

If $\dim L = 3$, then $P = I_3$, which is a trivial rotation. So suppose that $\dim L < 3$. We show that $\dim L = 1$. By way of contradiction, suppose $\dim L = 2$. Then $\dim L^{\perp} = 1$. Select any nonzero vector $\mathbf{w}$ in $L^{\perp}$. Then $\{\mathbf{w}\}$ is a basis for $L^{\perp}$. Since $P\mathbf{w}$ is in $L^{\perp}$, there is a scalar λ such that $P\mathbf{w} = \lambda\mathbf{w}$. Thus $\mathbf{w}$ is an eigenvector of P corresponding to the eigenvalue λ. Since L is an eigenspace of P corresponding to the eigenvalue 1 and $\mathbf{w}$ is in $L^{\perp}$, $\lambda \neq 1$. By Exercise 25 of Section 6.4, $\lambda = \pm 1$, and hence $\lambda = -1$. Let $\{\mathbf{v}_1, \mathbf{v}_2\}$ be a basis for L. Then $\mathcal{S} = \{\mathbf{v}_1, \mathbf{v}_2, \mathbf{w}\}$ is a basis for $\mathcal{R}^3$. Let B be the 3×3 matrix $B = [\mathbf{v}_1 \quad \mathbf{v}_2 \quad \mathbf{w}]$. Then

$$B^{-1}PB = [T_P]_{\mathcal{B}} = \begin{bmatrix} 1 & 0 & 0 \\ 0 & 1 & 0 \\ 0 & 0 & -1 \end{bmatrix},$$

and hence

$$P = B\begin{bmatrix} 1 & 0 & 0 \\ 0 & 1 & 0 \\ 0 & 0 & -1 \end{bmatrix}B^{-1}.$$

Therefore

$$\det P = \det\left(B\begin{bmatrix} 1 & 0 & 0 \\ 0 & 1 & 0 \\ 0 & 0 & -1 \end{bmatrix}B^{-1}\right)$$

$$= \det B \cdot \det\begin{bmatrix} 1 & 0 & 0 \\ 0 & 1 & 0 \\ 0 & 0 & -1 \end{bmatrix} \cdot \det(B^{-1})$$

$$= (\det B)(-1)(\det B)^{-1} = -1$$

contrary to the assumption that $\det P = 1$. We conclude that $\dim L = 1$.

Thus $\dim L^{\perp} = 2$. Let $\{\mathbf{v}_1, \mathbf{v}_2\}$ be an orthonormal basis for $L^{\perp}$, and let $\mathbf{v}_3$ be a unit vector in L. Then $\mathcal{B} = \{\mathbf{v}_1, \mathbf{v}_2, \mathbf{v}_3\}$ is an orthonormal basis for $\mathcal{R}^3$, and $P\mathbf{v}_3 = \mathbf{v}_3$. Furthermore, $P\mathbf{v}_1$ and $P\mathbf{v}_2$ are in $L^{\perp}$. Thus there are scalars a, b, c, and d such that

$$P\mathbf{v}_1 = a\mathbf{v}_1 + b\mathbf{v}_2 \qquad \text{and} \qquad P\mathbf{v}_2 = c\mathbf{v}_1 + d\mathbf{v}_2.$$

Let $V = [\mathbf{v}_1 \quad \mathbf{v}_2 \quad \mathbf{v}_3]$. Then V is an orthogonal matrix, and

$$[T_P]_\mathcal{B} = V^{-1}PV = \begin{bmatrix} a & c & 0 \\ b & d & 0 \\ 0 & 0 & 1 \end{bmatrix}.$$

Let $A = \begin{bmatrix} a & c \\ b & d \end{bmatrix}$. Comparing the columns of A with the first two columns of the orthogonal matrix $V^{-1}PV$, we see that the columns of A are orthonormal and hence A is a 2×2 orthogonal matrix. Furthermore,

$$\det A = \det \begin{bmatrix} a & c & 0 \\ b & d & 0 \\ 0 & 0 & 1 \end{bmatrix}$$

$$= \det(V^{-1}PV)$$

$$= \det(V^{-1}) \cdot \det P \cdot \det V$$

$$= (\det V)^{-1} \det P \det V$$

$$= \det P$$

$$= 1,$$

and hence A is a rotation matrix by Theorem 6.11. Therefore there is an angle θ such that

$$A = \begin{bmatrix} \cos\theta & -\sin\theta \\ \sin\theta & \cos\theta \end{bmatrix},$$

and thus

$$V^{-1}PV = \begin{bmatrix} \cos\theta & -\sin\theta & 0 \\ \sin\theta & \cos\theta & 0 \\ 0 & 0 & 1 \end{bmatrix} = R_\theta.$$

Hence $P = VR_\theta V^{-1}$. It follows by (23) that P is a rotation matrix. ∎

Practice Problems

1. Find the result of rotating the vector $\begin{bmatrix} 1 \\ -1 \\ 2 \end{bmatrix}$ by an angle of $60°$ about the y-axis followed by a rotation of $90°$ about the x-axis.

2. Let
$$W = \text{Span}\left\{ \begin{bmatrix} 1 \\ 2 \\ 3 \end{bmatrix}, \begin{bmatrix} 2 \\ 3 \\ 4 \end{bmatrix} \right\}.$$

Suppose that R is a 3×3 matrix such that for any vector $\mathbf{w}$ in W, the vector $R\mathbf{w}$ is also in W. Describe the axis of rotation.

Exercises

1. Determine if the following statements are true or false.

   (a) Every 3×3 orthogonal matrix is a rotation matrix.
   (b) For any 3×3 orthogonal matrix P, if $|\det P| = 1$, then P is a rotation matrix.
   (c) Every 3×3 orthogonal matrix has 1 as an eigenvalue.
   (d) Every 3×3 orthogonal matrix has -1 as an eigenvalue.
   (e) If P is a 3×3 rotation matrix and $P \neq I_3$, then 1 is an eigenvalue of P with multiplicity 1.

   (f) Every 3×3 orthogonal matrix is diagonalizable.
   (g) If P and Q are 3×3 rotation matrices, then PQ^T is a rotation matrix.

In Exercises 2–7, find the matrix M such that for each vector $\mathbf{v}$ in $\mathcal{R}^3$, $M\mathbf{v}$ is the result of the sequence of rotations described.

2. Each vector $\mathbf{v}$ is first rotated by $90°$ about the y-axis, and the result is then rotated $90°$ about the x-axis.

3. Each vector $\mathbf{v}$ is first rotated by 90° about the x-axis, and the result is then rotated 90° about the y-axis.

4. Each vector $\mathbf{v}$ is first rotated by 45° about the z-axis, and the result is then rotated 90° about the y-axis.

5. Each vector $\mathbf{v}$ is first rotated by 45° about the z-axis, and the result is then rotated 90° about the x-axis.

6. Each vector $\mathbf{v}$ is first rotated by 90° about the x-axis, and the result is then rotated 45° about the z-axis.

7. Each vector $\mathbf{v}$ is first rotated by 30° about the y-axis, and the result is then rotated 30° about the x-axis.

8. Find the rotation matrix P that rotates $\mathcal{R}^3$ by an angle of 90° about the axis containing $\begin{bmatrix} 1 \\ -1 \\ 1 \end{bmatrix}$, with orientation determined by the unit vector $\dfrac{1}{\sqrt{3}} \begin{bmatrix} 1 \\ -1 \\ 1 \end{bmatrix}$.

9. Find the rotation matrix P that rotates $\mathcal{R}^3$ by an angle of 180° about the axis containing $\begin{bmatrix} 1 \\ 0 \\ 1 \end{bmatrix}$, with orientation determined by the unit vector $\dfrac{1}{\sqrt{2}} \begin{bmatrix} 1 \\ 0 \\ 1 \end{bmatrix}$.

10. Find the rotation matrix P that rotates $\mathcal{R}^3$ by an angle of 45° about the axis containing $\begin{bmatrix} 1 \\ 1 \\ 0 \end{bmatrix}$, with orientation determined by the unit vector $\dfrac{1}{\sqrt{2}} \begin{bmatrix} 1 \\ 1 \\ 0 \end{bmatrix}$.

11. Find the rotation matrix P that rotates $\mathcal{R}^3$ by an angle of 45° about the axis containing $\begin{bmatrix} 1 \\ 1 \\ 0 \end{bmatrix}$, with orientation determined by the unit vector $\dfrac{-1}{\sqrt{2}} \begin{bmatrix} 1 \\ 1 \\ 0 \end{bmatrix}$.

12. Find the rotation matrix P that rotates $\mathcal{R}^3$ by an angle of 30° about the axis containing $\begin{bmatrix} 1 \\ -1 \\ 0 \end{bmatrix}$, with orientation determined by the unit vector $\dfrac{-1}{\sqrt{2}} \begin{bmatrix} 1 \\ -1 \\ 0 \end{bmatrix}$.

13. Find the rotation matrix P that rotates $\mathcal{R}^3$ by an angle of 30° about the axis containing $\begin{bmatrix} 1 \\ -1 \\ 0 \end{bmatrix}$, with orientation determined by the unit vector $\dfrac{1}{\sqrt{2}} \begin{bmatrix} 1 \\ -1 \\ 0 \end{bmatrix}$.

In Exercises 14–19, a rotation matrix M is given. (a) Find a vector that spans the axis of rotation, and (b) find the cosine of the angle of rotation.

14. The matrix M of Exercise 2.

15. The matrix M of Exercise 3.

16. The matrix M of Exercise 4.

17. The matrix M of Exercise 5.

18. The matrix M of Exercise 6.

19. The matrix M of Exercise 7.

20. Let P be a 3×3 rotation matrix with axis of rotation L, and suppose that $P \neq I_3$. Prove that for any vector $\mathbf{v}$ in $\mathcal{R}^3$, if $\mathbf{v}$ is not in L, then $P\mathbf{v} \neq \mathbf{v}$. *Hint:* For $\mathbf{v}$ not in L, let $\mathbf{v} = \mathbf{w} + \mathbf{z}$, where $\mathbf{w}$ is in L and $\mathbf{z}$ is in $L^{\perp}$. Now consider $P\mathbf{v} = P(\mathbf{w} + \mathbf{z})$.

21. Show by a direct computation that $\det P_\theta = \det Q_\theta = \det R_\theta = 1$ for any angle θ.

22. Prove that if P is a 3×3 orthogonal matrix, then P^2 is a rotation matrix.

23. Suppose that P is the 3×3 rotation matrix that rotates $\mathcal{R}^3$ by an angle θ about an axis L with orientation determined by the unit vector $\mathbf{v}$ in L. Prove that P^T rotates $\mathcal{R}^3$ by the angle $-\theta$ about L with the orientation determined by $\mathbf{v}$.

Definition.   For a two-dimensional subspace W of $\mathcal{R}^3$, let $T_W \colon \mathcal{R}^3 \to \mathcal{R}^3$ be the function defined as follows. Let $\mathbf{v}$ be in $\mathcal{R}^3$. By Theorem 6.7, $\mathbf{v}$ has a unique representation as the sum $\mathbf{v} = \mathbf{w} + \mathbf{z}$, where $\mathbf{w}$ is in W and $\mathbf{z}$ is in $W^{\perp}$. Using this representation, we define

$$T_W(\mathbf{v}) = \mathbf{w} - \mathbf{z}.$$

The function T_W is called the **reflection of $\mathcal{R}^3$ about W**. We sometimes refer to T_W as a **reflection operator**.

Notice in Figure 6.25 that for a reflection operator T_W and a vector $\mathbf{v}$ in $\mathcal{R}^3$, the image $T_W(\mathbf{v})$ is the "mirror image" of $\mathbf{v}$, where we think of W as the mirror. Hence the term "reflection" is appropriate here.

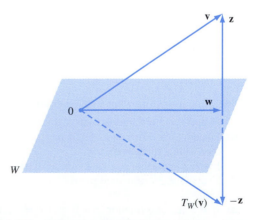

FIGURE 6.25   The reflection of $\mathcal{R}^3$ about a subspace W

24. Let Q be the standard matrix of the reflection of $\mathcal{R}^3$ about a two-dimensional subspace W. Prove that $Q = 2(P_W - I_3)$, where P_W is the orthogonal projection matrix for W.

25. Use Exercise 24 to prove that reflection operators are linear.

26. Let B be the standard matrix of a reflection operator, T_W. Prove that B is an orthogonal matrix, and hence T_W is an orthogonal operator.

27. Let T_W be the reflection of $\mathcal{R}^3$ about the two-dimensional subspace W.

    (a) Prove that 1 is an eigenvalue of T_W and that W is the corresponding eigenspace.
    (b) Prove that -1 is an eigenvalue with corresponding eigenspace $W^\perp$.

28. Prove the converse of Exercise 27: Let T be an orthogonal operator on $\mathcal{R}^3$ with eigenvalues 1 and -1 having corresponding multiplicities 2 and 1, respectively. Then T is a reflection operator.

In Exercises 29–32, find the standard matrix for the reflection of $\mathcal{R}^3$ about the given subspace W.

29. $W = \text{Span}\left\{ \begin{bmatrix} 1 \\ 2 \\ 3 \end{bmatrix}, \begin{bmatrix} 1 \\ 0 \\ -1 \end{bmatrix} \right\}$

30. $W = \text{Span}\left\{ \begin{bmatrix} 1 \\ 1 \\ 1 \end{bmatrix}, \begin{bmatrix} 1 \\ 1 \\ -1 \end{bmatrix} \right\}$

31. $W = \{(x, y, z): x + y + z = 0\}$
32. $W = \{(x, y, z): x + 2y - z = 0\}$

In Exercises 33 and 34, find the standard matrix for the reflection operator T such that $T(\mathbf{v}) = -\mathbf{v}$ for the given vector $\mathbf{v}$.

33. $\mathbf{v} = \begin{bmatrix} 1 \\ 2 \\ -1 \end{bmatrix}$ 34. $\mathbf{v} = \begin{bmatrix} -1 \\ 1 \\ 1 \end{bmatrix}$

35. Let T_W be a reflection operator with standard matrix B. Prove that $\det B = -1$.

36. Let B be the standard matrix of a reflection operator. Prove the following.

    (a) $B^2 = I_3$.
    (b) B is a symmetric matrix.

37. Find a 3×3 orthogonal matrix C such that $\det C = -1$ but C is not the standard matrix of a reflection operator. *Hint:* Multiply the standard matrix of a reflection operator by a rotation matrix whose axis of rotation is the eigenspace of the reflection operator corresponding to eigenvalue 1.

38. Let B and C be standard matrices of reflection operators. Prove that the product BC is a rotation matrix.

In Exercises 39–46:

(a) Determine if the given orthogonal matrix is a rotation matrix, the standard matrix of a reflection operator, or neither of these.

(b) If the matrix is a rotation matrix, find a vector that spans the axis of rotation. If the matrix is the standard matrix of a reflection operator, find a basis for the two-dimensional subspace about which $\mathcal{R}^3$ is reflected.

39. $\begin{bmatrix} 0 & 1 & 0 \\ -1 & 0 & 0 \\ 0 & 0 & -1 \end{bmatrix}$ 40. $\begin{bmatrix} 0 & 0 & 1 \\ 0 & 1 & 0 \\ 1 & 0 & 0 \end{bmatrix}$

41. $\begin{bmatrix} 1 & 0 & 0 \\ 0 & -1 & 0 \\ 0 & 0 & -1 \end{bmatrix}$ 42. $\begin{bmatrix} 0 & 0 & -1 \\ 0 & -1 & 0 \\ -1 & 0 & 0 \end{bmatrix}$

43. $\dfrac{1}{45} \begin{bmatrix} 35 & 28 & 4 \\ -20 & 29 & -28 \\ -20 & 20 & 35 \end{bmatrix}$ 44. $\dfrac{1}{9} \begin{bmatrix} 1 & -4 & 8 \\ -4 & 7 & 4 \\ 8 & 4 & 1 \end{bmatrix}$

45. $\begin{bmatrix} \frac{1}{\sqrt{2}} & 0 & \frac{1}{\sqrt{2}} \\ 0 & 1 & 0 \\ \frac{1}{\sqrt{2}} & 0 & \frac{-1}{\sqrt{2}} \end{bmatrix}$ 46. $\begin{bmatrix} \frac{1}{\sqrt{2}} & 0 & \frac{-1}{\sqrt{2}} \\ 0 & 1 & 0 \\ \frac{1}{\sqrt{2}} & 0 & \frac{1}{\sqrt{2}} \end{bmatrix}$

47. Let W be a two-dimensional subspace of $\mathcal{R}^3$, and suppose that $\{\mathbf{v}_1, \mathbf{v}_2\}$ is a basis for W. Let $\mathbf{v}_3$ be a nonzero vector that is orthogonal to both $\mathbf{v}_1$ and $\mathbf{v}_2$. Let

$$B = [\mathbf{v}_1 \quad \mathbf{v}_2 \quad \mathbf{v}_3] \quad \text{and} \quad C = [\mathbf{v}_1 \quad \mathbf{v}_2 \quad -\mathbf{v}_3].$$

Prove that CB^{-1} is the standard matrix for the reflection of $\mathcal{R}^3$ about W.

In Exercises 48 and 49, use a calculator with matrix capabilities or computer software such as MATLAB to find the axis of rotation and the angle of rotation, to the nearest degree, for the following rotation matrices.

48. $P_{22°} Q_{16°}$ 49. $R_{42°} P_{23°}$

Solutions to the Practice Problems

1. Let $\theta = 60°$ and $\phi = 90°$. Then the resulting vector is

$$P_\phi Q_\theta \begin{bmatrix} 1 \\ -1 \\ 2 \end{bmatrix} = \begin{bmatrix} 1 & 0 & 0 \\ 0 & \cos 90° & -\sin 90° \\ 0 & \sin 90° & \cos 90° \end{bmatrix}$$

$$\times \begin{bmatrix} \cos 60° & 0 & \cos 60° \\ 0 & 1 & 0 \\ -\sin 60° & 0 & \cos 60° \end{bmatrix} \begin{bmatrix} 1 \\ -1 \\ 2 \end{bmatrix}$$

$$= \begin{bmatrix} 1 & 0 & 0 \\ 0 & 0 & -1 \\ 0 & 1 & 0 \end{bmatrix} \begin{bmatrix} \frac{1}{2} & 0 & \frac{\sqrt{3}}{2} \\ 0 & 1 & 0 \\ -\frac{\sqrt{3}}{2} & 0 & \frac{1}{2} \end{bmatrix} \begin{bmatrix} 1 \\ -1 \\ 2 \end{bmatrix}$$

$$= \frac{1}{2} \begin{bmatrix} 1 + 2\sqrt{3} \\ -2 + \sqrt{3} \\ -2 \end{bmatrix}.$$

2. The axis of rotation is $W^{\perp}$, which is the solution space to the homogeneous system of linear equations

$$x_1 + 2x_2 + 3x_3 = 0$$
$$2x_1 + 3x_2 + 4x_3 = 0.$$

This solution space is spanned by the vector $\begin{bmatrix} 1 \\ -2 \\ 1 \end{bmatrix}$, and hence the axis of rotation is the line through the origin containing this vector.

Chapter 6 Review Exercises

1. Determine if the following statements are true or false.

   (a) The norm of a vector in $\mathcal{R}^n$ is a scalar.
   (b) The dot product of two vectors in $\mathcal{R}^n$ is a scalar.
   (c) The dot product of any two vectors is defined.
   (d) If the endpoint of a vector lies on a given line, then the vector equals the orthogonal projection of the vector on that line.
   (e) The distance between two vectors in $\mathcal{R}^n$ is the norm of their difference.
   (f) The orthogonal complement of the row space of a matrix equals the null space of the matrix.
   (g) If W is a subspace of $\mathcal{R}^n$, then every vector in $\mathcal{R}^n$ can be written uniquely as a sum of a vector in W and a vector in $W^{\perp}$.
   (h) Every orthonormal basis of a subspace is also an orthogonal basis of the subspace.
   (i) A subspace and its orthogonal complement have the same dimension.
   (j) An orthogonal projection matrix is never invertible.
   (k) If $\mathbf{w}$ is the closest vector in a subspace W of $\mathcal{R}^n$ to a vector $\mathbf{v}$ in $\mathcal{R}^n$, then $\mathbf{w}$ is the orthogonal projection of $\mathbf{v}$ onto W.
   (l) If $\mathbf{w}$ is the orthogonal projection of a vector $\mathbf{v}$ in $\mathcal{R}^n$ onto a subspace W of $\mathcal{R}^n$, then $\mathbf{w}$ is orthogonal to $\mathbf{v}$.
   (m) For a given set of data plotted in the xy-plane, the least-squares line is the unique line in the plane that minimizes the sum of squared distances from the data points to the line.
   (n) If the columns of an $n \times n$ matrix P are orthogonal, then P is an orthogonal matrix.
   (o) If the determinant of the standard matrix of a linear operator on $\mathcal{R}^2$ equals 1, then the linear operator is a rotation.
   (p) In $\mathcal{R}^2$, the composition of two reflections is a rotation.
   (q) In $\mathcal{R}^2$, the composition of two rotations is a rotation.
   (r) Every square matrix has a spectral decomposition.
   (s) If a matrix has a spectral decomposition, then the matrix must be symmetric.

In Exercises 2–5, two vectors $\mathbf{u}$ and $\mathbf{v}$ are given. In each exercise,

(a) compute the norm of each of the vectors,
(b) compute the distance d between the vectors,
(c) compute the dot product of the vectors, and
(d) determine if the vectors are orthogonal.

2. $\mathbf{u} = \begin{bmatrix} 2 \\ -3 \end{bmatrix}$, $\mathbf{v} = \begin{bmatrix} 4 \\ 1 \end{bmatrix}$ 3. $\mathbf{u} = \begin{bmatrix} 3 \\ -6 \end{bmatrix}$, $\mathbf{v} = \begin{bmatrix} 4 \\ 2 \end{bmatrix}$

4. $\mathbf{u} = \begin{bmatrix} 2 \\ -1 \\ 3 \end{bmatrix}$, $\mathbf{v} = \begin{bmatrix} 0 \\ 4 \\ 2 \end{bmatrix}$ 5. $\mathbf{u} = \begin{bmatrix} 1 \\ -1 \\ 2 \end{bmatrix}$, $\mathbf{v} = \begin{bmatrix} 2 \\ 4 \\ 1 \end{bmatrix}$

In Exercises 6 and 7, a vector $\mathbf{v}$ and a line $\mathcal{L}$ in $\mathcal{R}^2$ are given. Compute the orthogonal projection $\mathbf{w}$ of $\mathbf{v}$ on $\mathcal{L}$, and use it to compute the distance d from the endpoint of $\mathbf{v}$ to $\mathcal{L}$.

6. $\mathbf{v} = \begin{bmatrix} 3 \\ 5 \end{bmatrix}$, $y = 4x$ 7. $\mathbf{v} = \begin{bmatrix} 3 \\ 2 \end{bmatrix}$, $y = -2x$

In Exercises 8–11, suppose that $\mathbf{u}$, $\mathbf{v}$, and $\mathbf{w}$ are vectors in $\mathcal{R}^n$ such that $\|\mathbf{u}\| = 3$, $\|\mathbf{v}\| = 4$, $\|\mathbf{w}\| = 2$, $\mathbf{u} \cdot \mathbf{v} = -2$, $\mathbf{u} \cdot \mathbf{w} = 5$, and $\mathbf{v} \cdot \mathbf{w} = -3$.

8. Compute $\|-2\mathbf{u}\|$.   9. Compute $(2\mathbf{u} + 3\mathbf{v}) \cdot \mathbf{w}$.
10. Compute $\|3\mathbf{u} - 2\mathbf{w}\|^2$.   11. Compute $\|\mathbf{u} - \mathbf{v} + 3\mathbf{w}\|^2$.

In Exercises 12 and 13, determine if the given linearly independent set S is orthogonal. If the set is not orthogonal, apply the Gram–Schmidt process to S to find an orthogonal basis for the span of S.

12. $\left\{ \begin{bmatrix} 1 \\ 1 \\ 0 \end{bmatrix}, \begin{bmatrix} 2 \\ 0 \\ 1 \end{bmatrix}, \begin{bmatrix} 2 \\ 2 \\ 1 \end{bmatrix} \right\}$ 13. $\left\{ \begin{bmatrix} 1 \\ 1 \\ -1 \\ 0 \end{bmatrix}, \begin{bmatrix} 0 \\ 0 \\ 1 \\ 1 \end{bmatrix}, \begin{bmatrix} 1 \\ 2 \\ 0 \\ 1 \end{bmatrix} \right\}$

In Exercises 14 and 15, find a basis for $S^{\perp}$.

14. $S = \left\{ \begin{bmatrix} 2 \\ -1 \\ 3 \end{bmatrix} \right\}$ 15. $S = \left\{ \begin{bmatrix} 2 \\ 1 \\ -1 \\ 0 \end{bmatrix}, \begin{bmatrix} 3 \\ 4 \\ 2 \\ -2 \end{bmatrix} \right\}$

In Exercises 16 and 17, a vector $\mathbf{v}$ in $\mathcal{R}^n$ and an orthonormal basis S for a subspace W of $\mathcal{R}^n$ are given. Use S to obtain the unique vectors $\mathbf{w}$ in W and $\mathbf{z}$ in $W^{\perp}$ such that $\mathbf{v} = \mathbf{w} + \mathbf{z}$. Use your answer to find the distance from $\mathbf{v}$ to W.

16. $\mathbf{v} = \begin{bmatrix} 2 \\ 3 \end{bmatrix}$, $S = \left\{ \dfrac{1}{\sqrt{5}} \begin{bmatrix} 2 \\ 1 \end{bmatrix} \right\}$,

17. $\mathbf{v} = \begin{bmatrix} 1 \\ 2 \\ -3 \end{bmatrix}$, $S = \left\{ \dfrac{1}{\sqrt{5}} \begin{bmatrix} 1 \\ 2 \\ 0 \end{bmatrix}, \dfrac{1}{\sqrt{14}} \begin{bmatrix} -2 \\ 1 \\ 3 \end{bmatrix} \right\}$

In Exercises 18–21, a subspace W and a vector $\mathbf{v}$ are given. Find the orthogonal projection matrix P_W, and find the vector **w** in W that is closest to **v**.

18. $W = \text{Span}\left\{ \begin{bmatrix} 1 \\ -1 \\ 2 \end{bmatrix}, \begin{bmatrix} 1 \\ 0 \\ 1 \end{bmatrix} \right\}$, $\quad \mathbf{v} = \begin{bmatrix} 2 \\ -1 \\ 6 \end{bmatrix}$

19. $W = \text{Span}\left\{ \begin{bmatrix} 1 \\ 2 \\ 0 \\ -1 \end{bmatrix} \right\}$, $\quad \mathbf{v} = \begin{bmatrix} 2 \\ 1 \\ 3 \\ -8 \end{bmatrix}$

20. W is the solution set of

$$\begin{array}{c} x_1 + 2x_2 - x_3 = 0 \\ x_1 - x_2 - x_3 = 0 \end{array} \quad \text{and} \quad \mathbf{v} = \begin{bmatrix} 2 \\ 1 \\ 4 \end{bmatrix}$$

21. W is the orthogonal complement of the subspace of $\mathcal{R}^4$ spanned by

$$\left\{ \begin{bmatrix} 1 \\ -1 \\ 0 \\ 0 \end{bmatrix}, \begin{bmatrix} 1 \\ 0 \\ 1 \\ 0 \end{bmatrix} \right\} \quad \text{and} \quad \mathbf{v} = \begin{bmatrix} 2 \\ -1 \\ 1 \\ 2 \end{bmatrix}$$

22. Find the equation of the least-squares line for the data: $(1, 4), (2, 6), (3, 10), (4, 12), (5, 13)$.

23. An object is moving away from a fixed point at a constant speed v. At various times t, the distance d from the object to the point was measured. The results are listed in following table.

Time t (seconds)	Distance d (feet)
1	3.2
2	5.1
3	7.1
4	9.2
5	11.4

Assuming that d and t are related by the equation $d = vt + c$ for some constant c, use the method of least squares to estimate the speed of the object and the initial distance from the object to the fixed point (that is, at $t = 0$).

24. Use the method of least-squares to find the best quadratic fit for the data: $(1, 2), (2, 3), (3, 7), (4, 14), (5, 23)$.

In Exercises 25–28, determine if the given matrix is orthogonal.

25. $\begin{bmatrix} 0.7 & 0.3 \\ -0.3 & 0.7 \end{bmatrix}$

26. $\dfrac{1}{13} \begin{bmatrix} 5 & -12 \\ 12 & -5 \end{bmatrix}$

27. $\dfrac{1}{\sqrt{2}} \begin{bmatrix} 1 & 0 & 1 \\ 0 & \sqrt{2} & 0 \\ 1 & 0 & -1 \end{bmatrix}$

28. $\dfrac{1}{\sqrt{6}} \begin{bmatrix} \sqrt{2} & -\sqrt{3} & 1 \\ \sqrt{2} & \sqrt{3} & -2 \\ \sqrt{2} & 0 & 1 \end{bmatrix}$

In Exercises 29–32, determine if the given orthogonal matrix is the standard matrix of a rotation or a reflection. If the operator is a rotation, determine the angle of rotation. If the operator is a reflection, determine the equation of the line of reflection.

29. $\dfrac{1}{2} \begin{bmatrix} 1 & \sqrt{3} \\ -\sqrt{3} & 1 \end{bmatrix}$

30. $\dfrac{1}{2} \begin{bmatrix} 1 & -\sqrt{3} \\ \sqrt{3} & 1 \end{bmatrix}$

31. $\dfrac{1}{2} \begin{bmatrix} 1 & \sqrt{3} \\ \sqrt{3} & -1 \end{bmatrix}$

32. $\dfrac{1}{5} \begin{bmatrix} -3 & 4 \\ 4 & 3 \end{bmatrix}$

33. Let $T: \mathcal{R}^3 \to \mathcal{R}^3$ be defined by

$$T\left(\begin{bmatrix} x_1 \\ x_2 \\ x_3 \end{bmatrix} \right) = \begin{bmatrix} -x_2 \\ x_3 \\ x_1 \end{bmatrix}.$$

Prove that T is an orthogonal operator.

34. Suppose that $T: \mathcal{R}^2 \to \mathcal{R}^2$ is an orthogonal operator. Let $U: \mathcal{R}^2 \to \mathcal{R}^2$ be defined by

$$U\left(\begin{bmatrix} x_1 \\ x_2 \end{bmatrix} \right) = \frac{1}{\sqrt{2}} T\left(\begin{bmatrix} x_1 + x_2 \\ -x_1 + x_2 \end{bmatrix} \right).$$

(a) Prove that U is an orthogonal operator.
(b) Suppose that T is a rotation. Is U a rotation or a reflection?
(c) Suppose that T is a reflection. Is U a rotation or a reflection?

In Exercises 35 and 36, a symmetric matrix A is given. Find an orthonormal basis of eigenvectors of A and their corresponding eigenvalues. Use this information to obtain a spectral decomposition of A.

35. $A = \begin{bmatrix} 2 & 3 \\ 3 & 2 \end{bmatrix}$

36. $A = \begin{bmatrix} 6 & 2 & 0 \\ 2 & 9 & 0 \\ 0 & 0 & -9 \end{bmatrix}$

In Exercises 37 and 38, the equation of a conic section is given in xy-coordinates. Find the appropriate angle of rotation so that the equation may be written in $x'y'$-coordinates with no $x'y'$-term. Give the new equation and identify the type of conic section.

37. $x^2 + 6xy + y^2 - 16 = 0$

38. $3x^2 - 4xy + 3y^2 - 9 = 0$

39. Let W be a subspace of $\mathcal{R}^n$, and let Q be an $n \times n$ orthogonal matrix. Prove that $Q^T P_W Q = P_Z$, where $Z = \{Q^T \mathbf{w}: \mathbf{w} \text{ is in } W\}$.

40. Prove that rank $P_W = \dim W$ for any subspace W of $\mathcal{R}^n$.
Hint: Apply Exercises 28 and 29 of Section 6.3.

CHAPTER 7

Vector Spaces

Up to this point in our development of linear algebra, we have accumulated a rich body of facts about vectors in $\mathcal{R}^n$ and linear transformations acting on these vectors. In this chapter, we consider other mathematical systems that share many of the formal properties of $\mathcal{R}^n$. For example, consider differentiable functions, which are encountered in the study of calculus. These functions can be added and multiplied by scalars. Derivatives and integrals transform these functions in such a way that addition and multiplication by scalars are preserved, much as linear transformations preserve the corresponding operations on vectors. Because of these similarities, we could reformulate such notions as *linear combinations, linear independence,* and *linear transformation* in the context of differentiable functions.

Transplanting these concepts to the context of *function* gives us a way to analyze functions using the tools that we have developed and mastered. As a dramatic example of this, we see how to use the concept of *orthogonal projection* and *the closest vector property* to devise a method of approximating a given function by polynomials or by sines and cosines.

In order to study functions and other mathematical systems that share the appropriate formal properties of vectors, we extend the concepts we have developed in our studies of $\mathcal{R}^n$ to the more general context of *vector spaces* so that the study of $\mathcal{R}^n$ becomes a special case of the theory of vector spaces. In the general theory, a vector space is defined as any mathematical system that satisfies certain axioms (properties). The general theorems about vector spaces are then deduced from these axioms. Once it is shown that a system satisfies these axioms, it follows immediately that all of these theorems apply.

7.1 Vector Spaces and Their Subspaces

The operations of addition and multiplication by scalars that are central to the study of vectors in $\mathcal{R}^n$ have their analogs in other mathematical systems. For example, we have seen that any two $m \times n$ matrices can be added to produce another $m \times n$ matrix, and that any $m \times n$ matrix can be multiplied by a scalar to produce another $m \times n$ matrix. So it may make sense to start from scratch and study matrices in the same way that we have studied vectors in $\mathcal{R}^n$. If we were to do this, we would introduce definitions such as *subspace* and *dimension* that mimic the corresponding definitions we introduced in our study of $\mathcal{R}^n$, but this time we would formulate these definitions in the context of matrices.

Similarly, real-valued functions defined on $\mathcal{R}$ can be added to produce a new real-valued function defined on $\mathcal{R}$. Furthermore, such functions can be multiplied by scalars. So again, we have the opportunity to start at the beginning and develop a theory of functions imitating, as much as possible, the definitions and theorems developed in the context of vectors in $\mathcal{R}^n$.

In fact, there are many mathematical systems in which the notions of addition and multiplication by scalars are defined. It would be impractical to develop a body of theory using the formal properties of these operations for each such system as we have done for $\mathcal{R}^n$. For this reason, we resort to a "general theory" of *vector spaces* that can be applied to each of these systems. In this theory, a vector space is defined to be any mathematical system that satisfies certain prescribed axioms. The general theorems about vector spaces are then deduced from these axioms. Once it is shown that a particular mathematical system satisfies these axioms, it follows immediately that all of these theorems apply to that system.

We begin with the formal definition of a vector space. The reader should compare the axioms below to Theorem 1.1.

Definitions.   A (real) **vector space** is a set V on which two operations called **vector addition** and **scalar multiplication** are defined so that for any elements **u** and **v** in V and any scalar a, the sum $\mathbf{u} + \mathbf{v}$ and the scalar multiple $a\mathbf{u}$ are uniquely defined elements of V such that the following axioms hold.

Axioms of a Vector Space

1. $\mathbf{u} + \mathbf{v} = \mathbf{v} + \mathbf{u}$ for all **u** and **v** in V. (commutative law of vector addition)
2. $(\mathbf{u} + \mathbf{v}) + \mathbf{w} = \mathbf{u} + (\mathbf{v} + \mathbf{w})$ for all **u**, **v**, and **w** in V. (associative law of vector addition)
3. There exists an element **0** in V such that $\mathbf{u} + \mathbf{0} = \mathbf{u}$ for every **u** in V.
4. For each element **u** in V, there exists an element $-\mathbf{u}$ in V such that $\mathbf{u} + (-\mathbf{u}) = \mathbf{0}$.
5. $1\mathbf{u} = \mathbf{u}$ for every **u** in V.
6. $(ab)\mathbf{u} = a(b\mathbf{u})$ for all scalars a and b and every **u** in V.
7. $a(\mathbf{u} + \mathbf{v}) = a\mathbf{u} + a\mathbf{v}$ for every scalar a and all **u** and **v** in V.
8. $(a + b)\mathbf{u} = a\mathbf{u} + b\mathbf{u}$ for all scalars a and b and every **u** in V.

The elements of a vector space are called **vectors**. The vector **0** of axioms 3 and 4 is called the **zero vector**. We will show that it is unique; that is, there cannot be two distinct vectors in a vector space that both satisfy axiom 3. For any vector **u** in a vector space V, the vector $-\mathbf{u}$ is called the **additive inverse** of **u**. We will also see that the additive inverse of a vector in a vector space is unique. In view of the commutative law of vector addition (axiom 1), the zero vector must satisfy $\mathbf{0} + \mathbf{u} = \mathbf{u}$ for every **u** in V. Likewise for each element **u** in V its additive inverse satisfies $(-\mathbf{u}) + \mathbf{u} = \mathbf{0}$.

By applying Theorem 1.1 to $n \times 1$ matrices, we see that $\mathcal{R}^n$ is a vector space under the operations of addition and scalar multiplication defined in

Chapter 1. It also follows from the definition of subspace that any subspace of $\mathcal{R}^n$ is also a vector space in its own right. We are already quite familiar with these examples.

Function Spaces

Among the most important vector spaces are those consisting of functions. Such vector spaces are called **function spaces**. The area of modern mathematics called *functional analysis* is devoted to the study of function spaces.

Fix a nonempty set S, and let $\mathcal{F}(S)$ denote the set of all functions from S to $\mathcal{R}$. For any function f in $\mathcal{F}(S)$ and any element t in S, the image $f(t)$ of t under f is a real number. For functions f and g in $\mathcal{F}(S)$, the **sum** $f + g$ and the **scalar multiple** af are functions in $\mathcal{F}(S)$ defined by

$$(f + g)(t) = f(t) + g(t) \qquad \text{and} \qquad (af)(t) = a(f(t))$$

for all $t \in S$.

For example, suppose that S is the set $\mathcal{R}$ of real numbers and f and g are the functions from $\mathcal{R}$ to $\mathcal{R}$ defined by $f(t) = t^2 - t$ for all t in $\mathcal{R}$ and $g(t) = 2t + 1$ for all t in $\mathcal{R}$. Then

$$(f + g)(t) = f(t) + g(t) = t^2 - t + 2t + 1 = t^2 + t + 1$$

for all t in $\mathcal{R}$. Also, if $a = 3$, then the scalar multiple $af = 3f$ is defined by

$$(3f)(t) = 3f(t) = 3(t^2 - t)$$

for all t in $\mathcal{R}$.

Next we define the **zero function 0** in $\mathcal{F}(S)$ by $\mathbf{0}(t) = 0$ for all t in S. This function serves as the zero vector for axiom 3 of a vector space. Finally, for any $f \in \mathcal{F}(S)$, the function $-f \in \mathcal{F}(S)$ is defined by $(-f)(t) = -f(t)$ for all t in S. For example, if $S = \mathcal{R}$, and $f(t) = t - 1$ for all t in $\mathcal{R}$, then $-f(t) = 1 - t$ for all t in $\mathcal{R}$. For any function f, the function $-f$ serves as the additive inverse of f for axiom 4 of a vector space.

To prove that $\mathcal{F}(S)$ is a vector space, we must verify the eight axioms of a vector space. In the context of $\mathcal{F}(S)$, each axiom contains an equation involving functions. For this purpose, we use the definition that two functions f and g in $\mathcal{F}(S)$ are **equal** if $f(t) = g(t)$ for all t in S. We use this definition to verify that the equations in each axiom are true.

Theorem 7.1

For any nonempty set S, $\mathcal{F}(S)$ with the operations defined previously is a vector space.

Proof To prove this theorem, we must verify that the given definitions of addition and scalar multiplication satisfy the eight axioms of a vector space. We verify axioms 1, 3, and 7, leaving the verification of the other axioms as exercises.

Axiom 1. Let f and g be functions in $\mathcal{F}(S)$. Then for any t in S

$$\begin{aligned}(f + g)(t) &= f(t) + g(t) &&\text{(definition of sum of functions)}\\ &= g(t) + f(t) &&\text{(commutative law of addition for real numbers)}\\ &= (g + f)(t). &&\text{(definition of sum of functions)}\end{aligned}$$

Therefore, $f + g = g + f$, and axiom 1 is verified.

Axiom 3. Let f be any function in $\mathcal{F}(S)$. Then for any t in S

$$
\begin{aligned}
(f + \mathbf{0})(t) &= f(t) + \mathbf{0}(t) && \text{(definition of sum of functions)} \\
&= f(t) + 0 && \text{(definition of } \mathbf{0}) \\
&= f(t).
\end{aligned}
$$

Therefore, $f + \mathbf{0} = f$, and hence axiom 3 is verified.

Axiom 7. Let f and g be functions in $\mathcal{F}(S)$, and let a be a scalar. Then for any t in S,

$$
\begin{aligned}
(a(f + g))(t) &= a((f + g)(t)) && \text{(definition of scalar multiplication)} \\
&= a(f(t) + g(t)) && \text{(definition of sum of functions)} \\
&= a(f(t)) + a(g(t)) && \text{(distributive law for real numbers)} \\
&= (af)(t) + (ag)(t) && \text{(definition of scalar multiplication)} \\
&= (af + ag)(t). && \text{(definition of sum of functions)}
\end{aligned}
$$

Therefore, $a(f + g) = af + ag$, which verifies axiom 7.   ❑

Other Examples of Vector Spaces

In what follows, we briefly consider three examples of vector spaces. The notation and terminology introduced in these examples is used in other examples in this chapter.

Example 1

For fixed positive integers m and n, let $\mathcal{M}_{m \times n}$ denote the set of all $m \times n$ matrices. Then as a direct consequence of Theorem 1.1, $\mathcal{M}_{m \times n}$ is a vector space under the operations of matrix addition and multiplication of a matrix by a scalar. In this case, the $m \times n$ zero matrix plays the role of the zero vector.   ○

Example 2

For fixed positive integers m and n, let $\mathcal{L}(\mathcal{R}^n, \mathcal{R}^m)$ denote the set of all linear transformations from $\mathcal{R}^n$ to $\mathcal{R}^m$. It can be shown that $\mathcal{L}(\mathcal{R}^n, \mathcal{R}^m)$ is a vector space under the operations of addition of linear transformations and the product of a linear transformation by a scalar, as defined in the Exercises of Section 2.6. The zero transformation, T_0, plays the role of the zero vector, and for any transformation T, we have that $(-1)T$ is the additive inverse, $-T$.

The proof that $\mathcal{L}(\mathcal{R}^n, \mathcal{R}^m)$ is a vector space is very similar to the proof that $\mathcal{F}(S)$ is a vector space because both of these are function spaces. For this reason, the proof is left as an exercise.   ○

Example 3

Let $\mathcal{P}$ denote the set of all polynomials

$$
p(x) = a_0 + a_1 x + \cdots + a_n x^n,
$$

where n is a nonnegative integer and $a_0, a_1, \ldots, a_n$ are real numbers. For each i, a_i is called the **coefficient** of x^i. We usually write x^i in place of $1x^i$ and $-a_i x^i$ in place of $(-a_i)x^i$. Furthermore, if $a_i = 0$, we often omit the term $a_i x^i$ entirely. The unique polynomial $p(x)$ with only zero coefficients is called the **zero polynomial**. The **degree** of a nonzero polynomial $p(x)$ is defined to be the largest exponent of x that appears in the representation

$$
p(x) = a_0 + a_1 x + \cdots + a_n x^n
$$

with a nonzero coefficient. Two nonzero polynomials $p(x)$ and $q(x)$ are called **equal** if they have the same degree and equal corresponding coefficients. That is, if

$$p(x) = a_0 + a_1 x + \cdots + a_n x^n \qquad \text{and} \qquad q(x) = b_0 + b_1 x + \cdots + b_m x^m,$$

then $p(x)$ and $q(x)$ are equal if and only if $m = n$ and $a_i = b_i$ for $i = 0,$ $1, \ldots, n$. Notice that if two distinct polynomials $p(x)$ and $q(x)$ have different degrees, say m and n, respectively, with $m < n$, then we may still represent $q(x)$ in the form

$$q(x) = b_0 + b_1 x + \cdots + b_n x^n$$

by simply requiring that $b_i = 0$ for all $i > m$. With this in mind, for any polynomials

$$p(x) = a_0 + a_1 x + \cdots + a_n x^n \qquad \text{and} \qquad q(x) = b_0 + b_1 x + \cdots + b_n x^n,$$

(not necessarily of the same degree) and any scalar a, we define the **sum** $p(x) + q(x)$ and the **scalar multiple** $ap(x)$ by

$$p(x) + q(x) = (a_0 + b_0) + (a_1 + b_1)x + \cdots + (a_n + b_n)x^n$$

and

$$ap(x) = (a \cdot a_0) + (a \cdot a_1)x + \cdots + (a \cdot a_n)x^n.$$

For example, $(1 - x + 2x^2) + (3 + 2x) = 4 + x + 2x^2$ and $4(3 + 2x) = 12 + 8x$. We can also define the **additive inverse** of a polynomial $p(x)$ by $-p(x) = (-1)p(x)$. With these definitions, it can be shown that $\mathcal{P}$ is a vector space with respect to the operations just defined. The zero polynomial serves the role of the zero vector. We relegate the details to the exercises.   ○

Properties of Vector Spaces

The following results are deduced entirely from the axioms of a vector space. They are therefore valid for all of the examples of vector spaces we have considered.

Theorem 7.2
Let V be a vector space. Then for any $\mathbf{u}$, $\mathbf{v}$, and $\mathbf{w}$ in V, and any scalar a, the following are true.

   (a) If $\mathbf{u} + \mathbf{v} = \mathbf{w} + \mathbf{v}$, then $\mathbf{u} = \mathbf{w}$. (right cancellation law)
   (b) If $\mathbf{u} + \mathbf{v} = \mathbf{u} + \mathbf{w}$, then $\mathbf{v} = \mathbf{w}$. (left cancellation law)
   (c) The zero vector $\mathbf{0}$ is unique, that is, it is the only vector in V that satisfies axiom 3.
   (d) Each vector in V has exactly one additive inverse.
   (e) $0\mathbf{v} = \mathbf{0}$.
   (f) $a\mathbf{0} = \mathbf{0}$.
   (g) $(-1)\mathbf{v} = -\mathbf{v}$.
   (h) $(-a)\mathbf{v} = a(-\mathbf{v}) = -(a\mathbf{v})$.

Proof We prove (a), (c), (e), and (g), leaving the proofs of (b), (d), (f), and (h) as exercises.

(a) Suppose that $\mathbf{u} + \mathbf{v} = \mathbf{w} + \mathbf{v}$. Then

$$
\begin{aligned}
\mathbf{u} &= \mathbf{u} + \mathbf{0} && \text{(by axiom 3)} \\
&= \mathbf{u} + (\mathbf{v} + (-\mathbf{v})) && \text{(by axiom 4)} \\
&= (\mathbf{u} + \mathbf{v}) + (-\mathbf{v}) && \text{(by axiom 2)} \\
&= (\mathbf{w} + \mathbf{v}) + (-\mathbf{v}) && \\
&= \mathbf{w} + (\mathbf{v} + (-\mathbf{v})) && \text{(by axiom 2)} \\
&= \mathbf{w} + \mathbf{0} && \text{(by axiom 4)} \\
&= \mathbf{w}. && \text{(by axiom 3)}
\end{aligned}
$$

(c) Suppose $\mathbf{0}'$ is also a zero vector for V. Then

$$\mathbf{0}' + \mathbf{0} = \mathbf{0}' = \mathbf{0}' + \mathbf{0} = \mathbf{0} + \mathbf{0}'$$

and hence $\mathbf{0}' = \mathbf{0}$ by the right cancellation law. It follows that the zero vector is unique.

(e) For any vector $\mathbf{v}$,

$$
\begin{aligned}
0\mathbf{v} + 0\mathbf{v} &= (0 + 0)\mathbf{v} && \text{(by axiom 8)} \\
&= 0\mathbf{v} && \text{(property of 0)} \\
&= \mathbf{0} + 0\mathbf{v}. && \text{(by axioms 3 and 1)}
\end{aligned}
$$

So $0\mathbf{v} + 0\mathbf{v} = \mathbf{0} + 0\mathbf{v}$, and hence $0\mathbf{v} = \mathbf{0}$ by the right cancellation law.

(g) For any vector $\mathbf{v}$,

$$
\begin{aligned}
\mathbf{v} + (-1)\mathbf{v} &= (1)\mathbf{v} + (-1)\mathbf{v} && \text{(by axiom 5)} \\
&= (1 + (-1))\mathbf{v} && \text{(by axiom 8)} \\
&= 0\mathbf{v} && \\
&= \mathbf{0}. && \text{(by (e))}
\end{aligned}
$$

Therefore $(-1)\mathbf{v}$ is an additive inverse of $\mathbf{v}$. But by (d), additive inverses are unique, and hence $(-1)\mathbf{v} = -\mathbf{v}$. ❑

Subspaces

As is the case for $\mathcal{R}^n$, vector spaces have *subspaces*.

Definition.   A subset W of a vector space V is called a **subspace** of V if W satisfies the following three properties.

1. The zero vector of V is in W.

2. Whenever $\mathbf{u}$ and $\mathbf{v}$ belong to W, then $\mathbf{u} + \mathbf{v}$ belongs to W. (In this case, we say that W is *closed under (vector) addition*.)

3. Whenever $\mathbf{u}$ belongs to W and c is a scalar, then $c\mathbf{u}$ belongs to W. (In this case, we say that W is *closed under scalar multiplication*.)

It is a simple matter to verify that if V is a vector space, then V is a subspace of itself. In fact, V is the largest subspace of V. Moreover, the set $\{\mathbf{0}\}$ is also a subspace of V. This subspace is called the **zero subspace**, and is the smallest subspace of V. A subspace of a vector space other than $\{\mathbf{0}\}$ is called a **nonzero subspace**.

Example 4

Let S be a nonempty set. Fix an element s_0 in S, and let W be the subset of $\mathcal{F}(S)$ consisting of all real-valued functions f on S such that $f(s_0) = 0$. Clearly the zero function lies in W. For any functions f and g in S and any scalar a,

$$(f + g)(s_0) = f(s_0) + g(s_0) = 0 + 0 = 0,$$

and

$$(af)(s_0) = af(s_0) = a \cdot 0 = 0.$$

Hence $f + g$ and af are in W. We conclude that W is closed under these operations. Therefore W is a subspace of V. ○

Example 5

Let n be a positive integer, and let W be the set of all $n \times n$ matrices with trace equal to zero. (Recall that the trace of a square matrix equals the sum of its diagonal entries). Show that W is a subspace of $\mathcal{M}_{n\times n}$.

Solution

Since the $n \times n$ zero matrix has trace equal to zero, it lies in W. Suppose that A and B are matrices in W. Then by Exercise 46 of Section 1.1,

$$\text{trace } (A + B) = \text{trace } A + \text{trace } B = 0 + 0 = 0,$$

and for any scalar c,

$$\text{trace } (cA) = c \cdot \text{trace } A = c \cdot 0 = 0.$$

Therefore $A + B$ and cA are in W. We conclude that W is closed under vector addition and scalar multiplication. Therefore W is a subspace of $\mathcal{M}_{n\times n}$. ○

If W is a subspace of a vector space, then W satisfies all of the axioms in the definition of *vector space,* and hence W is itself a vector space (see Exercise 47). This provides a simpler way to prove that certain sets are vector spaces, namely, by verifying that they are actually subspaces of a known vector space. The next several examples demonstrate this technique.

Example 6

Let $\mathsf{C}(\mathcal{R})$ denote the set of all continuous real-valued functions on $\mathcal{R}$. Then $\mathsf{C}(\mathcal{R})$ is a subset of $\mathcal{F}(\mathcal{R})$, the vector space of all real-valued functions defined on $\mathcal{R}$. Since the zero function is a continuous function, the sum of continuous functions is a continuous function, and any scalar multiple of a continuous function is a continuous function, it follows that $\mathsf{C}(\mathcal{R})$ is a subspace of $\mathcal{F}(\mathcal{R})$. In particular, $\mathsf{C}(\mathcal{R})$ is a vector space. ○

Example 7

Recall the vector space $\mathcal{P}$ of all polynomials considered in Example 3. Let n be a nonnegative integer, and let $\mathcal{P}_n$ denote the subset of $\mathcal{P}$ consisting of the zero polynomial and all polynomials of degree less than or equal to n. Since the sum of two polynomials of degree less than or equal to n is the zero polynomial or has degree less than or equal to n, and a scalar multiple of a polynomial of degree less than or equal to n is either the zero polynomial or a polynomial of degree less than or equal to n, it is clear that $\mathcal{P}_n$ is closed under both addition and scalar multiplication. Therefore $\mathcal{P}_n$ is a subspace of $\mathcal{P}$ and hence is a vector space. ○

Practice Problems

1. Let W be the set of all 2×2 matrices of the form $\begin{bmatrix} a & a+b \\ b & 0 \end{bmatrix}$. Prove that W is a subspace of $\mathcal{M}_{2\times2}$.

2. Let S be the set $\mathcal{R}^2$ with addition and scalar multiplication defined as follows:

$$(a, b) \oplus (c, d) = (a+c, 0) \quad \text{and} \quad k \odot (a, b) = (ka, kb)$$

for all (a, b) and (c, d) in $\mathcal{R}^2$. Show that S is *not* a vector space.

Exercises

1. Determine if the following statements are true or false.

   (a) Every vector space has a zero vector.
   (b) A vector space may have more than one zero vector.
   (c) In any vector space, $a\mathbf{v} = \mathbf{0}$ implies that $\mathbf{v} = \mathbf{0}$.
   (d) $\mathcal{R}^n$ is a vector space for every positive integer n.
   (e) Only polynomials of the same degree can be added.
   (f) The set of polynomials of degree n is a subspace of the vector space of all polynomials.
   (g) Two polynomials of the same degree are equal if and only if they have equal corresponding coefficients.
   (h) The set of all $m \times n$ matrices with the usual definitions of matrix addition and scalar multiplication is a vector space.
   (i) The zero vector of $\mathcal{F}(S)$ is the function that assigns 0 as the image of f to every element of S.
   (j) Two functions in $\mathcal{F}(S)$ are equal if and only if they assign equal values to each element of S.
   (k) If V is a vector space and W is a subspace of V, then W is a vector space under the same operations that are defined on V.
   (l) The empty set is a subspace of every vector space.
   (m) If V is a nonzero vector space, then V contains a subspace other than itself.

In Exercises 2–6, verify that the set V is a vector space with respect to the indicated operations.

2. $V = \mathcal{P}$, as defined in Example 3.

3. Fix a nonempty set S and a positive integer n. Let V denote the set of all functions from S to $\mathcal{R}^n$. For any functions f and g and any scalar c, define the sum $f + g$ and the product cf by

$$(f + g)(s) = f(s) + g(s) \quad \text{and} \quad (cf)(s) = cf(s)$$

   for all s in S.

4. Let V be the set of all positive real numbers. Define the operation of *vector addition* by $\mathbf{u} \oplus \mathbf{v} = \mathbf{u} \cdot \mathbf{v}$, the usual product of these two elements as real numbers. Define the product of the scalar c and an element $\mathbf{v}$ of V by $c \odot \mathbf{v} = \mathbf{v}^c$, that is, the real number $\mathbf{v}$ raised to the power c.

5. Let V be the set of all 2×2 matrices of the form $\begin{bmatrix} a & 2a \\ b & -b \end{bmatrix}$, where a and b are any real numbers. Addition and multiplication by scalars are defined in the usual way for matrices.

6. Let V be the set of all functions $f: \mathcal{R} \to \mathcal{R}$ for which $f(t) = 0$ whenever $t < 0$. Addition of functions and multiplication by scalars are defined as in $\mathcal{F}(\mathcal{R})$.

In Exercises 7–11, show that the set S is not a vector space.

7. S is the set of functions f in $\mathcal{F}(\mathcal{R})$ such that $f(1) = 2$.

8. S is $\mathcal{R}^2$, with addition and scalar multiplication defined as follows:

$$(a, b) \oplus (c, d) = (a + c, b + d)$$

   and

$$k \odot (a, b) = (ka, b)$$

   for all (a, b) in $\mathcal{R}^2$.

9. S is $\mathcal{R}^2$, with addition and scalar multiplication defined as follows:

$$(a, b) \oplus (c, d) = (a + c, b + d)$$

   and

$$k \odot (a, b) = (k^2 a, k^2 b)$$

   for all (a, b) in $\mathcal{R}^2$.

10. S is $\mathcal{R}^2$, with addition and scalar multiplication defined as follows:

$$(a, b) \oplus (c, d) = (ac, bd) \quad \text{and} \quad k \odot (a, b) = (ka, kb)$$

   for all (a, b) in $\mathcal{R}^2$.

11. S is $\mathcal{R}^2$, with addition and scalar multiplication defined as follows:

$$(a, b) \oplus (c, d) = (ac, 0) \quad \text{and} \quad k \odot (a, b) = (ka, 0)$$

   for all (a, b) in $\mathcal{R}^2$.

12. Verify axiom 2 for $\mathcal{F}(S)$.   13. Verify axiom 4 for $\mathcal{F}(S)$.
14. Verify axiom 5 for $\mathcal{F}(S)$.   15. Verify axiom 6 for $\mathcal{F}(S)$.
16. Verify axiom 8 for $\mathcal{F}(S)$.

17. Verify the axioms of a vector space for $\mathcal{L}(\mathcal{R}^n, \mathcal{R}^m)$ in Example 2.

18. Prove Theorem 7.2(b).   19. Prove Theorem 7.2(d).

20. Prove Theorem 7.2(f).   21. Prove Theorem 7.2(h).

22. Prove that, for any vector $\mathbf{v}$ in a vector space, $-(-\mathbf{v}) = \mathbf{v}$.

23. Prove that, for any vectors $\mathbf{u}$ and $\mathbf{v}$ in a vector space, $-(\mathbf{u} + \mathbf{v}) = (-\mathbf{u}) + (-\mathbf{v})$.

24. Let $\mathbf{u}$ and $\mathbf{v}$ be vectors in a vector space, and suppose that $c\mathbf{u} = c\mathbf{v}$ for some scalar $c \neq 0$. Prove that $\mathbf{u} = \mathbf{v}$.

25. Prove that $(-c)(-\mathbf{v}) = c\mathbf{v}$ for any vector $\mathbf{v}$ in a vector space and any scalar c.

In Exercises 26–31, determine whether or not the set V is a subspace of the vector space $\mathcal{M}_{n \times n}$. Justify your answer.

26. V is the set of all $n \times n$ symmetric matrices.

27. V is the set of all $n \times n$ matrices with determinant equal to 0.

28. V is the set of all $n \times n$ matrices A such that $A^2 = A$.

29. Let B be a fixed $n \times n$ matrix. V is the set of all $n \times n$ matrices A such that $AB = BA$.

30. V is the set of all 2×2 matrices of the form $\begin{bmatrix} a & 2a \\ 0 & b \end{bmatrix}$, and $n = 2$.

31. V is the set of all $n \times n$ skew-symmetric matrices.

In Exercises 32–35, determine whether or not the set V is a subspace of the vector space $\mathcal{P}$. Justify your answer.

32. V is the set consisting of the zero polynomial and all polynomials of the form $c_0 + c_1 x + \cdots + c_m x^m$ with $c_k = 0$ if k is odd.

33. V is the set consisting of the zero polynomial and all polynomials of the form $c_0 + c_1 x + \cdots + c_m x^m$ with $c_k \neq 0$ if k is even.

34. V is the set consisting of the zero polynomial and all polynomials of the form $c_0 + c_1 x + \cdots + c_m x^m$ with $c_i \geq 0$ for all i.

35. V is the set consisting of the zero polynomial and all polynomials of the form $c_0 + c_1 x + \cdots + c_m x^m$ such that $c_0 + c_1 = 0$.

In Exercises 36–38, determine whether or not the set V is a subspace of the vector space $\mathcal{F}(S)$, where S is a fixed nonempty set. Justify your answer.

36. Let S' be a nonempty subset of S, and let V be the set of all functions f in $\mathcal{F}(S)$ such that $f(s) = 0$ for all s in S'.

37. Let $\{s_1, s_2, \ldots, s_n\}$ be a subset of S, and let V be the set of all functions in $\mathcal{F}(S)$ such that
$$f(s_1) + f(s_2) + \cdots + f(s_n) = 0.$$

38. Let s_1 and s_2 be elements of S, and let V be the set of all functions in $\mathcal{F}(S)$ such that $f(s_1) \cdot f(s_2) = 0$.

39. Fix a nonzero vector $\mathbf{v}$ in $\mathcal{R}^n$. Let V be the set of all linear operators T on $\mathcal{R}^n$ such that $T(\mathbf{v}) = \mathbf{0}$. Prove that V is a subspace of $\mathcal{L}(\mathcal{R}^n, \mathcal{R}^n)$.

40. Let W be the set of all differentiable functions from $\mathcal{R}$ to $\mathcal{R}$. Prove that W is a subspace of $\mathcal{F}(R)$.

41. Let S be the subset of the subspace W in Exercise 40 consisting of the functions f such that $f' = f$. Show that S is a subspace of W.

42. A function f in $\mathcal{F}(\mathcal{R})$ is called an **even function** if $f(t) = f(-t)$ for all t in $\mathcal{R}$ and is called an **odd function** if $f(-t) = -f(t)$ for all t in $\mathcal{R}$.

    (a) Show that the subset of all even functions is a subspace of $\mathcal{F}(\mathcal{R})$.

    (b) Show that the subset of all odd functions is a subspace of $\mathcal{F}(\mathcal{R})$.

43. Let V be the set of all continuous real-valued functions on the closed interval $[0, 1]$.

    (a) Show that V is a subspace of $\mathcal{F}([0, 1])$.

    (b) Let W be the subset of V defined by
    $$W = \left\{ f \in V : \int_0^1 f(t)\, dt = 0 \right\}.$$

    Prove that W is a subspace of V.

44. Suppose that W_1 and W_2 are subspaces of a vector space V. Prove that their intersection $W_1 \cap W_2$ is also a subspace of V.

45. Suppose that W_1 and W_2 are subspaces of a vector space V. Define
    $$W = \{\mathbf{w}_1 + \mathbf{w}_2 : \mathbf{w}_1 \text{ is in } W_1 \text{ and } \mathbf{w}_2 \text{ is in } W_2\}.$$

    Prove that W is a subspace of V.

46. Let W be a subset of a vector space V. Prove that W is a subspace of V if and only if the following conditions hold.

    (i) $\mathbf{0}$ is in W.
    (ii) $a\mathbf{w}_1 + \mathbf{w}_2$ is in W whenever $\mathbf{w}_1$ and $\mathbf{w}_2$ are in W and a is a scalar.

47. Suppose that W is a subspace of a vector space V. Prove that W satisfies the axioms in the definition of *vector space* and hence W is itself a vector space.

48. Suppose that W is a subset of a vector space V such that W is a vector space under the same operations that are defined on V. Prove that W is a subspace of V. (Most of the proof is routine; the part of the proof that is less than routine is to show that the zero vector of V is also the zero vector of W.)

Solutions to the Practice Problems

1. (i) Clearly W is a subset of $\mathcal{M}_{2\times2}$ that contains the 2×2 zero matrix.

(ii) Suppose that

$$A = \begin{bmatrix} a_1 & a_1 + b_1 \\ b_1 & 0 \end{bmatrix} \quad \text{and}$$

$$B = \begin{bmatrix} a_2 & a_2 + b_2 \\ b_2 & 0 \end{bmatrix}$$

are in W. Then

$$A + B = \begin{bmatrix} a_1 & a_1 + b_1 \\ b_1 & 0 \end{bmatrix} + \begin{bmatrix} a_2 & a_2 + b_2 \\ b_2 & 0 \end{bmatrix}$$

$$= \begin{bmatrix} a_1 + a_2 & (a_1 + b_1) + (a_2 + b_2) \\ b_1 + b_2 & 0 \end{bmatrix}$$

$$= \begin{bmatrix} a_1 + a_2 & (a_1 + a_2) + (b_1 + b_2) \\ b_1 + b_2 & 0 \end{bmatrix},$$

which is clearly in $\mathcal{M}_{2\times2}$. So W is closed under vector addition.

(iii) For any scalar c, we have

$$cA = c \begin{bmatrix} a_1 & a_1 + b_1 \\ b_1 & 0 \end{bmatrix} = \begin{bmatrix} ca_1 & c(a_1 + b_1) \\ cb_1 & c \cdot 0 \end{bmatrix}$$

$$= \begin{bmatrix} ca_1 & ca_1 + cb_1 \\ cb_1 & 0 \end{bmatrix},$$

which is clearly in $\mathcal{M}_{2\times2}$. So W is closed under scalar multiplication.

2. Suppose that (z, w) is an element of S. Then

$$(1, 1) \oplus (z, w) = (1 + z, 0) \neq (1, 1).$$

So Axiom 3 fails.

7.2   Dimension and Isomorphism

In this section, we reexamine the concepts of linear combinations, linear dependence and independence, basis, and dimension in the more general context of vector spaces.

A careful examination of Sections 1.2 and 1.7 and Chapter 4 reveals that the proofs of results dealing with these concepts for $\mathcal{R}^n$ can, with little or no modification, be adapted to establish the corresponding results for general vector spaces. This should come as no surprise, because most of these proofs are based on, or are consequences of, the items listed in Theorem 1.1. Remember that these items include the axioms for vector spaces. Consequently, we assume some of these results in this chapter.

Linear Combinations and Spanning Sets

In Chapter 4, we used spanning sets as a convenient way to describe subspaces. The notion of a spanning set is built on the definition of a linear combination of vectors given in Section 1.2. This definition of a linear combination can be applied to vectors in an arbitrary vector space. However, in contrast to subspaces of $\mathcal{R}^n$, there are important vector spaces with subspaces that have no finite spanning sets. Hence it is necessary to extend the definition of a linear combination to include vectors of an infinite set.

Definition.   A vector $\mathbf{v}$ is a **linear combination** of the vectors of a (possibly infinite) subset S of a vector space V if there exist vectors $\mathbf{v}_1, \mathbf{v}_2, \ldots, \mathbf{v}_n$ in S and scalars $c_1, c_2, \ldots, c_n$ such that

$$\mathbf{v} = c_1\mathbf{v}_1 + c_2\mathbf{v}_2 + \cdots + c_n\mathbf{v}_n.$$

The scalars are called the **coefficients** of the linear combination.

We consider examples of linear combinations of vectors of both finite and infinite sets.

Example 1   In the vector space of 2×2 matrices,

$$\begin{bmatrix} -1 & 8 \\ 2 & -2 \end{bmatrix} = 2 \begin{bmatrix} 1 & 3 \\ 1 & -1 \end{bmatrix} + (-1) \begin{bmatrix} 4 & 0 \\ 1 & 1 \end{bmatrix} + (1) \begin{bmatrix} 1 & 2 \\ 1 & 1 \end{bmatrix}.$$

Hence $\begin{bmatrix} -1 & 8 \\ 2 & -2 \end{bmatrix}$ is a linear combination of the matrices

$$\begin{bmatrix} 1 & 3 \\ 1 & -1 \end{bmatrix}, \qquad \begin{bmatrix} 4 & 0 \\ 1 & 1 \end{bmatrix}, \qquad \text{and} \qquad \begin{bmatrix} 1 & 2 \\ 1 & 1 \end{bmatrix},$$

with coefficients 2, −1, and 1.   ○

Example 2   Let $S = \{1, x, x^2, x^3\}$, which is a subset of the vector space $\mathcal{P}$ of all polynomials. Then the polynomial $f(x) = 2 + 3x - x^2$ is a linear combination of the vectors in S because there are scalars, namely, 2, 3, −1, and 0 such that

$$f(x) = (2)1 + (3)x + (-1)x^2 + (0)x^3.$$

In fact, the zero polynomial and any polynomial of degree less than or equal to 3 is a linear combination of the vectors in S. That is, the set of all linear combinations of the vectors in S is equal to $\mathcal{P}_3$, which is a subspace of $\mathcal{P}$.   ○

Example 3   Let S be the set of all real-valued functions

$$S = \{1, \sin t, \cos^2 t, \sin^2 t\},$$

which is a subset of $\mathcal{F}(R)$. Observe that the function $\cos 2t$ is a linear combination of the vectors in S because

$$\cos 2t = \cos^2 t - \sin^2 t$$

$$= (0)1 + (0)\sin t + (1)\cos^2 t + (-1)\sin^2 t.$$ ○

Example 4   Let

$$S = \{1, x, x^2, \ldots, x^n, \ldots\}$$

be the set of all monomials, which is an infinite subset of $\mathcal{P}$. Then the polynomial $p(x) = 3 - 4x^2 + 5x^4$ is a linear combination of the vectors in S because it is a linear combination of a finite number of vectors in S, namely, 1, x^2, and x^4. In fact, any polynomial

$$p(x) = a_0 + a_1 x + \cdots + a_n x^n$$

is a linear combination of the vectors in S because it is a linear combination of the vectors $1, x, \ldots, x^n$ in S.   ○

Now that we have extended the definition of *linear combination* to include infinite sets, we are ready to reintroduce the definition of *span*.

Definition. The **span** of a nonempty subset S of a vector space V is the set of all linear combinations of vectors in S. This set is denoted Span S.

By remarks made in Example 2,

$$\text{Span }\{1, x, x^2, x^3\} = \mathcal{P}_3,$$

and by the remarks above,

$$\text{Span }\{1, x, \ldots, x^n, \ldots\} = \mathcal{P}.$$

Example 5   Describe the span of the subset

$$S = \left\{ \begin{bmatrix} 1 & 0 \\ 0 & -1 \end{bmatrix}, \begin{bmatrix} 0 & 1 \\ 0 & 0 \end{bmatrix}, \begin{bmatrix} 0 & 0 \\ 1 & 0 \end{bmatrix} \right\}.$$

Solution   Consider any matrix A in Span S. Then there exist scalars a, b, and c, such that

$$A = a \begin{bmatrix} 1 & 0 \\ 0 & -1 \end{bmatrix} + b \begin{bmatrix} 0 & 1 \\ 0 & 0 \end{bmatrix} + c \begin{bmatrix} 0 & 0 \\ 1 & 0 \end{bmatrix} = \begin{bmatrix} a & b \\ c & -a \end{bmatrix}.$$

Notice that trace $A = a + (-a) = 0$. Conversely, suppose that $A = \begin{bmatrix} a & b \\ c & d \end{bmatrix}$ is a matrix in $\mathcal{M}_{2 \times 2}$ such that trace $A = 0$. Then $a + d = 0$, and hence $d = -a$. It follows that $A = \begin{bmatrix} a & b \\ c & -a \end{bmatrix}$, which is in Span S. Therefore Span S is the subset of all 2×2 matrices with trace equal to zero. Since this set was proved to be a subspace of $\mathcal{M}_{2 \times 2}$ in Example 5 of Section 7.1, Span S is a subspace of $\mathcal{M}_{2 \times 2}$. ○

In the examples above, we have anticipated the following result, which is an extension of Theorem 4.1 (page 202) to vector spaces. We omit the proof, which is similar to the proof of Theorem 4.1.

Theorem 7.3
The span of a nonempty subset of a vector space V is a subspace of V.

This result gives us a convenient way to define vector spaces. One example of this is the vector space of *trigonometric polynomials* on $[0, 2\pi]$, which we study in later sections. In this example, we consider all functions to have the domain $[0, 2\pi]$. Then we define the **space of trigonometric polynomials**, denoted by $\mathcal{T}[0, 2\pi]$, to be the subspace of $\mathcal{F}([0, 2\pi])$ defined by

$$\mathcal{T}[0, 2\pi] = \text{Span }\{1, \cos t, \sin t, \cos 2t, \sin 2t, \ldots, \cos nt, \sin nt, \ldots\}.$$

Linear Dependence, Linear Independence, and Bases

The related concepts of *linear dependence* and *linear independence* for general vector spaces are similar to the corresponding concepts for $\mathcal{R}^n$, except that we must allow for infinite sets. The definitions for finite sets are the same as for $\mathcal{R}^n$ (see page 69).

Definitions.   An infinite subset S of a vector space V is **linearly dependent** if some finite subset of S is linearly dependent. An infinite set S is **linearly independent** if S is not linearly dependent; that is, if every finite subset of S is linearly independent.

Example 6   The subset $S = \{x^2 - 3x + 2, 3x^2 - 5x, 2x - 3\}$ of $\mathcal{P}_2$ is linearly dependent because

$$3(x^2 - 3x + 2) + (-1)(3x^2 - 5x) + 2(2x - 3) = \mathbf{0},$$

where $\mathbf{0}$ is the zero polynomial. As with linearly dependent subsets of $\mathcal{R}^n$, S is linearly dependent because we are able to represent the zero vector as a linear combination of the vectors of S using at least one nonzero coefficient.   ○

Example 7   The infinite subset $\{1, x, x^2, \ldots, x^n, \ldots\}$ of the vector space $\mathcal{P}$ is linearly independent. To understand why, first observe that the only polynomial that is zero is the one with all coefficients zero. So any linear combination of vectors of a finite subset of S yields the zero polynomial if and only if all of the coefficients are zeros.   ○

Example 8   In Example 5, we noted that the set

$$S = \left\{ \begin{bmatrix} 1 & 0 \\ 0 & -1 \end{bmatrix}, \begin{bmatrix} 0 & 1 \\ 0 & 0 \end{bmatrix}, \begin{bmatrix} 0 & 0 \\ 1 & 0 \end{bmatrix} \right\}$$

spans the subspace of 2×2 matrices with trace equal to zero. We now show that this set is linearly independent. Consider any scalars a, b, and c such that

$$a \begin{bmatrix} 1 & 0 \\ 0 & -1 \end{bmatrix} + b \begin{bmatrix} 0 & 1 \\ 0 & 0 \end{bmatrix} + c \begin{bmatrix} 0 & 0 \\ 1 & 0 \end{bmatrix} = O,$$

where O is the zero matrix of $\mathcal{M}_{2 \times 2}$. In Example 5, we observed that this linear combination equals $\begin{bmatrix} a & b \\ c & -a \end{bmatrix}$, and hence

$$\begin{bmatrix} a & b \\ c & -a \end{bmatrix} = \begin{bmatrix} 0 & 0 \\ 0 & 0 \end{bmatrix}.$$

We equate corresponding entries to obtain $a = 0$, $b = 0$, and $c = 0$. It follows that S is linearly independent.   ○

Example 9   Let $S = \{e^t, e^{2t}, e^{3t}\}$. We show that S is a linearly independent subset of $\mathcal{F}(R)$. Consider any scalars a, b, and c such that

$$ae^t + be^{2t} + ce^{3t} = \mathbf{0},$$

where $\mathbf{0}$ is the zero function. We differentiate both sides of this equation twice to obtain

$$ae^t + 2be^{2t} + 3ce^{3t} = \mathbf{0}$$

and

$$ae^t + 4be^{2t} + 9ce^{3t} = \mathbf{0}.$$

Now we substitute $t = 0$ into these three equations to obtain the homogeneous system of linear equations

$$\begin{aligned} a + b + c &= 0 \\ a + 2b + 3c &= 0 \\ a + 4b + 9c &= 0 \, . \end{aligned}$$

It is easy to show that this system has only the trivial solution $a = b = c = 0$, and hence S is linearly independent.   ○

As for a subspace of $\mathcal{R}^n$, a subset S of a vector space V is a **basis** for V if S is linearly independent and spans V. Thus we see that the set S in Examples 5 and 8 is a basis for the subspace of 2×2 matrices with trace equal to zero. Furthermore, we know that the set $\{1, x, x^2, \ldots, x^n, \ldots\}$ is a basis for $\mathcal{P}$ because in Examples 4 and 7 we have seen that this set is linearly independent and spans $\mathcal{P}$. Thus, in contrast to the subspaces of $\mathcal{R}^n$, the vector space $\mathcal{P}$ has an infinite basis. The spanning set for the previously defined space of trigonometric polynomials, $\mathcal{T}[0, 2\pi]$, is also linearly independent (see Section 7.4). This is another example of an infinite basis for a vector space.

In Chapter 4, it was shown that any two bases for the same subspace of $\mathcal{R}^n$ contain the same number of vectors. The preceding observations lead to three questions about bases of vector spaces. We list these questions together with their answers.

1. Is it possible for a vector space to have both an infinite and a finite basis? The answer is *no*.

2. If a vector space V has a finite basis, must any two bases for V have the same number of vectors (as is the case for subspaces of $\mathcal{R}^n$)? The answer is *yes*.

3. Does every vector space have a basis? If a certain mildly controversial axiom of set theory (called the *axiom of choice*) is accepted, then the answer is *yes*.

We justify the first two answers in this section. The justification for the third answer is beyond the scope of this book (see [4: 55–58]).

To facilitate our study, we introduce a special type of transformation called an *isomorphism*. As is the case for functions from $\mathcal{R}^n$ to $\mathcal{R}^m$, a **linear transformation** $T: V \to W$, where V and W are vector spaces, is a function such that for any vectors $\mathbf{u}$ and $\mathbf{v}$ in V and any scalar c

$$T(\mathbf{u} + \mathbf{v}) = T(\mathbf{u}) + T(\mathbf{v}) \qquad \text{and} \qquad T(c\mathbf{u}) = cT(\mathbf{u}).$$

If T is also one-to-one and onto, we call T an **isomorphism**. We say that V is **isomorphic** to W if there exists an isomorphism from V to W. Any function that is both one-to-one and onto has an inverse. If T is an isomorphism, then its inverse, T^{-1}, is also linear and hence is an isomorphism (see Exercise 50). Therefore, if V is isomorphic to W, then W is isomorphic to V. For this reason, we simply say that V and W are **isomorphic**.

Because isomorphisms are linear, one-to-one, and onto, they "transfer" properties of V to W. A good illustration of this characteristic is given in

Theorem 7.4. Its proof is virtually the same as the proof of Exercise 65 in Section 2.7, and hence is omitted.

Theorem 7.4

Let V and W be vector spaces, $T: V \to W$ be an isomorphism, and $\{\mathbf{v}_1, \mathbf{v}_2, \ldots, \mathbf{v}_k\}$ be a linearly independent subset of V. Then the set of images $\{T(\mathbf{v}_1), T(\mathbf{v}_2), \ldots, T(\mathbf{v}_k)\}$ is a linearly independent subset of W.

We now show that if a vector space V has a finite basis of n vectors, then an isomorphism can be constructed from V to $\mathcal{R}^n$. We can then apply Theorem 7.4 to justify the answers to the first two questions that we posed. Suppose that $\mathcal{B} = \{\mathbf{v}_1, \mathbf{v}_2, \ldots, \mathbf{v}_n\}$ is such a basis. Then as is the case for any basis for a subspace of $\mathcal{R}^n$, a vector $\mathbf{v}$ in V can be represented as a linear combination of the vectors in $\mathcal{B}$ in exactly one way. (This is Theorem 4.8 on page 230.) The proof of Theorem 4.8 can be used word-for-word to prove the same assertion in the more general context of vector spaces. This enables us to define a mapping $\Phi_{\mathcal{B}}: V \to \mathcal{R}^n$ as follows. Consider any vector $\mathbf{v}$ in V, and suppose that the unique representation of $\mathbf{v}$ as a linear combination of the vectors of $\mathcal{B}$ is given by

$$\mathbf{v} = a_1 \mathbf{v}_1 + a_2 \mathbf{v}_2 + \cdots + a_n \mathbf{v}_n.$$

We define

$$\Phi_{\mathcal{B}}(\mathbf{v}) = \begin{bmatrix} a_1 \\ a_2 \\ \vdots \\ a_n \end{bmatrix}.$$

Then $\Phi_{\mathcal{B}}$ is one-to-one because the representation of a vector in V as a linear combination of the vectors of $\mathcal{B}$ is unique. Furthermore, any scalars $a_1, a_2, \ldots, a_n$ are the coefficients of some linear combination of the vectors of $\mathcal{B}$, and hence $\Phi_{\mathcal{B}}$ is onto. We leave as an exercise the straightforward proof that $\Phi_{\mathcal{B}}$ is linear (see Exercise 56). Thus we have the following result.

If V is a vector space with a basis $\mathcal{B} = \{\mathbf{v}_1, \mathbf{v}_2, \ldots, \mathbf{v}_n\}$, then the mapping $\Phi_{\mathcal{B}}: V \to \mathcal{R}^n$ defined by

$$\Phi_{\mathcal{B}}(a_1 \mathbf{v}_1 + a_2 \mathbf{v}_2 + \cdots + a_n \mathbf{v}_n) = \begin{bmatrix} a_1 \\ a_2 \\ \vdots \\ a_n \end{bmatrix}$$

is an isomorphism. Therefore if V has a basis of n vectors, then V is isomorphic to $\mathcal{R}^n$.

In view of this result, a vector space with a finite basis has the same vector space structure as $\mathcal{R}^n$. It follows that we can answer questions about the number of vectors in a basis for a vector space by comparing this vector space to $\mathcal{R}^n$.

Theorem 7.5

Let V be a vector space with a finite basis. Then every basis for V is finite and contains the same number of vectors.

Proof Let $\mathcal{B} = \{\mathbf{v}_1, \mathbf{v}_2, \ldots, \mathbf{v}_n\}$ be a finite basis for V and $\Phi_\mathcal{B}: V \to \mathcal{R}^n$ be the isomorphism defined above. Let $\mathcal{A}$ be any basis for V. First suppose that $\mathcal{A}$, which could be finite or infinite, contains more vectors than $\mathcal{B}$. Then there exists a subset $S = \{\mathbf{w}_1, \mathbf{w}_2, \ldots, \mathbf{w}_{n+1}\}$ of $\mathcal{A}$ consisting of $n + 1$ distinct vectors. By Exercise 43 of Section 1.7, which also applies to general vector spaces, S is linearly independent, and hence $\{\Phi_\mathcal{A}(\mathbf{w}_1), \Phi_\mathcal{A}(\mathbf{w}_2), \ldots, \Phi_\mathcal{A}(\mathbf{w}_{n+1})\}$ is a linearly independent subset of $\mathcal{R}^n$ by Theorem 7.4. But this is a contradiction because a linearly independent subset of $\mathcal{R}^n$ contains at most n vectors. It follows that $\mathcal{A}$ is finite and contains at most n vectors. Let m denote the number of vectors in $\mathcal{A}$. Then $m \leq n$. We can now apply the argument given above, but with the roles of $\mathcal{B}$ and $\mathcal{A}$ reversed, to deduce that $n \leq m$. Therefore, $m = n$, and we conclude that any two bases for V contain the same number of vectors. $\quad\square$

According to Theorem 7.5, vector spaces are of two types. One type consists of the zero vector space and those vector spaces that have a finite basis. These vector spaces are called **finite-dimensional**. The preceding boxed result shows that a vector space having a finite basis containing exactly n elements is isomorphic to $\mathcal{R}^n$. Moreover, every basis for such a vector space V must contain exactly n elements. In such a case, we say that n is the **dimension** of V and denote the dimension by dim V. The other type of vector space, which is not finite-dimensional, is called **infinite-dimensional**. It can be shown that every infinite-dimensional vector space contains an infinite linearly independent set (see Exercise 55). In fact, every infinite-dimensional vector space contains an infinite basis (see [4]).

Furthermore, dimension is preserved under isomorphism. That is, if V and W are isomorphic vector spaces and V is finite-dimensional, then W is finite-dimensional and the two vector spaces have the same dimension. On the other hand, if one of the vector spaces is infinite-dimensional, then so is the other (see Exercise 57).

We conclude this section with several examples.

Example 10

Recall the vector space $\mathcal{P}_n$ in Example 7 of Section 7.1. It is easy to see that the set $\mathcal{B} = \{1, x, x^2, \ldots, x^n\}$ is a linearly independent subset of $\mathcal{P}_n$. Furthermore, any polynomial $p(x)$ of degree at most n can be expressed as a linear combination of the polynomials in $\mathcal{B}$; for example,

$$p(x) = a_0 1 + a_1 x + \cdots + a_n x^n.$$

Therefore $\mathcal{B}$ spans $\mathcal{P}_n$, and we conclude that $\mathcal{B}$ is a basis for $\mathcal{P}_n$. Since $\mathcal{B}$ contains $n + 1$ polynomials, $\mathcal{P}_n$ is finite-dimensional with dimension equal to $n + 1$; that is, dim $\mathcal{P}_n = n + 1$.

Notice that $\Phi_\mathcal{B}: \mathcal{P}_n \to \mathcal{R}^{n+1}$ defined by

$$\Phi_\mathcal{B}(a_0 + a_1 x + \cdots + a_n x^n) = \begin{bmatrix} a_0 \\ a_1 \\ \vdots \\ a_n \end{bmatrix}$$

is an isomorphism.

Example 11

Recall $\mathcal{M}_{m \times n}$, the vector space of $m \times n$ matrices in Example 1 of Section 7.1. For each i and j, $1 \le i \le m$, and $1 \le j \le n$, let E_{ij} denote the $m \times n$ matrix whose (i, j)-entry is 1, and whose other entries are all 0. Let S be the set of all E_{ij}. Then S is a subset of $\mathcal{M}_{m \times n}$ consisting of mn matrices. Furthermore, any $m \times n$ matrix A can be written as a linear combination of the vectors in S, where a_{ij} is the coefficient of E_{ij} for all i and j. Finally, observe that if O is written as a linear combination of the matrices in S, then all of the coefficients must be zeros, and hence S is linearly independent. It follows that S is a basis for $\mathcal{M}_{m \times n}$, and hence this space has dimension mn, that is, $\dim \mathcal{M}_{m \times n} = mn$.

As an example, suppose $m = n = 2$. Then

$$\left\{ \begin{bmatrix} 1 & 0 \\ 0 & 0 \end{bmatrix}, \begin{bmatrix} 0 & 1 \\ 0 & 0 \end{bmatrix}, \begin{bmatrix} 0 & 0 \\ 1 & 0 \end{bmatrix}, \begin{bmatrix} 0 & 0 \\ 0 & 1 \end{bmatrix} \right\}$$

is a basis for $\mathcal{M}_{2 \times 2}$, and any 2×2 matrix A can be written as

$$A = \begin{bmatrix} a_{11} & a_{12} \\ a_{21} & a_{22} \end{bmatrix} = a_{11} \begin{bmatrix} 1 & 0 \\ 0 & 0 \end{bmatrix} + a_{12} \begin{bmatrix} 0 & 1 \\ 0 & 0 \end{bmatrix} + a_{21} \begin{bmatrix} 0 & 0 \\ 1 & 0 \end{bmatrix} + a_{22} \begin{bmatrix} 0 & 0 \\ 0 & 1 \end{bmatrix}.$$

Example 12

Recall the vector space $\mathcal{L}(\mathcal{R}^n, \mathcal{R}^m)$ of linear transformations defined in Example 2 of Section 7.1. Let $U \colon \mathcal{M}_{m \times n} \to \mathcal{L}(\mathcal{R}^n, \mathcal{R}^m)$ be defined by $U(A) = T_A$, where T_A is the matrix transformation induced by A, as defined in Section 2.6. Then for any $m \times n$ matrices A and B, we have

$$U(A + B) = T_{A+B} = T_A + T_B = U(A) + U(B).$$

Similarly, $U(cA) = c\,U(A)$ for any scalar c. By Theorem 2.11 on page 156, U is both one-to-one and onto. It follows that U is an isomorphism. By Exercise 44, isomorphisms preserve dimension. Thus, since $\mathcal{M}_{m \times n}$ is isomorphic to $\mathcal{L}(\mathcal{R}^n, \mathcal{R}^m)$ and $\mathcal{M}_{m \times n}$ has dimension mn, it follows that $\mathcal{L}(\mathcal{R}^n, \mathcal{R}^m)$ is a finite-dimensional vector space of dimension mn.

Example 13

Let $a_0, a_1, \ldots, a_n$ be $n + 1$ distinct real numbers. For $i = 0, 1, \ldots, n$, let $p_i(x)$ be the polynomial defined by

$$p_i(x) = \frac{(x - a_0)(x - a_1) \cdots (x - a_{i-1})(x - a_{i+1}) \cdots (x - a_n)}{(a_i - a_0)(a_i - a_1) \cdots (a_i - a_{i-1})(a_i - a_{i+1}) \cdots (a_i - a_n)}.$$

Then for $j = 0, 1, \ldots, n$, we have

$$p_i(a_j) = \begin{cases} 0 & \text{if } i \ne j \\ 1 & \text{if } i = j. \end{cases}$$

The polynomials defined above are called the **Lagrange interpolating polynomials** (associated with $a_0, a_1, \ldots, a_n$).[1] They are each of degree n and hence lie in $\mathcal{P}_n$. We can use Lagrange interpolating polynomials to construct a

---

[1] Joseph Louis Lagrange (1736–1813) was among the most important mathematicians and physical scientists of his time. Among his most significant accomplishments was the development of the calculus of variations, which he applied to problems in celestial mechanics. His 1788 treatise on analytical mechanics summarized the principal results in mechanics and demonstrated the importance of mathematics to mechanics. Lagrange also made important contributions to number theory, the theory of equations, and the foundations of calculus (by emphasizing functions and the use of Taylor series).

polynomial of degree less than or equal to n with specified values at $a_0, a_1, \ldots,$ a_n. That is, for any numbers $c_0, c_1, \ldots, c_n$, not necessarily distinct, there is a polynomial $p(x)$ in $\mathcal{P}_n$ such that $p(a_i) = c_i$ for $0 \leq i \leq n$. The required polynomial $p(x)$ can be obtained as a linear combination of the Lagrange interpolating polynomials with the c_i as the corresponding coefficients. In fact, if

$$p(x) = c_0 p_0(x) + c_1 p_1(x) + \cdots + c_n p_n(x),$$

then, for each i,

$$p(a_i) = c_0 p_0(a_i) + c_1 p_1(a_i) + \cdots + c_i p_i(a_i) + \cdots + c_n p_n(a_i)$$

$$= c_0 \cdot 0 + c_1 \cdot 0 + \cdots + c_i \cdot 1 + \cdots + c_n \cdot 0$$

$$= c_i. \tag{1}$$

To illustrate this method, we find a polynomial $p(x)$ in $\mathcal{P}_2$ such that $p(1) = 3$, $p(2) = 1$, and $p(4) = -1$. That is, we want a quadratic polynomial whose graph passes through the points $(1, 3)$, $(2, 1)$, and $(4, -1)$. Let $a_0 = 1$, $a_1 = 2$, and $a_2 = 4$. Then the Lagrange interpolating polynomials are

$$p_0(x) = \frac{(x - a_1)(x - a_2)}{(a_0 - a_1)(a_0 - a_2)} = \frac{(x - 2)(x - 4)}{(1 - 2)(1 - 4)} = \frac{1}{3}(x^2 - 6x + 8)$$

$$p_1(x) = \frac{(x - a_0)(x - a_2)}{(a_1 - a_0)(a_1 - a_2)} = \frac{(x - 1)(x - 4)}{(2 - 1)(2 - 4)} = -\frac{1}{2}(x^2 - 5x + 4)$$

$$p_2(x) = \frac{(x - a_0)(x - a_1)}{(a_2 - a_0)(a_2 - a_1)} = \frac{(x - 1)(x - 2)}{(4 - 1)(4 - 2)} = \frac{1}{6}(x^2 - 3x + 2).$$

Thus

$$p(x) = (3)p_0(x) + (1)p_1(x) + (-1)p_2(x)$$

$$= \frac{3}{3}(x^2 - 6x + 8) - \frac{1}{2}(x^2 - 5x + 4) - \frac{1}{6}(x^2 - 3x + 2)$$

$$= \frac{1}{3}x^2 - 3x + \frac{17}{3}.$$

In general, let $\mathcal{A} = \{p_0(x), p_1(x), \ldots, p_n(x)\}$ be the set of Lagrange interpolating polynomials associated with $a_0, a_1, \ldots, a_n$. We can apply (1) to show that $\mathcal{A}$ is linearly independent. For suppose that some linear combination

$$c_0 p_0(x) + c_1 p_1(x) + \cdots + c_n p_n(x) = \mathbf{0}$$

of the polynomials in $\mathcal{A}$ is equal to $\mathbf{0}$, the zero polynomial. Then for each i, $c_i = \mathbf{0}(a_i) = 0$ by (1), thus establishing that $\mathcal{A}$ is linearly independent. Furthermore, we can use Theorem 7.4 to prove that $\mathcal{A}$ is a basis for $\mathcal{P}_n$. Let $\mathcal{B}$ be the basis for $\mathcal{P}_n$ given in Example 10. Since $\Phi_\mathcal{B} \colon \mathcal{P}_n \to \mathcal{R}^{n+1}$ is an isomorphism and $\mathcal{A}$ is linearly independent, it follows by Theorem 7.4 that the set

$$\{\Phi_\mathcal{B}(p_0(x)), \Phi_\mathcal{B}(p_1(x)), \ldots, \Phi_\mathcal{B}(p_n(x))\}$$

is a linearly independent subset of $\mathcal{R}^{n+1}$. Because this set consists of $n + 1$ vectors, it is a basis for $\mathcal{R}^{n+1}$ by Theorem 4.5 on page 215. Thus, since $\Phi_\mathcal{B}$ is an isomorphism, $\mathcal{A}$ is a basis for $\mathcal{P}_n$ by Exercise 44. ○

In the example above, we used an isomorphism to compare a vector space with $\mathcal{R}^n$. This technique can be applied to obtain various properties about finite-dimensional vector spaces from the corresponding properties about $\mathcal{R}^n$ (see Exercises 51–54).

Practice Problems

1. Determine if the set

$$S = \left\{ \begin{bmatrix} 1 & 1 \\ 1 & 0 \end{bmatrix}, \begin{bmatrix} 0 & 0 \\ 1 & 1 \end{bmatrix}, \begin{bmatrix} 0 & 2 \\ 0 & -1 \end{bmatrix}, \begin{bmatrix} 0 & 1 \\ 1 & 0 \end{bmatrix} \right\}$$

is a linearly independent subset of $\mathcal{M}_{2\times 2}$.

2. Use the reasoning in Example 13 to prove that the set S in problem 1 is a basis for $\mathcal{M}_{2\times 2}$.

3. Let B be a 2×2 matrix, and define $T: \mathcal{M}_{2\times 2} \to \mathcal{M}_{2\times 2}$ by $T(A) = BA$.

   (a) Determine if T is linear.

   (b) If $B = \begin{bmatrix} 1 & 1 \\ 0 & 0 \end{bmatrix}$, determine if T is an isomorphism.

Exercises

1. Determine if the following statements are true or false.

   (a) The zero vector is a linear combination of any nonempty set of vectors.

   (b) The span of any nonempty subset of a vector space is a subspace of the vector space.

   (c) If a set is infinite, it cannot be linearly independent.

   (d) Every vector space has a finite basis.

   (e) The dimension of the vector space $\mathcal{P}_n$ equals n.

   (f) Every subspace of an infinite-dimensional vector space is infinite-dimensional.

   (g) It is possible for a vector space to have both an infinite basis and a finite basis.

   (h) Every isomorphism is linear and one-to-one.

   (i) Every nonzero finite-dimensional vector space is isomorphic to $\mathcal{R}^n$ for some n.

In Exercises 2–5, determine if the given matrix is a linear combination of the vectors in the set

$$\left\{ \begin{bmatrix} 1 & 2 & 1 \\ 0 & 0 & 0 \end{bmatrix}, \begin{bmatrix} 0 & 0 & 0 \\ 1 & 1 & 1 \end{bmatrix}, \begin{bmatrix} 1 & 0 & 1 \\ 1 & 2 & 3 \end{bmatrix} \right\}.$$

2. $\begin{bmatrix} 1 & 2 & 1 \\ 1 & 1 & 1 \end{bmatrix}$ 3. $\begin{bmatrix} 2 & 2 & 2 \\ 2 & 3 & 4 \end{bmatrix}$

4. $\begin{bmatrix} 2 & 2 & 2 \\ 2 & 2 & 2 \end{bmatrix}$ 5. $\begin{bmatrix} 2 & 2 & 2 \\ 1 & 1 & 1 \end{bmatrix}$

6. Let $S = \{x^3 - 2x^2 - 5x - 3, 3x^3 - 5x^2 - 4x - 9\}$, and $f(x) = 2x^3 - 2x^2 + 12x - 6$. Prove that $f(x)$ is a linear combination of the vectors in S.

7. Let $S = \{x^3 - 2x^2 - 5x - 3, 3x^3 - 5x^2 - 4x - 9\}$, and $f(x) = 3x^3 - 2x^2 + 7x + 8$. Prove that $f(x)$ is *not* a linear combination of the vectors in S.

8. Prove that Span $\{1 + x, 1 - x, 1 + x^2, 1 - x^2\} = \mathcal{P}_2$.

In Exercises 9–12, determine if the given subset of $\mathcal{M}_{2\times 2}$ is linearly independent or linearly dependent.

9. $\left\{ \begin{bmatrix} 1 & 2 \\ 3 & 1 \end{bmatrix}, \begin{bmatrix} 1 & -5 \\ -4 & 0 \end{bmatrix}, \begin{bmatrix} 3 & -1 \\ 2 & 2 \end{bmatrix} \right\}$

10. $\left\{ \begin{bmatrix} 1 & 2 \\ 3 & 1 \end{bmatrix}, \begin{bmatrix} 1 & -1 \\ 0 & 1 \end{bmatrix}, \begin{bmatrix} 1 & 0 \\ 1 & 1 \end{bmatrix} \right\}$

11. $\left\{ \begin{bmatrix} 1 & 2 \\ 2 & 1 \end{bmatrix}, \begin{bmatrix} 4 & 3 \\ -1 & 0 \end{bmatrix}, \begin{bmatrix} 12 & 9 \\ -3 & 0 \end{bmatrix} \right\}$

12. $\left\{ \begin{bmatrix} 1 & 2 \\ 2 & 1 \end{bmatrix}, \begin{bmatrix} 1 & 3 \\ 3 & 1 \end{bmatrix}, \begin{bmatrix} 1 & 2 \\ 3 & 1 \end{bmatrix} \right\}$

13. Prove that $\{1 + x, 1 - x, 1 + x^2, 1 - x^2\}$ is a linearly dependent subset of $\mathcal{P}_2$.

In Exercises 14–16, determine if the given subset of $\mathcal{P}$ is linearly independent or linearly dependent.

14. $\{x^2 - 2x + 5, 2x^2 - 4x + 10\}$

15. $\{x^2 - 2x + 5, 2x^2 - 5x + 10, x^2\}$

16. $\{x^3 + 4x^2 - 2x + 3, x^3 + 6x^2 - x + 4,$
    $3x^3 + 8x^2 - 8x + 7\}$

In Exercises 17–22, determine if the given subset of $\mathcal{F}(\mathcal{R})$ is linearly independent or linearly dependent.

17. $\{t, t\sin t\}$ 18. $\{t, t\sin t, e^{2t}\}$

19. $\{\sin t, \sin^2 t, \cos^2 t, 1\}$ 20. $\{\sin t, e^{-t}, e^t\}$

21. $\{e^t, e^{2t}, \dots, e^{nt}, \dots\}$ 22. $\{\cos^2 t, \sin^2 t, \cos 2t\}$

23. Let N be the set of positive integers, and let f, g, and h be the functions in $\mathcal{F}(N)$ defined by $f(n) = n + 1$, $g(n) = 1$, and $h(n) = 2n - 1$. Determine if the set $\{f, g, h\}$ is linearly independent. Justify your answer.

24. Use Lagrange interpolating polynomials to find the quadratic polynomial whose graph passes through the points $(1, 8)$, $(2, 5)$, and $(3, -4)$.

25. Use Lagrange interpolating polynomials to find the quadratic polynomial whose graph passes through the points $(0, 1)$, $(1, 0)$, and $(2, 3)$.

26. Let N denote the set of positive integers, and let V be the subset of $\mathcal{F}(N)$ consisting of the functions which are zero except at finitely many elements of N. For each n in N, let $f_n \colon N \to \mathcal{R}$ be defined by

$$f_n(k) = \begin{cases} 0 & \text{for } k \neq n \\ 1 & \text{for } k = n. \end{cases}$$

   (a) Show that V is a subspace of $\mathcal{F}(N)$.

(b) Show that $S = \{f_1, f_2, \ldots, f_n, \ldots\}$ is a linearly independent subset of V.

(c) Show that S in (b) is a basis for V.

In Exercises 27–30, find a basis for the subspace W of the vector space V.

27. Let W be the subspace of symmetric 3×3 matrices and $V = \mathcal{M}_{3 \times 3}$.

28. Let W be the subspace of skew-symmetric 3×3 matrices and $V = \mathcal{M}_{3 \times 3}$.

29. Let W be the subspace of 2×2 matrices with trace equal to 0 and $V = \mathcal{M}_{2 \times 2}$.

30. Let W be the subspace of $V = \mathcal{P}_n$ consisting of polynomials $p(x)$ for which $p(0) = 0$.

31. Let S be a subset of $\mathcal{P}_n$ consisting of exactly one polynomial of degree k for $k = 0, 1, \ldots, n$. Prove that S is a basis for $\mathcal{P}_n$.

In Exercises 32–37, determine if the given transformation T is linear. If T is linear, determine if it is an isomorphism. Justify your conclusions.

32. $T: \mathcal{M}_{n \times n} \to \mathcal{R}$ defined by $T(A) = \det A$

33. $T: \mathcal{P} \to \mathcal{P}$ defined by $T(f)(x) = xf(x)$

34. $T: \mathcal{P}_2 \to \mathcal{R}^3$ defined by $T(f) = \begin{bmatrix} f(0) \\ f(1) \\ f(2) \end{bmatrix}$

35. $T: \mathcal{P} \to \mathcal{P}$ defined by $T(f)(x) = (f(x))^2$

36. $T: \mathcal{M}_{2 \times 2} \to \mathcal{M}_{2 \times 2}$ defined by $T(A) = \begin{bmatrix} 1 & 1 \\ 1 & 1 \end{bmatrix} A$ for all A in $\mathcal{M}_{2 \times 2}$

37. $T: \mathcal{F}(\mathcal{R}) \to \mathcal{F}(\mathcal{R})$ defined by $T(f)(x) = f(x+1)$ for all f in $\mathcal{F}(\mathcal{R})$

The following definitions and notation apply to Exercises 38–43.

Definitions. An $n \times n$ matrix is called a **magic square of order n** if the sum of the entries in each row, the sum of the entries in each column, the sum of the diagonal entries, and the sum of the entries on the secondary diagonal are all equal. (The entries on the secondary diagonal are the $(1, n)$-entry, the $(2, n-1)$-entry, $\ldots$, the $(n, 1)$-entry.) This common value is called the **sum** of the magic square.

For example, the 3×3 matrix

$$\begin{bmatrix} 4 & 9 & 2 \\ 3 & 5 & 7 \\ 8 & 1 & 6 \end{bmatrix}$$

is a magic square of order 3 with sum equal to 15.

  Let V_n denote the set of all magic squares of order n, and let W_n denote the subset of V_n consisting of magic squares

with sum equal to 0.

38. (a) Show that V_n is a subspace of $\mathcal{M}_{n \times n}$.

  (b) Show that W_n is a subspace of V_n.

39. For each positive integer n, let C_n be the $n \times n$ matrix all of whose entries are equal to 1.

  (a) Prove that C_n is in V_n.

  (b) Prove that, for any positive integer n and magic square A in V_n, if A has sum s, then there is a unique magic square B in W_n such that $A = B + sC_n$.

40. Prove that W_3 has dimension equal to 7.

41. Prove that V_3 has dimension equal to 8.

42. Use the results of Exercise 39 to prove that, for any positive integer n, $\dim V_n = \dim W_n + 1$.

43. Prove that, for any $n \geq 3$, W_n has dimension equal to $2n + 1$, and hence by Exercise 42, V_n has dimension equal to $2n + 2$. *Hint:* You can identify $\mathcal{M}_{n \times n}$ with $\mathcal{R}^{n^2}$, and then analyze the description of W_n as the solution space of a system of homogeneous equations.

44. Let V and W be isomorphic finite-dimensional vector spaces, and suppose that $T: V \to W$ is an isomorphism. Prove that if $\{\mathbf{v}_1, \mathbf{v}_2, \ldots, \mathbf{v}_n\}$ is a basis for V, then the set of images $\{T(\mathbf{v}_1), T(\mathbf{v}_2), \ldots, T(\mathbf{v}_n)\}$ is a basis for W.

45. Use Exercise 44 and Example 12 to find a basis for $\mathcal{L}(\mathcal{R}^n, \mathcal{R}^m)$.

46. Let $S = \{s_1, s_2, \ldots, s_n\}$ be a set consisting of n elements, and let $T: \mathcal{F}(S) \to \mathcal{R}^n$ be defined by

$$T(f) = \begin{bmatrix} f(s_1) \\ f(s_2) \\ \vdots \\ f(s_n) \end{bmatrix}$$

for every f in $\mathcal{F}(S)$. Prove that T is an isomorphism.

47. Show that the vector space V in Exercise 26 is isomorphic to $\mathcal{P}$. *Hint:* Choose $T: V \to \mathcal{P}$ to be the transformation that maps a function f to the polynomial having $f(i+1)$ as the coefficient of x^i.

48. Let V be the vector space of all 2×2 matrices with trace equal to 0. Prove that V is isomorphic to $\mathcal{P}_2$ by constructing an isomorphism from V to $\mathcal{P}_2$. Verify your answer.

49. Let V be the subset of $\mathcal{P}_4$ of polynomials of the form $ax^4 + bx^2 + c$, where a, b, and c are scalars.

  (a) Prove that V is a subspace of $\mathcal{P}_4$.

  (b) Prove that V is isomorphic to $\mathcal{P}_2$ by constructing an isomorphism from V to $\mathcal{P}_2$. Verify your answer.

50.[2] Let V and W be isomorphic vector spaces, and suppose that $T: V \to W$ is an isomorphism. Prove that the inverse $T^{-1}: W \to V$ is also an isomorphism. *Hint (to prove that T^{-1} preserves vector sums):* Let $\mathbf{u}$ and

---

[2]The result of this exercise is used in Section 7.3 (page 424).

v be in W, and suppose $\mathbf{u}' = T^{-1}(\mathbf{u})$ and $\mathbf{v}' = T^{-1}(\mathbf{v})$. Then $\mathbf{u} = T(\mathbf{u}')$ and $\mathbf{v} = T(\mathbf{v}')$. Use this fact to compute $T^{-1}(\mathbf{u} + \mathbf{v})$.

In Exercises 51–54, use an isomorphism from V to $\mathcal{R}^n$ to prove the result.

51. Let n be a positive integer. Suppose that V is a vector space such that any subset of V consisting of more than n vectors is linearly dependent, and some linearly independent subset contains n vectors. Prove that any linearly independent subset of V consisting of n vectors is a basis for V, and hence dim $V = n$.

52. Let V be a finite-dimensional vector space of dimension $n \geq 1$.

    (a) Prove that any subset of V containing more than n vectors is linearly dependent.

    (b) Prove that any linearly independent subset of V consisting of n vectors is a basis for V.

53. Let V be a vector space of dimension $n \geq 1$, and suppose that $\mathcal{S}$ is a finite subset of V that spans V. Prove the following statements.

    (a) $\mathcal{S}$ contains at least n vectors.

    (b) If $\mathcal{S}$ consists of exactly n vectors, then $\mathcal{S}$ is a basis for V.

54. Let V be a finite-dimensional vector space and W be a subspace of V. Prove the following statements.

    (a) W is finite-dimensional, and dim $W \leq$ dim V.

    (b) If dim $W =$ dim V, then $W = V$.

55. Let V be an infinite-dimensional vector space. Prove that V contains an infinite linearly independent set. *Hint:* Choose a nonzero vector $\mathbf{v}_1$ in V. Next, choose a vector $\mathbf{v}_2$ not in the span of $\{\mathbf{v}_1\}$. Show that this process can be continued to obtain an infinite subset $\{\mathbf{v}_1, \mathbf{v}_2, \ldots, \mathbf{v}_n, \ldots\}$ of V such that for any n, $\mathbf{v}_{n+1}$ is not in the span of $\{\mathbf{v}_1, \mathbf{v}_2, \ldots, \mathbf{v}_n\}$. Now show that this infinite set is linearly independent.

56. Prove that if $\mathcal{B}$ is a basis for a vector space that contains exactly n vectors, then $\Phi_{\mathcal{B}}: V \to \mathcal{R}^n$ is a linear transformation.

57. Suppose that V and W are isomorphic vector spaces. Prove that if V is infinite-dimensional, then W is infinite-dimensional.

58. Suppose that T and U are invertible linear operators on a vector space. Prove that TU is invertible and $(TU)^{-1} = U^{-1}T^{-1}$.

59.[3] Prove the following extension of Exercise 58. If $T_1, T_2, \ldots, T_k$ are invertible linear operators on a vector space, then the composition $T_1 T_2 \cdots T_k$ is invertible and

$$(T_1 T_2 \cdots T_k)^{-1} = (T_k)^{-1}(T_{k-1})^{-1} \cdots (T_1)^{-1}.$$

---

[3]The results of this exercise are used in Section 7.3 (page 424).

The following definitions are used in Exercises 60–64.

    For vector spaces V and W, let $\mathcal{L}(V, W)$ denote the set of all linear transformations $T: V \to W$. For T and U in $\mathcal{L}(V, W)$ and any scalar c, define $T + U: V \to W$ and $cT: V \to W$ by

$$(T + U)(\mathbf{x}) = T(\mathbf{x}) + U(\mathbf{x}) \qquad \text{and} \qquad (cT)(\mathbf{x}) = cT(\mathbf{x})$$

for all $\mathbf{x}$ in V.

60. Let V and W be vector spaces, let T and U be in $\mathcal{L}(V, W)$, and let c be a scalar.

    (a) Prove that for any T and U in $\mathcal{L}(V, W)$, $T + U$ is a linear transformation.

    (b) Prove that for any T in $\mathcal{L}(V, W)$, cT is a linear transformation for any scalar c.

    (c) Prove that $\mathcal{L}(V, W)$ is a vector space under these operations.

    (d) Describe the zero vector of this vector space.

61. Let V and W be finite-dimensional vector spaces. Prove that dim $\mathcal{L}(V, W) = (\dim V) \cdot (\dim W)$.

62. Let n be a positive integer. For $0 \leq i \leq n$, let $T_i: \mathcal{P}_n \to \mathcal{R}$ be defined by $T_i(f(x)) = f(i)$. Prove that T_i is linear for all i and that $\{T_0, T_1, \ldots, T_n\}$ is a basis for $\mathcal{L}(\mathcal{P}_n, \mathcal{R})$. *Hint:* For each i, let $p_i(x)$ be the ith Lagrange interpolating polynomial associated with $0, 1, \ldots, n$. Show that for all i and j,

$$T_i(p_j(x)) = \begin{cases} 0 & \text{if } i \neq j \\ 1 & \text{if } i = j. \end{cases}$$

Use this to show that $\{T_0, T_1, \ldots, T_n\}$ is linearly independent. Now apply Exercises 61 and 52.

63. Apply Exercise 62 to prove that for any positive integer n and any scalars a and b, there exist unique scalars $c_0, c_1, \ldots, c_n$ such that

$$\int_a^b f(x)\,dx = c_0 f(0) + c_1 f(1) + \cdots + c_n f(n)$$

for every polynomial $f(x)$ in $\mathcal{P}_n$.

64. (a) Derive Simpson's rule: For any polynomial $f(x)$ in $\mathcal{P}_2$ and any scalars $a < b$,

$$\int_a^b f(x)\,dx = \frac{b - a}{6}\left[f(a) + 4f\left(\frac{a + b}{2}\right) + f(b)\right].$$

    (b) Verify that Simpson's rule is valid for the polynomial x^3, and use this fact to justify that Simpson's rule is valid for every polynomial in $\mathcal{P}_3$.

In Exercises 65–69, use a calculator with matrix capabilities or computer software such as MATLAB *to solve the problem.*

In Exercises 65–68, determine if the given set is linearly dependent. In the case that the set is linearly dependent, write some vector in the set as a linear combination of the others.

65. $\{1 + x - x^2 + 3x^3 - x^4, 2 + 5x - x^3 + x^4,$
$3x + 2x^2 + 7x^4, 4 - x^2 + x^3 - x^4\}$

66. $\{2 + 5x - 2x^2 + 3x^3 + x^4, 3 + 3x - x^2 + x^3 + x^4,$
$$6 - 3x + 2x^2 - 5x^3 + x^4, 2 - x + x^2 + x^4\}$$

67. $\left\{ \begin{bmatrix} 0.97 & -1.12 \\ 1.82 & 2.13 \end{bmatrix}, \begin{bmatrix} 1.14 & 2.01 \\ 1.01 & 3.21 \end{bmatrix}, \right.$
$$\left. \begin{bmatrix} -0.63 & 7.38 \\ -3.44 & 0.03 \end{bmatrix}, \begin{bmatrix} 2.12 & -1.21 \\ 0.07 & -1.32 \end{bmatrix} \right\}$$

68. $\left\{ \begin{bmatrix} 1.23 & -0.41 \\ 2.57 & 3.13 \end{bmatrix}, \begin{bmatrix} 2.71 & 1.40 \\ -5.23 & 2.71 \end{bmatrix}, \right.$
$$\left. \begin{bmatrix} 3.13 & 1.10 \\ 2.12 & -1.11 \end{bmatrix}, \begin{bmatrix} 8.18 & 2.15 \\ -1.21 & 4.12 \end{bmatrix} \right\}$$

69. In view of Exercise 63, find the scalars $c_0, c_1, c_2, c_3,$ and c_4 such that
$$\int_0^1 f(x)\, dx = c_0 f(0) + c_1 f(1) + c_2 f(2)$$
$$+ c_3 f(3) + c_4 f(4)$$
for every polynomial $f(x)$ in P_4. *Hint:* Apply this equation to $1, x, x^2, x^3,$ and x^4 to obtain a system of five linear equations in five variables.

Solutions to the Practice Problems

1. We suppose that

$$c_1 \begin{bmatrix} 1 & 1 \\ 1 & 0 \end{bmatrix} + c_2 \begin{bmatrix} 0 & 0 \\ 1 & 1 \end{bmatrix} + c_3 \begin{bmatrix} 0 & 2 \\ 0 & -1 \end{bmatrix} + c_4 \begin{bmatrix} 0 & 1 \\ 1 & 0 \end{bmatrix} = \begin{bmatrix} 0 & 0 \\ 0 & 0 \end{bmatrix}$$

for scalars $c_1, c_2, c_3,$ and c_4. This equation may be rewritten as

$$\begin{bmatrix} c_1 & c_1 + 2c_3 + c_4 \\ c_1 + c_2 + c_4 & c_2 - c_3 \end{bmatrix} = \begin{bmatrix} 0 & 0 \\ 0 & 0 \end{bmatrix}.$$

The last matrix equation is equivalent to the system

$$\begin{aligned} c_1 & & & = 0 \\ c_1 & & + 2c_3 + c_4 &= 0 \\ c_1 + c_2 & & + c_4 &= 0 \\ & c_2 - & c_3 & = 0. \end{aligned}$$

After we apply Gaussian elimination to this system, we discover that its only solution is $c_1 = c_2 = c_3 = c_4 = 0$. Therefore S is linearly independent.

2. From Example 11, we know that dim $\mathcal{M}_{2 \times 2} = 4$. So let B be a basis for $\mathcal{M}_{2 \times 2}$ with four vectors. Then

$\Phi_B \colon \mathcal{M}_{2 \times 2} \to \mathcal{R}^4$ is an isomorphism. Because S is linearly independent, the set of images of the vectors in S under Φ_B is also linearly independent, and hence is a basis for $\mathcal{R}^4$. As in Example 11, we conclude that S is a basis for $\mathcal{M}_{2 \times 2}$.

3. (a) Let C and D be vectors in $\mathcal{M}_{2 \times 2}$, and let k be a scalar. Then

$$T(C + D) = B(C + D) = BC + BD = T(C) + T(D)$$

and

$$T(kC) = B(kC) = k(BC) = kT(C).$$

So T is linear.

(b) For any matrix $C = \begin{bmatrix} a & b \\ c & d \end{bmatrix}$, we have

$$T(C) = \begin{bmatrix} 1 & 1 \\ 0 & 0 \end{bmatrix} \begin{bmatrix} a & b \\ c & d \end{bmatrix} = \begin{bmatrix} a + c & b + d \\ 0 & 0 \end{bmatrix}.$$

We see that it is impossible for any matrix with a nonzero $(2, 2)$-entry to be in the range of T. So T is not onto, and hence T is not an isomorphism.

7.3 Linear Transformations and Matrix Representations

In this section, we study linear transformations and *linear operators* in depth. As in the case of $\mathcal{R}^n$, a linear transformation $T \colon V \to V$, where V is a vector space, is called a **linear operator** on V, and the **null space** of the linear operator T on the vector space V is the set $\{\mathbf{v} \colon T(\mathbf{v}) = \mathbf{0}\}$. Later in this section, for a finite-dimensional vector space, we show how to identify a linear operator with a matrix, as was done in the context of $\mathcal{R}^n$. This is important because matrices give us computational tools that can be used in the study of linear operators.

We begin with examples of linear transformations and linear operators that arise quite naturally in the study of linear algebra and related fields.

Example 1 (**Differential operators**) Let C^∞ denote the subset of $\mathcal{F}(\mathcal{R})$ consisting of those functions that have derivatives of all orders. That is, a function $f \colon \mathcal{R} \to \mathcal{R}$ lies in C^∞ if the nth derivative of f exists for every positive integer n. Theorems from calculus imply that C^∞ is a subspace of $\mathcal{F}(\mathcal{R})$ (see

Exercise 42). Consider the mapping $D: C^\infty \to C^\infty$ that takes a function into its derivative, that is, $D(f) = f'$ for all f in C^∞. From elementary properties of the derivative, it follows that

$$D(f + g) = (f + g)' = f' + g' = D(f) + D(g)$$

for all functions f and g in C^∞. Similarly, $D(cf) = cD(f)$ for every scalar c. It follows that D is a linear operator on C^∞.

We may compose D with itself any number of times to obtain new linear operators. As with matrix multiplication, we use exponent notation. For example, D^3 is the linear operator that takes a function into its third derivative, that is, $D^3(f) = f'''$ for all f in C^∞. Furthermore, we can multiply these operators by scalars and add them as defined on page 417 to obtain a linear operator of the form

$$a_0 I + a_1 D + a_2 D^2 + \cdots + a_n D^n, \tag{2}$$

where I is the identity operator on C^∞. Such a mapping is an example of a *differential operator*. For example, the differential operator $2I + 3D - D^2$ acts on the function $f(t) = t^3$ as

$$(2I + 3D - D^2)(t^3) = 2t^3 + 9t^2 - 6t.$$

More generally, a **differential operator** is a function from C^∞ to C^∞ of the form (2), where the a_i are functions in C^∞. All differential operators are linear. These operators are important because many differential equations can be expressed in the context of differential operators, just as systems of linear equations can be expressed in the context of matrices. For example, the differential equation

$$t^2 y'' - (\sin t) y' + 3y = \mathbf{0}$$

can be rewritten using the notation of differential operators as

$$(t^2 D^2 - (\sin t)D + 3I)y = \mathbf{0}.$$

Thus the solution set of the differential equation is the same as the null space of the differential operator $t^2 D^2 - (\sin t)D + 3I$, which is the set of functions in C^∞ that have $\mathbf{0}$ as their images under this differential operator.   ○

Example 2   Let $C([a, b])$ denote the set of all continuous real-valued functions defined on the closed interval $[a, b]$. It can be shown that $C([a, b])$ is a subspace of $\mathcal{F}([a, b])$, the set of real-valued functions on $[a, b]$ (see Exercise 43). Any function f in $C([a, b])$ is integrable. Let $T: C([a, b]) \to \mathcal{R}$ be the mapping that takes a function f into its integral,

$$T(f) = \int_a^b f(t)\, dt.$$

The linearity of T follows from the elementary properties of integration. For example, for any f and g in $C([a, b])$,

$$T(f + g) = \int_a^b (f + g)(t)\, dt$$

$$= \int_a^b (f(t) + g(t))\, dt$$

$$= \int_a^b f(t)\, dt + \int_a^b g(t)\, dt$$

$$= T(f) + T(g).$$

Similarly, $T(cf) = cT(f)$ for every scalar c. Thus T is a linear transformation. ○

Example 3

Let $U: \mathcal{M}_{m \times n} \to \mathcal{M}_{n \times m}$ be the mapping that takes a matrix into its transpose, that is, $U(A) = A^T$. The linearity of U is a consequence of Theorem 1.1 on page 5. Since $(A^T)^T = A$, we have that U is one-to-one and onto. Therefore U is an isomorphism. ○

We now extend the definitions of *eigenvalue, eigenvector*, and *eigenspace*, as given for $\mathcal{R}^n$ in Chapter 5, to all vector spaces.

Let T be a linear operator on a vector space V. A nonzero vector $\mathbf{v}$ in V is called an **eigenvector** of T if there is a scalar λ such that $T(\mathbf{v}) = \lambda\mathbf{v}$. The scalar λ is called the **eigenvalue** of T corresponding to $\mathbf{v}$. If λ is an eigenvalue of T, then the set of all vectors $\mathbf{v}$ in V such that $T(\mathbf{v}) = \lambda\mathbf{v}$ is called the **eigenspace** of T corresponding to λ. As in $\mathcal{R}^n$, the eigenspace corresponding to an eigenvalue of T is a subspace of V consisting of the zero vector and all of the eigenvectors of T corresponding to the eigenvalue (see Exercise 38).

Example 4

Let $D: \mathsf{C}^\infty \to \mathsf{C}^\infty$ be the linear operator defined in Example 1. Let λ be a scalar, and let f be the exponential function $f(t) = e^{\lambda t}$. Then

$$D(f)(t) = (e^{\lambda t})' = \lambda e^{\lambda t} = \lambda f(t).$$

Thus f is an eigenvector of D, and λ is the eigenvalue corresponding to f. Since λ was chosen arbitrarily, we see that every scalar is an eigenvalue of D. Therefore D has infinitely many eigenvalues, in contrast to linear operators on $\mathcal{R}^n$. ○

Example 5

Show that the solution set to the differential equation

$$y'' + 4y = \mathbf{0}$$

coincides with the eigenspace of the differential operator D^2 corresponding to the eigenvalue $\lambda = -4$.

Solution

First observe that the solutions to this differential equation lie in C^∞. For suppose that f is a solution. Then f must be at least twice differentiable, and since $f'' = -4f$, we have that f'' is also twice differentiable. We can now differentiate both sides of this equation twice to obtain that $f'''' = -4f''$, from which we can infer that the fourth derivative of f is twice differentiable. Repetition of this argument leads to the conclusion that f has derivatives of any order. Since $y'' = D^2 y$, we can rewrite the given differential equation as

$$D^2 y = -4y.$$

But this last equation asserts that y is in the eigenspace of D^2 corresponding to the eigenvalue -4. Thus this eigenspace and the solution set to the differential equation are the same.

The functions $\sin 2t$ and $\cos 2t$ are solutions to this differential equation. Notice that they are also eigenvectors of D^2 corresponding to the eigenvalue $\lambda = -4$. ○

Example 6

Let n be a positive integer, and let $U \colon \mathcal{M}_{n \times n} \to \mathcal{M}_{n \times n}$ be the linear operator that takes a matrix into its transpose, that is, $U(A) = A^T$. We saw in Example 3 that U is an isomorphism. Recall that a matrix A in $\mathcal{M}_{n \times n}$ is symmetric if and only if $A^T = A$, that is, if and only if $U(A) = A$. Thus $\lambda = 1$ is an eigenvalue of U, and the eigenspace corresponding to $\lambda = 1$ is precisely the subset of symmetric matrices. Recall that an $n \times n$ matrix A is *skew-symmetric* if $A^T = -A$. Since there exist nonzero skew-symmetric matrices, for example (for $n = 2$)

$$\begin{bmatrix} 0 & 1 \\ -1 & 0 \end{bmatrix},$$

it follows immediately that $\lambda = -1$ is also an eigenvalue of U, and the corresponding eigenspace is the subset of skew-symmetric matrices. It can be shown that 1 and -1 are the only eigenvalues of U (see Exercise 11).   ○

Matrix Representations

We can use a *matrix representation* of a linear operator on a finite-dimensional vector space to compute the eigenvalues and eigenvectors of the operator, as well as its inverse (if such exists). For this purpose, we extend the definitions of *coordinate vector* and *matrix representation* given in Sections 4.4 and 4.5, respectively, to the general context of vector spaces.

Definition.   Let $\mathcal{B} = \{\mathbf{v}_1, \mathbf{v}_2, \ldots, \mathbf{v}_n\}$ be a basis for a vector space V, and let $\mathbf{v}$ be a vector in V. Then $\mathbf{v}$ can be represented as a unique linear combination

$$\mathbf{v} = a_1 \mathbf{v}_1 + a_2 \mathbf{v}_2 + \cdots + a_n \mathbf{v}_n$$

of vectors in $\mathcal{B}$. The vector

$$[\mathbf{v}]_{\mathcal{B}} = \begin{bmatrix} a_1 \\ a_2 \\ \vdots \\ a_n \end{bmatrix}$$

is called the **coordinate vector of v relative to** $\mathcal{B}$.

In the context of the definition above, notice that $[\mathbf{v}]_{\mathcal{B}} = \Phi_{\mathcal{B}}(\mathbf{v})$, where $\Phi_{\mathcal{B}}$ is as defined in Section 7.2. Since $\Phi_{\mathcal{B}}$ is a linear transformation, it follows that for any vectors $\mathbf{u}$ and $\mathbf{v}$ in V and any scalar c

$$[\mathbf{u} + \mathbf{v}]_{\mathcal{B}} = \Phi_{\mathcal{B}}(\mathbf{u} + \mathbf{v}) = \Phi_{\mathcal{B}}(\mathbf{u}) + \Phi_{\mathcal{B}}(\mathbf{v}) = [\mathbf{u}]_{\mathcal{B}} + [\mathbf{v}]_{\mathcal{B}} \qquad (3)$$

and

$$[c\mathbf{v}]_{\mathcal{B}} = \Phi_{\mathcal{B}}(c\mathbf{v}) = c\Phi_{\mathcal{B}}(\mathbf{v}) = c[\mathbf{v}]_{\mathcal{B}}. \qquad (4)$$

Example 7

Let $\mathcal{B} = \{1, x, \ldots, x^n\}$, which is a basis for $\mathcal{P}_n$. For any polynomial $p(x) = a_0 + a_1 x + \cdots + a_n x^n$, we have

$$[p(x)]_{\mathcal{B}} = \Phi_{\mathcal{B}}(p(x)) = \begin{bmatrix} a_1 \\ a_2 \\ \vdots \\ a_n \end{bmatrix}.$$

○

A coordinate vector, which depends on the choice of basis, is a concrete representation of a vector in a vector space. We use coordinate vectors to define a *matrix representation* of a linear operator, and hence a matrix representation of a linear operator also depends on the choice of basis.

Definition.   Let T be a linear operator on a finite-dimensional vector space V, and let $\mathcal{B} = \{\mathbf{v}_1, \mathbf{v}_2, \ldots, \mathbf{v}_n\}$ be a basis for V. The $n \times n$ matrix whose jth column is $[T(\mathbf{v}_j)]_\mathcal{B}$ is called the **matrix representation of T with respect to $\mathcal{B}$** and is denoted by $[T]_\mathcal{B}$. That is,

$$[T]_\mathcal{B} = [\, [T(\mathbf{v}_1)]_\mathcal{B} \quad [T(\mathbf{v}_2)]_\mathcal{B} \quad \cdots \quad [T(\mathbf{v}_n)]_\mathcal{B} \,].$$

Example 8   Let D be the linear operator on $\mathcal{P}_2$ that takes a polynomial into its derivative; that is, $D(p(x)) = p'(x)$ for every polynomial $p(x)$ in $\mathcal{P}_2$. Let $\mathcal{B} = \{1, x, x^2\}$, which is a basis for $\mathcal{P}_2$, and $A = [T]_\mathcal{B}$. Then

$$\mathbf{a}_1 = [D(1)]_\mathcal{B} = [\mathbf{0}]_\mathcal{B} = \begin{bmatrix} 0 \\ 0 \\ 0 \end{bmatrix}$$

$$\mathbf{a}_2 = [D(x)]_\mathcal{B} = [1]_\mathcal{B} = \begin{bmatrix} 1 \\ 0 \\ 0 \end{bmatrix}$$

and

$$\mathbf{a}_3 = [D(x^2)]_\mathcal{B} = [2x]_\mathcal{B} = \begin{bmatrix} 0 \\ 2 \\ 0 \end{bmatrix}.$$

Therefore

$$[T]_\mathcal{B} = [\mathbf{a}_1 \quad \mathbf{a}_2 \quad \mathbf{a}_3] = \begin{bmatrix} 0 & 1 & 0 \\ 0 & 0 & 2 \\ 0 & 0 & 0 \end{bmatrix}.$$

Example 9   Let $\mathcal{B} = \{e^t \cos t, \ e^t \sin t\}$, which is a subset of C^∞, and let $V = \operatorname{Span} \mathcal{B}$. It can be shown that $\mathcal{B}$ is linearly independent and hence is a basis for V. Let D be the linear operator on V defined by $D(f) = f'$ for all f in V. Then

$$D(e^t \cos t) = (1)e^t \cos t + (-1)e^t \sin t$$

and

$$D(e^t \sin t) = (1)e^t \cos t + (1)e^t \sin t.$$

Therefore

$$[D]_\mathcal{B} = \begin{bmatrix} 1 & 1 \\ -1 & 1 \end{bmatrix}.$$

The following result, which is the generalization of Theorem 4.11 on page 243, enables us to describe a linear operator on a finite-dimensional vector space in terms of matrix multiplication. We use this description to obtain information about a linear operator by making use of what we already know about matrices.

Theorem 7.6

Let T be a linear operator on a finite-dimensional vector space V with basis $\mathcal{B}$. Then for any vector $\mathbf{v}$ in V,

$$[T(\mathbf{v})]_\mathcal{B} = [T]_\mathcal{B}[\mathbf{v}]_\mathcal{B}. \tag{5}$$

Proof Let $\mathcal{B} = \{\mathbf{v}_1, \mathbf{v}_2, \ldots, \mathbf{v}_n\}$. Now consider any vector

$$\mathbf{v} = a_1\mathbf{v}_1 + a_2\mathbf{v}_2 + \cdots + a_n\mathbf{v}_n$$

in V. Then

$$
\begin{aligned}
[T(\mathbf{v})]_\mathcal{B} &= [T(a_1\mathbf{v}_1 + a_2\mathbf{v}_2 + \cdots + a_n\mathbf{v}_n)]_\mathcal{B} \\
&= [a_1 T(\mathbf{v}_1) + a_2 T(\mathbf{v}_2) + \cdots + a_n T(\mathbf{v}_n)]_\mathcal{B} \\
&= a_1[T(\mathbf{v}_1)]_\mathcal{B} + a_2[T(\mathbf{v}_2)]_\mathcal{B} + \cdots + a_n[T(\mathbf{v}_n)]_\mathcal{B} \quad \text{(by (3) and (4))} \\
&= [\,[T(\mathbf{v}_1)]_\mathcal{B} \quad [T(\mathbf{v}_2)]_\mathcal{B} \quad \cdots \quad [T(\mathbf{v}_n)]_\mathcal{B}\,] \begin{bmatrix} a_1 \\ a_2 \\ \vdots \\ a_n \end{bmatrix} \\
&= [T]_\mathcal{B}[\mathbf{v}]_\mathcal{B}.
\end{aligned}
$$

Example 10   Let D and $\mathcal{B}$ be as in Example 8. Consider the polynomial $p(x) = 5 - 4x + 3x^2$. Then

$$[p(x)]_\mathcal{B} = \begin{bmatrix} 5 \\ -4 \\ 3 \end{bmatrix}.$$

By Theorem 7.6,

$$
\begin{aligned}
[p'(x)]_\mathcal{B} = [D(p(x))]_\mathcal{B} &= [D]_\mathcal{B}[p(x)]_\mathcal{B} \\
&= \begin{bmatrix} 0 & 1 & 0 \\ 0 & 0 & 2 \\ 0 & 0 & 0 \end{bmatrix} \begin{bmatrix} 5 \\ -4 \\ 3 \end{bmatrix} \\
&= \begin{bmatrix} -4 \\ 6 \\ 0 \end{bmatrix}.
\end{aligned}
$$

The preceding vector is the coordinate vector for the polynomial $-4 + 6x$, which is the derivative of $p(x)$. Thus we can compute a derivative by taking a matrix–vector product.   ○

It is useful to understand Theorem 7.6 from the viewpoint of linear transformations. The isomorphism $\Phi_\mathcal{B}$ defined in Section 7.2 is related to coordinate vectors by

$$\Phi_\mathcal{B}(\mathbf{v}) = [\mathbf{v}]_\mathcal{B},$$

where V is a finite-dimensional vector space and $\mathcal{B}$ is a basis for V. We can reformulate (5) using $\Phi_\mathcal{B}$ in place of coordinate vectors. To simplify the notation, let $A = [T]_\mathcal{B}$, and recall that T_A is the matrix transformation induced by A. With this notation, (5) of Theorem 7.6 can be rewritten

$$\Phi_\mathcal{B} T(\mathbf{v}) = T_A \Phi_\mathcal{B}(\mathbf{v}).$$

Finally, since this equation is valid for every vector $\mathbf{v}$ in V, we can equate the transformations themselves to obtain

$$\Phi_{\mathcal{B}}T = T_A\Phi_{\mathcal{B}}. \tag{6}$$

Equation (6) gives us a visual way of understanding what Theorem 7.6 tells us. In Figure 7.1, we see two paths, labeled (1) and (2), that map a vector $\mathbf{v}$ in V to a vector in $\mathcal{R}^n$. The path labeled (1) maps $\mathbf{v}$ into $T_A\Phi_{\mathcal{B}}(\mathbf{v})$, and the path labeled (2) maps $\mathbf{v}$ into $\Phi_{\mathcal{B}}T(\mathbf{v})$. By (6), the two images are equal, and hence the two paths yield the same result.

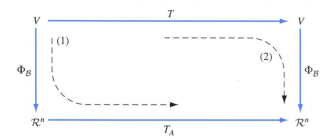

FIGURE 7.1

Our next task is to apply (6) to derive a test for invertibility of a linear operator on a finite-dimensional vector space, and to obtain a method for computing the inverse of an invertible operator.

Let T be a linear operator on a finite-dimensional vector space V with basis $\mathcal{B}$, and let $A = [T]_{\mathcal{B}}$. Since $\Phi_{\mathcal{B}}$ is an isomorphism, it has an inverse $(\Phi_{\mathcal{B}})^{-1}$, which is a linear transformation by Exercise 50 of Section 7.2. By (6),

$$T_A = T_A\Phi_{\mathcal{B}}(\Phi_{\mathcal{B}})^{-1} = \Phi_{\mathcal{B}}T(\Phi_{\mathcal{B}})^{-1}. \tag{7}$$

Thus by Exercise 59 of Section 7.2, T_A is invertible if and only if T is invertible, and hence A is invertible if and only if T is invertible.

Next assume that T, and hence A, is invertible, and let $C = [T^{-1}]_{\mathcal{B}}$. Apply (7) to T^{-1} and C in place of T and A to obtain

$$T_C = \Phi_{\mathcal{B}}T^{-1}(\Phi_{\mathcal{B}})^{-1}. \tag{8}$$

Take the inverse of both sides of (7) and apply Exercise 59 of Section 7.2 to obtain

$$T_{A^{-1}} = (T_A)^{-1} = \left(\Phi_{\mathcal{B}}T(\Phi_{\mathcal{B}})^{-1}\right)^{-1} = \Phi_{\mathcal{B}}T^{-1}(\Phi_{\mathcal{B}})^{-1}. \tag{9}$$

Finally, compare (8) and (9) to obtain that $T_C = T_{A^{-1}}$, and hence, $C = A^{-1}$.

We summarize these results in the following.

The Matrix Representation of the Inverse of an Invertible Linear Operator

Let T be a linear operator on a finite-dimensional vector space V with basis $\mathcal{B}$, and let $A = [T]_{\mathcal{B}}$. Then the following statements are true.

(a) T is invertible if and only if A is invertible.

(b) If T is invertible, then $[T^{-1}]_{\mathcal{B}} = A^{-1}$.

Example 11   Find an antiderivative of $e^t \sin t$.

Solution   Let $D, \mathcal{B}$, and V be as in Example 9. In the context of V, the derivative operator D is invertible because

$$[D]_{\mathcal{B}} = \begin{bmatrix} 1 & 1 \\ -1 & 1 \end{bmatrix}$$

is invertible. So our goal is to compute $D^{-1}(e^t \sin t)$. Since

$$[D^{-1}]_{\mathcal{B}} = ([D]_{\mathcal{B}})^{-1} = \begin{bmatrix} 1 & 1 \\ -1 & 1 \end{bmatrix}^{-1} = \begin{bmatrix} \frac{1}{2} & -\frac{1}{2} \\ \frac{1}{2} & \frac{1}{2} \end{bmatrix},$$

and

$$[e^t \sin t]_{\mathcal{B}} = \begin{bmatrix} 0 \\ 1 \end{bmatrix},$$

it follows that

$$[D^{-1}(e^t \sin t)]_{\mathcal{B}} = \begin{bmatrix} \frac{1}{2} & -\frac{1}{2} \\ \frac{1}{2} & \frac{1}{2} \end{bmatrix} \begin{bmatrix} 0 \\ 1 \end{bmatrix} = \begin{bmatrix} -\frac{1}{2} \\ \frac{1}{2} \end{bmatrix}.$$

Thus $D^{-1}(e^t \sin t) = (-\frac{1}{2})e^t \cos t + (\frac{1}{2})e^t \sin t$.   ○

Finally, we apply Theorem 7.6 to analyze the eigenvalues and eigenvectors of a linear operator on a finite-dimensional vector space. Let T be a linear operator on a finite-dimensional vector space V with basis $\mathcal{B}$, and let $A = [T]_{\mathcal{B}}$.

Suppose that $\mathbf{v}$ is an eigenvector of T with corresponding eigenvalue λ. Observe that $\mathbf{v} \neq \mathbf{0}$, and hence $\Phi_{\mathcal{B}}(\mathbf{v}) \neq \mathbf{0}$ since $\Phi_{\mathcal{B}}$ is an isomorphism. Furthermore, by (6),

$$A(\Phi_{\mathcal{B}}(\mathbf{v})) = T_A \Phi_{\mathcal{B}}(\mathbf{v}) = \Phi_{\mathcal{B}}(T(\mathbf{v})) = \Phi_{\mathcal{B}}(\lambda \mathbf{v}) = \lambda \Phi_{\mathcal{B}}(\mathbf{v}).$$

Therefore $\Phi_{\mathcal{B}}(\mathbf{v})$ is an eigenvector of A with corresponding eigenvalue λ.

Conversely, suppose that $\mathbf{w}$ is an eigenvector of A with corresponding eigenvalue λ. Let $\mathbf{v} = (\Phi_{\mathcal{B}})^{-1}(\mathbf{w})$. Then $\Phi_{\mathcal{B}}(\mathbf{v}) = \mathbf{w}$, and

$$A(\Phi_{\mathcal{B}}(\mathbf{v})) = A\mathbf{w} = \lambda \mathbf{w} = \lambda \Phi_{\mathcal{B}}(\mathbf{v}).$$

Therefore, by (6),

$$\Phi_{\mathcal{B}}(T(\mathbf{v})) = T_A \Phi_{\mathcal{B}}(\mathbf{v}) = A(\Phi_{\mathcal{B}}(\mathbf{v})) = \lambda \Phi_{\mathcal{B}}(\mathbf{v}) = \Phi_{\mathcal{B}}(\lambda \mathbf{v}).$$

Hence $T(\mathbf{v}) = \lambda \mathbf{v}$ because $\Phi_{\mathcal{B}}$ is an isomorphism. It follows that $\mathbf{v}$ is an eigenvector of T with corresponding eigenvalue λ.

We summarize these results in the following, replacing $\Phi_{\mathcal{B}}(\mathbf{v})$ by $[\mathbf{v}]_{\mathcal{B}}$.

Eigenvalues and Eigenvectors of a Matrix Representation of a Linear Operator

Let T be a linear operator on a finite-dimensional vector space V with basis $\mathcal{B}$, and let $A = [T]_{\mathcal{B}}$. Then a vector $\mathbf{v}$ in V is an eigenvector of T with corresponding eigenvalue λ if and only if $[\mathbf{v}]_{\mathcal{B}}$ is an eigenvector of A with corresponding eigenvalue λ.

Example 12   Let $T \colon \mathcal{P}_2 \to \mathcal{P}_2$ be defined by

$$T(p(x)) = p(0) + 3p(1)x + p(2)x^2$$

for every polynomial $p(x)$ in $\mathcal{P}_2$. For example, if $p(x) = 2 + x - 2x^2$, then $p(0) = 2$, $p(1) = 1$, and $p(2) = -4$. Therefore $T(p(x)) = 2 + 3x - 4x^2$. It can be shown that T is linear. We wish to find the eigenvalues and eigenvectors of T. Let $\mathcal{B} = \{1, x, x^2\}$, which is a basis for $\mathcal{P}_2$, and let $A = [T]_{\mathcal{B}}$. Since

$$T(1) = 1 + 3x + x^2,$$

$$T(x) = 0 + 3x + 2x^2,$$

and

$$T(x^2) = 0 + 3x + 4x^2,$$

we have

$$A = \begin{bmatrix} 1 & 0 & 0 \\ 3 & 3 & 3 \\ 1 & 2 & 4 \end{bmatrix}.$$

It can be shown that A has the characteristic polynomial $(-1)(t-1)^2(t-6)$, and so the eigenvalues of A, and hence those of T, are 1 and 6. We now determine the eigenvectors of T.

Eigenvectors corresponding to the eigenvalue 1: The eigenspace of A corresponding to $\lambda = 1$ is spanned by $\begin{bmatrix} 0 \\ -3 \\ 2 \end{bmatrix}$. Since this vector is the coordinate vector of the polynomial $p(x) = -3x + 2x^2$, the nonzero multiples of $p(x)$ are the eigenvectors of T corresponding to the eigenvalue 1.

Eigenvectors corresponding to the eigenvalue 6: The eigenspace of A corresponding to $\lambda = 6$ is spanned by $\begin{bmatrix} 0 \\ 1 \\ 1 \end{bmatrix}$. Since this vector is the coordinate vector of the polynomial $q(x) = x + x^2$, the nonzero multiples of $q(x)$ are the eigenvectors of T corresponding to the eigenvalue 6. ○

Example 13   Let $U \colon \mathcal{M}_{2 \times 2} \to \mathcal{M}_{2 \times 2}$ be the linear operator defined by $U(A) = A^T$, which is a special case of Example 6. We saw in Example 6 that 1 and -1 are eigenvalues of U. We now show that these are the only eigenvalues of U. Let

$$\mathcal{B} = \left\{ \begin{bmatrix} 1 & 0 \\ 0 & 0 \end{bmatrix}, \begin{bmatrix} 0 & 1 \\ 0 & 0 \end{bmatrix}, \begin{bmatrix} 0 & 0 \\ 1 & 0 \end{bmatrix}, \begin{bmatrix} 0 & 0 \\ 0 & 1 \end{bmatrix} \right\},$$

which is a basis for $\mathcal{M}_{2 \times 2}$. We leave it to the reader to verify that

$$[U]_{\mathcal{B}} = \begin{bmatrix} 1 & 0 & 0 & 0 \\ 0 & 0 & 1 & 0 \\ 0 & 1 & 0 & 0 \\ 0 & 0 & 0 & 1 \end{bmatrix},$$

which has the characteristic polynomial $(t - 1)^3(t + 1)$. Therefore 1 and -1 are the only eigenvalues of A, and we conclude that 1 and -1 are the only eigenvalues of U.

Practice Problems

1. Define $T: \mathcal{P}_2 \to \mathcal{P}_2$ by $T(p(x)) = (x+1)p'(x) + p(x)$.
   (a) Prove that T is linear.
   (b) Construct a matrix representation of T with respect to the basis $\mathcal{B} = \{1, \, x, \, x^2\}$ for $\mathcal{P}_2$.

2. Let $p(x) = 2 - 3x + 5x^2$. Use the linear transformation T in problem 1 to compute $T(p(x))$, first by using the rule for T, and second by using Theorem 7.6.

3. Let T be the linear transformation in problem 1.
   (a) Find the eigenvalues of T.
   (b) Show that T is invertible, and find a rule for its inverse.

Exercises

1. Determine if the following statements are true or false.

   (a) Every linear operator has an eigenvalue.
   (b) Every linear operator can be represented by a matrix.
   (c) Every linear operator on a nonzero finite-dimensional vector space can be represented by a matrix.
   (d) For any positive integer n, taking the derivative of a polynomial in $\mathcal{P}_n$ can be accomplished by matrix multiplication.
   (e) The inverse of a linear operator on a finite-dimensional vector space can be found by computing the inverse of a matrix.
   (f) The definite integral can be considered to be a linear transformation from $C([a, b])$ to the real numbers.
   (g) It is possible for a matrix to be an eigenvector of a linear operator.

In Exercises 2–6, show that the given function T is a linear transformation.

2. $T: \mathsf{C}^\infty \to \mathsf{C}^\infty$ is the differential operator defined by

$$T(f)(t) = (t^2 D^2 - (\sin t)D + 3I)f(t).$$

3. $T: \mathcal{M}_{n \times n} \to \mathcal{R}$ is defined by $T(A) = \text{trace } A$ for all A in $\mathcal{M}_{n \times n}$.

4. Fix vectors $\mathbf{u}_1, \mathbf{u}_2, \ldots, \mathbf{u}_{n-1}$ in $\mathcal{R}^n$, and define $T: \mathcal{R}^n \to \mathcal{R}$ by

$$T(\mathbf{v}) = \det [\mathbf{v} \quad \mathbf{u}_1 \quad \mathbf{u}_2 \quad \cdots \quad \mathbf{u}_{n-1}]$$

   for all $\mathbf{v}$ in $\mathcal{R}^n$.

5. Let S be a nonempty set and s an element of S. Define $T: \mathcal{F}(S) \to \mathcal{R}$ by $T(f) = f(s)$ for all f in $\mathcal{F}(S)$.

6. Fix a matrix C in $\mathcal{M}_{m \times n}$, and define $T: \mathcal{M}_{m \times n} \to \mathcal{M}_{n \times n}$ by $T(A) = A^T C$ for all A in $\mathcal{M}_{m \times n}$.

7. Fix a nonzero vector $\mathbf{u}$ in $\mathcal{R}^n$, and let $T: \mathcal{M}_{m \times n} \to \mathcal{R}^m$ be defined by $T(A) = A\mathbf{u}$.

   (a) Show that T is linear.
   (b) Show that T is onto. *Hint:* Suppose that u_i, the ith component of $\mathbf{u}$, is not 0. Let $\mathbf{v}$ be any vector in $\mathcal{R}^m$. Let A be the $m \times n$ matrix whose ith column is $(1/u_i)\mathbf{v}$ and whose other columns are zero.

8. Let D be the differential operator on $\mathcal{P}_2$.

   (a) Find the eigenvalues of D.
   (b) Find a basis for each of the corresponding eigenspaces.

9. Let T be the differential operator $D^2 + D$.

   (a) Show that 1 and e^{-t} lie in the null space of T.
   (b) Show that, for any real number a, the function e^{at} is an eigenvector of T corresponding to the eigenvalue $a^2 + a$.

10. For every integer $n > 1$, give an example of a nonzero $n \times n$ skew-symmetric matrix.

11. Let $U: \mathcal{M}_{n \times n} \to \mathcal{M}_{n \times n}$ be the linear operator defined by $U(A) = A^T$ for every $n \times n$ matrix A. Prove that 1 and -1 are the only eigenvalues of A. *Hint:* Suppose that A is a nonzero $n \times n$ matrix and λ is a scalar such that $A^T = \lambda A$. Take the transpose of both sides of this equation, and show that $A = \lambda^2 A$.

12. Let N denote the set of positive integers, and let $E: \mathcal{F}(N) \to \mathcal{F}(N)$ be defined by $E(f)(n) = f(n + 1)$.

   (a) Prove that E is a linear operator on $\mathcal{F}(N)$.
   (b) Since a sequence of real numbers is a function from N to $\mathcal{R}$, we may identify $\mathcal{F}(N)$ with the space of sequences. Recall the Fibonacci sequences defined in Section 5.5. Prove that a nonzero sequence f is a Fibonacci sequence if and only if f is an eigenvector of $E^2 - E$ with corresponding eigenvalue 1.

13. For the basis

$$\mathcal{B} = \left\{ \begin{bmatrix} 1 & 0 \\ 0 & 0 \end{bmatrix}, \begin{bmatrix} 0 & 0 \\ 1 & 0 \end{bmatrix}, \begin{bmatrix} 0 & 0 \\ 0 & 1 \end{bmatrix}, \begin{bmatrix} 0 & 1 \\ 0 & 0 \end{bmatrix} \right\}$$

for $\mathcal{M}_{2\times 2}$ and the matrix $A = \begin{bmatrix} 1 & 2 \\ 3 & 4 \end{bmatrix}$, find the coordinate vector of A relative to $\mathcal{B}$.

14. For the basis $\mathcal{B} = \{x^2, x, 1\}$ of $\mathcal{P}_2$, find the coordinate vector of the polynomial $2 + x - 3x^2$ relative to $\mathcal{B}$.

15. For the basis $\mathcal{B} = \{\cos^2 t, \sin^2 t, \sin t \cos t\}$ of the vector space $V = \operatorname{Span} \mathcal{B}$, find the coordinate vector of the function $f(t) = \sin 2t - \cos 2t$ relative to $\mathcal{B}$.

16. Let $\mathcal{B} = \{\mathbf{v}_1, \mathbf{v}_2, \ldots, \mathbf{v}_n\}$ be a basis for a vector space V. Show that, for any j, we have $[\mathbf{v}_j]_\mathcal{B} = \mathbf{e}_j$, where $\mathbf{e}_j$ is the jth standard vector in $\mathcal{R}^n$.

In Exercises 17–20, find the matrix representation $[T]_\mathcal{B}$, where T is a linear operator on the vector space V and $\mathcal{B}$ is a basis for V.

17. $\mathcal{B} = \{e^t, e^{2t}, e^{3t}\}$, $V = \operatorname{Span} \mathcal{B}$, and $T = D$, the differential operator.

18. $\mathcal{B} = \{e^t, te^t, t^2 e^t\}$, $V = \operatorname{Span} \mathcal{B}$, and $T = D$, the differential operator.

19. $\mathcal{B} = \{1, x, x^2\}$, $V = \mathcal{P}_2$, and $T(p(x)) = p(0) + 3p(1)x + p(2)x^2$ for all $p(x)$ in $\mathcal{P}_2$.

20. $V = \mathcal{M}_{2\times 2}$,

$$\mathcal{B} = \left\{ \begin{bmatrix} 1 & 0 \\ 0 & 0 \end{bmatrix}, \begin{bmatrix} 0 & 1 \\ 0 & 0 \end{bmatrix}, \begin{bmatrix} 0 & 0 \\ 1 & 0 \end{bmatrix}, \begin{bmatrix} 0 & 0 \\ 0 & 1 \end{bmatrix} \right\},$$

and $T(A) = \begin{bmatrix} 1 & 2 \\ 3 & 2 \end{bmatrix} A$ for all A in $\mathcal{M}_{2\times 2}$.

21. Use the technique in Example 10 to find the derivatives of the following polynomials.

(a) $p(x) = 6 - 4x^2$
(b) $p(x) = 2 + 3x + 5x^2$
(c) $p(x) = x^3$

22. Use the technique in Example 11 to find an antiderivative of $e^t \cos t$.

23. Let $\mathcal{B} = \{e^t, te^t, t^2 e^t\}$ be a basis for the subspace V of C^∞. Use the method in Example 11 to find antiderivatives of the following functions.

(a) te^t
(b) $t^2 e^t$
(c) $3e^t - 4te^t + 2t^2 e^t$

In Exercises 24–27, find the eigenvalues of T and a basis for each of the corresponding eigenspaces.

24. Let T be the linear operator in Exercise 18.

25. Let T be the linear operator in Exercise 17.

26. Let T be the linear operator in Exercise 20.

27. Let T be the linear operator in Exercise 19.

28. Let $\mathcal{B}$ be the basis for $\mathcal{M}_{2\times 2}$ given in Example 13, and let $T: \mathcal{M}_{2\times 2} \to \mathcal{M}_{2\times 2}$ be defined by

$$T\left(\begin{bmatrix} a & b \\ c & d \end{bmatrix} \right) = \begin{bmatrix} b & a+c \\ 0 & d \end{bmatrix}.$$

(a) Prove that T is linear.
(b) Determine the matrix representation $[T]_\mathcal{B}$.
(c) Find the eigenvalues of T.
(d) Find a basis for each eigenspace.

29. Fix a matrix B in $\mathcal{M}_{2\times 2}$, and let T be the function on $\mathcal{M}_{2\times 2}$ defined by $T(A) = (\operatorname{trace} A)B$.

(a) Prove that T is linear.
(b) Suppose that $\mathcal{B}$ is the basis for $\mathcal{M}_{2\times 2}$ given in Example 13, and that $B = \begin{bmatrix} 1 & 2 \\ 3 & 4 \end{bmatrix}$. Determine $[T]_\mathcal{B}$.
(c) Prove that if A is a nonzero matrix whose trace is zero, then A is an eigenvector of T.
(d) Prove that if A is an eigenvector of T with a corresponding nonzero eigenvalue, then A is a scalar multiple of B.

30. Let B be an $n \times n$ matrix and $T: \mathcal{M}_{n\times n} \to \mathcal{M}_{n\times n}$ be the function defined by $T(A) = BA$ for all A in $\mathcal{M}_{n\times n}$.

(a) Prove that T is linear.
(b) Prove that T is invertible if and only if B is invertible.
(c) Prove that a nonzero $n \times n$ matrix C is an eigenvector of T corresponding to eigenvalue λ if and only if λ is an eigenvalue of B and each column of C lies in the eigenspace of B corresponding to λ.

For Exercises 31–36, we define the differential operator T_a for any scalar a by $T_a = D - aI$.

31. Prove that, for any scalar b, $T_a(e^{bt}) = (b - a)e^{bt}$.

32. Use Exercise 31 to prove that $\{e^{b_1 t}, e^{b_2 t}\}$ is linearly independent if $b_1 \neq b_2$, by applying T_{b_1} to both sides of the equation $c_1 e^{b_1 t} + c_2 e^{b_2 t} = 0$, where c_1 and c_2 are any scalars.

33. Prove that $\{e^{b_1 t}, e^{b_2 t}, \ldots, e^{b_k t}\}$ is linearly independent if the b_i are distinct. This generalizes Exercise 32.

34. Consider the differential equation $y' + 3y = 0$.

(a) Express the differential equation in the form $T_a(y) = \mathbf{0}$ for some scalar a.
(b) Use (a) and Exercise 31 to show that e^{-3t} is a solution to the differential equation.

35. For any scalars a and b, prove that $T_a T_b = T_b T_a$; that is, $T_a T_b(f) = T_b T_a(f)$ for every f in C^∞.

36. Consider the differential equation $y'' + 2y - 8 = 0$.

(a) Express the differential equation in the form $T_b T_a(y) = \mathbf{0}$ for some scalars a and b.
(b) Use (a) and Exercises 35 and 31 to show that e^{2t} and e^{-4t} are solutions to the differential equation.
(c) Assume that the dimension of the solution set of the differential equation equals 2, and prove that every solution to the differential equation is of the form $ce^{2t} + de^{-4t}$, where c and d are scalars.
(d) Use (c) to find a solution to the differential equation that satisfies $y(0) = 1$ and $y'(0) = 2$.

37. Let V be a finite-dimensional vector space with basis $\mathcal{B}$. Prove that, for any linear operators T and U on V,
$$[UT]_\mathcal{B} = [U]_\mathcal{B}[T]_\mathcal{B}.$$
Hint: Apply Theorem 7.6 to $(UT)\mathbf{v}$ and $U(T(\mathbf{v}))$, where $\mathbf{v}$ is an arbitrary vector in V.

38. Let T be a linear operator on a vector space V, and suppose that λ is an eigenvalue of T. Prove that the eigenspace of T corresponding to λ is a subspace of V and consists of the zero vector and the eigenvectors of T corresponding to λ.

39. Let $T: V \to W$ be a linear tranformation between vector spaces V and W. Prove that the null space of T is a subspace of V.

40. Let $T: V \to W$ be a linear tranformation between vector spaces V and W. Prove that the range of T, the set of all images $T(\mathbf{v})$ for $\mathbf{v}$ in V, is a subspace of W.

41. Let $T: V \to W$ be a linear tranformation between finite-dimensional vector spaces V and W. Prove that the sum of the dimensions of the null space and the range of T equals dim V.

42. Recall the set C^∞ defined in Example 1.
   (a) Prove that C^∞ is a subspace of $\mathcal{F}(\mathcal{R})$.
   (b) Let $T: C^\infty \to C^\infty$ be defined by $T(f)(t) = e^t f''(t)$ for all t in $\mathcal{R}$. Prove that T is linear.

43. Recall the set $C([a, b])$ defined in Example 2.
   (a) Prove that $C([a, b])$ is a subspace of $\mathcal{F}([a, b])$.

(b) Let $T: C([a, b]) \to C([a, b])$ be defined by
$$T(f)(x) = \int_a^x f(t)\, dt \qquad \text{for } a \le x \le b.$$
Prove that T is linear and one-to-one.

In Exercises 44 and 45, use a calculator with matrix capabilities or computer software such as MATLAB *to solve the problem.*

44. Let T be the linear operator on $\mathcal{P}_3$ defined by
$$T(f(x)) = f(x) + f'(x) + f''(x) + f(0) + f(2)x^2$$
for all $f(x)$ in $\mathcal{P}_3$.
   (a) Determine the eigenvalues of T.
   (b) Find a basis for $\mathcal{P}_3$ consisting of eigenvectors of T.
   (c) For $f(x) = a_0 + a_1 x + a_2 x^2 + a_3 x^3$, find $T^{-1}(f(x))$.

45. Let T be the linear operator on $\mathcal{M}_{2\times 2}$ defined by
$$T(A) = \begin{bmatrix} 1 & 2 \\ 3 & 4 \end{bmatrix} A + 3A^T$$
for all A in $\mathcal{M}_{2\times 2}$.
   (a) Determine the eigenvalues of T.
   (b) Find a basis for $\mathcal{M}_{2\times 2}$ consisting of eigenvectors of T.
   (c) For $A = \begin{bmatrix} a & b \\ c & d \end{bmatrix}$, find $T^{-1}(A)$.

Solutions to the Practice Problems

1. (a) Let $q(x)$ and $r(x)$ be polynomials in $\mathcal{P}_2$ and let c be a scalar. Then
$$T(q(x) + r(x)) = (x + 1)(q(x) + r(x))'$$
$$+ (q(x) + r(x))$$
$$= (x + 1)(q'(x) + r'(x))$$
$$+ (q(x) + r(x))$$
$$= (x + 1)q'(x) + (x + 1)r'(x)$$
$$+ q(x) + r(x)$$
$$= ((x + 1)q'(x) + q(x))$$
$$+ ((x + 1)r'(x) + r(x))$$
$$= T(q(x)) + T(r(x))$$
and
$$T(cq(x)) = (x + 1)(cq(x))' + cq(x)$$
$$= c((x + 1)q'(x) + q(x))$$
$$= cT(q(x)).$$
So T is linear.

(b) We have
$$T(1) = (x + 1)(0) + 1 = 1,$$
$$T(x) = (x + 1)(1) + x = 1 + 2x,$$
and
$$T(x^2) = (x + 1)(2x) + x^2 = 2x + 3x^2.$$
So
$$[T(1)]_\mathcal{B} = \begin{bmatrix} 1 \\ 0 \\ 0 \end{bmatrix}, \quad [T(x)]_\mathcal{B} = \begin{bmatrix} 1 \\ 2 \\ 0 \end{bmatrix}, \quad \text{and}$$
$$[T(x^2)]_\mathcal{B} = \begin{bmatrix} 0 \\ 2 \\ 3 \end{bmatrix}.$$
Therefore
$$[T]_\mathcal{B} = \begin{bmatrix} 1 & 1 & 0 \\ 0 & 2 & 2 \\ 0 & 0 & 3 \end{bmatrix}.$$

2. Using the rule for T, we have
$$T(p(x)) = T(2 - 3x + 5x^2)$$
$$= (x + 1)(-3 + 10x) + (2 - 3x + 5x^2)$$
$$= -1 + 4x + 15x^2.$$

Using Theorem 7.6, we have

$$[T(2 - 3x + 5x^2)]_\mathcal{B} = [T]_\mathcal{B}[2 - 3x + 5x^2]_\mathcal{B}$$

$$= \begin{bmatrix} 1 & 1 & 0 \\ 0 & 2 & 2 \\ 0 & 0 & 3 \end{bmatrix} \begin{bmatrix} 2 \\ -3 \\ 5 \end{bmatrix} = \begin{bmatrix} -1 \\ 4 \\ 15 \end{bmatrix},$$

and hence $T(p(x)) = -1 + 4x + 15x^2$.

3. (a) The eigenvalues of T are the same as the eigenvalues of $[T]_\mathcal{B}$. But $[T]_\mathcal{B}$ is an upper triangular matrix, so its eigenvalues are its diagonal entries, that is, 1, 2, and 3.

   (b) Because 0 is not an eigenvalue of T, we have that T is invertible. To find the rule for T^{-1}, we can use the

result that $[T^{-1}]_\mathcal{B} = [T]_\mathcal{B}^{-1}$. So

$$[T^{-1}(a + bx + cx^2)]_\mathcal{B} = [T^{-1}]_\mathcal{B} \begin{bmatrix} a \\ b \\ c \end{bmatrix} = [T]_\mathcal{B}^{-1} \begin{bmatrix} a \\ b \\ c \end{bmatrix}$$

$$= \begin{bmatrix} 1 & -\frac{1}{2} & \frac{1}{3} \\ 0 & \frac{1}{2} & -\frac{1}{3} \\ 0 & 0 & \frac{1}{3} \end{bmatrix} \begin{bmatrix} a \\ b \\ c \end{bmatrix} = \begin{bmatrix} a - \frac{1}{2}b + \frac{1}{3}c \\ \frac{1}{2}b - \frac{1}{3}c \\ \frac{1}{3}c \end{bmatrix}.$$

Therefore

$$T^{-1}(a + bx + cx^2)$$

$$= \left(a - \frac{1}{2}b + \frac{1}{3}c\right) + \left(\frac{1}{2}b - \frac{1}{3}c\right)x + \left(\frac{1}{3}c\right)x^2.$$

7.4 Inner Product Spaces

The dot product introduced in Chapter 6 provides a stronger link between vectors and matrices and the geometry of $\mathcal{R}^n$. For example, we saw how the concept of dot product leads to deep and elegant results about symmetric matrices.

In certain vector spaces, especially function spaces, there are scalar-valued products, called *inner products*, that share the important formal properties of dot products. These inner products allow us to define and apply such concepts as *distance* and *orthogonality* to vector spaces.

Definition.   Let V be a vector space. An **inner product** on V is a real-valued function that assigns to any pair of vectors **u** and **v** a scalar, denoted $\langle \mathbf{u}, \mathbf{v} \rangle$, such that, for any vectors **u**, **v**, and **w** in V and any scalar a, the following axioms hold.

Axioms of an Inner Product

1. $\langle \mathbf{u}, \mathbf{u} \rangle > 0$ if $\mathbf{u} \neq \mathbf{0}$
2. $\langle \mathbf{u}, \mathbf{v} \rangle = \langle \mathbf{v}, \mathbf{u} \rangle$
3. $\langle \mathbf{u} + \mathbf{v}, \mathbf{w} \rangle = \langle \mathbf{u}, \mathbf{w} \rangle + \langle \mathbf{v}, \mathbf{w} \rangle$
4. $\langle a\mathbf{u}, \mathbf{v} \rangle = a \langle \mathbf{u}, \mathbf{v} \rangle$

Suppose that $\langle \mathbf{u}, \mathbf{v} \rangle$ is an inner product on a vector space V. For any scalar $r > 0$, defining $\langle\langle \mathbf{u}, \mathbf{v} \rangle\rangle$ by $\langle\langle \mathbf{u}, \mathbf{v} \rangle\rangle = r \langle \mathbf{u}, \mathbf{v} \rangle$ gives another inner product on V. Thus there can be infinitely many different inner products on a vector space. A vector space endowed with a particular inner product is called an **inner product space**.

The dot product for $\mathcal{R}^n$ is a familiar example of an inner product, where $\langle \mathbf{u}, \mathbf{v} \rangle = \mathbf{u} \cdot \mathbf{v}$ for **u** and **v** in $\mathcal{R}^n$. Notice that the axioms of an inner product are proved in Theorem 6.1 on page 313.

As is the case for vector spaces, many facts about dot products are valid for inner products. Often a proof of a result for inner product spaces requires little or no modification of the proof for the dot product on $\mathcal{R}^n$. What follows are additional examples of inner products.

Example 1

Let $C([a, b])$ denote the vector space of continuous real-valued functions defined on the closed interval $[a, b]$, which was described in Example 2 of Section 7.3. For f and g in $C([a, b])$, let

$$\langle f, g \rangle = \int_a^b f(t)g(t)\, dt.$$

This definition determines an inner product on $C([a, b])$.

To verify axiom 1, let f be any nonzero function in $C([a, b])$. Then f^2 is continuous and nonnegative and, furthermore, $f^2(t_0) > 0$ for some t_0 in $C([a, b])$. Since f^2 is continuous, there is an interval containing t_0 of length $r > 0$, and there is a number $p > 0$ such that $f^2(t) > p$ on that interval. Therefore,

$$\langle f, f \rangle = \int_a^b f^2(t)\, dt \geq r \cdot p > 0.$$

To verify axiom 2, let f and g be functions in $C([a, b])$. Then

$$\langle f, g \rangle = \int_a^b f(t)g(t)\, dt = \int_a^b g(t)f(t)\, dt = \langle g, f \rangle.$$

We leave the verifications of axioms 3 and 4 as exercises.　　　　　　　○

Example 2

Recall the definition of *trace* given in Exercise 46 of Section 1.1. For A and B in $\mathcal{M}_{n \times n}$, the vector space of $n \times n$ matrices, define

$$\langle A, B \rangle = \text{trace}\,(AB^T).$$

This definition determines an inner product, called the **Frobenius inner product**[4] on $\mathcal{M}_{n \times n}$.

To verify axiom 1, let A be any nonzero matrix, and let $C = AA^T$. Then

$$\langle A, A \rangle = \text{trace}\,(AA^T) = \text{trace}\, C$$

$$= c_{11} + c_{22} + \cdots + c_{nn}.$$

Furthermore, for each i,

$$c_{ii} = a_{i1}^2 + a_{i2}^2 + \cdots + a_{in}^2.$$

It follows that $\langle A, A \rangle$ is the sum of squares of all of the entries of A. Since $A \neq O$, it follows that $(a_{ij})^2 > 0$ for some i and j, and hence $\langle A, A \rangle > 0$.

To verify axiom 2, let A and B be matrices in V. Then by Exercise 46 of Section 1.1 and Theorem 2.2 on page 88,

$$\langle A, B \rangle = \text{trace}\,(AB^T)$$

$$= \text{trace}\,(AB^T)^T$$

$$= \text{trace}\,(BA^T)$$

$$= \langle B, A \rangle.$$

---

[4]Ferdinand Georg Frobenius (1849–1917) was a German mathematician best known for his work in group theory. His research combined results from the theory of algebraic equations, number theory, and geometry. His representation theory for finite groups made important contributions to quantum mechanics.

We leave the verifications of axioms 3 and 4 as exercises.

It can be shown that the Frobenius inner product of two $n \times n$ matrices is simply the sum of the products of their corresponding entries (see Exercises 33 and 34). Thus the Frobenius inner product looks like an ordinary inner product in $\mathcal{R}^n$, except that the components are entries in a matrix. For example,

$$\left\langle \begin{bmatrix} 1 & 2 \\ 3 & 4 \end{bmatrix}, \begin{bmatrix} 5 & 6 \\ 7 & 8 \end{bmatrix} \right\rangle = 1 \cdot 5 + 2 \cdot 6 + 3 \cdot 7 + 4 \cdot 8 = 70.$$

As with the dot product on $\mathcal{R}^n$, we can define the *length* of a vector in an inner product space. For any vector $\mathbf{v}$ in an inner product space V, the **norm** or **length** of $\mathbf{v}$, denoted $\|\mathbf{v}\|$, is defined by

$$\|\mathbf{v}\| = \sqrt{\langle \mathbf{v}, \mathbf{v} \rangle}.$$

The **distance** between two vectors $\mathbf{u}$ and $\mathbf{v}$ in V is defined in the usual way as $\|\mathbf{u} - \mathbf{v}\|$.

As stated earlier, many of the elementary properties of dot products are also valid for all inner products. In particular, all of the items of Theorem 6.1 are valid for inner product spaces. For example, the analog of Theorem 6.1(d) for inner products follows from axioms 2 and 3 of inner products. For let $\mathbf{u}, \mathbf{v}$, and $\mathbf{w}$ be vectors in an inner product space. Then

$$\begin{aligned} \langle \mathbf{u}, \mathbf{v} + \mathbf{w} \rangle &= \langle \mathbf{v} + \mathbf{w}, \mathbf{u} \rangle && \text{(by axiom 2)} \\ &= \langle \mathbf{v}, \mathbf{u} \rangle + \langle \mathbf{w}, \mathbf{u} \rangle && \text{(by axiom 3)} \\ &= \langle \mathbf{u}, \mathbf{v} \rangle + \langle \mathbf{u}, \mathbf{w} \rangle. && \text{(by axiom 2)} \end{aligned}$$

The Cauchy–Schwarz inequality (Theorem 6.3 on page 316) and the triangle inequality (Theorem 6.4 on page 317) are valid for all inner product spaces because their proofs are also based on the items of Theorem 6.1 that correspond to the axioms of an inner product. Thus, in Example 1, we can obtain an inequality about the integrals of functions in $C([a, b])$ by applying the Cauchy–Schwartz inequality and squaring both sides:

$$\left[\int_a^b f(t)g(t)\,dt \right]^2 \leq \left[\int_a^b f^2(t)\,dt \right] \left[\int_a^b g^2(t)\,dt \right],$$

where f and g are continuous functions on the closed interval $[a, b]$.

Orthogonality and the Gram–Schmidt Process

Let V be an inner product space. As in Chapter 6, two vectors $\mathbf{u}$ and $\mathbf{v}$ in V are called **orthogonal** if $\langle \mathbf{u}, \mathbf{v} \rangle = 0$, and a subset S of V is **orthogonal** if any two distinct vectors in S are orthogonal. Again a vector $\mathbf{u}$ in V is called a **unit vector** if $\|\mathbf{u}\| = 1$. An orthogonal subset S of V is called **orthonormal** if S is an orthogonal set and every vector in S is a unit vector. Any nonzero vector $\mathbf{v}$ in an inner product space can be replaced by the unit vector $\dfrac{1}{\|\mathbf{v}\|}\mathbf{v}$, which is a scalar multiple of $\mathbf{v}$, called its **normalized vector**. If S is an orthogonal set of nonzero vectors, then replacing every vector in S by its normalized vector results in an orthonormal set that spans the same subspace as S.

It was shown in Section 6.2 that any finite orthogonal set of nonzero vectors is linearly independent (Theorem 6.5 on page 322). This result is valid for any inner product space, and the proof is identical. Furthermore, we can show that this result is also valid for infinite orthogonal sets (see Exercise 36).

Example 3   Let $f(t) = \sin 3t$ and $g(t) = \cos 2t$ be defined on the closed interval $[0, 2\pi]$. Then f and g are functions in the inner product space $C([0, 2\pi])$ of Example 1. Show that f and g are orthogonal.

Solution   We apply the trigonometric identity

$$\sin \alpha \cos \beta = \frac{1}{2}[\sin(\alpha + \beta) + \sin(\alpha - \beta)]$$

for $\alpha = 3t$ and $\beta = 2t$ to obtain

$$\langle f, g \rangle = \int_0^{2\pi} \sin 3t \cos 2t \, dt$$

$$= \frac{1}{2} \int_0^{2\pi} [\sin 5t + \sin t] \, dt$$

$$= \frac{1}{2} \left[-\frac{1}{5} \cos 5t - \cos t \right] \Big|_0^{2\pi}$$

$$= 0,$$

and hence f and g are orthogonal.

Example 4   Recall the vector space of trigonometric polynomials $T[0, 2\pi]$ defined on page 408. This space is defined as the span of a set S of trigonometric functions defined on $[0, 2\pi]$, namely

$$S = \{1, \cos t, \sin t, \cos 2t, \sin 2t, \ldots, \cos nt, \sin nt, \ldots\}.$$

We assert that S is an orthogonal set. To prove this, we must show that any two distinct functions f and g in S are orthogonal. If $f(t) = 1$ and $g(t) = \cos nt$ for some positive integer n, then

$$\langle f, g \rangle = \int_0^{2\pi} \cos nt \, dt = \frac{1}{n} \sin nt \Big|_0^{2\pi} = 0.$$

In a similar manner, if $f(t) = \sin mt$ and $g(t) = 1$, then $\langle f, g \rangle = 0$.

If $f(t) = \sin mt$ and $g(t) = \cos nt$ for positive integers m and n, we can apply the trigonometric identity in Example 3 (we omit the details) to obtain that $\langle f, g \rangle = 0$. The other two cases are treated in the exercises (see Exercises 18 and 19).

Since S is orthogonal and its members are nonzero, S is linearly independent and hence is a basis for $T[0, 2\pi]$.

A computational method for converting a linearly independent subset of $\mathcal{R}^n$ into an orthogonal set, the Gram–Schmidt process, was introduced in

Section 6.2 (see Theorem 6.6 on page 323). The Gram–Schmidt process is also valid for any inner product space, and its justification is identical to the proof of Theorem 6.6. We can use the Gram–Schmidt process to replace an arbitrary basis for a finite-dimensional inner product space with an orthogonal or an orthonormal basis. It follows that *every finite-dimensional inner product space has an orthonormal basis.*

Example 5   Define an inner product on $\mathcal{P}_2$ by

$$\langle f(x), g(x) \rangle = \int_{-1}^{1} f(t)g(t)\, dt$$

for all polynomials $f(x)$ and $g(x)$ in $\mathcal{P}_2$. (It can be verified that this does indeed define an inner product for $\mathcal{P}_2$. See, for example, the argument in Example 1.) Use the Gram–Schmidt process to convert the basis $\{1, x, x^2\}$ into an orthogonal basis for $\mathcal{P}_2$. Then normalize the vectors of this orthogonal basis to obtain an orthonormal basis for $\mathcal{P}_2$.

Solution   To facilitate the following computations, we use the notation of Theorem 6.6. Let $\mathbf{u}_1 = \mathbf{v}_1 = 1$, $\mathbf{u}_2 = x$, and $\mathbf{u}_3 = x^2$. Then

$$\mathbf{v}_2 = \mathbf{u}_2 - \frac{\langle \mathbf{u}_2, \mathbf{v}_1 \rangle}{\|\mathbf{v}_1\|^2} \mathbf{v}_1$$

$$= x - \frac{\displaystyle\int_{-1}^{1} t \cdot 1\, dt}{\displaystyle\int_{-1}^{1} 1^2\, dt} \quad (1)$$

$$= x - 0 \cdot 1$$

$$= x$$

and

$$\mathbf{v}_3 = \mathbf{u}_3 - \frac{\langle \mathbf{u}_3, \mathbf{v}_1 \rangle}{\|\mathbf{v}_1\|^2} \mathbf{v}_1 - \frac{\langle \mathbf{u}_3, \mathbf{v}_2 \rangle}{\|\mathbf{v}_2\|^2} \mathbf{v}_2$$

$$= x^2 - \frac{\displaystyle\int_{-1}^{1} t^2 \cdot 1\, dt}{\displaystyle\int_{-1}^{1} 1^2\, dt}\, (1) - \frac{\displaystyle\int_{-1}^{1} t^2 \cdot t\, dt}{\displaystyle\int_{-1}^{1} t^2\, dt}\, (x)$$

$$= x^2 - \frac{\frac{2}{3}}{2} \cdot 1 - 0 \cdot x$$

$$= x^2 - \frac{1}{3}.$$

Thus the set $\{1, x, x^2 - \frac{1}{3}\}$ is an orthogonal basis for $\mathcal{P}_2$.

Next we normalize each vector in this set to obtain an orthonormal basis for $\mathcal{P}_2$. Since

$$\|\mathbf{v}_1\| = \sqrt{\int_{-1}^{1} 1^2\, dx} = \sqrt{2},$$

$$\|\mathbf{v}_2\| = \sqrt{\int_{-1}^{1} x^2\, dx} = \sqrt{\frac{2}{3}},$$

and

$$\|\mathbf{v}_3\| = \sqrt{\int_{-1}^{1} \left(x^2 - \frac{1}{3}\right)^2 dx} = \sqrt{\frac{8}{45}},$$

the desired orthonormal basis for $\mathcal{P}_2$ is

$$\left\{\frac{1}{\|\mathbf{v}_1\|}\mathbf{v}_1, \frac{1}{\|\mathbf{v}_2\|}\mathbf{v}_2, \frac{1}{\|\mathbf{v}_3\|}\mathbf{v}_3\right\} = \left\{\frac{1}{\sqrt{2}}, \sqrt{\frac{3}{2}}\,x, \sqrt{\frac{45}{8}}\left(x^2 - \frac{1}{3}\right)\right\}.$$

The example above can be extended to $\mathcal{P}_n$ for any positive integer n by using the same inner product and choosing $\{1, x, \ldots, x^n\}$ as the initial basis. As the Gram–Schmidt process is applied to polynomials of higher degree, the polynomials of lower degree remain unchanged. Thus we obtain an infinite sequence of polynomials $p_0(x), p_1(x), \ldots, p_n(x), \ldots$ such that, for any n, the first $n + 1$ polynomials in the sequence form an orthonormal basis for $\mathcal{P}_n$. These polynomials, called the **normalized Legendre polynomials**,[5] form an orthonormal basis for the infinite-dimensional space $\mathcal{P}$. They have applications to differential equations, statistics, and numerical analysis. In Example 5, we computed the first three normalized Legendre polynomials.

Orthogonal Projections and Least-Squares Approximation

Suppose that V is an inner product space and that W is a finite-dimensional subspace of V. Select an orthonormal basis $\mathcal{B} = \{\mathbf{v}_1, \mathbf{v}_2, \ldots, \mathbf{v}_n\}$ for W. Because the proof of Theorem 6.7 on page 329 applies directly to this context, we assume the result here. Thus, for any vector $\mathbf{v}$ in V, there exist unique vectors $\mathbf{w}$ in W and $\mathbf{z}$ in $W^{\perp}$ such that $\mathbf{v} = \mathbf{w} + \mathbf{z}$. Furthermore,

$$\mathbf{w} = \langle \mathbf{v}, \mathbf{v}_1 \rangle \mathbf{v}_1 + \langle \mathbf{v}, \mathbf{v}_2 \rangle \mathbf{v}_2 + \cdots + \langle \mathbf{v}, \mathbf{v}_n \rangle \mathbf{v}_n. \tag{10}$$

The vector $\mathbf{w}$ is called the **orthogonal projection** of $\mathbf{v}$ onto W. In 10, the representation for $\mathbf{w}$ is independent of the selection of the orthonormal basis $\mathcal{B}$ because the orthogonal projection $\mathbf{w}$ is unique.

Of particular interest to us is the *closest vector property* of orthogonal projections, which is stated and justified on page 331. In the notation above, *among all the vectors in W, the vector closest to $\mathbf{v}$ is $\mathbf{w}$, the orthogonal projection of $\mathbf{v}$ onto W.*

---

[5]Adrien Marie Legendre (1752–1833) was a French mathematician who taught at the Ècole Militaire and the Ècole Normale in Paris. He is best known for his writing on elliptic functions but also produced important results in number theory such as the law of quadratic reciprocity. His paper *Nouvelles Méthodes pour la détermination des orbites des comètes* contained the first mention of the method of least squares.

If the inner product space V is a function space, then the closest vector property can be used to express the best approximation to a function in V as a linear combination of some specified finite set of functions. Here "best" means nearest, as measured by the distance between two vectors in V. Since the inner product of two functions in a function space is usually defined by an integral of the product of the functions, a distance usually involves an integral of a square of a function. For this reason, the orthogonal projection, which serves as the approximation to a function, is called the **least-squares approximation** of the function. We illustrate the technique in the following examples.

Example 6   Find the least-squares approximation to the function $f(x) = \sqrt[3]{x}$, with domain restricted to $[-1, 1]$, as a polynomial of degree less than or equal to 2.

Solution   The function f described above lies in the function space of Example 1 with $a = -1$ and $b = 1$, namely, $C([-1, 1])$. Considering polynomials as functions and restricting their domains to $[-1, 1]$, we may view $\mathcal{P}_2$ as a finite-dimensional subspace of $C([-1, 1])$. Then the required least-squares approximation to f is the orthogonal projection of f onto $\mathcal{P}_2$. We apply 10 with $\mathbf{w} = f$, and with the $\mathbf{v}_i$ as the vectors of the orthonormal basis for $\mathcal{P}_2$ obtained in Example 5. Thus we set

$$\mathbf{v}_1 = \frac{1}{\sqrt{2}}, \qquad \mathbf{v}_2 = \sqrt{\frac{3}{2}}\, x, \qquad \text{and} \qquad \mathbf{v}_3 = \sqrt{\frac{5}{2}}\left(x^2 - \frac{1}{3}\right)$$

to obtain

$$\mathbf{w} = \langle \mathbf{v}, \mathbf{v}_1 \rangle\, \mathbf{v}_1 + \langle \mathbf{v}, \mathbf{v}_2 \rangle\, \mathbf{v}_2 + \langle \mathbf{v}, \mathbf{v}_3 \rangle\, \mathbf{v}_3$$

$$= \left(\int_{-1}^{1} \sqrt[3]{x} \cdot \frac{1}{\sqrt{2}}\, dx \right) \frac{1}{\sqrt{2}} + \left(\int_{-1}^{1} \sqrt[3]{x} \cdot \sqrt{\frac{3}{2}}\, x\, dx \right) \sqrt{\frac{3}{2}}\, x$$

$$+ \left(\int_{-1}^{1} \sqrt[3]{x} \cdot \sqrt{\frac{45}{8}}\left(x^2 - \frac{1}{3}\right) dx \right) \sqrt{\frac{45}{8}}\left(x^2 - \frac{1}{3}\right)$$

$$= 0 \cdot \frac{1}{\sqrt{2}} + \frac{6}{7}\sqrt{\frac{3}{2}}\sqrt{\frac{3}{2}}\, x + 0 \cdot \sqrt{\frac{45}{8}}\left(x^2 - \frac{1}{3}\right)$$

$$= \frac{9}{7}x.$$

Thus the function $g(x) = \frac{9}{7}x$, which is the orthogonal projection of $f(x) = \sqrt[3]{x}$ onto $\mathcal{P}_2$ with respect to the inner product used here, is the least-squares approximation of f as a linear combination of 1, x, and x^2. This means that for any polynomial $p(x)$ of degree less than or equal to 2, if $p(x) \neq g(x)$, then

$$\|f - p\| > \|f - g\|$$

(see Figure 7.2).   ○

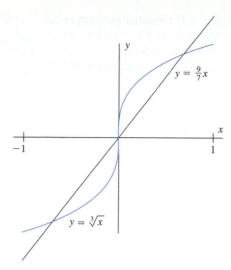

FIGURE 7.2

Approximation by Trigonometric Polynomials

A function $y = f(t)$ is called **periodic of period** p if $f(t + p) = f(t)$ for all t. Periodic functions are used to model phenomena that regularly repeat. An example of this is a vibration that generates a sound at a fixed pitch or frequency. The frequency is the number of vibrations per unit time, usually measured in seconds. In this case, we can define a function f of time t so that $f(t)$ is the relative pressure caused by the sound at time t on some fixed location, such as on the diaphragm of a microphone. For example, a musical instrument that plays a sustained note of middle C vibrates at the rate of 256 cycles per second. The length of one cycle, therefore, is $1/256$ of a second. So the function f associated with this sound has a period of $1/256$ seconds, that is, $f(t + \frac{1}{256}) = f(t)$ for all t.

We can use orthogonal projections to obtain least-squares approximations of periodic functions using trigonometric polynomials. Suppose that $y = f(t)$ is a continuous periodic function. To simplify matters, we adjust the units of t so that f has period 2π. Then we may regard f and all trigonometric polynomials as continuous functions on $[0, 2\pi]$, that is, as vectors in $C([0, 2\pi])$. The least squares approximations of f of interest are the orthogonal projections of f onto particular finite-dimensional subspaces of trigonometric polynomials.

For each positive integer n, let

$$S_n = \{1, \cos t, \sin t, \cos 2t, \sin 2t, \ldots, \cos nt, \sin nt\}.$$

Then S_n spans a finite-dimensional subspace of trigonometric polynomials, which we denote by W_n. Furthermore, S_n is an orthogonal set, as we saw in Example 4. We can normalize each function in S_n to obtain an orthonormal basis for W_n, which can be used to compute the orthogonal projection of f on W_n. For this purpose, we compute the norms of the functions in S_n:

$$\|1\| = \sqrt{\int_0^{2\pi} 1 \, dt} = \sqrt{2\pi},$$

and for each positive integer k,

$$\| \cos kt \| = \sqrt{\int_0^{2\pi} \cos^2 kt \, dt}$$

$$= \sqrt{\frac{1}{2} \int_0^{2\pi} (1 + \cos 2kt) \, dt}$$

$$= \sqrt{\frac{1}{2} \left(t + \frac{1}{2k} \sin 2kt \right) \Bigg|_0^{2\pi}}$$

$$= \sqrt{\pi}.$$

Similarly, $\| \sin kt \| = \sqrt{\pi}$ for every positive integer k. If we divide each function in $\mathcal{S}_n$ by its norm, we obtain the orthonormal basis

$$\mathcal{B}_n = \left\{ \frac{1}{\sqrt{2\pi}}, \frac{1}{\sqrt{\pi}} \cos t, \frac{1}{\sqrt{\pi}} \sin t, \frac{1}{\sqrt{\pi}} \cos 2t, \frac{1}{\sqrt{\pi}} \sin 2t, \dots, \right.$$

$$\left. \frac{1}{\sqrt{\pi}} \cos nt, \frac{1}{\sqrt{\pi}} \sin nt \right\}$$

for W_n. We can use $\mathcal{B}_n$ to compute the orthogonal projection of f onto W_n, in order to obtain the least-squares approximation as described earlier in this section. The following example illustrates this approach.

Example 7   Consider a sound at a fixed frequency that causes fluctuations in pressure at a fixed point in space which are described by a graph in the shape of sawteeth (that is, like the edge of a saw—see Figure 7.3).

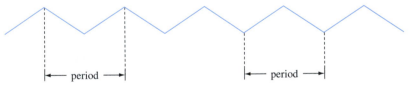

FIGURE 7.3
Sawtooth tone.

To simplify our computations, we adjust the units of time and relative pressure so that the function describing the relative pressure has a period of 2π and varies between 1 and -1. Furthermore, we select $t = 0$ at a maximum value of the relative pressure. Thus we obtain a function $y = f(t)$ defined on $[0, 2\pi]$ by

$$f(t) = \begin{cases} 1 - \dfrac{2}{\pi} t & \text{if } 0 \le t \le \pi \\[2mm] \dfrac{2}{\pi} t - 3 & \text{if } \pi \le t \le 2\pi \end{cases}$$

(see Figure 7.4).

For each positive integer n, let f_n be the orthogonal projection of f onto W_n. We can compute f_n using (10) in conjunction with the orthonormal

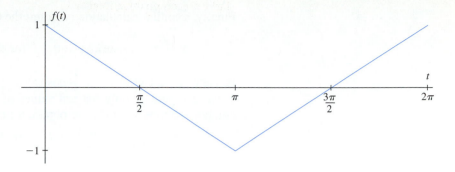

FIGURE 7.4

One period of the sawtooth tone.

basis $\mathcal{B}_n$:

$$f_n = \left\langle f, \frac{1}{\sqrt{2\pi}} \right\rangle \frac{1}{\sqrt{2\pi}} +$$

$$\left\langle f, \frac{1}{\sqrt{\pi}} \cos t \right\rangle \frac{1}{\sqrt{\pi}} \cos t + \left\langle f, \frac{1}{\sqrt{\pi}} \sin t \right\rangle \frac{1}{\sqrt{\pi}} \sin t + \cdots$$

$$+ \left\langle f, \frac{1}{\sqrt{\pi}} \cos nt \right\rangle \frac{1}{\sqrt{\pi}} \cos nt + \left\langle f, \frac{1}{\sqrt{\pi}} \sin nt \right\rangle \frac{1}{\sqrt{\pi}} \sin nt. \quad (11)$$

To find f_n, we must calculate the inner products in (11). First,

$$\left\langle f, \frac{1}{\sqrt{2\pi}} \right\rangle = \frac{1}{\sqrt{2\pi}} \int_0^{\pi} \left(1 - \frac{2}{\pi} t \right) dt + \frac{1}{\sqrt{2\pi}} \int_{\pi}^{2\pi} \left(\frac{2}{\pi} t - 3 \right) dt$$

$$= 0 + 0$$

$$= 0.$$

Next, for each positive integer k, we use integration by parts to compute

$$\left\langle f, \frac{1}{\sqrt{\pi}} \cos kt \right\rangle = \frac{1}{\sqrt{\pi}} \int_0^{\pi} \left(1 - \frac{2}{\pi} t \right) \cos kt \, dt$$

$$+ \frac{1}{\sqrt{\pi}} \int_{\pi}^{2\pi} \left(\frac{2}{\pi} t - 3 \right) \cos kt \, dt$$

$$= \frac{1}{\sqrt{\pi}} \left[\frac{-2(-1)^k}{\pi k^2} + \frac{2}{\pi k^2} \right] + \frac{1}{\sqrt{\pi}} \left[\frac{2}{\pi k^2} - \frac{-2(-1)^k}{\pi k^2} \right]$$

$$= \frac{4}{\pi \sqrt{\pi} k^2} (1 - (-1)^k)$$

$$= \begin{cases} \dfrac{4}{\pi \sqrt{\pi} k^2} & \text{if } k \text{ is odd} \\ 0 & \text{if } k \text{ is even.} \end{cases}$$

Finally, a similar calculation (we omit the details) shows that

$$\left\langle f, \frac{1}{\sqrt{\pi}} \sin kt \right\rangle = 0 \qquad \text{for every positive integer } k.$$

In view of the fact that $\left\langle f, \frac{1}{\sqrt{\pi}} \cos kt \right\rangle = \left\langle f, \frac{1}{\sqrt{\pi}} \sin kt \right\rangle = 0$ for even integers k, we compute f_n only for odd values of n. Substituting the inner products computed above into (11), we obtain, for every odd positive integer n,

$$f_n(t) = \frac{8}{\pi^2} \left[\frac{\cos t}{1^2} + \frac{\cos 3t}{3^2} + \cdots + \frac{\cos nt}{n^2} \right].$$

Figure 7.5 allows us to compare the graphs of f with three least-squares approximations, f_1, f_5 and f_{15}, obtained by taking orthogonal projections of f onto W_1, W_5, and W_{15}, respectively. Notice that as n increases, the graph of f_n more closely approximates the graph of f.

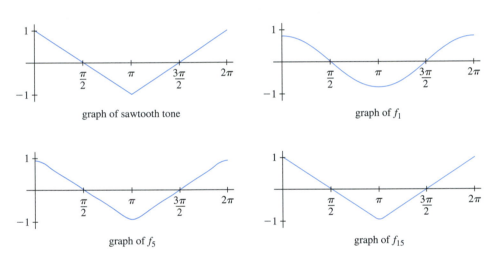

graph of sawtooth tone

graph of f_1

graph of f_5

graph of f_{15}

FIGURE 7.5
f and three least squares approximations of f.

Simple electronic circuits can be designed to generate alternating currents described by functions of the form $\cos kt$ and $\sin kt$. These currents can be combined with a simple direct current (corresponding to a constant function) to produce a current that describes any chosen trigonometric polynomial. This current can be fed to an audio amplifier to produce an audible tone that approximates a given tone, such as the sawtooth tone of Example 7. Electronic devices called synthesizers do exactly this. Least-squares approximations of musical tones produced on different instruments such as violins and clarinets can be computed, and synthesizers can then use this information to produce sounds that convincingly mimic these instruments.

An area of mathematics called *Fourier analysis*[6] is concerned with the study of periodic functions, including many that are not continuous, and their approximations by trigonometric polynomials.

---

[6] Jean Baptiste Joseph Fourier (1768–1830) was a French mathematician and professor of analysis at the Ècole Polytechnique in Paris. His *Théorie Analytique de la Chaleur* (1822) was an important contribution to physics (to problems involving the radiation of heat) and to mathematics (to the development of what are now called Fourier or trigonometric series). In 1798, he was appointed governor of Lower Egypt by Napoleon.

Practice Problems

1. Let $f(t) = t$ and $g(t) = t^2$ be vectors in the inner product space $C([0, 1])$ defined in Example 1. Compute the following quantities.

   (a) $\|f\|^2$
   (b) $\langle f, g \rangle$

2. Let $A = \begin{bmatrix} 2 & 1 \\ 0 & 3 \end{bmatrix}$ and $B = \begin{bmatrix} 1 & 1 \\ 2 & 0 \end{bmatrix}$. Use the Frobenius inner product defined in Example 2 to compute the fol-

   lowing quantities.

   (a) $\|A\|^2$
   (b) $\langle A, B \rangle$

3. Let W be the subset of $\mathcal{M}_{2\times2}$ consisting of the 2×2 matrices A such that trace $A = 0$, and let $B = \begin{bmatrix} 1 & 2 \\ 3 & 5 \end{bmatrix}$. Find the matrix in W that is closest to B, where distance is defined using the Frobenius inner product on $\mathcal{M}_{2\times2}$.

Exercises

1. Determine if the following statements are true or false.

   (a) The inner product of two vectors in an inner product space is a vector in the same inner product space.
   (b) An inner product is a real-valued function on the set of ordered pairs of vectors in a vector space.
   (c) An inner product on a vector space V is a linear operator on V.
   (d) There can be at most one inner product on a vector space.
   (e) Every nonzero finite-dimensional inner product space has an orthonormal basis.
   (f) Every orthogonal set in an inner product space is linearly independent.
   (g) Every orthonormal set in an inner product space is linearly independent.
   (h) It is possible to define an inner product on the set of $n \times n$ matrices.
   (i) The dot product is a special case of an inner product.
   (j) The definite integral can be used to define an inner product on $\mathcal{P}_2$.
   (k) The indefinite integral can be used to define an inner product on $\mathcal{P}_2$.
   (l) In an inner product space, the orthogonal projection of a vector $\mathbf{v}$ onto a finite-dimensional subspace W is the vector in W that is closest to $\mathbf{v}$.

In Exercises 2 and 3, use the inner product defined in Example 1 for $C([1, 2])$ to compute $\langle f, g \rangle$.

2. $f(t) = t^2$, $g(t) = \frac{1}{t}$
3. $f(t) = t$, $g(t) = e^t$

In Exercises 4 and 5, use the Frobenius inner product for $\mathcal{M}_{2\times2}$ defined in Example 2 to compute $\langle A, B \rangle$.

4. $A = \begin{bmatrix} 1 & 0 \\ 0 & 2 \end{bmatrix}$, $B = \begin{bmatrix} 2 & 3 \\ 1 & 0 \end{bmatrix}$

5. $B = \begin{bmatrix} 1 & -1 \\ 2 & 3 \end{bmatrix}$, $B = \begin{bmatrix} 2 & 4 \\ 1 & 0 \end{bmatrix}$

In Exercises 6 and 7, use the inner product for $\mathcal{P}_2$ defined in Example 5 to compute $\langle f(x), g(x) \rangle$.

6. $f(x) = x$, $g(x) = 2x + 1$
7. $f(x) = x^2 - 2$, $g(x) = 3x + 5$
8. Let V be a finite-dimensional vector space and $\mathcal{B}$ be a basis for V. For $\mathbf{u}$ and $\mathbf{v}$ in V, define

   $$\langle \mathbf{u}, \mathbf{v} \rangle = [\mathbf{u}]_\mathcal{B} \cdot [\mathbf{v}]_\mathcal{B}.$$

   Prove that this rule defines an inner product on V.
9. Let A be an $n \times n$ positive definite matrix (as defined in the Exercises of Section 6.5). For $\mathbf{u}$ and $\mathbf{v}$ in $\mathcal{R}^n$, define

   $$\langle \mathbf{u}, \mathbf{v} \rangle = (A\mathbf{u}) \cdot \mathbf{v}.$$

   Prove that this rule defines an inner product on $\mathcal{R}^n$.
10. Let A be an $n \times n$ invertible matrix. For $\mathbf{u}$ and $\mathbf{v}$ in $\mathcal{R}^n$, define

    $$\langle \mathbf{u}, \mathbf{v} \rangle = (A\mathbf{u}) \cdot (A\mathbf{v}).$$

    Prove that this rule defines an inner product on $\mathcal{R}^n$.
11. In Example 1, verify axioms 3 and 4 of the definition of inner product.
12. In Example 2, verify axioms 3 and 4 of the definition of inner product.

In Exercises 13–17, a vector space V and a rule are given. Determine if the rule defines an inner product on V. Justify your answer.

13. $V = \mathcal{R}^n$, $\langle \mathbf{u}, \mathbf{v} \rangle = 2(\mathbf{u} \cdot \mathbf{v})$
14. $V = \mathcal{R}^n$, $\langle \mathbf{u}, \mathbf{v} \rangle = |\mathbf{u} \cdot \mathbf{v}|$
15. Let $V = C([0, 2])$, and

    $$\langle f, g \rangle = \int_0^1 f(t)g(t)\, dt$$

    for all f and g in V. (Note that the limits of integration are not 0 and 2.)

16. $V = \mathcal{R}^2$, $D = \begin{bmatrix} 3 & 0 \\ 0 & 2 \end{bmatrix}$, and $\langle \mathbf{u}, \mathbf{v} \rangle = (D\mathbf{u}) \cdot \mathbf{v}$.
17. Let V be any vector space on which two inner products $\langle \mathbf{u}, \mathbf{v} \rangle_1$ and $\langle \mathbf{u}, \mathbf{v} \rangle_2$ are defined for $\mathbf{u}$ and $\mathbf{v}$ in V. Define

$\langle \mathbf{u}, \mathbf{v} \rangle$ by

$$\langle \mathbf{u}, \mathbf{v} \rangle = \langle \mathbf{u}, \mathbf{v} \rangle_1 + \langle \mathbf{u}, \mathbf{v} \rangle_2 .$$

18. Use the inner product in Example 4 to prove that $\sin mt$ and $\sin nt$ are orthogonal for any two distinct integers m and n. *Hint:* Use the trigonometric identity

$$\sin a \sin b = \frac{\cos(a+b) - \cos(a-b)}{2} .$$

19. Use the inner product in Example 4 to prove that $\cos mt$ and $\cos nt$ are orthogonal for any two distinct integers m and n. *Hint:* Use the trigonometric identity

$$\cos a \cos b = \frac{\cos(a+b) + \cos(a-b)}{2} .$$

20. (a) Use the methods of Example 5 to obtain $p_3(x)$, the normalized Legendre polynomial of degree 3.
    (b) Use the result of (a) to find the least-squares approximation of $\sqrt[3]{x}$ on $[-1, 1]$ as a polynomial of degree less than or equal to 3.

21. Find an orthogonal basis for the subspace $C([0, 1])$ (defined in Example 1) that is spanned by $\{1, e^t, e^{-t}\}$.

22. Suppose that $\langle \mathbf{u}, \mathbf{v} \rangle$ is an inner product for a vector space V. For any scalar $r > 0$, define $\langle\langle \mathbf{u}, \mathbf{v} \rangle\rangle = r \langle \mathbf{u}, \mathbf{v} \rangle$.

    (a) Prove that $\langle\langle \mathbf{u}, \mathbf{v} \rangle\rangle$ is an inner product on V.
    (b) Why is $\langle\langle \mathbf{u}, \mathbf{v} \rangle\rangle$ not an inner product if $r \leq 0$?

In Exercises 23–30, let $\mathbf{u}$, $\mathbf{v}$, and $\mathbf{w}$ be vectors in an inner product space V, and let c be a scalar.

23. Prove that $\|\mathbf{v}\| = 0$ if and only if $\mathbf{v} = \mathbf{0}$.

24. Prove that $\|c\mathbf{v}\| = |c| \|\mathbf{v}\|$.

25. Prove that $\langle \mathbf{0}, \mathbf{u} \rangle = \langle \mathbf{u}, \mathbf{0} \rangle = 0$.

26. Prove that $\langle \mathbf{u} - \mathbf{w}, \mathbf{v} \rangle = \langle \mathbf{u}, \mathbf{v} \rangle - \langle \mathbf{w}, \mathbf{v} \rangle$.

27. Prove that $\langle \mathbf{v}, \mathbf{u} - \mathbf{w} \rangle = \langle \mathbf{v}, \mathbf{u} \rangle - \langle \mathbf{v}, \mathbf{w} \rangle$.

28. Prove that $\langle \mathbf{u}, c\mathbf{v} \rangle = c \langle \mathbf{u}, \mathbf{v} \rangle$.

29. Prove that if $\langle \mathbf{u}, \mathbf{w} \rangle = 0$ for all $\mathbf{u}$ in V, then $\mathbf{w} = \mathbf{0}$.

30. Prove that if $\langle \mathbf{u}, \mathbf{v} \rangle = \langle \mathbf{u}, \mathbf{w} \rangle$ for all $\mathbf{u}$ in V, then $\mathbf{v} = \mathbf{w}$.

31. Let V be a finite-dimensional inner product space, and suppose that $\mathcal{B}$ is an orthonormal basis for V. Prove that, for any vectors $\mathbf{u}$ and $\mathbf{v}$ in V,

$$\langle \mathbf{u}, \mathbf{v} \rangle = [\mathbf{u}]_\mathcal{B} \cdot [\mathbf{v}]_\mathcal{B}.$$

32. Prove that if A is an $n \times n$ symmetric matrix and B is an $n \times n$ skew-symmetric matrix, then A and B are orthogonal with respect to the Frobenius inner product defined in Example 2.

33. Prove that if A and B are 2×2 matrices, then the Frobenius inner product $\langle A, B \rangle$ can be computed as

$$\langle A, B \rangle = a_{11}b_{11} + a_{12}b_{12} + a_{21}b_{21} + a_{22}b_{22}.$$

34. Extend Exercise 33 to the general case. That is, prove that if A and B are $n \times n$ matrices, then the Frobenius inner product $\langle A, B \rangle$ can be computed as

$$\langle A, B \rangle = a_{11}b_{11} + a_{12}b_{12} + \cdots + a_{nn}b_{nn}.$$

35. Consider the inner product space $\mathcal{M}_{2 \times 2}$ with the Frobenius inner product.

    (a) Find an orthonormal basis for the subspace of 2×2 symmetric matrices.
    (b) Use (a) to find the 2×2 symmetric matrix that is closest to

$$\begin{bmatrix} 1 & 2 \\ 4 & 8 \end{bmatrix}.$$

36. Prove that if $\mathcal{B}$ is an infinite orthogonal subset of nonzero vectors in an inner product space V, then $\mathcal{B}$ is a linearly independent subset of V.

37. Prove that if $\{\mathbf{u}, \mathbf{v}\}$ is a linearly dependent subset of an inner product space, then $\langle \mathbf{u}, \mathbf{v} \rangle^2 = \langle \mathbf{u}, \mathbf{u} \rangle \langle \mathbf{v}, \mathbf{v} \rangle$.

38. Prove the converse of Exercise 37: If $\mathbf{u}$ and $\mathbf{v}$ are vectors in an inner product space and $\langle \mathbf{u}, \mathbf{v} \rangle^2 = \langle \mathbf{u}, \mathbf{u} \rangle \langle \mathbf{v}, \mathbf{v} \rangle$, then $\{\mathbf{u}, \mathbf{v}\}$ is a linearly dependent set. *Hint:* Suppose that $\mathbf{u}$ and $\mathbf{v}$ are nonzero vectors. Show that

$$\left\| \mathbf{v} - \frac{\langle \mathbf{u}, \mathbf{v} \rangle}{\langle \mathbf{u}, \mathbf{u} \rangle} \mathbf{u} \right\| = 0.$$

39. Let V be an inner product space and $\mathbf{u}$ be a vector in V. Define $F_\mathbf{u} : V \to \mathcal{R}$ by

$$F_\mathbf{u}(\mathbf{v}) = \langle \mathbf{v}, \mathbf{u} \rangle$$

for all $\mathbf{v}$ in V. Prove that $F_\mathbf{u}$ is a linear transformation.

40. Prove the converse of Exercise 39 for finite-dimensional inner product spaces: If V is a finite-dimensional inner product space and $T : V \to \mathcal{R}$ is a linear transformation, then there exists a unique vector $\mathbf{u}$ in V such that $T = F_\mathbf{u}$. *Hint:* Let $\{\mathbf{v}_1, \mathbf{v}_2, \ldots, \mathbf{v}_n\}$ be an orthonormal basis for V, and let

$$\mathbf{u} = T(\mathbf{v}_1)\mathbf{v}_1 + T(\mathbf{v}_2)\mathbf{v}_2 + \cdots + T(\mathbf{v}_n)\mathbf{v}_n.$$

41. (a) Prove that $B^T B$ is positive definite for any invertible matrix B.
    (b) Use (a) and Exercise 31 to prove the converse of Exercise 9: For any inner product on $\mathcal{R}^n$, there exists a positive definite matrix A such that

$$\langle \mathbf{u}, \mathbf{v} \rangle = (A\mathbf{u}) \cdot \mathbf{v}$$

for all vectors $\mathbf{u}$ and $\mathbf{v}$ in $\mathcal{R}^n$.

The following definitions are used in Exercises 42 and 43.

Definitions.   A linear transformation $T: V \to W$ is called a **linear isometry** if T is an isomorphism and $\langle T(\mathbf{u}), T(\mathbf{v}) \rangle = \langle \mathbf{u}, \mathbf{v} \rangle$ for every $\mathbf{u}$ and $\mathbf{v}$ in V. The inner product spaces V and W are called **isometric** if there exists a linear isometry from V to W.

42. Let V, W, and Z be inner product spaces. Prove the following statements.

    (a) V is isometric to itself.
    (b) If V is isometric to W, then W is isometric to V.
    (c) If V is isometric to W and W is isometric to Z, then V is isometric to Z.

43. Prove that, for any n-dimensional inner product space V and any orthonormal basis $\mathcal{B}$ of V, the linear transformation $\Phi_{\mathcal{B}}: V \to \mathcal{R}^n$ defined by $\Phi_{\mathcal{B}}(\mathbf{v}) = [\mathbf{v}]_{\mathcal{B}}$ is a linear isometry. Thus every n-dimensional inner product space is isometric to $\mathcal{R}^n$.

44. Let $\{\mathbf{w}_1, \mathbf{w}_2, \dots, \mathbf{w}_n\}$ be an orthonormal basis for a subspace W of an inner product space. Prove that, for any vector $\mathbf{v}$ in W,

$$\mathbf{v} = \langle \mathbf{v}, \mathbf{w}_1 \rangle \, \mathbf{w}_1 + \langle \mathbf{v}, \mathbf{w}_2 \rangle \, \mathbf{w}_2 + \cdots + \langle \mathbf{v}, \mathbf{w}_n \rangle \, \mathbf{w}_n.$$

45. Let $\{\mathbf{w}_1, \mathbf{w}_2, \dots, \mathbf{w}_n\}$ be an orthonormal basis for a subspace W of an inner product space. Prove that, for any vectors $\mathbf{u}$ and $\mathbf{v}$ in W, we have

$$\mathbf{u} + \mathbf{v} = (\langle \mathbf{u}, \mathbf{w}_1 \rangle + \langle \mathbf{v}, \mathbf{w}_1 \rangle)\mathbf{w}_1 + \cdots$$
$$+ (\langle \mathbf{u}, \mathbf{w}_n \rangle + \langle \mathbf{v}, \mathbf{w}_n \rangle)\mathbf{w}_n.$$

46. Let $\{\mathbf{w}_1, \mathbf{w}_2, \dots, \mathbf{w}_n\}$ be an orthonormal basis for a subspace W of an inner product space. Prove that, for any

vectors $\mathbf{u}$ and $\mathbf{v}$ in W, we have

$$\langle \mathbf{u}, \mathbf{v} \rangle = \langle \mathbf{u}, \mathbf{w}_1 \rangle \, \langle \mathbf{v}, \mathbf{w}_1 \rangle + \langle \mathbf{u}, \mathbf{w}_2 \rangle \, \langle \mathbf{v}, \mathbf{w}_2 \rangle + \cdots$$
$$+ \langle \mathbf{u}, \mathbf{w}_n \rangle \, \langle \mathbf{v}, \mathbf{w}_n \rangle.$$

47. Let W be the one-dimensional subspace of $\mathcal{M}_{n \times n}$ spanned by I_n. That is, W is the set of all $n \times n$ scalar matrices. Prove that for any $n \times n$ matrix A, the matrix in W that is nearest to A is $\left(\dfrac{\text{trace } A}{n} \right) I_n$, where distance is defined using the Frobenius inner product on $\mathcal{M}_{n \times n}$.

In Exercise 48, use a calculator with matrix capabilities or computer software such as MATLAB *to solve the problem.*

48. Let

$$A = \begin{bmatrix} 25 & 24 & 23 & 22 & 21 \\ 20 & 19 & 18 & 17 & 16 \\ 15 & 14 & 13 & 12 & 11 \\ 10 & 9 & 8 & 7 & 6 \\ 5 & 4 & 3 & 2 & 1 \end{bmatrix}$$

and

$$B = \begin{bmatrix} 1 & 2 & 3 & 4 & 5 \\ 6 & 7 & 8 & 9 & 10 \\ 11 & 12 & 13 & 14 & 15 \\ 16 & 17 & 18 & 19 & 20 \\ 21 & 22 & 23 & 24 & 25 \end{bmatrix}.$$

Find the matrix in the one-dimensional space spanned by A that is nearest to B, where distance is defined using the Frobenius inner product on $\mathcal{M}_{5 \times 5}$.

Solutions to the Practice Problems

1. (a) $\|f\|^2 = \langle f, f \rangle = \int_0^1 f(t) \cdot f(t) \, dt = \int_0^1 t^2 \, dt$
    $= \frac{1}{3} t^3 \big|_0^1 = \frac{1}{3}.$
   (b) $\langle f, g \rangle = \int_0^1 t \cdot t^2 \, dt = \int_0^1 t^3 \, dt = \frac{1}{4} t^4 \big|_0^1 = \frac{1}{4}.$

2. (a) $\|A\|^2 = \text{trace}\,(AA^T) = \text{trace} \left(\begin{bmatrix} 2 & 1 \\ 0 & 3 \end{bmatrix} \begin{bmatrix} 2 & 1 \\ 0 & 3 \end{bmatrix}^T \right)$

    $= \text{trace} \left(\begin{bmatrix} 2 & 1 \\ 0 & 3 \end{bmatrix} \begin{bmatrix} 2 & 0 \\ 1 & 3 \end{bmatrix} \right)$

    $= \text{trace} \left(\begin{bmatrix} 5 & 3 \\ 3 & 9 \end{bmatrix} \right) = 14.$

   (b) $\langle A, B \rangle = \text{trace}\,(AB^T) = \text{trace} \left(\begin{bmatrix} 2 & 1 \\ 0 & 3 \end{bmatrix} \begin{bmatrix} 1 & 1 \\ 2 & 0 \end{bmatrix}^T \right)$

    $= \text{trace} \left(\begin{bmatrix} 2 & 1 \\ 0 & 3 \end{bmatrix} \begin{bmatrix} 1 & 2 \\ 1 & 0 \end{bmatrix} \right) = \text{trace} \left(\begin{bmatrix} 3 & 4 \\ 3 & 0 \end{bmatrix} \right)$

    $= 3.$

3. The set W is a subspace of $\mathcal{M}_{2 \times 2}$ by Example 5 of Section 7.1, and hence the desired matrix A is the orthogonal projection of B onto W. Since

$$\{A_1, A_2, A_3\} = \left\{ \frac{1}{\sqrt{2}} \begin{bmatrix} 1 & 0 \\ 0 & -1 \end{bmatrix}, \begin{bmatrix} 0 & 1 \\ 0 & 0 \end{bmatrix}, \begin{bmatrix} 0 & 0 \\ 1 & 0 \end{bmatrix} \right\}$$

is an orthonormal basis for W, we can apply 10 to obtain the orthogonal projection. For this purpose, observe that

$$\langle B, A_1 \rangle = \text{trace}\left(BA_1^T\right)$$

$$= \text{trace}\left(\begin{bmatrix} 1 & 2 \\ 3 & 5 \end{bmatrix} \frac{1}{\sqrt{2}} \begin{bmatrix} 1 & 0 \\ 0 & -1 \end{bmatrix}^T\right) = -\frac{4}{\sqrt{2}}.$$

Similarly,

$$\langle B, A_2 \rangle = 2 \quad \text{and} \quad \langle B, A_3 \rangle = 3.$$

Therefore

$$A = \langle B, A_1 \rangle A_1 + \langle B, A_2 \rangle A_2 + \langle B, A_3 \rangle A_3$$

$$= -\frac{4}{\sqrt{2}}\left(\frac{1}{\sqrt{2}}\right)\begin{bmatrix} 1 & 0 \\ 0 & -1 \end{bmatrix} + 2\begin{bmatrix} 0 & 1 \\ 0 & 0 \end{bmatrix} + 3\begin{bmatrix} 0 & 0 \\ 1 & 0 \end{bmatrix}$$

$$= \begin{bmatrix} -2 & 2 \\ 3 & 2 \end{bmatrix}.$$

Chapter 7 Review Exercises

1. Determine if the following statements are true or false.
   (a) Every subspace of a vector space is a subset of $\mathcal{R}^n$ for some integer n.
   (b) Every $m \times n$ matrix is a vector in the vector space $M_{m \times n}$.
   (c) $\dim M_{m \times n} = m + n$.
   (d) A matrix representation of a linear operator on $M_{m \times n}$ is an $m \times n$ matrix.
   (e) The Frobenius inner product of two matrices is a scalar.
   (f) Suppose that $\mathbf{u}$, $\mathbf{v}$, and $\mathbf{w}$ are vectors in an inner product space. If $\mathbf{u}$ is orthogonal to $\mathbf{v}$ and $\mathbf{v}$ is orthogonal to $\mathbf{w}$, then $\mathbf{u}$ is orthogonal to $\mathbf{w}$.
   (g) Suppose that $\mathbf{u}$, $\mathbf{v}$, and $\mathbf{w}$ are vectors in an inner product space. If $\mathbf{u}$ is orthogonal to both $\mathbf{v}$ and $\mathbf{w}$, then $\mathbf{u}$ is orthogonal to $\mathbf{v} + \mathbf{w}$.

In Exercises 2–5, determine if the given set V is a vector space with respect to the indicated operations. Justify your conclusions.

2. V is the set of all sequences $\{a_n\}$ of real numbers. For any sequences $\{a_n\}$ and $\{b_n\}$ in V and any scalar c, define the sum $\{a_n\} + \{b_n\}$ and the product $c\{a_n\}$ by

$$\{a_n\} + \{b_n\} = \{a_n + b_n\} \quad \text{and} \quad c\{a_n\} = \{ca_n\}.$$

3. V is the set of all real numbers with vector addition, $\oplus$, and scalar multiplication, $\odot$, defined by

$$a \oplus b = a + b + ab \quad \text{and} \quad c \odot a = ca,$$

   where a and b are in V and c is any scalar.

4. V is the set of all 2×2 matrices with vector addition, $\oplus$, and scalar multiplication, $\odot$, defined by

$$A \oplus B = A + B \quad \text{and} \quad t \odot \begin{bmatrix} a & b \\ c & d \end{bmatrix} = \begin{bmatrix} ta & tb \\ c & d \end{bmatrix}$$

   for all 2×2 matrices A and B and scalars t.

5. V is the set of all functions from $\mathcal{R}$ to $\mathcal{R}$ such that $f(x) > 0$ for all x in $\mathcal{R}$. Vector addition, $\oplus$, and scalar

multiplication, $\odot$, are defined by

$$(f \oplus g)(x) = f(x)g(x) \quad \text{and} \quad (c \odot f)(x) = [f(x)]^c$$

for all f and g in V, x in $\mathcal{R}$, and scalars c.

In Exercises 6–9, determine if the given subset W is a subspace of the vector space V. Justify your conclusion.

6. $V = \mathcal{F}(\mathcal{R})$, and W is the set of all functions f in V such that $f(x) \geq 0$ for all x in $\mathcal{R}$.

7. $V = \mathcal{P}$, and W is the set consisting of the zero polynomial and all polynomials of even degree.

8. Fix a nonzero vector $\mathbf{v}$ in $\mathcal{R}^n$, and let W be the set of all $n \times n$ matrices A such that $\mathbf{v}$ is an eigenvector of A, and $V = M_{n \times n}$.

9. Fix a nonzero scalar λ, and let W be the set of all $n \times n$ matrices A such that λ is an eigenvalue of A, and $V = M_{n \times n}$.

In Exercises 10–13, determine if the given matrix is a linear combination of the matrices in the set

$$\left\{\begin{bmatrix} 1 & 2 \\ 1 & -1 \end{bmatrix}, \begin{bmatrix} 0 & 1 \\ 2 & 0 \end{bmatrix}, \begin{bmatrix} -1 & 3 \\ 1 & 1 \end{bmatrix}\right\}.$$

10. $\begin{bmatrix} 1 & 10 \\ 9 & -1 \end{bmatrix}$ 11. $\begin{bmatrix} 2 & 8 \\ 1 & -5 \end{bmatrix}$ 12. $\begin{bmatrix} 3 & 1 \\ -2 & -4 \end{bmatrix}$

13. $\begin{bmatrix} 4 & 1 \\ -2 & -4 \end{bmatrix}$

In Exercises 14 and 15, let $\mathcal{S}$ be the following subset of $\mathcal{P}$:

$$\mathcal{S} = \{x^3 - x^2 + x + 1, 3x^2 + x + 2, x - 1\}.$$

14. Show that the polynomial $x^3 + 2x^2 + 5$ is a linear combination of the polynomials in $\mathcal{S}$.

15. Find a constant c so that $f(x) = 2x^3 + x^2 + 2x + c$ is a linear combination of the polynomials in $\mathcal{S}$.

In Exercises 16 and 17, find a basis for the subspace W of the vector space V. Then find the dimension of W.

16. $V = \mathcal{M}_{2\times2}$ and

$$W = \left\{ A \text{ in } V: \begin{bmatrix} 1 & 2 \\ 1 & 2 \end{bmatrix} A = \begin{bmatrix} 0 & 0 \\ 0 & 0 \end{bmatrix} \right\}.$$

17. $V = \mathcal{P}_3$ and

$$W = \{f(x) \text{ in } V: f(0) + f'(0) + f''(0) = 0\}.$$

In Exercises 18–21, determine if the given function T is linear. If T is linear, determine if it is an isomorphism.

18. $T: \mathcal{R}^3 \to \mathcal{P}_2$ defined by

$$T\left(\begin{bmatrix} a \\ b \\ c \end{bmatrix}\right) = (a + b) + (a - b)x + cx^2$$

19. $T: \mathcal{M}_{2\times2} \to \mathcal{R}$ defined by $T(A) = \text{trace } A^2$

20. $T: \mathcal{R}^3 \to \mathcal{M}_{2\times2}$ defined by

$$T\left(\begin{bmatrix} a \\ b \\ c \end{bmatrix}\right) = \begin{bmatrix} a & b \\ c & a+b+c \end{bmatrix}$$

21. $T: \mathcal{P}_2 \to \mathcal{R}^3$ defined by

$$T(f(x)) = \begin{bmatrix} f(0) \\ f'(0) \\ \int_0^1 f(t)\,dt \end{bmatrix}$$

In Exercises 22–25, a vector space V, a basis B for V, and a linear operator T on V are given. Find $[T]_B$.

22. $V = \mathcal{P}_2$, $T(p(x)) = p(1) + 2p'(1)x - p''(1)x^2$ for all $p(x)$ in V, and $\mathcal{B} = \{1, x, x^2\}$

23. $V = \text{Span } \mathcal{B}$, where $\mathcal{B} = \{e^{at} \cos bt, e^{at} \sin bt\}$ for some nonzero scalars a and b, and $T = D$

24. $V = \text{Span } \mathcal{B}$, where $\mathcal{B} = \{e^t \cos t, e^t \sin t\}$, and $T = D^2 + 2D$

25. $V = \mathcal{M}_{2\times2}$,

$$\mathcal{B} = \left\{ \begin{bmatrix} 1 & 0 \\ 0 & 0 \end{bmatrix}, \begin{bmatrix} 0 & 1 \\ 0 & 0 \end{bmatrix}, \begin{bmatrix} 0 & 0 \\ 1 & 0 \end{bmatrix}, \begin{bmatrix} 0 & 0 \\ 0 & 1 \end{bmatrix} \right\},$$

and T is defined by $T(A) = 2A + A^T$ for all A in V

26. Find an expression for $T^{-1}(a + bx + cx^2)$, where T is the linear operator in Exercise 22.

27. Find an expression for $T^{-1}(c_1 e^{at} \cos bt + c_2 e^{at} \sin bt)$, where T is the linear operator in Exercise 23.

28. Find an expression for $T^{-1}(c_1 e^t \cos t + c_2 e^t \sin t)$, where T is the linear operator in Exercise 24.

29. Find an expression for

$$T^{-1}\left(\begin{bmatrix} a & b \\ c & d \end{bmatrix}\right),$$

where T is the linear operator in Exercise 25.

30. Find the eigenvalues and a basis for each eigenspace of the linear operator in Exercise 22.

31. Find the eigenvalues and a basis for each eigenspace of the linear operator in Exercise 23.

32. Find the eigenvalues and a basis for each eigenspace of the linear operator in Exercise 24.

33. Find the eigenvalues and a basis for each eigenspace of the linear operator in Exercise 25.

In Exercises 34–37, $V = \mathcal{M}_{2\times2}$ with the Frobenius inner product, and W is the subspace of V defined by

$$W = \left\{ A \text{ in } \mathcal{M}_{2\times2}: \text{trace}\left(\begin{bmatrix} 0 & 1 \\ 1 & 0 \end{bmatrix} A\right) = 0 \right\}.$$

34. Find $\langle A, B \rangle$, for $A = \begin{bmatrix} 1 & 2 \\ -1 & 3 \end{bmatrix}$ and $B = \begin{bmatrix} 2 & -1 \\ 1 & 1 \end{bmatrix}$.

35. Find a basis for the subspace of V consisting of all matrices that are orthogonal to $\begin{bmatrix} 1 & 3 \\ 4 & 2 \end{bmatrix}$.

36. Find an orthonormal basis for W.

37. Find the orthogonal projection of $\begin{bmatrix} 2 & 5 \\ 9 & -3 \end{bmatrix}$ onto W.

In Exercises 38 and 39, let $V = C([0, 1])$ with the inner product defined by

$$\langle f, g \rangle = \int_0^1 f(t)g(t)\,dt,$$

and let W be the subspace of V consisting of all polynomial functions of degree less than or equal to 2 with domain restricted to [0, 1].

38. Let f and g be the functions in V defined by $f(t) = \cos 2\pi t$ and $g(t) = \sin 2\pi t$. Prove that f and g are orthogonal.

39. Find an orthonormal basis for W.

40. Determine the orthogonal projection of the function $f(t) = t$ onto W without doing any computations. Now compute the orthogonal projection of f to verify your answer.

41. Find the orthogonal projection of the function $f(t) = \sqrt{t}$ onto W.

42. Prove that for any $n \times n$ matrix A, $\langle A, I_n \rangle = \text{trace } A$, where the inner product is the Frobenius inner product.

Appendix: Complex Numbers

Throughout this book the word *scalar* has been used almost interchangeably with *real number*. However, many of the results that have been established can be reformulated to hold when scalars are allowed to be *complex numbers*.

Definitions. A **complex number** z is an expression of the form

$$z = a + bi,$$

where a and b are real numbers. The real numbers a and b are called the **real part** and the **imaginary part** of z, respectively. We denote the set of all complex numbers by $\mathcal{C}$.

Thus $z = 3 + (-2)i = 3 - 2i$ is a complex number. The real part of z is 3, and the imaginary part is -2. When the imaginary part of a complex number is 0, we identify the number with its real part. Thus $4 + 0i$ is identified with the real number 4. In this way, $\mathcal{R}$ may be regarded as a subset of $\mathcal{C}$.

Two complex numbers are called **equal** if their real parts are equal and their imaginary parts are equal. Thus two complex numbers $a + bi$ and $c + di$, where a, b, c, and d are real numbers, are equal if and only if $a = c$ and $b = d$.

The arithmetic operations on $\mathcal{R}$ can be extended to $\mathcal{C}$. The **sum** of two complex numbers $z = a + bi$ and $w = c + di$, where a, b, c, and d are real numbers, is defined by

$$z + w = (a + bi) + (c + di) = (a + c) + (b + d)i,$$

and their **product** is defined by

$$zw = (a + bi)(c + di) = (ac - bd) + (bc + ad)i.$$

Example 1 Compute the sum and product of $z = 2 + 3i$ and $w = 4 - 5i$.

Solution By definition

$$z + w = (2 + 3i) + (4 - 5i) = (2 + 4) + [3 + (-5)]i = 6 + (-2)i = 6 - 2i$$

447

""

and

$$zw = (2 + 3i)(4 - 5i) = [2(4) - 3(-5)] + [3(4) + 2(-5)]i = 23 + 2i.$$

Note that the complex number $i = 0 + 1i$ has the property that

$$i^2 = (0 + 1i)(0 + 1i) = [0(0) - 1(1)] + [1(0) + 0(1)]i = -1 + 0i = -1.$$

This provides an easy method for multiplying complex numbers: Multiply the numbers as though they were algebraic expressions, and then replace i^2 by -1. Thus the computation in Example 1 can be performed as follows.

$$zw = (2 + 3i)(4 - 5i)$$
$$= 8 + (12 - 10)i - 15i^2$$
$$= 8 + 2i - 15(-1)$$
$$= 23 + 2i$$

The sum and product of complex numbers share many of the same properties as sums and products of real numbers. In particular, the following theorem can be proved.

Theorem 1
For all complex numbers x, y, and z, the following are true.

(a) $x + y = y + x$.   (commutativity of addition)
(b) $xy = yx$.   (commutativity of multiplication)
(c) $x + (y + z) = (x + y) + z$.   (associativity of addition)
(d) $x(yz) = (xy)z$.   (associativity of multiplication)
(e) $0 + x = x$.   (0 is an identity element for addition)
(f) $1 \cdot x = x$.   (1 is an identity element for multiplication)
(g) $x + (-1)x = 0$.   (existence of additive inverses)
(h) If $x \neq 0$, there is a u in $\mathcal{C}$
such that $xu = 1$.   (existence of multiplicative inverses)
(i) $x(y + z) = xy + xz$.   (distributive property)

As usual, the **difference** of complex numbers z and w is defined by $z - w = z + (-1)w$. Thus

$$(2 + 3i) - (4 - 5i) = (2 + 3i) + (-4 + 5i) = -2 + 8i.$$

Because of Theorem 1(h), it is also possible to define division for complex numbers. In order to obtain an efficient method for computing the quotient of complex numbers, we need the following concept.

Definition. The **(complex) conjugate** of the complex number $z = a + bi$, where a and b are real numbers, is the complex number $a - bi$. It is denoted $\bar{z}$.

Thus the conjugate of $z = 4 - 3i$ is $\bar{z} = 4 - (-3)i = 4 + 3i$. The following result lists some useful properties of conjugates.

Theorem 2

For all complex numbers z and w, the following are true.

(a) $\overline{\overline{z}} = z$.
(b) $\overline{z + w} = \overline{z} + \overline{w}$.
(c) $\overline{zw} = \overline{z}\,\overline{w}$.
(d) z is a real number if and only if $z = \overline{z}$.

Complex numbers can be visualized as the vectors in a plane with two axes, which are called the **real axis** and the **imaginary axis** (see Figure 1). In this interpretation, the sum of complex numbers $z = a + bi$ and $w = c + di$, where $a, b, c,$ and d are real numbers, corresponds to the sum of the vectors

$$\begin{bmatrix} a \\ b \end{bmatrix} \quad \text{and} \quad \begin{bmatrix} c \\ d \end{bmatrix}$$

in $\mathcal{R}^2$. The **absolute value** or (**modulus**) of z, denoted $|z|$, corresponds to the length of a vector in $\mathcal{R}^2$ and is defined as the nonnegative real number

$$|z| = \sqrt{a^2 + b^2}.$$

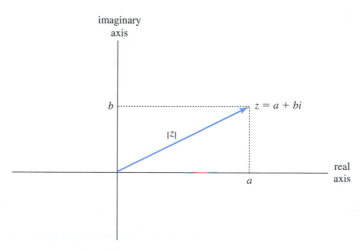

FIGURE 1

The following properties of absolute value are easy to verify.

Theorem 3

For all complex numbers z and w, the following are true.

(a) $|z| \geq 0$, and $|z| = 0$ if and only if $z = 0$.
(b) $z\overline{z} = |z|^2$.
(c) $|zw| = |z||w|$.
(d) $|z + w| \leq |z| + |w|$.

Note that Theorem 3(b) tells us that the product of a complex number and its conjugate is a real number. This fact provides an easy method for evaluating the quotient of two complex numbers. For suppose that $z = a + bi$ and $w = c + di$, where $a, b, c,$ and d are real numbers and $w \neq 0$. We wish to represent $\dfrac{z}{w}$ in the form $r + si$, where r and s are real numbers. Since $w\overline{w} = |w|^2$ is real, we can obtain such a representation by multiplying the numerator and denominator of $\dfrac{z}{w}$ by the conjugate of the denominator, as follows:

$$\frac{z}{w} = \frac{z\,\overline{w}}{w\,\overline{w}} = \frac{z\overline{w}}{|w|^2} = \frac{(a+bi)\cdot(c-di)}{c^2+d^2} = \frac{ac+bd}{c^2+d^2} + \frac{bc-ad}{c^2+d^2}i.$$

Example 2 Compute $\dfrac{9+8i}{2-i}$.

Solution Multiplying the numerator and denominator of the given expression by the conjugate of the denominator, we obtain

$$\frac{9+8i}{2-i} = \frac{9+8i}{2-i}\cdot\frac{2+i}{2+i} = \frac{10+25i}{5} = 2+5i.$$

It is possible to define the expression e^z when z is a complex number in a manner that reduces to the familiar case when z is real. (Here e is the base of the natural logarithm.) If $z = a + bi$, where a and b are real numbers, we define

$$e^z = e^{a+ib} = e^a(\cos b + i \sin b),$$

where b is in radians. This definition is called *Euler's formula*.[7] Notice that if $b = 0$, so that z is real, this expression reduces to e^a. Moreover, with this definition, all of the familiar exponent properties are preserved. For example, the equations

$$e^z e^w = e^{z+w} \qquad \text{and} \qquad \frac{e^z}{e^w} = e^{z-w}$$

are true for all complex numbers z and w.

Recall that there are polynomials with real coefficients that have no real roots, for example, $t^2 + 1$. The principal reason for the importance of the complex number system is the following result due to Gauss,[8] which shows that such an occurrence is impossible in $\mathcal{C}$.

The Fundamental Theorem of Algebra
Any polynomial of positive degree with complex coefficients has a (complex) root.

An important consequence of this result is that every polynomial of positive degree with complex coefficients can be factored into a product of linear factors. For example, the polynomial $t^3 - 2t^2 + t - 2$ can be factored as

$$t^3 - 2t^2 + t - 2 = (t-2)(t^2+1) = (t-2)(t+i)(t-i).$$

This fact is useful in our discussion of eigenvalues in Chapter 5.

---

[7]Leonhard Euler (1707–1783), a Swiss mathematician, wrote more than 500 books and papers during his lifetime. He is responsible for much of our present-day mathematical notation, including the symbols e (for the base of the natural logarithm) and i (for the complex number whose square is -1).

[8]The German mathematician Karl Friedrich Gauss (1777–1885) is regarded by many as the greatest mathematician of all time. Although the fundamental theorem of algebra had been previously stated by others, Gauss gave the first successful proof in his doctoral thesis at the University of Helmstädt.

References

[1] Abbot, Stephen D. and Matt Richey. Take a Walk on the Boardwalk. *The College Mathematics Journal* 28 (1997): 162–71.

[2] Ash, Robert and Richard Bishop. Monopoly as a Markov process. *Mathematics Magazine* 45 (1972): 26–29.

[3] Bourne, Larry S. Physical Adjustment Process and Land Use Succession: A Conceptual Review and Central City Example. *Economic Geography* 47 (1971): 1–15.

[4] Friedberg, Stephen H., Arnold J. Insel, and Lawrence E. Spence. *Linear Algebra*. 3d ed. Upper Saddle River, NJ: Prentice-Hall, 1997.

[5] Gabriel, K. R. and J. Neumann. A Markov chain model for daily rainfall occurrence at Tel Aviv. *Quarterly Journal of the Royal Meteorological Society* 88 (1962): 90–95.

[6] Hampton, P. Regional Economic Development in New Zealand. *Journal of Regional Science* 8 (1968): 41–51.

[7] Hunter, Albert. Community Change: A Stochastic Analysis of Chicago's Local Communities, 1930–1960. *American Journal of Sociology* 79 (January 1974): 923–47.

Answers to Odd-Numbered Exercises

CHAPTER 1

Section 1.1

1. (a) T   (b) T   (c) T   (d) F   (e) F
   (f) T   (g) F

3. $\begin{bmatrix} 8 & -4 & 20 \\ 12 & 16 & 4 \end{bmatrix}$ 5. $\begin{bmatrix} 6 & -4 & 24 \\ 8 & 10 & -4 \end{bmatrix}$ 7. $\begin{bmatrix} 2 & 4 \\ 0 & 6 \\ -4 & 8 \end{bmatrix}$

9. -2 11. $\begin{bmatrix} 3 \\ 0 \\ 2\pi \end{bmatrix}$ 13. $\begin{bmatrix} 2 \\ 2e \end{bmatrix}$

15. $[2 \quad -3 \quad 0.4]$ 17. $\begin{bmatrix} 150 \\ 150\sqrt{3} \\ 10 \end{bmatrix}$

19. (a) $\begin{bmatrix} 150\sqrt{2} + 50 \\ 150\sqrt{2} \end{bmatrix}$ (b) $50\sqrt{37 + 6\sqrt{2}} \approx 337.21$

35. $\begin{bmatrix} 2 & 5 \\ 5 & 8 \end{bmatrix}$ and $\begin{bmatrix} 2 & 5 & 6 \\ 5 & 7 & 8 \\ 6 & 8 & 4 \end{bmatrix}$

41. No. Consider $\begin{bmatrix} 2 & 5 & 6 \\ 5 & 7 & 8 \\ 6 & 8 & 4 \end{bmatrix}$ and $\begin{bmatrix} 2 & 6 \\ 5 & 8 \end{bmatrix}$.

43. They must equal 0.

Section 1.2

1. (a) T   (b) F   (c) T   (d) T   (e) T
   (f) F   (g) F

3. $\begin{bmatrix} 12 \\ 14 \end{bmatrix}$ 5. $\begin{bmatrix} 9 \\ 0 \\ 10 \end{bmatrix}$ 7. $\begin{bmatrix} a \\ b \end{bmatrix}$ 9. $\begin{bmatrix} 22 \\ 5 \end{bmatrix}$

11. $\begin{bmatrix} sa \\ tb \\ uc \end{bmatrix}$ 13. $\begin{bmatrix} 21 \\ 13 \end{bmatrix}$

15. $\dfrac{1}{2}\begin{bmatrix} \sqrt{2} & -\sqrt{2} \\ \sqrt{2} & \sqrt{2} \end{bmatrix}, \dfrac{1}{2}\begin{bmatrix} -\sqrt{2} \\ \sqrt{2} \end{bmatrix}$

17. $\dfrac{1}{2}\begin{bmatrix} 1 & -\sqrt{3} \\ \sqrt{3} & 1 \end{bmatrix}, \dfrac{1}{2}\begin{bmatrix} 3 - \sqrt{3} \\ 3\sqrt{3} + 1 \end{bmatrix}$

19. $\dfrac{1}{2}\begin{bmatrix} -\sqrt{3} & 1 \\ -1 & -\sqrt{3} \end{bmatrix}, \dfrac{1}{2}\begin{bmatrix} \sqrt{3} - 3 \\ 3\sqrt{3} + 1 \end{bmatrix}$

23. $\begin{bmatrix} 1 \\ 1 \end{bmatrix} = (1)\begin{bmatrix} 1 \\ 0 \end{bmatrix} + (1)\begin{bmatrix} 0 \\ 1 \end{bmatrix}$

25. not possible   27. not possible

29. $\begin{bmatrix} -1 \\ 11 \end{bmatrix} = 3\begin{bmatrix} 1 \\ 3 \end{bmatrix} - 2\begin{bmatrix} 2 \\ -1 \end{bmatrix}$

31. $\begin{bmatrix} 3 \\ 8 \end{bmatrix} = 7\begin{bmatrix} 1 \\ 2 \end{bmatrix} - 2\begin{bmatrix} 2 \\ 3 \end{bmatrix} + 0\begin{bmatrix} -2 \\ -5 \end{bmatrix}$

33. (a) 349,000 in the city and 351,000 in the suburbs
     (b) 307,180 in the city and 392,820 in the suburbs

37. $B = \begin{bmatrix} 1 & 0 \\ 0 & -1 \end{bmatrix}$

55. (a) $\begin{bmatrix} 24.6 \\ 45.0 \\ 26.0 \\ -41.4 \end{bmatrix}$ (b) $\begin{bmatrix} 134.1 \\ 44.4 \\ 7.6 \\ 104.8 \end{bmatrix}$ (c) $\begin{bmatrix} 128.4 \\ 80.6 \\ 63.5 \\ 25.8 \end{bmatrix}$

     (d) $\begin{bmatrix} 653.09 \\ 399.77 \\ 528.23 \\ -394.52 \end{bmatrix}$

Section 1.3

1. (a) F   (b) F   (c) T   (d) F   (e) T   (f) T
   (g) F   (h) T   (i) T   (j) F   (k) T

3. (a) $\begin{bmatrix} 0 & -1 & 2 \\ 1 & 3 & 0 \end{bmatrix}$ (b) $\begin{bmatrix} 0 & -1 & 2 & 0 \\ 1 & 3 & 0 & -1 \end{bmatrix}$

5. (a) $\begin{bmatrix} 1 & 2 \\ -1 & 3 \\ -3 & 4 \end{bmatrix}$ (b) $\begin{bmatrix} 1 & 2 & 3 \\ -1 & 3 & 2 \\ -3 & 4 & 1 \end{bmatrix}$

7. $\begin{bmatrix} 0 & 2 & -4 & 4 & 2 \\ -2 & 6 & 3 & -1 & 1 \\ 1 & -1 & 0 & 2 & -3 \end{bmatrix}$

9. $\begin{bmatrix} 1 & -1 & 0 & 2 & -3 \\ 0 & 4 & 3 & 3 & -5 \\ 0 & 2 & -4 & 4 & 2 \end{bmatrix}$

11. $\begin{bmatrix} 1 & -1 & 0 & 2 & -3 \\ -2 & 6 & 3 & -1 & 1 \\ 0 & 1 & -2 & 2 & 1 \end{bmatrix}$

13. $\begin{bmatrix} -2 & 4 & 0 \\ -1 & 1 & -1 \\ 2 & -4 & 6 \\ -3 & 2 & 1 \end{bmatrix}$ **15.** $\begin{bmatrix} 1 & -2 & 0 \\ -1 & 1 & -1 \\ 0 & 0 & 6 \\ -3 & 2 & 1 \end{bmatrix}$

17. $\begin{bmatrix} 1 & -2 & 0 \\ 2 & -4 & 6 \\ -1 & 1 & -1 \\ -3 & 2 & 1 \end{bmatrix}$

19. yes   **21.** no

23. $x_1 = 2 + x_2$ **25.** $x_1 = 6 + 2x_2$
   x_2 free x_2 free

27. not consistent

29. $x_1 = 4 + 2x_2$ **31.** $x_1 = \quad 3x_4$
   x_2 free $x_2 = \quad 4x_4$
   $x_3 = 3$ $x_3 = -5x_4$
                          x_4 free

33. x_1 free
   $x_2 = -3$, $\begin{bmatrix} x_1 \\ x_2 \\ x_3 \\ x_4 \end{bmatrix} = x_1 \begin{bmatrix} 1 \\ 0 \\ 0 \\ 0 \end{bmatrix} + \begin{bmatrix} 0 \\ -3 \\ -4 \\ 5 \end{bmatrix}$
   $x_3 = -4$
   $x_4 = \quad 5$

35. $x_1 = 6 - 3x_2 + 2x_4$
   x_2 free   ,
   $x_3 = 7 - 4x_4$
   x_4 free

$\begin{bmatrix} x_1 \\ x_2 \\ x_3 \\ x_4 \end{bmatrix} = \begin{bmatrix} 6 \\ 0 \\ 7 \\ 0 \end{bmatrix} + x_2 \begin{bmatrix} -3 \\ 1 \\ 0 \\ 0 \end{bmatrix} + x_4 \begin{bmatrix} 2 \\ 0 \\ -4 \\ 1 \end{bmatrix}$

37. not consistent   **39.** $n - k$ **45.** 7

Section 1.4

1. (a) T   (b) F   (c) T   (d) T   (e) T   (f) T
    (g) F   (h) F   (i) T   (j) T   (k) T   (l) F
    (m) F

3. $x_1 = -2 - 3x_2$ **5.** $x_1 = 4$ **7.** not consistent
   x_2 free $x_2 = 5$

9. $x_1 = -1 + 2x_2$ **11.** $x_1 = \quad 1 + 2x_3$
   x_2 free $x_2 = -2 - \quad x_3$
   $x_3 = \quad 2$ x_3 free
                           $x_4 = -3$

13. $x_1 = -4 - 3x_2 + x_4$ **15.** not consistent
   x_2 free
   $x_3 = \quad 3 - 2x_4$
   x_4 free

17. $x_1 = -2 + \quad x_5$
   x_2 free
   $x_3 = \quad 3 - 3x_5$
   $x_4 = -1 - 2x_5$
   x_5 free

19. -12 **21.** $r \neq 0$ **23.** no r

25. (a) $r = 2, s \neq 15$ (b) $r \neq 2$ (c) $r = 2, s = 15$

27. (a) $r = 3$ (b) $r \neq \pm 3$ (c) $r = -3$

29. 3, 1   **31.** 2, 3

33. (a) 10, 20, and 25 days, respectively   (b) no

35. (a) 15 units   (b) no

37. $2x^2 - 5x + 7$ **39.** $4x^2 - 7x + 2$

41. It is e_3.   **43.** the $m \times n$ zero matrix   **45.** 4

47. 3

49. the minimum of m and n **51.** no

63. I_{m+n} **65.** $x_1 = \quad 2.32 + 0.32x_5$ **67.** 3, 2
                       $x_2 = -6.44 + 0.56x_5$
                       $x_3 = \quad 0.72 - 0.28x_5$
                       $x_4 = \quad 5.92 + 0.92x_5$
                       x_5 free

69. 4, 1

Section 1.5

1. (a) F   (b) T   (c) T   (d) F   (e) T

3. $11 million   **5.** services   **7.** entertainment

9. $16.1 million of agriculture, $17.8 million of manufacturing, $18 million of services, and $10.1 million of entertainment

11. $13.9 million of agriculture, $22.2 million of manufacturing, $12 million of services, and $9.9 million of entertainment

13. (a) $15.5 million of transportation, $1.5 million of food, and $9 million of oil
    (b) $128 million of transportation, $160 million of food, and $128 million of oil

15. (a) $\begin{bmatrix} .1 & .4 \\ .3 & .2 \end{bmatrix}$

    (b) $34 million of electricity and $22 million of oil
    (c) $128 million of electricity and $138 million of oil

17. (a) $49 million of finance, $10 million of goods, and $18 million of services
    (b) $75 million of finance, $125 million of goods, and $100 million of services
    (c) $75 million of finance, $104 million of goods, and $114 million of services

21. $I_1 = 9, I_2 = 4, I_3 = 5$

23. $I_1 = 21, I_2 = 3, I_3 = 18$ **25.** $v = 12$

Section 1.6

1. (a) T (b) T (c) T (d) F (e) T (f) T
   (g) T (h) F (i) F (j) F
3. yes   5. no   7. yes   9. 3   11. −6
13. no   15. yes
17. yes   19. no   21. yes   23. no   25. no
27. yes

29. $\left\{\begin{bmatrix}1\\3\end{bmatrix},\begin{bmatrix}0\\1\end{bmatrix}\right\}$ 31. $\left\{\begin{bmatrix}1\\0\\-1\end{bmatrix},\begin{bmatrix}0\\1\\0\end{bmatrix}\right\}$

33. $\left\{\begin{bmatrix}1\\-2\\1\end{bmatrix}\right\}$ 35. $\left\{\begin{bmatrix}-1\\0\\1\end{bmatrix},\begin{bmatrix}0\\1\\2\end{bmatrix}\right\}$

37. (a) 2 (b) infinitely many   45. no   51. yes
53. no

Section 1.7

1. (a) T (b) F (c) F (d) T (e) T (f) T
   (g) F (h) T (i) F (j) F (k) F (l) T
   (m) T
3. yes   5. no

7. $\left\{\begin{bmatrix}1\\-2\\3\end{bmatrix}\right\}$ 9. $\left\{\begin{bmatrix}-3\\2\\0\end{bmatrix},\begin{bmatrix}1\\6\\0\end{bmatrix}\right\}$

11. no   13. yes   15. yes   17. no

19. $-3\begin{bmatrix}-1\\1\\2\end{bmatrix}=\begin{bmatrix}3\\-3\\-6\end{bmatrix}$ 21. $5\begin{bmatrix}0\\1\\1\end{bmatrix}+4\begin{bmatrix}1\\0\\-1\end{bmatrix}=\begin{bmatrix}4\\5\\1\end{bmatrix}$

23. all real numbers   25. −2

27. $\begin{bmatrix}x_1\\x_2\\x_3\end{bmatrix}=x_2\begin{bmatrix}4\\1\\0\end{bmatrix}+x_3\begin{bmatrix}-2\\0\\1\end{bmatrix}$

29. $\begin{bmatrix}x_1\\x_2\\x_3\\x_4\end{bmatrix}=x_2\begin{bmatrix}-3\\1\\0\\0\end{bmatrix}+x_4\begin{bmatrix}-2\\0\\6\\1\end{bmatrix}$

31. $\begin{bmatrix}x_1\\x_2\\x_3\\x_4\end{bmatrix}=x_3\begin{bmatrix}-4\\3\\1\\0\end{bmatrix}+x_4\begin{bmatrix}2\\-5\\0\\1\end{bmatrix}$

33. $\begin{bmatrix}x_1\\x_2\\x_3\\x_4\\x_5\\x_6\end{bmatrix}=x_2\begin{bmatrix}0\\1\\0\\0\\0\\0\end{bmatrix}+x_4\begin{bmatrix}-1\\0\\2\\1\\0\\0\end{bmatrix}+x_6\begin{bmatrix}-3\\0\\-1\\0\\0\\1\end{bmatrix}$

35. $A=\begin{bmatrix}1&0\\0&1\end{bmatrix}$

53. The set is linearly dependent, and $\mathbf{v}_5 = 2\mathbf{v}_1 - \mathbf{v}_3 + \mathbf{v}_4$, where $\mathbf{v}_j$ is the jth vector in the set.
55. The set is linearly independent.

Chapter 1 Review

1. (a) F (b) T (c) T (d) T (e) T (f) T
   (g) T (h) F (i) F (j) T (k) T (l) T
   (m) F (n) T (o) T (p) F (q) F
3. (a) There is at most one solution.
   (b) There is at least one solution.

5. $\begin{bmatrix}3&2\\-2&7\\4&3\end{bmatrix}$

7. undefined because A has 2 columns and D^T has 3 rows

9. $\begin{bmatrix}3\\3\end{bmatrix}$

11. undefined because C^T and D do not have the same number of columns
13. The components are the average values of sales for all stores during January of last year for produce, meats, dairy, and processed foods, respectively.

15. $\begin{bmatrix}0\\-4\\3\\-2\end{bmatrix}$ 17. $\dfrac{1}{2}\begin{bmatrix}2-\sqrt{3}\\-2\sqrt{3}-1\end{bmatrix}$

19. $\mathbf{v}=(-1)\begin{bmatrix}-1\\5\\2\end{bmatrix}+3\begin{bmatrix}1\\3\\4\end{bmatrix}+1\begin{bmatrix}1\\-1\\1\end{bmatrix}$

21. $\mathbf{v}$ is not in the span of $\mathcal{S}$.

23. $x_1 = 1 - 2x_2 + x_3$ 25. inconsistent
    x_2 free
    x_3 free

27. $x_1 = 7 - 5x_3 - 4x_4$
    $x_2 = 5 + 3x_3 + 3x_4$
    x_3 free
    x_4 free

29. The rank is 1, and the nullity is 4.
31. The rank is 3, and the nullity is 2.
33. 20 of the first pack, 10 of the second pack, 40 of the third pack
35. yes   37. no   39. yes   41. yes   43. no
45. linearly independent   47. linearly dependent

49. $\begin{bmatrix}3\\3\\8\end{bmatrix}=2\begin{bmatrix}1\\2\\3\end{bmatrix}+1\begin{bmatrix}1\\-1\\2\end{bmatrix}$

51. $\begin{bmatrix} 1 \\ -1 \\ 1 \\ -1 \end{bmatrix} = 2 \begin{bmatrix} 1 \\ 0 \\ 1 \\ 0 \end{bmatrix} + (-1) \begin{bmatrix} 1 \\ 1 \\ 1 \\ 1 \end{bmatrix}$

53. $\begin{bmatrix} x_1 \\ x_2 \\ x_3 \end{bmatrix} = x_3 \begin{bmatrix} -3 \\ 2 \\ 1 \end{bmatrix}$ **55.** $\begin{bmatrix} x_1 \\ x_2 \\ x_3 \\ x_4 \end{bmatrix} = x_4 \begin{bmatrix} -2 \\ 5 \\ 0 \\ 1 \end{bmatrix}$

CHAPTER 2

Section 2.1

1. (a) F   (b) T   (c) F   (d) F   (e) T
(f) T   (g) F

3. AB is defined and has size 2×2.

5. Let A be any 2×3 matrix and B be any 3×4 matrix.

7. $C\mathbf{y} = \begin{bmatrix} 22 \\ -18 \end{bmatrix}$ **9.** $\mathbf{xz} = \begin{bmatrix} 14 & -2 \\ 21 & -3 \end{bmatrix}$

11. $AC\mathbf{x}$ is undefined.

13. $AB = \begin{bmatrix} 5 & 0 \\ 25 & 20 \end{bmatrix}$ **15.** $BC = \begin{bmatrix} 29 & 56 & 23 \\ 7 & 8 & 9 \end{bmatrix}$

17. CB^T is undefined.

19. $A^3 = \begin{bmatrix} -35 & -30 \\ 45 & 10 \end{bmatrix}$ **21.** C^2 is undefined.

25. -2 **27.** 24   **29.** $\begin{bmatrix} -4 \\ -9 \\ -2 \end{bmatrix}$ **31.** $[7 \quad 5 \quad 10]$

33. $AB = \begin{bmatrix} -1 & 0 \\ -2 & 0 \\ 3 & 0 \end{bmatrix} + \begin{bmatrix} 8 & 2 \\ -4 & -1 \\ -8 & -2 \end{bmatrix} + \begin{bmatrix} 9 & -6 \\ 12 & -8 \\ 0 & 0 \end{bmatrix}$

35. $\begin{bmatrix} 1 & 1 & 2 & 1 \\ 1 & 0 & 1 & -1 \\ 0 & 1 & -1 & 1 \end{bmatrix}$ **37.** $\begin{bmatrix} 3 & 6 \\ 9 & 12 \\ 2 & 4 \\ 6 & 8 \end{bmatrix}$

39. $\begin{bmatrix} A_{60°} & A_{70°} \\ A_{70°} & A_{80°} \end{bmatrix}$ **41.** (a) $B = \begin{bmatrix} .70 & .95 \\ .30 & .05 \end{bmatrix}$

43. (a)

	Today	
	Hot	Bag
	lunch	lunch
Next day Hot lunch	.3	.4
Bag lunch	.7	.6

$A = \begin{bmatrix} .3 & .4 \\ .7 & .6 \end{bmatrix}$

(b) $A^3 \begin{bmatrix} u_1 \\ u_2 \end{bmatrix} = \begin{bmatrix} 109.1 \\ 190.9 \end{bmatrix}$. Approximately 109 students will buy hot lunches and 191 students will bring bag lunches three school days from today.

(c) $A^{100} \begin{bmatrix} u_1 \\ u_2 \end{bmatrix} = \begin{bmatrix} 109.0909 \\ 190.9091 \end{bmatrix}$ (rounded to four places after the decimal)

59. (a), (b), and (c) have the same answer, namely,
$\begin{bmatrix} -1 & 0 \\ 0 & -1 \end{bmatrix}$.

61. (c) $A^k = \begin{bmatrix} B^k & * \\ 0 & D^k \end{bmatrix}$, where $*$ represents some 2×2 matrix.

63. (a) The population of the city is 205,688. The population of the suburbs is 994,332.
(b) The population of the city is 200,015. The population of the suburbs is 999,985.

Section 2.2

1. (a) F   (b) F   (c) T   (d) F   (e) T

3. (a) all of them
(b) 0 from the first and 1 from the second
(c) $\begin{bmatrix} a \\ b \end{bmatrix}$ in even-numbered years and $\begin{bmatrix} b \\ a \end{bmatrix}$ in odd-numbered years

5. (a) $\begin{bmatrix} 0 & 2 & 1 \\ q & 0 & 0 \\ 0 & .5 & 0 \end{bmatrix}$
(b) The population grows without bound.
(c) The population approaches $\mathbf{0}$.
(d) $q = .4$, $\begin{bmatrix} 400 \\ 160 \\ 80 \end{bmatrix}$
(e) Over time, it approaches $\begin{bmatrix} 450 \\ 180 \\ 90 \end{bmatrix}$.
(f) $q = .4$
(g) $\mathbf{x} = x_3 \begin{bmatrix} 5 \\ 2 \\ 1 \end{bmatrix}$. The stable distributions have this form.

7. (a) $\begin{bmatrix} 0 & 2 & b \\ .2 & 0 & 0 \\ 0 & .5 & 0 \end{bmatrix}$ (b) The population approaches $\mathbf{0}$.
(c) The population grows without bound.
(d) $b = 6$, $\begin{bmatrix} 1600 \\ 320 \\ 160 \end{bmatrix}$
(e) Over time, it approaches $\begin{bmatrix} 1500 \\ 300 \\ 150 \end{bmatrix}$.
(f) $b = 6$

(g) $\begin{bmatrix} p_1 \\ p_2 \\ p_3 \end{bmatrix} = c \begin{bmatrix} 10 \\ 2 \\ 1 \end{bmatrix}$ where $c = \dfrac{1}{26}(p_1 + 5p_2 + 6p_3)$.

9. $\begin{bmatrix} 0.644 & 0.628 \\ 0.356 & 0.372 \end{bmatrix}$

11. (a) There are no nonstop flights from any of the cities 1, 2, and 3 to the cities 4 and 5, and vice versa.

(b) $A^2 = \begin{bmatrix} B^2 & O_1 \\ O_2 & C^2 \end{bmatrix}$, $A^3 = \begin{bmatrix} B^3 & O_1 \\ O_2 & C^2 \end{bmatrix}$, and

$A^k = \begin{bmatrix} B^k & O_1 \\ O_2 & C^2 \end{bmatrix}$.

(c) There are no flights with any number of layovers from any of the cities 1, 2, and 3 to the cities 4 and 5, or vice versa.

13. (a) 1 and 2, 1 and 4, 2 and 3, 3 and 4

(c) $\begin{bmatrix} 0 & 1 & 0 & 1 \\ 1 & 0 & 1 & 0 \\ 0 & 1 & 0 & 1 \\ 1 & 0 & 1 & 0 \end{bmatrix}$, yes

15. (c) There is one student who wants to take both course 1 and course 2, and there are three students who want to take course 1 and course 9.

(d) For each i, the ith diagonal entry of AA^T represents the number of students who take course i.

17. (a)

k	Sun	Noble	Honored	Stinkard
1	100	300	500	7700
2	100	400	800	7300
3	100	500	1200	6800

(b)

k	Sun	Noble	Honored	Stinkard
9	100	1100	5700	1700
10	100	1200	6800	500
11	100	1300	8000	−800

Section 2.3

1. (a) F   (b) T   (c) T   (d) F   (e) T   (f) T

3. no   5. yes

7. $(AB^T)^{-1} = \begin{bmatrix} 5 & 7 & 3 \\ -3 & -4 & -1 \\ 12 & 7 & 12 \end{bmatrix}$ 11. $\begin{bmatrix} 1 & 0 & 0 \\ 2 & 1 & 0 \\ 0 & 0 & 1 \end{bmatrix}$

13. $\begin{bmatrix} 1 & 0 & 0 & 0 \\ 0 & .25 & 0 & 0 \\ 0 & 0 & 1 & 0 \\ 0 & 0 & 0 & 1 \end{bmatrix}$

15. $E = \begin{bmatrix} 1 & 0 & 0 \\ 0 & 0 & 1 \\ 0 & 1 & 0 \end{bmatrix}$ 29. $A = \begin{bmatrix} 3 & 2 & 7 \\ -1 & 5 & 9 \end{bmatrix}$

31. $A = \begin{bmatrix} -1 & 1 & 1 & 4 & 13 \\ 2 & -2 & -1 & 1 & 3 \\ -1 & 1 & 0 & 3 & 8 \end{bmatrix}$

33. $A = \begin{bmatrix} 1 & 2 & 0 & -1 \\ 0 & 0 & 1 & 1 \\ 0 & 0 & 0 & 0 \end{bmatrix}$

45. (a) $A^{-1} = \begin{bmatrix} -7 & 2 & 3 & -2 \\ 5 & -1 & -2 & 1 \\ 1 & 0 & 0 & 1 \\ -3 & 1 & 1 & -1 \end{bmatrix}$

(b) $B^{-1} = \begin{bmatrix} 3 & 2 & -7 & -2 \\ -2 & -1 & 5 & 1 \\ 0 & 0 & 1 & 1 \\ 1 & 1 & -3 & -1 \end{bmatrix}$

and

$C^{-1} = \begin{bmatrix} -7 & -2 & 3 & 2 \\ 5 & 1 & -2 & -1 \\ 1 & 1 & 0 & 0 \\ -3 & -1 & 1 & 1 \end{bmatrix}$

(c) B^{-1} can be obtained by interchanging columns 1 and 3 of A^{-1}, and C^{-1} can be obtained by interchanging columns 2 and 4 of A^{-1}.

(d) B^{-1} can be obtained by interchanging columns i and j of A^{-1}.

47. (b) $(A^2)^{-1} = (A^{-1})^2 = \begin{bmatrix} 113 & -22 & -10 & -13 \\ -62 & 13 & 6 & 6 \\ -22 & 4 & 3 & 2 \\ 7 & -2 & -1 & 0 \end{bmatrix}$

49. (b) $A^{-1} = \begin{bmatrix} 0.2870 & 0.3912 & 0.2400 & -0.0848 \\ -0.2993 & -0.3759 & -0.1471 & 0.2094 \\ -0.2339 & -0.4576 & -0.3095 & 0.2114 \\ -0.1369 & -0.3289 & -0.1287 & 0.0582 \end{bmatrix}$,

$B^{-1} = \begin{bmatrix} -0.0384 & 0.0042 & 0.0558 & 0.0740 \\ 0.0813 & 0.0483 & -0.0412 & -0.0335 \\ -0.0188 & -0.0112 & -0.0674 & 0.0846 \\ 0.0784 & -0.0615 & 0.0070 & 0.0426 \end{bmatrix}$,

and

$C^{-1} = \begin{bmatrix} -0.0556 & -0.1429 & -0.1032 & 0.1746 \\ 0.0444 & -0.1714 & 0.0540 & 0.0317 \\ 0.0028 & -0.1714 & -0.2377 & 0.1984 \\ 0.0306 & 0.2571 & 0.0996 & -0.1032 \end{bmatrix}$

(c)

$-A^{-1}BC^{-1} = \begin{bmatrix} 0.0569 & 1.4822 & -0.3997 & -0.6180 \\ 0.0183 & -1.2210 & 0.5182 & 0.3251 \\ -0.0902 & -1.5286 & 0.5630 & 0.5010 \\ -0.0607 & -1.3683 & 0.9012 & 0.6247 \end{bmatrix}$

Section 2.4

1. (a) F   (b) T   (c) T   (d) T   (e) T
   (f) F   (g) T

3. $\begin{bmatrix} -2 & 3 \\ 1 & -1 \end{bmatrix}$ **5.** not invertible

7. $\dfrac{1}{3}\begin{bmatrix} -7 & 2 & 3 \\ -6 & 0 & 3 \\ 8 & -1 & -3 \end{bmatrix}$ **9.** not invertible

11. $A^{-1}B = \begin{bmatrix} -1 & 3 & -4 \\ 1 & -2 & 3 \end{bmatrix}$

13. $A^{-1}B = \begin{bmatrix} -1 & -4 & 7 & -7 \\ 2 & 6 & -6 & 10 \end{bmatrix}$

15. $A^{-1}B = \begin{bmatrix} 1.0 & -0.5 & 1.5 & 1.0 \\ 6.0 & 12.5 & -11.5 & 12.0 \\ -2.0 & -5.5 & 5.5 & -5.0 \end{bmatrix}$

17. $R = \begin{bmatrix} 1 & 0 & -1 \\ 0 & 1 & -3 \end{bmatrix}$, $P = \begin{bmatrix} -1 & -1 \\ -2 & -1 \end{bmatrix}$

19. $R = \begin{bmatrix} 1 & 0 & -2 & -1 \\ 0 & 1 & 1 & -1 \\ 0 & 0 & 0 & 0 \end{bmatrix}$. One possibility is

$P = \begin{bmatrix} -1 & 0 & 0 \\ 0 & 1 & 0 \\ 2 & -3 & 1 \end{bmatrix}$.

21. $R = \begin{bmatrix} 1 & 0 & 0 & 0 \\ 0 & 1 & 0 & 0 \\ 0 & 0 & 1 & 0 \\ 0 & 0 & 0 & 1 \end{bmatrix}$, $P = \begin{bmatrix} -4 & -15 & -8 & 1 \\ 1 & 4 & 2 & 0 \\ 1 & 3 & 2 & 0 \\ -4 & -13 & -7 & -1 \end{bmatrix}$

25. (a) $\begin{bmatrix} -1 & -3 \\ 2 & 5 \end{bmatrix}\begin{bmatrix} x_1 \\ x_2 \end{bmatrix} = \begin{bmatrix} -6 \\ 4 \end{bmatrix}$

(c) $\begin{bmatrix} x_1 \\ x_2 \end{bmatrix} = A^{-1}\mathbf{b} = \begin{bmatrix} -18 \\ 8 \end{bmatrix}$

27. (a) $\begin{bmatrix} -1 & 0 & 1 \\ 1 & 2 & -2 \\ 2 & -1 & 1 \end{bmatrix}\begin{bmatrix} x_1 \\ x_2 \\ x_3 \end{bmatrix} = \begin{bmatrix} -4 \\ 3 \\ 1 \end{bmatrix}$

(c) $\begin{bmatrix} x_1 \\ x_2 \\ x_3 \end{bmatrix} = A^{-1}\mathbf{b} = \begin{bmatrix} 1 \\ -2 \\ -3 \end{bmatrix}$

29. (b) $A^{-1} = A^{k-1}$

33. no   **39.** $x_1 = -3 + x_3$

$x_2 = 4 - 2x_3$

x_3 free

41. \$2 million of electricity and \$4.5 million of oil

43. \$12.5 million of finance, \$15 million of goods, and \$65 million of services

53. The reduced row echelon form of A is I_4.

55. rank $A = 4$

Section 2.5

1. (a) F   (b) T

3. $L = \begin{bmatrix} 1 & 0 & 0 \\ 3 & 1 & 0 \\ -1 & 1 & 1 \end{bmatrix}$, $U = \begin{bmatrix} 2 & 3 & 4 \\ 0 & -1 & -2 \\ 0 & 0 & 3 \end{bmatrix}$

5. $L = \begin{bmatrix} 1 & 0 & 0 & 0 \\ -1 & 1 & 0 & 0 \\ 2 & -1 & 1 & 0 \\ 1 & 3 & 0 & 1 \end{bmatrix}$,

$U = \begin{bmatrix} -1 & 2 & 1 & -1 & 3 \\ 0 & -2 & 1 & 4 & -2 \\ 0 & 0 & -2 & 1 & -1 \\ 0 & 0 & 0 & 0 & 1 \end{bmatrix}$

7. $\begin{bmatrix} x_1 \\ x_2 \\ x_3 \end{bmatrix} = \begin{bmatrix} 2 \\ -1 \\ 0 \end{bmatrix}$ **9.** $\begin{bmatrix} x_1 \\ x_2 \\ x_3 \\ x_4 \\ x_5 \end{bmatrix} = \begin{bmatrix} 1 \\ -3 \\ 2 \\ 0 \\ 4 \end{bmatrix} + x_4\begin{bmatrix} 4 \\ \frac{9}{4} \\ \frac{1}{2} \\ 1 \\ 0 \end{bmatrix}$

15. $P = \begin{bmatrix} 1 & 0 & 0 \\ 0 & 0 & 1 \\ 0 & 1 & 0 \end{bmatrix}$, $L = \begin{bmatrix} 1 & 0 & 0 \\ -1 & 1 & 0 \\ 2 & 0 & 1 \end{bmatrix}$, and

$U = \begin{bmatrix} 1 & -1 & 3 \\ 0 & 1 & 2 \\ 0 & 0 & -1 \end{bmatrix}$

17. $P = \begin{bmatrix} 1 & 0 & 0 & 0 \\ 0 & 0 & 0 & 1 \\ 0 & 1 & 0 & 0 \\ 0 & 0 & 1 & 0 \end{bmatrix}$, $L = \begin{bmatrix} 1 & 0 & 0 & 0 \\ 2 & 1 & 0 & 0 \\ 2 & 0 & 1 & 0 \\ 3 & -4 & 0 & 1 \end{bmatrix}$, and

$U = \begin{bmatrix} 1 & 2 & 1 & -1 \\ 0 & 1 & 1 & 2 \\ 0 & 0 & -1 & 3 \\ 0 & 0 & 0 & 9 \end{bmatrix}$

19. $\begin{bmatrix} x_1 \\ x_2 \\ x_3 \end{bmatrix} = \begin{bmatrix} -2 \\ 1 \\ 3 \end{bmatrix}$ **21.** $\begin{bmatrix} x_1 \\ x_2 \\ x_3 \\ x_4 \end{bmatrix} = \begin{bmatrix} -3 \\ 2 \\ 1 \\ -1 \end{bmatrix}$ **25.** $m(2n-1)p$

27. $L = \begin{bmatrix} 1 & 0 & 0 & 0 & 0 \\ -1 & 1 & 0 & 0 & 0 \\ 2 & 3 & 1 & 0 & 0 \\ 2 & -3 & 2 & 1 & 0 \\ 2 & 0 & 1 & -1 & 1 \end{bmatrix}$, $U = \begin{bmatrix} 2 & -1 & 3 & 2 & 1 \\ 0 & 1 & 2 & 3 & 5 \\ 0 & 0 & 3 & -1 & 2 \\ 0 & 0 & 0 & 1 & 8 \\ 0 & 0 & 0 & 0 & 13 \end{bmatrix}$

29. $P = \begin{bmatrix} 0 & 1 & 0 & 0 & 0 \\ 1 & 0 & 0 & 0 & 0 \\ 0 & 0 & 1 & 0 & 0 \\ 0 & 0 & 0 & 1 & 0 \\ 0 & 0 & 0 & 0 & 1 \end{bmatrix}$, $L = \begin{bmatrix} 1.0 & 0 & 0 & 0 & 0 \\ 0.0 & 1 & 0 & 0 & 0 \\ 0.5 & 2 & 1 & 0 & 0 \\ -0.5 & -1 & -3 & 1 & 0 \\ 1.5 & 7 & 9 & -9 & 1 \end{bmatrix}$,

and

$$U = \begin{bmatrix} 2 & -2 & -1.0 & 3.0 & 4 \\ 0 & 1 & 2.0 & -1.0 & 1 \\ 0 & 0 & -1.5 & -0.5 & -2 \\ 0 & 0 & 0.0 & -1.0 & -2 \\ 0 & 0 & 0.0 & 0.0 & -9 \end{bmatrix}$$

Section 2.6

1. (a) F   (b) T   (c) F   (d) T   (e) F
   (f) T   (g) F   (h) T

3. The domain is $\mathcal{R}^3$ and the codomain is $\mathcal{R}^2$.

5. $\begin{bmatrix} 11 \\ 8 \end{bmatrix}$ **7.** $\begin{bmatrix} 8 \\ -6 \\ 11 \end{bmatrix}$ **9.** $\begin{bmatrix} 2 \\ 4 \end{bmatrix}$ and $\begin{bmatrix} 1 \\ -2 \end{bmatrix}$

11. $n = 2, m = 3$

13. $\begin{bmatrix} 4 \\ -8 \\ 12 \end{bmatrix}$ and $\begin{bmatrix} -1 \\ 2 \\ -3 \end{bmatrix}$

15. $T\left(\begin{bmatrix} x_1 \\ x_2 \end{bmatrix}\right) = \begin{bmatrix} 12x_1 + 5x_2 \\ 3x_1 + x_2 \end{bmatrix}$

17. $\begin{bmatrix} 0 & 1 \\ 1 & 1 \end{bmatrix}$ **19.** $\begin{bmatrix} 1 & 1 & 1 \\ 2 & 0 & 0 \end{bmatrix}$

21. $\begin{bmatrix} 1 & 0 & 0 \\ 0 & 1 & 0 \\ 0 & 0 & 1 \end{bmatrix}$ **23.** $\begin{bmatrix} 4 & 0 & 0 \\ 0 & 4 & 0 \\ 0 & 0 & 4 \end{bmatrix}$ **25.** $\begin{bmatrix} 2a + 4b \\ 3a + b \end{bmatrix}$

27. linear   **29.** not linear   **31.** linear

33. not linear

43. (b) $\begin{bmatrix} 1 & 0 \\ 0 & 0 \end{bmatrix}$

45. $T = T_A$ for $A = \begin{bmatrix} -1 & 0 \\ 0 & 1 \end{bmatrix}$ (b) $\mathcal{R}^2$

47. (b) $\mathcal{R}^2$ **51.** Both are **v**.

57. The given vector is in the range of T.

Section 2.7

1. (a) T   (b) F   (c) F   (d) T   (e) T
   (f) T   (g) F   (h) T

3. $\left\{ \begin{bmatrix} 2 \\ 4 \end{bmatrix}, \begin{bmatrix} 3 \\ 5 \end{bmatrix} \right\}$ **5.** $\left\{ \begin{bmatrix} 0 \\ 2 \\ 1 \end{bmatrix}, \begin{bmatrix} 3 \\ -1 \\ 1 \end{bmatrix} \right\}$

7. $\left\{ \begin{bmatrix} 2 \\ 2 \\ 4 \end{bmatrix}, \begin{bmatrix} 1 \\ 2 \\ 1 \end{bmatrix}, \begin{bmatrix} 1 \\ 3 \\ 0 \end{bmatrix} \right\}$ **9.** $\left\{ \begin{bmatrix} 0 \\ 0 \end{bmatrix} \right\}$

11. $\left\{ \begin{bmatrix} 1 \\ 0 \end{bmatrix} \right\}$ **13.** $\{\mathbf{0}\}$, one-to-one

15. $\left\{ \begin{bmatrix} 0 \\ -1 \\ 1 \end{bmatrix} \right\}$, not one-to-one

17. $\left\{ \begin{bmatrix} 1 \\ -1 \\ 1 \end{bmatrix} \right\}$, not one-to-one

19. $\{\mathbf{0}\}$, one-to-one   **21.** $\{\mathbf{e}_2\}$, not one-to-one

23. $\left\{ \begin{bmatrix} 1 \\ -3 \\ 1 \\ 0 \end{bmatrix}, \begin{bmatrix} 3 \\ -5 \\ 0 \\ 1 \end{bmatrix} \right\}$, not one-to-one

25. $\begin{bmatrix} 2 & 3 \\ 4 & 5 \end{bmatrix}$, one-to-one   **27.** $\begin{bmatrix} 0 & 3 \\ 2 & -1 \\ 1 & 1 \end{bmatrix}$, one-to-one

29. $\begin{bmatrix} 2 & 3 \\ 4 & 5 \end{bmatrix}$, onto   **31.** $\begin{bmatrix} 0 & 3 \\ 2 & -1 \\ 1 & 1 \end{bmatrix}$, not onto

33. (a) $\{\mathbf{0}\}$ (b) one-to-one   (c) $\mathcal{R}^2$ (d) yes

35. (a) Span $\{\mathbf{e}_1\}$ (b) no   (c) Span $\{\mathbf{e}_2\}$ (d) no

37. (a) Span $\{\mathbf{e}_3\}$ (b) no   (c) Span $\{\mathbf{e}_1, \mathbf{e}_2\}$
   (d) no

39. (a) one-to-one   (b) onto

41. The domain and codomain are $\mathcal{R}^2$. The rule is

$$UT\left(\begin{bmatrix} x_1 \\ x_2 \end{bmatrix}\right) = \begin{bmatrix} 16x_1 + 4x_2 \\ 4x_1 - 8x_2 \end{bmatrix}.$$

43. $A = \begin{bmatrix} 1 & 1 \\ 1 & -3 \\ 4 & 0 \end{bmatrix}$ and $B = \begin{bmatrix} 1 & -1 & 4 \\ 1 & 3 & 0 \end{bmatrix}$

45. The domain and codomain are $\mathcal{R}^3$. The rule is

$$TU\left(\begin{bmatrix} x_1 \\ x_2 \\ x_3 \end{bmatrix}\right) = \begin{bmatrix} 2x_1 + 2x_2 + 4x_3 \\ -2x_1 - 10x_2 + 4x_3 \\ 4x_1 - 4x_2 + 16x_3 \end{bmatrix}.$$

47. $\begin{bmatrix} 2 & 2 & 4 \\ -2 & -10 & 4 \\ 4 & -4 & 16 \end{bmatrix}$ **49.** $\begin{bmatrix} -1 & 5 \\ 15 & -5 \end{bmatrix}$ **51.** $\begin{bmatrix} -1 & 5 \\ 15 & -5 \end{bmatrix}$

53. $\begin{bmatrix} 2 & 9 \\ 6 & -8 \end{bmatrix}$

55. $T^{-1}\left(\begin{bmatrix} x_1 \\ x_2 \end{bmatrix}\right) = \begin{bmatrix} \dfrac{1}{3}x_1 + \dfrac{1}{3}x_2 \\ -\dfrac{1}{3}x_1 + \dfrac{2}{3}x_2 \end{bmatrix}$

57. $T^{-1}\left(\begin{bmatrix} x_1 \\ x_2 \\ x_3 \end{bmatrix}\right) = \begin{bmatrix} 2x_1 + x_2 - x_3 \\ -9x_1 - 2x_2 + 5x_3 \\ 4x_1 + x_2 - 2x_3 \end{bmatrix}$

59. yes

67. (a) $A = \begin{bmatrix} 1 & 3 & -2 & 1 \\ 3 & 0 & 4 & 1 \\ 2 & -1 & 0 & 2 \\ 0 & 0 & 1 & 1 \end{bmatrix}$ and

$B = \begin{bmatrix} 0 & 1 & 0 & -3 \\ 2 & 0 & 1 & -1 \\ 1 & -2 & 0 & 4 \\ 0 & 5 & 1 & 0 \end{bmatrix}$.

(b) $AB = \begin{bmatrix} 4 & 10 & 4 & -14 \\ 4 & 0 & 1 & 7 \\ -2 & 12 & 1 & -5 \\ 1 & 3 & 1 & 4 \end{bmatrix}$

(c) $TU\left(\begin{bmatrix} x_1 \\ x_2 \\ x_3 \\ x_4 \end{bmatrix}\right) = \begin{bmatrix} 4x_1 + 10x_2 + 4x_3 - 14x_4 \\ 4x_1 + x_3 + 7x_4 \\ -2x_1 + 12x_2 + x_3 - 5x_4 \\ x_1 + 3x_2 + x_3 + 4x_4 \end{bmatrix}$

Chapter 2 Review

1. (a) T (b) F (c) F (d) T (e) F (f) F
(g) T (h) T (i) F (j) T (k) F (l) T
(m) F (n) T (o) T (p) F (q) T (r) F
(s) F (t) T

3. (a) BA is defined if and only if $q = m$.   (b) $p \times n$

5. $\begin{bmatrix} 64 & -4 \\ 32 & -2 \end{bmatrix}$ **7.** $\begin{bmatrix} 2 \\ 29 \\ 4 \end{bmatrix}$ **9.** incompatible dimensions

11. $\frac{1}{6}\begin{bmatrix} 5 & 10 \\ 2 & 4 \end{bmatrix}$ **13.** $\begin{bmatrix} 30 \\ 42 \end{bmatrix}$

15. incompatible dimensions

17. $\begin{bmatrix} 1 \\ 3 \end{bmatrix} + \begin{bmatrix} 7 \\ 4 \end{bmatrix} = \begin{bmatrix} 8 \\ 7 \end{bmatrix}$ **19.** $\frac{1}{50}\begin{bmatrix} 22 & 14 & -2 \\ -42 & -2 & 11 \\ -5 & -10 & 5 \end{bmatrix}$

23. $\begin{bmatrix} -2 \\ 7 \end{bmatrix}$

25. $\begin{bmatrix} 3 & 6 & 2 & 2 & 2 \\ 5 & 10 & 0 & -1 & -11 \\ 2 & 4 & -1 & 3 & -4 \end{bmatrix}$

27. The range and codomain both equal $\mathcal{R}^3$.

29. $\begin{bmatrix} 20 \\ -2 \\ 2 \end{bmatrix}$ **31.** $\begin{bmatrix} 2 & 0 & -1 \\ 4 & 0 & 0 \end{bmatrix}$

33. The standard matrix is $\begin{bmatrix} 4 & 1 \\ 3 & 2 \end{bmatrix}$.

35. linear   **37.** linear   **39.** $\left\{ \begin{bmatrix} 1 \\ 0 \end{bmatrix}, \begin{bmatrix} 2 \\ 1 \end{bmatrix}, \begin{bmatrix} 0 \\ -1 \end{bmatrix} \right\}$

41. $\left\{ \begin{bmatrix} -2 \\ 1 \\ 1 \end{bmatrix} \right\}$, T is not one-to-one.

43. $\begin{bmatrix} 1 & 1 \\ 0 & 0 \\ 2 & -1 \end{bmatrix}$; the columns are linearly independent, so T is one-to-one.

45. $\begin{bmatrix} 3 & -1 \\ 0 & 1 \\ 1 & 1 \end{bmatrix}$; the rank is 2, so T is not onto.

47. $\begin{bmatrix} 5 & -1 & 4 \\ 1 & 1 & -1 \\ 3 & 1 & 0 \end{bmatrix}$ **49.** $\begin{bmatrix} 5 & -1 & 4 \\ 1 & 1 & -1 \\ 3 & 1 & 0 \end{bmatrix}$

51. $\begin{bmatrix} 7 & -1 \\ 1 & -2 \end{bmatrix}$

53. $T^{-1}\left(\begin{bmatrix} x_1 \\ x_2 \end{bmatrix}\right) = \frac{1}{5}\begin{bmatrix} 3x_1 - 2x_2 \\ x_1 + x_2 \end{bmatrix}$

CHAPTER 3

Section 3.1

1. (a) F (b) F (c) T (d) F (e) F

3. -25 **5.** 0   **7.** 16   **9.** -30 **11.** 19

13. -2 **15.** 20   **17.** 2   **19.** 60   **21.** 180

23. -147 **25.** -24 **27.** 31   **29.** 0   **31.** 2

35. 2   **47.** $\frac{1}{2}|\det [\mathbf{u} \ \mathbf{v}]|$ **49.** (c) no **51.** (c) yes

Section 3.2

1. (a) F (b) T (c) F (d) T (e) F (f) T
(g) F (h) F (i) T (j) F

3. 19   **5.** -60 **7.** -15 **9.** 30

11. -20 **13.** -3 **15.** 18   **17.** -95

19. -8 **21.** -6 and 2 **23.** 5   **25.** -14

27. -5 and 3

29. $\begin{bmatrix} x_1 \\ x_2 \end{bmatrix} = \begin{bmatrix} -15.0 \\ 10.5 \end{bmatrix}$ **31.** $\begin{bmatrix} x_1 \\ x_2 \end{bmatrix} = \begin{bmatrix} 11 \\ -6 \end{bmatrix}$

33. $\begin{bmatrix} x_1 \\ x_2 \\ x_3 \end{bmatrix} = \begin{bmatrix} 2 \\ 3 \\ -2 \end{bmatrix}$

35. $\begin{bmatrix} x_1 \\ x_2 \\ x_3 \end{bmatrix} = \begin{bmatrix} -0.4 \\ 1.8 \\ -2.4 \end{bmatrix}$ **37.** Take $k = 2$ and $A = I_2$.

53. (a)

$$A \longrightarrow \begin{bmatrix} 2.4 & 3.0 & -6 & -9 \\ 0.0 & -3.0 & -2 & -5 \\ -4.8 & 6.3 & 4 & -2 \\ 9.6 & 1.5 & 5 & 9 \end{bmatrix}$$

$$\longrightarrow \begin{bmatrix} 2.4 & 3.0 & -6 & 9 \\ 0.0 & -3.0 & -2 & -5 \\ 0.0 & 12.3 & -8 & 16 \\ 0.0 & -10.5 & 29 & -27 \end{bmatrix}$$

$$\longrightarrow \begin{bmatrix} 2.4 & 3 & -6.0 & 9.0 \\ 0.0 & -3 & -2.0 & -5.0 \\ 0.0 & 0 & -16.2 & -4.5 \\ 0.0 & 0 & 36.0 & -9.5 \end{bmatrix}$$

$$\longrightarrow \begin{bmatrix} 2.4 & 3 & -6.0 & 9.0 \\ 0.0 & -3 & -2.0 & -5.0 \\ 0.0 & 0 & -16.2 & -4.5 \\ 0.0 & 0 & 0.0 & -19.5 \end{bmatrix}$$

(b) 2274.48

55. $\begin{bmatrix} 13 & -8 & -3 & 6 \\ -10 & -16 & -10 & -12 \\ -17 & 8 & -1 & 2 \\ -12 & 0 & -12 & -8 \end{bmatrix}$

Chapter 3 Review

1. (a) F (b) F (c) T (d) F (e) T (f) F
  (g) F (h) T (i) F (j) F

3. 5 **5.** -3 **7.** $2(-3) + 1(-1) + 3(1)$

9. $1(7) + (-1)5 + 2(-3)$

11. 0 **13.** 3 **15.** -3 and 4 **17.** -3 **19.** 25

21. $x_1 = 2.1, x_2 = 0.8$ **23.** 5 **25.** 40 **27.** 5

29. 20 **31.** $\det B = 0$ or $\det B = 1$

CHAPTER 4

Section 4.1

1. (a) T (b) F (c) F (d) T (e) T (f) F
  (g) F (h) F (i) T (j) T (k) F

3. $\left\{ \begin{bmatrix} 4 \\ -1 \end{bmatrix} \right\}$ **5.** $\left\{ \begin{bmatrix} -1 \\ 2 \\ 1 \end{bmatrix}, \begin{bmatrix} 1 \\ -1 \\ 3 \end{bmatrix} \right\}$

7. $\left\{ \begin{bmatrix} -1 \\ 0 \\ 0 \\ 3 \end{bmatrix}, \begin{bmatrix} 1 \\ 4 \\ 0 \\ 0 \end{bmatrix}, \begin{bmatrix} 0 \\ -3 \\ 0 \\ -1 \end{bmatrix} \right\}$

9. $\left\{ \begin{bmatrix} 0 \\ 3 \\ 1 \\ -1 \end{bmatrix}, \begin{bmatrix} 2 \\ 1 \\ -4 \\ 2 \end{bmatrix}, \begin{bmatrix} -5 \\ -2 \\ 3 \\ 0 \end{bmatrix} \right\}$

11. $\begin{bmatrix} 1 \\ 0 \end{bmatrix}$ and $\begin{bmatrix} 0 \\ 1 \end{bmatrix}$ are in the set, but $\begin{bmatrix} 1 \\ 0 \end{bmatrix} + \begin{bmatrix} 0 \\ 1 \end{bmatrix}$ is not.

13. $\begin{bmatrix} 0 \\ 0 \\ 0 \end{bmatrix}$ is not in the set.

15. $\begin{bmatrix} 1 \\ 0 \\ -1 \end{bmatrix}$ is in the set, but $(-2) \begin{bmatrix} 1 \\ 0 \\ -1 \end{bmatrix}$ is not.

17. $\begin{bmatrix} 6 \\ 2 \\ 3 \end{bmatrix}$ is in the set, but $(-1) \begin{bmatrix} 6 \\ 2 \\ 3 \end{bmatrix}$ is not. **19.** yes

21. no **23.** no **25.** yes **27.** $\left\{ \begin{bmatrix} 7 \\ 5 \\ 1 \end{bmatrix} \right\}$

29. $\left\{ \begin{bmatrix} -1 \\ 1 \\ 1 \\ 0 \\ 0 \end{bmatrix}, \begin{bmatrix} 1 \\ 0 \\ 0 \\ -1 \\ 1 \end{bmatrix} \right\}$

31. $\left\{ \begin{bmatrix} -1 \\ 1 \end{bmatrix}, \begin{bmatrix} 1 \\ -2 \end{bmatrix} \right\}$ **33.** $\left\{ \begin{bmatrix} 1 \\ 3 \\ 0 \end{bmatrix}, \begin{bmatrix} 1 \\ 2 \\ -1 \end{bmatrix}, \begin{bmatrix} 0 \\ 1 \\ 1 \end{bmatrix}, \begin{bmatrix} 2 \\ 6 \\ -1 \end{bmatrix} \right\}$

35. $\mathcal{R}^n$, the zero subspace of $\mathcal{R}^m$, the zero subspace of $\mathcal{R}^n$

37. no **39.** $\begin{bmatrix} 1 & -1 \\ -1 & 1 \end{bmatrix}$

45. $\{1, 2, -1\}, \left\{ \begin{bmatrix} -2 \\ 1 \\ 0 \end{bmatrix}, \begin{bmatrix} 1 \\ 0 \\ 1 \end{bmatrix} \right\}$

47. $\left\{ \begin{bmatrix} 1 \\ 1 \\ 1 \\ 0 \end{bmatrix}, \begin{bmatrix} 1 \\ -1 \\ 0 \\ 1 \end{bmatrix} \right\}, \left\{ \begin{bmatrix} 0 \\ 0 \end{bmatrix} \right\}$

49. $\left\{ \begin{bmatrix} 1 \\ 0 \\ 2 \end{bmatrix}, \begin{bmatrix} 1 \\ 0 \\ 0 \end{bmatrix}, \begin{bmatrix} -1 \\ 0 \\ -1 \end{bmatrix} \right\}, \left\{ \begin{bmatrix} 1 \\ 1 \\ 2 \end{bmatrix} \right\}$

61. (a) yes (b) no **63.** (a) yes (b) no

Section 4.2

1. (a) F (b) T (c) F (d) F (e) T (f) T
  (g) T (h) F (i) F (j) T

3. No linearly independent subset of $\mathcal{R}^3$ can contain more than 3 vectors.

5. $\left\{ \begin{bmatrix} 1 \\ -2 \end{bmatrix} \right\}$ **7.** $\left\{ \begin{bmatrix} 3 \\ 1 \\ 0 \end{bmatrix}, \begin{bmatrix} -5 \\ 0 \\ 1 \end{bmatrix} \right\}$ **9.** $\left\{ \begin{bmatrix} 1 \\ 2 \\ 1 \end{bmatrix}, \begin{bmatrix} 2 \\ 1 \\ 3 \end{bmatrix} \right\}$

11. $\left\{ \begin{bmatrix} 1 \\ -1 \\ 3 \end{bmatrix}, \begin{bmatrix} 0 \\ -1 \\ 1 \end{bmatrix}, \begin{bmatrix} 1 \\ -2 \\ 0 \end{bmatrix} \right\}$

13. $\left\{ \begin{bmatrix} 1 \\ 0 \\ -1 \\ 2 \end{bmatrix}, \begin{bmatrix} 1 \\ 1 \\ -2 \\ 1 \end{bmatrix}, \begin{bmatrix} 0 \\ 1 \\ -1 \\ 2 \end{bmatrix} \right\}$

15. (a) $\left\{ \begin{bmatrix} 1 \\ -1 \\ 2 \end{bmatrix}, \begin{bmatrix} 0 \\ 1 \\ 3 \end{bmatrix} \right\}$ (b) $\left\{ \begin{bmatrix} 2 \\ 1 \\ 0 \\ 0 \end{bmatrix}, \begin{bmatrix} -2 \\ 0 \\ 1 \\ 1 \end{bmatrix} \right\}$

17. (a) $\left\{ \begin{bmatrix} -1 \\ 2 \\ 1 \\ 0 \end{bmatrix}, \begin{bmatrix} 1 \\ 0 \\ -1 \\ 1 \end{bmatrix}, \begin{bmatrix} 2 \\ -5 \\ -1 \\ -2 \end{bmatrix} \right\}$ (b) $\left\{ \begin{bmatrix} -4 \\ -4 \\ -1 \\ 1 \end{bmatrix} \right\}$

19. (a) $\left\{ \begin{bmatrix} 1 \\ 2 \\ 1 \end{bmatrix}, \begin{bmatrix} 2 \\ 3 \\ 2 \end{bmatrix}, \begin{bmatrix} 1 \\ 3 \\ 4 \end{bmatrix} \right\}$ (b) The null space of T is $\{\mathbf{0}\}$.

21. (a) $\left\{ \begin{bmatrix} 1 \\ 2 \\ 1 \end{bmatrix}, \begin{bmatrix} -2 \\ -5 \\ -3 \end{bmatrix} \right\}$ (b) $\left\{ \begin{bmatrix} -3 \\ -1 \\ 1 \\ 0 \end{bmatrix}, \begin{bmatrix} 1 \\ 1 \\ 0 \\ 1 \end{bmatrix} \right\}$

23. (a) $\left\{ \begin{bmatrix} 1 \\ 2 \\ 0 \\ 3 \end{bmatrix}, \begin{bmatrix} 1 \\ 1 \\ 0 \\ 1 \end{bmatrix} \right\}$ (b) $\left\{ \begin{bmatrix} 1 \\ -3 \\ 1 \\ 0 \end{bmatrix}, \begin{bmatrix} -1 \\ 2 \\ 0 \\ 1 \end{bmatrix} \right\}$

25. (a) $\left\{ \begin{bmatrix} 1 \\ 3 \\ 7 \end{bmatrix}, \begin{bmatrix} 2 \\ 1 \\ 4 \end{bmatrix} \right\}$ (b) $\left\{ \begin{bmatrix} 1 \\ -2 \\ 1 \\ 0 \\ 0 \end{bmatrix}, \begin{bmatrix} 0 \\ 0 \\ 0 \\ 1 \\ 0 \end{bmatrix}, \begin{bmatrix} 2 \\ -3 \\ 0 \\ 0 \\ 1 \end{bmatrix} \right\}$

31. 0   **33.** $n - 2$

43. $\left\{ \begin{bmatrix} 2 \\ 3 \\ 0 \end{bmatrix}, \begin{bmatrix} 1 \\ 0 \\ 0 \end{bmatrix}, \begin{bmatrix} 0 \\ 0 \\ 1 \end{bmatrix} \right\}$ **45.** $\left\{ \begin{bmatrix} 0 \\ 2 \\ 1 \\ 0 \end{bmatrix}, \begin{bmatrix} 1 \\ 1 \\ 0 \\ 0 \end{bmatrix}, \begin{bmatrix} -1 \\ 0 \\ 0 \\ 1 \end{bmatrix} \right\}$

47. (c) No, $\mathcal{S}$ is not a subset of V.

49. (a) $\left\{ \begin{bmatrix} 0.1 \\ 0.7 \\ -0.5 \end{bmatrix}, \begin{bmatrix} 0.2 \\ 0.9 \\ 0.5 \end{bmatrix}, \begin{bmatrix} 0.5 \\ -0.5 \\ -0.5 \end{bmatrix} \right\}$

(b) $\left\{ \begin{bmatrix} 1.2 \\ -2.3 \\ 1.0 \\ 0.0 \\ 0.0 \end{bmatrix}, \begin{bmatrix} -1.4 \\ 2.9 \\ 0.0 \\ -0.7 \\ 1.0 \end{bmatrix} \right\}$

Section 4.3

1. (a) F   (b) T   (c) T   (d) F   (e) F   (f) T
(g) T   (h) F   (i) T

3. 2

5. (a) 2   (b) 2   (c) 2   (d) 1

7. (a) 3   (b) 2   (c) 3   (d) 0

9. (a) 1   (b) 3   (c) 1   (d) 0

11. (a) 2   (b) 1   (c) 2   (d) 0

13. (a) 2   (b) 2   (c) 2   (d) 1

15. (a) 2   (b) 1   (c) 2   (d) 2

17. (a) 2   (b) 0   one-to-one and onto

19. (a) 1   (b) 2   neither one-to-one nor onto

21. (a) 2   (b) 0   one-to-one, not onto

23. (a) 2   (b) 1   onto, not one-to-one

33. (a) $\left\{ \begin{bmatrix} 1 \\ 0 \\ 6 \\ 0 \end{bmatrix}, \begin{bmatrix} 0 \\ 1 \\ -4 \\ 1 \end{bmatrix} \right\}, \left\{ \begin{bmatrix} -6 \\ 4 \\ 1 \\ 0 \end{bmatrix}, \begin{bmatrix} 0 \\ -1 \\ 0 \\ 1 \end{bmatrix} \right\}$

43. Take $V = \text{Span}\{\mathbf{e}_1, \mathbf{e}_2\}$ and $W = \text{Span}\{\mathbf{e}_4, \mathbf{e}_5\}$.

49. (a) $\begin{bmatrix} 1 & 2 & 0 & 0 \\ -1 & 1 & 0 & 0 \\ 1 & 0 & 0 & 0 \\ 0 & 1 & 0 & 0 \end{bmatrix}$

51. (a) No, the first vector in $\mathcal{A}_1$ is not in W.   (b) yes
(c) $[\mathbf{e}_1 \ \mathbf{e}_2 \ \mathbf{e}_3]$, $[\mathbf{e}_1 \ \mathbf{e}_2 \ \mathbf{e}_3]$, $[\mathbf{e}_1 \ \mathbf{e}_2 \ \mathbf{e}_3]$,

$\begin{bmatrix} 1 & 0 & 0 & -.4 & -.2 \\ 0 & 1 & 0 & .8 & .4 \\ 0 & 0 & 1 & -.2 & -.6 \end{bmatrix}, \begin{bmatrix} 1 & 0 & 0 & -.4 & -.2 \\ 0 & 1 & 0 & .8 & .4 \\ 0 & 0 & 1 & -.2 & -.5 \end{bmatrix},$

$\begin{bmatrix} 1 & 0 & 0 & -.4 & -.2 \\ 0 & 1 & 0 & .8 & .4 \\ 0 & 0 & 1 & -.2 & -.6 \end{bmatrix}$

Section 4.4

1. (a) F   (b) T   (c) T   (d) T   (e) T

3. $\begin{bmatrix} -5 \\ 11 \end{bmatrix}$ **5.** $\begin{bmatrix} -3 \\ 8 \end{bmatrix}$ **7.** $\begin{bmatrix} 4 \\ 5 \\ 4 \end{bmatrix}$ **9.** $\begin{bmatrix} -7 \\ -3 \\ 2 \end{bmatrix}$

11. (b) $\begin{bmatrix} 5 \\ -3 \end{bmatrix}$ **13.** (b) $\begin{bmatrix} 3 \\ 0 \\ -1 \end{bmatrix}$ **15.** $\begin{bmatrix} -5 \\ -1 \end{bmatrix}$ **17.** $\begin{bmatrix} 7 \\ 2 \end{bmatrix}$

19. $\begin{bmatrix} 0 \\ -1 \\ 3 \end{bmatrix}$ **21.** $\begin{bmatrix} -5 \\ 1 \\ 2 \end{bmatrix}$ **23.** $(a+2b)\mathbf{b}_1 + (a+3b)\mathbf{b}_2 = \mathbf{u}$

25. (b) $\begin{bmatrix} -3 & 2 \\ 2 & -1 \end{bmatrix}$ (c) $A = B^{-1}$

27. (b) $\begin{bmatrix} 1 & 0 & 1 \\ 1 & 1 & 3 \\ 0 & -1 & -1 \end{bmatrix}$ (c) $A = B^{-1}$

29. $x' = \dfrac{\sqrt{3}}{2}x + \dfrac{1}{2}y$ **31.** $x' = -\dfrac{\sqrt{2}}{2}x + \dfrac{\sqrt{2}}{2}y$

$y' = -\dfrac{1}{2}x + \dfrac{\sqrt{3}}{2}y$ $y' = -\dfrac{\sqrt{2}}{2}x - \dfrac{\sqrt{2}}{2}y$

33. $x' = -x + y + 2z$ **35.** $x' = \ \ x - \ y + \ z$
$y' = 2x - y - 2z$ $y' = -3x + 4y - 2z$
$z' = \ \ x - y - \ z$ $z' = \ \ x - 2y + \ z$

37. $x = \dfrac{1}{2}x' - \dfrac{\sqrt{3}}{2}y'$

$\quad\,\, y = \dfrac{\sqrt{3}}{2}x' + \dfrac{1}{2}y'$

39. $x = -\dfrac{\sqrt{2}}{2}x' - \dfrac{\sqrt{2}}{2}y'$

$\quad\,\, y = \dfrac{\sqrt{2}}{2}x' - \dfrac{\sqrt{2}}{2}y'$

41. $x = x' - y'$

$\quad\,\, y = 3x' + y' - z'$

$\quad\,\, z = \qquad\; y' + z'$

43. $x = \;\; x' - y' - z'$

$\quad\,\, y = -x' + 3y' + z'$

$\quad\,\, z = \;\; x' + 2y' + z'$

45. $73x^2 + 18\sqrt{3}xy + 91y^2 = 1600$

47. $8x^2 - 34xy + 8y^2 = 225$

49. $2(x')^2 - 5(y')^2 = 10$ **51.** $4(x')^2 + 3(y')^2 = 12$

53. $\begin{bmatrix} \frac{a_1}{c_1} \\ \frac{a_2}{c_2} \\ \vdots \\ \frac{a_n}{c_n} \end{bmatrix}$ **55.** $\begin{bmatrix} a_1 \\ a_2 - a_1 \\ \vdots \\ a_n - a_1 \end{bmatrix}$ **57.** no

67. (b) $\begin{bmatrix} 29 \\ 44 \\ -52 \\ 33 \\ 39 \end{bmatrix}$ **69.** $\begin{bmatrix} 0 \\ 2 \\ -2 \\ 2 \\ 1 \end{bmatrix}$

Section 4.5

1. (a) T   (b) F   (c) T   (d) F   (e) T
(f) T

3. $\begin{bmatrix} 1 & -3 \\ 4 & 0 \end{bmatrix}$ **5.** $\begin{bmatrix} 0 & 2 & 3 \\ -5 & 0 & 0 \\ 4 & -7 & 1 \end{bmatrix}$

7. (a) $\begin{bmatrix} 0 & 3 \\ 1 & 0 \end{bmatrix}$ (b) $\begin{bmatrix} -1 & 2 \\ 1 & 1 \end{bmatrix}$

(c) $T\left(\begin{bmatrix} x_1 \\ x_2 \end{bmatrix}\right) = \begin{bmatrix} -x_1 + 2x_2 \\ x_1 + x_2 \end{bmatrix}$

9. (a) $\begin{bmatrix} 0 & 0 & 1 \\ -1 & 0 & 2 \\ 0 & 2 & 0 \end{bmatrix}$ (b) $\begin{bmatrix} -1 & 2 & 1 \\ 0 & 2 & -1 \\ 1 & 0 & -1 \end{bmatrix}$

(c) $T\left(\begin{bmatrix} x_1 \\ x_2 \\ x_3 \end{bmatrix}\right) = \begin{bmatrix} -x_1 + 2x_2 + x_3 \\ 2x_2 - x_3 \\ x_1 - x_3 \end{bmatrix}$

11. $\begin{bmatrix} 1 & 1 \\ 3 & 0 \end{bmatrix}$ **13.** $\begin{bmatrix} 1 & 2 \\ 1 & 1 \end{bmatrix}$ **15.** $\begin{bmatrix} 10 & 19 & 16 \\ -5 & -8 & -8 \\ 2 & 2 & 3 \end{bmatrix}$

17. $\begin{bmatrix} 0 & -19 & 28 \\ 3 & 34 & -47 \\ 3 & 23 & -31 \end{bmatrix}$ **19.** $\begin{bmatrix} -10 & -12 & -9 & 1 \\ 20 & 26 & 20 & -7 \\ -10 & -15 & -12 & 7 \\ 7 & 7 & 5 & 1 \end{bmatrix}$

21. $\begin{bmatrix} 10 & -19 \\ 3 & -4 \end{bmatrix}$ **23.** $\begin{bmatrix} 2 & 5 & 10 \\ -6 & 1 & -7 \\ 2 & -2 & 0 \end{bmatrix}$

25. $9\mathbf{b}_1 + 12\mathbf{b}_2$ **27.** $-6\mathbf{b}_1 - 5\mathbf{b}_2 + 25\mathbf{b}_3$ **29.** I_n

31. $T\left(\begin{bmatrix} x_1 \\ x_2 \end{bmatrix}\right) = \begin{bmatrix} .8x_1 + .6x_2 \\ .6x_1 - .8x_2 \end{bmatrix}$

33. $T\left(\begin{bmatrix} x_1 \\ x_2 \end{bmatrix}\right) = \begin{bmatrix} -.6x_1 - .8x_2 \\ -.8x_1 + .6x_2 \end{bmatrix}$

35. $U\left(\begin{bmatrix} x_1 \\ x_2 \end{bmatrix}\right) = \begin{bmatrix} .5x_1 + .5x_2 \\ .5x_1 + .5x_2 \end{bmatrix}$

37. $U\left(\begin{bmatrix} x_1 \\ x_2 \end{bmatrix}\right) = \begin{bmatrix} .1x_1 - .3x_2 \\ -.3x_1 + .9x_2 \end{bmatrix}$

39. (b) $\begin{bmatrix} 1 & 0 & 0 \\ 0 & 1 & 0 \\ 0 & 0 & -1 \end{bmatrix}$ (c) $\dfrac{1}{7}\begin{bmatrix} 6 & -2 & 3 \\ -2 & 3 & 6 \\ 3 & 6 & -2 \end{bmatrix}$

(d) $T\left(\begin{bmatrix} x_1 \\ x_2 \\ x_3 \end{bmatrix}\right) = \dfrac{1}{7}\begin{bmatrix} 6x_1 - 2x_2 + 3x_3 \\ -2x_1 + 3x_2 + 6x_3 \\ 3x_1 + 6x_2 - 2x_3 \end{bmatrix}$

41. (a) $\begin{bmatrix} 1 & 0 & 0 \\ 0 & 1 & 0 \\ 0 & 0 & 0 \end{bmatrix}$ (b) $\dfrac{1}{14}\begin{bmatrix} 13 & -2 & 3 \\ -2 & 10 & 6 \\ 3 & 6 & 5 \end{bmatrix}$

(c) $U\left(\begin{bmatrix} x_1 \\ x_2 \\ x_3 \end{bmatrix}\right) = \dfrac{1}{14}\begin{bmatrix} 13x_1 - 2x_2 + 3x_3 \\ -2x_1 + 10x_2 + 6x_3 \\ 3x_1 + 6x_2 + 5x_3 \end{bmatrix}$

53. (a) $\begin{bmatrix} 11 & 5 & 13 & 1 \\ -2 & 0 & -5 & -3 \\ -8 & -3 & -9 & 0 \\ 6 & 1 & 8 & 1 \end{bmatrix}, \begin{bmatrix} -5 & 10 & -38 & -31 \\ 2 & -3 & 9 & 6 \\ 6 & -10 & 27 & 17 \\ -4 & 7 & -25 & -19 \end{bmatrix},$

$\begin{bmatrix} 43 & 58 & -21 & -66 \\ -8 & -11 & 8 & 17 \\ -28 & -34 & 21 & 53 \\ 28 & 36 & -14 & -44 \end{bmatrix}$

55. (a) $\begin{bmatrix} 0 & 0 & 0 & 1 \\ 1 & 0 & 0 & 0 \\ 0 & 1 & 0 & 0 \\ 0 & 0 & 1 & 0 \end{bmatrix},$

$T\left(\begin{bmatrix} x_1 \\ x_2 \\ x_3 \\ x_4 \end{bmatrix}\right) = \begin{bmatrix} 8x_1 - 4x_2 + 3x_3 + x_4 \\ -11x_1 + 7x_2 - 4x_3 - 2x_4 \\ -35x_1 + 20x_2 - 13x_3 - 5x_4 \\ -9x_1 + 4x_2 - 3x_3 - 2x_4 \end{bmatrix}$

57. $[T^{-1}]_{\mathcal{B}} = ([T]_{\mathcal{B}})^{-1}$

Chapter 4 Review

1. (a) T   (b) T   (c) F   (d) F   (e) F   (f) T
(g) T   (h) F   (i) F   (j) F   (k) T   (l) T
(m) T   (n) T   (o) T   (p) T   (q) F   (r) T
(s) F   (t) F   (u) T   (v) T   (w) F   (x) T

3. (a) There are at least k vectors in a spanning set for V.
   (b) There are at most k vectors in a linearly independent subset of V.

5. No, $\begin{bmatrix} 1 \\ 0 \\ -1 \\ 0 \end{bmatrix}$ and $\begin{bmatrix} 1 \\ 0 \\ 1 \\ 0 \end{bmatrix}$ are in the set, but their sum is not.

7. (a) $\left\{ \begin{bmatrix} -3 \\ 2 \\ 1 \end{bmatrix} \right\}$ (b) $\left\{ \begin{bmatrix} 1 \\ -1 \\ 2 \\ 1 \end{bmatrix}, \begin{bmatrix} 2 \\ -1 \\ 1 \\ 4 \end{bmatrix} \right\}$

   (c) $\{[1 \quad 0 \quad 3], [0 \quad 1 \quad -2]\}$

9. (a) $\left\{ \begin{bmatrix} 0 \\ -1 \\ 1 \\ 2 \end{bmatrix}, \begin{bmatrix} 1 \\ 3 \\ -4 \\ -1 \end{bmatrix}, \begin{bmatrix} -2 \\ 1 \\ 1 \\ 3 \end{bmatrix} \right\}$

   (b) The null space of T is $\{\mathbf{0}\}$.

11. The given set is a linearly independent subset of the null space that contains two vectors.

13. (b) $\begin{bmatrix} -1 \\ -2 \\ 5 \end{bmatrix}$ (c) $\begin{bmatrix} 1 \\ -8 \\ -6 \end{bmatrix}$

15. (a) $\begin{bmatrix} -17 & 1 \\ -10 & 1 \end{bmatrix}$ (b) $\begin{bmatrix} -7 & -5 \\ -14 & -9 \end{bmatrix}$

   (c) $T\left(\begin{bmatrix} x_1 \\ x_2 \end{bmatrix} \right) = \begin{bmatrix} -7x_1 - 5x_2 \\ -14x_1 - 9x_2 \end{bmatrix}$

17. $T\left(\begin{bmatrix} x_1 \\ x_2 \\ x_3 \end{bmatrix} \right) = \begin{bmatrix} x_1 + 6x_2 - 5x_3 \\ -4x_1 + 4x_2 + 5x_3 \\ -x_1 + 3x_2 + x_3 \end{bmatrix}$

19. $21x^2 - 10\sqrt{3}xy + 31y^2 = 144$

21. $50(x')^2 + 8(y')^2 = 200$

23. $T\left(\begin{bmatrix} x_1 \\ x_2 \end{bmatrix} \right) = \dfrac{1}{13} \begin{bmatrix} -5x_1 - 12x_2 \\ -12x_1 + 5x_2 \end{bmatrix}$

25. (a) $m \le k$ (b) Nothing can be said about m and k.
    (c) $m \ge k$

CHAPTER 5

Section 5.1

1. (a) F   (b) F   (c) T   (d) T   (e) T   (f) F
   (g) T   (h) T   (i) F

3. 3   5. -2 7. -3 9. -4 11. 2

13. $\left\{ \begin{bmatrix} -1 \\ 1 \end{bmatrix} \right\}$ 15. $\left\{ \begin{bmatrix} -3 \\ 1 \end{bmatrix} \right\}$ 17. $\left\{ \begin{bmatrix} -1 \\ 1 \\ 0 \end{bmatrix} \right\}$

19. $\left\{ \begin{bmatrix} -2 \\ -1 \\ 1 \end{bmatrix} \right\}$ 21. $\left\{ \begin{bmatrix} -1 \\ 3 \\ 0 \end{bmatrix}, \begin{bmatrix} 2 \\ 0 \\ 3 \end{bmatrix} \right\}$

23. $\left\{ \begin{bmatrix} 1 \\ 1 \\ 0 \end{bmatrix}, \begin{bmatrix} 1 \\ 0 \\ 1 \end{bmatrix} \right\}$

25. 6   27. -3 29. 5

31. $\left\{ \begin{bmatrix} 2 \\ 3 \end{bmatrix} \right\}$ 33. $\left\{ \begin{bmatrix} -1 \\ 1 \\ 0 \end{bmatrix}, \begin{bmatrix} -3 \\ 0 \\ 1 \end{bmatrix} \right\}$ 35. $\left\{ \begin{bmatrix} 1 \\ -2 \\ 2 \end{bmatrix} \right\}$

37. The only eigenvalue is 1; its eigenspace is $\mathcal{R}^n$.

41. Null A

47. Either $\mathbf{v} = \mathbf{0}$ or $\mathbf{v}$ is an eigenvector of A.   53. no

57. yes, $\begin{bmatrix} -1 \\ 1 \\ -2 \\ 1 \end{bmatrix}, \begin{bmatrix} 2 \\ 0 \\ 3 \\ 3 \end{bmatrix}, \begin{bmatrix} 1 \\ -1 \\ 2 \\ 0 \end{bmatrix}, \begin{bmatrix} 0 \\ -1 \\ 0 \\ 1 \end{bmatrix}$

Section 5.2

1. (a) F   (b) T   (c) T   (d) F   (e) F   (f) F
   (g) F   (h) T   (i) F   (j) T   (k) F   (l) F

3. (a) $(t-5)^3(t+9)$
   (b) $(t-5)^3(t+9)$, $(t-5)^2(t+9)^2$, $(t-5)(t+9)^3$
   (c) $(t-5)^2(t+9)^2$, $(t-5)^3(t+9)$

5. $5, \left\{ \begin{bmatrix} -3 \\ 2 \end{bmatrix} \right\}, 6, \left\{ \begin{bmatrix} -1 \\ 1 \end{bmatrix} \right\}$ 7. $0, \left\{ \begin{bmatrix} 3 \\ 5 \end{bmatrix} \right\}, -1, \left\{ \begin{bmatrix} 2 \\ 3 \end{bmatrix} \right\}$

9. $-3, \left\{ \begin{bmatrix} 1 \\ 1 \\ 1 \end{bmatrix} \right\}, 2, \left\{ \begin{bmatrix} 1 \\ 0 \\ 1 \end{bmatrix} \right\}$

11. $6, \left\{ \begin{bmatrix} 1 \\ -1 \\ 1 \end{bmatrix} \right\}, -2, \left\{ \begin{bmatrix} 1 \\ 2 \\ 0 \end{bmatrix}, \begin{bmatrix} 1 \\ 0 \\ 2 \end{bmatrix} \right\}$

13. $-3, \left\{ \begin{bmatrix} -1 \\ 1 \\ 1 \end{bmatrix} \right\}, -2, \left\{ \begin{bmatrix} -1 \\ 1 \\ 0 \end{bmatrix} \right\}, 1, \left\{ \begin{bmatrix} 1 \\ 0 \\ 1 \end{bmatrix} \right\}$

15. $3, \left\{ \begin{bmatrix} 1 \\ 1 \\ 0 \\ 0 \end{bmatrix} \right\}, 4, \left\{ \begin{bmatrix} 0 \\ 1 \\ 0 \\ 1 \end{bmatrix} \right\}, -1, \left\{ \begin{bmatrix} 0 \\ 1 \\ 1 \\ 0 \end{bmatrix}, \begin{bmatrix} -1 \\ 1 \\ 0 \\ 1 \end{bmatrix} \right\}$

17. $-4, \left\{ \begin{bmatrix} 1 \\ 2 \end{bmatrix} \right\}, 1, \left\{ \begin{bmatrix} 1 \\ 1 \end{bmatrix} \right\}$

19. $3, \left\{ \begin{bmatrix} -2 \\ 3 \end{bmatrix} \right\}, 5, \left\{ \begin{bmatrix} -1 \\ 2 \end{bmatrix} \right\}$ 21. $-3, \left\{ \begin{bmatrix} 1 \\ 0 \\ 1 \end{bmatrix} \right\}, 1, \left\{ \begin{bmatrix} 1 \\ 0 \\ 2 \end{bmatrix} \right\}$

23. $-1, \left\{ \begin{bmatrix} 1 \\ 1 \\ 0 \end{bmatrix}, \begin{bmatrix} 1 \\ 0 \\ 1 \end{bmatrix} \right\}, 5, \left\{ \begin{bmatrix} 1 \\ 0 \\ 2 \end{bmatrix} \right\}$

25. $-6, \left\{ \begin{bmatrix} -1 \\ 1 \\ 1 \end{bmatrix} \right\}, -2, \left\{ \begin{bmatrix} 1 \\ 1 \\ 1 \end{bmatrix} \right\}, 4, \left\{ \begin{bmatrix} 0 \\ 1 \\ 0 \end{bmatrix} \right\}$

27. $-1, \left\{ \begin{bmatrix} -1 \\ 0 \\ 1 \\ 0 \end{bmatrix} \right\}, 2, \left\{ \begin{bmatrix} 0 \\ -1 \\ 1 \\ 0 \end{bmatrix}, \begin{bmatrix} 1 \\ 0 \\ 0 \\ 1 \end{bmatrix} \right\}$

29. $5, \left\{ \begin{bmatrix} 1 \\ 1 \end{bmatrix} \right\}, 7, \left\{ \begin{bmatrix} 3 \\ 4 \end{bmatrix} \right\}$

31. $-2, \left\{ \begin{bmatrix} 1 \\ 1 \\ 0 \end{bmatrix} \right\}, 4, \left\{ \begin{bmatrix} 1 \\ 0 \\ 1 \end{bmatrix} \right\}$

33. $1, \left\{ \begin{bmatrix} 1 \\ 2 \\ 3 \end{bmatrix} \right\}, 2, \left\{ \begin{bmatrix} -2 \\ 1 \\ 0 \end{bmatrix}, \begin{bmatrix} 2 \\ 0 \\ 1 \end{bmatrix} \right\}$

35. $-3, \left\{ \begin{bmatrix} 1 \\ 1 \end{bmatrix} \right\}, -2, \left\{ \begin{bmatrix} 1 \\ 2 \end{bmatrix} \right\}$ **37.** $-3, \left\{ \begin{bmatrix} 1 \\ 1 \\ 0 \end{bmatrix} \right\}, 2, \left\{ \begin{bmatrix} 0 \\ 0 \\ 1 \end{bmatrix} \right\}$

39. $-3, \left\{ \begin{bmatrix} 1 \\ 2 \\ 3 \end{bmatrix} \right\}, 1, \left\{ \begin{bmatrix} 0 \\ 1 \\ 0 \end{bmatrix}, \begin{bmatrix} 0 \\ 0 \\ 1 \end{bmatrix} \right\}$

43. c is not an eigenvalue of A.

47. (a) $\left\{ \begin{bmatrix} -1 \\ 1 \end{bmatrix} \right\}, \left\{ \begin{bmatrix} -2 \\ 1 \end{bmatrix} \right\}$,   (b) $\left\{ \begin{bmatrix} -1 \\ 1 \end{bmatrix} \right\}, \left\{ \begin{bmatrix} -2 \\ 1 \end{bmatrix} \right\}$

(c) $\left\{ \begin{bmatrix} -1 \\ 1 \end{bmatrix} \right\}, \left\{ \begin{bmatrix} -2 \\ 1 \end{bmatrix} \right\}$

(d) $\mathbf{v}$ is an eigenvector of B if and only if $\mathbf{v}$ is an eigenvector of cB.

(e) λ is an eigenvalue of B if and only if $c\lambda$ is an eigenvalue of cB.

49. (a) $(t - 6)(t - 7)$

(b) The characteristic polynomials of B and B^T are equal.

(c) The eigenvalues of B and B^T are the same.

(d) no

53. $-t^3 + \dfrac{23}{15}t^2 - \dfrac{127}{720}t + \dfrac{1}{2160}$ **55.** $\begin{bmatrix} 0 & 0 & 0 & 5 \\ 1 & 0 & 0 & -7 \\ 0 & 1 & 0 & -23 \\ 0 & 0 & 1 & 11 \end{bmatrix}$

57. (a) $3, -0.5$ (b) $\dfrac{1}{3}, \begin{bmatrix} 1 \\ 1 \end{bmatrix}, -2, \begin{bmatrix} 1 \\ 2 \end{bmatrix}$

Section 5.3

1. (a) F   (b) T   (c) T   (d) T   (e) F   (f) F
(g) F   (h) T   (i) F   (j) F   (k) T   (l) F

3. (a) if the eigenspace corresponding to -1 is two-dimensional

(b) if the eigenspace corresponding to -1 is one-dimensional

5. (a) if the eigenspace corresponding to -3 is four-dimensional

(b) if the eigenspace corresponding to -3 is not four-dimensional

7. (a) $-(t - 4)^2(t - 5)(t - 8)^2$

(b) There is insufficient information because the dimensions of W_1 and W_2 are not given. Therefore, the multiplicities of the eigenvalues 4 and 8 are not determined.

(c) $-(t - 4)(t - 5)^2(t - 8)^2$

9. $P = \begin{bmatrix} -2 & -3 \\ 1 & 1 \end{bmatrix}, D = \begin{bmatrix} 4 & 0 \\ 0 & 5 \end{bmatrix}$

11. The eigenspace corresponding to 2 is one-dimensional.

13. $P = \begin{bmatrix} 0 & -2 & -1 \\ 1 & 3 & 1 \\ 1 & 2 & 1 \end{bmatrix}, D = \begin{bmatrix} -5 & 0 & 0 \\ 0 & 2 & 0 \\ 0 & 0 & 3 \end{bmatrix}$

15. There is only one real eigenvalue, and its multiplicity is one.

17. $P = \begin{bmatrix} -1 & -1 & 1 \\ 4 & 1 & 0 \\ 2 & 0 & 1 \end{bmatrix}, D = \begin{bmatrix} 5 & 0 & 0 \\ 0 & 3 & 0 \\ 0 & 0 & 3 \end{bmatrix}$

19. $P = \begin{bmatrix} 0 & 1 & 0 & 0 \\ 0 & 0 & 1 & 0 \\ 1 & 0 & 0 & 1 \\ 0 & 1 & 1 & 1 \end{bmatrix}, D = \begin{bmatrix} 4 & 0 & 0 & 0 \\ 0 & -1 & 0 & 0 \\ 0 & 0 & -1 & 0 \\ 0 & 0 & 0 & -1 \end{bmatrix}$

21. The eigenspace corresponding to 1 is one-dimensional.

23. $P = \begin{bmatrix} -2 & -3 \\ 1 & 2 \end{bmatrix}, D = \begin{bmatrix} 3 & 0 \\ 0 & 2 \end{bmatrix}$

25. $P = \begin{bmatrix} 1 & -1 & -1 \\ 0 & 1 & 1 \\ 0 & 0 & 5 \end{bmatrix}, D = \begin{bmatrix} -1 & 0 & 0 \\ 0 & -3 & 0 \\ 0 & 0 & 2 \end{bmatrix}$

27. The eigenspace corresponding to 1 is one-dimensional.

29. $\begin{bmatrix} -4^k + 2 \cdot 3^k & 2 \cdot 4^k - 2 \cdot 3^k \\ -4^k + 3^k & 2 \cdot 4^k - 3^k \end{bmatrix}$

31. $\begin{bmatrix} -2 \cdot 2^k + 3 \cdot 3^k & -6 \cdot 2^k + 6 \cdot 3^k \\ 2^k - 3^k & 3 \cdot 2^k - 2 \cdot 3^k \end{bmatrix}$

33. $\begin{bmatrix} -5^k + 2 & -2 \cdot 5^k + 2 & 0 \\ 5^k - 1 & 2 \cdot 5^k - 1 & 0 \\ 0 & 0 & 5^k \end{bmatrix}$

35. 3   **37.** all c **39.** no c **41.** -2 and -1

43. 2   **45.** $\begin{bmatrix} -7 & 4 \\ -12 & 9 \end{bmatrix}$ **47.** $\begin{bmatrix} 0 & 0 \\ 0 & 1 \end{bmatrix}$ and $\begin{bmatrix} 0 & -1 \\ 0 & -1 \end{bmatrix}$

59. $\det A$

61. $P = \begin{bmatrix} -3 & -1 & -8 & -1 \\ -1 & -1 & -1 & -2 \\ 2 & 0 & 3 & 0 \\ 0 & 2 & 0 & 3 \end{bmatrix}$,

$D = \begin{bmatrix} -1 & 0 & 0 & 0 \\ 0 & -1 & 0 & 0 \\ 0 & 0 & -2 & 0 \\ 0 & 0 & 0 & -2 \end{bmatrix}$

63. The eigenspace corresponding to 1 has dimension 2.

Section 5.4

1. (a) F　(b) F　(c) T　(d) F　(e) F
　(f) F　(g) F

3. $\begin{bmatrix} 0 & 0 & 3 \\ 0 & -2 & 0 \\ -4 & 0 & 0 \end{bmatrix}$, no　5. $\begin{bmatrix} 2 & 0 & 0 \\ 0 & -1 & 0 \\ 0 & 0 & -3 \end{bmatrix}$, yes

7. There are no real eigenvalues.　9. $\left\{ \begin{bmatrix} 1 \\ 2 \end{bmatrix}, \begin{bmatrix} 1 \\ 1 \end{bmatrix} \right\}$

11. $\left\{ \begin{bmatrix} 1 \\ -1 \\ 1 \end{bmatrix}, \begin{bmatrix} 0 \\ -1 \\ 1 \end{bmatrix}, \begin{bmatrix} 0 \\ 0 \\ 1 \end{bmatrix} \right\}$

13. The eigenspace corresponding to -1 is one-dimensional.

15. $\left\{ \begin{bmatrix} 1 \\ 1 \\ 0 \end{bmatrix}, \begin{bmatrix} 0 \\ 1 \\ 1 \end{bmatrix}, \begin{bmatrix} 1 \\ 0 \\ 0 \end{bmatrix} \right\}$　17. $\left\{ \begin{bmatrix} 1 \\ 0 \\ 1 \\ 0 \end{bmatrix}, \begin{bmatrix} -1 \\ 2 \\ 0 \\ 0 \end{bmatrix}, \begin{bmatrix} 1 \\ 0 \\ 2 \\ 0 \end{bmatrix}, \begin{bmatrix} -1 \\ 0 \\ 0 \\ 2 \end{bmatrix} \right\}$

19. T has no real eigenvalues.　21. $\left\{ \begin{bmatrix} 1 \\ 1 \end{bmatrix}, \begin{bmatrix} -3 \\ 4 \end{bmatrix} \right\}$

23. $\left\{ \begin{bmatrix} 0 \\ 1 \\ 0 \end{bmatrix}, \begin{bmatrix} -1 \\ 0 \\ 1 \end{bmatrix}, \begin{bmatrix} 0 \\ 1 \\ 1 \end{bmatrix} \right\}$

25. The eigenspace corresponding to 1 is two-dimensional.

27. 7　29. all scalars c　31. -3 and -1

33. all scalars c

35. no scalars c

37. $U\left(\begin{bmatrix} x_1 \\ x_2 \\ x_3 \end{bmatrix} \right) = \frac{1}{3} \begin{bmatrix} 2x_1 - x_2 - x_3 \\ -x_1 + 2x_2 - x_3 \\ -x_1 - x_2 + 2x_3 \end{bmatrix}$

39. $U\left(\begin{bmatrix} x_1 \\ x_2 \\ x_3 \end{bmatrix} \right) = \frac{1}{6} \begin{bmatrix} 5x_1 - 2x_2 + x_3 \\ -2x_1 + 2x_2 + 2x_3 \\ x_1 - 2x_2 + 5x_3 \end{bmatrix}$

41. $U\left(\begin{bmatrix} x_1 \\ x_2 \\ x_3 \end{bmatrix} \right) = \frac{1}{9} \begin{bmatrix} 5x_1 - 4x_2 - 2x_3 \\ -4x_1 + 5x_2 - 2x_3 \\ -2x_1 - 2x_2 + 8x_3 \end{bmatrix}$

43. $T\left(\begin{bmatrix} x_1 \\ x_2 \\ x_3 \end{bmatrix} \right) = \frac{1}{3} \begin{bmatrix} x_1 - 2x_2 - 2x_3 \\ -2x_1 + x_2 - 2x_3 \\ -2x_1 - 2x_2 + x_3 \end{bmatrix}$

45. $T\left(\begin{bmatrix} x_1 \\ x_2 \\ x_3 \end{bmatrix} \right) = \frac{1}{3} \begin{bmatrix} 2x_1 - 2x_2 + x_3 \\ -2x_1 - x_2 + 2x_3 \\ x_1 + 2x_2 + 2x_3 \end{bmatrix}$

47. $T\left(\begin{bmatrix} x_1 \\ x_2 \\ x_3 \end{bmatrix} \right) = \frac{1}{9} \begin{bmatrix} x_1 - 8x_2 - 4x_3 \\ -8x_1 + x_2 - 4x_3 \\ -4x_1 - 4x_2 + 7x_3 \end{bmatrix}$

55. $\left\{ \begin{bmatrix} 2 \\ -2 \\ -4 \\ 3 \\ 0 \end{bmatrix}, \begin{bmatrix} -1 \\ 1 \\ 2 \\ 0 \\ 3 \end{bmatrix}, \begin{bmatrix} -1 \\ 1 \\ -3 \\ 2 \\ 0 \end{bmatrix}, \begin{bmatrix} 1 \\ 1 \\ 3 \\ 0 \\ 0 \end{bmatrix}, \begin{bmatrix} 1 \\ 0 \\ 1 \\ 0 \\ 0 \end{bmatrix} \right\}$

Section 5.5

1. (a) F　(b) F　(c) T　(d) F　(e) T
　(f) F　(g) F　(h) T

3. no　5. yes　7. $\begin{bmatrix} .75 \\ .25 \end{bmatrix}$　9. $\begin{bmatrix} .25 \\ .25 \\ .50 \end{bmatrix}$　11. $\frac{1}{6} \begin{bmatrix} 1 \\ 3 \\ 2 \end{bmatrix}$

13. (a) $\begin{bmatrix} .25 & .5 \\ .75 & .5 \end{bmatrix}$　(b) .375　(c) .6

15. (a) $\begin{bmatrix} .7 & .1 & .1 \\ .1 & .6 & .1 \\ .2 & .3 & .8 \end{bmatrix}$　(b) .6　(c) .33

　(d) .25 buy brand A, .20 buy brand B, and .55 buy brand C

17. (a) $\begin{bmatrix} .6 & \frac{16}{123} \\ .4 & \frac{107}{123} \end{bmatrix}$　(b) $\frac{107}{123}$　(c) about .809

　(d) about .676　(e) about .245

19. (a) .05　(b) .1　(c) .3　(d) $\begin{bmatrix} .6 \\ .3 \\ .1 \end{bmatrix}$

21. (a) $\mathbf{u}$　(b) 1 is an eigenvalue of A^T.
　(d) 1 is an eigenvalue of A.

27. $y_1 = -ae^{-3t} + 2be^{4t}$
　$y_2 = 3ae^{-3t} + be^{4t}$

29. $y_1 = -2ae^{-4t} - be^{-2t}$
　$y_2 = 3ae^{-4t} + be^{-2t}$

31. $y_1 = -ce^{2t}$
　$y_2 = -ae^{-t} + be^{2t}$
　$y_3 = ae^{-t} + ce^{2t}$

33. $y_1 = 10e^{-t} + 5e^{3t}$
　$y_2 = -20e^{-t} + 10e^{3t}$

35. $y_1 = -3e^{4t} + 5e^{6t}$
　$y_2 = 6e^{4t} - 5e^{6t}$

37. $y_1 = 4e^{-t} + 5e^t - 9e^{2t}$
　$y_2 = 5e^t - 3e^{2t}$
　$y_3 = 4e^{-t} - 3e^{2t}$

39. $y = 3e^{-t} - 2e^t + e^{2t}$

41. $y = e^{-t}(a \cos \sqrt{3}t + b \sin \sqrt{3}t)$

43. (a) $y_1 = 100e^{-2t} + 800e^t$, $y_2 = 100e^{-2t} + 200e^t$
　(b) 2188 and 557 at time 1, 5913 and 1480 at time 2, and 16069 and 4017 at time 3
　(c) .25, no

47. $r_n = 3 \cdot 2^n + 4(-1)^n$; $r_6 = 196$

49. $r_n = \left(\frac{2}{5}\right) 4^n + \left(\frac{3}{5}\right)(-1)^n$; $r_6 = 1639$

51. (a) $r_0 = 1, r_1 = 2, r_2 = 7, r_3 = 20$
　(b) $r_n = 2r_{n-1} + 3r_{n-2}$
　(c) $r_n = \left(\frac{3}{4}\right) 3^n + \left(\frac{1}{4}\right)(-1)^n$

53. $\begin{bmatrix} r_{n+3} \\ r_{n+2} \\ r_{n+1} \end{bmatrix} = \begin{bmatrix} 4 & -2 & 5 \\ 1 & 0 & 0 \\ 0 & 1 & 0 \end{bmatrix} \begin{bmatrix} r_{n+2} \\ r_{n+1} \\ r_n \end{bmatrix}$

61. $y_1 = 8e^t - 6e^{-.8t} - 2e^{-.1t} + e^{.3t}$
$\; y_2 = -4e^t \phantom{- 6e^{-.8t}} - 2e^{-.1t} + 2e^{.3t}$
$\; y_3 = -4e^t + 6e^{-.8t}$
$\; y_4 = 8e^t - 6e^{-.8t} + 4e^{-.1t} - 3e^{.3t}$

Chapter 5 Review

1. (a) T (b) F (c) T (d) T (e) T (f) T
(g) F (h) F (i) F (j) F (k) T (l) T
(m) F (n) T (o) T (p) T

3. $\left\{ \begin{bmatrix} -3 \\ 2 \end{bmatrix} \right\}$ for 1 and $\left\{ \begin{bmatrix} -2 \\ 1 \end{bmatrix} \right\}$ for 2

5. $\left\{ \begin{bmatrix} -1 \\ 1 \\ 0 \end{bmatrix} \right\}$ for -2 and $\left\{ \begin{bmatrix} 0 \\ 1 \\ 0 \end{bmatrix} \right\}$ for -1

7. $P = \begin{bmatrix} 2 & 1 \\ 1 & 3 \end{bmatrix}$ and $D = \begin{bmatrix} 2 & 0 \\ 0 & 7 \end{bmatrix}$

9. The eigenspace corresponding to -1 has dimension 1.

11. $\left\{ \begin{bmatrix} -2 \\ 1 \end{bmatrix}, \begin{bmatrix} -1 \\ 4 \end{bmatrix} \right\}$ **13.** $\left\{ \begin{bmatrix} -1 \\ 0 \\ 1 \end{bmatrix}, \begin{bmatrix} 1 \\ 1 \\ 0 \end{bmatrix}, \begin{bmatrix} 0 \\ 0 \\ 1 \end{bmatrix} \right\}$

15. none   **17.** -2 and 2

19. $\begin{bmatrix} (-1)^{k+1} + 2^{k+1} & 2(-1)^k - 2^{k+1} \\ (-1)^{k+1} + 2^k & 2(-1)^k - 2^k \end{bmatrix}$

21. $\left\{ \begin{bmatrix} -1 \\ 1 \\ 0 \end{bmatrix}, \begin{bmatrix} -1 \\ 0 \\ 1 \end{bmatrix}, \begin{bmatrix} -1 \\ 0 \\ 2 \end{bmatrix} \right\}$

23. The eigenvalue a has multiplicity 2, but its eigenspace has dimension 1.

CHAPTER 6

Section 6.1

1. (a) T (b) F (c) F (d) F (e) F (f) T
(g) T

3. $\|\mathbf{u}\| = \sqrt{2}$, $\|\mathbf{v}\| = \sqrt{5}$, and $d = \sqrt{5}$
5. $\|\mathbf{u}\| = \sqrt{11}$, $\|\mathbf{v}\| = \sqrt{5}$, and $d = \sqrt{14}$
7. $\|\mathbf{u}\| = \sqrt{7}$, $\|\mathbf{v}\| = \sqrt{15}$, and $d = \sqrt{26}$
9. 1, no   **11.** 0, yes   **13.** -2, no
15. $\|\mathbf{u}\|^2 = 13$, $\|\mathbf{v}\|^2 = 0$, $\|\mathbf{u} + \mathbf{v}\|^2 = 13$
17. $\|\mathbf{u}\|^2 = 14$, $\|\mathbf{v}\|^2 = 138$, $\|\mathbf{u} + \mathbf{v}\|^2 = 152$
19. $\|\mathbf{u}\| = \sqrt{20}$, $\|\mathbf{v}\| = \sqrt{10}$, $\|\mathbf{u} + \mathbf{v}\| = \sqrt{50}$
21. $\|\mathbf{u}\| = \sqrt{14}$, $\|\mathbf{v}\| = \sqrt{17}$, $\|\mathbf{u} + \mathbf{v}\| = \sqrt{53}$
23. $\|\mathbf{u}\| = \sqrt{17}$, $\|\mathbf{v}\| = 2$, $\mathbf{u} \cdot \mathbf{v} = -2$

25. $\|\mathbf{u}\| = \sqrt{21}$, $\|\mathbf{v}\| = \sqrt{6}$, $\mathbf{u} \cdot \mathbf{v} = 5$
27. $\mathbf{w} = \dfrac{1}{2} \begin{bmatrix} -1 \\ 1 \end{bmatrix}$ and $d = \dfrac{7\sqrt{2}}{2}$

29. $\mathbf{w} = \begin{bmatrix} 0.7 \\ 2.1 \end{bmatrix}$ and $d = 1.1\sqrt{10}$

31. -3 **33.** 11   **35.** 441

Section 6.2

1. (a) F (b) T (c) F (d) F (e) T
(f) T

3. no   **5.** yes

7. (a) $\left\{ \begin{bmatrix} 1 \\ 1 \\ 1 \end{bmatrix}, \begin{bmatrix} 3 \\ -3 \\ 0 \end{bmatrix} \right\}$ (b) $\left\{ \dfrac{1}{\sqrt{3}} \begin{bmatrix} 1 \\ 1 \\ 1 \end{bmatrix}, \dfrac{1}{\sqrt{2}} \begin{bmatrix} 1 \\ -1 \\ 0 \end{bmatrix} \right\}$

9. (a) $\left\{ \begin{bmatrix} 0 \\ 1 \\ 1 \\ 1 \end{bmatrix}, \dfrac{1}{3} \begin{bmatrix} 3 \\ -2 \\ 1 \\ 1 \end{bmatrix}, \dfrac{1}{5} \begin{bmatrix} 3 \\ 3 \\ -4 \\ 1 \end{bmatrix} \right\}$

(b) $\left\{ \dfrac{1}{\sqrt{3}} \begin{bmatrix} 0 \\ 1 \\ 1 \\ 1 \end{bmatrix}, \dfrac{1}{\sqrt{15}} \begin{bmatrix} 3 \\ -2 \\ 1 \\ 1 \end{bmatrix}, \dfrac{1}{\sqrt{35}} \begin{bmatrix} 3 \\ 3 \\ -4 \\ 1 \end{bmatrix} \right\}$

11. (a) $\left\{ \begin{bmatrix} 1 \\ 0 \\ -1 \\ 1 \end{bmatrix}, \begin{bmatrix} 1 \\ 1 \\ 0 \\ -1 \end{bmatrix}, \begin{bmatrix} 2 \\ -1 \\ 3 \\ 1 \end{bmatrix} \right\}$

(b) $\left\{ \dfrac{1}{\sqrt{3}} \begin{bmatrix} 1 \\ 0 \\ -1 \\ 1 \end{bmatrix}, \dfrac{1}{\sqrt{3}} \begin{bmatrix} 1 \\ 1 \\ 0 \\ -1 \end{bmatrix}, \dfrac{1}{\sqrt{15}} \begin{bmatrix} 2 \\ -1 \\ 3 \\ 1 \end{bmatrix} \right\}$

13. $\mathbf{v} = 2 \begin{bmatrix} 2 \\ 1 \end{bmatrix} + 3 \begin{bmatrix} -1 \\ 2 \end{bmatrix}$

15. $\mathbf{v} = \dfrac{5}{2} \begin{bmatrix} 1 \\ 0 \\ 1 \end{bmatrix} + \dfrac{3}{6} \begin{bmatrix} 1 \\ 2 \\ -1 \end{bmatrix} + 0 \begin{bmatrix} 1 \\ -1 \\ -1 \end{bmatrix}$

17. $\left\{ \begin{bmatrix} 1 \\ 1 \\ 0 \end{bmatrix}, \begin{bmatrix} -2 \\ 0 \\ 1 \end{bmatrix} \right\}$ **19.** $\left\{ \begin{bmatrix} -5 \\ -2 \\ 1 \\ 0 \end{bmatrix}, \begin{bmatrix} -3 \\ -1 \\ 0 \\ 1 \end{bmatrix} \right\}$

21. (a) $\mathbf{w} = \begin{bmatrix} -1 \\ 1 \end{bmatrix}$ and $\mathbf{z} = \begin{bmatrix} 2 \\ 2 \end{bmatrix}$ (b) $\begin{bmatrix} -1 \\ 1 \end{bmatrix}$ (c) $\sqrt{8}$

23. (a) $\mathbf{w} = \mathbf{v}$ and $\mathbf{z} = \mathbf{0}$ (b) $\begin{bmatrix} 1 \\ 4 \\ -1 \end{bmatrix}$ (c) 0

25. $\mathbf{w} = \begin{bmatrix} -6 \\ 8 \end{bmatrix}$, $\mathbf{z} = \begin{bmatrix} -4 \\ -3 \end{bmatrix}$, $\begin{bmatrix} -6 \\ 8 \end{bmatrix}$, and 5

27. $\mathbf{w} = \dfrac{1}{3} \begin{bmatrix} -1 \\ 2 \\ 1 \end{bmatrix}$, $\mathbf{z} = \dfrac{4}{3} \begin{bmatrix} 1 \\ 1 \\ -1 \end{bmatrix}$, $\dfrac{4}{3} \begin{bmatrix} 1 \\ 1 \\ -1 \end{bmatrix}$, and $\dfrac{4}{\sqrt{3}}$

Section 6.3

1. (a) T   (b) F   (c) T   (d) F   (e) T

3. $\dfrac{1}{3}\begin{bmatrix} 1 & -1 & 1 & 0 \\ -1 & 1 & -1 & 0 \\ 1 & -1 & 1 & 0 \\ 0 & 0 & 0 & 0 \end{bmatrix}$ 5. $\dfrac{1}{2}\begin{bmatrix} 2 & 0 & 0 & 0 \\ 0 & 1 & 0 & 1 \\ 0 & 0 & 0 & 0 \\ 0 & 1 & 0 & 1 \end{bmatrix}$

7. $\dfrac{1}{6}\begin{bmatrix} 5 & -2 & 1 \\ -2 & 2 & 2 \\ 1 & 2 & 5 \end{bmatrix}$

9. $\dfrac{1}{42}\begin{bmatrix} 25 & -20 & 5 \\ -20 & 16 & -4 \\ 5 & -4 & 1 \end{bmatrix}$ 11. $\dfrac{1}{11}\begin{bmatrix} 6 & -2 & -1 & -5 \\ -2 & 8 & 4 & -2 \\ -1 & 4 & 2 & -1 \\ -5 & -2 & -1 & 6 \end{bmatrix}$

13. $\dfrac{2}{3}\begin{bmatrix} 1 \\ 2 \\ 1 \end{bmatrix}, \dfrac{5\sqrt{3}}{3}$ 15. $\dfrac{1}{3}\begin{bmatrix} 3 \\ 4 \\ 1 \\ 1 \end{bmatrix}, 1$

17. $y = 13.5 + x$ 19. $y = 3.2 + 1.6x$

21. $y = 44 - 3x$

23. $y = -6.35 + 2.1x$. So the estimate for the spring constant is $b = 2.1$.

25. $y = 1.42 + 0.49x + 0.38x^2 + 0.73x^3$

39. $\dfrac{1}{6}\begin{bmatrix} 4 & 2 & 0 & -2 \\ 2 & 3 & 2 & 1 \\ 0 & 2 & 2 & 2 \\ -2 & 1 & 2 & 3 \end{bmatrix}$ 41. $\dfrac{1}{3}\begin{bmatrix} 2 & -1 & 0 & 1 \\ -1 & 1 & -1 & 0 \\ 0 & -1 & 2 & -1 \\ 1 & 0 & -1 & 1 \end{bmatrix}$

45. (rounded to four places after the decimal)

$$P_W = \begin{bmatrix} 0.7201 & 0.0001 & -0.1845 & -0.3943 & -0.1098 \\ 0.0001 & 0.4915 & 0.4391 & -0.1547 & -0.1823 \\ -0.1845 & 0.4391 & 0.4993 & -0.1263 & 0.0850 \\ -0.3943 & -0.1547 & -0.1263 & 0.3975 & -0.2102 \\ -0.1098 & -0.1823 & 0.0850 & -0.2102 & 0.8915 \end{bmatrix}$$

The distance equals 3.4418.

Section 6.4

1. (a) T   (b) F   (c) F   (d) T   (e) T   (f) T
   (g) F   (h) F

3. no   5. yes   7. no

9. a reflection, $y = (\sqrt{2} - 1)x$ 11. a rotation, $\theta = 30°$

13. a reflection, $y = \frac{2}{3}x$ 15. a rotation, $\theta = 270°$

17. (b) The only eigenvalue is $\lambda = 1$, and the corresponding eigenspace is spanned by $\{\mathbf{e}_3\}$.

37. $Q = \begin{bmatrix} 1 & 0 \\ 0 & -1 \end{bmatrix}$ and $\mathbf{b} = \begin{bmatrix} 1 \\ 4 \end{bmatrix}$

45. $\begin{bmatrix} 0.7833 & 0.6217 \\ 0.6217 & -0.7833 \end{bmatrix}$ (rounded to four places after the decimal)

47. $231°$

Section 6.5

1. (a) T   (b) F   (c) F   (d) F   (e) T
   (f) T   (g) F   (h) T

3. (a) $\begin{bmatrix} 1 & -6 \\ -6 & -4 \end{bmatrix}$ (b) about $56.3°$

   (c) $x = \dfrac{2}{\sqrt{13}}x' - \dfrac{3}{\sqrt{13}}y'$

   $y = \dfrac{3}{\sqrt{13}}x' + \dfrac{2}{\sqrt{13}}y'$

   (d) $-8(x')^2 + 5(y')^2 = 40$ (e) a hyperbola

5. (a) $\begin{bmatrix} 5 & 2 \\ 2 & 5 \end{bmatrix}$ (b) $45°$ (c) $x = \dfrac{1}{\sqrt{2}}x' - \dfrac{1}{\sqrt{2}}y'$

   $y = \dfrac{1}{\sqrt{2}}x' + \dfrac{1}{\sqrt{2}}y'$

   (d) $7(x')^2 + 3(y')^2 = 9$ (e) an ellipse

7. (a) $\begin{bmatrix} 1 & 2 \\ 2 & 1 \end{bmatrix}$ (b) $45°$ (c) $x = \dfrac{1}{\sqrt{2}}x' - \dfrac{1}{\sqrt{2}}y'$

   $y = \dfrac{1}{\sqrt{2}}x' + \dfrac{1}{\sqrt{2}}y'$

   (d) $3(x')^2 - (y')^2 = 7$ (e) a hyperbola

9. (a) $\begin{bmatrix} 2 & -6 \\ -6 & -7 \end{bmatrix}$ (b) about $63.4°$

   (c) $x = \dfrac{1}{\sqrt{5}}x' - \dfrac{2}{\sqrt{5}}y'$

   $y = \dfrac{2}{\sqrt{5}}x' + \dfrac{1}{\sqrt{5}}y'$

   (d) $-10(x')^2 + 5(y')^2 = 200$ (e) a hyperbola

11. (a) $\begin{bmatrix} 1 & 1 \\ 1 & 1 \end{bmatrix}$ (b) $45°$ (c) $x = \dfrac{1}{\sqrt{2}}x' - \dfrac{1}{\sqrt{2}}y'$

    $y = \dfrac{1}{\sqrt{2}}x' + \dfrac{1}{\sqrt{2}}y'$

    (d) $2\sqrt{2}(x')^2 + 9x' - 7y' = 0$ (e) a parabola

13. $\left\{ \dfrac{1}{\sqrt{2}}\begin{bmatrix} 1 \\ -1 \end{bmatrix}, \dfrac{1}{\sqrt{2}}\begin{bmatrix} 1 \\ 1 \end{bmatrix} \right\}$, 2 and 4,

    $A = 2\begin{bmatrix} 0.5 & -0.5 \\ -0.5 & 0.5 \end{bmatrix} + 4\begin{bmatrix} 0.5 & 0.5 \\ 0.5 & 0.5 \end{bmatrix}$

15. $\left\{ \dfrac{1}{\sqrt{2}}\begin{bmatrix} 1 \\ 1 \end{bmatrix}, \dfrac{1}{\sqrt{2}}\begin{bmatrix} 1 \\ -1 \end{bmatrix} \right\}$, 3 and -1,

    $A = 3\begin{bmatrix} 0.5 & 0.5 \\ 0.5 & 0.5 \end{bmatrix} + (-1)\begin{bmatrix} 0.5 & -0.5 \\ -0.5 & 0.5 \end{bmatrix}$

17. $\left\{ \dfrac{1}{3}\begin{bmatrix} -1 \\ -2 \\ 2 \end{bmatrix}, \dfrac{1}{3}\begin{bmatrix} 2 \\ 1 \\ 2 \end{bmatrix}, \dfrac{1}{3}\begin{bmatrix} -2 \\ 2 \\ 1 \end{bmatrix} \right\}$, 3, 6, and 0,

$$A = 3 \begin{bmatrix} \frac{1}{9} & \frac{2}{9} & -\frac{2}{9} \\ \frac{2}{9} & \frac{4}{9} & -\frac{4}{9} \\ -\frac{2}{9} & -\frac{4}{9} & \frac{2}{9} \end{bmatrix} + 6 \begin{bmatrix} \frac{4}{9} & \frac{2}{9} & \frac{4}{9} \\ \frac{2}{9} & \frac{1}{9} & \frac{2}{9} \\ \frac{4}{9} & \frac{2}{9} & \frac{4}{9} \end{bmatrix}$$

$$+ 0 \begin{bmatrix} \frac{4}{9} & -\frac{4}{9} & -\frac{2}{9} \\ -\frac{4}{9} & \frac{4}{9} & \frac{2}{9} \\ -\frac{2}{9} & \frac{2}{9} & \frac{1}{9} \end{bmatrix}$$

19. $\left\{ \begin{bmatrix} 1 \\ 0 \\ 0 \end{bmatrix}, \frac{1}{\sqrt{5}} \begin{bmatrix} 0 \\ -2 \\ 1 \end{bmatrix}, \frac{1}{\sqrt{5}} \begin{bmatrix} 0 \\ 1 \\ 2 \end{bmatrix} \right\}$ $-1, -1,$ and $4,$

$$A = (-1) \begin{bmatrix} 1 & 0 & 0 \\ 0 & 0 & 0 \\ 0 & 0 & 0 \end{bmatrix} + (-1) \begin{bmatrix} 0 & 0 & 0 \\ 0 & .8 & -.4 \\ 0 & -.4 & .2 \end{bmatrix}$$

$$+ 4 \begin{bmatrix} 0 & 0 & 0 \\ 0 & .2 & .4 \\ 0 & .4 & .8 \end{bmatrix}$$

21. $2\begin{bmatrix} 1 & 0 \\ 0 & 0 \end{bmatrix} + 2\begin{bmatrix} 0 & 0 \\ 0 & 1 \end{bmatrix}$ and $2\begin{bmatrix} .5 & .5 \\ .5 & .5 \end{bmatrix} + 2\begin{bmatrix} .5 & -.5 \\ -.5 & .5 \end{bmatrix}$

Section 6.6

1. (a) F (b) T (c) F (d) T (e) F

3. $\begin{bmatrix} \frac{1}{3} & \frac{2}{\sqrt{5}} & \frac{2}{3\sqrt{5}} \\ \frac{2}{3} & \frac{-1}{\sqrt{5}} & \frac{4}{3\sqrt{5}} \\ \frac{2}{3} & 0 & \frac{-5}{3\sqrt{5}} \end{bmatrix} \begin{bmatrix} 3 \\ 0 \\ 0 \end{bmatrix} [1]^T$

5. $\begin{bmatrix} \frac{3}{\sqrt{35}} & \frac{1}{\sqrt{10}} & \frac{-3}{\sqrt{14}} \\ \frac{-1}{\sqrt{35}} & \frac{3}{\sqrt{10}} & \frac{1}{\sqrt{14}} \\ \frac{5}{\sqrt{35}} & 0 & \frac{2}{\sqrt{14}} \end{bmatrix} \begin{bmatrix} \sqrt{7} & 0 \\ 0 & \sqrt{2} \\ 0 & 0 \end{bmatrix} \begin{bmatrix} \frac{1}{\sqrt{5}} & \frac{2}{\sqrt{5}} \\ \frac{2}{\sqrt{5}} & \frac{-1}{\sqrt{5}} \end{bmatrix}^T$

7. $\begin{bmatrix} \frac{1}{\sqrt{2}} & \frac{1}{\sqrt{2}} \\ \frac{-1}{\sqrt{2}} & \frac{1}{\sqrt{2}} \end{bmatrix} \begin{bmatrix} 2 & 0 & 0 \\ 0 & \sqrt{2} & 0 \end{bmatrix} \begin{bmatrix} 0 & 1 & 0 \\ \frac{1}{\sqrt{2}} & 0 & \frac{1}{\sqrt{2}} \\ \frac{1}{\sqrt{2}} & 0 & \frac{-1}{\sqrt{2}} \end{bmatrix}^T$

9. $\begin{bmatrix} \frac{1}{\sqrt{6}} & \frac{5}{\sqrt{30}} & 0 \\ \frac{2}{\sqrt{6}} & \frac{-2}{\sqrt{30}} & \frac{1}{\sqrt{5}} \\ \frac{1}{\sqrt{6}} & \frac{-1}{\sqrt{30}} & \frac{-2}{\sqrt{5}} \end{bmatrix} \begin{bmatrix} \sqrt{6} & 0 & 0 \\ 0 & \sqrt{6} & 0 \\ 0 & 0 & 1 \end{bmatrix} \begin{bmatrix} 1 & 0 & 0 \\ 0 & \frac{1}{\sqrt{5}} & \frac{2}{\sqrt{5}} \\ 0 & \frac{2}{\sqrt{5}} & \frac{-1}{\sqrt{5}} \end{bmatrix}^T$

11. $\begin{bmatrix} \frac{2}{\sqrt{5}} & \frac{1}{\sqrt{5}} & 0 \\ \frac{1}{\sqrt{5}} & \frac{-2}{\sqrt{5}} & 0 \\ 0 & 0 & 1 \end{bmatrix} \begin{bmatrix} \sqrt{60} & 0 & 0 & 0 \\ 0 & \sqrt{15} & 0 & 0 \\ 0 & 0 & 0 & 0 \end{bmatrix} \begin{bmatrix} \frac{1}{\sqrt{3}} & \frac{-1}{\sqrt{3}} & \frac{1}{\sqrt{6}} & \frac{1}{\sqrt{6}} \\ \frac{1}{\sqrt{3}} & \frac{1}{\sqrt{3}} & \frac{1}{\sqrt{6}} & \frac{-1}{\sqrt{6}} \\ \frac{1}{\sqrt{3}} & 0 & \frac{-2}{\sqrt{6}} & 0 \\ 0 & \frac{1}{\sqrt{3}} & 0 & \frac{2}{\sqrt{6}} \end{bmatrix}^T$

13. $\begin{bmatrix} \frac{1}{\sqrt{3}} & \frac{2}{\sqrt{6}} & 0 \\ \frac{1}{\sqrt{3}} & \frac{-1}{\sqrt{6}} & \frac{1}{\sqrt{2}} \\ \frac{-1}{\sqrt{3}} & \frac{1}{\sqrt{6}} & \frac{1}{\sqrt{2}} \end{bmatrix} \begin{bmatrix} \sqrt{21} & 0 & 0 & 0 \\ 0 & \sqrt{18} & 0 & 0 \\ 0 & 0 & 0 & 0 \end{bmatrix}$

$$\times \begin{bmatrix} \frac{1}{\sqrt{7}} & \frac{1}{\sqrt{3}} & \frac{1}{\sqrt{11}} & \frac{1}{\sqrt{2}} \\ \frac{2}{\sqrt{7}} & \frac{-1}{\sqrt{3}} & \frac{1}{\sqrt{11}} & 0 \\ \frac{1}{\sqrt{7}} & 0 & \frac{-3}{\sqrt{11}} & 0 \\ \frac{1}{\sqrt{7}} & \frac{1}{\sqrt{3}} & 0 & \frac{-1}{\sqrt{2}} \end{bmatrix}^T$$

15. In the accompanying figure:
$$\mathbf{u}_1 = \frac{1}{\sqrt{2}} \begin{bmatrix} 1 \\ -1 \end{bmatrix}, \mathbf{u}_2 = \frac{1}{\sqrt{2}} \begin{bmatrix} 1 \\ 1 \end{bmatrix},$$
$$OP = 2\sqrt{2}, \quad \text{and} \quad OQ = \sqrt{2}.$$

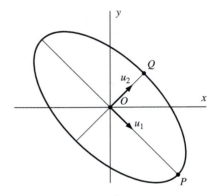

Figure for Exercise 15

17. $\begin{bmatrix} 1 \\ 1 \end{bmatrix}$ **19.** $\begin{bmatrix} 3 \\ 1 \\ 1 \end{bmatrix}$ **21.** $\frac{1}{7}\begin{bmatrix} 12 \\ 3 \end{bmatrix}$ **23.** $\frac{5}{6}\begin{bmatrix} 3 \\ 3 \\ 2 \end{bmatrix}$

25. $\frac{1}{9}[1 \ 2 \ 2]$ **27.** $\frac{1}{14}\begin{bmatrix} 4 & 8 & 2 \\ 1 & -5 & 4 \end{bmatrix}$ **29.** $\frac{1}{4}\begin{bmatrix} 2 & 2 \\ 1 & -1 \\ 1 & -1 \end{bmatrix}$

31. $\frac{1}{9}\begin{bmatrix} 1 & 2 & 2 \\ 2 & 4 & 4 \\ 2 & 4 & 4 \end{bmatrix}$ **33.** $\frac{1}{14}\begin{bmatrix} 5 & 3 & 6 \\ 3 & 13 & -2 \\ 6 & -2 & 10 \end{bmatrix}$

51. (rounded to four places after the decimal)
$$U = \begin{bmatrix} 0.5836 & 0.7289 & -0.3579 \\ 0.7531 & -0.6507 & -0.0970 \\ 0.3036 & 0.2129 & 0.9287 \end{bmatrix},$$

$$\Sigma = \begin{bmatrix} 5.9073 & 0 & 0 & 0 \\ 0 & 2.2688 & 0 & 0 \\ 0 & 0 & 1.7194 & 0 \end{bmatrix},$$

$$V = \begin{bmatrix} 0.3024 & -0.3462 & -0.8612 & -0.2170 \\ 0.0701 & 0.9293 & -0.3599 & 0.0434 \\ 0.2777 & 0.1283 & 0.2755 & -0.9113 \\ 0.9091 & 0.0043 & 0.2300 & 0.3472 \end{bmatrix},$$

and

$$A^\dagger = \begin{bmatrix} 0.0979 & 0.1864 & -0.4821 \\ 0.3804 & -0.2373 & -0.1036 \\ 0.0113 & -0.0169 & 0.1751 \\ 0.0433 & 0.1017 & 0.1714 \end{bmatrix}$$

Section 6.7

1. (a) F   (b) F   (c) F   (d) F   (e) T
   (f) F   (g) T

3. $\begin{bmatrix} 0 & 1 & 0 \\ 0 & 0 & -1 \\ -1 & 0 & 0 \end{bmatrix}$ 5. $\dfrac{1}{\sqrt{2}}\begin{bmatrix} 1 & -1 & 0 \\ 0 & 0 & -\sqrt{2} \\ 1 & 1 & 0 \end{bmatrix}$

7. $\dfrac{1}{4}\begin{bmatrix} 2\sqrt{3} & 0 & 2 \\ 1 & 2\sqrt{3} & -\sqrt{3} \\ -\sqrt{3} & 2 & 3 \end{bmatrix}$

9. $\begin{bmatrix} 0 & 0 & 1 \\ 0 & -1 & 0 \\ 1 & 0 & 0 \end{bmatrix}$ 11. $\dfrac{1}{2\sqrt{2}}\begin{bmatrix} \sqrt{2}+1 & \sqrt{2}-1 & -\sqrt{2} \\ \sqrt{2}-1 & \sqrt{2}+1 & \sqrt{2} \\ \sqrt{2} & -\sqrt{2} & 2 \end{bmatrix}$

13. $\dfrac{1}{4}\begin{bmatrix} \sqrt{3}+2 & \sqrt{3}-2 & -\sqrt{2} \\ \sqrt{3}-2 & \sqrt{3}+2 & -\sqrt{2} \\ \sqrt{2} & \sqrt{2} & 2\sqrt{3} \end{bmatrix}$

15. (a) $\begin{bmatrix} 1 \\ 1 \\ -1 \end{bmatrix}$ (b) $-\dfrac{1}{2}$ 17. (a) $\begin{bmatrix} \sqrt{2}+1 \\ -1 \\ 1 \end{bmatrix}$ (b) $\dfrac{1-\sqrt{2}}{2\sqrt{2}}$

19. (a) $\begin{bmatrix} 1 \\ 1 \\ 2-\sqrt{3} \end{bmatrix}$ (b) $\dfrac{4\sqrt{3}-1}{16}$

29. $\dfrac{1}{3}\begin{bmatrix} 2 & 2 & -1 \\ 2 & -1 & 2 \\ -1 & 2 & 2 \end{bmatrix}$ 31. $\dfrac{1}{3}\begin{bmatrix} 1 & -2 & -2 \\ -2 & 1 & -2 \\ -2 & -2 & 1 \end{bmatrix}$

33. $\dfrac{1}{3}\begin{bmatrix} 2 & -2 & 1 \\ -2 & -1 & 2 \\ 1 & 2 & 2 \end{bmatrix}$

39. (a) neither   41. (a) a reflection (b) $\{e_2, e_3\}$

43. (a) a rotation (b) $\begin{bmatrix} 2 \\ 1 \\ -2 \end{bmatrix}$

45. (a) a reflection   (b) $\left\{\begin{bmatrix} 1 \\ 0 \\ \sqrt{2}-1 \end{bmatrix}, \begin{bmatrix} 0 \\ 1 \\ 0 \end{bmatrix}\right\}$

49. (rounded to four places after the decimal)

Span $\left\{\begin{bmatrix} .4609 \\ .1769 \\ .8696 \end{bmatrix}\right\}$,   48°

Chapter 6 Review

1. (a) T   (b) T   (c) F   (d) T   (e) T   (f) T
   (g) T   (h) T   (i) F   (j) F   (k) T   (l) F
   (m) F   (n) F   (o) F   (p) T   (q) T   (r) F
   (s) T

3. (a) $\|u\| = \sqrt{45}$, $\|v\| = \sqrt{20}$ (b) $d = \sqrt{65}$
   (c) $u \cdot v = 0$ (d) u and v are orthogonal.

5. (a) $\|u\| = \sqrt{6}$, $\|v\| = \sqrt{21}$ (b) $d = \sqrt{27}$
   (c) $u \cdot v = 0$ (d) u and v are orthogonal.

7. $w = \dfrac{1}{5}\begin{bmatrix} -1 \\ 2 \end{bmatrix}$, $d = 3.5777$ 9. 1   11. 113

13. $\left\{\begin{bmatrix} 1 \\ 1 \\ -1 \\ 0 \end{bmatrix}, \dfrac{1}{3}\begin{bmatrix} 1 \\ 1 \\ 2 \\ 3 \end{bmatrix}, \dfrac{1}{5}\begin{bmatrix} -2 \\ 3 \\ 1 \\ -1 \end{bmatrix}\right\}$

15. $\left\{\begin{bmatrix} 6 \\ -7 \\ 5 \\ 0 \end{bmatrix}, \begin{bmatrix} -2 \\ 4 \\ 0 \\ 5 \end{bmatrix}\right\}$

17. $w = \dfrac{1}{14}\begin{bmatrix} 32 \\ 19 \\ -27 \end{bmatrix}$ and $z = \dfrac{1}{14}\begin{bmatrix} -18 \\ 9 \\ -15 \end{bmatrix}$

19. $P_W = \dfrac{1}{6}\begin{bmatrix} 1 & 2 & 0 & -1 \\ 2 & 4 & 0 & -2 \\ 0 & 0 & 0 & 0 \\ -1 & -2 & 0 & 1 \end{bmatrix}$ and $w = \begin{bmatrix} 2 \\ 4 \\ 0 \\ -2 \end{bmatrix}$

21. $P_W = \dfrac{1}{3}\begin{bmatrix} 2 & -1 & 1 & 0 \\ -1 & 2 & 1 & 0 \\ 1 & 1 & 2 & 0 \\ 0 & 0 & 0 & 0 \end{bmatrix}$ and $w = \begin{bmatrix} 2 \\ -1 \\ 1 \\ 0 \end{bmatrix}$

23. $v \approx 2.05$ and $c \approx 1.05$

25. no   27. yes   29. a rotation, $\theta = -60°$

31. a reflection, $y = \dfrac{1}{\sqrt{3}}x$

35. $\left\{\dfrac{1}{\sqrt{2}}\begin{bmatrix} -1 \\ 1 \end{bmatrix}, \dfrac{1}{\sqrt{2}}\begin{bmatrix} 1 \\ 1 \end{bmatrix}\right\}$, -1 and 5,

$A = (-1)\begin{bmatrix} 0.5 & -0.5 \\ -0.5 & 0.5 \end{bmatrix} + 5\begin{bmatrix} 0.5 & 0.5 \\ 0.5 & 0.5 \end{bmatrix}$

37. $45°$, $\dfrac{(x')^2}{4} - \dfrac{(y')^2}{8} = 1$,   a hyperbola

CHAPTER 7

Section 7.1

1. (a) T   (b) F   (c) F   (d) T   (e) F   (f) F
   (g) T   (h) T   (i) T   (j) T   (k) T   (l) F
   (m) T

27. no   29. yes   31. yes   33. no   35. yes

37. yes

Section 7.2

1. (a) T   (b) T   (c) F   (d) F   (e) F   (f) F
   (g) F   (h) T   (i) T

3. yes   5. no   9. no   11. no

15. linearly independent

17. linearly independent   19. linearly dependent

21. linearly independent

23. linearly dependent   25. $p(x) = 2x^2 - 3x + 1$

27. $\left\{ \begin{bmatrix} 1 & 0 & 0 \\ 0 & 0 & 0 \\ 0 & 0 & 0 \end{bmatrix}, \begin{bmatrix} 0 & 0 & 0 \\ 0 & 1 & 0 \\ 0 & 0 & 0 \end{bmatrix}, \begin{bmatrix} 0 & 0 & 0 \\ 0 & 0 & 0 \\ 0 & 0 & 1 \end{bmatrix}, \begin{bmatrix} 0 & 1 & 0 \\ 1 & 0 & 0 \\ 0 & 0 & 0 \end{bmatrix}, \right.$
$\left. \begin{bmatrix} 0 & 0 & 1 \\ 0 & 0 & 0 \\ 1 & 0 & 0 \end{bmatrix}, \begin{bmatrix} 0 & 0 & 0 \\ 0 & 0 & 1 \\ 0 & 1 & 0 \end{bmatrix} \right\}$

29. $\left\{ \begin{bmatrix} 1 & 0 \\ 0 & -1 \end{bmatrix}, \begin{bmatrix} 0 & 1 \\ 0 & 0 \end{bmatrix}, \begin{bmatrix} 0 & 0 \\ 1 & 0 \end{bmatrix} \right\}$

33. linear, not an isomorphism   35. not linear

37. linear, an isomorphism

65. The set is linearly independent.

67. The set is linearly dependent, and $M_3 = (-3)M_1 + 2M_2$, where M_j is the jth matrix in the set.

69. (rounded to four places after the decimal) $c_0 = 0.3486$, $c_1 = 0.8972$, $c_2 = -0.3667$, $c_3 = 0.1472$, $c_4 = -0.0264$

Section 7.3

1. (a) F   (b) F   (c) T   (d) T   (e) T
   (f) T   (g) T

13. $[A]_\mathcal{B} = \begin{bmatrix} 1 \\ 3 \\ 4 \\ 2 \end{bmatrix}$

15. $[\sin 2t - \cos 2t]_\mathcal{B}$

$= [2\sin t \cos t - \cos^2 t + \sin^2 t]_\mathcal{B} = \begin{bmatrix} -1 \\ 1 \\ 2 \end{bmatrix}$

17. $\begin{bmatrix} 1 & 0 & 0 \\ 0 & 2 & 0 \\ 0 & 0 & 3 \end{bmatrix}$ 19. $\begin{bmatrix} 1 & 0 & 0 \\ 3 & 3 & 3 \\ 1 & 2 & 4 \end{bmatrix}$

21. (a) $-8x$ (b) $3 + 10x$ (c) $3x^2$

23. (a) $-e^t + te^t$ (b) $2e^t - 2te^t + t^2 e^t$
    (c) $11e^t - 8te^t + 2t^2 e^t$

25. 1, 2, 3, $\{e^t\}$, $\{e^{2t}\}$, $\{e^{3t}\}$

27. 1, 6, $\{3x - 2x^2\}$, $\{x + x^2\}$ 29. (b) $\begin{bmatrix} 1 & 0 & 0 & 1 \\ 2 & 0 & 0 & 2 \\ 3 & 0 & 0 & 3 \\ 4 & 0 & 0 & 4 \end{bmatrix}$

45. (rounded to four places after the decimal)
   (a) $-1.6533, 2.6277, 6.6533, 8.3723$

   (b) $\left\{ \begin{bmatrix} -0.1827 & -0.7905 \\ 0.5164 & 0.2740 \end{bmatrix}, \begin{bmatrix} 0.6799 & -0.4655 \\ -0.4655 & 0.3201 \end{bmatrix}, \right.$

   $\left. \begin{bmatrix} 0.4454 & 0.0772 \\ 0.5909 & -0.6681 \end{bmatrix}, \begin{bmatrix} 0.1730 & 0.3783 \\ 0.3783 & 0.8270 \end{bmatrix} \right\}$

   (c) $\begin{bmatrix} 0.2438a - 0.1736b & -0.2603a - 0.2893b \\ +0.0083c + 0.0496d & +0.3471c + 0.0826d \\ & \\ 0.0124a + 0.3471b & -0.1116a + 0.1240b \\ -0.0165c - 0.0992d & -0.1488c + 0.1074d \end{bmatrix}$

Section 7.4

1. (a) F   (b) T   (c) F   (d) F   (e) T   (f) F
   (g) T   (h) T   (i) T   (j) T   (k) F   (l) T

3. e^2 5. 0   7. $-\dfrac{50}{3}$ 13. yes   15. no

17. yes

21. $\left\{ 1, e^t - e + 1, e^{-t} + \dfrac{-3e^2 + 6e - 5}{e(e-3)} \right.$

$\left. + \dfrac{2(e^2 - e + 1)}{e(e-3)(e-1)} e^t \right\}$

35. (a) $\left\{ \begin{bmatrix} 1 & 0 \\ 0 & 0 \end{bmatrix}, \dfrac{1}{\sqrt{2}} \begin{bmatrix} 0 & 1 \\ 1 & 0 \end{bmatrix}, \begin{bmatrix} 0 & 0 \\ 0 & 1 \end{bmatrix} \right\}$ (b) $\begin{bmatrix} 1 & 3 \\ 3 & 8 \end{bmatrix}$

Chapter 7 Review

1. (a) F   (b) T   (c) F   (d) F   (e) T
   (f) F   (g) T

3. no   5. yes   7. yes   9. no   11. no

13. yes   15. $c = 5$

17. $\{-1 + x, -2 + x^2, x^3\}$, dim $W = 3$ 19. not linear

21. linear, an isomorphism   23. $\begin{bmatrix} a & b \\ -b & a \end{bmatrix}$

25. $\begin{bmatrix} 3 & 0 & 0 & 0 \\ 0 & 2 & 1 & 0 \\ 0 & 1 & 2 & 0 \\ 0 & 0 & 0 & 3 \end{bmatrix}$

27. $\dfrac{1}{a^2 + b^2}(ac_1 - bc_2)e^{at} \cos bt$

$+ \dfrac{1}{a^2 + b^2}(bc_1 + ac_2)e^{at} \sin bt$

29. $\dfrac{1}{3}\begin{bmatrix} a & 2b - c \\ -b + 2c & d \end{bmatrix}$

31. T has no (real) eigenvalues.

33. 1 and 3, with corresponding bases $\left\{ \begin{bmatrix} 0 & 1 \\ -1 & 0 \end{bmatrix} \right\}$

and $\left\{ \begin{bmatrix} 1 & 0 \\ 0 & 0 \end{bmatrix}, \begin{bmatrix} 0 & 1 \\ 1 & 0 \end{bmatrix}, \begin{bmatrix} 0 & 0 \\ 0 & 1 \end{bmatrix} \right\}$, respectively

35. $\left\{ \begin{bmatrix} -3 & 1 \\ 0 & 0 \end{bmatrix}, \begin{bmatrix} -4 & 0 \\ 1 & 0 \end{bmatrix}, \begin{bmatrix} -2 & 0 \\ 0 & 1 \end{bmatrix} \right\}$ **37.** $\begin{bmatrix} 2 & -2 \\ 2 & -3 \end{bmatrix}$

39. $\left\{ 1, \sqrt{3}(2x - 1), \sqrt{\dfrac{5}{19}}(12x^2 - 6x - 1) \right\}$

41. $\dfrac{6}{35} + \dfrac{8}{35}x + \dfrac{8}{7}x^2$

Index

A

Addition, *see* Sum, 3
Additive inverse
    of a matrix, 4
    of a polynomial, 401
    of a vector, 398
Adjacency matrix, 104
Adjoint, classical, 195
Algorithm
    for computing $A^{-1}B$, 130
    Gaussian elimination, 36–40
    Gram-Schmidt process, 323
    for matrix diagonalization, 277
    for matrix inversion, 128
    for solving a system of linear
        equations, 32
Anthropology, 106
Area of a parallelogram, 181
Associated quadratic form, 358
Associative law for matrix
    multiplication, 88
Augmented matrix of a system of linear
    equations, 26
Averaging class size, 318–319
Axis of rotation, 383

B

$\mathcal{B}$-coordinate vector, 231
$\mathcal{B}$-matrix, 240
Back substitution, 140
Backward pass, 40
Basic variables, 29
Basis, 211
    for the column space of a matrix, 211
    of eigenvectors, 276, 284
    left singular vectors, 366
    for the null space of a matrix, 221
    ordered, 231
    orthogonal, 322
    orthonormal, 325
    right singular vectors, 366
    for the row space of a matrix, 223
    standard, 211
Best fit, line of, 336
Binomial formula for matrices, 111

Block, 91
Block multiplication, 92
Block problem, 299
Bunyakovsky, Viktor Yakovlevich, 316

C

Cauchy, Augustin-Louis, 175, 316
Cauchy-Schwarz inequality, 316, 432
Cayley, Arthur, 282
Cayley-Hamilton theorem, 282, 362
Characteristic equation
    of a linear operator, 261
    of a matrix, 259
Characteristic polynomial
    of a linear operator, 261
    of a matrix, 259
City-suburb application, 18, 89
Classical adjoint, 195
Clique, 111
Closest vector property, 331, 435
Closure, 199, 402
Codomain of a function, 151
Coefficient matrix of a system of linear
    equations, 26
Coefficients of a linear combination, 12
Coefficients of a linear equation, 23
Coefficients of a polynomial, 400
Cofactor, 177
Cofactor expansion
    along a column, 190
    along the first row, 177
    along the ith row, 178
Column of a matrix, 6
Column space of a matrix, 204
    basis, 211
    dimension, 222
Components of a vector, 5
Composite of functions, 165
Computer graphics, 378–383
Conic sections, 357–360
    ellipse, 228, 237, 357–360, 362
    hyperbola, 234–235, 237, 362
Consistent system of linear equations, 24
Constant term of a linear equation, 23

Consumption matrix, *see* Input-output
    matrix, 50
Contraction, 157, 159
Coordinate vector, 231, 421
Cramer's rule, 191
Cramer, Gabriel, 175, 191
Cube root of a matrix, 282
Current flow, 54–55

D

Damping force, 298
Degree of a polynomial, 400
Demand vector, 53
Determinant
    1×1 matrix, 176
    2×2 matrix, 175
    $n \times n$ matrix, 177
    area of a parallelogram, 181
    cofactor expansion along a column, 190
    cofactor expansion along the first
        row, 177
    cofactor expansion along the ith
        row, 178
    Cramer's rule, 191
    and elementary row operations, 185–187
    lower triangular matrix, 180
    nilpotent matrix, 195
    orthogonal matrix, 195
    properties, 188
    skew-symmetric matrix, 195
    upper triangular matrix, 180
    Vandermonde matrix, 195
    volume of a parallelepiped, 182
Deterministic relationship, 335
Diagonal entry of a matrix, 89
Diagonal matrix, 10, 89
Diagonal of a matrix, 89
Diagonalizable linear operator, 284
Diagonalizable matrix, 271, 355
Diagonalization of a matrix, 277
Diagonals of a parallelogram, 315
Difference equation, 299–302
    initial conditions, 301
    kth-order homogeneous, 301
    nonhomogeneous, 306

Differential equations
  initial conditions, 294
  system of, 294–299
Differential operator, 419
Dilation, 159
Dimension, 213, 412
  column space of a matrix, 222
  null space of a matrix, 222
  row space of a matrix, 224
Dimension theorem, 227
Diplomatic relations, 103
Distance
  between vectors, 312
  from a point to a line, 315
  from a vector to a subspace, 331
Distributive laws for matrix
  multiplication, 88
Domain, 151
Dot product, 313

E
Eigenspace
  of a linear operator, 254, 420
  of a matrix, 254
Eigenvalue
  of a linear operator, 252, 420
  of a matrix, 252
Eigenvector
  of a linear operator, 252, 420
  of a matrix, 252
Electrical network, 54–55
Elementary column operation, 125
Elementary matrix, 117
Elementary row operation, 27
Ellipse, 228, 237, 357–360, 362
Entry of a matrix, 2
Entry of a matrix product, 87
Equality
  of matrices, 2
  of polynomials, 401
Equivalent systems of linear equations, 24
Error sum of squares, 336
Euler's formula, 298
Even function, 405
Extension principle, 212

F
Fibonacci sequence, 301, 427
Finite-dimensional vector space, 412
Flop, 143
Flop count, 143
  Gaussian elimination, 143
  inverse of a matrix, 143
  LU decomposition, 143
  matrix product, 149
  matrix-vector product, 149

solving a system using an *LU*
  decomposition, 143
Focal point, 381
Forward pass, 40
Forward substitution, 140
Fourier analysis, 440
Fourier, Jean Baptiste Joseph, 440
Free variables, 30
Frobenius inner product, 431
Frobenius, Ferdinand Georg, 431
Function, 151
  composite, 165
  even, 405
  inverse, 167
  invertible, 167
  odd, 405
  one-to-one, 163
  onto, 161
Function space, 399

G
Gaussian elimination, 36
  flop count, 143
  for matrix inversion, 128–129
General solution to a system of linear
  equations, 30
Geometry
  area of a parallelogram, 181
  contraction, 157
  dilation, 159
  perspective, 381–383
  projections, 152, 159, 246, 286, 315,
    331, 340
  reflections, 159, 239, 246, 392
  rigid motion, 350–352
  rotations, 20, 84, 378–381, 383–391
  shear, 153
  translation, 351
  volume of a parallelepiped, 182
Gibbs, Josiah Willard, 7
Gram, Jorgen P., 323
Gram-Schmidt process, 323, 433
Gross production vector, 52

H
Hamilton, William Rowan, 282
Harmonic motion, 298–299
Heat loss, 302
Heaviside, Oliver, 7
Homogeneous difference equation, 301
Homogeneous system of linear
  equations, 72
  parametric representation of
    solutions, 73
Hooke's law, 343
Hyperbola, 234–235, 237, 362

I
Identity matrix, 17
Identity transformation, 154
Image, 151
Inconsistent system of linear equations, 24
Infinite-dimensional vector space, 412
Initial conditions
  difference equation, 301
  system of differential equations, 294
Inner product, 430
  axioms, 430
  Frobenius, 431
Inner product space, 430
  isometric, 443
Input-output matrix, 50, 131
Inverse of a function, 167
Inverse of a linear transformation, 416
Inverse of a matrix, 113
  flop count, 143
Invertible function, 167
Invertible matrix, 113
Isometric inner product spaces, 443
Isomorphic vector spaces, 410
Isomorphism, 410

J
Jordan canonical form, 297

K
Kirchoff's current law, 55
Kirchoff's voltage law, 54
Kirchoff, Gustav Robert, 54

L
Lagrange interpolating polynomial, 413
Lagrange, Joseph Louis, 413
Law of cosines, 321
Leading entry of a matrix in reduced row
  echelon form, 35
Least squares approximation of a
  function, 436
Least squares line, 336
Least squares problem, 335–339, 371–375
Left singular vectors of a matrix, 366
Legendre polynomials, normalized, 435
Legendre, Adrien Marie, 336, 435
Length of a vector, *see* Norm, 311
Leontief input–output model, 50
Leontief input-output model, 131
Leontief, Wassily, 50
Leslie matrix, 101
Line of best fit, 336
Linear combination, 12, 406
  and linearly dependent sets, 73
  coefficients, 12, 406
Linear correspondence property, 120
Linear equation, 23

Linear isometry, 443
Linear operator, *see also* Linear
    transformation, 239, 418
    characteristic equation, 261
    characteristic polynomial, 261
    diagonalizable, 284
    differential operator, 419
    eigenspace, 254, 420
    eigenvalue, 252, 420
    eigenvector, 252, 420
    matrix representation, 240, 422
    null space, 418
    orthogonal, 345
    orthogonal on $\mathcal{R}^2$, 348–350
    orthogonal projection, 286
    orthogonal projection of $\mathcal{R}^2$ on a
        line, 246
    orthogonal projection of $\mathcal{R}^3$ on a
        plane, 246
    reflection of $\mathcal{R}^2$ about a line, 239,
        348–350
    reflection of $\mathcal{R}^3$ about a plane, 246, 392
    rotation, 345, 348–350
    square root, 290
Linear relationship among the columns of a
    matrix, 119
Linear transformation, *see also* Linear
    operator *and* Matrix, 154, 410
    contraction, 157, 159
    dilation, 159
    identity, 154
    inverse, 416
    isometry, 443
    null space, 163, 418
    scalar multiple, 159, 417
    shear transformation, 153
    standard matrix, 156
    sum, 159, 417
    zero, 154
Linearly dependent set, 69, 409
Linearly independent set, 69–76, 409
Lower triangular matrix, 137, 180
LU decomposition, 137
    flop count, 143
    used to solve a system of linear
        equations, 140–143

M
Magic square, 416
Markov chain, 292
Marriage laws, 106
Mathematical model
    clique, 111
    current flow, 54–55
    diplomatic relations, 103
    electrical network, 54–55

    harmonic motion, 298–299
    input-output model, 50, 131
    Markov chain, 292
    marriage laws, 106
    predator-prey, 295
    rainfall in Tel Aviv, 304
    scheduling, 105
    traffic flow, 101
Matrix, *see also* Linear operator *and* Linear
    transformation, 2
    addition, 3
    adjacency, 104
    associative law of multiplication, 88
    augmented, 26
    binomial formula, 111
    block, 91
    block multiplication, 92
    characteristic equation, 259
    characteristic polynomial, 259
    classical adjoint, 195
    coefficient, 23
    column, 6
    column space, 204
    consumption, *see* Input-output
        matrix, 50
    cube root, 282
    determinant, 177
    diagonal, 10, 89
    diagonalizable, 271
    diagonalization, 277
    distributive laws for multiplication, 88
    eigenspace, 254
    eigenvalue, 252
    eigenvector, 252
    elementary, 117
    entry, 2
    entry of a matrix product, 87
    equality, 2
    identity, 17
    input-output, 50, 131
    inverse, 113
    invertible, 113
    Jordan canonical form, 297
    left singular vectors, 366
    Leslie, 101
    linear relationship among columns, 119
    lower triangular, 137, 180
    Moore-Penrose generalized inverse, *see*
        Pseudoinverse, 375
    nilpotent, 195, 258
    null space, 203
    nullity, 42
    orthogonal, 195, 345
    orthogonal projection, 340
    partitioned, 91
    permutation, 144

    positive definite, 362, 441, 442
    positive semidefinite, 362
    powers, 89, 270
    product, 85
    pseudoinverse, 375
    rank, 42, 71, 97, 124, 132
    reduced row echelon form, 28, 32, 121
    representation of a linear operator, 240,
        422
    right singular vectors, 366
    rotation, 20, 84, 378–381,
        383–391
    row, 6
    row echelon form, 28
    row space, 206
    scalar, 135
    scalar multiple, 4
    similar, 134, 242, 261
    singular value, 366
    singular value decomposition, 370
    size, 2
    skew-symmetric, 10, 195, 421, 442
    spectral decomposition, 360
    square, 2
    stochastic, 18
    subtraction, 4
    symmetric, 10, 90, 355–364, 377, 421,
        442
    trace, 10, 97, 106, 282
    transition, 292
    transpose, 3, 116
    unit lower triangular, 137
    upper triangular, 97, 137, 180
    Vandermonde, 195
    zero, 4
Matrix product
    flop count, 149
Matrix transformation, 152
Matrix-vector product, 16
    flop count, 149
Method of least squares, 336
Moore-Penrose generalized inverse, *see*
    Pseudoinverse, 375
Multiplicity of an eigenvalue, 262
Multiplier, 138

N
n-tuple, 6
Natchez Indians, 106
Net production vector, 52
Nilpotent matrix, 195, 258
Nonhomogeneous difference equation, 306
Nonzero subspace, 201, 402
Norm of a vector, 311, 432
Normalized Legendre polynomials, 435
Normalized vector, 432

Null space, 418
   basis, 221
   dimension, 222
   of a linear transformation, 163
   of a matrix, 203
Nullity of a matrix, 42

O

Odd function, 405
Ohm, Georg Simon, 54
Ohm's law, 54
One-to-one function, 163
Onto function, 161
Operations count, *see also* Flop count, 143
Operations count, computing a determinant by cofactor expansion, 179
Ordered basis, 231
Orientation, 384
Orthogonal basis, 322
Orthogonal complement of a subset, 327
Orthogonal matrix, 195, 345
Orthogonal operator, 345
Orthogonal operator on $\mathcal{R}^2$, 348–350
Orthogonal projection
   of $\mathcal{R}^2$ on a line, 246
   of $\mathcal{R}^2$ on the x-axis, 159
   of $\mathcal{R}^3$ on a plane, 246
   of $\mathcal{R}^3$ on the xy-plane, 152
   of $\mathcal{R}^3$ onto the yz-plane, 159
Orthogonal projection matrix, 340
Orthogonal projection of a vector, 331, 435
Orthogonal projection of a vector on a line, 315
Orthogonal projection operator, 286
Orthogonal set, 321, 432
Orthogonal vectors, 313, 432
Orthonormal basis, 325
Orthonormal set, 432
Outer product, 94
Overdetermined system of linear equations, 48

P

Parallel vectors, 8
Parallelepiped volume, 182
Parallelogram
   area, 181
   determined by vectors, 7, 181
   diagonals, 315
Parallelogram law, 7, 320
Parametric representation, 73
Parseval's identity, 333
Partition of a matrix, 91
Partitioned matrix, 91
Partitioning, 91
Period of a function, 437

Periodic function, 437
Permutation matrix, 144
Perpendicular, 313
Perspective, 381–383
Perspective projection, 381
Pivot column, 35
Pivot position, 35
Population distribution, 99
Positive definite matrix, 362, 441, 442
Positive semidefinite matrix, 362
Powers of a matrix, 89, 270
Predator-prey models, 295, 305
Probabilistic relationship, 335
Probability vector, 6
Product
   of matrices, 85
   of a scalar and a linear transformation, 159, 417
   of a scalar and a matrix, 4
   of a scalar and a vector, 398
Projection, *see* Orthogonal projection, 152
Pseudoinverse, 375
Pythagorean theorem, 315

Q

Quadratic form, 358

R

Rabbit problem, 301
Range of a function, 151
Rank of a matrix, 42, 71, 97, 124, 132
Recurrence relation, *see* Difference equation, 299
Reduced row echelon form, 28, 32, 121
Reduction principle, 212
Reflection
   of $\mathcal{R}^2$ about a line, 239
   of $\mathcal{R}^2$ about the y-axis, 159
   of $\mathcal{R}^3$ about a plane, 246, 392
   of $\mathcal{R}^3$ about the xy-plane, 159
   on $\mathcal{R}^2$ about a line, 348–350
Regular Markov chain, 292
Representation of a linear operator by a matrix, 240
Rhombus, 315
Right singular vectors of a matrix, 366
Rigid motion, 350–352
Rotation matrix, 20, 84, 378–381, 383–391
Rotation operator, 345, 348–350
Roundoff errors, 45
Row echelon form, 28
Row of a matrix, 6
Row space of a matrix, 206
   basis, 223
   dimension, 224

S

Scalar, 2
Scalar matrix, 135
Scalar multiple
   of a function, 399
   of a linear transformation, 159, 417
   of a matrix, 4
   of a vector, 6
Scheduling, 105
Schmidt, Erhard, 323
Schwarz, Amandus, 316
Shear transformation, 153
Similar matrices, 134, 242, 261
Singular value, 366
Singular value decomposition of a matrix, 370
Size of a matrix, 2
Skew-symmetric matrix, 10, 195, 421, 442
Solution set of a system of linear equations, 24
Solution to a system of linear equations, 24
Span, 60–65, 75, 408
Spanning set, 63
Spectral decomposition of a matrix, 360
Spring constant, 343
Square matrix, 2
Square root of a linear operator, 290
Standard basis, 211
Standard matrix of a linear transformation, 156
Standard vectors, 15
Steady-state vector, 293
Stochastic matrix, 18
Stochastic relationship, 335
Student enrollments, 318
Submatrix, 2
Subspace, 199, 402
   basis, 211
   column space of a matrix, 204
   dimension of, 213
   nonzero, 201, 402
   null space of a linear transformation, 163
   null space of a matrix, 203
   of $\mathcal{R}^n$, 199
   row space of a matrix, 206
   span, 202–203
   spanning set, 63
   trigonometric polynomials, 408, 433, 437–440
   of a vector space, 402
   zero, 201, 402

Subtraction
   of matrices, 4
Sum
   of functions, 399
   of linear transformations,
      159, 417
   of matrices, 3
   of polynomials, 401
   of vectors, 6, 398
Sylvester, James Joseph, 2
Symmetric matrix, 10, 90, 355–364, 377,
   421, 442
   diagonalization, 355
   positive definite, 362
   positive semidefinite, 362
   quadratic form, 358
   spectral decomposition, 360
System of differential equations,
   294–299
System of linear equations, 23
   augmented matrix, 26
   coefficient matrix, 23
   consistent, 24
   general solution, 30
   homogeneous, 72
   inconsistent, 24
   overdetermined, 48
   parametric representation, 73
   solution, 24
   solution set, 24
   underdetermined, 48

T
Tel Aviv, 304
Test for consistency, 44
Trace of a matrix, 10, 97, 106, 282
Traffic flow, 101
Transition matrix, 292
Translation, 351
Transpose of a matrix, 3, 116
Triangle inequality, 317
Trigonometric polynomials, subspace of,
   408, 433, 437–440

U
Underdetermined system of linear
   equations, 48
Uniqueness of the reduced row echelon
   form, 121
Unit lower triangular matrix, 137
Unit vector, 325, 432
Upper triangular matrix, 97, 137, 180

V
Value-added vector, 58
Vandermonde matrix, 195
Variables of a linear equation, 23
Vector, *see also* Matrix, 5, 398
   addition, 6
   additive inverse, 398
   components, 5
   coordinate, 231, 421
   dot product, 313
   gross production, 52
   length, *see* norm, 311
   net production, 52
   norm, 311, 432
   normalized, 432
   orthogonal, 313, 432
   orthogonal projection of, 331, 435
   parallel, 8
   population distribution, 99
   probability, 6
   scalar multiple, 6
   standard, 15
   steady-state, 293
   unit, 325, 432
   value-added, 58
   zero, 6, 398
Vector space, 398
   dimension, 412
   finite-dimensional, 412
   infinite-dimensional, 412
   isomorphic, 410
   isomorphism, 410
   of linear transformations, 400
   of matrices, 400
   of polynomials, 400
Volume of a parallelepiped, 182

Z
Zero function, 399
Zero matrix, 4
Zero polynomial, 400
Zero subspace, 201, 402
Zero transformation, 154
Zero vector, 6, 398